Handbook of Bioenergy Crops

Handbook of Bioenergy Crops

A Complete Reference to Species, Development and Applications

N. El Bassam

earthscan
from Routledge

First published by Earthscan in the UK and USA in 2010

For a full list of publications please contact:
Earthscan
2 Park Square, Milton Park, Abingdon, Oxfordshire OX14 4RN
711 Third Avenue, New York, NY 10017

First issued in paperback 2015

Earthscan is an imprint of the Taylor & Francis Group, an informa business

ISBN 13: 978-1-138-97571-2 (pbk)
ISBN 13: 978-1-84407-854-7 (hbk)

Typeset by Domex e-Data, India
Cover design by Yvonne Booth

A catalogue record for this book is available from the British Library

Library of Congress Cataloging-in-Publication Data

El Bassam, Nasir.
  Bioenergy crops : a development guide and species reference / N. El Bassam. -- 1st ed.
      p. cm.
  Includes bibliographical references and index.
  ISBN 978-1-84407-854-7 (hardback : alk. paper) 1. Energy crops. I. Title.
  SB288.E43 2009
  333.95'39--dc22

                                                          2009012294

Contents

PART I

PART II

PART III

List of Figures and Tables

Figures

Tables

List of Acronyms and Abbreviations

ABE	acetone-butanol-ethanol (fermentation)
AEAM	Association of European Automotive Manufacturers
BIG/CC	biomass integrated gasifier/combined cycle
BIG/GT	biomass integrated gasifier/gas turbine
BSRS	BioSaline Research Station
BTL	biomass to liquid
BTX	benzene, toluene and xylene
CAM	crassulacean acid metabolism
CFB	circulating fluidized bed (gasification)
CHP	combined heat and power
DM	dry matter
DME	dimethyl ether
DOE	US Department of Energy
DW	dry weight
EC	electrical conductivity
EDP	energy demand planning
EPA	Environmental Protection Agency
ETBE	ethyl tertiary butyl ether
EthOH	ethanol
FAO	UN Food and Agriculture Organization
FCV	(hydrogen) fuel cell vehicle
FT	Fischer-Tropsch
GHG	greenhouse gas
HHV	high heat value
HTU	high temperature upgraded
ICEV	internal combustion engine vehicle
ICRISAT	International Crops Research Institute for the Semi-Arid Tropics
IEA	International Energy Agency
IEF	integrated energy farm
IFEED	Internationales Forschungszentrum für Erneuerbare Energien (International Research Centre for Renewable Energy)
IFPRI	International Food Policy Research Institute
IREF	integrated renewable energy farm
LDV	light-duty vehicle
LHV	low heat value
MDGs	(UN) millennium development goals
MeOH	methanol
MTBE	methyl tertiary butyl ether
MWe	megawatts (electric)
MWth	megawatts (thermal)
NPP	net primary production
ODT	oven dried tonne
OECD	Organisation for Economic Co-operation and Development
ORC	organic rankine cycle

PAH	polycyclic aromatic hydrocarbon
PAR	photosynthetically active radiation
PEM	proton exchange membrane (fuel cell)
PME	plant methyl ester
PPP	public-private partnership
R&D	research and development
RD&D	research, development and demonstration
RME	rape methyl ester
SHSS	super hybrid sweet sorghums
SRC	short rotation coppice
SRF	short rotation forestry
TER	total energy radiation
TS	total solids
UFOP	Union for the Promotion of Oil
UNESCO	United Nations Educational, Scientific and Cultural Organization
VS	volatile solids
WTW	'well-to-wheel'

Foreword

Energy is directly related to the most critical economic and social issues which affect sustainable development: mobility, job creation, income levels and access to social services, gender disparity, population growth, food production, climate change, environmental quality, industry, communications, regional and global security issues. Many of the crises on our planet arise from the desire to secure supplies of raw materials, particularly energy sources, at low prices. The International Energy Agency (IEA) forecasts that world primary energy demand will grow by 1.6 per cent per year on average in 2006–2030, from 11,730Mtoe to just over 17,010Mtoe – an increase of 45 per cent.

Two-thirds of the new demand will come from developing nations, with China accounting for 30 per cent. Without adequate attention to the critical importance of energy to all these aspects, the global, social, economic and environmental goals of sustainability cannot be achieved. Indeed, the magnitude of change needed is immense, fundamental and directly related to the energy produced and consumed nationally and internationally. Today, it is estimated that more than 2 billion people worldwide lack access to modern energy resources.

Current approaches to energy are unsustainable and non-renewable. Today, the world's energy supply is largely based on fossil fuels and nuclear power. These sources of energy will not last forever and have proven to be contributors to our environmental problems and the world cannot indefinitely continue to base its life on the consumption of finite energy resources. In less than three centuries since the industrial revolution, mankind has already burned roughly half of the fossil fuels that accumulated under the Earth's surface over hundreds of millions of years. Nuclear power is also based on a limited resource (uranium) and the use of nuclear power creates such incalculable risks that nuclear power plants cannot be insured. Although some of the fossil energy resources might last a little longer than predicted, especially if additional reserves are discovered, the main problem of 'scarcity' will remain, and this represents the greatest challenge to humanity.

Renewable energy offers our planet a chance to reduce carbon emissions, clean the air, and put our civilization on a more sustainable footing. Renewable sources of energy are an essential part of an overall strategy of sustainable development. They help reduce dependence on energy imports, thereby ensuring a sustainable supply and climate protection. Furthermore, renewable energy sources can help improve the competitiveness of industries over the long run and have a positive impact on regional development and employment. Renewable energies will provide a more diversified, balanced and stable pool of energy sources. With rapid growth in Brazil, China and India, and continued growth in the rest of the world, it is no longer a question of when we will incorporate various renewable energy sources more aggressively into the mix, but how fast?

The Scientific Steering Committee of the International Scientific Congress on Climate Change, Copenhagen, published a press release in March 2009 entitled: 'New renewable to power 40 per cent of global electricity demand by 2050'. But other organizations, such as the Energy Watch Group at Stanford University, Greenpeace and others, have shown that renewable energies can provide 100 per cent of the global energy supply – even beyond electricity – and that this can well be achieved by the year 2050. Some countries such as Denmark, China, Germany, Spain and India have already demonstrated the impressive pace of transition which can be achieved in renewable energy deployment, if the right policies and frameworks are in place. Also the new US President has made clear his determination to massively increase renewable energy in the US, giving strong and clear signals to the world.

Moving away from food crops, biomass energy has now become one of the most dynamic aspects of the modern global energy market with governments around the world seeking to adapt existing renewable policy and regulatory structures to encourage the development of biomass to energy projects. Plantation owners, refiners and biofuel producers are integrating biomass as part of the value chain to achieve energy efficiency and cost savings through the production of utilities, solid, gaseous and liquid fuels and other by-products. Biomass provides the fuel-switching feedstock option to the power and heat generation sector, answering to the high cost of electricity especially for those with large energy needs and the rural areas. Lignocellulosic conversion of biomass such as corn (maize) stover, crop residues and wood to cellulosic ethanol is promoted globally as the answer to the transportation fuels market, complementing the demand for gasoline.

Of all renewable energy sources, generally the largest contribution, especially in the short and medium range, is expected to come from biomass. Fuels derived from energy crops are not only potentially renewable, but are also sufficiently similar in origin to the fossil fuels (which also began as biomass) to provide direct substitution. They are storable, transportable, available and affordable and can be converted into a wide variety of energy carriers using existing and novel conversion technologies, and thus have the potential to be significant new sources of energy into the twenty-first century.

REN21 (2007) estimates the current bioenergy supply to be about 13.3 per cent of the total primary energy demand of 50EJ/year of which 7–10EJ/year is used in industrial countries and 40–45EJ/year is used in developing countries. China and India are the largest biomass energy producers worldwide. While most biomass electricity production occurs in OECD countries, several developing countries, especially India, Brazil, other Latin American/Caribbean and African countries generate large amounts of electricity from combustion of bagasse from sugar alcohol production. Denmark, Finland, Sweden and the Baltic countries provide substantial shares (5–50 per cent) of district heating fuel. Among developing countries, small-scale power and heat production from agricultural waste is common, for example from rice or coconut husks. Biomass pellets have become more common, with about 6 million tonnes consumed in Europe in 2005, about half for residential heating and half for power generation – often in small-scale combined heat and power (CHP) plants. The main European countries employing pellets are Austria, Belgium, Denmark, Germany, Italy, The Netherlands and Sweden. Although a global division of biomass consumption for heating versus power is not available, in Europe two-thirds of biomass is used for heating. For the transport fuels sector, production of fuel ethanol for vehicles reached 39 billion litres in 2006, an 18 per cent increase from 2005. Most of the increased production occurred in the United States, with significant increases also in Brazil, France, Germany and Spain. The United States became the leading fuel ethanol producer in 2006, producing over 18 billion litres and jumping ahead of long-standing leader Brazil. US production increased by 20 per cent as dozens of new production plants came on line. Even so, production of ethanol in the United States could not keep up with demand during 2006, and ethanol imports increased sixfold, with about 2.3 billion litres imported in 2006. By 2007, most gasoline sold in the country was being blended with some share of ethanol. Biodiesel production jumped 50 per cent in 2006, to over 6 billion litres globally. Half of world biodiesel production continued to be in Germany. Significant production increases also took place in Italy and the United States (where production more than tripled). In Europe, supported by new policies, biodiesel gained broader acceptance and market.

The potential of bioenergy crops is huge. More than 450,000 plant species have been identified worldwide; approximately 3000 of these are used by humans as sources of food, energy and other feedstock. About 300 plant species have been domesticated as crops for agriculture; of these, 60 species are of major importance. It will be a vital task to increase the number of plant species that could be grown to produce fuel feedstock. This can be done by identifying, screening, adapting and breeding, biotechnology and genetic engineering a part of the large potential of the flora available on our planet. The introduction of new crops to agriculture would lead to an improvement in the biological and environmental condition of soils, water, vegetation and landscapes, and increasing biodiversity. They can be converted by first-generation biofuels technologies (combustion, ethanol from sugar and starch crops, biodiesel from oil crops) or the second-generation biofuels (hydrolysis or synthesis of lignocellulosic crops and their residues to liquid or gaseous fuels such as hydrogen).

A large potential and opportunity could be expected from the production of fuels from microalgae extraction which I would characterize as the third generation. They can produce at least ten times more biomass than land crops per unit area; they can be grown in saline, brackish or waste waters in ponds or in bioreactors and can be converted to oil, ethanol or hydrogen. Biomass is:

- storable;
- transportable;
- convertible;
- available and affordable; and
- always with positive energy balance.

Of all options, biomass represents a large and sustainable alternative to substitute fossil and nuclear fuels in the mid-term timescale in a 'win–win' strategy.

This book deals with various aspects related to the potential of bioenergy crops that can be grown on plantations for production of fuel feedstock, and with appropriate upgrading and conversion technologies, along with their environmental, economic and social dimensions.

Most grateful thanks are due to Ms Marcia Schlichting who did the most arduous and time-consuming work of preparing the manuscript. I would also like to thank S. G. Agong, M. Assefa, C. Baldelli, R. E. Behl, D. G. Christian, L. Dajue, C. D. Dalianis, J. Decker, H. W. Elbersen, J. Fernández, P. Goosse, A. K. Gupta, K. A. Malik, T. A. Mohammed, Y. Nitta, V. Petríková, M. Satin, H. K. Were, Q. Xi and B. Zitouni for their efforts in providing additional information.

I wish and hope that this book will contribute to enlighten and understand the vital economic and social roles of biomass to meet the growing demand for energy and to face the present and future challenges of limited fossil fuel reserves, global climate change and financial crises.

Professor Dr Nasir El Bassam
International Research Centre for Renewable Energy
Sievershausen, January 2010

Part I

1

Global Energy Production, Consumption and Potentials of Biomass

Structure of energy production and supply

The global energy system currently relies mainly on hydrocarbons such as oil, gas and coal, which together provide nearly 80 per cent of energy resources. Traditional biomass – such as wood and dung – accounts for 11 per cent, and nuclear for 6 per cent, whilst all renewable sources combined contribute just 3 per cent. Energy resources, with the exception of nuclear, are ultimately derived from the sun. Non-renewable resources such as coal, oil and gas are the result of a process that takes millions of years to convert sunlight into hydrocarbons. Renewable energy sources convert solar radiation, the rotation of the Earth and geothermal energy into usable energy in a far shorter time. In the International Energy Agency (IEA) Reference Scenario (OECD, 2008), world primary energy demand grows by 1.6 per cent per year on average during 2006–2030, from 472EJ (11,730Mtoe) to just over 714EJ (17,010Mtoe). Due to continuing strong economic growth, China and India account for just over half of the increase in world primary energy demand between 2006 and 2030. Middle Eastern countries strengthen their position as an important demand centre, contributing a further 11 per cent to incremental world demand. Collectively, non-OECD countries account for 87 per cent of the increase. As a result, their share of world primary energy demand rises from 51 per cent to 62 per cent. Their energy consumption overtook that of the Organisation for Economic Co-operation and Development (OECD) in 2005 (IEA, 2006a).

Current global energy final supplies are dominated by fossil fuels (388EJ per year), with much smaller contributions from nuclear power (26EJ) and hydropower (28EJ). Biomass provides about 45 ± 10EJ, making it by far the most important renewable energy source used.

Renewable energy supplies 18 per cent of the world's final energy consumption, counting traditional biomass, large hydropower, and 'new' renewable (small hydro, modern biomass, wind, solar, geothermal and biofuels), see Figure 1.1. Traditional biomass, primarily for cooking and heating, represents about 13 per cent and is growing slowly or even declining in some regions as biomass is used more efficiently or replaced by more modern energy forms.

Global potential of biomass

On average, in the industrialized countries biomass contributes less than 10 per cent to total energy supplies, but in developing countries the proportion is as high as 20–30 per cent. In a number of countries, biomass supplies 50–90 per cent of the total energy demand. A considerable part of this biomass use is, however, non-commercial and relates to cooking and space heating, generally by the poorer part of the population. Part of this use is commercial: the household fuelwood in industrialized countries and charcoal and firewood in urban and industrial areas in developing countries, but there are very limited data on the size of those markets. An estimated 9 ± 6EJ are included in this category (WEA, 2000, 2004).

Modern bioenergy (commercial energy production from biomass for industry, power generation or transport fuels) makes a lower, but still very significant contribution (some 7EJ per year in 2000), and this share is growing. It is estimated that by 2000, 40GW of

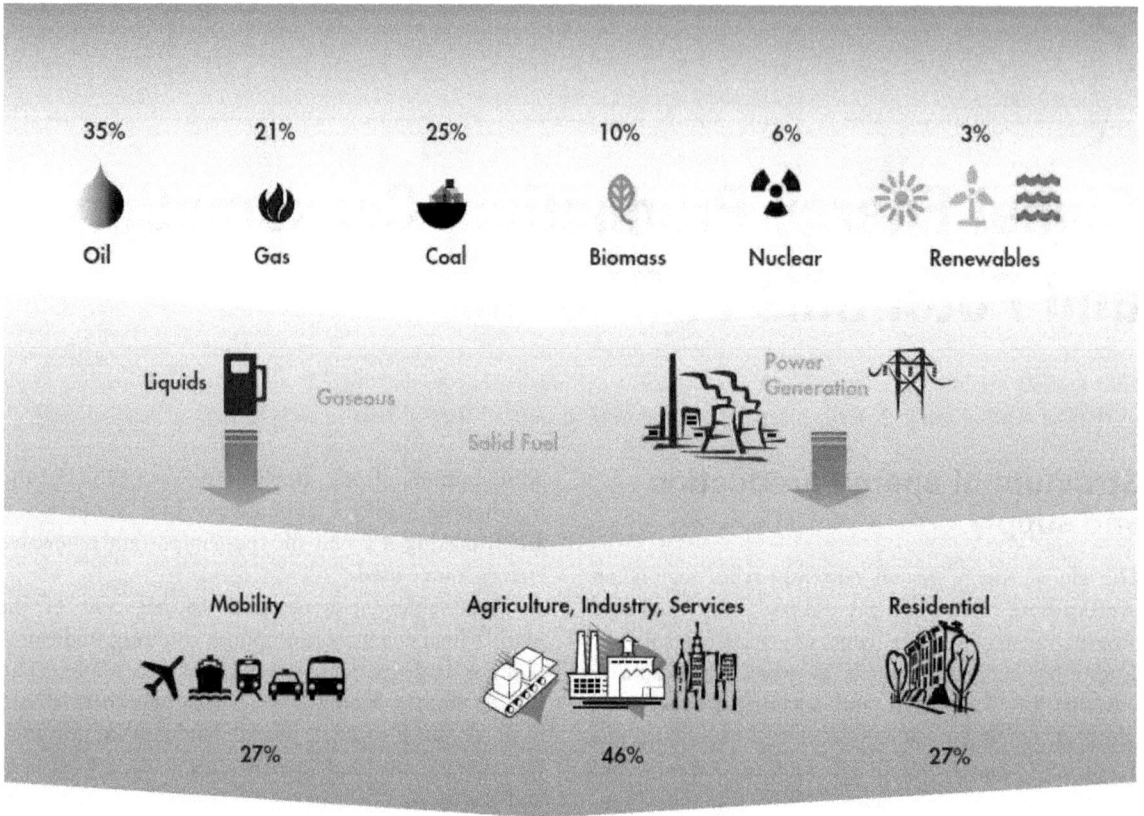

Source: Shell Technology BV, www-static.shell.com/static/innovation/downloads/information/technology-futures.pdf, page 13

Figure 1.1 Primary energy resources, supply and utilization

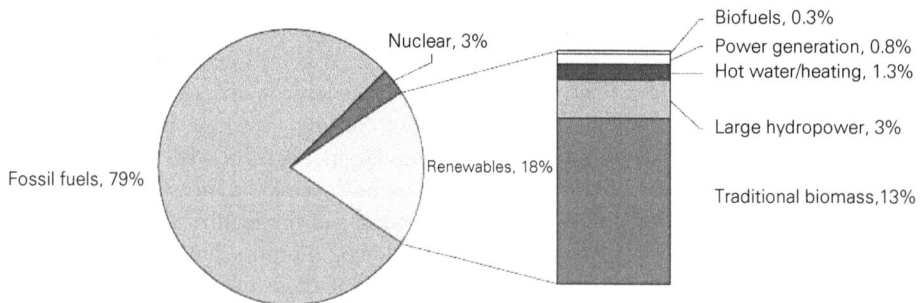

Source: REN21 (2007)

Figure 1.2 Global final energy consumption, 2006

biomass-based electricity production capacity was installed worldwide (producing 0.6EJ electricity per year) and 200GW of heat production capacity (2.5EJ heat per year) (WEA, 2000).

Biomass combustion is responsible for over 90 per cent of the current production of secondary energy carriers from biomass. Combustion for domestic use (heating, cooking), waste incineration, use of process residues in industries, and state-of-the-art furnace and boiler designs for efficient power generation all play their role in specific contexts and markets. Biofuels, mainly ethanol produced from sugar cane and surpluses of corn (maize) and cereals, and to a far lesser extent biodiesel from oilseed crops, represent a modest 1.5EJ (about 1.5 per cent) of transport fuel use worldwide. Global interest in transport biofuels is growing, particularly in Europe, Brazil, North America and Asia (most notably Japan, China and India) (WEA, 2000, 2004; IEA, 2007). Global ethanol production has more than doubled since 2000, while production of biodiesel, starting from a much smaller base, has expanded nearly threefold. In contrast, crude oil production has increased by only 7 per cent since 2000 (WorldWatch Institute, 2007).

The potential contribution of bioenergy to the world energy demand of some 467EJ per year (2004) may be increased considerably compared to the current 45–55EJ. A range of 200–400EJ per year in biomass harvested for energy production may be expected during this century. Assuming expected average conversion efficiencies, this would result in 130–260EJ per year of transport fuels or 100–200EJ per year of electricity.

Various biomass resource categories can be considered: residues from forestry and agriculture, various organic waste streams and, most importantly, the possibilities for dedicated biomass production on land of different categories, for example grass production on pasture land, wood plantations and sugar cane on arable land, and low productivity afforestation schemes for marginal and degraded lands.

The potential for energy crops depends largely on land availability considering that worldwide a growing demand for food has to be met, combined with environmental protection, sustainable management of soils and water reserves, and a variety of other sustainability requirements. Given that a major part of the future biomass resource availability for energy and materials depends on these complex and related factors, it is not possible to present the future biomass potential in one simple figure. Table 1.1 provides a synthesis of analyses of the longer-term potential of biomass resource availability on a global scale. Also, a number of uncertainties are highlighted that can affect biomass availability. These estimates are sensitive to assumptions about crop yields and the amount of land that could be made available for the production of biomass for energy uses, including biofuels.

Focusing on the more average estimates of biomass resource potentials, energy farming on current agricultural (arable and pasture) land could, with projected technological progress, contribute 100–300EJ annually, without jeopardizing the world's future food supply. A significant part of this potential (around 200EJ in 2050) for biomass production may be developed at low production costs in the range of €2/GJ assuming this land is used for perennial crops (Hoogwijk et al, 2005b; WEA, 2000). Another 100EJ could be produced with lower productivity and higher costs, from biomass on marginal and degraded lands. Regenerating such lands requires more upfront investment, but competition with other land uses is less of an issue and other benefits (such as soil restoration, and improved water retention functions) may be obtained, which could partly compensate for biomass production costs. Combined and using the more average potential estimates, organic wastes and residues could possibly supply another 40–170EJ, with uncertain contributions from forest residues and potentially a significant role for organic waste, especially when biomaterials are used on a larger scale. In total, the bioenergy potential could amount to 400EJ per year during this century. This is comparable to the total current fossil energy use of 388EJ. Key to the introduction of biomass production in the suggested orders of magnitude is the rationalization of agriculture, especially in developing countries. There is room for considerably higher land-use efficiencies that can more than compensate for the growing demand for food (Smeets et al, 2007). The development and deployment of perennial crops (in particular in developing countries) is of key importance for bioenergy in the long run. Regional efforts are needed to deploy biomass production and supply systems adapted to local conditions, e.g. for specific agricultural, climatic and socio-economic conditions. Table 1.2 demonstrates an assumption including the possible available area for bioenergy crops by 2050.

Table 1.1 Overview of the global potential of biomass for energy (EJ per year) to 2050 for a number of categories and the main preconditions and assumptions that determine these potentials

Biomass category	Main assumptions and remarks	Energy potential in biomass up to 2050
Energy farming on current agricultural land	Potential land surplus: 0–4Gha (average: 1–2Gha). A large surplus requires structural adaptation towards more efficient agricultural production systems. When this is not feasible, the bioenergy potential could be reduced to zero. On average higher yields are likely because of better soil quality: 8–12 dry tonne/ha/yr* is assumed.	0–700EJ (more average development: 100–300EJ)
Biomass production on marginal lands	On a global scale a maximum land surface of 1.7Gha could be involved. Low productivity of 2–5 dry tonne/ha/yr.* The net supplies could be low due to poor economics or competition with food production.	<60–110EJ
Residues from agriculture	Potential depends on yield/product ratios and the total agricultural land area as well as type of production system. Extensive production systems require re-use of residues for maintaining soil fertility. Intensive systems allow for higher utilization rates of residues.	15–70EJ
Forest residues	The sustainable energy potential of the world's forests is unclear – some natural forests are protected. Low value: includes limitations with respect to logistics and strict standards for removal of forest material. High value: technical potential. Figures include processing residues.	30–150 EJ
Dung	Use of dried dung. Low estimate based on global current use. High estimate: technical potential. Utilization (collection) in the longer term is uncertain	5–55EJ
Organic wastes	Estimate on basis of literature values. Strongly dependent on economic development, consumption and the use of biomaterials. Figures include the organic fraction of MSW and waste wood. Higher values possible by more intensive use of biomaterials.	5–50EJ
Combined potential	Most pessimistic scenario: no land available for energy farming; only utilization of residues. Most optimistic scenario: intensive agriculture concentrated on the better quality soils. In parentheses: average potential in a world aiming for large-scale deployment of bioenergy.	40–1100EJ (200–400EJ)

* Heating value: 19GJ/tonne dry matter.

Sources: Berndes et al (2003); Hoogwijk et al (2005a); Smeets et al (2007)

Table 1.2 Projection of technical energy potential from power crops grown by 2050

	Africa	China	Latin America	Industrialized	All regions
Available area for biomass production in 2050 (Gha)	0.484	0.033	0.665	0.100	1.28
Maximum additional asset of energy from biomass (EJ/year)	145	21	200	30	396
Total (including traditional biomass 45EJ/year) (IPPC, 2001)					441

Source: El Bassam, N. (2004a)

Perspectives of bioenergy crops

The following facts demonstrate the huge potentials of biomass which could boost the production of food and bioenergy feedstock:

- Annual primary biomass production: 220 billion tonnes dry matter (t DM), 4500EJ = 10 times world primary energy consumption. Biomass used for food: 800 millions t DM = 0.4 per cent of primary biomass production.
- Biomass currently supplies 14 per cent of the worldwide energy consumption and 1 per cent of global transportation fuel supply. The level varies from 90 per cent in countries such as Nepal, 45 per cent in India, 28 per cent in China and Brazil, with conversion efficiency of less than 10 per cent. The potential of improving this efficiency through novel technologies is very high.
- Annual food production corresponds to 140 per cent of the needs of world population.
- The potential biomass productivity is the result of interactions between their genetic make up, the environments and the external inputs. The optimization of these three factors is the key issue for a successful introduction of power crops.
- Microalgae have the potential to achieve high levels of photosynthetic efficiency. If laboratory production can be effectively scaled up to commercial quantities, levels of up to 200t/ha/yr may be obtained.
- Large areas of surplus agricultural land in the EU, Eastern Europe and the US could become significant biomass-producing areas (>200 million ha). There is also a huge potential in Latin America.
- Developments in power and heat generation as well as in car technologies are leading to a significant reduction in biomass fuel consumption and land areas.
- The average efficiency of photosynthesis is around 1 per cent. An increase in this efficiency (through plant breeding and genetic engineering) would have spectacular effects in biomass productivity: successful transformation of C_4-mechanism crops (e.g. maize) to C_3 crops (rice). New achievement in accelerating cell division opens opportunities to speed up the growing seasons, resulting in several harvests per year and an overall increase in biomass.
- Bioenergy crops may also be produced on good quality agricultural and pasture lands without jeopardizing the world's food and feed supply if agricultural land-use efficiency is increased, especially in developing regions. Revenues from biomass and biomass-derived products could provide a key lever for rural development and enhanced agricultural production.

2
Bioenergy Crops versus Food Crops

Impact of bioenergy on food security and food sovereignty

The issue of biofuel quotas is a bone of contention with some environmental and humanitarian groups, which say over-zealous biofuel production could cause massive deforestation and lead to food shortages. The United Nations Special Rapporteur on the Right to Food, Jean Ziegler, in 2007 warned: 'the conversion of arable land for plants used for green fuel had led to an explosion of agricultural prices which was punishing poor countries forced to import their food at a greater cost. Using land for biofuels would result in "massacres" ... It's a total disaster for those who are starving' (Lederer, 2007). This is one of various statements on the possible negative effects of large-scale fuel generation from biomass on food availability, food prices and food security. Recently, Brazilian President Lula de Silva declared: 'I am offended when people point their fingers at clean biofuels – those fingers that are besmirched with coal and fossil fuels.'

Some facts: only 2 per cent of arable land worldwide is used for bioenergy development, while 30 per cent of arable land, crucial for growing food crops, lies fallow. Those numbers alone demonstrate that land used to grow biofuel raw materials is not the main cause for global hunger. What it does tell us is the fact that farmers in developing countries do not have the financial resources to pay for seed, and thus cannot utilize these available acres for food production. Globally, 15 million ha are planted in coffee and tea, crops not known to alleviate hunger.

The UN Food and Agriculture Organization (FAO) and the International Food Policy Research Institute (IFPRI) have both published damning reports in recent months in which biofuels are portrayed as the main culprit for the 2007 and 2008 crop price hikes.

Both organizations argue that governments should (urgently) review their biofuel policy because of the devastating effect biofuels production is having on food prices and increased world hunger. These reports were published at a time when prices for maize and wheat were at levels lower than what they were at the beginning of 2007. In early 2008, prices went as high as nearly €250/tonne for wheat and maize, and prices for cereals have now reached such a level that the European Commission is restarting its policy of intervention. Biofuels have been made a scapegoat for rising commodity prices but price hikes are common in agricultural markets due to a combination of relatively inelastic demand and volatile supply. Historical data shows that real (inflation-adjusted) world wheat prices were 15 per cent higher in 1995 and 1996 than the 2007 price spike. Moreover, EU production of bioethanol from wheat only began in earnest in 2003. Therefore there must be a number of factors affecting commodity prices, some of which are cyclical and some of which are structural in nature. On the one hand, among structural factors are the growing demand from emerging economies, the historically low levels of investment in agriculture and agricultural research which have slowed down productivity, the rising biofuels production and higher oil prices. In particular, the Sustainable Development Commission suggests that an increase in oil price from US$50 (€38) to US$100 a barrel could cause a 13 per cent increase in production costs in commodity prices for crops and 3–5 per cent for livestock products. Cyclical factors also affect commodity prices, such as adverse weather conditions resulting in bad harvests in key production areas of the world, limited international commodity trade due to the imposition of export restrictions in various countries and, it seems now above all, speculative investment in agricultural commodity markets. In summary, growing demand and sluggish productivity growth led to the change from a surplus to a shortage era and set the stage for commodity price increases. When weather and crop disease shocks hit commodity markets in 2006 and 2007, stocks of many agricultural commodities were already low, thus exacerbating the price impacts. The policy actions of some countries to isolate their domestic markets through export restraints made the situation even worse, particularly for rice. Rising biofuel production has had a very modest impact on commodity prices.

In Germany, the Association for Bioenergy has set an achievable formula for the year 2020: 10 per cent bioenergy in the electric power sector, 10 per cent in the heating sectors, and 12 per cent for automotive fuel emissions. Those goals and many others for the future are achievable without conflicting with crops for human consumption. A Swiss Study in 2007 showed that if the available biomass is transformed into energy in an efficient and environmentally friendly manner, while at the same time consumption is reduced and energy efficiency increased, these alternative energy carriers can, together with other forms of renewable energy, play a role in our future energy supply that should not be neglected.

Brazil has 340 million ha of arable land, of which only a fifth is under the plough. This calculation excludes the Amazon forests and other environmentally sensitive areas (Brazil's total land mass is 850 million ha). Due to sugar cane's very high productivity, using approximately 3.4 million ha, or only 1 per cent of its arable land, Brazil now produces enough ethanol to power approximately 50 per cent of its passenger vehicles. This is a remarkable achievement. Indeed, competitive ethanol prices are helping to keep petrol prices in check, as the latter struggles to preserve its market share.

The International Research Centre for Renewable Energy (IFEED) is carrying out a research study on biomass biofuel production, land availability and food security in developed and developing countries and their impact on climate, economic and social constraints. The study considers short-, mid- and long-term issues of biomass productivity, conversion and vehicle technologies.

Biomass supply implies a complex analysis of the local natural and agro-environmental conditions. Perspectives of increasing and improving biomass productivity, via plant breeding, gene- and biotechnologies, and optimizing management practices of conventional and new crops, as well as new species including algae and micro-organisms, improvement of conversion technologies and engine efficiencies, are considerable. The primary results indicate that on a long-term timescale, the generation of up to 10 per cent of the agricultural cultivated areas (in Brazil the current land under sugar cane cultivation for sugar and ethanol production amounts to only 2 per cent) would have positive effects on the income of the farmers, poverty alleviation in developing countries, the mitigation of greenhouse gas (GHG) emissions and the environment as well as on food security.

Food production is closely correlated with energy availability and energy supply. The lack of energy in farming systems in rural areas in developing countries is one of the key factors of low productivity, poverty and food shortage (Figure 2.1).

Bioenergy and food prices

The sharp decline of grain prices from recent peaks in commodity prices is a further indication that bioenergy plantations are not the real reason for the food commodity price increase at the beginning of 2008. Oilseed prices have also fallen sharply (Figure 2.2). Palm oil prices averaged less than US$480 a tonne in November, down from US$1250 a tonne in March 2008. Similar declines took place in most edible oils (soya bean oil dropped from US$1475 a tonne to US$835, and coconut oil from US$1470 a tonne to US$705 over the period). The weakening of edible oil

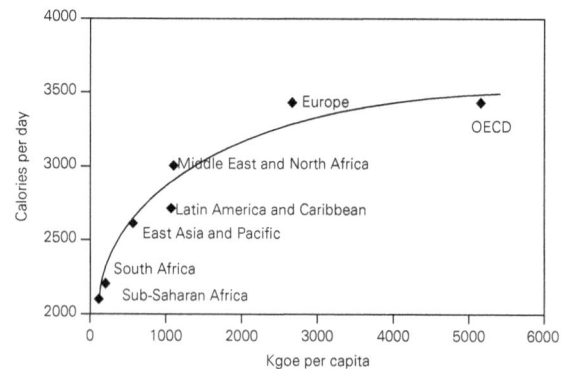

Source: El Bassam (2002b)

Figure 2.1 Correlation between energy input and availability of food as calorie supply

Price changes of major agricultural commodities

Prices of most food commodities have fallen sharply

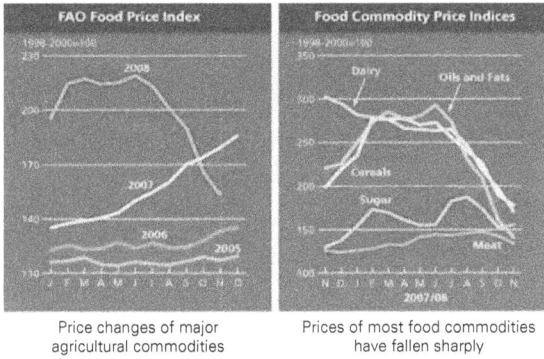

Source: FAO (2008b)

Figure 2.2 Food price indices

a. Soybeans vs. Crude Oil Prices

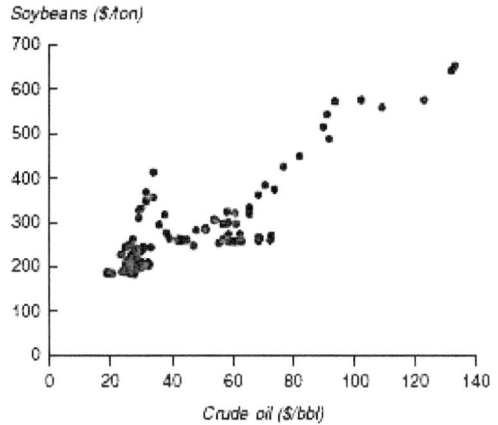

b. Wheat vs. Crude Oil Prices

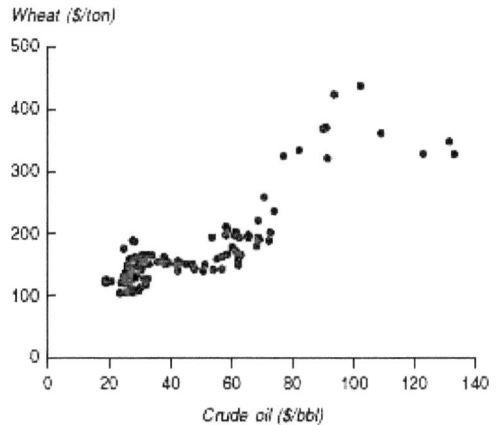

c. Maize vs. Crude Oil Prices

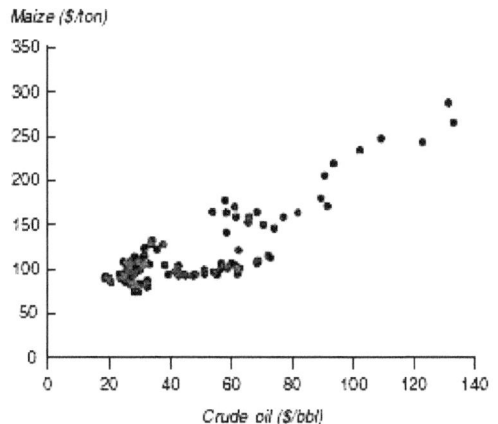

Source: World Bank (2009)

Figure 2.3 Correlation between oil and food prices

prices reflects not only slowing economic growth but also improved supplies (World Bank, 2009).

Over the past 30 years, agricultural productivity has improved much faster than demand; as a result, agricultural output has increased rapidly even as the share of agricultural workers in total employment has steadily declined and prices have fallen (World Bank, 2009). Oil prices are having a direct impact on food prices (Figure 2.3). The OECD member states are spending US$1 billion every day supporting their farmers, resulting in the export of cheap food commodities to developing countries, which inhibits the development of their agriculture, decreasing the income of poor farmers and accelerating depopulation of the rural areas. Currently, nearly 4 million ha in the EU is not used for food production due to the overproduction set-aside policy. The set-aside policy ended in January 2009. The food surplus in the EU countries cannot reach the poor people in developing countries and contribute towards solving their hunger problems.

The UN Energy (2005) paper 'The energy challenge for achieving the Millennium Development Goals', pointed out that available energy services fail to meet the needs of the world's poor, with 2.4 billion people relying on traditional biomass for their energy needs and 1.6 billion not having any access to electricity. The basic commitments to poor people cannot be met without a far more focused approach to energy services.

World hunger is not caused by the lack of food, but by poverty and the lack of adequate income. Eliminating high tariffs on biofuels, combating agricultural subsidies and creating a global biofuels market will certainly help to create new opportunities for developing countries, contributing towards the creation of new jobs and helping to combat rural poverty. Bioelectricity obtained from cogeneration can help to provide electricity to remote rural areas and to further promote economic development.

Food prices have been skyrocketing worldwide due to high oil prices, changing diets, urbanization, expanding populations, flawed trade policies, extreme weather conditions and speculation. Nevertheless, adequate rules should be developed to make sure that only environmentally friendly biofuels are counted towards the overall target. In this context, the ability of modern bioenergy to provide energy services for the poor, implications for agro-industrial development, job creation, health and gender, food and energy security,

trade, foreign exchange balances, climate change and impacts on biodiversity and natural resource management should always be verified and considered. It should be clearly stated that deforestation or shifting large areas for the growing of energy crops is not only unnecessary but also cannot be tolerated.

Internet resources

www.bioenergy-lamnet.org
www.fao.org
http://go.worldbank.org/Y3FILKN180
www.ieabioenergy.com
www.ifpri.org
www.ren21.net

3

Transportation Biofuels

Transport and mobility are now high on many agendas as countries and regions across the world seek to increase mobility and to lessen transport's impact on the environment and the climate. As the world scrambles to meet its increasing energy demands, the search for renewable energy sources is more and more desperate. Energy experts predict a 35 per cent increase in worldwide petroleum demand by 2025. Biofuels are poised to meet the world's increased energy demands. With the number of vehicles in the world projected to rise from 700 million today to over 2 billion by 2050, efforts are intensifying to find ways to meet this fast-growing demand while at the same time minimizing the sector's greenhouse gas emissions. The future strategies of clean transportation depend on the availability of fuels, appropriate engine technologies and their impact on climate and the environment. The transition from fossil fuels to alternative and renewable fuels is already taking place, along with new engines, and will continue to grow. Leading car and vehicle manufacturers have contributed towards creating the necessary technologies and continue to do so. Progress has been achieved in the implementation of legislative structures, directories, improvement of engine efficiencies, fuel generation and marketing strategies in Japan, the US, the EU and other parts of the world.

Automotive biofuels

Liquid and gaseous transportation fuels derived from biomass reduce the dependence on crude oil imports and therefore increase the stability of national fuel markets. Most bioenergy systems generate significantly less greenhouse gas emissions than do fossil fuels and can even be greenhouse gas neutral if efficient methods for biofuels production are developed. Technologies to produce the first-generation liquid and gaseous fuels, such as ethanol from starch and sugar, and biodiesel from vegetable oils, are commercially installed. However, fossil fuel replacement is limited due to feedstock availability. Second-generation liquid transportation fuel utilizes more abundant biomass such as agricultural and forestry residues. Technologies to convert lignocellulosic biomass into liquid fuels are available, but have not yet been applied to large-scale production.

For the transport fuels sector, production of fuel ethanol for vehicles reached 39 billion litres in 2006, an 18 per cent increase from 2005. Most of the increased production occurred in the United States, with significant increases also in Brazil, France, Germany, and Spain. The United States became the leading fuel ethanol producer in 2006, producing over 18 billion litres and jumping ahead of long-standing leader Brazil. US production increased by 20 per cent as dozens of new production plants came on line. Even so, production of ethanol in the United States could not keep up with demand during 2006, and ethanol imports increased sixfold, with about 2.3 billion litres imported in 2006. By 2007, most gasoline sold in the country was being blended with some share of ethanol as a substitute oxygenator for the chemical compound methyl tertiary butyl ether (MTBE), which more and more states have banned due to environmental concerns. Brazilian ethanol production increased to almost 18 billion litres in 2006, nearly half the world's total.

All fuelling stations in Brazil sell both pure ethanol and gasohol, a 25 per cent ethanol/75 per cent gasoline blend. Demand for ethanol fuels, compared to gasoline, was very strong in 2007, due to the introduction of so-called 'flexible-fuel' cars by automakers in Brazil over the past several years. Such cars are able to use either blend and have been widely embraced by drivers, with an 85 per cent share of all auto sales in Brazil. In recent years, significant global

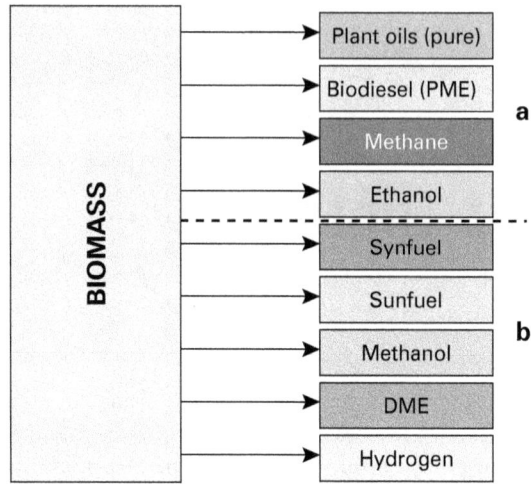

Source: El Bassam (2007b)

Figure 3.1 (a) First; and (b) second generation of biofuels

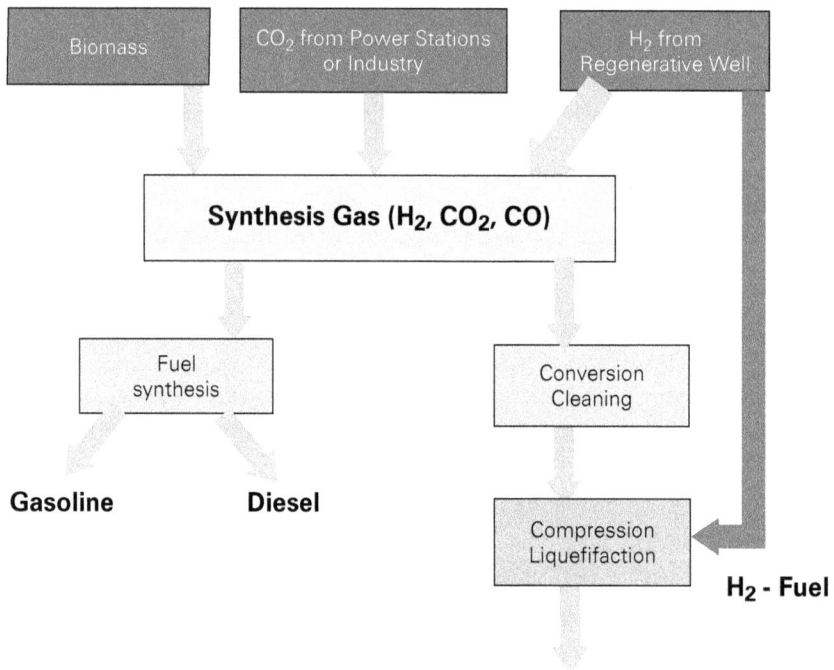

Source: Volkswagen AG, GroupResearch

Figure 3.2 Pathways of innovative synthetic biofuels

trade in fuel ethanol has emerged, with Brazil being the leading exporter.

Other countries producing fuel ethanol include Australia, Canada, China, Colombia, the Dominican Republic, France, Germany, India, Jamaica, Malawi, Poland, South Africa, Spain, Sweden, Thailand and Zambia. Biodiesel production jumped 50 per cent in 2006, to over 6 billion litres globally. Half of world biodiesel production continued to be in Germany. Significant production increases also took place in Italy and the United States (where production more than tripled). In Europe, supported by new policies, biodiesel gained broader acceptance and market share. Aggressive expansion of biodiesel production was also occurring in Southeast Asia (Malaysia, Indonesia, Singapore and China), Latin America (Argentina and Brazil), and Southeast Europe (Romania and Serbia). Malaysia's ambition is to capture 10 per cent of the global biodiesel market by 2010 based on its palm oil plantations. Indonesia also planned to expand its palm oil plantations by 1.5 million ha by 2008, to reach 7 million ha in total, as part of a biofuels expansion programme that includes US$100 million in subsidies for palm oil and other agro-fuels like soya and maize. Other biodiesel producers are Austria, Belgium, the Czech Republic, Denmark, France and the United Kingdom. Seventeen characteristics and costs of the most common renewable energy applications are shown in Table 3.1.

Many of these costs are still higher than conventional energy technologies. (Typical wholesale power generation costs from conventional fuels are in the range US$0.4–US$0.8 per kilowatt hour (kWh) for new base-load power, but can be higher for peak power and higher still for off-grid diesel generators.) Higher costs and other market barriers mean that most renewables continue to require policy support. However, economic competitiveness is not static. The costs of many renewables technologies have been declining significantly with technology improvements and market maturity. At the same time, some conventional technology costs are declining (for

Table 3.1 Biofuels blending mandates

Country	Mandate
Australia	E2 in New South Wales, increasing to E10 by 2011; E5 in Queensland by 2010
Argentina	E5 and B5 by 2010
Bolivia	B2.5 by 2007 and B20 by 2015
Brazil	E22 to E25 existing (slight variation over time); B2 by 2008 and B5 by 2013
Canada	E5 by 2010 and B2 by 2012; E7.5 in Saskatchewan and Manitoba; E5 by 2007 in Ontario
China	E10 in 9 provinces
Colombia	E10 existing; B5 by 2008
Dominican Republic	E15 and B2 by 2015
Germany	E2 and B4.4 by 2007; B5.75 by 2010
India	E10 in 13 states/territories
Italy	E1 and B1
Malaysia	B5 by 2008
New Zealand	3.4 per cent total biofuels by 2012 (ethanol or biodiesel or combination)
Paraguay	B1 by 2007, B3 by 2008, and B5 by 2009
Peru	B5 and E7.8 by 2010 nationally; starting regionally by 2006 (ethanol) and 2008 (biodiesel)
Philippines	B1 and E5 by 2008; B2 and E10 by 2011
South Africa	E8–E10 and B2–B5 (proposed)
Thailand	E10 by 2007; 3 per cent biodiesel share by 2011
United Kingdom	E2.5/B2.5 by 2008; E5/B5 by 2010
United States	Nationally, 130 billion litres/year by 2022 (36 billion gallons); E10 in Iowa, Hawaii, Missouri and Montana; E20 in Minnesota; B5 in New Mexico; E2 and B2 in Louisiana and Washington State; Pennsylvania 3.4 billion litres/year biofuels by 2017 (0.9 billion gallons)
Uruguay	E5 by 2014; B2 from 2008–2011 and B5 by 2012

Note: E2 is a blend containing 2 per cent ethanol; B5 a blend containing 5 per cent biodiesel, etc.

Source: REN21 (2007)

example with improvements in gas turbine technology), while others are increasing due to rising fuel costs and environmental requirements, among many factors. Future cost competitiveness also relates to uncertain future fossil fuel prices and future carbon-related policies. Table 3.1 shows binding obligations on fuel suppliers in some countries. There are more countries with future indicative targets. Gasoline blenders have steadily used more ethanol and biodiesel in their blends than the stated minimum standard, because increasing oil prices meant that ethanol is an attractive fuel in its own right.

The German Federal Research Centre for Agriculture has demonstrated the future energy farming systems which depend on non-food crops. Tall grasses, i.e. miscanthus, *Arundo donax*, bamboo, eucalyptus, acacia, salicornia, and sweet and fibre sorghum are some of the most promising candidates for future energy crops. They have naturally high productivity, the majority are perennial and need less chemicals and water. They can be converted to a wide variety of biofuels (Figure 3.3).

Production and technology in biofuels have evolved considerably in the last 30 years. Today, approximately 90 per cent of new cars sold in Brazil are flex-fuel vehicles, which allow consumers to choose at the pump between petrol, ethanol or any combination of the two. Thanks to biofuels and to the remarkable achievements of Petrobràs, Brazil is no longer dependent on oil imports and is in a much better position to weather what some leaders and analysts are a calling the 'third oil shock'.

Source: El Bassam (2007b)

Figure 3.3 Integrated bioenergy plantations, Braunschweig, Germany

Major car companies such as General Motors, Volkswagen and Toyota intensified their efforts to improve all-round efficiencies and to develop alternative engines for various fuels and purposes. General Motors Corp. introduced 14 new or significantly revised power trains in the 2008 model year – including five 1.0–2.0 litre small-displacement engine variants – with a focus on saving fuel and improving performance in GM's cars and trucks. For 2008, GM's power-train line-up included hybrids, clean diesels and fuel-saving technologies such as active fuel management, direct injection, variable valve timing, six-speed transmissions and flex-fuel options for consumers.

Aviation biofuels

Contributed by Jeffrey Decker, Oshkosh, WI, USA

Commercial and private aviation is growing rapidly around the world, but the increasing cost of fuel and pressure to reduce emissions are leading the industry to look at alternative sources of energy. More attention and resources than ever before are directed to finding a replacement to petroleum, and investment dollars are also at record amounts.

And the need for new fuels will only increase. According to Airports Council International, if current trends continue, airline ticket sales will double by 2025, reaching 9 billion annually. Today, fuel accounts for about 30 per cent of the airline industry's operating costs, compared with about 10 per cent 5 years ago. The cost of jet fuel is forcing the aviation world to look seriously at energy efficiency and new sources of fuel.

In 2005, US airlines alone spent more than US$33 billion on fuel, but governments are taking a hit, too. The largest fuel customer in the world is the US military, at 8 billion gallons per year. The US air force spent US$5 billion on fuel in 2006 and is currently leading tests of synthetic jet fuel converted from coal and natural gas to liquids using an 80-year-old process called Fischer-Tropsch. Officials say this synthetic fuel ought to provide half of the air force's fuel by 2016.

Boeing is at the lead of the new Sustainable Aviation Fuel Users Group, which includes Air France, Air New Zealand, All Nippon Airways, Cargolux, Continental Airlines, Gulf Air, Japan Airlines, KLM, SAS and Virgin Atlantic Airways. Together they claim to use 15–20 per cent of the jet fuel consumed in

Source: El Bassam (2007b)

Figure 3.4 Total flex technology (alcohol or gasoline), Brazil

commercial flights. Along with engine-maker Honeywell and its technology developer Universal Oil Products, the group will research renewable fuel sources with the aim of cutting emissions and reducing fossil fuel dependence.

The steps on that path are regularly outlined in a series of high-profile demonstration flights that began in early 2008 and will certainly continue for years as new fuels come on line.

In February 2008, Virgin operated a Boeing 747 between London Heathrow and Amsterdam using a 20 per cent blend of coconut and palm oil to power one of four GE CF6-80C2 engines. Those feedstocks won't be right for large-scale flights, Virgin stated, but did prove the concept and earn it some positive attention.

Airbus used a synthetic 40 per cent blend derived from the gas-to-liquids process in one of the Rolls-Royce Trent 900 engines of an Airbus A380 in February 2008. Airbus, Honeywell Aerospace, US carrier Jetblue Airways, International Aero Engines (IAE) and Honeywell's UOP formed their own pact to develop second-generation biofuels to provide as much as 30 per cent of world jet fuel demand by 2030.

British Airways is teamed with Rolls-Royce to test several alternative fuels and then pick four, with each supplier asked to provide up to 60,000 litres for testing on one of the airline's RB211 engines. Boeing and Japan Airlines (JAL), with the close cooperation of Pratt & Whitney, are planning a demo flight in 2009. A still-unnamed second-generation biofuel will be blended with jet fuel and tested in one of the four engines of a JAL Boeing 747-300 aircraft equipped with Pratt & Whitney JT9D engines. Japan Airlines will provide the aeroplane and staff for the one-hour flight from an airport in Japan scheduled for 31 March 2009. The flight will be the first biofuel demonstration by an Asian carrier and the first using Pratt & Whitney engines.

The world's first commercial aviation test flight powered by the sustainable second-generation biofuel jatropha has been successfully completed in Auckland. Air New Zealand conducted a biofuel test flight in December 2008 using a 747-400 aircraft, with *Jatropha curcas*-based fuel refined in the US. More than a dozen key performance tests were undertaken in the two-hour test flight which took off at 11.30 am (New Zealand time) on Tuesday 30 December from Auckland International Airport. A biofuel blend of 50:50 jatropha and Jet A1 fuel was used to power one of the Air New Zealand Boeing 747-400's Rolls-Royce RB211 engines. By 2013, Air New Zealand expects to use at least 1 million barrels of environmentally sustainable fuel annually, meeting at least 10 per cent of its total annual needs. (Flug Revue)

In June 2008 Air New Zealand, Continental, Virgin Atlantic Airways, Boeing and UOP became the first wave of aviation-related members to join the newly formed Algal Biomass Organization.

Boeing and Japan Airlines (JAL) were the first to use camelina as a feedstock on 30 January 2009 at Haneda Airport, Tokyo. It was also the first biofuel flight to use Pratt & Whitney engines, with one of four engines on the 747-300 aircraft running on a 50 per cent blend of three second-generation biofuel feedstocks. The biofuel portion was 84 per cent camelina-based, with less than 16 per cent from jatropha and less than 1 per cent from algae. Camelina, also known as false flax, is an energy crop, with a high oil content and an ability to grow in rotation with wheat and other cereal crops. The crop is mostly grown in more moderate climates such as the northern plains of the US. It can be grown in dry areas, poor soil and at high altitudes. The camelina to be used in the JAL demo flight was sourced by Sustainable Oils, a US-based provider of renewable, environmentally clean, and high-value camelina-based fuels. Terasol Energy sourced and provided the jatropha oil, and the algae oil was provided by Sapphire Energy. 'Prior to take-off, we will run the No. 3 engine (middle right) using the fuel blend to confirm everything operates normally. In the air, we will check the engine's performance during normal and non-normal flight operations, which will include quick accelerations and decelerations, and engine shutdown and restart,' JAL environmental affairs VP Yasunori Abe explains. Once the flight has been completed, data recorded on the aircraft will be analysed by Pratt & Whitney and Boeing engineers. Several of the engine readings will be used to determine if equivalent engine performance was seen from the biofuel blend compared to typical Jet A1 fuel.

'There is significant interest across multiple sectors in the potential of algae as an energy source, and nowhere is that more evident than in aviation,' said Billy Glover, ABO co-chair and managing director of Environmental Strategy for Boeing Commercial Airplanes. Dr Max Shauck proclaims ethanol as 'the best fuel there is'. In 1989, Shauck, director of the Institute for Air Science at Baylor University in Waco, Texas, flew a single-propeller prototype Velocity with his wife, Grazia Zanin, across the Atlantic Ocean. In

Source: Arpingstone http://commons.wikimedia.org/wiki/File:Airnz. b747.zk-nbt.arp.jpg

Figure 3.5 Air New Zealand biofuel test flight (747-400)

Brazil, where one-third of the transportation fuel is now ethanol, airframe manufacturer Embraer got the world's first certification for a production aeroplane for use of 100 per cent ethanol – the Ipanema.

Brazil's Centro Tecnico Aerospacial aviation authority worked with Embraer for one and a half years to approve the certification. The Ipanema crop duster first flew in 1970 and its most recent version was introduced in 2004 with approval for ethanol use. More than 60 are burning 100 per cent ethanol today, with the Lycoming IO-540-K1J5 engines delivered with a special fuel system designed by Embraer. More than 100 kits to convert earlier Ipanemas to run on ethanol have been sold. The range is 40 per cent lower than with AvGas, though, and the engines come set for a 40 per cent richer fuel feed. Even Shauck concedes ethanol won't work in jets because its energy density and specific energy are too low. Range and payload would be greatly limited, and, at just 12°C, its flash point would present significant safety dangers.

A bonus is the doubling of time between engine overhaul. 'There is less vibration when you use ethanol as a fuel. That's a function of the larger range of flammability. All of the fuel is consumed in the initial spark,' Shauck says.

An August, 2003 report titled 'The potential for renewable energy sources in aviation' by the Imperial College of London (Saynor et al, 2003) has received much more attention lately than upon its release, says

Dr Ausilio Bauen, one of the three authors. 'It's strange,' he remarks, 'It was not picked up on for a long time, except for the airline industry, which had some knowledge of it.' It examined the feasibility of biodiesel, methanol, ethanol, Fischer-Tropsch synthetics, nuclear, hydrogen and liquefied biomethane. The authors concluded that the cost of producing biodiesel for aircraft would be between US\$33.50 and US\$52.60/GJ (£20.90–£32.80/GJ). That figure includes distribution, necessary chemicals and the conversion process. Raising crops is 75 per cent of the cost. The figure doesn't include benefits from selling the waste flakes, which could have value to farmers and food producers. The 2003 rate for kerosene was US\$4.6/GJ (£2.90/GJ). Costs may come down, said Dr Bauen, especially as more attention and investment dollars go into renewable energy.

For Fischer-Tropsch (FT) fuels, the report examines miscanthus, reed canarygrass and herbaceous perennials and states, 'No particular environmental problems are envisaged with the production of biomass-derived FT kerosene, unless ligno-cellulosic FT plantations replace ecologically important land use, for e.g. native forests (Saynor et al, 2003).'

Operating either the air-fed or the more expensive oxygen-fed facilities is inexpensive after initial capital investments, though costs are highly dependent on electricity prices. Considering transportation, harvesting and other expenses, the report concludes FT can cost as little as US\$5.8/GJ (£3.60/GJ), compared with the 2003 price of US\$4.6/GJ (£2.88/GJ).

Assuming one-third of the United Kingdom's arable land was available to produce FT feedstock, The UK's annual production would be 51PJ (1×10^{15} Joules) and the world could produce 22,800PJ. FT synthetic fuel has fewer particulates and is virtually sulphur-free. That's an emissions advantage, but it also contributes to poor lubricity.

Coal-to-liquids' likely successor

The US Air Force is leading testing of the first successor to petroleum-based jet fuel, or kerosene, which is certain to be coal- and natural gas-based through the Fischer-Tropsch process. A leap for acceptance came with the much-anticipated approval of a 100 per cent synthetic fuel on 9 April 2008, as Sasol of South Africa again leads the way in coal-to-liquids and gas-to-liquids technology for aircraft.

The approval covers jet fuel produced at Sasol's Synfuels facility in Secunda, South Africa. Johan Botha, general manager of Sasol's product applications, says the 100 per cent synthetic fuel will become part of normal Jet A1 supply to the market. 'Once the product is certified, it loses its identity as synthetic fuel and becomes Jet A1. No special arrangements are necessary to supply or use the fuel,' Botha states. Vivian Stockman of the Ohio Valley Environmental Coalition wishes coal-to-liquids wasn't on the fast track to widespread use. She points out that the National Coal Association predicts mining activity will double. 'The coal field communities, we believe, cannot bear a doubling of mining where people already suffer from poison water, poison air and annihilated ecosystems,' she says. Environmentalists also worry about drawdown from the five barrels of water needed to produce a barrel of fuel.

The industry and its regulators have declared any new fuels will be greener, or at least no dirtier, than petroleum. A major new player in fuels development should soon see its first major certification goal realized in a bold new approval process for 50 per cent blends of synthetic jet fuel. If all goes well, CAAFI (The Commercial Aviation Alternative Fuels Initiative) will shepherd certification of 100 per cent synthetic

Source: J. Decker.

Figure 3.6 Brazilian ethanol-powered crop duster Ipanema

kerosene blends by 2010 and similar biofuels approvals for jets as early as 2013. CAAFI pools the resources of its formidable sponsors and stakeholders from across the aviation industry. The Aerospace Industries Association (AIA), the Air Transport Association (ATA), Airports Council International – North America (ACI-NA), and the Office of Environment and Energy of the US Federal Aviation Administration (FAA) are CAAFI's sponsors. They've brought dozens of stakeholders into the fold, including representatives of eight other countries, to advance alternative fuels for aviation.

Presently, the few individual alternative fuels in use have been approved individually, says CAAFI executive director Richard Altman, but soon any producer could be welcome to meet the new standards on fuel for US skies. The spark for this change came in 2006 when the US Air Force (one of seven CAAFI US government stakeholder entities) issued a Request for Information (RFI) before its high-profile B-52 engine tests with 50 per cent synthetic kerosene made by the Fischer-Tropsch process. Currently the certification and qualification panel is focused on drop-in fuels, says Mark Rumizen, FAA fuels specialist and CAAFI Certification team leader. 'That is, it's chemically identical to fuel made from petroleum.'

Solar hydrogen

In 2001 NASA's Helios achieved an altitude of nearly 97,000 feet (29,600m), putting solar power back in the headlines and breaking ground for unmanned solar-powered vehicles. The first and possibly greatest solar flight was by inventor Paul MacCready, when he flew his Solar Challenger across the English Channel in 1981. Before his death in August 2007, MacCready was chairman of Aerovironment, which finds energy solutions for numerous transportation needs, including unmanned aerial vehicles (UAVs). 'We've given up using solar cells for it,' he said. 'As you go to high northern latitudes there

isn't enough sunlight to do the job, so we now are using hydrogen stored as liquid, because that can provide power to fly at 65,000 feet [19,800m] for a week or two weeks, and that's long enough to meet our needs.'

Hydrogen is not a good fuel to substitute for the fuel that airliners use, MacCready explained. 'It has good energy, a lot of energy per pound, but you carry up so few pounds in a large volume, even when it's liquid.' The Fuel Cell Demonstrator Airplane of Boeing Phantom Works has been flying since 2007 after five years of development. 'For the takeoff and climb we use power from the auxiliary batteries. On cruise we fly on power from the fuel cells,' says Jonay Mosquera, the aeronautical engineer in charge of layout, moulding, and perfecting calculations.

Mosquera says, 'I'd guess in 10–15 years the technology should be ready to be implemented with endurance for conventional products.' For now, 'This project is a symbol of the investment and research the Boeing company is doing to create environmentally friendly aircraft.' While Mosquera says his is the only manned fuel cell-powered aircraft, several commercially available unmanned aerial vehicles are powered by hydrogen.

Once new engines and designs start to gain acceptance, the Advisory Council for Aeronautical Research in Europe (ACARE) sees vast changes taking hold. The Out of the Box project established for ACARE by the Dutch aeronautical research institute NIVR is pondering flying saucers and permanent flotillas (Decker, 2007a, 2007b, 2008a, 2008b).

Internet resources

Flug Revue, 'Air New Zealand flies 747-400 with Biofuel', 4 Jan 2009, www.flugrevue.de/de//zivilluftfahrt/air-new-zealand-flies-747-400-with-biofuel.6551.htm
www.purdue.edu/dp/energy/pdf/Harrison08-30-06.pdf
www.ren21.net

4

Primary Biomass Productivity, Current Yield Potentials, Water and Land Availability

Basic elements of biomass accumulation

Bioenergy from crops is a renewable and sustainable source of energy that may be close to carbon neutral, although greenhouse gas mitigation potential and carbon footprints of these crops may vary widely. Solar energy drives the growth of green plants, in the process of photosynthesis, resulting in the production of simple and complex carbohydrates. These carbohydrates can be used as feedstock in a plethora of thermochemical, biological and gasification processes, producing a whole range of energy sources. It has been estimated that some two billion people rely on biomass for primary energy of cooking and space heating. This is mostly in the developing countries of Africa, Asia and Latin America, and wood fuels account for some 14 per cent of global primary energy (OECD and FAO, 2007).

Nearly all of the energy that we use today has been collected and stored by the process of photosynthesis. The amount of carbon fixed each year by autotrophic plants, in the form of biomass, is roughly 200 billion tonnes, or approximately ten times the energy equivalent used yearly in the world. Of these 200 billion tonnes, 800 million (0.4 per cent) are used to feed the human population.

The fundamental process of biomass accumulation within the context of energy is based on photosynthesis. The green plant is the only organism able to absorb solar energy with the help of the pigment chlorophyll, converting this energy into the chemical energy of organic compounds with the aid of carbon dioxide and water (Figure 4.1). In addition to carbon, hydrogen and oxygen, plants also incorporate nitrogen and sulphur into organic matter by means of light reactions.

The photosynthetic process also requires other nutrients for optimum functioning, and an adequate temperature at which to take place. The efficiency of photosynthesis, expressed as the ratio between chemical energy fixed by plants and the energy contained in light rays falling on plants, is less than 1 per cent. A small increase in this efficiency would have spectacular effects in view of the scale of the process involved. However, what has been defined as 'photosynthetic efficiency' is actually the sum of the individual effectiveness of a large number of complex and related processes taking place both within the plants and in their environment.

The solar energy that hits every hectare globally averages out to be the amount found at 40° latitude where every hectare receives 1.47×10^{13} calories of total energy radiation (TER). If this energy were completely converted into the chemical energy of carbohydrates we would expect a yield of more than 2000 tonnes per hectare. But the photosynthetically active radiation (PAR) amounts to 43 per cent of the total solar radiation on the earth's surface.

In the case of a crop yielding 10 tonnes of dry matter per hectare per year, located at 40° latitude, the photosynthetic efficiency would be 0.27 per cent for TER or 0.63 per cent for PAR. As PAR is about 43 per cent of TER, the upper limit of efficiency for the absorption of solar radiation is 6.8 per cent of the total radiation, or 15.8 per cent with respect to PAR.

Considering the total amount of radiation incident on certain areas, and keeping in mind that the calorific content of dry matter is about 4000 kilocalories per kilogram (= 4kcal/g), a theoretical upper production limit of about 250 tonnes of dry matter per hectare could be considered for an area having an amount of incident radiation similar to that at 40° latitude. This

Solar Radiation Energy

$$6 CO_2 + 6 H_2O \xrightarrow[\text{chlorophyll (flora)}]{} 6 O_2 + C_6H_{12}O_6 \xrightarrow{\text{combustion}} 6 CO_2 + 6 H_2O + ENERGY + ASH$$

carbon dioxide   water   oxygen   glucose

nutrients

Biomass
sugars, oils, starches, cellulose, lignin, protein

Heat Power

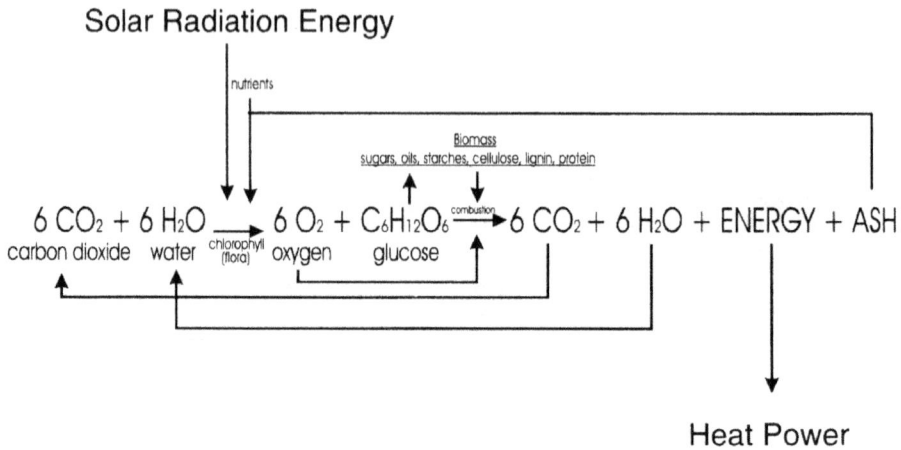

Figure 4.1 The almost closed cycle of energy production from biomass

production value has not yet been attained with any crop, because the vegetation cycle and high growth and solar utilization rates cannot be achieved and maintained continuously throughout the entire year.

From laboratory studies, we know that the photosynthetic apparatus of green plants works with an efficiency of more than 30 per cent. Why this discrepancy? What we really measure when we determine the harvested biomass of our fields is the difference between the efficiencies of photosynthesis and respiration. Furthermore, we have to consider that the growing season for main crops comprises just a fraction of the year, and that mutual shading from the leaves, unfavourable water content of the soil, air humidity and too low or too high temperatures all keep the yield below the theoretical limit. Besides this, the solar spectrum contains broad regions (wavelengths below 400nm and above 700nm) that are almost photosynthetically inactive. Another important factor is the availability of adequate resources of water and nutrients during the growth cycle. There are three main groups of determinants that affect photosynthesis: environmental factors, the genetic structure of genotype and population, and external inputs (Figure 4.2). Their influence on the plant's photosynthesis controls the final biomass yield.

Photosynthetic pathways

The two major pathways of photosynthesis are the C_3 pathway and the C_4 pathway. A third, less common pathway is the crassulacean acid metabolism (CAM)

pathway. In the C_3 pathway, the first product of photosynthesis is a 3-carbon organic acid (3-phosphoglyceric acid), whereas in the C_4 pathway, the first products are 4-carbon organic acids (malate and aspartate). In general, the C_3 assimilation pathway is adapted to operate at optimal rates under conditions of low temperature (15–20°C). At a given radiation level, the C_3 species have lower rates of CO_2 exchange than C_4 species, which are adapted to operate at optimal rates under conditions of higher temperature (30–35°C) and have higher rates of CO_2 exchange. Furthermore, C_4 species have maximum rates of photosynthesis in the range of 70–100mg $CO_2/dm^2/h$ with light saturation at 1.0–1.4cal/cm²/min total radiation, while C_3 species have maximum rates of photosynthesis in the range 15–30mg $CO_2/dm^2/h$ with light saturation at 0.2–0.6cal/cm²/min.

One additional group of species has evolved and adapted to operate under xerophytic conditions. These species have the CAM. Although the biochemistry of photosynthesis in the CAM species has several features in common with that in the C_4 species, particularly the synthesis of 4-carbon organic acids, CAM species have some unique features that are not observed in C_3 or C_4 species. These include capturing light energy during daytime and fixing CO_2 during the night and, consequently, very high water use efficiencies. There are two CAM species of agricultural importance: pineapple and sisal.

The C_3, C_4 and CAM crops can be roughly classified into five groups (Table 4.1), based on the differences between crop species in their photosynthetic pathways

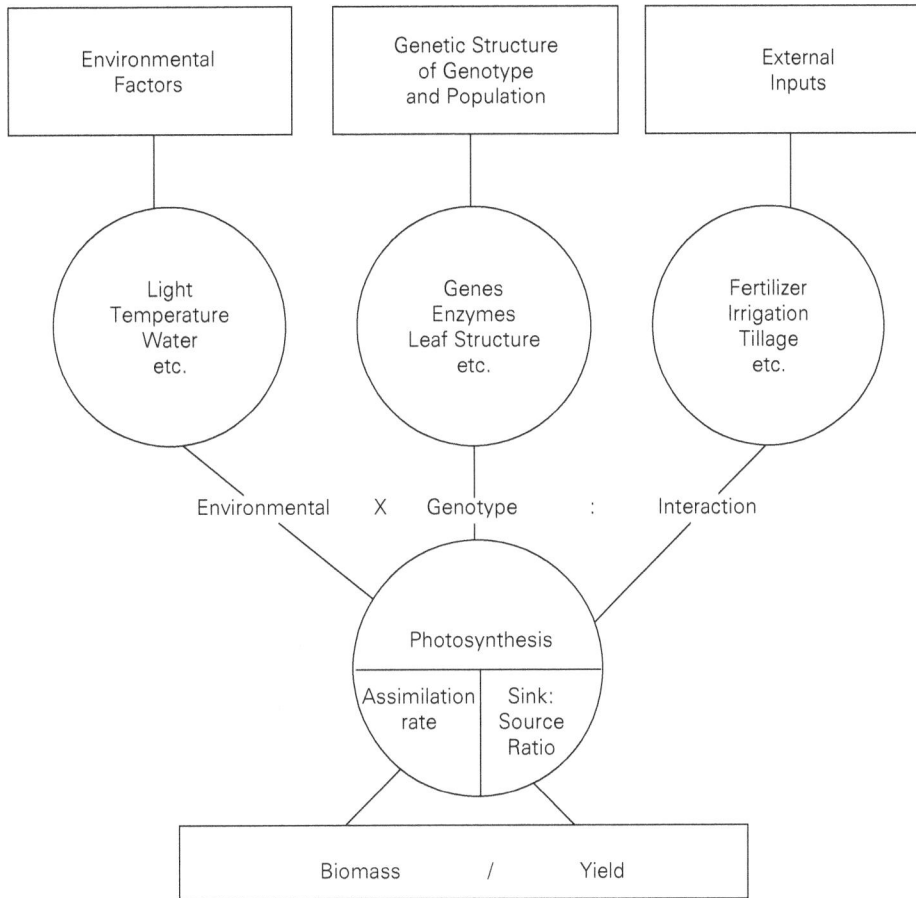

Source: El Bassam (1990)

Figure 4.2 Determinants of potential biomass production

and the response of photosynthesis to radiation and temperature.

C$_4$ plant species tend to be more efficient in water utilization because they do not have to open their stomata so much for the inward diffusion of CO$_2$. Thus transpiration is reduced. Figure 4.3 demonstrates that, although irrigation generally leads to higher yields, the C$_4$ crops miscanthus and maize had higher dry matter yields under conditions of non-irrigation than did the C$_3$ crop rye. The corollary is that C$_3$ species are usually more efficient net photosynthesizers at lower temperatures than C$_4$ species. Both metabolic groups are capable of substantial climatic adaptation (Clayton and Renovize, 1986). *Miscanthus* spp. and *Spartina* spp.

are both C$_4$ species and more efficient photosynthesizers than a wide variety of C$_3$ grasses. This reflects the fact that there are possibilities for the genetic selection of C$_4$ plants with improved cold tolerance. The C$_4$ species may utilize nitrogen with greater efficiency.

An estimate of the upper threshold for converting incident solar energy into biomass production for C$_3$ and C$_4$ species was made by Hall et al (1993). For C$_4$ plants it is 100 per cent × 0.50 × 0.80 × 0.28 × 0.60 = 6.7 per cent, where:

- 0.50 represents the 50 per cent of light that is photosynthetically active radiation (PAR), at wavelengths between 400 and 700nm;

Table 4.1 Some physiological characteristics according to photosynthesis type

Photosynthesis characteristics	Crop group				
	1	2	3	4	5
Photosynthesis pathway	C_3	C_3	C_4	C_4	CAM
Radiation intensity at max. photosynthesis (cal/cm²/min)	0.2–0.6	0.3–0.8	1.0–1.4	1.0–1.4	0.6–1.4
Max. net rate of CO_2 exchange at light saturation (mg/dm²/h)	20–30	40–50	70–100	70–100	20–50
Temperature response of photosynthesis: optimum temp.	15–20°C	5–30°C	25–30°C	10–35°C	30–35°C
operative temp.	15–45°C	20–30°C	10–35°C	25–35°C	10–45°C
Max. crop growth rate (g/m²/day)	20–30	30–40	30–60	40–60	20–30
Water use efficiency (g/g)	400–800	300–700	150–300	150–350	50–200
Crop species	field mustard, potato, oat, rye, rape, sugar beet, wheat, sunflower, olive, barley, lentil, linseed	groundnut, rice, soya bean, sesame, tobacco, sunflower, castor, safflower, kenaf, sweet potato, bananas, coconut, cassava, *Arundo donax*	Japanese barnyard millet, sorghum, maize, sugar cane	millet, sorghum, maize, miscanthus, *Spartina*, *Panicum virgatum*	sisal, pineapple

Source: FAO (1980); revised El Bassam (1995)

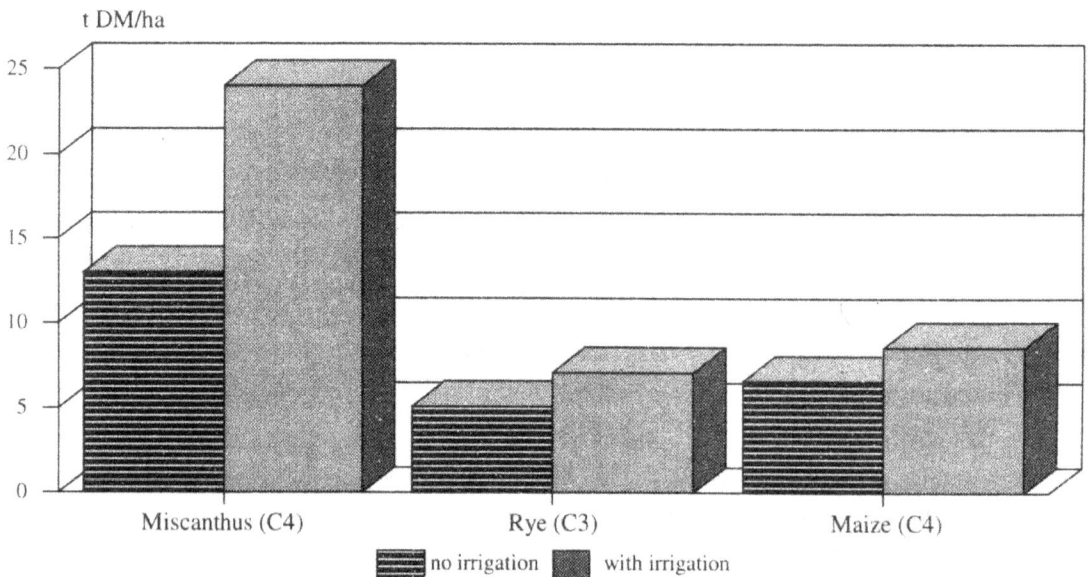

Figure 4.3 Production of biomass of the C_4 crops miscanthus and maize and the C_3 crop rye under two water supply conditions

- 0.80 represents the roughly 80 per cent of PAR captured by photosynthetically active compounds; the rest is reflected, transmitted, and absorbed by non-photosynthesizing leaves;
- 0.28 represents 28 per cent, the theoretical maximum energy efficiency of converting the effectively absorbed PAR to glucose; and
- 0.60 represents the roughly 60 per cent energy stored in photosynthesis remaining after 40 per cent is consumed during dark respiration to sustain plant metabolic processes.

For C_3 plants the equivalent figure is 6.7 per cent × 0.70 × 0.70 = 3.3 per cent. There are two additional losses for these plants: C_3 plants lose about 30 per cent of the already fixed CO_2 during photorespiration; and C_3 plants become light-saturated at lower light intensities than C_4 plants, so that C_3 plants are unable to utilize perhaps 30 per cent of the light absorbed by photosynthetically active compounds.

These figures can now be used to calculate the potential dry matter yield for each biomass crop. However, the following information is also required:

- length of growing season;
- average daily solar radiation in the region;
- energy content of the feedstocks;
- partitioning of the biomass above and below ground; and
- availability of water and nutrients.

The final biomass accumulation is then determined by the genetic structure of the genotype and population.

Plants with the C_4 photosynthetic pathway have a maximum possible potential in temperate climates of 55t/ha, compared to 33t/ha for temperate C_3 crops (Bassham, 1980). The potentials could be much higher for warmer and more humid climates. Samson and Chen (1995) estimated that the maximum photosynthetic efficiency of a biomass plantation studied in Canada would be 64.8 oven dried tonnes per hectare (ODT/ha) for switchgrass (C_4) and 33.6ODT/ha for willow (C_3).

The growth and productivity of crops depends on their genetic potential. How far this potential is realized is closely related to the environmental factors dominating in the region and to the external inputs. Yields achieved in field experiments do not represent the physiological limits of the present cultivars, but

only demonstrate that portion of the genetic potential that is realized by the optimal utilization of present means of cultivation and levels of inputs (Dambroth and El Bassam, 1990). Yield and response of cultivars to environmental conditions and inputs is under genetic control, and therefore an improved response is accessible via screening, selection and breeding.

Climate and local soil properties are the most important constraints influencing the biomass productivity within a specific region. Temperature, precipitation and global radiation are the few classical climatic factors that influence, to a great extent, the regional distribution of various plant species.

Breeding and selection, through both natural and human agencies, have changed the temperature response of photosynthesis in some C_3 and C_4 species. Consequently, there are C_3 species (such as cotton and groundnut) whose optimal temperature is in a medium to high range (25–30°C), and there are C_4 species (such as maize and sorghum), where, for temperate and tropical highland cultivars, the optimal temperature is in a low to medium range (20–30°C).

The potential biomass resource base

The earth's land surface is 13 bn ha. Arable land currently in use is estimated at 1.6 bn ha. Total arable land suitable for rainfed crop cultivation is estimated at 4 bn ha. The area of pasture land is about twice the area of crop land. (OECD and FAO, 2007)

According to the IEA about 14 million ha of arable land are currently used for the production of modern biofuels (IEA, 2006a). Biofuels currently take less than 20 million ha worldwide, compared with the 5000 million ha (sum of arable and pasture lands) that are used worldwide for food and feed. Bio-saline agriculture may bring into play large areas of saline wasteland. According to the FAO there is little evidence to suggest that global land scarcities lie ahead. Between the early 1960s and the late 1990s, world cropland grew by only 11 per cent, while world population almost doubled. As a result, cropland needed per person fell by 40 per cent, from 0.43ha to only 0.26ha, whereas world population has almost doubled. Yet, over this same period, nutrition levels improved considerably and the real price of food declined. This trend is explained by a spectacular increase in land productivity which reduced land use per unit of output by 56 per cent over the same period. Although an estimated 4 billion ha are considered suitable for rain-fed crop production, only some 1.6 billion is currently in use. This is not to say that regional shortages cannot occur as a result of energy crop cultivation, particularly as these surpluses are not evenly distributed globally, nor that in some regions competing claims on water resources and soil nutrients need not be reconciled. The actual claim on arable land resources implied by the large-scale cultivation of energy crops depends to a large extent on the crop choice and the efficiency of the entire energy conversion route from crop to drop. This is illustrated by the figures in Table 4.2. It is important to stress that when lignocellulose is the feedstock of choice, production is not constrained to arable land, but amounts to the sum of residues and production from degraded/marginal lands not used for current food production. Ultimately, this will be the preferred option in most cases. The US National Agricultural Biotechnology Council estimates that with 10 billion people, the required arable land for food would be 2.6–1.2 billion ha (low/high yield) whereas available arable land will shrink to 1.1–2.1 billion ha (NABC, 2000).

Table 4.2 illustrates the indicative ranges for biomass yield and subsequent fuel production per hectare per year for different cropping systems in different settings. Starch and sugar crops require conversion via fermentation to ethanol, and oil crops to biodiesel via esterification (commercial technology at present). The woody and grass crops require either hydrolysis technology followed by conversion to ethanol or gasification to syngas to produce synthetic fuel (neither are yet commercial conversion routes).

Water scarcity can be a serious, highly political matter, mostly locally and regionally specific, and mainly a matter of price. The lack of national water and water-pricing policies and legislation to promote efficiencies are a great barrier, in semi-arid and arid areas, to agriculture in general and hence to water use in biomass production. Adequate agricultural management techniques exist in many cases, and research is under way to promote higher efficiencies through specific temporal application (water the plant root, not the soil, and water the plant when the fruit is developing, not before) such as drip irrigation and so-called deficit irrigation that reduce water use even more. Dramatic improvements in recycling runoff and percolated water can be achieved, so that the only water consumption is in evapotranspiration. In some cases, biomass crops can contribute to improved water management and retention functions, especially in reforestation schemes. If the resource base for biofuels is expanded to cellulosic feedstocks such as grasses and fast-growing trees such as willow or eucalyptus, the whole picture changes dramatically. Not only is the net energy yield per hectare considerably higher (Table 4.2), but these crops also open up a much larger land base for production compared to annual crops.

Table 4.2 Indicative ranges for biomass yield and subsequent fuel production per hectare

Crop	Biomass yield (ODT/ha/yr)	Energy yield in fuel (GJ/ha/yr)
Wheat	4–5	~50
Corn	5–6	~60
Sugar beet	9–10	~110
Soya bean	1–2	~20
Sugar cane	5–20	~180
Palm oil	10–15	~160
Jatropha	5–6	~60
SRC temperate climate	10–15	100–180
SRC tropical climate	15–30	170–350
Energy grasses good conditions	10–20	170–230
Perennials marginal/degraded lands	3–10	30–120

SRC = short rotation coppice

Source: Faaij (2008)

Table 4.3 Overview and evaluation of selected biomass potential studies

Study	Subject	Biomass potential	Evaluation
Fischer et al, 2005	Assessment of eco-physiological biomass yields	Central and Eastern Europe, North and Central Asia; EC (poplar, willow, miscanthus); TP	*Strong*: detailed differentiation of land suitability for biomass production of specific crops on a grid cell level (0.5 degree) *Weak*: not considering interlinkages with food, energy, economy, biodiversity and water demands
Hoogwijk et al, 2005a, b	Integrated assessment based on Intergovernmental Panel on Climate Change (IPCC) Special Report on Emissions Scenarios (SRES) scenarios	Global, EC (short rotation crops); TP	*Strong*: integrated assessment considering food, energy material demands including a scenario analyses based; analyses of different categories of land (e.g. marginal, abandoned) *Weak*: crop yields not modelled detailed for different species and management systems
Hoogwijk et al, 2004	Cost-supply curves of biomass based on integrated assessment	Global; EC (short rotation crops); TP, EP (as cost-supply curve)	*Strong*: establishes a global cost-supply curve for biomass based on integrated assessment *Weak*: linkage land/energy prices not regarded
Obersteiner et al, 2006	Biomass supply from afforestation/ reforestation activities	Global; F (incl. short rotation); EP	*Strong*: modelling of economic potential by comparing net present value of agriculture and forestry on grid-cell level *Weak*: yields of forestry production not dependent on different technology levels
Perlack et al, 2005	Biomass supply study based on outlook studies from agriculture and forestry	USA; EC, F, FR, AR, SR, TR; TP	*Strong*: detailed inclusion of possible advances in agricultural production systems (incl. genetic manipulation) *Weak*: no integrated assessment, e.g. demands for food and materials not modelled
Rokityanski et al, 2007	Analysis of land-use change mitigation options; methods similar to Obersteiner et al, 2006	Global; F (incl. short rotation); EP	*Strong*: policy analysis of stimulating land-use options including carbon prices *Weak*: agricultural land not included
Smeets et al, 2007	Bottom-up assessment of bioenergy potentials	Global; EC, F, AR, FR. SR, TR; TP	*Strong*: detailed bottom-up information on agricultural production systems incl. animal production *Weak*: yield data for crops only regionally modelled
Wolf et al, 2003	Bottom-up assessment of bioenergy potentials mainly analysing food supplies	Global; EC, TP	*Strong*: various scenarios on production systems and demand showing a large range of potentials *Weak*: yields of energy crops not specified for different species and land types

Biomass: EC = energy crops, F = forestry production, FR = primary forest residues, AR = primary agricultural residues, SR = secondary residues, TR = tertiary residues. Potentials: TP = technical potential, EP = economic potential

Source: Faaij (2008)

The size of the biomass resource potentials and subsequent degree of utilization depend on numerous factors. Some of those factors are (largely) beyond policy control. (Smeets and Faaij, 2007). Examples are population growth and food demand. Factors that can be more strongly influenced by policy are development and commercialization of key technologies (such as conversion technology for producing fuels from lignocellulosic biomass and perennial cropping systems), e.g. by means of targeted RD&D strategies. Other areas are:

• Sustainability criteria, as currently defined by various governments and market parties.
• Regimes for trade of biomass and biofuels and adoption of sustainability criteria (typically to be addressed in the international arena, for example via the World Trade Organization).

- Infrastructure: investment in infrastructure (agriculture, transport and conversion) is still an important factor in further deployment of bioenergy.
- Modernization of agriculture: particularly in Europe, the Common Agricultural Policy and related subsidy instruments allow for targeted developments of both conventional agriculture and second-generation bioenergy production. Such sustainable developments are, however, crucial for many developing countries and are a matter for national governments, international collaboration and various UN bodies (such as the FAO).
- Nature conservation: policies and targets for biodiversity protection determine to what extent nature reserves are protected and expanded and set standards for management of other lands.
- Regeneration of degraded lands (and required preconditions), is generally not attractive for market parties and requires government policies to be realized.

Conversion of lignocellulosic biomass requires the availability of commercial conversion processes for such feedstocks and the market and supply infrastructure needed for the second-generation biofuel production capacity. Apart from the matter of what land bioenergy crops should be produced on, and hence apart from the direct competition with crops for food and feed, a new area of research opens up with the use of new species and varieties. In terms of plants, the search is on for species that cannot be consumed by humans or animals. Experience with species such as jatropha, which is exclusively grown for energy content, are promising. Many of these grow under harsh conditions on soils that are unsuitable for most food or feed crops. The challenge will be to reduce water use to a minimum so that they fit marginal conditions. Today, many of these so-called wild species are often very low in yield (a maximum of 1.5 tonnes oil/ha/yr on poor soils) and major breeding efforts are still needed to increase their productivity. In general, the same applies as for any market crop such as cotton: if these new crops are grown on previously cleared land, especially in a rotation with food crops, they could provide useful cash revenue to farmers and would complement food production rather than displace it. The use of species adapted to marginal conditions in combination with improved husbandry methods in grazing systems of small ruminants, now often one of the lowest-yielding agricultural systems in the world, would allow the freeing up of this land to grow crops like *Jatropha* spp. or for the harvesting of grasses. In fact, innovations in protein production (e.g. via algae, new fish farm concepts, even biorefining certain crops directly, single cell proteins) may lead to substantial efficiency increases in protein production and thus reduced land demand compared to the base projections. Land shortage, in other words, is unlikely to be the major long-term factor in the biomass for food or fuel debate. Low productivity in agriculture in many regions has resulted in unsustainable land use, erosion and loss of soils, deforestation and poverty. Increased productivity over time as a result of better farm management, new technologies, improved varieties, energy-related capital investment and capacity building would gradually increase the intensity of land use so that sufficient land becomes available to meet the growing demand for food, feed, fibre and biofuel production (Faaij, 2008).

5

Harvesting, Logistics and Delivery of Biomass

Harvesting biomass to accomplish the goals of fuel reduction and supplying material for energy production is a new practice in Minnesota. Fuel reduction prescriptions need to be adjusted to address operational challenges, and planning and coordination concerns. Once biomass harvest is identified as a management option, incorporating an early understanding of production logistics into harvest plans and prescriptions can reduce fuel management and biomass production costs. Site prescriptions, distance to market, size and efficiency of operations, and equipment all influence the economic viability of biomass harvests as a tool to manage bioenergy crops. Environmental effects of biomass removal on soils, wildlife habitats, and other natural features should be considered. Under the right combination of these circumstances, biomass harvest can reduce crop management costs (Arnosti et al, 2008).

The availability of adequate logistic systems, which include harvesting, recovery, compacting, transport, upgrading and storage, represents a basic requirement for the utilization of energy crops as feedstocks for industrial and energy purposes.

Each conversion technology has specific requirements concerning dry matter content, shape, size, and particle consistency of the raw material. The logistics of the raw materials provide the tool for establishing an effective link between agricultural production systems and industrial activities.

Mechanization of harvest, transport and storage is defined by the methods and processing of the primary products and the need for year-round availability. Since semi-finished products can be created with various properties and qualities, and can be used for various final products, chain management is needed. This also relates to possible storage techniques and conditions during harvest (in winter and spring), and therefore to the workability and timeliness aspects of harvest and storage.

Harvesting of agricultural products is always associated with the handling, transportation and processing of large volumes. This is particularly true of the harvest of biofuels from annually yielding herbaceous field crops, such as energy cereals, miscanthus, straw and hay. Because the total dry matter contributes to the energetic yield of these crops, the annual turnover can be extremely high. Moreover, cost-effective combustion units have to exceed a certain size; thus the area for biofuel production and supply has to be relatively large and increases the need for long-distance transportation in order to meet the demand for biofuel. This reveals one of the main problems for biofuels: while the mass-related energetic density of solid biomass varies only a little, the required volume for a single unit of fuel equivalent can easily vary by a factor of ten, depending on the method of harvesting or processing (Figure 5.1). In the case of the storage of chopped biomass, the volume demand can be reduced tenfold by the application of a high-density compression technology. Thus, the development of appropriate harvesting and compacting systems promises to be extremely beneficial in view of logistical improvements. Such systems appear to be essential for a large-scale introduction of biofuels to the energy market (Hartmann, 1994).

Harvesting systems and machinery

Biomass can be harvested in different ways. From the harvesting standpoint, the most important characteristics of energy crops are moisture content and hardness of the stem. There are now several existing harvesting systems, and newly developed or modified harvesting machines are being tested. These harvesting systems follow two main procedures:

- Multi-phase procedure (several machines must be used):
  – cutting (mowing);
  – pick-up, compacting and baling.
- Single-phase procedure (one machine is used):
  – chopping line;
  – baling line;
  – bundling line;
  – pelleting line.

Multi-phase procedure

For the multi-phase procedure, existing harvesting machines such as mowers and balers can be employed. A rotary or double knife mower can be used to cut the whole crop, which must then be put in swathes before baling. After swathing, the following baler picks up almost all of the harvested material.

The numerous types of baling machines produce rectangular bales, round bales or compact rolls. The big round baler and the big rectangular baler have demonstrated good results for compacting: dry matter density in these high pressure bales is about 120kg/m³ (see Figure 5.2).

The Welger CRP 400 compact round baler produces very highly compacted bales of 0.40m diameter and varying in length from 0.50m to 2.50m. The dry matter density of the compact roll is about 350kg/m³. This high compaction technology helps to

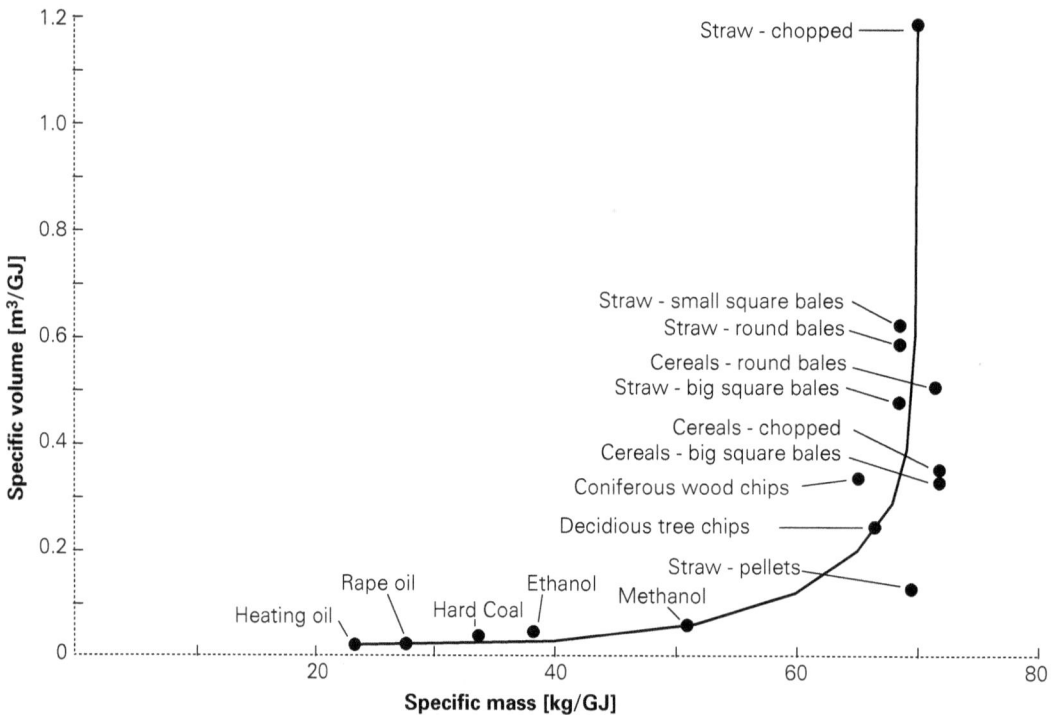

Source: Hartmann (1994)

Figure 5.1 Mass- and volume-related energy density of biofuels

Source: Welger

Figure 5.2 Big round baler

Source: Dajue (2008)

Figure 5.3 Whole crop chopping machine and collecting and transportation tracks

minimize transport and storage: the volume demand of a compact baler is about ten times lower than that of chopped material (Hartmann, 1994). The bulk density can obviously be reduced. Thus, harvesting with a compact roller baler can increase harvesting and transportation efficiency in comparison to other types of balers.

Single-phase procedure

Chopping line

Two different systems can be used for the single-phase chopping line. One is the chopping system normally used for short rotation forestry, which consists of a

mowing unit in front of a tractor in combination with a trailer. Depending on the cutting unit, this system can be applied in stands with or without a special row spacing. An alternative approach uses the chop forage harvester designed for harvesting maize. In most cases, a row-independent mowing attachment is necessary, because the planted rows are not distinguishable with increasing age of the stand.

A chopping machine developed by the Claas company has two heavy mowing devices, which also enable cutting and chopping with one machine. A

Source: Claas

Figure 5.4 Harvesting devices for large bioenergy crop fields

pneumatic conveyor feeds the chopped biomass to a trailer. One disadvantage of this system is that a strong wind can adversely influence the recovery of the biomass.

Depending on the cut length of the material, the densities in dry matter mass of the chopped product are between 70 and 95kg/m³. This low dry matter density is another disadvantage of the chopping system compared to the baling line: the chopped material needs very high transport and storage capacities; and transportation efficiency, especially over long distances, is extremely low.

Baling line

The single-phase baling line process is based on a type of machine developed by several companies (Figure 5.5). This full plant harvester combines mowing, pick-up, compaction and baling in one pass, and produces a highly compacted square bale. The full harvester combines the advantages of the single-phase procedure with the baling line, and is characterized by very low biomass losses during the whole harvesting process.

Bundling line

A bundling line needs a special machine and should be used only when the whole stem of the plant is necessary for further processing. The harvesting method is based on methods developed for reed culture. One harvesting machine developed by Agostini consists of a mowing unit, binding equipment and a transport/deposit unit attached to the three-point linkage of a tractor. The crop is cut with a cutter bar and transported via the binding unit to the side. The density of the bundles is approximately 140kg/m³, the weight 9kg and the bundle diameter 0.2m. The machine is a low-cost and comprehensive unit.

Pelleting line

The Haimer company's newly developed 'Biotruck' (Figure 5.6) means that a single-phase pelleting line is now an option. This machine combines mowing, chopping and pelleting in one procedure on the field. After mowing and chopping, the material is predried using the thermal energy of the engine. The raw material is compacted, pressed and pelleted without additional bonding agents. The result of the process is a corrugated plate with a length of 30–100mm. The single pellet density ranges from 850 to 1000kg/m³, and the bulk density is about 300–500kg/m³. The Biotruck has a capacity of 3–8t/h. Figure 5.7 shows various forms of pellets.

Pelleting offers an excellent opportunity for the reduction of bulk density and hence transportation and storage requirements. In addition, pellets are generally easier to handle than chopped material or bales. If the approach is further utilized, there are good prospects for the automatic handling of such pellets.

Source: John Deere

Figure 5.5 Biomass combine harvester

Source: Haimer

Figure 5.6 Biotruck for harvesting and pelleting

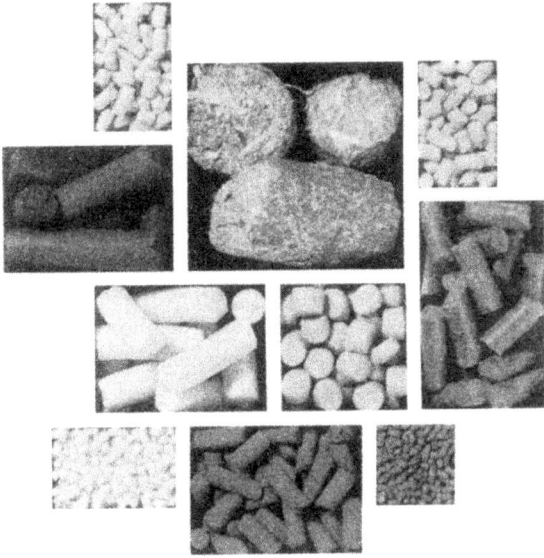

Source: Kahl (1987)

Figure 5.7 Different types of pellets and briquettes

Upgrading and storage

For the year-round delivery of biomass, on-farm storage is necessary. Experience with crop harvests has shown that the moisture content at spring harvest is usually between 20 and 40 per cent, so some kind of drying process is required before storage. It is important to preserve the quality of the product. Depending on the harvest method, the material can be stored chopped (at various different lengths), baled, bundled or pelleted.

Chopped material

The maximum moisture content for the storage of chopped material is 25 per cent, though for a safe storage over a longer period (such as one year) the moisture content must be 18 per cent or lower. When harvest conditions are poor, and the material contains more than 25 per cent moisture, the chopped material can be stored in any storage facility where ventilation from a floor system is possible. Tests show that daily ventilation for 1.5 hours with ambient air reduces the moisture content sufficiently. The cheapest way to store chopped material is in outdoor piles covered with a plastic that allows vapours to pass through.

Experiments with outside storage piles have given different results concerning the moisture content of the plant material. One study compared two piles: one had a ventilation channel and was covered with normal plastic; the other was only covered by a net to prevent the material from blowing away. After five months, the covered pile had a moisture content ranging from 15 per cent in the centre to 10 per cent in the top layer. The open pile showed an increase in moisture content, with contents ranging from 24 per cent in the centre to 64 per cent in the top layer. The results indicate that a covered and non-mechanically ventilated pile is preferable. However, the open pile in this experiment was very small, so storage in a bigger open pile could be possible. The characteristics of the outside layer, such as its thickness and behaviour, are subjects for further research.

Another storage method is to ensilage the fresh biomass, a method commonly used for grass, maize and hemp. The silage could then be pressed so that the solid phase provides a feedstock with more than 50 per cent dry matter content; the liquid would be fermented to produce biogas.

Bales

The compaction rates and density produced by the baler are very important parameters for the storage of bales. Two possible storage methods are field (open) storage, and under-cover storage (with or without a drying system).

The moisture content of the bales at harvest time is an important factor when deciding which type of storage should be implemented in relation to the weather conditions at the site. Large bales can be stored with up to 25 per cent moisture, and are difficult to dry. Bales with a high moisture content (about 40 per cent) at the time of harvest will be prone to mildew, especially under field storage conditions. Compact rolls with a dry matter density of $350kg/m^3$ cannot be further dried, so the moisture content of the material has to be lower than 25 per cent. On the other hand, a higher compaction density will enable storage of the bales under field conditions without large losses: in rainy weather only the first 10cm layer will absorb water, and this layer protects the other, deeper layers. Storage under cover is a favourite method, because the sides are open and allow air to circulate and dry the bales.

Bundles

Bundles can be stored outside. In an experiment to compare two different kinds of bundles, one bundle was tightly bound and the other had a loose binding. There were two further variations for each kind of bundle: covered with plastic, and uncovered. The results showed a great difference in the months of July and September between the covered and uncovered bundles. The difference between loose and tight binding is greater when the bundles are not covered. The water that has infiltrated the tight bundle cannot easily evaporate or trickle through. Thus the outdoor storage of bundles is quite feasible, though a plastic cover against rain prevents fluctuations in moisture content.

Pellets

The advantages of high-density compaction systems for biofuel treatment become obvious when the logistical improvements for long-distance transportation to a combustion unit is considered. However, advanced big square bales and small-sized compact roller bales (400mm diameter) present an interesting advantage, since their drawbacks compared to pelletized fuels are rather marginal, a position that is confirmed by an examination of the aggregate energy demand and total costs. However, there is a strong indication that pellets are associated with a series of benefits for combustion, such as the reduced risk of breakdown. The interactions of biofuel treatment and combustion performance are still awaiting systematic investigation.

High-density biofuel feedstocks (compact bales or pellets) have two important advantages: they need much less space for storage and transport; and their recovery capacity (tonnes of dry matter per manpower hour) is about four times higher than that of chopped feedstocks – see Figure 5.8 (Schön and Strehler, 1992).

Chips

Biofuel from short rotation *Salix* is harvested during the winter when the plants are in dormancy and the dry matter content is about 50 per cent. The harvested, chipped material is either burned directly after

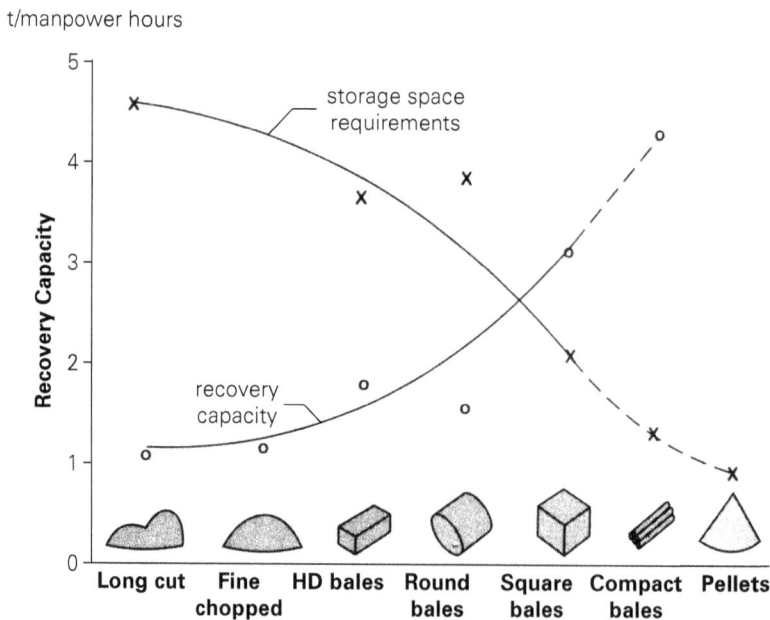

Source: Schön and Strehler (1992)

Figure 5.8 The recovery capacity and storage space requirements for different harvest products as biofuel feedstocks

harvesting or dried in the field for shorter or longer periods before it is chipped and burned. This material is used only at large plants where heat recovery is utilized.

Machines for harvesting coppice have been developed along two principal lines (Danfors, 1994):

1 Direct harvest the crop, chop it up and transport it to a heating plant where the material is stored for only a few days before it is burned. There may be intermediate storage in the field if the transport capacity is too low, or if the fuel cannot be received at the heating plant for some reason.
2 Harvest whole stems with the other type of harvester (Figure 5.9), and store the material in large heaps outside the coppice plantation. The dry matter content decreases gradually during the spring and summer. The material is chipped and delivered at a convenient time to suit demand.

Figure 5.10 shows the land requirements for biomass power plants calculated by taking the power plant's capacity and the yield of the crop to determine the plant's land requirement in hectares. This requirement is correlated with net land use to discover the radius of the land area around the power plant. The example in the figure uses a 20MW power plant with a 10t

Source: Warren Gretz, NREL

Figure 5.9 Harvesting poplar for fibre and fuel

(oven dry)/ha yield. This plant would require 7000ha which for 10 per cent land use will require land within a 14km radius of the power plant. This assumes 45 per cent efficiency, 80 per cent load and 18GJ/ODT.

Internet resources

www.claas.de
www.sustainable-agro.com

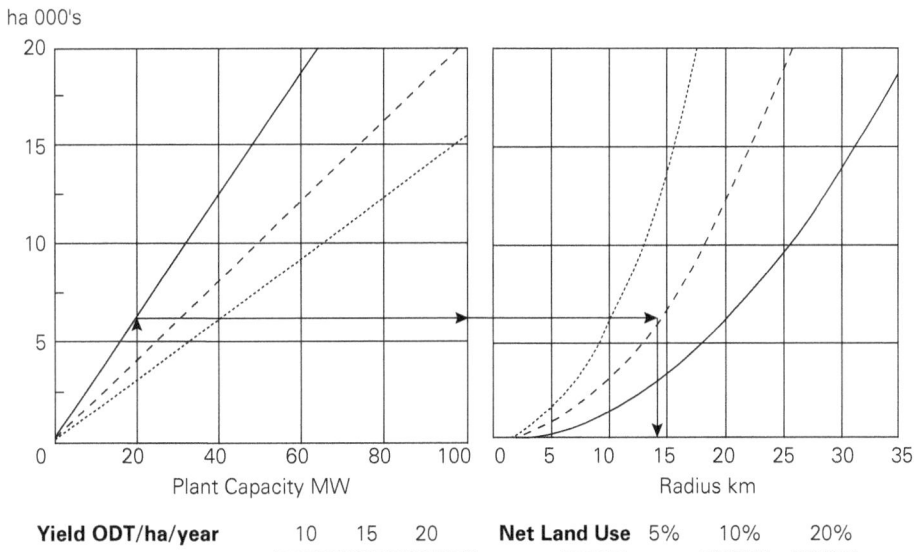

(Hall D., 1994)

Figure 5.10 Land requirements for biomass power plants

6

Technical Overview: Feedstocks, Types of Biofuels and Conversion Technologies

Agriculture and energy have always been tied by close links, but the nature and strength of the relationship have changed over time. Agriculture has always been a source of energy, and energy is a major input in modern agricultural production. Demand for agricultural feedstocks for biofuels will be a significant factor for agricultural markets and for world agriculture over the next decade and perhaps beyond (FAO, 2008c).

A range of biomass resources are used to generate electricity and heat through combustion. Sources include various forms of waste, such as residues from agro-industries, post-harvest residues left on the fields, animal manure, wood wastes from forestry and industry, residues from food and paper industries, municipal solid wastes, sewage sludge, and biogas from the digestion of agricultural and other organic wastes. Dedicated energy crops, such as short rotation perennials (e.g. eucalyptus, poplar and willow) and grasses (e.g. miscanthus and switchgrass) are also used.

Research and development (R&D) work in various industrialized countries currently focuses on technologies that are relevant to their needs and future markets. It is important to note that the same technologies can often be employed with minimal modifications in less developed areas of the world. Hence renewable energy will play an increasing role in future cooperation schemes between developed and developing countries and in export markets for several regions.

Grassi and Bridgewater (1992) indicated that the full range of traditional and modern liquid fuels can be manufactured from biomass by thermochemical conversion and synthesis or upgrading of the products. Advanced gasification for the production of electricity through integrated power generation cycles has a significant role to play in the short term, and is being developed in many parts of the world including Europe, and North and South America. Longer-term objectives are the production of hydrogen, methanol, ammonia and transport fuels through advanced gasification and synthesis.

Biological conversion technologies use microbial and enzymatic processes to produce sugars that can later be developed into alcohol and other solvents of interest to the fuel and chemical industries. Yeast-based fermentation, for example, has yielded ethanol from sugar or starch crops. Solid and liquid wastes can be used to produce methane through anaerobic digestion. Figure 6.1 identifies different forms of biomass with high oil, sugar or starch contents, and technologies used to convert them into biofuels.

Other conversion technologies concentrate on the conversion of biomass high in lignocellulose. Figure 6.2 demonstrates the variety of conversion technologies that can be used on high-lignocellulose biomass to create a range of biofuels.

This chapter provides additional details that demonstrate both current and future possibilities for producing biofuels, and their conversion technologies.

Table 6.1 Biofuel yields for different feedstocks and countries

Crop	Global/national estimates	Biofuel	Crop yield (tonnes/ha)	Conversion efficiency (litres/tonne)	Biofuel yield (litres/ha)
Sugar beet	Global	Ethanol	46.0	110	5060
Sugar cane	Global	Ethanol	65.0	70	4550
Cassava	Global	Ethanol	12.0	180	2070
Maize	Global	Ethanol	4.9	400	1960
Rice	Global	Ethanol	4.2	430	1806
Wheat	Global	Ethanol	2.8	340	952
Sorghum	Global	Ethanol	1.3	380	494
Sugar cane	Brazil	Ethanol	73.5	74.5	5476
Sugar cane	India	Ethanol	60.7	74.5	4522
Oil palm	Malaysia	Biodiesel	20.6	230	4736
Oil palm	Indonesia	Biodiesel	17.8	230	4092
Maize	United States of America	Ethanol	9.4	399	3751
Maize	China	Ethanol	5.0	399	1995
Cassava	Brazil	Ethanol	13.6	137	1863
Cassava	Nigeria	Ethanol	10.8	137	1480
Soya bean	United States of America	Biodiesel	2.7	205	552
Soya bean	Brazil	Biodiesel	2.4	205	491

Source: FAO (2008c)

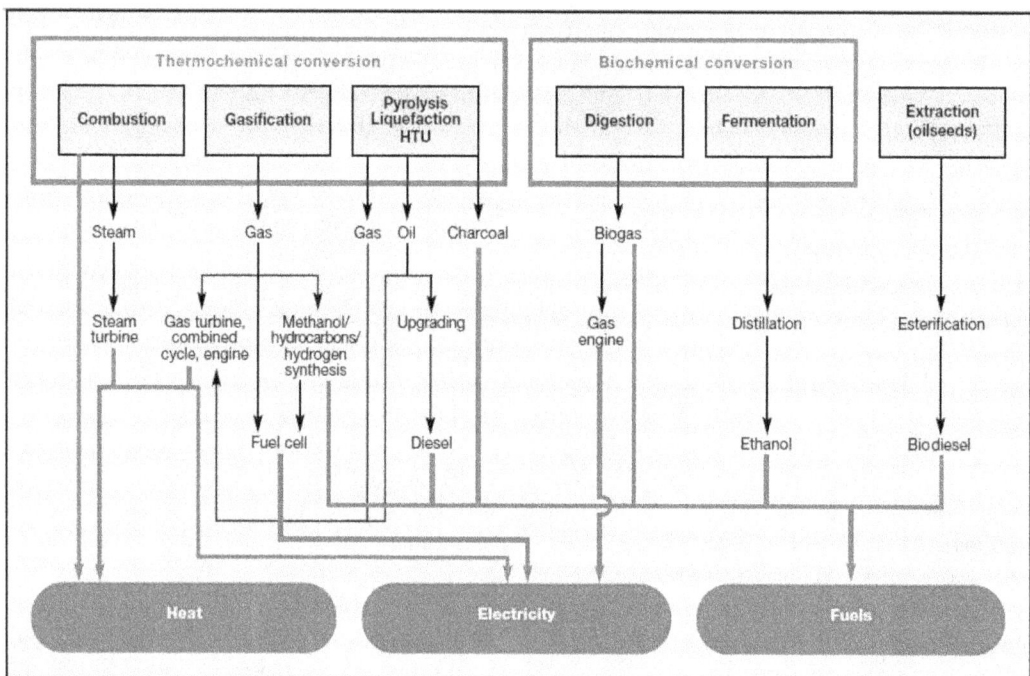

Source: WEA (2000)

Figure 6.1 Main conversion options for biomass to secondary energy carriers

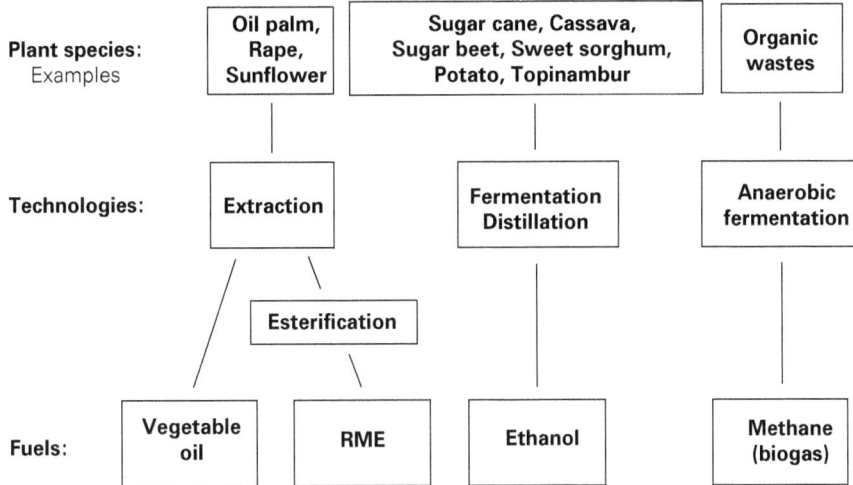

Source: El Bassam (1993)

Figure 6.2 Conversion possibilities for biomass rich in oils, sugars and starch

Ethanol

Ethanol (ethyl alcohol) is obtained by sugar fermentation (sugar cane), or by starch hydrolysis or cellulose degradation followed by sugar fermentation, and then by subsequent distillation.

Obtaining alcohol from vegetable raw materials has a long tradition in agriculture. The fermentation of sugar derived from agricultural crops using yeast to give alcohol, followed by distillation, is a well-established commercial technology. Alcohol can also be produced efficiently from starch crops (wheat, maize, potato, cassava, etc.). The glucose produced by the hydrolysis of starch can also be fermented to alcohol. The goal-directed use of cellulose-containing biomass from agriculturally utilized species for producing alcohol has not yet been practised on a large-scale engineering level. Production has been confined to the use of wood, residues and waste materials. Table 6.2 lists the ethanol production of several sugar- or starch-producing plants.

Technologies that can produce fuel ethanol from lignocellulose are already available (Ingram et al, 1995). The challenge today is to assemble these technologies into a commercial demonstration plant.

Many wood wastes and residues are available as feed materials, while high-yielding woody crops will also provide lignocellulose at lower prices than agriculture crops.

Approaches for the improvement of acetone-butanol-ethanol (ABE) fermentation have focused on the development of hyper-amylolytic and hyper-cellulolytic clostridial strains with an improved potential for biomass-to-butanol conversion. The developed strains produce approximately 60 per cent more butanol than the parental wild type strain (Blaschek, 1995). These findings suggest that the developed strains are stable and offer a significant economic advantage for use in commercial fermentation processes for producing butanol.

Bioethanol can be used as a fuel in different ways: directly in special or modified engines as in Brazil or in Sweden, blended with gasoline at various percentages, or after transformation into ethyl tertiary butyl ether (ETBE). In the first case it does not need to be dehydrated, whereas blends and ETBE production need anhydrous ethanol. Moreover, ETBE production requires neutral alcohol with a very low level of impurities. ETBE is an ether resulting from a catalytic reaction between bioethanol and isobutylene, just as

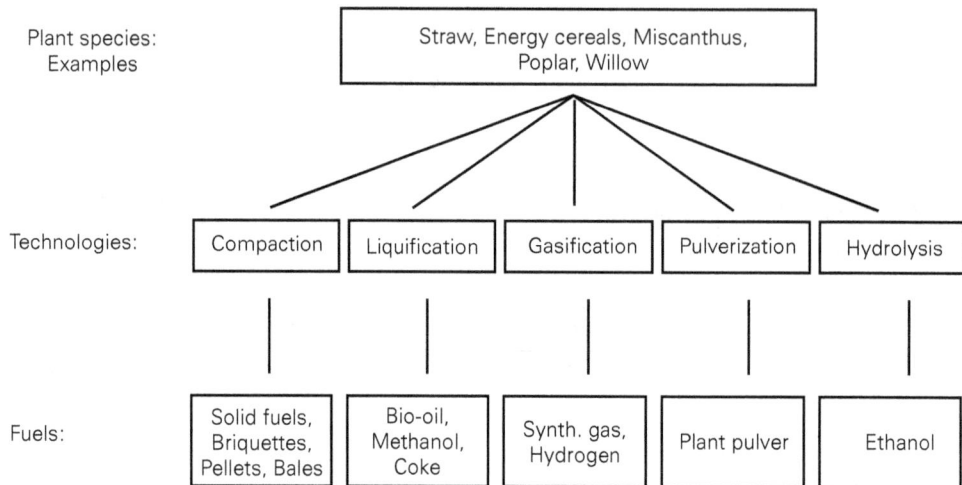

Source: El Bassam (1993)

Figure 6.3 Conversion possibilities for lignocellulose biomass

Table 6.2 Ethanol yields of some sugar- or starch-producing plants

Source	Ethanol yield (litres per tonne biomass)
Molasses	270
Sugar cane	70
Sweet sorghum	80
Corn	370
Sweet potato	125
Cassava	180
Babassu	80
Wood	160

MTBE is processed from methanol. MTBE is the most commonly known fuel ether, used in European gasoline since 1973 and still generally used throughout the world to improve gasoline. Even the aviation industry has become involved in using ethanol as fuel (Figure 6.4). Bioethanol can also be the alcohol used for the production of biodiesel, thus giving ethyl esters instead of methyl esters; the other constituent is vegetable oil from rape, sunflower, soya beans or other sources.

European regulations enable direct blending of bioethanol up to 5 per cent and of ETBE up to 15 per cent with other oxygenated products, according to a Directive dated 5 December 1985 concerning crude oil savings by utilization of substitute fuels. On the other hand, the European Union's objective, as expressed in its Altener programme, for example, is to let biofuels take a 5 per cent market share of fuel consumption in the year 2005. In fact, oil companies and car manufacturers in Europe do not like bioethanol blends, primarily because of their lack of water tolerance and their volatility and the need to put a label on the pumps to inform the consumer.

However, oil companies and car manufacturers are interested in ETBE, which is very similar and slightly better than the MTBE they are used to. ETBE has a

Source: Aviation Sciences

Figure 6.4 An ethanol-powered plane

lower Reid vapour pressure (RVP) than MTBE, and a higher octane value and lower oxygen content. Moreover, most MTBE producers could readily convert their plants to make ETBE after some process and equipment modifications.

ETBE- and MTBE-blended gasolines have several benefits:

- the oxygen content reduces carbon monoxide emissions;
- the lower RVP lessens the pollution that forms ozone; and
- the high octane value reduces the need for hydrocarbon-based aromatic octane enhancers like benzene, which is carcinogenic.

This growing interest in one product but not the other brought bioethanol producers into discussions with MTBE producers, and a partnership was initiated. There were only two of them in France: ELF and ARCO. ELF made a first test of ETBE production over five days in 1990 at its refinery in Feyzin near the town of Lyon, and a second test during a one-month period in 1992. The result was conclusive, so the company started ETBE production on an industrial scale in 1995.

Oxygenated fuel feedstocks (ETBE and MTBE) are considered as convenient options for reaching the octane value requirements of gasoline fuels. Oxygenated

compounds offer considerable benefits if they are mixed with gasoline: reduction in the amount of carbon monoxide emissions and unburned hydrocarbons. There is a growing demand for oxygenates in Europe, as well as in the USA as a result of the Clean Air Act Amendments of 1990.

Bioethanol can be used in an undiluted form or mixed with motor gasoline. 'Gasohol' is the term used in the USA to describe a maize-based mixture of gasoline (90 per cent) and ethyl alcohol (10 per cent); it should not be confused with gasoil, an oil product used to fuel diesel engines. Although engines need to be adapted to use bioethanol at 100 per cent, unadapted engines can run the mixed fuels. Bioethanol can be used to substitute for MTBE and is added to unleaded fuel to increase octane ratings. In Europe, the preferred percentage – as recommended by the Association of European Automotive Manufacturers (AEAM) – is a 5 per cent ethanol or 15 per cent ETBE mix with gasoline.

Oils

Vegetable oils and fats, in contrast to the simple carbohydrate building blocks glucose and fructose, exhibit a number of modifications to the molecule's structure and are also more valuable regarding energy. From the point of view of plant breeding and plant cultivation of energy crops, it is high oil yields per hectare rather than quality aspects that are of primary importance.

The use of vegetable oils as fuels is not new. Since ancient times, oils have also been used as a material for burning and for lighting (the oil lamp). The inventor Rudolph Diesel used peanut oil to fuel one of his engines at the Paris Exhibition of 1900, and wrote in 1912 that: 'The use of vegetable oils for engine fuels may seem insignificant today, but such oils may become, in the course of time, as important as petroleum and the coal tar products of the present time' (Nitske and Wilson, 1965). Now, biodiesel is increasingly used as a transportation fuel (Figure 6.5).

Worldwide, there are more than 280 plant species with more or less considerable oil contents in their seeds, fruits, tubers and/or roots. The oil can be extracted by compressing the seeds (sunflower, rapeseed, etc.) mechanically with screw presses and expellers, with or without preheating. Pressing with

Source: UFOP

Figure 6.5 Biodiesel fuel station

preheating allows for the removal of up to 95 per cent of the oil, whereas without preheating the amount is considerably less. Extraction by the use of solvents is more efficient: it removes nearly 99 per cent of the oil from the seeds, but it has a greater energy consumption than the expeller. Table 6.3 looks at the seed and oil yields of various crops. Table 6.4 shows the oil and fatty acid composition of various crops.

Raw vegetable oils must be refined prior to their utilization in engines or transesterification. The most widespread idea is that fuel must be adapted to present-day diesel engines and not vice versa. In order to meet the requirements of diesel engines, vegetable oils can be modified into vegetable oil esters (transesterification). The transesterification procedure includes the production of methyl esters – RME (rape methyl ester) or sunflower methyl ester – and glycerol (glycerine) by the processing of plant oils (triglycerides), alcohol (methanol) and catalysts (aqueous sodium hydroxide or potassium hydroxide). Table 6.5 provides a comparison of the fuel properties of three of the most common vegetable oils, their corresponding methyl esters, and diesel fuel. Table 6.6 lists the properties of some additional vegetable oils.

Cracking is another option for modifying the triglyceride molecule of the oil (Pernkopf, 1984). However, the cracking products are very irregular and more suitable for gasoline substitution; the process has to be conducted on a large scale; costs are considerable;

and conversion losses are also significant. These negative aspects together with the much lower efficiency of gasoline engines make the cracking process of minor interest.

The Veba process is yet another procedure for converting the triglyceride molecule. During the refining of mineral oil to form the different conventional fuels (gasoline, diesel, propane, butane, etc.), up to 20 per cent rapeseed oil is added to the vacuum distillate. The molecules are cracked and the mixture is treated with hydrogen. The generated fuel molecules are not different from conventional fuel molecules. Advantages of the Veba process are that no glycerine is produced as a by-product, and that the generated fuel is not different from standardized fuels so there is no need for a special distribution system and handling. But there are also disadvantages: the high consumption of valuable hydrogen, and reductions in biodegradability (Vellguth, 1991).

Pure plant oils, particularly when refined and deslimed, can be used in pre-chamber (indirect injected, like the Deutz engine described below) and swirl-chamber (Elsbett) diesel engines as pure plant oil or in a mixture with diesel (Schrottmaier et al, 1988). Pure plant oil cannot be used in direct injection diesel engines, which are used in standard tractors and cars, because engine coking occurs after some hours of operation. The addition of a small proportion of plant oils to diesel fuel is possible for all engine types, but will also lead to increased deposits in the engines in the long term (Ruiz-Altisent, 1994).

The 'Elsbett' engine is a recently developed diesel engine variant with a 'duothermic combustion system' that uses a special turbulence swirl chamber and runs on pure vegetable oils. Another engine, developed by Deutz, uses the principle of turbulence with indirect injection and can run on purified plant oils. Its consumption is about 6 per cent higher than that of other diesel engines, but it has proved to be robust and reliable.

Other sources (Löhner, 1963; van Basshuysen et al, 1989) indicate that the consumption level of indirect injection swirl chamber diesel engines is 10–20 per cent higher in relation to comparable diesel engines with direct injection. Another engine, still in development, is the John Deere 'Wankelmotor', which works with multi-type fuels, including vegetable oils.

The transesterification of vegetable oils permits the wide utilization of these oils in existing engines, either

Table 6.3 Oilseed yields and oil contents of various plant species

Botanical name	Common name	Seed or oil yield (kg/ha)		% oil
Ablemoschus moschatus Medik	Ambrette	836–1693	seed	
Aleurites fordii Hemsl.	Tung-oil tree		seed	
Aleurites montana (Lour.) Wils.	Mu-oil tree	5500	oil	
Anacardium occidentale L.	Cashew	1000	seed	
Arachis hypogaea L.	Peanut	5000	seed	35–55
Brassica juncea (L.) Czern	Mustard, green	1166	seed	30–38
Brassica napus L.	Rape	3000	seed	40–45
Brassica nigra (L.) Koch	Mustard, black	1100	seed	30–35
Brassica rapa L.	Turnip	1000	seed	
Calophyllum inophyllum	Calophyllum			50–73
Camellia japonica	Tsubaki		seed	66
Cannabis sativa L.	Hemp	1500	seed	25–30
Carthamus tinctorius L.	Safflower	4500	seed	35–45
Ceiba pentandra	Kapok			20–25
Cier arietinum L.	Chickpea	2000	seed	
Citrullus colocynthis (L.) Schrad.	Colocynth	6700	seed	47
Cocos nucifera L.	Coconut	1000	copra	62.5
Coronilla varia L.	Crownvetch	500	seed	
Crambe abyssinica Hochst ex R.E. Fries	Crambe	5000	seed	25–33
Croton tiglium L.	Croton, purging	900	seed	50–55
Cucurbita foetidissima HBK	Buffalo gourd	3000	seed	24–34
Cyamopsis tetragonoloba (L.) Taub	Guar	2000	seed	
Elaeis guineensis Jacq.	African oil palm	2200	oil	55 (of kernel))
Glycine max (L.) Merr.	Soya bean	3100	seed	13–25
Gossypium hirsutum L.	Cotton	900	seed	16
Guizotia abyssinica (L. f.) Cass.	Niger seed	600	seed	25–35
Helianthus annuus L.	Sunflower	3700	seed	40–50
Juglans nigra L.	Black walnut	7500	seed	60
Juglans regia L.	Persian walnut	7500	seed	63–67
Lens culinaris Medik.	Lentil	1700	seed	
Lesquerella spp.	Lesquerella	1121	seed	11–39
Limnanthes bakeri J. T. Howell	Meadowfoam, Baker's	400	seed	24–30
Limnanthes douglasii R. Br.	Meadowfoam, Douglas'	1900	seed	24–30
Linum usitatissimum	Flax	650	seed	34
Lotus corniculatus L.	Trefoil, birdsfoot	600	seed	
Lupinus albus L.	Lupin, white	1000	seed	
Macadamia spp.	Macadamia nut	4000	seed	15–20 (of nut)
Medicago sativa	Alfalfa seed	800	seed	8–11
Mucuna deeringiana (Bort) Merr.	Velvetbean	2000	seed	
Oryza sativa	Rice bran	800	seed	15–20
Pachyrhizus erosus (L.) Urb.	Jicama	600	seed	
Pachyrhizus tuberosus (Lam.) Spreng	Ajipo	600	seed	
Papaver somniferum L.	Poppy, opium	900	seed	45–50
Perilla frutescens (L.) Britt.	Perilla	1500	seed	35–45
Phaseolus acutifolius A. Gray	Tepary bean	1700	seed	
Phaseolus coccineus L.	Scarlet runner bean	1700	seed	
Phaseolus lunatus L.	Lima bean	2500	seed	
Pisum sativum L.	Garden pea	1800	seed	
Prunus dulcis (Mill.) D. A. Webb	Almond	3000	seed	50–55
Ricinus communis L.	Castorbean	5000	seed	35–55
Sambucus conodensis L.	Elderberry			22–28
Sapium sebiferum (L.) Roxb.	Chinese tallow tree	14000	seed	19
Sesamum indicum L.	Sesame	1000	seed	45–50

Table 6.3 Oilseed yields and oil contents of various plant species (*Cont'd*)

Botanical name	Common name	Seed or oil yield	(kg/ha)	% oil
Simmondsia chinensis (Link)				
C. Schneid	Jojoba	2250	seed	43–56
Sinapis alba L.	Mustard, white	8000	seed	25–30
Stokesia laevis (Hill) Green	Stokes aster	1121	seed	27–44
Stylosanthes humilis HBK	Townsville style	1200	seed	
Telfairia pedata (Sm. ex Sims.) Hook	Zanzibar olivine	2000	seed	35
Theobroma cacao L.	Cacao	3300	seed	35–50
Trigonella foenum-graecum L.	Fenugreek	3000	seed	
Vigna angularis (Willd.)				
Ohwi & Ohashi	Adzuki bean	1100	seed	
Vigna radiata (L.) Wilczek var. *rodiato*	Mung bean	1100	seed	
Vigna umbelata (Thunb.)				
Ohwi & Ohashi	Rice bean	200	seed	
Vigna unguiculata (L.) Walp.	Cow pea	2500	seed	
Vitis vinifera	Grape seed oil			6–21

Source: Peterson (1985)

Table 6.4 Oil and fatty acid contents of various plant species

Species	Family	Oil content (%)	Fatty acid composition (wt.%)						
			16:0	18:0	18:1	18:2	18:3	18:4	others
Asphodelus tenuifolius	Liliaceae	22.0	7.6	16.0	12.8	78.0			
Calotropis procera	Asclepidaceae	26.0	14.3	19.3	62.7	3.1			0.6% 16:1
Carthamus oxycantha	Compositae	26.3	8.5	2.4	17.3	71.3	0.5		
Celastrus paniculatus	Celastraceae	50.6	24.1	5.6	16.6	18.3	29.9		2.2% 12:0, 0.9% 10:0
Chrozophora plicata	Euphorbiaceae	26.0	4.1	21.4	13.4	59.3	1.8		
Chrozophora rottleri	Euphorbiaceae	24.0	5.6	18.2	18.9	56.4			0.9% 14:0, 0.2% 12:0
Cisternum divernum	Solanaceae	25.0	28.0	6.2	6.4	40.0			19.3% 14:0
Cynoglossum zeylanicum	Boraginaceae	28.0	11.8	2.8	33.5	20.0	10.5	2.7	0.4% 14:0
Echium italicum	Boraginaceae	24.0	8.4	4.4	36.8	18.6	21.1	5.7	4.9% 20:0
Euphorbia genicelate	Euphorbiaceae	20.6	31.5	8.7	13.9	24.3	21.6		
Evolvulus nummulari	Convolvulaceae	47.0	36.3	14.4	18.4	20.3	7.1		3.5% 14:0
Fimbristylis quinqueangularis	Cyperaceae	23.6	5.5	3.6	21.4	52.9	3.2		3.5% 14:0
Gelonium multiflorum	Euphorbiaceae	21.0	12.8	4.5	17.7	15.2	47.8		1.6% 12:0

Table 6.4 Oil and fatty acid contents of various plant species (*Cont'd*)

Species	Family	Oil content (%)	Fatty acid composition (wt.%)						
			16:0	18:0	18:1	18:2	18:3	18:4	others
Justicia adhatoda	Acanthaceae	26.2	5.5	0.6	80.1	5.2	3.3	5.4	
Lepidum sativum	Cruciferae	29.0	8.9	0.2	33.7	29.1	10.1		0.4% 14:0, 16.3% 20:0
Martynia diandra	Pedaliaceae	48.0	48.2	0.5	37.0	0.8	9.2		1.7% 14:0, 2.5% 12:0
Ochna squarrosa	Ochnaceae	23.4	73.5	1.5	14.1	10.9			
Passiflora foetida	Passifloraceae	21.2	13.3	4.1	17.2	65.4			
Pogestemon plactranthoides	Labiatae	32.0	22.8	trace	3.2	70.0			
Ruellia tuberosa	Acanthaceae	24.0	30.3	7.2	17.5	32.5			3.8% 14:0, 3.6% 12:0, 5.1% 10:0
Salvia plebeia	Labiatae	23.0	7.8	3.3	13.4	43.9	28.5		3.1% 14:0
Sesamum indicum	Pedaliaceae	28.6	10.5	2.3	42.3	41.8	2.3		0.8% 12:0
Solanum sisymbrifolium	Solanaceae	20.1	14.0	2.9	22.7	54.9	3.8		1.7% 14:0
Tabernaemontana coronaria	Apocyanaceae	42.6	21.3	0.5	50.5	23.1	4.6		
Tragia involucrata	Euphorbiaceae	27.0	10.6	2.4	15.8	61.7	7.3		2.4% 14:0
Xantrium strumarium	Compositae	36.0	20.9	2.6	54.3	22.2			

Source: Osman and Ahmad (1981)

Table 6.5 Fuel properties of vegetable oils and methyl esters in comparison with diesel fuel

Fuel property	Diesel fuel	Sunflower oil	Sunflower methyl ester	Rape oil	Rape methyl ester	Linseed oil	Linseed methyl ester
Specific gravity (kg/dm³)	0.835	0.924	0.88	0.916	0.88	0.932	0.896
Viscosity (cSt):							
at 20°C	5.1	65.8		77.8	7.5	50.5	8.4
at 50°C	2.6	34.9	4.22	25.7	3.8		
Heat of combustion:							
Gross (MJ/litre)	38.4	36.5	35.3	37.2		36.9	35.6
Net(MJ/litre)	35.4	34.1	33.0	34.3	33.1		
Cetane number	>45	33	45–51	44–51	52–56		
Carbon residue (%)	0.15	0.42	0.05	0.25	0.02		0.22
Sulphur (%)	0.29	0.01	0.01	0.002	0.002	0.35	0.24

Source: Ortiz-Cañavate (1994)

Table 6.6 Physical-chemical properties of some additional vegetable oils

	Groundnut	Copra	Cotton	Palm oil	Soya
Density (kg/litre) (20°C)	0.914	0.915 (30°C)	0.915		0.916
Viscosity (cSt):					
20.0°C	88.5		69.9		
50.0°C	29.0		24.8	28.6	
100.0°C		6.1	8.4	8.3	7.6
Melting point	(−3)–0	20–28	(−4)–0	23–27	(−29)–(−12)
Composition (weight %):					
C	77.35	73.4	77.7	76.4	78.3
H	11.8	11.9	11.7	11.7	11.3
O	10.9	14.7	10.6	11.5	10.3
Formula	$CH_{1.83}O_{0.11}$	$CH_{1.95}O_{0.15}$	$CH_{1.81}O_{0.10}$	$CH_{1.84}O_{0.11}$	$CH_{1.73}O_{0.10}$
Cetane number	39–41	40–2	35–40	38–40	36–9
Heat of combustion (MJ/kg)	36.7	37.4	36.8	36.5	36.8
Heat of combustion (MJ/litre)	33.5	34.2	33.7		33.7

Source: Guibet (1988)

as a 100 per cent substitute or in blends with mineral diesel oil. Although vegetable oil esters have good potential, are well suited for mixture with or replacement of diesel fuels, and are effective in eliminating injection problems in direct injection diesel engines, their use still creates problems.

Solid biofuels

There are four basic groups of plant species rich in lignin and cellulose that are suitable for conversion into solid biofuels (as bales, briquettes, pellets, chips, powder, etc.):

1 annual plant species such as cereals, pseudocereals, hemp, kenaf, maize, rapeseed, mustard, sunflower, and reed canarygrass (whole plants);
2 perennial species harvested annually, such as miscanthus and other reeds;
3 fast-growing tree varieties like poplar, aspen or willow with a perennial harvest rhythm (short rotation or cutting cycle), short rotation coppice (SRC); and
4 tree species with a long rotation cycle.

The raw materials can be used directly after mechanical treatment and compaction, or are converted to other types of biofuels (Figure 6.6). The lignocellulose plant species offer the greatest potential within the array of worldwide biomass feedstocks. The following basic processes are of major importance for the conversion of lignocellulose biomass into fuel suitable for electricity production: direct combustion of biomass to produce high-grade heat; advanced gasification to produce fuel gas of medium heating value; and flash pyrolysis to produce bio-oil, with the possibility of upgrading to give hydrocarbons similar to those in mineral crude oils.

The production of methanol or hydrogen from woody biomass feedstocks (such as lignocellulose biomass woodchips from fast-growing trees, *Miscanthus* spp. or *Arundo donax*) via processes that begin with thermochemical gasification can provide considerably more useful energy per hectare than the production of ethanol from starch or sugar crops and vegetable oils like RME.

Source: El Bassam

Figure 6.6 The 'bio-pipelines' of the future

However, methanol and hydrogen derived from biomass are likely to be much more costly than conventional hydrocarbon fuels, unless oil prices rise to a level that is far higher than expected prices in coming decades.

The most promising thermochemical conversion technology of lignocellulose raw materials currently available seems to be the production of pyrolytic oil or 'bio-oil' (sometimes called bio-crude oil). This liquid can be produced by flash or fast pyrolysis processes at up to 80 per cent weight yield. It has a heating value of about half that of conventional fossil fuels, but can be stored, transported and utilized in many circumstances where conventional liquid fuels are used, such as boilers, kilns and possibly turbines. It can be readily upgraded by hydrotreating or zeolite cracking into hydrocarbons for more demanding combustion applications such as gas turbines, and further refined into gasoline and diesel for use as transport fuels. In addition, there is considerable unexploited potential for the extraction and recovery of specialized chemicals.

The technologies for producing bio-oil are evolving rapidly with improving process performance, larger yields and better quality products. Catalytic upgrading and extraction also show considerable promise and potential for transport fuels and chemicals; this is at a much earlier stage of development with more fundamental research under way at several laboratories. The utilization of the products is of major importance for the industrial realization of these technologies, and this is being investigated by several laboratories and companies. The economic viability of these processes is very promising in the medium term, and their integration into conventional energy systems presents no major problems.

Charcoal is manufactured by traditional slow pyrolysis processes, and is also produced as a by-product from flash pyrolysis. It can be used industrially as a solid speciality fuel/reductant, for liquid slurry fuels, or for the manufacture of activated charcoal.

To provide a better understanding of the fossil fuels (gasoline and diesel) and the biofuels (ethanol and methanol), Table 6.7 makes a closer comparison of these four important fuels. Some physical and chemical properties of biofuel feedstocks are provided in the subsequent tables. Table 6.8 provides a comparison of the energy contents of the most common biofuels and fossil fuels. Table 6.9 primarily compares the elemental constituents of several fossil fuels and a variety of crops. The biofuels have an average energy content that is approximately equal to that of brown coal (20MJ/kg).

There is no expected environmental impact or any disadvantage created by mixing some percentage of alcohol with diesel or gasoline fuel. There are no legal constraints at present.

Wet biofuels

Biomass which can be harvested only in wet conditions and cannot be dried directly in the field can be upgraded as a fuelstock through the technique of ensiling (Figure 6.7): wet biomass → ensiling → pressing (mechanical water reduction) → fuel combustion (Scheffer, 1998). After the reduction of water content in the ensiled biomass by mechanical pressing, the biomass can be used as fuel for thermal power generation. Harvested crops are preserved in a clamp silo, which has a solid ground plate with drainage furrows and a collector for water.

Water reduction by mechanical pressing processes silage into fuel with a minimum dry matter content of 60 per cent. The addition of straw, crops and hay leads to a further increase of dry matter content which, with less pressing, results in a dry matter content of only 50 per cent.

Table 6.7 Some properties of ethanol, methanol, gasoline and diesel fuel

Property	Ethanol	Methanol	Gasoline	Diesel
Boiling temperature (°C)	78.30	64.5	99.2	140.0
Density (15°C)	0.79	0.77	0.70–0.78	0.83–0.88
Heating value (MJ/kg)	26.90	21.3	43.7	42.7
Vaporization heat (KJ/kg)	842.00	1100.0	300.0	
Stoichiometric ratio	9.00	6.5	15.1	14.5
Octane number (MON)	106.00	105.0	79.0–98	

Table 6.8 Heating values of selected fuels

Fuel	Heating value (MJ/kg)
Biogas	61.0
Brown coal	20.0
Cellulose	15.0
Diesel oil	42.0
Energy cereals	15.0
Ethanol	26.9
Gasoline	46.0
Hydrogen	144.0
Lignin	28.0
Methanol	19.5
Miscanthus	17.0
Pit coal	32.0
Rape oil	40.0
Straw	14.5
Wood	17.0

Table 6.9 Energy (MJ/kg) and chemical content (per cent) of different biofuel feedstocks

Feedstock	Heating value (HHV, MJ/kg)	Ash	Volatile constituents	Water	C	H	N	S	O	P	K	Mg
Bamboo	15.85	3.98	67.89	10.4								
Tree bark	19.5	3.2	76		52	5	0.4	0.05	39	2.3	6.8	2.8
Barley, full plant	17.6	3.7	77.8	14.1	46.1	6.63	1.24	0.11	42	7.6	15.4	2.5
Brown coal	27	7.6	55		68.4	5.5	1.8	1.3	15			
Coal	27–30	5	25–33	8	65–80	5	1.5–2	0.5–1.5	2			
Eucalyptus	19.35	0.52	82.55		48.33	5.89	0.15	0.01	45			
Heating oil	41	0.03	99		85	11–13	0.5	0.5–1.0				
Maize silage	17.1	5.5	75.7		47.3	7.54	1.85	0.43	39			
Maize straw	16.8	5.3	75–81		45.6	5.4	0.3	0.04	43	2.2	21.8	4.3
Meadow hay	16.9	5.8	80.5		45	6.2	I	0.08	44	1.7	8	1.8
Miscanthus spp.	18	3.7	71.2	12.5	45–50	5.7	0.4	0.3	42	3	23.7	3.3
Pseudo acacia	19.5		80									
Poplar	18.7	0.3	84		50.9	6.6	0.2	0.02	42	3.5	10.4	3.3
Rape straw	17	6.5	78.7	48.3	6.3	0.7	0.2	38				
Reed	16.3	3–5	70–80		45–50	5.5–6.5	1.0	0.1–0.3				
Sorghum, full plant	16.9	6.3	2840	6.2	44	5.6	2	0.05	38			
Triticale straw	17.4	4.9	80		47	6.1	0.4	0.1	41	2.9	16.3	1.5
Turf, dry	14.7	0.4–9.0	60–65									
Wheat, full plant	16.99	3.6	78.9	16.5	46.5	6.84	1.71	0.13	41	5.8	14.5	2
Wheat straw	17.1	5.3	79.6		46.7	6.3	0.4	0.01	41	3.1	17	1.5

HHV = high heat value

For the same dry matter content, fuels with a water content of 40 per cent have a 7 per cent lower heating value than dry straw of 15 per cent water content. This energy difference results from the evaporation of the higher water content during combustion, though most of the energy evaporation can be reclaimed by steam condensation.

Besides a reduction of water content, pressing produces an additional qualitative improvement in the fuel. Up to 50 per cent of the plant's nitrogen and 40–80 per cent of the other minerals are transported in the effluent by pressing. This reduces NO_x emission caused by nitrogen, the formation of dioxin by chlorine and damage from corrosion, whereas the reduction of the potassium

Source: G.M. Prine

Figure 6.7 Ensilaging of biomass as a fuel feedstock

Table 6.10 Content of dry matter and nutritive material (in dry matter) before and after pressing of winter rye silage, harvested during the milky stage (per cent)

	Input	Output	Relative content in fuels
Dry matter	28	55	
Nitrogen	1.25	1.25	47
Phosphorus	0.28	0.28	28
Potassium	1.68	0.53	32
Magnesium	0.10	0.03	10
Sodium	0.01	<0.01	
Chloride	0.57	0.57	12

Source: Scheffer (1998)

content increases the melting point of the ash. Table 6.10 shows as an example the result of a pressing experiment with rye silage. The proportions of the elements that remain in the output are far below the maximum limits set by the technical staff of energy plants (Scheffer, 1998).

Apart from nitrogen, all the elements important for crop nutrition are contained in the effluent and in the ashes produced by combustion. A partly closed nutrient cycle can be achieved when the effluent and the ashes are utilized in the field.

Biogas

Biogas consists of similar proportions of methane and carbon dioxide and is produced by the anaerobic fermentation of wet organic feedstocks in a process called 'biomethanization'. Biomethanization has a certain significance for the disposal of organic residues and waste products in the processing of agricultural products and in animal husbandry. Here, environmentally relevant aspects are of primary concern.

The cultivation of plants with the goal of producing biogas is hardly ever practised, though there are a number of green plants that are suitable in their fresh or in their ensiled form for biogas production. Through biomethanization, a broad spectrum of green plants could be used for energy because, in contrast to combustion, the raw material could be used in its natural moist state. This cannot be put into practice, however, because a number of fundamental conditions are not met. These will not be considered here.

The plant species most suitable for the production of biogas are those that are rich in easily degradable carbohydrates, such as sugar and protein matter. According to investigations by Zauner and Küntzel (1986), the methane yields from maize, reed canarygrass and perennial ryegrass after ensilaging and fermentation were identical.

Raw materials from lignocellulose-containing plant species are hardly suitable for the production of biogas. Biogas production from green plants has turned out to be very complicated, and the control of the fermentation process is very expensive and awkward. Besides this, for an acceptable yield of biogas from vegetable raw materials, production on a continuous, long-term basis and a homogeneous substrate would be indispensable. The biogas yields from several plant species after 20 days retention time are given in Table 6.11.

Table 6.11 Biogas yield from fresh green plant materials, 20 days retention time

Plant material	Biogas yield (m³/t VS*)
Alfalfa	440–630
Clover	430–450
Grass	520–640
Jerusalem artichoke	480–590
Maize plant	530–750
Sugar beet leaves	490–510
Sweet sorghum	640–670

* Volatile solids
Source: Weiland (1997)

Conversion of biofuels to heat, power and electricity

Production of heat and electricity dominate current bioenergy use. At present, the main growth markets for bioenergy are the European Union, North America, Central and Eastern Europe, and Southeast Asia (Thailand, Malaysia and Indonesia), especially with respect to efficient power generation from biomass wastes and residues, and for biofuels. Two key industrial sectors for application of state-of-the-art biomass combustion (and potentially gasification) technology for power generation are the paper and pulp sector and cane-based sugar industry. Power generation from biomass by advanced combustion technology and co-firing schemes is a growth market worldwide.

Mature, efficient and reliable technology is available to turn biomass into power. In various markets the average scale of biomass combustion schemes rapidly increases due to improved availability of biomass resources and economies of scale of conversion technology. Competitive performance compared to fossil fuels is possible where lower-cost residues are available particularly in co-firing schemes, where investment costs can be minimal. Specific (national) policies such as carbon taxes or renewable energy support can accelerate this development. Gasification technology (integrated with gas turbines/ combined cycles) offers even better perspectives for power generation from biomass in the medium term and can make power generation from energy crops competitive in many areas in the world once this technology has been proven on a commercial scale. Gasification, in particular larger-scale circulating fluidized bed (CFB) concepts, also offers excellent possibilities for co-firing schemes (EREC, 2007).

Heat

Biomass is the renewable heat source for small-, medium- and large-scale solutions. Pellets, chips and various by-products from agriculture and forestry deliver the feedstock for bioheat. Pellets in particular offer possibilities for high energy density and standard fuels to be used in automatic systems, offering convenience for the final users. The construction of new plants to produce pellets, the installation of millions of burners/boilers/stoves and appropriate logistical solutions to serve consumers should result in a significant growth of the pellets market. Stoves and boilers operated with chips, wood pellets and wood logs have been optimized in recent years with respect to efficiency and emissions. However, more can be achieved in this area. In particular, further improvements regarding fuel handling, automatic control and maintenance requirements are necessary. Rural areas present a significant market development potential for the application of those systems. There is a growing interest in district heating plants, which currently are run mainly by energy companies, and sometimes by farmers' cooperatives for small-scale systems. The systems applied so far generally use forestry and wood processing residues but the application of agro-residues will be an important issue in the coming years (EREC, 2007).

Direct combustion of biomass is the established technology for converting biomass to heat at commercial scales. Hot combustion gases are produced when the solid biomass is burned under controlled conditions. The gases are often used directly for product drying, but more commonly the hot gases are channelled through a heat exchanger to produce hot air, hot water or steam.

The type of combuster most often implemented uses a grate to support a bed of fuel and to mix a controlled amount of combustion air. Usually, the grates move in such a way that the biomass can be added at one end and burned in a fuel bed which moves progressively down the grate to an ash removal system at the other end. More sophisticated designs permit the overall combustion process to be divided into its three main stages – drying, ignition and combustion of volatile constituents, and burnout of char – with separate control of conditions for each activity being possible. A fixed grate may be used for low-ash fuels, in which case the fuel charge is input by a spreader-stoker that maintains an even bed and fuel distribution, and hence optimum combustion conditions.

Grates are well proven and reliable, and can tolerate a varied range in fuel quality (moisture content and particle size). They have also been shown to be controllable and efficient. The goal of reducing emissions is one of the driving forces behind current developments. This goal has also driven the development of fluidized bed technology, which is the main alternative to grate-based systems.

Table 6.12 Overview of current and projected performance data for the main conversion routes of biomass to power and heat, and summary of technology status and deployment

Conversion option	Typical capacity	Net efficiency (LHV basis)	Investment cost ranges (€/kW)	Status and deployment
Biogas production via anaerobic digestion	Up to several MWe	10–15% electrical (assuming on-site production of electricity)		Well-established technology. Widely applied for homogeneous wet organic waste streams and wastewater. To a lesser extent used for heterogeneous wet wastes such as organic domestic wastes.
Landfill gas production	Generally several hundred kWe	As above		Very attractive GHG mitigation option. Widely applied and, in general, part of waste treatment policies of many countries.
Combustion for heat	Residential: 5–50kWth Industrial: 1–5MWth	Low for classic fireplaces, up to 70–90% for modern furnaces	~100/kWth for logwood stoves, 300–800/kWth for automatic furnaces, 300–700/kWth for larger furnaces	Classic firewood use still widely deployed, but not growing. Replacement by modern heating systems (i.e., automated, flue-gas cleaning, pellet firing) in e.g. Austria, Sweden, Germany ongoing for years.
Combined heat and power	0.1–1MWe 1–20MWe	60–90% (overall) 80–100% (overall)	3500 (Stirling) 2700 (ORC) 2500–3000 (Steam turbine)	Stirling engines, steam screw type engines, steam engines, and organic Rankine cycle (ORC) processes are in demonstration for small-scale applications between 10kW and 1MWe. Steam turbine-based systems 1–10MWe are widely deployed throughout the world.
Combustion for power generation	20– >100MWe	20–40% (electrical)	2500–1600	Well-established technology, especially deployed in Scandinavia and North America; various advanced concepts using fluid bed technology giving high efficiency, low costs and high flexibility. Commercially deployed waste to energy (incineration) has higher capital costs and lower (average) efficiency.
Co-combustion of biomass with coal	Typically 5–100MWe at existing coal-fired stations. Higher for new multifuel power plants.	30–40% (electrical)	100–1000 + costs of existing power station (depending on biomass fuel + co-firing configuration)	Widely deployed in various countries, now mainly using direct combustion in combination with biomass fuels that are relatively clean. Biomass that is more contaminated and/or difficult to grind can be indirectly co-fired, e.g. using gasification processes. Interest in larger biomass co-firing shares and utilization of more advanced options is increasing.
Gasification for heat production	Typically hundreds kWth	80–90% (overall)	Several hundred/ kWth, depending on capacity	Commercially available and deployed; but total contribution to energy production to date limited.

Table 6.12 Overview of current and projected performance data for the main conversion routes of biomass to power and heat, and summary of technology status and deployment (*Cont'd*)

Conversion option	Typical capacity	Net efficiency (LHV basis)	Investment cost ranges (€/kW)	Status and deployment
Gasification/CHP using gas engines	0.1–1MWe	15–30% (electrical) 60–80% (overall)	1000–3000 (depends on configuration)	Various systems on the market. Deployment limited due to relatively high costs, critical operational demands, and fuel quality.
Gasification using combined cycles for electricity (BIG/CC)	30–200MWe	40–50% (or higher; electrical)	5000–3500 (demos) 2000–1000 (longer term, larger scale)	Demonstration phase at 5–10MWe range obtained. Rapid development in the 1990s has stalled in recent years. First-generation concepts prove capital intensive.
Pyrolysis for production of bio-oil	10 tonnes/hr in the shorter term up to 100 tonnes/hr in the longer term	60–70% bio-oil/feedstock and 85% for oil + char	Scale and biomass supply dependent; Approx 700/kWth input for a 10MWth input unit	Commercial technology available. Bio-oil is used for power production in gas turbines, gas engines, for chemicals and precursors, direct production of transport fuels, as well as for transporting energy over longer distances.

Due to the variability of technological designs and conditions assumed, all costs are indicative.
Source: DOE (1998); Dornburg and Faaij (2001); van Loo and Koppejan (2002); Knoef (2005)

A fluidized bed involves the fuel being burned in a bed of inert material. Air blown through the fluidized bed suspends the inert material and is used to combust or partially combust the biomass fuel. In the case of partial combustion, additional secondary air is added downstream of the bed to allow total combustion. At any given time, there is a large ratio of inert material to fuel in the fluidized bed itself and the bed is in continuous motion because of its suspended state. The process thus promotes rapid fuel–air mixing and rapid heat transfer between bed and fuel during ignition and combustion, minimizing the impact of variations in fuel quality. The bed and overbed temperatures can be controlled to maintain optimum combustion conditions, maximizing combustion efficiency and reducing emissions. Powerful fans are therefore needed to maintain fluidization of the bed, which increases the auxiliary power required compared to grate systems and may make it impossible to realize all of the potentially available efficiency improvements (Dumbleton, 1997).

Combined Heat and Power (CHP)

Significant improvement in efficiencies can be achieved by installing systems that generate both useful power and heat (cogeneration plants have a typical overall annual efficiency of 80–90 per cent). CHP is generally the most profitable choice for power production with biomass if heat, as hot water or as process steam, is needed. The increased efficiencies reduce both fuel input and overall greenhouse gas emissions compared to separate systems for power and heat, and also realize improved economics for power generation where expensive natural gas and other fuels are displaced. The technology for medium-scale CHP from 400kW to 4MW is now commercially available in the form of organic Rankine cycle (ORC) systems or steam turbine systems. The first commercially available units for small-scale CHP (1–10kW) are just arriving on the market, a breakthrough for the gasification of biomass in the size between 100 and 500kW might occur in a few years (EREC, 2007).

Electricity

The use of biomass for power generation has increased over recent years mainly due to the implementation of a favourable European and national political framework. In the EU-25 electricity generation from biomass (solid biomass, biogas and biodegradable fraction of municipal solid waste) grew by 19 per cent in 2004 and 23 per cent in 2005. However, most biomass power plants operating today are characterized by low boiler and thermal plant efficiencies and such plants are still costly to build. The main challenge therefore is to develop more efficient lower-cost systems. Advanced biomass-based systems for power generation require fuel upgrading, combustion and cycle improvement, and better flue-gas treatment. Future technologies have to provide superior environmental protection at lower cost by combining sophisticated biomass preparation, combustion and conversion processes with post-combustion clean-up. Such systems include fluidized bed combustion, biomass-integrated gasification, and biomass externally fired gas turbines.

Steam technology

At present, the generation of electricity from biomass uses one of the combustion systems discussed above. This process consists of creating steam and then using the steam to power an engine or turbine for generating electricity. Even though the production of steam by combustion of biomass is efficient, the conversion of steam to electricity is much less efficient. Where the production of electricity is to be maximized, the steam engine or turbine will exhaust into a vacuum condenser and conversion efficiencies are likely to be in the 5–10 per cent range for plants of less than 1MWe, 10–20 per cent for plants of 1 to 5MWe and 15–30 per cent for plants of 5 to 25MWe. Low-temperature heat (<50°C) is usually available from the condenser, though this is insufficient for most applications so it is normally wasted by dispersal into the atmosphere or a local waterway. The average conversion efficiency of steam plants in the USA is approximately 18 per cent. The USA has installed 7000MWe of wood-fired plants since 1979. Figure 6.8 shows the 50MWe McNeil power plant in Burlington, Vermont.

Where there is a need for heat as well as electricity – for example, for the processing of biomass products such as the kilning of wood, or the processing of sugar, palm oil, etc. – the plant can be arranged to provide high-temperature steam. This is achieved by taking some steam directly from the boiler, by extracting partially expanded steam from a turbine designed for this purpose, or by arranging for the steam engine or turbine to produce exhaust steam at the required temperature.

All three options significantly reduce the amount of electricity available from the plant, though the overall energy efficiency may be much higher – 50–80 per cent being common (Dumbleton, 1997). See Table 6.13 at

Source: Warren Gretz, NREL

Figure 6.8 The wood-fired McNeil electric generating facility in Burlington, Vermont, USA

the end of this chapter for a summary of power ranges and efficiencies of different technologies.

Bioelectricity, particularly in the form of combined heat and power (CHP) is in the mainstream of current technological trends. Aside from the large number of small individual heating systems, more than 1000MW of bioelectricity are currently produced annually by European utilities through conventional or advanced combustion (Palz, 1995). Biomass, mainly in the form of industrial and agricultural residues and municipal solid waste, is presently used to generate electricity with conventional steam turbine power systems. The USA has an installed biomass (not just wood) electricity generating capacity of more than 8000MW (Williams, 1995). Although the power plants are small – typically 20MW or less – and relatively capital intensive and energy inefficient, they can provide cost-competitive power where low-cost biomass is available, especially in combined heat and power applications.

However, the use of this technology will not expand considerably in the future because low-cost biomass supplies are limited. Less capital-intensive and more energy-efficient technologies are needed to make the more abundant, but more costly, biomass sources usable. This involves the production of biomass grown on plantations dedicated to energy crops, which will be competitive particularly for power-only applications (Williams, 1995).

Higher efficiency and lower unit capital costs can be realized with cycles involving gas turbines. Present developmental efforts are focused on biomass integrated gasifier/gas turbine (BIG/GT) cycles (Williams and Larson, 1993).

In all of the cases so far described, steam technology can be considered as sound and well proven, though expensive. Electricity and CHP systems using biomass and steam technology can be competitive with electricity produced from fossil fuels in places where biomass residues are available at low or no cost, but electricity prices will not be competitive if the biomass fuel has to be purchased at market prices. In this case, biomass-to-electricity systems need other reasons for their utilization, and electricity price structures that recognize these conditions. Steam technology will continue to be an acceptable alternative for biomass-to-electricity plants where these price structures exist. However, the environment and other benefits of using biomass are not maximized and continuous price support is needed, which are unacceptable factors in many instances.

The cost-effectiveness of electricity generation from biomass can be improved if conversion efficiencies increase and capital costs decrease. Raising conversion efficiencies also helps to maximize environmental benefits and associated environmental tax credits by decreasing the general dependency on fossil fuels. Because steam technology is in essence fully developed, there is unfortunately limited scope for finding improvements. New conversion technologies are therefore important: gasification and pyrolysis are two relatively new technologies that are close to commercialization (Dumbleton, 1997).

Gasification

Gasification is a thermochemical process in which carbonaceous feedstocks are partially oxidized by heating at temperatures as high as 1200°C to produce a stable fuel gas (Dumbleton, 1997). By an exothermic chemical reaction with oxygen, a proportion of the carbon in the biomass fuel is converted to gas. The normal producer gas chemical reactions occur when the remainder of the biomass fuel is subjected to high temperatures in an oxygen-depleted atmosphere. The resulting fuel gas is mainly carbon monoxide, hydrogen and methane, with small amounts of higher hydrocarbons such as ethane and ethylene. These combustible gases are unfortunately reduced in quantity by carbon dioxide and nitrogen if air is used as the source of oxygen. Because carbon dioxide and nitrogen have no heating value, the heating value of

the final fuel gas mixture is low: 4–6MJ/Nm^3. This is equivalent to only 10–40 per cent of the value for natural gas, usually 32MJ/Nm^3, for which the current engines and turbines were created. The low heating value also makes pipeline transportation of the gas inappropriate. Unwanted by-products such as char particles (unconverted carbon), tars, oils and ash are also present in small amounts. These could damage the engines and turbines and therefore must be removed or processed into additional fuel gas.

If oxygen is incorporated instead of air, the heating value of the gas is improved to 10–15MJ/Nm^3. This improvement would permit unmodified engines and turbines to be used to generate electricity. However, the cost of oxygen production and the potential dangers associated with its utilization mean that the use of oxygen-blown gasifiers is not a favoured option (Dumbleton, 1997).

Biomass may also be gasified under pressure, though it is not known if this will prove to be more cost-effective. Pressurized gasifiers are more expensive but will be smaller than the present normal gasifiers. The pressurized gasifier system would be more efficient overall because tars will be more completely converted, sensible heat will be retained, and the need for fuel gas compression will be eliminated. However, pressure seals and the need to purge the feed system with inert gas will make the fuel feed system for pressurized gasifiers much more complicated. Such factors increase the capital and operating costs of pressurized gasifier plant, though a demonstration plant at Vernamo in Sweden uses this technology so it will be possible to evaluate these factors in the near future.

Direct combustion of the hot fuel gas from the gasifier in a boiler or furnace may only be possible for heating applications. The condensation of the tars that are cracked and burned in the combustor is prevented by maintaining high temperatures. If the biomass fuel or gasification process leads to the production of dust (ash), it may well be possible to remove this simply, using a hot gas cyclone. Some combustion systems will be able to withstand a limited dust loading with no gas clean-up necessary. Biomass gasification for combustion is often found in the pulp and paper industry, where waste products are utilized as fuel. There are also a number of Bioneer gasifiers on district heating schemes in Finland, though it is not clear whether this approach is more cost-effective than combustion.

The quality of the gas must be improved before it can be used in combustion engines or turbines, and the gas may have to be cooled to intermediate, if not low, temperatures because of temperature limitations in the engine's or turbine's fuel control systems. Reducing the gas temperature will increase the volumetric heating value of the gas, but it will also increase the condensation of tars making the gas even less suitable for engine and turbines. A gas-cleaning system will be essential in this case, and may consist of cyclones, filters or wet scrubbers. Wet scrubbers are particularly effective: in a single operation, they reduce the gas temperature and also capture tars (which are water soluble) and inert dust in the form of ash and mineral contaminants. However, a contaminated liquid waste stream is produced. This casts a shadow on the idea of biomass fuels and the concept of sustainability being clean alternatives.

Fuels with ashes that have low ash softening and melting temperatures are inappropriate because of the high temperatures used during the gasification process. Numerous annual crops and their residues fall into this category (Dumbleton, 1997).

Advanced gasification for the production of electricity through integrated power generation cycles has a significant short-term potential, and is being developed in many parts of the world including Europe, and North and South America. Both atmospheric and pressurized air gasification technologies are being promoted that are close-coupled to a gas turbine. There are still problems to be resolved in interfacing the gasifier and turbine, and meeting the gas quality requirements of the turbine fuel gas, but both areas are receiving substantial support. The production of hydrogen, methanol, ammonia and transport fuels through advanced gasification and synthesis are longer-term objectives that rely on larger-capacity plants that can take advantage of economies of scale.

Pyrolysis

Pyrolysis is the thermal degradation of carbonaceous material in the absence of air or oxygen. Temperatures in the 350–800°C range are most often used. Gas, liquid and solid products (char/coke) are always produced in pyrolysis reactions, but the amounts of each can be influenced by controlling the reaction temperatures and residence time. The output of the desired product can be maximized by careful control of the reaction conditions. Because of this, heat is commonly added to the reaction indirectly.

Charcoals for barbecues and industrial processes are the most common products of present pyrolysis. An overall efficiency of 35 per cent by weight can be achieved by maximizing the output of the solid product through the implementation of long reaction times (hours to days) and low temperatures (350°C). Heat for the process in traditional kilns can be produced by burning the gas and liquid by-products.

Higher temperatures and shorter residence times are used to produce pyrolysis oils. Optimum conditions are approximately 500°C with reaction times of 0.5 to 2 seconds. The necessary short reaction time is achieved by rapid quenching of the fuel, which prevents further chemical reactions from taking place. Preventing additional chemical reactions also allows higher molecular weight molecules to survive. The fast heating and cooling requirements of the raw material creates process control problems, which are most often dealt with by fine milling the feedstock to less than a few millimetres in size.

Conversion efficiency to liquid can be up to 85 per cent by weight. Some gas is produced, and is commonly utilized in the fuel-heating process. Solid char is also produced. The char remains in the pyrolysis oil, and must be removed before the oil can be used in combustion engines or turbines; it too can be used in the fuel-heating process. The liquid product is a highly oxygenated hydrocarbon with a high water content, which is chemically and physically unstable. This instability may create difficulties in all aspects having to do with the storage and use of this product. Detrimental chemical reactions within the liquid could also limit the maximum practical fuel storage time.

Figure 6.10 shows an example of the technology of flash or fast pyrolysis being utilized in a double screw reactor. This form of low-temperature pyrolysis (450–550°C) is characterized by a reaction time of under one second. The main products produced are gas, bio-oil and coke. Figure 6.11 demonstrates the mass balance for this technology with respect to the initial weight of the biomass, using miscanthus as an example. The energy balance can also be determined with respect to the heating value of the fraction: Figure 6.11 shows the energy balance obtained if the drying step is regarded as a separate process (Shakir, 1996).

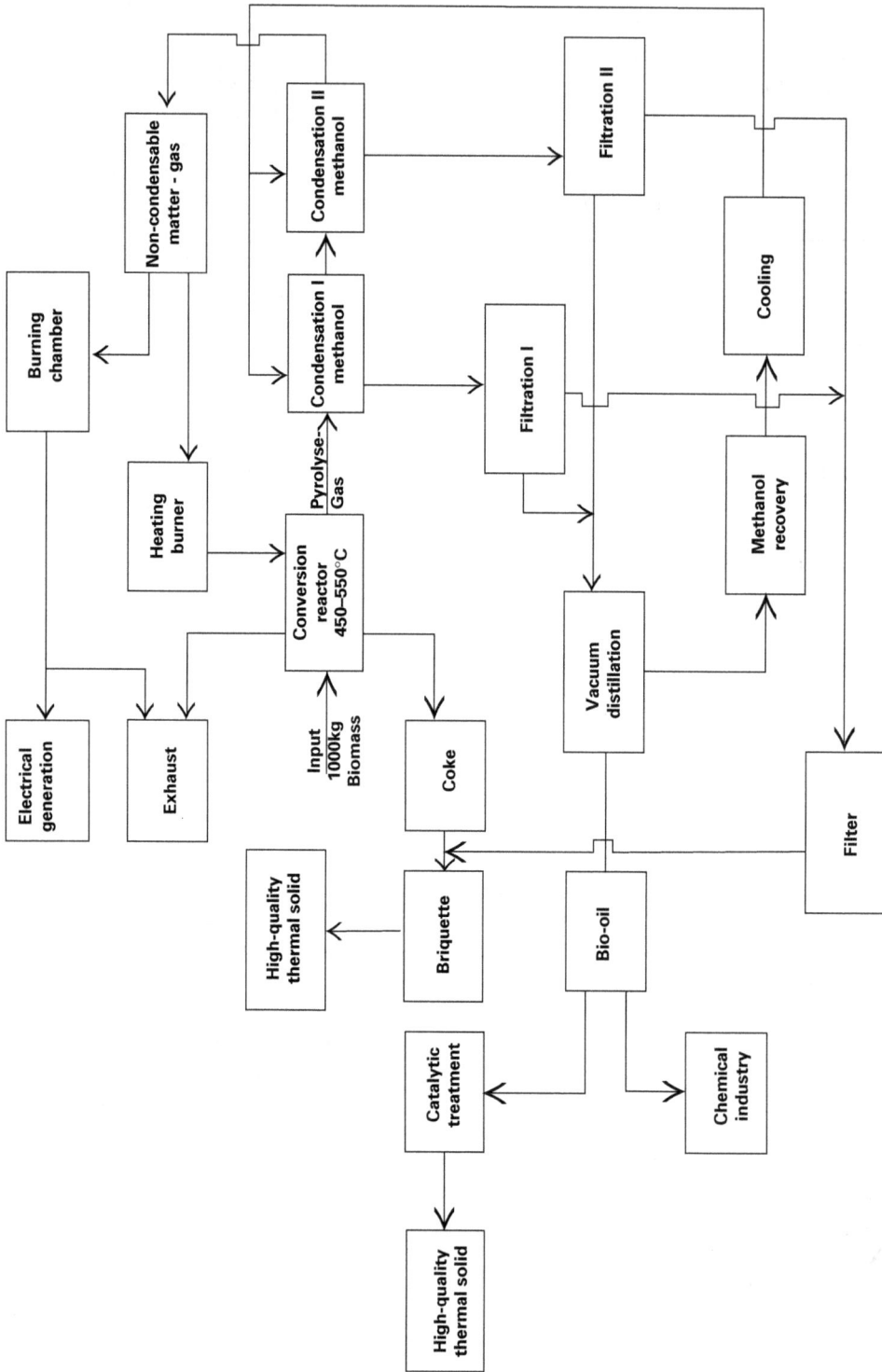

Figure 6.9 Flow diagram of flash or fast pyrolysis biomass conversion

Source: Shakir (1996)

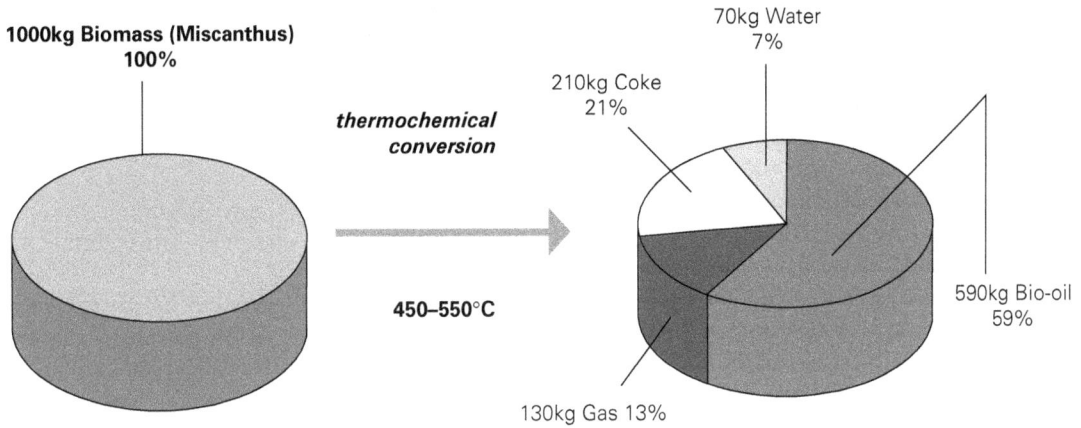

Source: Shakir (1996)

Figure 6.10 Mass balance of miscanthus biomass conversion using flash or fast pyrolysis

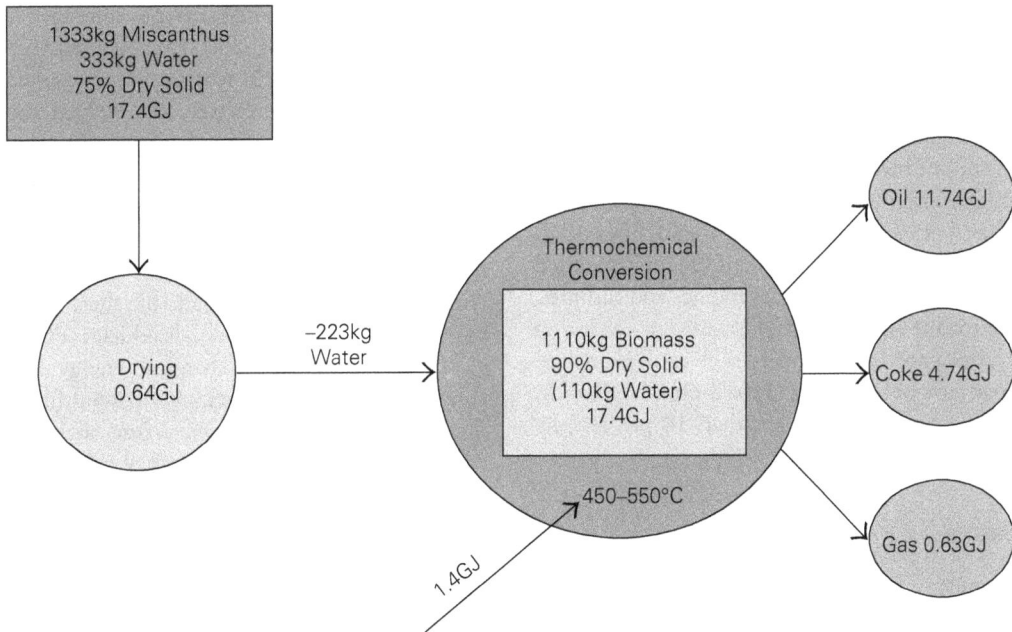

Source: Shakir (1996)

Figure 6.11 Energy balance of miscanthus biomass conversion using flash or fast pyrolysis
with pre-conversion drying

The production of pyrolysis oil from biomass fuels has a number of potential advantages over gasification and combustion. The pyrolysis process can be separated from the final energy conversion process by both space and time. Pyrolysis oil could also be transported from a central production site, close to the biomass source, to one or more energy conversion plants situated in the most appropriate location(s). It is also theoretically possible to upgrade the crude pyrolysis oil into more familiar petroleum products, though the technology for this is yet to be used commercially. Because the moderate process temperatures are not likely to create ash melting or softening, and alkali metals are likely to be retained in the char, pyrolysis is the favoured technology for annual crops and their residues.

Successful trials have shown that pyrolysis oil can be used as a substitute for heavy fuel oil in boilers. Some modification of burners is required, and care must be taken to limit pyrolysis oil temperature in storage tanks and heated fuel feed lines because of the oil's instability. There have also been limited trials of pyrolysis oil as a substitute for diesel fuel in diesel engines, and the results have been sufficiently encouraging for full-scale trials to be planned. The diesel engine needs to be adapted to include a pilot injector ignition system for pyrolysis oil because it will not self-ignite.

Pyrolysis reactions produce a much greater proportion of gas, with correspondingly less liquid and solid products, at temperatures as high as 700–800°C. Gas yields can be greater than 80 per cent by weight. The gas has a moderate heating value, 15–20MJ/Nm^3, which is enough to allow various types of combustion engines and turbines to use the fuel gas without modification. Liquids, in the form of tars, usually make up less than 5 per cent of the fuel by weight, but as with gasification, the tars must be removed or converted before the gas fuel can be used in engines and turbines. The solid residue consists of highly active dry char that tends to attract any contaminants, such as alkali metals, which may have been present in the biomass feedstock. The solid fraction is usually less than 10 per cent by weight of the fuel (Dumbleton, 1997).

Methanol

Biomethanol is a fuelwood-based methyl alcohol obtained by the destructive distillation of wood or by gas production through gasification. Biomethanol is a substitute product for synthetic methanol and is extensively used in the chemical industry (methanol is generally manufactured from natural gas and, to a lesser extent, from coal).

Bio-oil

Bio-oil is directly produced from biomass by fast pyrolysis. The oil has some particular characteristics but has been successfully combusted for both heat and electricity generation. The largest plant built today produces 2 tonnes per hour, but plants for 4 and 6t/h (equivalent to 6–10MWe) are at an advanced stage of planning. In the long term, pyrolysis oil could be used as a transportation fuel.

Fuel cells

The different fuel cell types can be divided into two groups: low-temperature, or first-generation (alkaline fuel cell, solid polymer fuel cell and phosphoric acid fuel cell); and high-temperature, or second-generation (molten carbonate fuel cell and solid oxide fuel cell). Low-temperature cells have been commercially demonstrated, but are restricted in their fuel supply and are not readily integrated into combined heat and power applications. The high-temperature cells can use a wide variety of fuels through internal reforming techniques, have a high efficiency, and can be integrated into combined heat and power systems. Although fuel cell capital costs are falling, increasing environmental restrictions are required before fuel cells will begin to replace conventional technologies in either the power generation or the transport sectors (Williams and Campbell, 1994).

Methanol (MeOH) and hydrogen (H$_2$) derived from biomass have the potential to make major contributions to transport fuel requirements by competitively addressing all of these challenges, especially when used in fuel cell vehicles (FCVs).

In a fuel cell, the chemical energy of fuel is converted directly into electricity without first burning the fuel to generate heat to run a heat engine. The fuel cell offers a quantum leap in energy efficiency and the virtual elimination of air pollution without the use of emission control devices. Dramatic technological advances, particularly for the proton exchange membrane (PEM) fuel cell, focused attention on this technology for motor vehicles. The FCV has the potential to compete with the petroleum-fuelled internal combustion engine vehicle (ICEV) on both cost and performance grounds (AGTD, 1994; Williams, 1993, 1994), while effectively addressing air quality, energy security and global warming concerns.

One of the newest markets being looked at for bioethanol uses is fuel cells. Electrochemical fuel cells convert the chemical energy of bioethanol directly into electrical energy to provide a clean and highly efficient energy source. Bioethanol is one of the most ideal fuels for a fuel cell. Besides the fact that it comes from renewable resources, highly purified bioethanol can solve the major problem of membrane contamination and catalyst deactivation within the fuel cell, which limits its life expectancy. Extensive research activities ensure that bioethanol remains among the most desirable fuels for fuel cells, delivering all the benefits that the bioethanol fuel cell technologies promise.

Ballard Power Systems, Inc., of Vancouver, Canada, introduced a prototype PEM fuel cell bus in 1993 and planned to introduce PEM fuel cell buses on a commercial basis beginning in 1998. In April 1994, Germany's Daimler Benz introduced a prototype PEM fuel cell light-duty vehicle (a van) and announced plans to develop the technology for commercial automotive applications. In the USA, the FCV was a leading candidate technology for accelerated development under the 'Partnership for a New Generation of Vehicles', a joint venture launched in September 1993 between the Clinton Administration and the US automobile industry. The partnership's goal was to develop in a decade production-ready prototypes of advanced, low-polluting, safe cars that could be run on secure energy sources, especially renewable sources, that would have up to three times the fuel economy existing gasoline ICEVs of comparable performance, and that cost no more to own and operate (Williams et al, 1995).

The Stirling engine

The Stirling engine's name comes from its implementation of the Stirling cycle. Nitrogen or helium gas (Weber, 1987) is shuttled back and forth between the hot and cold ends of the machine by the displacer piston. The power piston, with attached permanent magnets, oscillates within the linear alternator to generate electricity. New developments (Sunpower Inc., 1997) indicate that the Stirling engine's generator efficiently converts rough biomass into electricity with intrinsic load matching capacity. A single engine produces 2.5kW of 60Hz, 120V alternating current power, and up to four engines may be used with a single burner. It is proposed for use at domestic and light industrial sites, and may be used for cogeneration, delivering both electricity and heat.

Positive features of the engine include the direct conversion of biomass heat into electricity, and an integrated alternator with greater than 90 per cent efficiency. Further development of these types of engines is necessary and will open significant future opportunities to produce electricity directly from biomass. Table 6.14 summarizes the power ranges and efficiencies of the biomass conversion technologies mentioned in this chapter.

Biofuel properties

Certain properties of biomass detrimentally affect its efficient conversion as a biofuel. The utilization of biomass for high-efficiency thermal electricity generation has been shown to be limited not by the properties of the organic component of biomass, but by the characteristics of the associated mineral matter at high temperatures (Overend, 1995). On a moisture- and ash-free basis, biomass has a relatively low heating value of 18.6GJ/t. This, however, would not limit its use in high-efficiency combustion systems since sufficiently high temperatures could be reached to achieve high Carnot cycle efficiencies. The high temperatures cannot be reached because of the fouling and slagging

Table 6.13 Power range and reported efficiencies of different technologies

Technology	Power range (MWe)	Typical overall efficiency (%)
Steam engine	0.025–2.0	16
Steam turbine (back pressure)	1–150	25
Steam turbine (extraction-condensation cycle)	5–800	35
Steam turbine (condensation cycle)	1–800	40
Gas engine	0.025–1.5	27
Gas turbine	1.0–200	35
Integrated gas and steam combined cycle	5–450	55
Stirling engine	0.0003–?	40
Fuel cell	0.005–?	70

Source: Grimm (1996)

properties of the minerals in the biomass. Mainly responsible is the potassium component, alone or in combination with silica. Under combustion conditions the potassium is mobilized at relatively low temperatures and can then foul heating transfer surfaces and corrode high-performance metals used in the high-temperature sections of burners and gas turbines (Overend, 1995).

Techniques that separate the mineral matter from the fuel components (carbon and hydrogen) at low temperatures can result in very high-efficiency combustion applications in combustors, gas turbines and diesel engines. Gasification and various types of clean-up systems, as well as pyrolysis techniques, are able to separate the minerals from the fuel component and provide the organic part of the biomass as a suitable fuel for high-efficiency energy conversion in gas turbines, fuel cells and other engines (Overend, 1995). The conversion pathways are shown in Figure 6.12.

Internet resources

www.erec.org
www.ipcc.ch

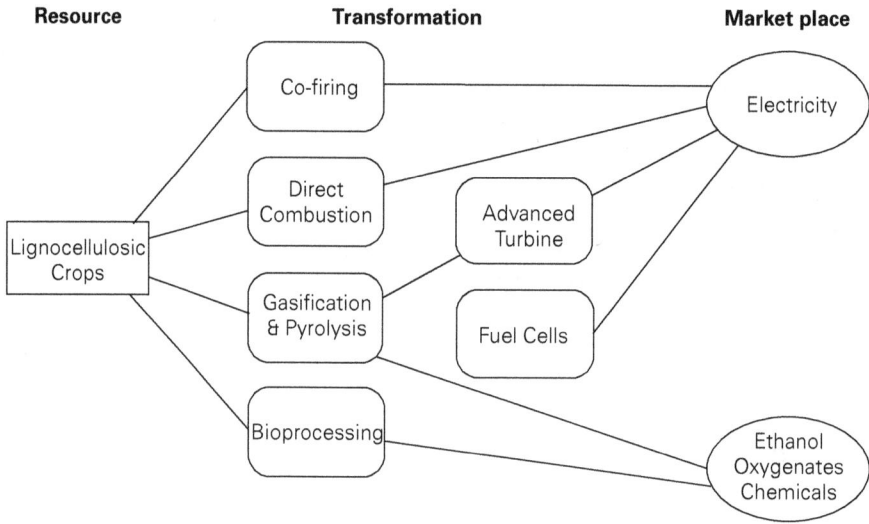

Source: Overend (1995)

Figure 6.12 Bioenergy pathways

7
Environmental Impacts

There is an increasing awareness that excessive use of fossil fuels may be causing serious damage to the environment and climate, contributing to global warming and possible negative effects on vegetation through acid rain. There is also major concern about the damage that could be caused by transportation accidents and the risks from nuclear power plants.

The potential contamination by mineral oils of the soil, water and air is considerable: one litre of mineral diesel contaminates one million litres of water. In cases of leakage and transport accidents, oils of fossil origin create extensive damage to flora and fauna in both terrestrial and aquatic environments.

Nuclear fuels are limited, but their residues are nearly unlimited in their radioactivity. In addition, the transportation, handling, installation, deposition and storage of radioactive materials need long-term monitoring and high inputs. Accidents in nuclear power plants like Chernobyl can be the cause of extensive disasters.

Biomass is a non-intermittent renewable energy source that can provide energy to be used for heating and cooling, electricity and transport. Biomass fuels can easily be stored, meeting both peak and baseline energy demands. Biomass can take different forms (solid, liquid or gaseous), and can directly replace coal, oil or natural gas, either fully or in blends of various percentages. Bioenergy is CO_2 neutral, as all carbon emitted by combustion has been taken up from the atmosphere by plants beforehand.

Wide utilization of plant raw materials for energy offers the chance to reorganize agricultural production towards an environmentally consistent system through increasing the number of plant species, reintroducing traditional crops and introducing new alternative crops. This will lead to the production of different energy feedstocks with greater outputs and lower environmental inputs. It will also lead to diversification, improving the appearance of the landscape, reducing the inputs, such as fertilizers, herbicides, fungicides and fuels, in crop management, and improving the microclimate through water use and recycling mechanisms. The introduction of energy crops allows a significant quantity of renewable fuels to be consumed and energy to be produced without markedly increasing the CO_2 content of the atmosphere. Crops can absorb CO_2 released from the combustion of biofuels and produce oxygen from water during photosynthesis, thus reducing the depletion of oxygen in the atmosphere.

Biofuels make a beneficial contribution to the problems of the 'greenhouse effect' and global climatic change in both industrialized and developing countries. There is no doubt that energy crops can be grown in ways that are environmentally desirable. It is possible to improve the land environmentally relative to present use through the production of biomass for energy. The environmental outcome closely depends on how the biomass is produced (Larson and Williams, 1995). Table 7.1 takes a closer look at the environmental benefits and constraints that are present when solid energy crops are produced and harvested.

Thermochemical conversion of vegetative feedstocks leads to the recovery of all mineral nutrients, except nitrogen, in the form of ash at the biomass conversion facility. The ash can then be returned to the fields as fertilizers. Fixed nitrogen can be restored in environmentally consistent procedures – that is, by improving the crop rotation systems through selection of plant species that fix nitrogen (legumes) to eliminate or to reduce the levels of the necessary mineral nitrogen fertilizers. Perennial lignocellulose energy crops need less nitrogen than annual food crops. Energy crops also

Table 7.1 Environmental benefits and constraints of production and harvest of some energy crops as solid feedstocks

| | Wood | | Straw | Hay | Miscanthus |
	Short rotation	Forest residues			
Environmental benefits	Little soil erosion risk; little soil compaction risk; area of relaxation for local people	Possible reduction of risk of disease and/or pest infestation	No need for additional production means	Very little soil erosion risk; very little soil compaction risk; little risk of nitrogen removal; very little risk of pesticide removal	Little soil erosion risk; little risk of nitrogen removal; very little risk of pesticide removal
Environmental constraints	Higher nutrient and pesticide demand compared to conventional forest; monoculture plantation	Nutrient extraction; reduction of humus supply	Reduction of humus supply		Medium soil compaction risk; monoculture plantation; water demand for irrigation possible

offer flexibility in dealing with erosion and chemical pollution from herbicide use. The erosion and herbicide pollution problems would be similar to those for annual food or feed crops. But the growing of cereals (whole crop) as biofuel feedstocks needs much less nitrogen – about one-third of that needed to produce grain cereals, in which most of the nitrogen is used to improve the nutritional value. In the case of whole cereal crops for energy, the food quality aspects are not valid.

Some perennial grasses (miscanthus and *Arundo donax*) can be harvested annually for decades after planting, and fast-growing trees (woods) that are harvested only every 2–5 years are replanted perhaps every 10–20 years. Erosion would be sharply reduced in both cases as compared to annual crops such as corn and soya bean. There would also be a reduction in the need for

fertilizers and herbicides, as shown in Table 7.2 (Larson and Williams, 1995).

Careful selection of crop species and good plantation design and management can be helpful in controlling pests and disease, and could thereby minimize or even eliminate the use of chemical pesticides.

Energy crops could replace monoculture food crops, and in many cases the shift would be to an ecologically more diversified landscape. Improving biodiversity on a regional basis might enable many species to migrate from one habitat to another and could help to address the aesthetic concerns sometimes expressed about intensive monocultures.

The use of energy crops on available land to substitute for fossil fuels would be an effective means of decreasing atmospheric carbon dioxide pollution,

Table 7.2 Fertilizer and herbicide application and soil erosion rates in the USA

Cropping system	N-P-K application rate (kg/ha/year)	Herbicide application rate (kg/ha/year)	Soil erosion rate (tonne/ha/year)
Annual crops			
Corn	135-60-80	3.06	21.8[a]
Soya bean	20[b]-45-70	1.83	40.9[a]
Perennial energy crops			
Herbaceous	50[c]-60-60	0.25	0.2
Short-rotation woods	60[c]-15-15	0.39	2.0

[a] Based on data collected in the early 1980s. New tillage practices used today may lower these values.

[b] The nitrogen input is inherently low for soya bean, a nitrogen-fixing crop.

[c] Not including nitrogen-fixing species.

Source: Hohenstein and Wright (1994)

because these crops can absorb CO_2 released from the combustion of biofuels.

The biodegradability of various biodiesel fuels in an aquatic environment, including rape methyl ester (RME), soyate methyl ester and blends of biodiesel and mineral diesel at different ratios, was determined by Zhang et al (1995). All of the biodiesel fuels are 'readily biodegraded' compounds according to Environmental Protection Agency (EPA) standards, and have a relatively high biodegradation rate in an aquatic environment. Biodiesel can promote and speed up the biodegradation of mineral diesel. The more biodiesel present in a biodiesel/mineral diesel mixture, the faster the degradation rate. The biodegradation pattern in a biodiesel/diesel mixture is that micro-organisms metabolize both biodiesel and diesel at the same time and at almost the same rate.

Measurements were carried out on buses and private vehicles to compare CO, CO_2, SO_x and NO_x emissions with respect to mineral diesel, biodiesel and a 30 per cent biodiesel blend. The results are given in Figure 7.1. CO_2 and SO_x emissions were eight times higher when mineral diesel was used. The emission of NO_x was 10 per cent higher from biodiesel (RME), but the NO_x emission was similar when a blend of 30 per cent biodiesel was used (ADEME, 1996).

The energy output/input ratio, a characteristic value for the energetic efficiency of production processes, for the main products was 2.14 for RME production. The ratio for RME was more than twice as high as the ratio for bioalcohol production from sugar beet, which was 1.04. Taking into account the energy of the by-products, the output/input ratio as a measure for the energy balance of all the involved

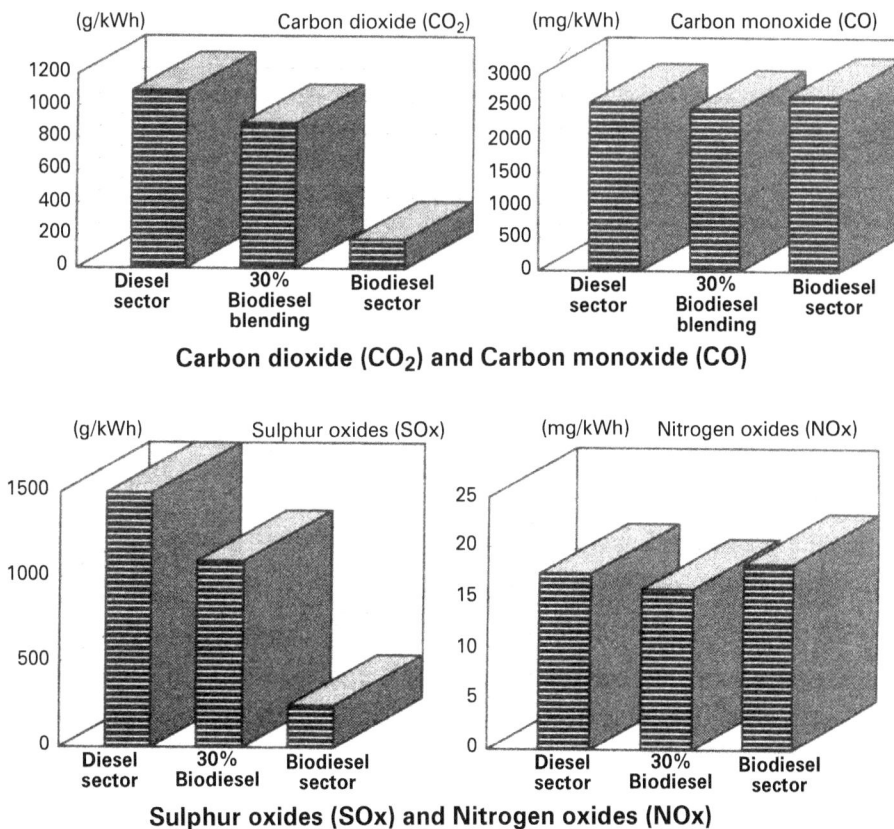

Carbon dioxide (CO_2) and Carbon monoxide (CO)

Sulphur oxides (SOx) and Nitrogen oxides (NOx)

Source: ADEME (1996)

Figure 7.1 Comparison of oxide emissions for diesel, biodiesel and blended fuels

production processes was about 1.63 for bioalcohol and 3.18 for RME production (Graef et al, 1994).

Energy plant species

If bioalcohol is used instead of gasoline in Otto cycle engines there is an improvement in knock resistance, and a reduction of carbon monoxide (CO) and hydrocarbons (HCs) in the exhaust gas. Running diesel engines on bioalcohol leads to reduction of smoke, nitrogen oxides (NO_x) and HC emissions. The use of rapeseed oil and RME in diesel engines instead of diesel fuel reduces the emission of carcinogens and polycyclic aromatic hydrocarbons (PAHs) in the exhaust gas (Krahl, 1993).

Biomass integrated gasifier/gas turbine (BIG/GT) power plants would be characterized by low levels of sulphur and particulate emissions. The low sulphur content of biomass makes it possible to avoid the costly capital equipment and operating cost penalties associated with sulphur removal. However, NO_x emissions could arise from nitrogen in the biofuels. These could be kept at low levels by growing biomass feedstocks with low nitrogen content and/or by selectively harvesting the portions of the biomass having high C/N ratios. Also, feedstocks from old cereal varieties with a low harvest index contain considerably less nitrogen than new high-yielding varieties.

The combustion of biomass (miscanthus) reduced the carbon dioxide (CO_2) emission by up to 90 per cent in a comparison with the combustion of fossil feedstocks (hard coal) (Lewandowski and Kicherer, 1995). Investigations have been carried out into the potential reductions of different gases emitted into the atmosphere where biofuels are used instead of fossil fuel in a thermal energy plant. Total emissions were determined using a base of 12.7MWh net heat capacity; the biofuel was straw and the fossil fuel brown coal. The results are summarized in Figure 7.2, and show that the bioenergy heat power plant in Schkölen, Germany, emits 28 per cent of the CO_2-equivalent gases emitted by the comparable brown coal plant.

Environmental impact

A significant amount of research is presently under way to develop and refine the use of fuel cell vehicles (FCVs). Williams et al (1995) explain that the fuel cell converts its fuel (hydrogen and methanol) directly into electricity, without burning. Through the use of indirectly heated gasifiers, biomass can be used to produce methanol and hydrogen for FCV fuels. When the biomass feedstock is grown sustainably, the use of these fuels in FCVs would result in extremely low levels of CO_2 emissions and very little or no air pollution,

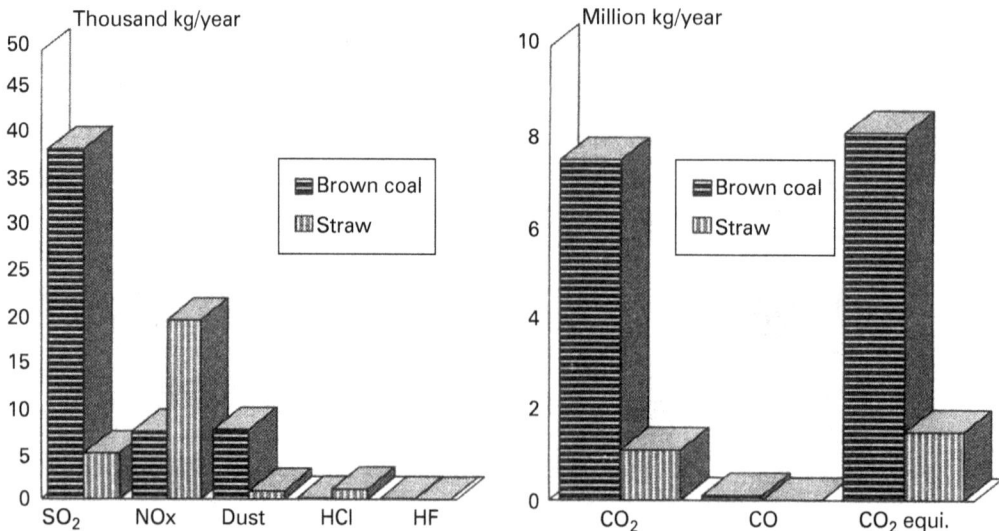

Source: Garbe (1995)

Figure 7.2 Emission of different gases at the energy plant in Schkölen

Table 7.3 Energy yield for alternative feedstock/conversion technologies

Option	Feedstock yield (dry tonnes/ha/year)	Transport fuel yield (GJ/ha/year)	Transport services yield[a] (10³ vehicle-km/ha/year)
Rape methyl ester (Netherlands, year 2000)	3.7 (rapeseed)	47	21 (ICEV)
EthOH from maize (USA)	7.2 (maize)	76	27 (ICEV)
EthOH from wheat (Netherlands, year 2000)	6.6 (wheat)	72	26 (ICEV)
EthOH from sugar beet (Netherlands, year 2000)	15.1 (sugar beet)	132	48 (ICEV)
EthOH from sugar cane (Brazil)	38.5 (cane stems)	111	40 (ICEV)
EthOH, enzymatic hydrolysis of wood (present technology)	15 (wood[b])	122	44 (ICEV)
EthOH, enzymatic hydrolysis of wood (improved technology)	15 (wood[b])	179	64 (ICEV)
MeOH, thermochemical gasification of wood	15 (wood[b])	177	64/133 (ICEV/FCV)
H_2, thermochemical gasification of wood	15 (wood[b])	213	84/189 (ICEV/FCV)

[a] Fuel economy assumed (in litres of gasoline-equivalent per 100km): 6.30 for rape methyl ester, 7.97 for ethanol (EthOH), 7.90 for methanol (MeOH), 7.31 for hydrogen used in ICEVs and 3.81 for methanol and 3.24 for hydrogen used in FCVs

[b] Including high-yielding lignocellulose crops and short rotation coppice.

Source: Williams et al (1995)

and would make a significant contribution to road transportation energy needs. Table 7.3 compares current and future energy yields from alternative feedstocks and conversion technologies for internal combustion engine vehicles (ICEVs) and FCVs. In the cases of MeOH and H_2, the FCVs offer more than double the transport services yield.

Figure 7.3 continues the comparison of alternative fuel feedstocks and traditional fuel feedstocks (natural gas and coal) with respect to carbon dioxide emissions.

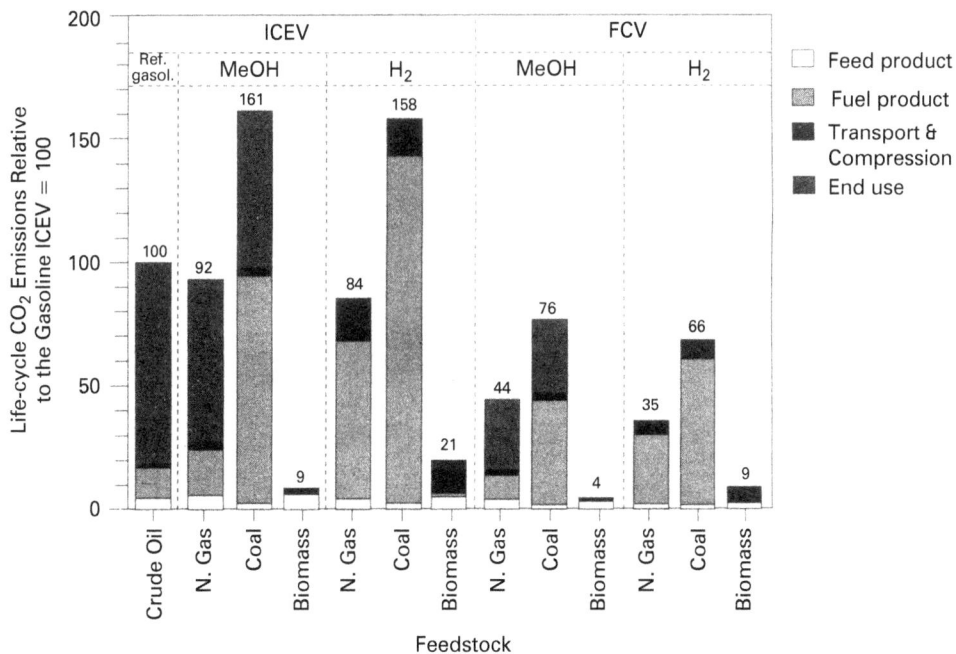

Source: Williams et al (1995)

Figure 7.3 Estimated life cycle of carbon dioxide emissions from methanol and hydrogen ICEVs and FCVs, with comparisons to emissions from a gasoline ICEV

The factors taken into consideration are feed production, fuel production, transport and compression, and end use. The biomass feedstocks, which in each case have considerably lower life-cycle carbon dioxide emissions than the traditional feedstocks, have most of their emissions coming from feed production and transport and compression. There are no end-use emissions, and fuel production emissions are virtually non-existent.

Production of biomass biofuel for transportation, along with the subsequent improvement in land usage, could lead to less dependence on oil for traditional transportation fuels. Biomass-derived methanol and hydrogen fuels would offer the additional advantages of providing a sustainable source of income for the rural areas of developing countries, and improving competition and price stability in world transportation fuel markets. These fuels bring increased land-use efficiency and decreased adverse environmental impacts when compared with conventional biofuels such as ethanol from maize. They can also be used in environmentally friendly and efficient FCVs.

It is important that biomass producers are aware of the variety of ways in which their biomass can be utilized and of the increasing need for methanol and hydrogen fuels to be used in transportation vehicles such as FCVs. In turn, FCV developers should be aware that the production of these fuels through sustainable biomass utilization makes their innovation and research efforts worthwhile.

Energy balances

Energy balance value alone is not meaningful in evaluating the benefit of ethanol or any other energy product; energy balance must be compared with that of the product it replaces. Compared to gasoline, any type of fuel ethanol substantially helps reduce fossil energy and petroleum use. Ethanol produced from corn can achieve moderate reductions in greenhouse gas emissions. Ethanol produced from 'cellulosic' plants, such as grass and weeds, can achieve much greater energy and greenhouse gas benefits.

The environmental advantages of biomass from energy plant species can be summarized as follows:

- They are a renewable resource that can be used to generate heat, electricity and transport fuels.
- The energy balance is positive – that is, more output energy than input energy, up to 25:1 (Figure 7.4).
- The ashes are recyclable.

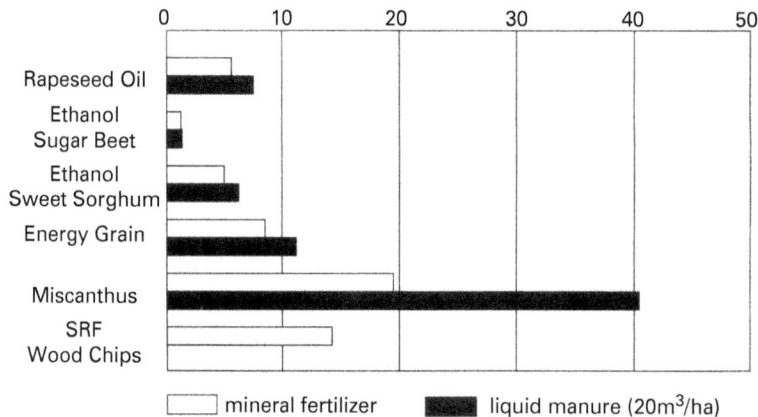

Source: Hartmann (1994)

Figure 7.4 Energy balance, output/input of different energy crops as a result of mineral or organic fertilizer application

- If they are cultivated and harvested properly, they make no or very little net contribution to global carbon dioxide.
- Other important environmental benefits, especially of the perennial plant species, are the protection and improvement of local watersheds and surrounding wildlife habitats, and reduction of the use of chemical pesticides and fertilizers.
- They create no major environmental risks through transportation, storage, processing and conversion.

Reductions in greenhouse gas emissions

Bioenergy can affect net carbon emissions in two main ways: (1) it provides energy that can displace fossil fuel energy, and (2) it can change the amount of carbon sequestered on land. The net carbon benefit depends on what would have happened otherwise – that is, both the amount and type of fossil fuel that would otherwise have been consumed and the land use that would otherwise have prevailed.

Most studies have found that producing first-generation biofuels from current feedstocks results in emission reductions in the range of 20–60 per cent relative to fossil fuels, provided the most efficient systems are used and carbon releases deriving from land-use change are excluded. Figure 7.5 shows estimated ranges of reduction in greenhouse gas emissions for a series of crops and locations, excluding the effects of land-use change. Brazil, which has long experience of producing ethanol from sugar cane, shows even greater reductions. Second-generation biofuels, although still insignificant at the commercial level, typically offer emission reductions in the order of 70–90 per cent, compared with fossil diesel and petrol, also excluding carbon releases related to land-use change.

Bioenergy crop systems can – if properly designed – yield significant benefits, both environmental and social. The right choice of biomass crops and production methods can lead to favourable carbon and energy balances and a net reduction in greenhouse gas emissions.

Internet resources

www.ifap.org
www.ifpri.org

Table 7.4 Approximate carbon emissions from sample bioenergy and fossil energy technologies for electricity generation

Fuel and technology	Generation efficiency	Grams of CO_2 per kWh
Diesel generator	20%	1320
Coal steam cycle	33%	1000
Natural gas combined cycle	45%	410
Biogas digester and diesel generator (with 15% diesel pilot fuel)	18%	220
Biomass steam cycle (biomass energy ratio[a] = 12)	22%	100
Biomass gasifier and gas turbine (biomass energy ratio[a] = 12)	35%	60

[a]The energy content of the biomass produced divided by the energy of the fossil fuel consumed to produce the biomass.

Source: IFPRI (2006)

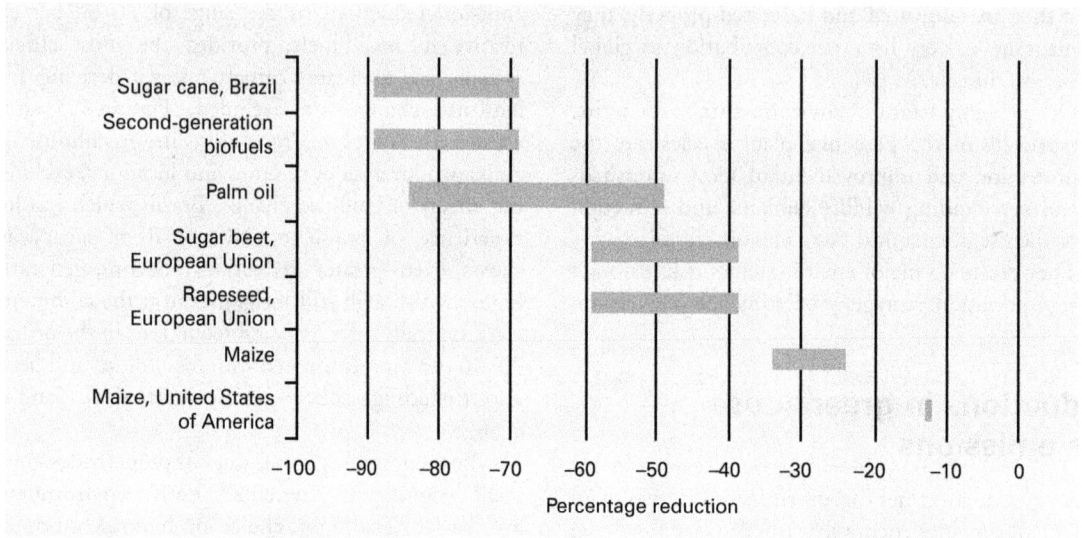

Note: Excludes the effects of land-use change.

Source: FAO (2008c)

Figure 7.5 Reductions in greenhouse gas emissions of selected biofuels relative to fossil fuels

Table 7.5 Ethanol production (L/ha) of various crops

Miscanthus	14,031
Switchgrass	10,757
Sweet Potatoes	10,000
Poplar Wood	9354
Sweet Sorghum	8419
Sugar Beet	6679
Sugar Cane	6192
Cassava	3835
Corn (maize)	3461
Wheat	2591

Source: El Bassam (2007b); USDA (2006)

8

Economic and Social Dimensions

Energy and development

> With oil prices in excess of US$60 a barrel, interest in bioenergy is running high. The energy needs of rapidly growing countries like China and India, together with unstable oil supplies, suggest that the days of cheap oil are over. Bioenergy offers an attractive alternative for many industrial and developing countries, but if its full potential is to be captured, then both the public and private sectors, working as partners, must make long-term commitments and investments in innovation. (IFPRI, 2006)

Fossil fuel utilization is highest in developed countries, where it is usually used for transportation and for the production of heat and electricity. Fossil fuels have a detrimental impact on the environment and our planet, and supplies are limited, so it is now time to look for a new source of energy that is environmentally friendly and that can be sustainably produced for future generations. The sustainable production of biomass is one of the most promising possibilities for moving the energy production trend away from fossil fuels.

Biomass utilization is most commonly found in developing countries, where it is usually the primary source of energy (making up approximately one-third of the total energy production). In addition, in some countries (Rwanda, Nepal and Tanzania) biomass utilization accounts for over 90 per cent of their energy production.

Economic development is closely correlated with the availability and utilization of modern energy sources. Also, the production and consumption of food is linked to the amount of energy used (Figure 8.1). In 1990, the per capita consumption of modern energy for 21 African countries was less than 100 kilograms of oil equivalent (kgoe) (World Bank, 1992). In many of these countries, daily per capita caloric supply is below 2000 calories. Food production is unlikely to increase without greater access to modern energy (FAO, 1995b).

According to Hohmeyer and Ottinger (1994), one of the main obstacles to the expansion and acceptance of renewable fuels and biomass energy technologies into world markets is that the markets do not acknowledge the adverse social and environmental costs and risks connected with the usage of fossil and nuclear fuels. Oil shipping accidents, such as the Exxon Valdez disaster (US$2.2 billion clean-up costs), cause devastating and long-lasting effects on the environment and wildlife, yet these disasters are becoming increasingly common occurrences. Nuclear accidents such as Three Mile Island (US$1 billion medical and clean-up costs) and Chernobyl still leave questions as to the effects on future generations (Miller, 1992). The costs of maintaining channels to fossil fuel sources through military means, along with the depletion of these non-renewable limited resources, should also be taken into consideration when promoting increased use of biomass fuels. Table 8.1 takes a comparative look at the social, economic and environmental aspects of the utilization of fossil, nuclear and biomass fuels.

The comparatively large amount of subsidies and support provided to conventional energy sources is another problem preventing biomass-derived fuels from playing a more substantial role in global energy supply (Hubbard, 1991). The amount of money spent by developed countries of the International Energy Agency (IEA) on R&D from 1988 to 1900 was US$73 billion for nuclear, $12 billion for coal, $11 billion for all renewables and $1 billion for biomass (OECD, 1991). Energy prices in numerous developing countries are subsidized by as much as 30–50 per cent.

Calories per day (1989)

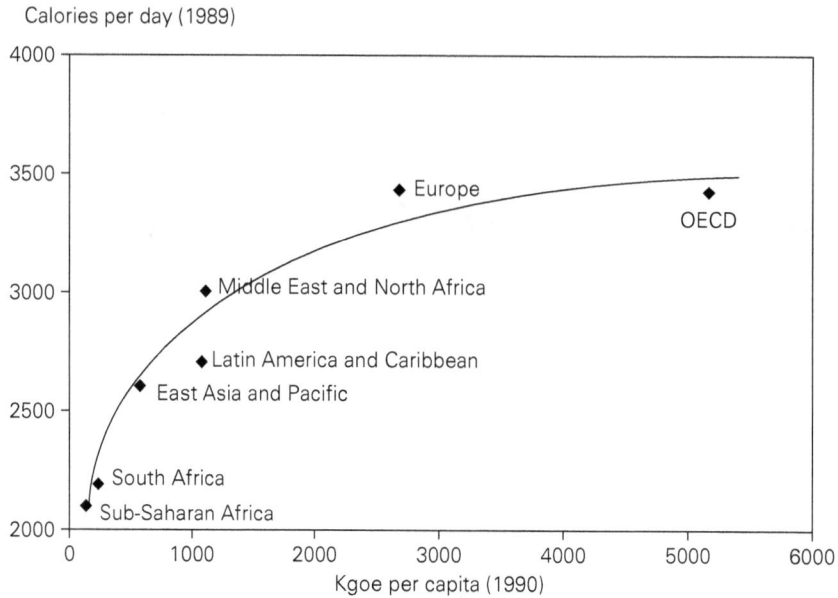

Source: El Bassam (2002b)

Figure 8.1 Energy consumption and calorie supply in different regions
of the world

Table 8.1 Social, economic and environmental aspects of utilization of different energy sources

	Energy fuels		
	Fossil	**Nuclear**	**Biomass**
Renewable	no	no	yes
CO_2 reduction	no	yes	yes
Reduction of heat emission	no	no	yes
Landscape	no	no	yes
Avoidance of large accident risks	no	no	yes
Excessive costs of environmental repair	yes	yes	no
Reduction of administrative costs	no	no	yes
Innovation	no	yes	yes
Creating new industrial jobs	no	yes	yes
Promoting decentralization of economic structure	no	no	yes
Promoting export	no	yes	yes
Increasing autonomous energy supply (industrialized countries)	no	yes	yes
Increasing autonomous energy supply (developing countries)	no	no	yes
Improving farmers' incomes	no	no	yes
Significant time of waste decay	yes	yes	no
Migration to urban areas	yes	yes	no
Favourable public opinion	no	no	yes
Avoidance of international conflicts and wars	no	no	yes
Gene deformation	no	yes	no

Source: Scheer (1993); El Bassam (1996)

Although biomass energy has become competitive with conventional energy sources in many areas, in order for biomass-derived fuels to compete more evenly it is necessary that subsidies and support be more equally distributed and guaranteed for longer time periods, and that environmental effects are given the recognition that they deserve.

A study was undertaken by Environmental Resources Management (ERM, 1995) to look at the economics of bioethanol, along with the effects of subsidies for helping increase the income of wheat and sugar beet farmers. The study divided the famers into 'base case' and 'favoured region' categories. The base case considered small farms that were not in traditional sugar beet farming territories, making up approximately 70 per cent of EU farmers. Here, yields were estimated at 49t/ha for sugar beet and 6t/ha for wheat; the set-aside subsidy was €270/ha. The favoured region took larger farms in traditional sugar beet farming territories, with bioethanol processing facilities in the region. The crop yields were estimated at 63t/ha for sugar beet and 7t/ha for wheat; the set-aside subsidy was €315/ha.

Figure 8.2 shows the results of the study in terms of farmers' net income in both the base case and favoured region for sugar beet and wheat production when the crops are grown as a food crop, as a bioethanol crop on set-aside land, and when the land is left fallow and the subsidy payment is collected. Only the wheat farmer is presently able to collect an additional set-aside subsidy for growing wheat for a non-food purpose.

The results of the study also showed the money that would be returned to the base case farmers because of a tax subsidy paid at the gasoline blending stage. For sugar beet, a €40/t tax would be forgone by the EU. At a crop price to the farmer of €23/t, this would contribute €2/t to the farmer's income. In the case of wheat, a €151/t tax would be forgone by the EU. At a crop price to the farmer of €28/t, along with a €45/t set-aside payment, this would contribute €2.5/t to the farmer's income.

The study concluded that it would not be very feasible for farmers in the EU to grow sugar beet or wheat crops for bioethanol production. In addition, no significant amount of money would be returned to the farmer because of a tax reduction on bioethanol blended fuels. The tax subsidy for bioethanol in motor fuels would not be sufficient to help support rural communities.

Medium- and long-term projects show an unavoidable increase in oil prices, which will make biofuels more competitive. Energy crops and biomass in general seem to have, in many instances, much lower external costs than fossil fuels, provided they are grown

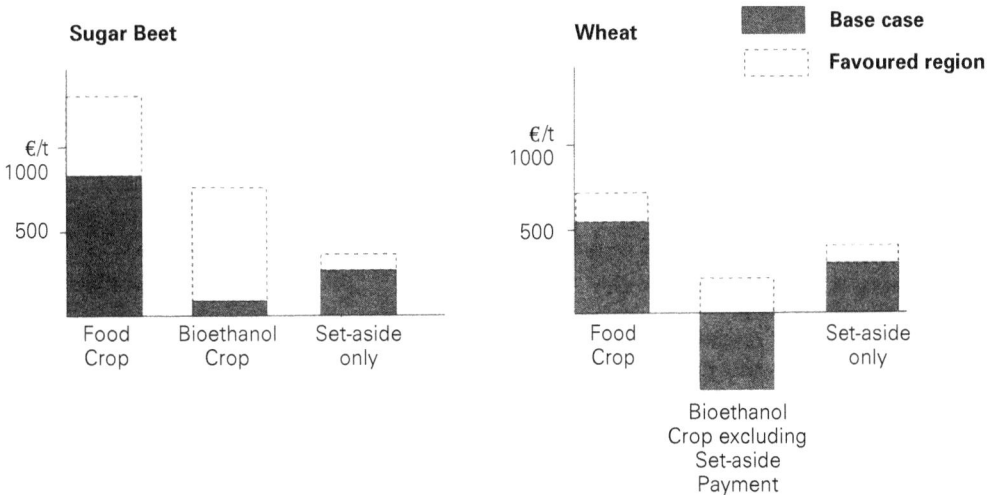

Source: ERM (1995)

Figure 8.2 Farmers' net incomes from food crops, biofuel crops and set-aside subsidies

and processed in an environmentally consistent way. It is unhelpful to argue the case for bioenergy with yesterday's technology without taking into account the innovation possibilities and the prices of today, along with the scarcity, the growing future demand for energy and the environmental damage and risks associated with fossil and nuclear energy feedstocks.

There are various concerns regarding the production of biomass as a fuel source, especially the establishment of large-scale crop production systems. These concerns include environmental impact, land availability, and possible conflicts with food production. The penetration of bioenergy into the market will require overcoming preconceived ideas of bioenergy within governments, the market and by the general public (Hall and House, 1995).

On the other hand, more energy will be available from a specific unit area as a result of:

- the introduction of naturally high-yielding plant species, such as *Miscanthus* spp., *Arundo donax,* etc.;
- the enhancement of the acreage yield of traditional crops through further developments in plant breeding, gene technology and biotechnology;
- improvements in the efficiency of engines and conversion technologies: in 1995, the major car companies in Europe agreed to develop 3 litre per 100km diesel car engines by the year 2000, which means more mileage per hectare.

Biotechnological advances in biomass conversion may initially alter the prevailing technical and economic conditions in comparison with the production of non-biomass fuels. Assessments of emerging technologies (using the biomass integrated gasifier as an example) suggest that power plants of medium scale (20–50MW) could achieve thermal efficiencies in excess of 40 per cent within a few years (eventually reaching 50 per cent or more) combined with capital costs well below those of comparable conventional biomass plants utilizing boiler/steam technology (Elliott and Booth, 1993).

The competitiveness of biomass utilization for biofuels will thus be considerably improved. Energy crops, as converters of solar energy and grown with environmentally sustainable methods, fulfil the sustainability criterion because they avoid the depletion of non-renewable fossil resources. This will impose positive impacts on soils (less erosion), landscapes, ecosystems and biodiversity, which should be of economic importance.

Although biomass feedstock prices are presently high, it would be possible for biomass methanol and hydrogen fuel prices to be competitive with those of coal. If the cost of fuel production from natural gas rises, as expected, it is estimated that by about the year 2010 biomass methanol and hydrogen fuels will be nearly competitive with natural gas. At that time, a tax of less than 2 per cent for using natural gas-derived fuels in fuel cell vehicles (FCVs) would be enough to make biomass-derived fuels more economical than natural gas.

In industrialized countries, the contribution of biofuels to energy supply varies. In Germany the percentage of biomass to total primary energy supply is less than 2 per cent, in Denmark it is 7 per cent, in the USA 4.5 per cent, in Austria 13 per cent and in Sweden almost 17 per cent. Sweden has developed a very effective taxation system for energy sources, and has implemented three environmental taxes: a sulphur tax, a CO_2 tax and an energy tax. All these led to increases in the prices of fossil energy resources in comparison to biofuels (Figure 8.3).

Jobs and employment

A study by Grassi (1997) looked at the impact of bioenergy development on job creation in the EU. The study came to the following conclusions:

- The renewable energy sector is labour intensive, and is thus a significant source of employment.
- Bioenergy, especially given its ability to penetrate all energy markets (power, heat, transport and chemicals) in the future, can offer the largest contribution to the EU, with an envisaged 12 per cent renewable energy target for the year 2010.
- At the EU level, the foreseen bioenergy contribution of about 113Mtoe/yr by the year 2020 should open up opportunities for about 1,500,000 new jobs.
- The total average investment cost for about 1 million new direct jobs (the cost of the indirect jobs will derive automatically from private investment) will be around €250 billion less than the estimated €345 billion of total subsidies paid

€/MWh

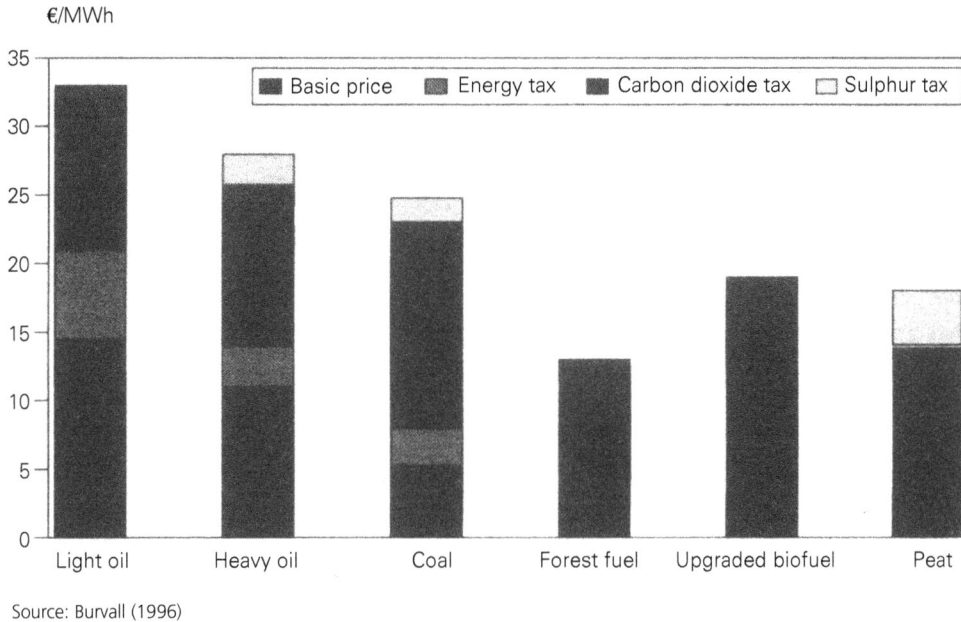

Source: Burvall (1996)

Figure 8.3 Energy prices for heating plants in Sweden, 1996

to 1.5 million unemployed people in our socially protected EU society during the 20-year minimum life of the investments (plants).

In 1997 the US Department of Energy (DOE, 1997) reported that economic activity in the USA associated with biomass supported about 66,000 jobs, with most of them in rural regions. It is predicted that by the year 2020, over 30,000MW of biomass power could be installed, with about 60 per cent of the fuel supplied from over 10 million acres of energy crops and the remainder from biomass residues. This would support over 260,000 US jobs and would substantially revitalize rural economies (Figure 8.4).

For farmers, biomass energy crops can be a profitable alternative that will complement, not compete with, existing crops and thus provide an additional source of income for the agricultural industry. It is envisioned that biomass energy crops will be grown on currently underutilized agricultural land.

In addition to rural jobs, expanded biomass power will create high-skill, high-value job opportunities for utility and power equipment vendors, power plant owners and operators, and agricultural equipment vendors.

Jobs

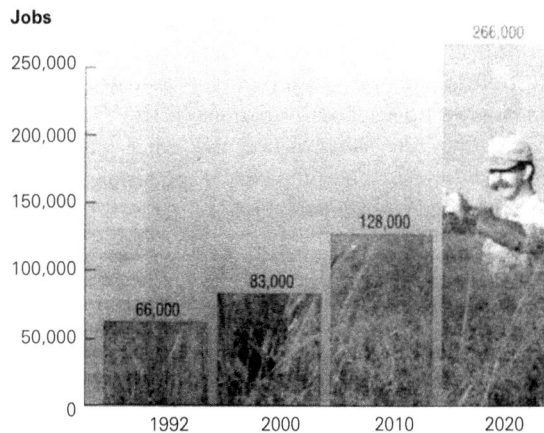

Source: DOE (1997).

Figure 8.4 Projected annual US employment impacts from biomass power

Biomass stoves

The most important energy service today in many developing countries is cooking. Traditional fuels – fuelwood, crop residues and dung – are the main cooking fuels in rural areas of these countries. In many urban areas, charcoal and coal are also used. More than half of the world's 2 billion poor people depend on these crude, polluting fuels for their cooking and other heating needs. Figure 8.5 shows the large amount of firewood, collected over great distances, necessary for fuelling the small and inefficient stoves. Figure 8.6 shows a slightly more modern and efficient stove, but there is still a large demand for significant improvement and distribution.

Higher incomes, and reliable access to fuel supplies, enable people to switch to modern stoves and cleaner fuels such as kerosene, liquefied petroleum gas (LPG) and electricity. This transition can be observed worldwide in various cultural traditions. These technologies are preferred for their convenience, comfort, cleanliness, ease of operation, speed, efficiency and other attributes. The efficiency, cost and performance of stoves generally increase as consumers shift progressively from wood stoves to charcoal, kerosene, LPG or gas, and electric stoves.

There can be a substantial reduction in both operating costs and energy use in going from traditional stoves using commercially purchased fuelwood to improved biomass, gas or kerosene stoves. There are also opportunities to substitute high-performance biomass stoves for traditional ones, or to substitute liquid or gas (fossil- or biomass-based) stoves for biomass stoves. Local variations in stove and fuel costs, availability, convenience and other attributes, and in consumer perceptions of stove performance, will then determine consumer choice.

Source: El Bassam

Figure 8.6 Improved type of biomass stove, northwest China

In rural areas, biomass is likely to be the cooking fuel for many years to come. This is exemplified by the common establishment of biofuel marketplaces on which many towns and villages are dependent throughout the world (Figure 8.7). Alternatively, particularly in urban areas, liquid- or gas-fuelled stoves offer the consumer greater convenience and performance at a reasonable cost.

From a national perspective, public policy can help to shift consumers towards the more economically and environmentally promising cooking technologies. In particular, improved biomass stoves are probably the most cost-effective option in the short to medium term, but require significant additional work to improve their performance.

Source: El Bassam

Figure 8.5 Extremely inefficient, traditional type of cooking stove (stove is on the right side of the photograph), northwest China

Source: Liese

Figure 8.7 Biofuel marketplace in Ethiopia

In the long term, the transition to high-quality liquid and gas fuels for cooking is inevitable. With this transition, substantial amounts of labour now expended to gather biomass fuels in rural areas could be freed; the time and attention needed to cook using biomass fuels could be substantially reduced; and household, local and regional air pollution from smoky biomass (or coal) fires could be largely eliminated. The use of biomass-derived liquid or gaseous fuels (such as ethanol, biogas and producer gas) for cooking and other advanced options is particularly relevant (UNDP, 1997).

The development, improvement and manufacture of biomass stoves and cookers will not only lead to improved efficiency of the energy conversion outputs and improve the environment, but will also provide large

opportunities for job creation both directly and indirectly throughout the world, especially in rural areas. In some cases, women in Africa and Asia need up to 8 hours and must go as far as 20km to collect firewood in order to cook one meal for the family (Figure 8.8). The growing of energy crops and utilization of plant residues, the development and production of efficient cookers, and the development of technologies for pelleting and briquetting of biomass will have great positive impacts on rural markets. The money that has to be spent on fossil energy fuels will stay in the region instead.

Investment in the energy supply sector

The present level of worldwide investment in the energy supply sector, US$450 billion per year, is projected to increase to perhaps US$750 billion per year by 2020, about half of which would be for the power sector. Such investment levels cannot be sustained by traditional sources of energy financing. Also, the multilateral financing agencies are able to provide only a small fraction of the capital needed. All of this undermines self-reliance, and leads to deals that reflect the high price of capital arising from the high financial risks involved.

The Belgian subsidiary of Germany-based ethanol producer CropEnergies, has started production at its next-generation bioethanol plant in Wanze, Belgium. The bioethanol plant, the largest in Belgium, will

Source: Liese

Figure 8.8 Women carrying firewood in Ethiopia

Source: Российская Биотопливная Ассоциация
www.bioethanol.ru/bioethanol/plants/

Figure 8.9 Bioethanol plant

produce up to 300,000m³ of bioethanol a year from wheat and sugar syrups. The manufacturing process, which uses biomass as a primary energy source, lowers carbon dioxide emissions by up to 70 per cent compared to fossil fuels.

With this, the company has been awarded half of the distributed production licences or 125,000m³ a year until 2013 with which the Belgian government promotes biofuels in the domestic market. The investment for this technology exceeds €250 million.

Besides the acquisition of the French alcohol producer Ryssen Alcools with an annual production capacity of 100,000m³ of bioethanol for fuel applications in June 2008, CropEnergies expanded the production capacities in Zeitz, Saxony-Anhalt, from 260,000m³ to 360,000m³ of bioethanol a year in June 2008, establishing its position as the largest bioethanol plant in Europe. These projects have almost tripled CropEnergies' annual production capacity to over 700,000m³ of bioethanol.

The dependence on fossil fuels has created a wide variety of problems for non-oil-producing developing countries, as well as for some industrialized countries and economies in transition. In over 30 countries, energy imports exceed 10 per cent of the value of all exports, a heavy burden on their balance of trade that often leads to debt problems. In about 20 developing countries, payments for oil imports exceed payments for external debt servicing. This is an important aspect of the energy–foreign-exchange nexus.

Advanced technology in bioenergy would facilitate decentralized rural electrification, and thereby promote rural development as well as considerably reduce the costs of imported fuels (UNDP, 1997).

Economic and social impact of bioenergy in developing countries

For developing countries with a large number of poor people reliant on agriculture, the first priority should be given to effective use of existing agricultural wastes for energy generation. This option has the least adverse impact on the poor and could provide additional revenue for poor rural communities. It requires, however, establishing effective revenue-sharing mechanisms that ensure that the higher revenues from the exploitation of agricultural wastes are shared in an equitable fashion and flow to all stakeholders,

including low-income farmers. It also requires enacting a legal and regulatory framework that allows for the development of modern agricultural waste-based bioenergy and that provides, among other incentives, access to the power grid and transport fuel market. In some cases, mechanisms for efficient centralization of agricultural wastes would need to be in place. Once developing countries have optimized the use of existing agricultural wastes for energy generation and put in place adequate revenue-sharing, regulatory and policy frameworks, they can consider the option of dedicated energy plantations, while carefully balancing any associated trade-offs between food security and energy generation. Fortunately, the technical, regulatory and policy expertise needed to promote an equitable agricultural waste energy industry also provides, in many cases, the skills needed to develop and nurture a sustainable dedicated energy plantation sector that does not adversely affect the poor or decrease food security.

Biomass is a primary source of energy for close to 2.4 billion people in developing countries. Easily available to many of the world's poor, biomass provides vital and affordable energy for cooking and space heating. Although widespread use of traditional and inefficient biomass energy in poor countries has been linked to indoor air pollution as well as to land degradation and attendant soil erosion, biomass-based industries are a significant source of jobs and income in poor rural areas with few other opportunities. The share of biomass energy in total energy consumption varies across developing countries, but generally the poorer the country, the greater its reliance on traditional biomass resources (see Figure 8.10). Biomass has considerable potential to become more important in total energy consumption, and this growth could have significant impacts, both positive and negative, on agriculture and the poor.

There are several ways to reduce the trade-offs between bioenergy crops and food production:

- Develop biomass crops that yield much higher amounts of energy per hectare or unit of water, thereby reducing the resource needs of bioenergy crops.
- Focus on food crops that generate by-products that can be used for bioenergy, and breed varieties that generate larger amounts of by-products.
- Develop and grow biomass in less-favoured areas rather than in prime agricultural lands – an

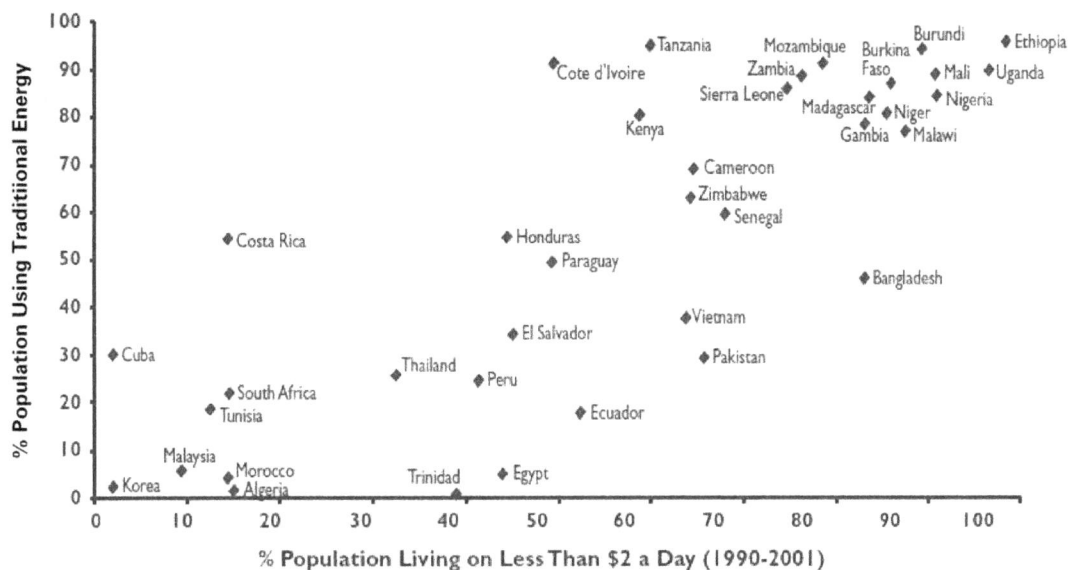

Source: IFPRI (2006)

Figure 8.10 Poverty and traditional energy use

approach that would benefit some of the poorest people. Second-generation technologies that enable cost-effective conversion of cellulose-rich biomass, like fast-growing trees, shrubs, and grasses that can grow in less fertile and low-rainfall areas, will greatly expand this option within the next 10–15 years.

• Invest in increasing the productivity of the food crops themselves, since this would free up additional land and water for the production of bioenergy crops.

• Remove barriers to international trade in biofuels.

The world has enough capacity to grow all the food that is needed as well as large amounts of biomass for energy use, but not in all countries and regions. Trade is a powerful way of spreading the benefits of this global capacity while enabling countries to focus on growing the kinds of food, feed or energy crops for which they are most competitive. Trade would also allow bioenergy production patterns to change in the most cost-effective ways as new second-generation technologies come on line. The benefits for the poor can also be enhanced by choosing appropriate scales and techniques for producing and processing biomass. So far, most attention has been given to large-scale production and processing of bioenergy for the market, which is often

the most cost-effective approach for private firms. This is because biomass crops lend themselves to economies of scale in growth and processing. Yet the scale benefits need to be balanced against the costs and energy loss of transporting biomass products, given their bulk and weight. This situation opens up opportunities for smaller-scale and rural-based production and processing, which would be much more beneficial for the poor than large-scale and urban-based processing. In many developing countries it may also be inappropriate to consolidate land into heavily mechanized farms for growing biomass. A better approach is to organize smallholders so that they can grow and market biomass crops to large processing firms. Small-scale processing of biomass to produce, for instance, electricity or biogas already helps meet local energy needs in rural areas in many developing countries, and these options can be expanded in the future. The agricultural research systems in developing countries have a key role to play in addressing these issues to make biofuels pro-poor. This is a promising area for public-private partnership in research. The Consultative Group on International Agricultural Research (CGIAR) could also play a key role in strengthening international knowledge and facilitating the exchange of information on pro-poor development of biofuels.

Nigeria-based Global Biofuels is establishing the first of a number of refineries for the production of bioethanol from sweet sorghum. Ethanol will be blended with petrol to make E10. Global Biofuels is planning to establish the project in nine Nigerian states. It expects to start production in December 2009, with an initial production capacity of 27 million litres a year. The company will also produce biodiesel from safflower (an oilseed crop used for food) to power industries and electric power generators.

The production of food and feed remains paramount for the farmers of IFAP [International Federation of Agricultural Producers]; however, biofuels represent a new market opportunity, help diversify risk and promote rural development. Biofuels are the best option currently available to bring down greenhouse gas emissions from the transport sector and thus to help mitigate climate change. With oil prices currently at record levels, biofuels also support fuel security.

Recently, biofuels have been blamed for soaring food prices. There are many factors behind the rise in food prices, including supply shortages due to poor weather conditions, and changes in eating habits which are generating strong demand. The proportion of agricultural land given over to producing biofuels in the world is very small: 1 per cent in Brazil, 1 per cent in Europe, 4 per cent in the United States of America, and so biofuel production is a marginal factor in the rise in food prices.

The misconceptions about biofuels are important to overcome for a farming community that has long suffered from low incomes. Bioenergy represents a good opportunity to boost rural economies and reduce poverty, provided this production complies with sustainability criteria. Sustainable biofuel production by family farmers is not a threat to food production. It is an opportunity to achieve profitability and to revive rural communities.

Development of biofuels depends on positive public policy frameworks and incentives such as mandatory targets for biofuel use and fiscal incentives that favour biofuels relative to fossil fuels until the industry matures. This is in the public interest when biofuels are produced from local sources since they create employment and wealth in the country. Governments should also provide investment incentives including: income tax credits for small biofuel producers, financing bioenergy plants,

increasing farmers' participation through matching grants, and reducing business risk for the adoption of new technologies. Support for research and development, particularly for small-scale technology and enhancing the energy potential of indigenous plants, is crucial.

Biofuels are not a miracle solution, but they offer significant income opportunities for farmers. If farmers are to benefit, careful long-term assessment of economic, environmental and social benefits and costs are required to identify real opportunities aimed at improving producers' incomes. Sound strategies, developed along with the different stakeholders, are needed to capture the potential environmental and economic benefits, including the setting up of a rational land-use policy, appropriate selection of crops and production areas, and protection of rights of farmers. Farmers' organizations need to push for the creation of the right incentive mechanisms that will allow their members to benefit from this new opportunity and generate complementary incomes.

Further research and development are needed in order to avoid competition between food and fuel uses of certain crops and also to get the right signals regarding the development of biofuel production worldwide. Therefore, bridging the knowledge gap on biofuels through information dissemination and capacity building programmes to support farmers in developing ownership of the value chain are of utmost importance. (IFAP, 2008)

Investments and market development

Biodiesel

Biodiesel feedstock markets worldwide are in transition from increasingly expensive first-generation feedstocks such as soya, rapeseed and palm oil, to alternative, lower-cost, non-food feedstocks. As a result, a surge in demand for alternative feedstocks is driving new growth opportunities in the sector.

'Biodiesel growth from non-food feedstocks is gaining traction around the world', said Thurmond (2008). 'For example, China recently set aside an area the size of England to produce jatropha and other non-food plants for biodiesel. India has up to 60 million ha of non-arable land available to produce jatropha, and intends to replace 20 per cent of diesel fuels with jatropha-based biodiesel. In Brazil and Africa, there are significant programs underway dedicated to producing non-food crops jatropha and castor for biodiesel.'

In the USA, the market for biodiesel is growing at a large rate – from 25 million gallons per year in 2004 to over 450 million gallons in 2007. The total biodiesel being sold in the US amounts to less than 1 per cent of all diesel consumption. In Europe, biodiesel represents 2–3 per cent of total transportation consumption and is targeted to reach 6 per cent by 2010. In China, India, Brazil and Europe, economic and environmental security concerns are leading to new government targets and incentives, aimed at reducing petroleum imports and increasing the consumption and production of renewable fuels. Europe, Brazil, China and India each have targets to replace 5 to 20 per cent of total diesel with biodiesel.

Ethanol

With its tropical climate, large landmass and abundant fertility, South America has plenty of potential for growth in biofuels. In the case of Brazil, with a solid and mature three decades of ethanol development already under its belt, it is a matter of adapting to maintain its leading position as a key producer of low-cost and sustainable biofuels and regain its crown from the US as the world's biggest ethanol producer. Peru and Colombia are progressing new biodiesel and ethanol projects at a steady pace while nations such as Ecuador are in the early stages of developing viable biofuels markets. According to figures from New Energy Finance, a research company focused on renewable energy, Latin America accounted for well over a third of total global investment in biofuels in 2007. 'What we saw in 2007 was US$19.5 billion (€15 billion) of investment globally in biofuels projects', says New Energy Finance's São Paulo-based analyst, Camila Ramos. 'Investment in bioenergy in Latin America was US$7.9 billion in 2007, and in the first half of 2008 alone investment in the region totalled US$5.1 billion. This is stronger than 2007 already, and the majority of the investment has been in Brazil and in developing the sugar cane sector.' In fact, while the overall rate of global investment in biofuels has slowed, Brazil topped the most attractive countries for biofuels investment in the first and second quarters of 2008, regaining that leading position from the US, which headed the 2007 list. Annual Brazilian ethanol production is forecast to reach

26 billion litres in 2008, and 38 billion litres by 2012. Brazil has some 400 plants, and uses close to 80 per cent of this production domestically, exporting a modest 15–20 per cent. Almost 90 per cent of the cars in Brazil are now flex-fuel cars taking an ethanol blend, and ethanol now accounts for 52 per cent of all fuel for passenger vehicles in Brazil, its use having surpassed petrol for the first time in March 2008. In Brazil, experience counts, and the country has had plenty of it. Efficiency levels are impressive. World Bank figures show that Brazil produces almost ten times more energy from sugar cane than the US produces from corn, and that while the US uses over 10 million ha to grow its ethanol-destined corn, Brazil uses just 3.6 million ha.

Driven by high crude oil prices, rising concerns over environmental pollution, and the consequent switch to alternative fuels, the world ethanol market is projected to reach 27.7 billion gallons by the year 2012. Stringent legislation of emission standards and governmental intervention, by means of subsidies and tax incentives, are expected to foster market growth in the medium to long run. Both developed and developing markets are expected to add to the market's growth in the future. China and India represent lucrative markets to develop.

The world ethanol market is strengthened by rising consumption patterns in end-use markets. Fuel ethanol is witnessing unprecedented interest, encouraged largely by the ban on MTBE in several countries and its resulting replacement by ethanol. Regulations imposed by most governments in the developed markets are additionally helping to increase demand for ethanol in fuels. In the US, the renewable fuels standard mandates the use of approximately 8 billion gallons of ethanol by the year 2012. The need to avoid painful rises in crude oil prices, reduce greenhouse gas emissions, and lower international dependence on oil is encouraging governmental intervention in fostering consumption of ethanol. For instance, consumption of ethanol will be upheld by government subsidies and incentives which are being offered at both the state and federal levels in most countries. The rising popularity of flexible fuel vehicles (FFVs) and oxydiesel, a blend of ethanol and diesel fuel, are also expected to bode well for the world ethanol market. Technology developments, which enable the blending of more than 10 per cent ethanol in gasoline, are expected to result in increased consumption of ethanol in fuels. Development of new technologies and emergence of new end-use applications are expected to bring in new growth opportunities. Ethanol production is aligned with the yields of sugar cane crops, given the fact that over 50 per cent of the world sugar cane produce is used in manufacturing ethanol.

As stated in a recent report published by Global Industry Analysts Inc. (2008), South America and the United States dominate the world ethanol market, together cornering over 66.5 per cent of total volume sales estimated in the year 2008. About 90 per cent of ethanol demanded in South America is consumed by Brazil. Ethanol consumption in Brazil is expected to reach 7.45 billion gallons by 2015. Sales of ethanol in Canada, one of the fastest-growing markets worldwide, are expected to rise by approximately 208.25 million gallons over the period 2008 to 2012.

Global growth is expected to be led by the use of ethanol in fuels. In the fuels end-use market, volume consumption of ethanol is projected to grow at double-digit rates, and rise by about 7597 million gallons over the period 2008 to 2012. Asia-Pacific dominates the global food and beverage end-use market with a 64.2 per cent share estimated in the year 2008. The solvent end-use market in the United States is projected to consume over 230 million gallons of ethanol by the year 2015. Europe, Germany and France collectively account for 35.5 per cent of the regional ethanol market as estimated in 2008.

Biogas

A total of 8900 biogas plants will be built in 2009 generating 2700MW of electricity worldwide. Globally, Asia is leading with 34 per cent of total capacity and Europe with 26 per cent, followed by NAFTA and others. Biogas from by-products and combinations is the most efficient form of renewable energy and therefore not affected by the financial crisis like other renewable energy technologies. Biogas production comes from at least three sources: agriculture wastes, sewage sludge and solid domestic wastes. Germany is a technology leader with about €700 million invested with 400 companies in 2006. The world market was at approximately €2 billion in 2006 and is expected to swell to in excess of €25 billion by 2020.

Biogas power plants are a combination of anaerobic digestion systems with associated electricity generators such as gas turbines or gas engines. The electricity they produce is classified as renewable or green energy and if sold into the national grid attracts subsidies. In the last 20 years, biogas use has been successful in wastewater treatment plants, industrial processing applications, landfill and the agricultural sector.

Algae

In the US and the EU, algae-based biodiesel ventures are growing in response to demands for clean fuels. Each of these endeavours clearly demonstrates increased public and private sector interest in non-food, second-generation markets.

Over US$300 million (€233 million) has been invested in algae so far in 2008. The initial findings from a new study, Algae 2020, identify three key trends or waves of investments now emerging in the path towards the commercialization of the algae biofuels industry. The first wave of algae investment is coming from public-private partnerships between governments, universities, research labs and private companies including DARPA, National Renewable Energy Laboratory (NREL) and the UK's Carbon Trust. These investments started late in 2006, and are growing in 2008 and beyond. Chevron, for example, has invested in and partnered with NREL to produce algae for biocrude, jet fuel and biodiesel. BP has invested in PPPs with the University of California at Berkeley and Arizona State University along with DARPA to advance the use of algae for jet fuel and biofuels. The US Department of Energy invested US$2.3 million in algae projects in 2008. Shell has invested in a PPP with Cellana, a joint venture with the Hawaiian Natural Energy Laboratory and HR Biopetroleum to produce algae for biofuels (Emerging Markets Online, 2008).

Internet resources

www.biofuels-news.com
www.biofuels-news.com/industry_news.php
www.emerging-markets.com
www.ifap.org
www.ifpri.org

9

Integrated Bioenergy Farms and Rural Settlements

Introduction

The world still continues to seek energy to satisfy its needs without giving due consideration to the consequent social, environmental, economic and security impacts. It is now clear that current approaches to energy are unsustainable. It is the responsibility of political institutions to ensure that the research and the development of technologies supporting sustainable systems be transferred to the end users. Scientists and individuals must bear the responsibility of understanding that the Earth is an integrated whole and must recognize the impact of our actions on the global environment, in order to ensure sustainability and avoid disorder in the natural life cycle. Sustainability in a regional and global context requires that demands are satisfied and risks overcome (El Bassam, 2004b) (Figure 9.1).

Current approaches to energy are unsustainable and non-renewable. Furthermore, energy is directly related to the most critical social issues which affect sustainable development: poverty, jobs, income levels, access to social services, gender disparity, population growth, agricultural production, climate change and environmental quality, and economic and security issues. Without adequate attention to the critical importance of energy to all these aspects, the global social, economical and environmental goals of sustainability cannot be achieved. Indeed, the magnitude of change needed is immense, fundamental and directly related to the energy produced and consumed nationally and internationally. The key challenge to realizing these targets is to overcome the lack of commitment and to develop the political will

to protect people and the natural resource base. Failure to take action will lead to continuing degradation of natural resources, increasing conflicts over scarce resources and widening gaps between rich and poor. We must act while we still have choices. Implementing sustainable energy strategies is one of the most important levers humankind has for creating a sustainable world. More than two billion people, mostly living in rural areas, have no access to modern energy sources. Food and fodder availability is very closely related to energy availability.

Integrated energy farms (IEFs) – a concept of the FAO

In order to meet challenges, future energy policies should put more emphasis on developing the potential of energy sources, which should form the foundation of future global energy structure. In this context, the FAO of the United Nations, in support of the Sustainable Rural Energy Network (SREN), has developed the concept for the optimization, evaluation and implementation of integrated renewable farms for rural communities under different climatic and environmental conditions (El Bassam, 1999).

Milestones

The IEF concept includes farms or decentralized living areas from which the daily necessities (water, food and energy) can be produced directly on site with minimal external energy inputs.

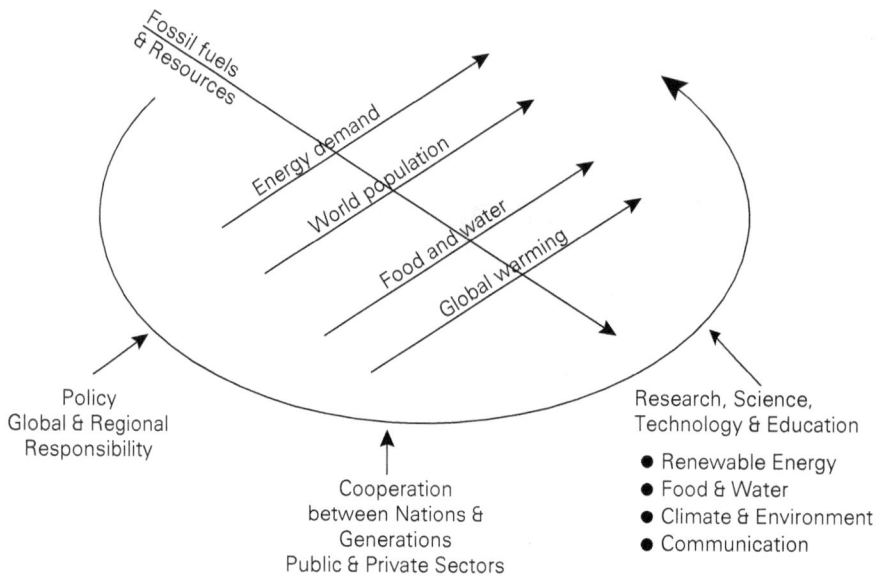

Source: El Bassam (2004b)

Figure 9.1 Sustainability in regional and global context demands, risks and measures

Energy production and consumption at the IEF has to be environmentally friendly, sustainable and ultimately based mainly on renewable energy sources. It includes a combination of different possibilities for non-polluting energy production, such as modern wind and solar electricity production, as well as the production of energy from biomass. It should seek to optimize energetic autonomy and an ecologically semi-closed system, while also providing socio-economic viability and giving due consideration to the newest concept of landscape and biodiversity management.

The concept considers the following needs of the rural population to improve their living conditions and raise their living standards and to improve the environment:

Heat

Heat can be generated from biomass or solar thermal to create both high-temperature steam and low-temperature heat for space heating, domestic and industrial hot water, pool heating, desalination, cooking and crop drying.

Electric power

Solar photovoltaics, solar thermal, biomass, wind and micro-hydro.

Water (drinking and irrigation)

Water is an essential resource for which there can be no substitute. Renewable energy can play a major role in supplying water in remote areas. Several systems could be adopted for this purpose:

- solar distillation;
- renewable energy-operated desalination units; and
- solar-, wind- and biomass-operated water pumping and distribution systems.

Lighting and cooling

In order to improve living standards and encourage the spread of education in rural areas, a supply of electricity is vital. Several systems could be adopted to generate electricity for lighting and cooling:

- solar systems (photovoltaic and solar thermal);
- wind energy systems; and
- biomass and biogas systems (engines, fuel cells and Stirling engines).

Cooking

Women in rural communities spend long hours collecting firewood and preparing food. There are other methods which are more efficient, healthy and environmentally benign. Among them are:

- solar cookers and ovens;
- biogas cooking systems;
- improved biomass stoves using briquettes and pellets; and
- plant oil and ethanol cookers.

Health and sanitation

To improve serious health problems among villagers, solar energy from photovoltaic, wind and biogas could be used to operate:

- refrigerators for vaccine and medicine storage;
- sterilizers for clinical items;
- wastewater treatment units; and
- ice making.

Communications

Communication systems are essential for rural development. The availability of these systems has a great impact on people's lives and can advance their development process more rapidly. Electricity can be generated from any renewable energy source to operate the basic communication needs such as radio,

television, weather information systems and mobile telephone.

Transportation

Improved transportation in rural areas and villages has a positive effect on the economic situation as well as on the social relations between the people of these areas. Several methods could be adapted for this purpose:

- solar electric vehicles;
- ethanol, plant oil fuel and hydrogen vehicles (engines and fuel cells); and
- traction animals.

Food and agriculture

In rural areas agriculture represents a major energy end use. Mechanization using renewable sources of energy can reduce the time spent in labour-intensive processes, freeing time for other income-producing activities. Renewable sources of energy can be applied to:

- soil preparation and harvesting;
- husking and milling of grain;
- crop drying and preservation; and
- textile processing.

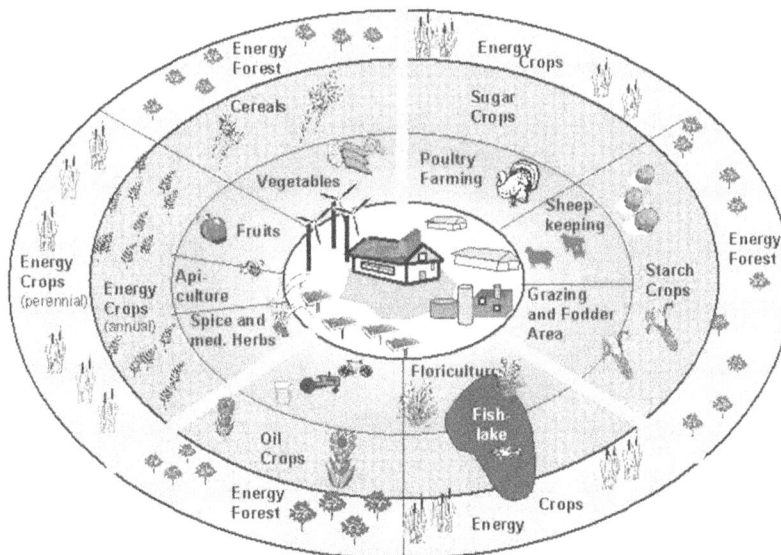

Source: El Bassam (2001a)

Figure 9.2 Basic energy requirements for rural populations

The concept of an integrated energy farm (IEF) or settlement includes four pathways:

1   economic and social pathway;
2   energy pathway;
3   food pathway; and
4   environmental pathway.

Basic data should be collected for the verification of an IEF. Various climatic constraints, water availability, soil conditions, infrastructure, availability of skills and technology, population structure, flora and fauna, common agricultural practices and economic, educational and administrative facilities in the region should be taken into consideration.

It is evident that throughout Europe, wind and biomass energies contribute the major share to the energy-mix, while in North Africa and the Sahara the main emphasis obviously lies with solar and wind energies. Equatorial regions offer great possibilities for solar as well as biomass energies and little share is expected from the wind source of energy in these regions. Under these assumptions, in southern Europe, the equatorial regions and North and Central Europe, a farm area of 4.8, 10 and 12 ha, respectively, would be needed for the cultivation of biomass for energy purposes. This would correspond to annual production of 36, 45 and 60 tonnes for the respective regions. In the North Africa and Sahara

regions, in addition to wind and solar energy, 14 tonnes of biomass from 1.2 per cent of the total area would be necessary for energy provisions. Projection steps are illustrated in Figure 9.4.

Moving ahead, in order to broaden the scope and seek the practical feasibility of such farms, the dependence of local inhabitants (end users) is to be integrated in this system. Roughly 500 persons (125 households) can be integrated in one farm or rural settlement unit. They have to be provided with food as well as energy. As a consequence, the estimated extra requirement of 1900MWh of heat and 600MWh of power has to be supplied from alternative sources. Under the assumption that the share of wind and solar energy in the complete energy provision remains at the same level, the production of 450 tonnes of dry biomass is needed to fulfil the demands of such farm units. For the production of this quantity of biomass, 20 per cent of farm area needs to be dedicated to cultivation. In southern Europe and the equatorial regions, 15 per cent of the land area should be made available for the provision of additional biomass.

Climatic conditions prevailing in a particular region are the major determinants of agricultural production. In addition to that, other factors like local and regional needs, availability of resources and other infrastructure facilities also determine the size and the product spectrum of the farmland. The same requisites also apply to an integrated renewable

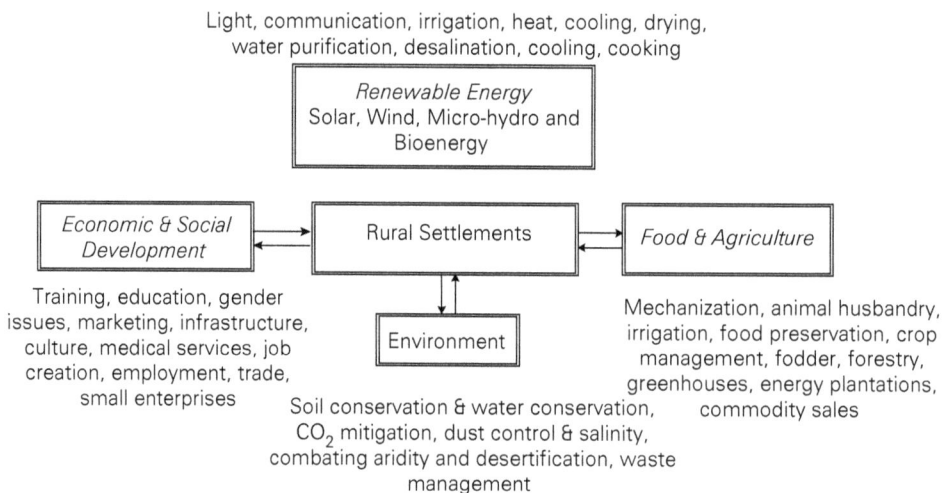

Figure 9.3 Pathways of the integrated energy farms (IEFs)

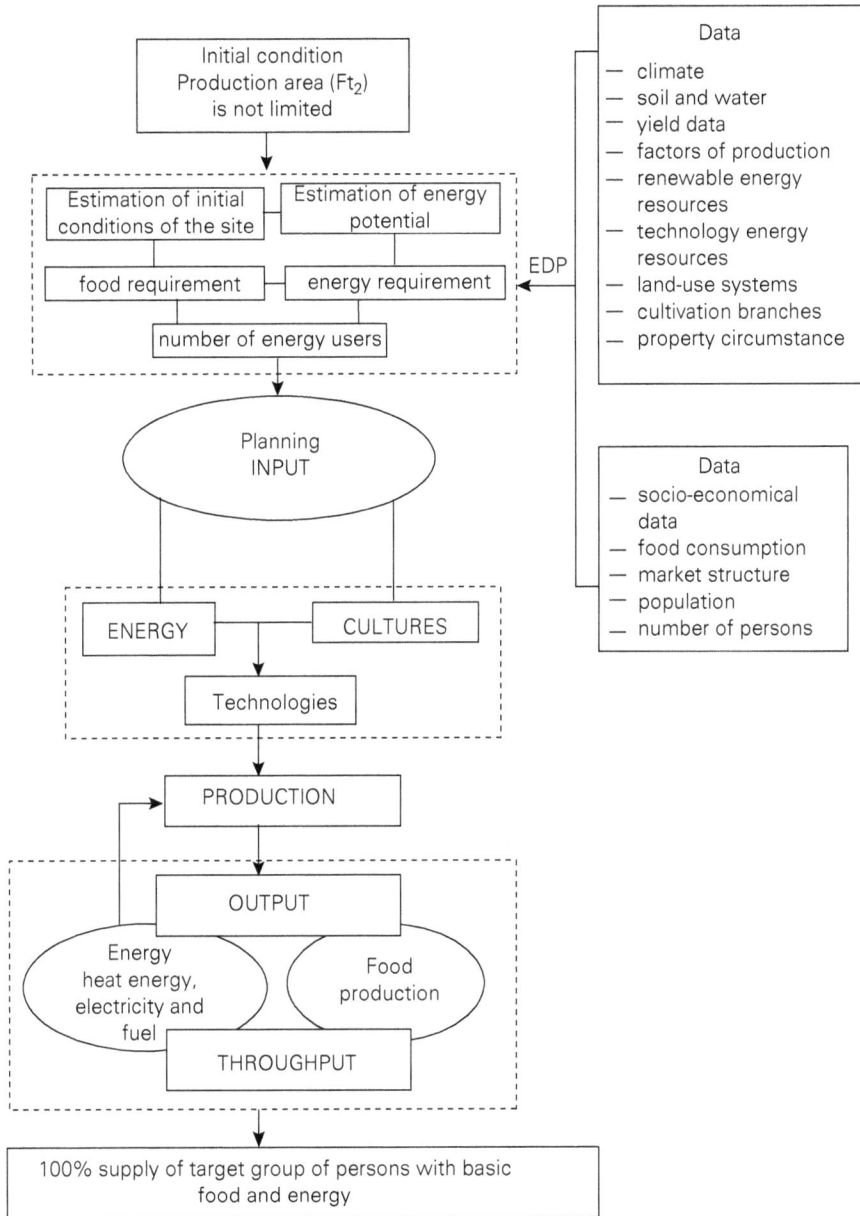

Figure 9.4 Projection steps of IEF

energy farm (IREF). The climate fundamentally determines the selection of plant species and their cultivation intensity for energy production on the farm. Moreover, climate also influences the production of energy-mix (consisting of biomass, wind and solar energies) essential at a given location; and the type of technology that can be installed also depends decisively on the climatic conditions of the locality in question. For example, cultivation of biomass for power generation is not advisable in arid areas. Instead, a larger share can be allocated to solar energy techniques in such areas. Likewise, coastal regions are ideal for wind power installations. More than 50 crops have been identified in different regions of the world to serve as sources for biofuels. Selected crops which can be grown under various climatic conditions are documented in Tables 9.1, 9.2 and 9.3.

Regional implementation

Verifying the implementation of the IREF in a practical sense at regional level, taking into consideration the climatic and soil conditions, planning work has been started at Dedelstorf (northern Germany). An area of 280ha has been earmarked for this farm, which would satisfy the food and energy demands of the 700 participants in the project. For the settlement purposes, old military buildings are being renovated. The main elements of heat and power generation will be solar generators and collectors, a wind generator, a biomass combined heat and power generator, a Stirling motor and a biogas plant.

The total energy to be provided amounts to 8000MWh heat and 2000MWh power energy. The cultivation of food and energy crops will be according to ecological guidelines. The energy plant species foreseen are: short rotation coppice, willow and

Table 9.1 Representative crops for temperate regions

• Cordgrass (*Spartina* spp.)	• Reed canarygrass (*Phalaris arundinacea*)
• Fibre sorghum (*Sorghum bicolor*)	• Rosin weed (*Silphium perfoliatum*)
• Giant knotweed (*Polygonum sachalinensis*)	• Safflower (*Carthamus tinctorius*)
• Hemp (*Cannabis sativa*)	• Soya bean (*Glycine max*)
• Kenaf (*Hibiscus cannabinus*)	• Sugar beet (*Beta vulgaris*)
• Linseed (*Linum usitatissimum*)	• Sunflower (*Helianthus annuus*)
• Miscanthus (*Miscanthus x giganteus*)	• Switchgrass (*Panicum virgatum*)
• Poplar (*Populus* spp.)	• Topinambur (*Helianthus tuberosus*)
• Rape (*Brassica napus*)	• Willow (*Salix* spp.)

Source: El Bassam (1996) and (1998b).

Table 9.2 Representative crops for arid and semi-arid regions

• Argan tree (*Argania spinosa*)	• Olive (*Olea europaea*)
• Broom (ginestra)(*Spartium junceum*)	• Poplar (*Populus* spp.)
• Cardoon (*Cynara cardunculus*)	• Rape (*Brassica Napus*)
• Date palm (*Phoenix dactylifera*)	• Safflower (*Carthamus tinctorius*)
• Eucalyptus (*Eucalyptus* spp.)	• Salicornia (*Salicornia bigelovii*)
• Giant reed (*Arundo donax*)	• Sesbania (*Sesbania* spp.)
• Groundnut (*Arachis hypogaea*)	• Soya bean (*Glycine max*)
• Jojoba (*Simmondsia chinensis*)	• Sweet sorghum (*Sorghum bicolor*)

Source: El Bassam (1996) and (1998b).

Table 9.3 Representative energy crops for tropical and subtropical regions

• Aleman grass (*Echinochloa polystachya*)	• Jatropha (*Jatropha curcas*)
• Babassu palm (*Orbignya oleifera*)	• Jute (*Crocorus* spp.)
• Bamboo (*Bambusa* spp.)	• Leucaena (*Leucaena leucoceohala*)
• Banana (*Musa x paradisiaca*)	• Neem tree (*Azadirachta indica*)
• Black locust (*Robinia pseudoacacia*)	• Oil palm (*Elaeis guineensis*)
• Brown beetle grass (*Leptochloa fusca*)	• Papaya (*Carica papaya*)
• Cassava (*Manihot esculenta*)	• Rubber tree (*Acacia senegal*)
• Castor oil plant (*Ricinus communis*)	• Sisal (*Agave sisalana*)
• Coconut palm (*Cocos nucifera*)	• Sorghum (*Sorghum bicolor*)
• Eucalyptus (*Eucalyptus* spp.)	• Soya bean (*Glycine max*)
	• Sugar cane (*Saccharum officinarum*)

Source: El Bassam (1996) and (1998b).

poplar, miscanthus, polygonum, sweet and fibre sorghum, switchgrass and reed canarygrass, and bamboo. Adequate food and fodder crops as well as animal husbandry is under implementation according to the needs of people and specific environmental conditions of the site. Energy supply will be produced by 90 per cent biomass, 7 per cent wind and 3 per cent solar resources, and the farm will act as a research centre for renewable energies – solar, wind and energy from biomass, as well as their configuration. Special emphasis is dedicated to optimization of energetic autonomy in decentralized living areas and to promote regional resource management (Figure 9.5).

Other IEFs had been projected in Bulgaria, Iran and in an ecological farm in Adolphshof, Germany.

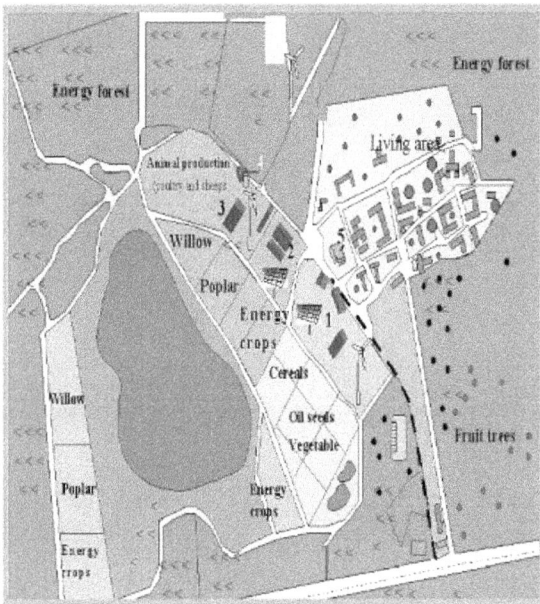

1 Thermal and Power unit (Biomass, Wind, Solar) 2 Pelleting, Oil mill, Ethanol unit
3 Animal husbandry 4 Biogas unit 5 Administration

Source: El Bassam (1999)

Figure 9.5 Project Integrated Renewable Energy Farm, Dedelstorf, Germany

Jühnde bioenergy village

This project was inspired by the IEF concept and other institutions. The village of Jühnde (750 inhabitants) is located in southern Lower Saxony in the middle of Germany. It was started in 2001 to become a 'bioenergy village'. With one-third of funds from the German Ministry BMELV and Lower Saxony it was possible to invest in such a project. The main emphasis is that the whole village is involved. More than 70 per cent of the households are connected to the hot water grid. The aim of the project is to convert biological material into electrical power and heat. A block-type thermal power station (or heat and power generator) run by biogas is now realized. For additional heating during winter a wood-fired heating system is implemented (IZNE, 2007).

The energy production process is as follows: under anaerobic conditions, micro-organisms engage in enzymatic digestion to create biogas. Biogas is obtained during the fermentation process of liquid manure and plant silage in an anaerobic digestion plant. The combustion of biogas in a combined heat and power generator (CHP) generates enough electricity for the entire village.

Biogas also generates heat as a by-product. This heat is mainly used to heat homes and other living spaces, replacing the conventional fossil fuels, oil and natural gas. A smaller portion of the generated heat is required to fuel the digestion process described above. The amount of heat generated cannot cover the high demand during winter months in Germany. During this period, an additional heating plant fuelled with wood chips is required. Rarely, on extremely cold days, peak demand necessitates a further boiler fuelled by oil or biodegradable diesel.

The distribution of heat energy to the 140 households in Jühnde (750 inhabitants) started in 2005. The project has produced in 2008 more than 10 million kWh electricity and is saving 3300 tonnes of CO_2 annually.

Sustainable cultivation of energy crops has been achieved according to the following measures:

- increased crop diversity;
- minimal soil erosion;
- restricted nitrate leaching;
- no pesticide treatments;
- fertilization with digestion residues and ashes (nutrient recycling); and
- high biomass yield/year because of multiple cropping.

Source: Nachwachsende Rohstoffe (p28)

Figure 9.6 Jühnde bioenergy village, Germany, with
the energy generation installations

Climate protection

The bioenergy project is contributing towards
reduction of CO_2 emissions by 3300 tonnes/year =
60 per cent CO_2 reduction/capita/year. It has already
reached the CO_2-reduction aims of the European
Union for 2005.

Internet resources

www.bioenergiedorf.de
www.oecd.org/agr/env
www.nachwachsenderohstoffe.de/presseservice/presse
mitteilungen/archiv/archiv-nachricht/archive/2008/
february/article/wege-zum-bioenergiedorf-1.html?
tx_ttnews[day]=28&cHash=15bb708976

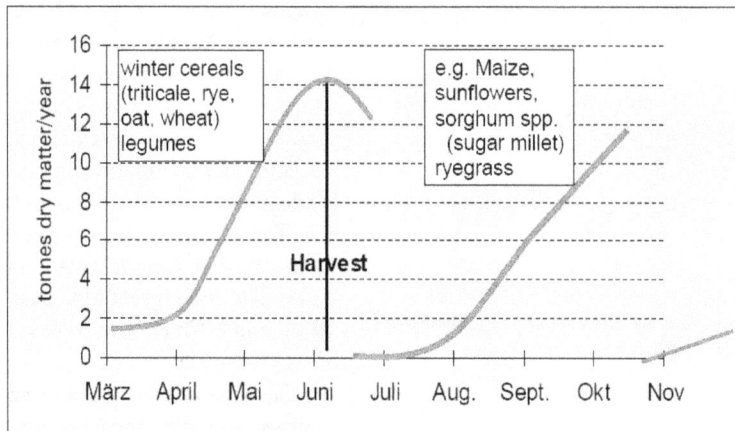

Source: IZNE (2005); Ruwisch and Sauer (2007)

Figure 9.7 Cultivation concept: various crops and double-cropping system

Liquid manure 6 animal farms (pig and cattle)	9000m³
Annual crops cultivated on farmland = 25% of the total farmland	300–330ha
Wood chips from forest = 10% of the annual growth	300 tonnes/year

Figure 9.8 Required biomass quantities

| Current energy con-
sumption in Jühnde
(800 inhabitants) | 2 million kWh electricity
4.5 million kWh heat energy |
| Bioenergy facilities
in Jühnde | • *Anaerobic digestion plant* with a
700kW$_{el}$ power station ~ 5 million kWh
• *Central heating plant* fed by wood chips:
550kW
• *2 heat storage facilities*
• *Boiler for peak load* in winter time (fuel
oil): 1600kW |

Figure 9.9 Heat and electricity consumption and
production of the village

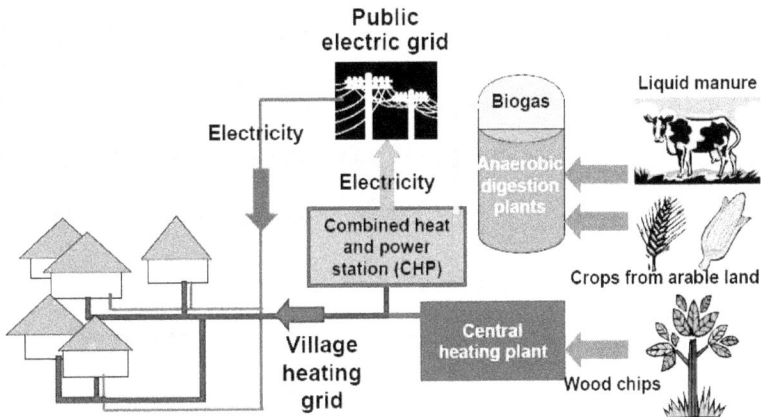

Figure 9.10 Technical concept of the project

Part II

The greatest service which can be rendered any country
is to add a useful plant to its culture

Thomas Jefferson (1821)

10
Energy Crops Guide

SCOPE AND DEFINITION

Scope

Soaring prices of fossil fuels, geopolitical issues and environmental pollution associated with fossil fuel use has led to worldwide interest in the production and use of biofuels. Both the developed and developing countries have generated a range of policies to encourage production of combustible fuels from plants, which triggered public and private investments in biofuel crop R&D and biofuels production (Reddy et al, 2008). The UN Millennium Development Goals (MDGs) provide a blueprint for improving livelihoods (by alleviating poverty), and preserving natural resources and the environment, with 2015 as target date. Though energy is the fuel of economic prosperity that is essential for alleviating poverty, nonetheless, diversifying crop uses, and identifying and introducing biofuel crops would lead to enhanced farmers' incomes, thereby contributing to eradicating extreme poverty (MDG 1) in rural areas. 'Energizing' the agriculture production chain is critical to achieving food security, considering the strong correlation between per capita energy consumption and crop yields in both developed and developing countries. Energy feedstocks produced from selected crops, among other renewable sources, provide sustainable and eco-friendly energy options that foster environmental sustainability (MDG 7) and offer opportunities to improve the income level of the developing world's smallholder subsistence farmers, who depend on agriculture for their livelihoods. However, not all crops offer equal environmental advantages. The crop, cultivar, production system and the processing technology are critical. Bioenergy R&D will lead to new local, regional and national public-private partnerships for development (MDG 8) (Reddy et al, 2008).

This part concentrates on options of crops and R&D interventions required to generate feedstocks to produce biofuels to meet projected demand without compromising food/fodder security.

Definition

Biomass

Biomass is the name given to any recent organic matter that has been derived from plants as a result of the photosynthetic conversion process. Biomass energy is derived from plant and animal material, such as wood from forests, residues from agricultural and forestry processes, and industrial, human or animal wastes. Biomass is not fossilized material (like oil, coal and gas) but fresh material that can grow again after having been harvested. During growth, plants use atmospheric carbon dioxide (CO_2) to build up their substance – if growth and use are balanced the use of biomass is carbon-neutral.

The energy value of biomass from plant matter originally comes from solar energy through photosynthesis. The chemical energy that is stored in plants (and in animals because they eat plants or other animals), or in the wastes that they produce, is known as bioenergy. When burning, biomass produces energy in the form of heat, and the carbon is reoxidized to carbon dioxide and released back into the atmosphere.

Of all the renewable sources of energy, biomass is unique in that it is effectively stored solar energy. Furthermore, it is the only renewable source of carbon, and is able to be processed into convenient solid, liquid and gaseous fuels.

Resources

Biomass resources that can be used for energy production cover a wide range of materials. The use of biomass energy can be separated into two categories:

Traditional biomass is generally confined to developing countries and small-scale uses. It includes fuelwood and charcoal for domestic use, rice husks, other plant residues and animal dung.

Modern biomass usually involves large-scale uses and is a substitute for conventional fossil fuel energy sources. It includes forest wood and agricultural residues, urban wastes, and biogas and biofuels from energy crops (such as plant oils and plants containing starch and sugar).

Applications

Bioenergy applications are extremely diverse, including heat supply, power generation and transport fuels. Biomass can be used directly (e.g. burning wood for heating and cooking) or indirectly by converting it into a liquid or gaseous fuel (e.g. ethanol from sugar crops or biogas from animal waste).

Traditional biomass use in open hearths for cooking and heating continues to be very important in developing countries due to a lack of alternatives. Elsewhere, burning wood in small systems such as stoves or open chimneys for heating has a long tradition. Processed wood wastes in the form of pellets or chips are now used in innovative heating systems.

Modern biomass is used to produce power and heat in large-scale facilities: solid biomass such as wood discards, yard wastes and straw can be burned in specially designed power plants, or together with coal in existing coal-fired plants. Biogas can be extracted in special facilities from, for instance, agricultural residues such as slurry. It can either be fed into gas networks, or used to produce power and heat.

Biofuel or biodiesel are generated from plant oils extracted from plants such as rape, sunflower or oil palm. Plants containing starch and sugar (such as potatoes and sugar beet) yield bioethanol after fermentation. Solid biomass can also be processed to make hydrogen or methanol.

Biomass use for energy can make a major contribution to climate protection and resource conservation, regardless of whether wastes or specially cultivated crops are used. It can also be used in all applications – power, heat and fuels. Using biogas, digester gas and landfill gas for energy is beneficial for the climate, but also presents opportunities for farmers who can enhance the value of their slurry and reduce odour emissions.

However, bioenergy production can have negative environmental impacts such as acidification, eutrophication or summer smog. The production of energy crops can also have negative impacts depending on what agricultural or forestry methods are used.

In each application, standards must therefore be set for the best possible methods of production – both for sustainable management of the landscapes and for maximum energy gain. The World Wide Fund for Nature is working across its own programmes – climate and energy, forestry, agriculture and fresh water – to determine such standards.

Bioenergy crops

Vegetative biomass consists of living plant species all around us. As they grow, plants store the sun's energy in their leaves, stems, bark, fruits, seeds and roots. Bioenergy crops are so diverse that they grow in virtually every part of the world. Energy crop species are understood to mean those annual and perennial species that can be cultivated to produce solid, liquid or gaseous energy feedstocks. The organic residues and wastes from the most widely diverse types of plant production, also used for producing energy, do not fall under this term but nevertheless represent a large potential.

All plant species that store primarily carbohydrates or oils are suitable for producing liquid energy sources. Cellulose, starch, sugar and inulin can be used to produce ethanol. Vegetable oils can be used as fuels. Parts of plants containing lignocellulose can provide energy directly as solid fuels or indirectly after conversion.

Obtaining alcohol from vegetable raw materials has a long tradition in agriculture. From an agricultural point of view, the species available for ethanol production are starch plant species, and cereals including maize and grain sorghum, as well as potato, topinambur, and the sugar crops (sugar beet, root chicory, sweet sorghum and sugar cane). Most of the crop species presently used to produce ethanol are given in Table 10.1. The goal-directed use of cellulose-containing biomass from agricultural crops for producing alcohol has not been practised on a large-scale engineering level, but the potential for the future is very large. According to a communication from Ingram et al (1995), the entire gasoline needs of the USA could be substituted in this way if systems developed at the laboratory scale could be implemented at large technical production scales.

Oil crops are well distributed throughout the world from north to south, but only a few of them have a high oil yield per unit area (tonnes per hectare) and this is a disadvantage in comparison to other fuel feedstocks like ethanol or solid biofuel. The main oil-yielding crops are rape, sunflower and soya bean. Table 10.2 lists several oil crops and their potentials. Most of them need further improvements in breeding and agricultural practices to improve the yield.

Table 10.1 Productivity of ethanol crops

Plant species	Yield (t/ha)	Sugar/starch (% FS)	Contents yield (t/ha)	Ethanol efficiency (l/ha)
Barley (*Hordeum vulgare*)	5.8	58.0	3.36	2150
Cassava, manioc (*Manihot esculenta*)	9.0	35.0	3.15	2900
Fodder beet (*Beta vulgaris* var. *rapacea*)	98.5	8.2	8.08	4923
Maize (*Zea mays*)	6.9	65.0	4.49	2874
Potato (*Solanum tuberosum*)	32.4	17.8	5.77	3693
Root chicory (*Cichorium intybus*)	35.0	16.0	5.60	3248
Sugar beet (*Beta vulgaris* var. *altissima*)	57.4	16.0	9.18	5600
Sugar cane (*Saccharum officinarum*)	80.0	10.0	8.00	5400
Sweet potato (*Ipomoea batatas*)	12.0	25.0	3.00	2400
Sweet sorghum (*Sorghum bicolor*)	90.0	10.0	9.00	5400
Topinambur (*Helianthus tuberosus*)	30.0	15.0	4.50	2610
Wheat (*Triticum aestivum*)	7.2	62.0	4.46	2854

Table 10.2 Productivity of oil crops

Common name	Botanical name	Seed yield (t/ha)	Oil content (%)	Oil yield (t/ha)
Abyssinian kale	*Crambe abyssinica*	2.0–3.5	30–45	0.74
Abyssinian mustard	*Brassica carinata*	1.1–3.0	23–40	0.26–1.2
Bird rape	*B. rapa* ssp. *oleifera*	1.0–2.5	38–48	0.38–1.2
Black mustard	*B. nigra*	0.5–2.0	24–38	0.12–0.76
Brown (Indian) mustard	*B. juncea*	1.5–3.3	30–40	0.72
Castor	*Ricinus communis*	1.2	50	0.60
Coconut palm	*Cocos nucifera*	4.17	36	1.5
Coriander	*Coriandrum sativum*	2.0–3.0	18–22	0.4–0.7
Cotton	*Gossypium* spp.	1.2	15–25	0.29
Dill	*Anethum graveolens*	1.0–1.5	16–20	0.16–0.30
Fennel	*Foeniculum vulgare*	1.0–2.0	10–12	0.08
Flax, linseed	*Linum usitatissimum*	1.8	30–48	0.70
Gold of pleasure	*Camelina sativa*	2.25	33–42	0.88
Groundnut	*Arachis hypogaea*	2.0	45–53	1.00
Hemp	*Cannabis sativa*	0.5–2.0	28–35	0.14–0.7
Jojoba	*Simmondsia chinensis*	2.1	48–56	1.01–1.18
Meadowfoam	*Limnanthes alba*	1.2	20–30	0.34
Nigerseed, ramtil	*Guizotia abyssinica*		35–45	
Oil palm	*Elaeis guineensis*	30	26	7.8
Oil radish	*Raphanus sativus* var. *oleiformis*	0.7–1.1	38–50	0.27–0.55
Oil squash	*Cucurbita pepo*	0.8–1.6	40–58	0.32–0.93
Olive	*Oleo* spp.	1–12.5	40	0.4–5.0
Opium poppy	*Papaver somniferum*	1.0–1.8	40–55	0.4–1.0
Penny flower	*Lunaria* spp.		30–40	
Pot marigold	*Calendula officinalis*	1.5	18–20	0.29
Rape	*Brassica napus* spp. *oleifera*	2–3.5	40–50	1.26
Rocket	*Eruca sativa*	0.9	24–3.5	0.27
Safflower	*Carthamus tinctorius*	1.8	18–50	0.63
Sesame	*Sesamum indicum*	0.5	50–60	0.25
Soya bean	*Glycine max*	2.1	18–24	0.38
Spurge	*Euphorbia lathyris*	1.5	48	0.72
Spurge	*Euphorbia lagascae*	0.6	46	0.28
Sunflower	*Helianthus annuus*	2.5–3.2	35–52	0.88–1.67
White mustard	*Sinapis alba* ssp. *alba*	1.5–2.5	22–42	0.64

Energy plant species

The number of plant species that can be utilized as solid biofuels is much higher than those usable for ethanol and oil production. The level of production of these crops is, besides the genetic potential, largely influenced by the availability of water and other external inputs, but the competition for available water between food crops and energy crops increases from northern to southern regions.

Depending on the time and methods of harvest and utilization, and the prevailing economics, energy plant species could include roots, tubers, stems, branches, leaves, fruits and seeds, or even whole plants.

The main goals of growing crops on plantations dedicated to energy crops can be summarized as follows:

- The growing of starch and sugar plant species to produce ethanol.
- The cultivation of oil crops as sources for biodiesel.
- The production of solid biomass to obtain heat and electricity, either directly through combustion or indirectly through conversion for use as fuels. Lignocellulose-rich raw materials can be used to produce fuels like methanol, biodiesel, synthetic gas and hydrogen (using thermal and thermochemical processes, by direct or indirect liquefaction or gasification) and ethanol (through hydrolysis and subsequent fermentation to produce).
- The cultivation of biomass to produce biogas.

The crop species discussed in this publication are not in all cases described in a uniform structure. This is due to the varied and unbalanced level of available information pertaining to the various crops.

Special emphasis will be placed on aspects related to identification of high-yielding energy plant species in which the production system should be sustainable and environmentally acceptable.

ALEMAN GRASS (CARIB GRASS) (*Echinochloa polystachya* (H. B. K.) Hitchc.)

Description

Echinochloa is a C_4 plant that occurs from Mexico to Argentina. It forms large monotypic stands on the

Source: Cyril Crusson, Plantes des rizières de Guyane, http://plantes-rizieres-guyane.cirad.fr/monocotyledones/poaceae/echinochloa_polystachya

Figure 10.1 Aleman grass

fertile floodplains along the white waters of the Amazon region. Each plant has a single unbranched stem, which varies little in diameter or bulk density along its length (Figure 10.1). The stem consists of internodes, which at maturity are of roughly equal length. It is a perennial grass that forms large monotypic stands on the fertile floodplains of white water rivers and lakes (white water because of the light coloured suspended sediment). In the wet tropics the C_3 plants normally dominate, but not in some communities on river and lake margins. There, monotypic stands of C_4 species are frequent, like *Cyperus papyrus* in Africa, and *Paspalum regens, Paspalum fasciculatum* and *Echinochloa polystachya* in South America.

The C_4 grass *Echinochloa polystachya* forms dense and extensive monotypic stands on the Varzea floodplains of the Amazon region and provides the most productive natural higher plant communities known. The seasonal cycle of growth of this plant is closely linked to the annual rise and fall of water level over the floodplain surface (Piedade et al, 1994).

Ecological requirements

The life cycle of plants that grow on the floodplains is regulated by the annual oscillation of the water level. The variation in water level shows a mean amplitude of 10m, but this may range from 6m in some years to 14m in others. *Echinochloa* is not a floating plant. The roots are fixed in the soil and the leaves reach out of the water. The stem has to be rigid and long enough to withstand the flowing and rising water. The C_4 plants benefit from their higher productivity: they can invest more assimilates into the stem. From November onwards the new plants grow steadily in length at approximately 1.1m per month. The crown reaches 1–2m outside of the water if the stems stand vertically. The increase in stem length is achieved by successive addition of internodes at a rate of seven per month.

When the water level drops, exposing the sediment surface in October, new shoots form at the nodes of the old stems, and root in the sediment. The old stems die and rot away, and each of the shoots becomes an individual plant. Each new plant normally remains as a single unbranched shoot. As the water level rises to cover the sediment in late November or December, the stems grow upwards, keeping pace with the steady rise

in water level. As each node becomes submerged its leaves die and adventitious roots form. If the water is highly turbid, submerged leaves cannot get enough light for photosynthesis. Photosynthesis is only possible if the leaves form a canopy above the water surface. The longevity of the leaves is only 34 ± 4 days. The leaves die and decay rapidly on submersion by the rising water.

New leaves are formed at the top of the stems with sufficient rapidity to maintain a dense canopy. As the water level drops the stems become increasingly exposed and bent, eventually collapsing onto the re-exposed sediment surface in October.

Propagation

The stands are perpetuated by vegetative propagation, but micropropagation is also feasible.

Production

The maximum standing of the crop at 80t/ha was determined at a site in central Amazon. The mean rate of dry matter production per unit of solar radiation intercepted by the stand was 2.3g/MJ, which is close to the considered maximum for C_4 plants.

From November on, the biomass increases steadily to peak in September with 6880g/m², or 8000g/m² if attached dead material is included. In the nine-month growth period from the December harvest until the September harvest, the total dry mass increases by 7000 ± 35g/m² with a daily growth rate of 25.9 ± 0.1g/m². The peak dry mass of 8000g/m² does not represent the total net primary production. If the dead leaves are taken into account 9930 ± 670g/m² are obtained, of which 9420 ± 660g/m² are constituted by the shoots. The accumulated net primary production rose steadily in each month from November, suggesting a complete absence of seasonal limitation of productivity.

The cultivation of this plant species for several years on dry land (in Germany as a perennial crop) showed that a biomass yield of up to 20 tonnes dry matter per ha is possible (El Bassam, 2001b).

Engle et al (2008) investigated whether rates of net primary production (NPP) and biomass turnover of floating grasses in a central Amazon floodplain lake (Lake Calado) are consistent with published evidence that CO_2 emissions from Amazon rivers and floodplains are largely supplied by carbon from C_4

plants. Ground-based measurements of species composition, plant growth rates, plant densities, and aerial biomass were combined with low altitude videography to estimate community NPP and compare expected versus observed biomass at monthly intervals during the aquatic growth phase (January to August). Principal species at the site were *Oryza perennis* (a C_3 grass), *Echinochloa polystachya* and *Paspalum repens* (both C_4 grasses). Monthly mean daily NPP of the mixed species community varied from 50 to 96g dry mass/m²/day, with a seasonal average (± 1SD) of 64 ± 12g dry mass/m²/day. Mean daily NPP (± 1SE) for *P. repens* and *E. polystachya* was 77 ± 3 and 34 ± 2g dry mass/m²/day, respectively. Monthly loss rates of combined above- and below-water biomass ranged from 31 to 75 per cent, and averaged 49 per cent. Organic carbon losses from aquatic grasses ranged from 30 to 34g C/m²/day from February to August. A regional extrapolation indicated that respiration of this carbon potentially accounts for about half (46 per cent) of annual CO_2 emissions from surface waters in the central Amazon, or about 44 per cent of gaseous carbon emissions, if methane flux is included.

Processing and utilization

Aleman grass is capable of very high productivity levels, up to 100 tonnes dry matter per hectare, which exceeds the productivity of almost all other plant species. The biomass is currently being used as an energy source for cooking. There are a wide range of possibilities to upgrade the biomass to produce solid biofuels and to convert them to ethanol, charcoal or bio-oil through hydrolysis, fermentation or pyrolysis.

Internet resources

http://plantes-rizieres-guyane.cirad.fr/

ALFALFA (*Medicago sativa L.*)

Description

Alfalfa (*Medicago sativa* L.), has the potential of being a leading crop in the production of cellulosic ethanol. Because of its high biomass production, perennial nature, its symbiotic relationship with specific bacteria for producing its own nitrogen and valuable co-products,

more alfalfa acres for biofuel production would reduce greenhouse gas emissions, protect water quality and improve the soil as a resource (Boateng et al, 2006).

Alfalfa, which originated near Iran, also has related forms and species found growing wild scattered over central Asia and into Siberia. Nowadays, alfalfa is grown worldwide (IFFS, 1995) and is one of the most important fodder crops. Alfalfa is a perennial herbaceous legume, reaching a height of 30–90cm. The plant is upright or ascending and highly branched. The flowers are bluish-purple to yellow in colour. They are a source for honey production (FAO, 1996a). Normally the leaves are trifoliate, but leaves with more than three leaflets are not uncommon. The content of proteins and the production of protein for a given area is higher than for non-legumes (Franke, 1985).

Ecological requirements

A wide range of soil and climatic conditions are suitable for alfalfa, but for good production certain conditions have to be fulfilled. The optimum temperature for growing is about 25°C. However, alfalfa has survived summer temperatures of 50°C in California. In the winter, alfalfa can resist temperatures down to –25°C: although the top growth may be killed by the frost, the plant will regenerate from the roots. Alfalfa needs an annual precipitation of 600 to 1200mm for optimal growth; 350mm is the minimum. The plant can be found at elevations of 4000m in Bolivia, but normally it is grown up to 2400m. The pH value of the soil should be nearly neutral, in the range 6.5 to 7.5. Alfalfa does not tolerate an acid soil, especially in the seedling state. A well-drained, deep and light soil is preferred. Because of the high production of alfalfa the soil fertility must be high. In warm climates, humidity should be low because hot and humid conditions promote diseases (FAO, 1996a).

Propagation

Alfalfa is propagated by seed. The seed is sown not deeper than 6–12mm into a firmly packed seedbed (Huffaker, 1994). Alfalfa can be sown either in the early spring or in late summer and autumn. The time of alfalfa seeding is influenced by precipitation, temperature, and cropping patterns. Spring seeding allows the first harvest during the seeding year, but weed control is usually required. Competition from weeds and unfavourable summer temperatures and moisture conditions can be avoided if alfalfa is sown in the late summer or autumn. The remaining time must allow adequate seedling development prior to the onset of winter to minimize loss of stand from winter injury. Spring seeding should be made early enough to allow the formation of good root systems before high temperature and low moisture conditions slow the growth rates. The seed must be in contact with moist soil. The seedlings are unable to emerge from the soil if they are planted too deep. The autotetraploid alfalfa originated from a diploid species, native to an area south of the Black and Caspian Seas (FAO, 1996b).

Many cultivars are available, offering specific characteristics for climatic and soil conditions and disease problems. The seedling forms a taproot that is later used for the storage of nutrients (IFFS, 1995). This taproot enables the plant to survive drought conditions, because it allows the plant to get water from deep soil layers.

Selection for disease resistance is important in alfalfa breeding. For alfalfa intended for use as an energy crop, specially adapted cultivars have to be developed. The proportion of the stems relative to the whole crop should be higher than for the fodder type (Elbersen, 1998).

Colombia Basin College (CBC) has begun an alfalfa breeding program for both standard varieties for dairy forage and germplasm for biomass types. One advanced variety was harvested for foundation seed production in 2007 while breeder seed was produced on eight more experimental varieties. Selections for nonlodging and biomass yield were made in the field and crossed in the greenhouse. Seven thousand progeny from this cross were planted in a spaced plant nursery. Selections for tall, upright growth habit suitable for infrequent cutting management were made from the nursery and are being crossed for potential new varieties targeted as feedstock for cellulosic ethanol production. Alfalfa varieties for the Columbia Basin, regardless of their use, will be selected for resistance to blue aphid, clover root curculio, *Verticillium* wilt, stem nematode and root knot nematode to enhance persistence (CBC, no date).

Crop management

A seeding rate of 13–17kg/ha planted in 15cm rows is usually sufficient for a good stand. Alfalfa is a perennial crop, lasting up to 6 years, but is usually used for 3–4 years. With increasing stand age the yield and stand density usually decrease due to competition between the plants, weed incursion, diseases and poor harvest management (Sheaffer et al, 1988). The growing period depends on the climate and ranges from 100 to 365 days per year. The seed needs to be inoculated with an effective strain of *Rhizobium meliloti* that requires a soil pH of at least 6.0. Lime-coated, inoculated seeds have been used with success on more acid soils (FAO, 1996a).

Alfalfa lives, like other legumes, in symbiosis with soil bacteria *(Rhizobium meliloti)* that are able to fix nitrogen from the air. The bacteria may be present in the soil, but to be sure that the plants are provided with an effective strain, the seed should be inoculated with a fresh (commercial) inoculum immediately prior to seeding. The freshness of the inoculum is more important than the rate. A complete fertilizer programme is essential to a long-lived stand. If the pH value of the soil is below 6.2, lime application is necessary. Nitrogen fertilizer is not required because of the fixation of atmospheric nitrogen by the bacteria in the nodules. Addition of nitrogen fertilizer will reduce the effectiveness of the natural nitrogen fixing mechanism (IFFS, 1995). Potassium (K), phosphorus (P) and boron (B) are most often limiting for alfalfa production (Barnes and Sheaffer, 1995). Potassium is especially important for winter survival and particularly for nitrogen fixation (Duke et al, 1980). In symbiosis with *Rhizobium meliloti*, the N_2 fixation ranges from 50 to 463kg/ha with an average of 200kg/ha (Vance et al, 1988).

More than 20 diseases are serious problems for alfalfa (the examples below are from the USA). These include fungal and bacterial wilts, leaf spots, crown and root rots, viruses and nematodes. Important wilts are bacterial wilt (*Corynebacterium insidiosum* (McCull.) H. L. Jens), fusarium wilt (*Fusarium oxysporum* Schlecht. f. sp. *medicaginis* (Weimer) Snyd. & Hans.), and verticillium wilt (*Verticillium alboatrum* Reinke & Berth). The most serious leaf spots are common leaf spot (*Pseudopeziza medicaginis* (Lib.) Sacc.), lepto leaf spot (*Leptosphaerulina briosianna* (Pollacci) J. H. Graham & Luttrell), stemphylium leaf spot (*Stemphylium botryosum* Wallr.), and summer blackstem (*Cercospora medicaginis* Ellis & Everh). Important crown and root rots include anthracnose (*Colletotrichum trifolii* Bain & Essary), *Aphanomyces* spp. root rot, spring blackstem (*Phoma medicaginis* Malbr. & Roum. var. *medicaginis*), phytophthora root rot, (*Phytophthora megasperma* Drechs.), rhizoctonia diseases (*Rhizoctonia solani* Kuehn.), and sclerotinia crown and stem rot (*Sclerotinia trifoliorum* sensu Kohn). Alfalfa mosaic (alfalfa mosaic virus complex), is the primary virus disease. Alfalfa stem nematode (*Ditylenchus dipsaci* (Kuhn) Filpjev), root-knot nematodes (*Meloidogyne* spp.), and root-lesion nematodes (*Pratylenchus* spp.) are the most prevalent nematode species on alfalfa. Resistant cultivars are available for most of the diseases and nematodes.

There are a number of insect pests on alfalfa (again the examples are from the USA). The insect pests that interfere with forage production include the potato leafhopper (*Empoasca fabae* (Harris)), the alfalfa weevil (*Hypera postica* (Gyll.)), the spotted alfalfa aphid, the pea aphid (*A. kondoi* Shinji), the alfalfa plant bug (*Adelphocoris lineolatus* (Goeze)), and the meadow spittlebug (*Philaenus spumarius* L.). The potato leafhopper is the most problematic pest and causes damage throughout most alfalfa-producing areas in the eastern and central USA. It causes yellowing of the foliage and stunting of stems. The damage results in significant losses in yield and forage quality, especially loss in carotene (IFFS, 1995).

Biological control of insect pests is possible in some cases. The aphids can be suppressed with significant success by the ladybird. There are experiments to control the alfalfa weevil with parasitic wasps (Bambara, 1994).

Alfalfa is an integral component of many crop rotations because of its ability to fix nitrogen, improve the soil structure and tilth, and to control weeds in subsequent crops (IFFS, 1995). The crop can be used as green manure to provide the soil with nitrogen and organic material. It can provide winter wheat as the following crop with 80–100 per cent of its needed nitrogen. If alfalfa is re-established immediately after a previous alfalfa crop, or if old alfalfa stands are thickened, autotoxicity effects are sometimes observed, in which case the alfalfa shows lower germination, lower production and poorer establishment. These effects are attributed to plant exudates and decomposition products. To avoid the effects, an interval of at least two or three weeks between ploughing or herbicide killing, respectively, and seeding, is necessary (IFFS, 1995).

The cropping plan for the supply of a planned biomass power plant in Minnesota is a seven-year cycle with 4–2–1 years of alfalfa, corn and soya beans.

Source: Gary D. Robson 2006 http://commons.wikimedia.org/wiki/File:Alfalfa_round_bales.jpg

Figure 10.2 Alfalfa field

Immediate environmental benefits as a result of this include increased yield from other rotation crops, reduced external inputs of nitrogen, reduced potential for nitrate leaching, reduced soil erosion, and improved soil tilth (Downing et al, 1996)

Production

Alfalfa stands can be cut several times per year. The dry matter yields maybe 10–20t/ha (FAO, 1996a). Production decreases with increasing age of the stand.

Processing and utilization

Alfalfa has the potential to be a significant contributor to a renewable energy future. In an alfalfa biomass energy production system, alfalfa forage would be separated into stems and leaves. The stems would be processed to produce liquid fuel (ethanol), and the leaves would be sold separately as a livestock feed. One of the advantages of alfalfa over other crops to produce biomass energy is this potential for a secondary income selling the leaves as an animal feed. Therefore, concentrations and yields of leaf protein and stem cell wall sugars (used to produce ethanol) are key plant traits in new alfalfa varieties developed for use in biofuel production systems (Lamb et al, 2007).

Alfalfa is not only usable fresh; it can be dried to make alfalfa hay or processed into silage, pellets or meal. The hay is pressed into bales and can be stored. The time of cutting determines the features of the hay.

If cut in the early bud stage, the plants have a high proportion of leaves that are rich in protein. If cut when the plants are mature, the stems constitute a larger proportion of the plant, the protein content decreases, and the fibre content increases; the total yield is also higher. The timing of mowing the stand has a profound influence on the productivity and the life of the alfalfa stand (Figure 10.3). The large taproot is a storage organ for food reserves that are needed to regenerate the plant in the spring or after cutting.

The maintenance of food reserves is necessary to keep the stand vigorous and productive. The quantity of reserves increases if the interval between the cuttings is prolonged to 35 days, after which it declines. Cutting more frequently than 28 days or continuous grazing will weaken the stand (IFFS, 1995).

Alfalfa is usually grown as a forage crop. It is rich in proteins, minerals and vitamins. The plant can be used fresh or dried to alfalfa hay. Research is under way to use alfalfa not only as a fodder plant, but also for energy purposes – for example, by using the protein-rich leaves as fodder and the coarse, fibrous stems as an energy source. In Minnesota, USA, the Northern States Power Company (NSP) had plans in 1996 to build a 75MW power plant that runs on biofuel made from alfalfa stems. This procedure takes advantage of the different characteristics of the plant fractions, because the stems are relatively low in nutrient value, whereas products made of the protein-rich leaf material have numerous market opportunities. If the stems are used for energy purposes, their value is higher than if they are used as fodder. After gasification, the gas drives a 50MW combustion turbine and the heat is used to produce superheated high pressure steam that drives a 29MW steam turbine. The net output of the plant is 75MW, and the plant efficiency is 40 per cent. The direct combustion of the stems is also possible and would make the construction of the power plant cheaper, but the efficiency would then be lower (Campbell, 1996).

The plant is dried before the stems and leaves are separated. The separation of leaf and fibrous material not only raises the value of the fodder but also removes the fuel-bond nitrogen. This lowers NO_x production in the turbine. Calculated energy balances have provided total system efficiency for the production of leaf meal and electricity from alfalfa of 1:3 – that is, for each unit of energy input, the system produces three units of energy output (Downing et al, 1996).

When alfalfa is grown strictly for biomass with two to three cuts, rather than the standard practice of four to five as cuts for producing dairy hay, total yield of alfalfa can increase as much as 42 per cent, and potential ethanol yield from stems doubles.

To maximize energy yield by reducing transportation costs, cellulosic biomass production and processing from alfalfa will need to be local. The fit is natural for the Columbia Basin when considering the high yield production by local growers. When located within 15 miles from a processing facility, the efficiency of energy production by alfalfa is two to three times that of corn grain or soya beans (Lamb et al, 2007).

Source: Moon (1997); photo: NSP

Figure 10.3 Alfalfa plantation, USA: leaves for livestock feed and stems for power

Internet resources

www.columbiabasin.edu
http://sunsite.unc.edu/london/orgfarm/biocontrol/
http://sunsite.unc.edu/london/orgfarm/cover-crops/
www.ussoil.net/

ALGAE (*Oleaginous* spp.)

Partially contributed by
P. Goosse and M. Satin

Every drop of oil on Earth comes from millions of years of build-up from algae and other natural residue – buried, compressed and eventually drilled – supplying our energy since the late 1800s. Now, consider that we're going to deplete, in less than 300 years, what took hundreds of millions of years to form. And with the inevitable global depletion of oil, alternative forms of energy are destined to emerge (Kertz, 2009).

Algae are plants that can absorb solar energy and carbon dioxide from the atmosphere, which they store as biomass. Globally, research is under way on microalgal cultivation in various laboratories at different scales. Microalgal biomass production has great potential to contribute to world energy supplies, and to control CO_2 emissions as the demand for energy increases. This technology makes productive use of arid and semi-arid lands and highly saline water, resources that are not suitable for agriculture and other biomass technologies (Brown, 1996).

Description

Microalgae comprise a vast group of photosynthetic, heterotrophic organisms which have an extraordinary potential for cultivation as energy crops. They can be cultivated under difficult agroclimatic conditions and are able to produce a wide range of commercially interesting by-products such as fats, oils, sugars and functional bioactive compounds. As a group, they are of particular interest in the development of future renewable energy scenarios. Certain microalgae are effective in the production of hydrogen and oxygen through the process of biophotolysis, while others naturally manufacture hydrocarbons suitable for direct use as high-energy liquid fuels. It is this last class that forms the subject here.

Once algae species have been grown, their harvesting and transportation costs are lower than with conventional crops, and their small size allows for a range of cost-effective processing options. They are easily studied under laboratory conditions and can effectively incorporate stable isotopes into their biomass, thus allowing effective genetic and metabolic research to be carried out over a much shorter period than with conventional plants.

Eukaryotic microalgae also represent a promising alternative renewable source of feedstock for biofuel production. With over 40,000 identified species, microalgae are one of the more diverse groups of organisms on Earth. They naturally produce large quantities of many biomaterials, including lipids/oil. Nature has had about 4 billion years to engineer strains with unique abilities to grow robustly in diverse environmental conditions and evolve unique metabolic characteristics such as intracellular lipid storage, with a growth potential an order of magnitude greater than terrestrial crop plants due to their extraordinarily efficient light and nutrient utilization. The exploitation of naturally occurring photosynthetic microalgae, collected and isolated over the past 25 years for production of renewable liquid fuels, initially in collaboration with the Aquatic Species Program of the DOE, and subsequently at Arizona State University, provides a green and renewable resource of feedstock biomass to meet increasing energy needs and especially the demand for liquid fuels.

Ecological requirements

Microalgae represent an immense range of genetic diversity and can exist as unicells, colonies and extended filaments. They are ubiquitously distributed throughout the biosphere and grow under the widest possible variety of conditions. Microalgae can be cultivated under aqueous conditions ranging from fresh water to situations of extreme salinity. They live in moist, black earth, in the desert sands and in all the conditions in between. Microalgae have been found living in clouds and have long been known to be essential components of coral reefs. This wide span of ecological requirements plays a significant role in determining the range of metabolic products they produce.

Propagation

Microalgae can be grown both in open culture systems such as ponds, lakes and canals, and in highly controlled closed culture systems similar to those used in commercial fermentation processes. Certain microalgae are very suitable for open system culture where the environmental conditions are very specific, such as high-salt or high-alkaline ponds, lakes or lagoons. The extreme nature of these environments severely limits the growth of competitive species, though other types of organisms may contaminate the culture. The advantages of such systems are that they generally require low investment, and are very cost-effective and easy to manage. Closed culture systems, on the other hand, require significantly higher investments and operating costs, but are independent of all variations in agroclimatic conditions and are very closely controlled for optimal performance and quality.

Open culture systems take advantage of natural sunlight and are totally subject to the vagaries of weather unless some form of shading system is utilized. Highly controlled closed systems use photobioreactors for phototrophic culture and conventional fermenters for heterotrophic growth.

The range of sophistication available for the two systems is very great, as is the associated investment. As an example, photobioreactors can vary from simple, externally illuminated glass jars to highly engineered fermenters saturated with light transmitting fibre optic filaments to ensure even lighting to all cells and infused with specific gas mixtures to control metabolism and growth rates. Certain new photobioreactors incorporate an α-type tubular design for greater cost-effectiveness and commercial efficiency (Lee et al, 1995).

Algae ponds

Algae grow rapidly, are rich in vegetable oil and can be cultivated in ponds of seawater, minimizing the use of fertile land and fresh water. Algae can double their mass several times a day and produce at least 15 times more oil per hectare than alternatives such as rape, palm soya or jatropha. Facilities can be built on coastal land unsuitable

for conventional agriculture. In the long term, algae cultivation facilities also have the potential to absorb waste carbon dioxide directly from industrial facilities such as power plants. Oil companies (e.g. Shell), the DOE and other institutions are intensifying research activities in this field. Shell and Hawaii-based algal biofuels company HR Biopetroleum have formed a joint venture to grow marine algae for conversion into biodiesel, and some companies have far surpassed the 15,000 gallon per acre accepted benchmark. In fact, one company can produce 180,000 gallons of biodiesel every year from just one acre of algae. That equates to about 4000 barrels, at a cost of US$25 per barrel or US$0.59 per gallon.

Algae bioreactors

Algal production systems have long been recognized as the most efficient means of producing biomass for food or fuel; they do not require arable land and therefore don't compete for space with existing crops. Over the same area, microalgae can produce 20–300 times more biodiesel than traditional crops (Table 10.3) and the remaining algal cake can still be useful for animal feed, fertilizer or other biofuel production systems. However, the initial set-up and maintenance of such systems has, to date, always proven to be cost-prohibitive for fuel.

In contrast to conventional crop plants, which yield a harvest once or twice a year, the microalgae have a life cycle of around 1–10 days depending on the process, with the result that multiple or continuous harvests with increased yield can be produced.

Source: Robert Henrikson, Ronore Enterprises, Inc http://www.spirulinasource.com/earthfoodch6c.html

Figure 10.4 A farm in southern Japan grows *Chlorella* in a circular pond, and New Ambadi farm in India grows *Spirulina* in canal-style ponds

Table 10.3 Comparison between crop efficiencies for biodiesel production

Plant Source	Biodiesel (l/ha/y)	Area required to match current global oil demand (million ha)	Area required as a percentage of global land mass
Soya bean	446	10,932	72.9
Rapeseed	119	4097	27.3
Mustard	1300	3750	25.0
Jatropha	1892	2577	17.2
Palm oil	5950	819	5.5
Algae (low)	45,000	108	0.7
Algae (high)	137,000	36	0.2

Screening and evaluation of naturally occurring algal strains that exhibit high growth rate and high oil content progresses to genetic improvement of selected strains for robustness in performance under diverse environmental and culture conditions, to large-scale photobioreactor design and optimization, to outdoor mass culture and downstream processing (i.e. harvesting, dewatering and drying), to algal oil extraction, pretreatment, and oil conversion to biodiesel and jet fuel, to systems/process scale-up analysis, and life cycle assessment.

Energy production

In the production of energy from microalgal biomass, two basic approaches are employed depending upon the particular organism and the hydrocarbons that they produce. The first is simply the biological conversion of nutrients into lipids or hydrocarbons. The second procedure entails the thermochemical liquefaction of algal biomass into usable hydrocarbons.

Anabolic production of lipids and hydrocarbons by microalgae

Lipids and hydrocarbons can normally be found throughout the microalgal cell mass. They occur as storage product inclusions in the cytoplasm and as functional components of various membranes. In some cases, they are excreted extracellularly into the microalgal colony matrix as almost pure oleaginous

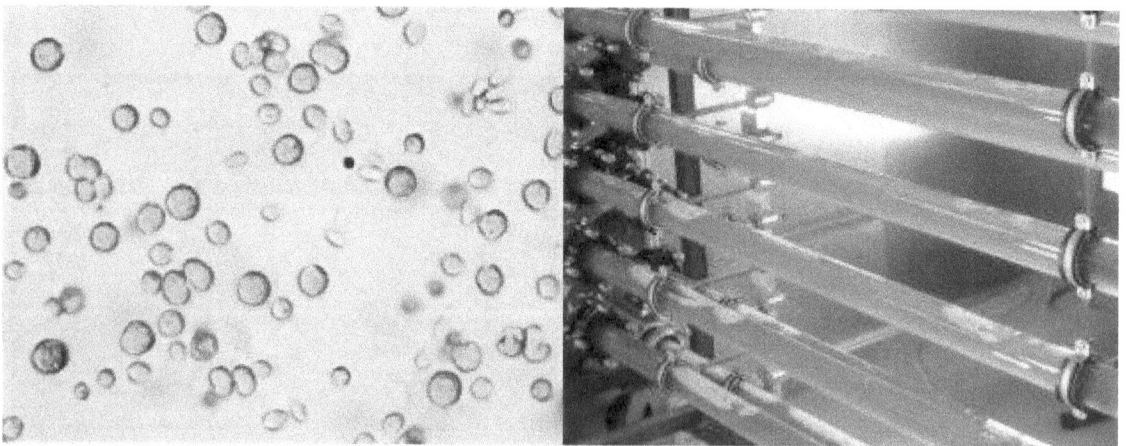

Source: Ben Hankamer and Clemens Posten, Solar Biofuels Consortium (www.solarbiofuels.org)

Figure 10.5 (a) *Chlamydomonas rheinhardtii* green algal cells. (b) Tubular photobioreactor

Source: Doug Frater, Global Green Solutions Vertigro Algae
Technologies LLC (USA) www.globalgreensolutionsinc.com

Figure 10.6 Closed loop continuous process algae
cultivation bioreactor system

globules. In certain cases, the lipid composition can be
regulated through the addition or restriction of certain
components in the diet. For example, nitrogen or
silicon starvation and other stress producers may
increase total lipid production.

The type and level of hydrocarbons produced is
often affected by environmental factors such as light,
temperature, ion concentration and pH. While it is
not uncommon to find levels of 20–40 per cent lipids
on a dry basis, on occasion the quantities of lipids
found in microalgae can be extraordinarily high. For
example, in one particular genus, *Botryococcus,* the
concentration of hydrocarbons in the dry matter may
exceed 90 per cent, under certain conditions (Largeau
et al, 1980).

Thermochemical liquefaction of microalgae

The convenience of microalgal harvesting and handling
makes it equally suitable for thermochemical
processing such as liquefaction. Following a process
reminiscent of the origin of petroleum products,
microalgae are converted into oily substances under the
influence of high temperature and high pressure. Yields
in the 30–40 per cent range of heavy-type oil can be
obtained in this manner (Kishimoto, 1997). Because of
the high levels of proteinaceous materials in the system,
nitrogen levels leading to NO_x formation have to be
carefully controlled. Yields close to 50 per cent of

Source: Christine Lambrakis/ASU Arizona State University http://biofuels.asv.edu/biomaterials.shtml

Figure 10.7 Demonstration field photobioreactors developed for algae feedstock production

liquid hydrocarbon have been obtained using a very high-temperature, high-pressure, catalysed hydrogenation process (Chin, 1979). Table 10.4 provides some examples of the lipid contents of various microalgae.

Processing and handling

In cases where the hydrocarbons are produced anabolically by the microalgae, direct extraction is the simplest and most effective way of obtaining products. This can be effected through the employment of solvents, through the direct expression of the liquid lipids, or a combination of both methods. The thermochemical liquefaction process often results in a heavy oily or tarry material which is then separated into different fractions by conventional catalytic cracking. As with hydrocarbons derived from other forms of renewable biomass, microalgal lipids can be converted into suitable gasoline and diesel fuels through transesterification.

Future prospects

The future of microalgal fuel production will be dependent upon its economy compared to other sources. Economies of production are steadily coming closer to fossil fuels and will continue to do so in the future. This phenomenon will accelerate as the extent of environmental degradation is factored into the cost of conventional fossil fuels.

Table 10.4 Lipid contents of different algae

Strain	% Lipid (on a dry basis)
Scenedesmus spp.	12–40
Chlomydomonas spp.	21
Clorello spp.	14–22
Spirogyra spp.	11–21
Dunoliello spp.	6–8
Eugleno spp.	14–20
Prymnesium spp.	22–38
Porphyridium spp.	9–14
Synechoccus spp.	11

Microalgae have the potential to achieve a greater level of photosynthetic efficiency than most other forms of plant life and are very amenable to genetic engineering of their photosynthetic apparatus for increased efficiency. If laboratory production can be effectively scaled up to commercial quantities, levels of up to 200t/ha/year may be obtained. Considering that the carbon source for microalgal hydrocarbon production is CO_2, the net production of CO_2 from fuel utilization will be nil. Whether the production focus is on hydrocarbons, hydrogen or useful energy products from waste conversion, the functional microbial characteristics of microalgae make them important components of all future renewable energy programmes.

Seeing microalgae farms as a means to reduce the effects of the greenhouse gas CO_2 changes the view of the economics of the process. Instead of requiring that microalgae-derived fuel be cost competitive with fossil fuels, the process economics must be compared with those of other technologies proposed to deal with CO_2 pollution.

The microalgal research and development effort couples the use of microalgae for biofuels production with environmental bioremediation. Microalgae naturally remove and recycle nutrients (such as nitrogen and phosphorus) from water and wastewater and carbon dioxide from flue-gases emitted from fossil fuel-fired power plants, providing an added environmental benefit. The integration of wastewater bioremediation and carbon sequestration with biofuel production in a novel field-scale bioreactor has been demonstrated. Although algal biomass residues derived from the oil extraction process can be used for animal feed or fertilizer, we are currently exploring, in collaboration with our industrial partners, the opportunity for using biomass residues to produce ethanol, and methane, and high-value biomaterials, such as biopolymers, carotenoids, and very long-chain polyunsaturated fatty acids. Collaboration with industrial partners to provide flue gas (APS), animal wastewater (United Dairymen of Arizona), commercial algal feedstock production capabilities (PetroAlgae), technical assistance with conversion of algae oil to biofuels (UOP and Honeywell Aerospace Division), and assistance with marketing of algal feedstock (Cargill) has either been initiated or is on-going. (ASU, no date)

In the US and the EU, algae-based biodiesel ventures are growing in response to demands for clean fuels. Each of these endeavours clearly demonstrates increased public and private sector interest in non-food, second-generation markets. Investments started late in 2006, and are growing in 2008 and beyond. Chevron, for example, has invested in and partnered with the National Renewable Energy Laboratory (NREL) to produce algae for biocrude, jet fuel and biodiesel. BP has invested in PPPs with the University of California at Berkeley and Arizona State University along with DARPA to advance the use of algae for jet fuel and biofuels. The US Department of Energy invested US$2.3 million in algae projects in 2008. Shell has invested in a PPP with Cellana, a joint venture with the Hawaiian Natural Energy Laboratory and HR Biopetroleum to produce algae for biofuels (Emerging Markets Online, 2008).

Over US$300 million (€233 million) has been invested in algae so far in 2008. The initial findings from a new study, Algae 2020, identify three key trends or waves of investments now emerging in the path towards the commercialization of the algae biofuels industry. The first wave is investment in public-private partnerships (PPPs).

Internet resources

http://biofuels.asu.edu/biomaterials.shtml
www.emerging-markets.com.
www.globalgreensolutionsinc.com/s/Vertigro.asp
www.solarbiofuels.org/
www.spirulinasource.com/earthfoodch6c.html
www.strategyr.com

ANNUAL RYEGRASS (*Lolium multiflorum* Lam.)

Description

Annual ryegrass has its origins in the Mediterranean region. It grows in semi-erect to erect clumps. Worldwide there are about a dozen species of ryegrass, including both annual and perennial plants. The stems grow to a height of 30 to 80cm, though the plant may grow as high as 1m or more. The leaf blades are 4–10cm wide, 21–24cm long, grooved and light green with a glossy underside. The inflorescence is a two-rowed lax ear. Annual ryegrass flowering occurs between June and August.

Ecological requirements

Temperate and subtropical climate regions are most appropriate for growing annual ryegrass. The plant prefers a fresh, nutrient-rich, medium loamy soil and thrives in humid climates with plenty of rainfall. A sufficient water supply is necessary for optimal yields. Annual ryegrass is known not to tolerate drought, wet soil and salt. Crop damage has been seen in areas with heavy snowfall and during cold winters.

Propagation

Annual ryegrass is propagated using seed.

Source: NC Cooperative Extension 2008 http://www.ces.ncsu.edu/wayne/graphics/eileen/forage/foragepic.html

Figure 10.8 Annual ryegrass

Table 10.5 Recommended fertilizer levels for annual ryegrass as used in Germany

Fertilizer	Amount (kg/ha)
Nitrogen (N)	300–350
Phosphorus (P$_2$O$_5$)	80–100
Potassium (K$_2$O)	400–420

Crop management

The annual ryegrass seed can be sown in the late summer (beginning of September), following the harvest of the previous crop, or in the spring. The amount of seed sown is approximately 35kg/ha. The recommended seeding date is 2 to 4 weeks before the average first frost date.

Table 10.5 shows the suggested fertilizer rates from Germany for annual ryegrass crops with more than one harvest per year. Nitrogen fertilizer is applied according to the individual growth stages of the plant and can be in the form of up to 7.5 per cent liquid manure. The fertilizer rates vary with soil composition. Among the diseases most damaging to annual ryegrass crops are mould (*Fusarium nivale*) and rusts (*Puccinia* spp.).

Insertion of annual ryegrass into the crop rotation is easy. Because annual ryegrass has not been shown to have negative effects on crops coming after it, and because no crops have been shown to have negative effects on annual ryegrass, there is currently no recommended crop rotation.

Annual ryegrass is a fast-growing, competitive winter annual cool-season grass grown nearly anywhere there is adequate available soil moisture. It tolerates wet, poorly drained soils, but not extensive periods of flooding. It could be seeded alone, with small grains, legumes (clover), or overseeded in warm-season pastures (Bermuda or bahia grass). Annual ryegrass is usually planted from September to early October at rate of 20 to 30lb/acre in pure stand or 10 to 15lb/acre in a mixture (small grains or overseeding Bermuda or bahia grass). Overseeding of warm-season grass should be done in November and it should be ensured that enough forage has been removed to allow a good soil–seed contact. Armyworms and rust could be a problem. Seeding depth should be between 0.5 and 1cm (msucares.com).

Production

Field tests at four locations in Germany where annual ryegrass was harvested once a year at the onset of flowering provided yields averaging between 7.1 and 11.8t/ha dry matter. The plant's dry matter content averaged about 33–35 per cent. The heating value averaged approximately 17MJ/kg.

Additional tests in Germany comparing different annual ryegrass varieties showed an average fresh matter yield of 105.5t/ha and an average dry matter yield of 18.5t/ha. The varieties with the highest fresh matter yield were Adrina and Lolita. Lolita was also the variety with the highest dry matter yield.

Forage yield depended on the harvest date and cultivar (Table 10.6). Cultivars that stood out for producing the most forage during early spring, more than 4000lb/acre by 1 April, included Attain, Big Boss, Diamond T, Ed, Fantastic, Flying A, ME4, ME94, Surrey II, TXR2006-T22 and WD-40. Ed, ME4, ME94 and WD-40 also produced good forage yields during this same time period last year. Cultivars that stood out for producing more forage during late spring (April harvest) included AM-4T, Attain, Barextra, Big Boss, Florida 4N, Hercules, ISI-LWD4, Jumbo, Marshall, MO 1, Surrey II, Tam TBO, TXR2006-T22 and Verdure (Table 10.6). When forage yield was examined over the whole season, differences among entries were mostly not significant. Forage yield among the top cultivars ranged from 6079 to 7420lb/acre (Table 10.6). Any entry producing more than 6000lb/acre across the season performed well (Guretzky and Norton, 2008).

Processing and utilization

Once the crop has reached maturity it may be harvested up to six times per year. This corresponds to about once every 4 weeks from the beginning of May through until the end of October.

Although annual ryegrass is known as a very valuable fodder plant, much of the present research on it has the goal of utilizing the entire plant as a biofuel for the direct production of heat and the indirect production of electricity.

Table 10.6 Dry matter forage yields of annual ryegrass cultivars at Ardmore, Oklahoma, harvested on 4 January, 14 March, 31 March, 29 April and 3 June

Cultivar [Source]	Harvest dates					
	4/1	14/3	31/3	29/4	3/6	Sum
	lb/acre					
AM-4T[Ampac]	984	1171	1558	2039	536	6288
Attain [Smith]	1107	1236	1662	1970	691	6666
Barextra [Barenbrug]	831	879	950	1856	1087	5605
Big Boss [Smith]	1319	1277	1470	1875	751	6692
Diamond T [Oregro]	1215	1580	1249	1714	815	6573
Ed [Smith]	1837	1863	1167	1386	622	6876
Fantastic [Ampac]	946	1728	1467	1827	976	6496
Florida 4N [DLF]	961	1298	1434	1986	543	6223
Flying A [Oregro]	1392	1367	1341	1443	923	6467
Hercules [Barenbrug]	696	1167	1283	2113	569	5828
ISI-LWD4 [DLF]	169	625	1842	2458	1304	6398
Jackson [Wax]	1104	1415	1321	1720	666	6228
Jumbo [Barenbrug]	104	971	1606	1939	1071	5691
Marshall [Wax]	1200	1236	1458	2102	680	6678
ME4 [Wax]	1154	1425	1624	1595	1213	7011
ME94 [Wax]	1242	1189	1715	1835	865	6846
M01 [DLF]	1019	751	1482	2000	706	5959
Surrey II [DLF]	1258	1743	1345	1888	982	7216
Tam 90 [TX AES]	741	1376	1401	1800	761	6079
Tam TBO [TX AES]	201	1147	1425	2153	1210	6137
Tetra Pro [TX AES]	678	957	1321	1476	863	5295
TXR2006-T22 [TX AES]	1293	1488	1501	2239	899	7420
Verdure [Smith]	1136	1281	1435	2232	688	6772
WD-40 [Oregro]	1565	1548	1296	1678	504	6593
WMN97 [Wax]	1141	1209	1089	1765	878	6083
Mean	1011	1277	1418	1884	832	6423
LSD	409	541	425	620	437	1397
P value	0.001	0.01	0.05	0.15	0.01	0.364
CV	25	25	18	20	32	13

Source: Guretzky and Norton (2008)

Internet resources

www.msucares.com/crops/forages/grasses/cool/annualryegrass.html

ARGAN TREE (IRONWOOD) (*Argania spinosa* (L.) Skeels; syn. *A. sideroxylon* Roem. & Schult)

Description

The ironwood is an evergreen olive-like tree native to Morocco. It reaches 4–10m in height, occasionally attaining 21m with a trunk diameter of up to 1m. The ironwood tree flourishes under very dry conditions, forming olive-like berries on its gnarled branches and thorny twigs. The berries contain one to three relatively small seeds (Gliese and Eitner, 1995). The kernels contain about 68 per cent oil. The timber is very hard, heavy and durable, and suitable for agricultural implements and building poles in addition to making a good charcoal. The tree can be planted as a shade tree and for soil conservation and windbreaks (FAO, 1996a). The argan tree is cultivated in those parts of Morocco where the olive tree cannot be grown (Franke, 1985).

Argan oil is produced from the kernels of the endemic argan tree, valued for its nutritive, cosmetic and numerous medicinal properties. The tree, a relict

Source: Luc Viatour GFDL/CC www.lucnix.be

Figure 10.9 Argan trees in Morocco

species from the tertiary age, is extremely well adapted to drought and other environmentally difficult conditions of south-western Morocco. The genus *Argania* once covered North Africa and is now endangered and under the protection of UNESCO. The argan tree grows wild in semi-desert soil, its deep root system helping to protect against soil erosion and the northern advance of the Sahara. This biosphere reserve, the Arganeraie Biosphere Reserve, covers a vast intramontane plain of more than 2,560,000 hectares, bordered by the High Atlas and Anti-Atlas Mountains and open to the Atlantic in the west. Argan oil remains one of the rarest oils in the world due the small and very specific growing area (Lybbert, 2007).

Ecological requirements

Argania spinosa can be found from sea level to an elevation of 1500m. The tree is drought resistant but not adapted to drifting sands. It can shed its foliage and remain in a state of dormancy for several years during prolonged droughts. An annual rainfall of 200–300mm is the optimum; the minimum of precipitation is 100mm and the maximum 400mm. The tree grows best at temperatures from 20–30°C; minimum and maximum temperatures are 10°C and 35°C respectively. The tree is killed by frost (−2°C). It needs a medium deep, well-drained soil with a pH value of 6.5–7.5. Bright light is required for growth (FAO, 1996a).

Production

The endosperm of the seed contains approximately 68 per cent oil. The oil is rich in oleic acid and edible

with a nutty taste. The tree may start to bear fruit after 5–6 years and reaches maximum production after about 60 years. These long-living trees may reach an age of 200–400 years.

The average yield of argan fruits per tree is 8kg. Global production of argan fruits is estimated to be about 350,000 tonnes, or about 50 billion fruits. The region of Essaouira produces between 1000 to 2000 tonnes of oil per year, corresponding very roughly to a local population of 60,000–120,000 trees, producing 142 million to 286 million fruits per year. Apart from oil, argan fruit, foliage and oil extraction by-products constitute valuable food resources for livestock. The hard, heavy and durable argan timber gives good charcoal and construction wood, and is also used to fashion ploughs, tool handles and household utensils. (Moussouris and Pierce, 2000)

The oil contains 80 per cent unsaturated fatty acids, is rich in essential fatty acids and is more resistant to oxidation than olive oil. Table 10.7 indicates the percentage of fatty acids in argan oil.

Argan fruit falls in July, when black and dry. Until that time, goats are kept out of the argan woodlands by wardens. Rights to collect the fruit are controlled by law and village traditions. The leftover nut is gathered after consumption by goats, but the oil produced from these nuts has an unpleasant taste, and is not used for human consumption (Nouaim, 2005).

Approximately 100kg of seeds yield 1 to 2kg of oil and 2kg of pressed 'cake' – a pasty by-product – plus 25kg of dried husk.

Processing and utilization

The oil is edible and is also used as fuel (for illumination) and in the production of soap (Gliese and Eitner, 1995). The wood can be used as fuelwood or be processed to charcoal. Although the tree's nuts provide husks for livestock and oil for cooking and medicinal purposes, many of the trees have been cut down for firewood and to clear land for other agricultural uses.

Internet resources

Wikipedia.rg, 'Argan, *Argania spinosa*', http://en.wikipedia. org/w/index.php?title=Argan&oldid= 260114382, accessed 26 December, 2008

Figure 10.10 Foliage, flowers and immature fruit of the argan tree near Agadir, Morocco

BABASSU PALM (BABAÇÚ) (*Orbignya oleifera* Burret)

Description

The babassu palm is a perennial evergreen palm that grows to a height of 15 to 20m. It has a slender trunk, and the crown is formed of 15–20 huge feathery leaves measuring up to 7m that stand at upright angles. It is native to the dry semi-deciduous forests of central Brazil (FAO, 1996a), where it grows wild in large populations. The babassu palm flowers and bears fruit over the whole year, but the yield is higher during the summer (Lennerts, 1984). In Brazil the fruit ripens from July to November. The male and female inflorescences with a length up to 1m hang down from

Table 10.7 Contents of fatty acids in Argan oil

Fatty Acid	Percentage
Palmitic	12.0%
Stearic	6.0%
Oleic	42.8%
Linoleic	36.8%
Linolenic	<0.5%

the leave axils. Each inflorescence carries between 200 and 600 oblong nuts the size of goose eggs, each weighing 80 to 250g and containing four to six kernels (Gliese and Eitner, 1995). The kernels are surrounded by a pulp and a hard woody shell, nearly 12mm thick, similar to a coconut shell. The fruit bunches weigh from 14–90kg; one to four fruit bunches are produced per year.

Related species that offer a commercial potential as oilseeds:

- *Orbygnia martiana* Barb. is the other main species of babassu found in Brazil. It grows in the wet forest areas of the Amazon basin. The preceding descriptions apply to this species, too.
- *Orbygnia cuatrecasana* A. Dugand grows in the Chocó region on the Pacific coast of Colombia. The seeds are edible and are used for the same purposes as coconuts.
- *Altalia speciosa* Mart. (formerly known as *Orbygnia speciosa* (Mart.) Barb.-Rodr.) grows in tropical South America, especially in Brazil. The fruits are smaller and contain fewer seeds than the previously mentioned species. The palm provides an excellent oil.
- *Orbygnia cohune* Mart., the cohune palm, grows in the lowlands of Yucatan/Mexico, Belize, Honduras and Guatemala. Each palm bears from 1000 to 2000 fruits per year, but only a small amount is harvested (NAS, 1975). The fruit flesh is edible and is also used as animal fodder (Rehm and Espig, 1991).

The seeds of palms like *Astrocaryum vulgare* Mart. and the *Scheelea* species and of the American oil palm *Elaeis oleifera* (H. B. K.) Cortes are collected in a similar way.

Ecological requirements

All of the species mentioned require tropical climates with warm temperatures, ample sunshine and moist soils. Babassu palms are most abundant along rivers and valley floors, but they also grow in semi-arid areas (Gliese and Eitner, 1995). An annual rainfall of 1200–1700mm is the optimum, but 700–3000mm is acceptable. The optimum temperature is 25–30°C.

The palm accepts a wide range of soil textures, but the soil should be well drained and not waterlogged

Figure 10.11 Babassu palm tree

(FAO, 1996a). The climate in the area in which the babassu palm is found is marked by short and irregular periods of heavy rainfall (Lennerts, 1984).

Propagation

The wild palms reproduce by seeds, which remain germinable for a long time. Clearings in babassu forests are quickly overgrown with babassu palms again. Because of the high number of seeds, the stock of babassu palms is rather dense. The germination has a special mechanism. First, a pregermination of the seed occurs: a 30–50cm long carrier is formed, growing vertically into the soil and carrying the embryo at the tip. There the embryo generates roots and the cotyledon (Lennerts, 1984).

For breeding, high-yielding varieties should be selected from the wild stands and collected in a germplasm bank. The reason for the great differences in yield in various regions should be investigated (NAS, 1975).

Crop management

Most babassu palms are not cultivated in plantations. Only a few plantations have been established. The fruits of the wild palm are collected. With arranged harvesting and cultivation of the palm in plantations, the yield could be increased (Franke, 1985). In wild stands the plant density is too high for good yields. There are some indications that 100 palms per ha will yield from 50 to 500kg of oil. The yield obtained from this heliotrophic plant is related directly to the available space. With increasing stocking density, the average number and weight of infructescences decrease. By merely considering parameters like the stocking density and age structure of babassu woods, enormous yield reserves could be opened by purposeful stocking alone (Ortmaier and Schmittinger, 1988). However, there is a complete disregard for domestication: babassu oil has been substituted by other oils, and the labour to harvest the babassu kernels competes with other activities (Lieberei et al, 1996).

Production

The babassu palm tree, unknown to most people in the West, grows in the wild on 18 million hectares (45 million acres) in the forest in the northeast Brazil, an area corresponding to roughlyhalf the size of Switzerland.

The babassu palm is slow to mature. It begins to yield after 8 years and reaches full production after 10–15 years (NAS, 1975). It is said to be productive for 200 years. The kernels contain 64–67 per cent fatty oil; this oil is similar to palm oil and coconut oil. The high-quality oil stores well. Oil yields peaking at about 100–200L/ha are possible (Gliese and Eitner, 1995). Various levels of yields are reported for individual trees. One palm may produce 1 tonne of nuts per year, which corresponds to 90kg of kernels. In cultivated plantations 1.5 tonnes of nuts per year can be achieved (NAS, 1975).

The yield potential has hardly been tapped: the stock of palms has a yield potential of 883,680 tonnes of kernels, but only 281,344 tonnes are collected and used for oil production (Ortmaier and Schmittinger, 1988). A lack of infrastructure and transport facilities and the use of wild, uncultivated babassu palms prevent higher yields. The fat-containing kernels make up less than 10 per cent of the fruit, while the share of the shells alone is more than 50 per cent. Table 10.8 shows the composition of fruits and kernels. Although the kernel has the most important properties for commercial use, it accounts for only 10 per cent of the fruit. Therefore, it is not usually economic to transport whole fruits to cracking and separating centres, since 90 per cent of the fruit has only incidental by-product value. Cracking and separating should be done locally.

source: www.britannica.com/EBchecked/topic/47367/babassu-palm

Figure 10.12 The nuts of a babassu tree

Table 10.8 The composition of babassu palm fruits and kernels

The fruits consist of:	The kernels contain:
10% fibrous husk	67% fat
24% fruit flesh	9% protein
58% hard nutshell	10.5% N-free extracts
7–8% kernels	6.5% fibre
	2% ash

Source: Lennerts, 1984

Processing and utilization

There is no well-organized harvest technique. In the wild stands, the ripe fruits fall off and are collected and brought to a collecting and storage point. After collection the fruit is dried to facilitate the removal of the shell from the kernel. If the kernels are not well dried, damage to the kernels initiates enzyme action and the oil may become rancid. The cleansed kernels can be stored or exported.

Babassu oil has a melting point between 21 and 31°C. It consists predominantly of glycerides of lauric acid and myristic acid (Franke, 1985). The oil is almost colourless and does not easily become rancid. The press cake contains up to 27 per cent protein and gives a valuable fodder (NAS, 1975). The profitable use of the babassu palm requires a practical utilization of the by-products. If the babassu palm is used as an energy plant, it delivers three forms of raw material: the oil, the fruit flesh and the nutshells. The difficulty is to develop an economic process to divide the kernels from the very hard shell. Along with the uses for nutrition, the oil is used for technical purposes (lubrication) and as fuel. For the production of 200 tonnes of oil, 2000–2500 tonnes of shells are needed. The shells are used directly as an excellent solid fuel. To a large extent they are used to produce charcoal (Rehm and Espig, 1991). The ripe fruit flesh can be milled and is used as a source of starch. In Brazil, research was conducted to produce ethanol from the fruit flesh for use in the Brazilian ethanol-as-fuel programme: 1 tonne of fruit flesh gives 80L ethanol (Wright, 1996).

Expedito Parente discovered that the nuts of the babassu – or more precisely the kernels within – can be turned into an oil that possesses energetic properties which make it an ideal feedstock for bio-jet fuel (Parente, 2007). There are certainly social and ecological arguments in favour of the babassu palm: its use does not compete for land with food production, nor does it imply deforestation as the palm grows in the wild and can thus be integrated in sustainable agroforestry systems. In February 2008, Babassu Palm oil was used in a blend with coconut oil and jet fuel to power one engine of a Virgin Atlantic Boeing 747 during a test flight.

BAMBOO (*Gramineae* (*Poaceae*), subfamily Bambusoideae)

Description

Bamboos are a group of giant woody grasses that make their natural habitat roughly between the 40° southern and northern latitudes, excluding Europe. Currently, more than 130 genera of woody bamboos and 25 grass bamboos have been named worldwide, though many other estimates place these numbers lower. These bamboos are made up of over 1300 species of the subfamily Bambusoideae that are distributed among tropical, subtropical and mild temperate zones of the world. Countries such as Bangladesh, China, India and Indonesia along with other Asian countries depend heavily on bamboo.

They are distributed among tropical, subtropical and mild temperate zones covering an area of more than 25 million hectares (Figure 10.13).

Bamboo plants, which are usually perennial, consist of an underground root system and rhizome mat from which rise erect aerial stems, referred to as culms. Culms are usually hollow, smooth and round with a brown or yellow-brown to yellow-green colour and consist mainly of cellulose, hemicellulose and lignin. They can have a diameter of up to 20cm or more and reach a full grown height of 10–40m in about 3–4 months. One genus, *Phyllostachys*, studied as a possible energy plant, grows to heights of 2–10m. Branches are often produced from lateral buds near the top of the culm once the culm reaches its full height. The branches are capable of secondary and tertiary branching, with the ultimate branches bearing leaves. The long fibres, which can be from 1.5 to 3.2mm in length, comprise 60–70 per cent of the culm's weight and are the characteristic that makes bamboo desirable for paper production. Bamboo provides a high biomass yield, is strong and has a similar amount of energy to wood.

Bamboos can be put into two categories with respect to their rhizome system: sympodial or monopodial. Sympodial bamboos grow in clumps and the rhizome's neck has no buds or roots. Monopodials, which grow spread out from each other, have roots and buds on the rhizomes, with axillary buds leading to the growth of new culms at intervals. Tropical bamboos are sympodial, while temperate bamboos can be of either category.

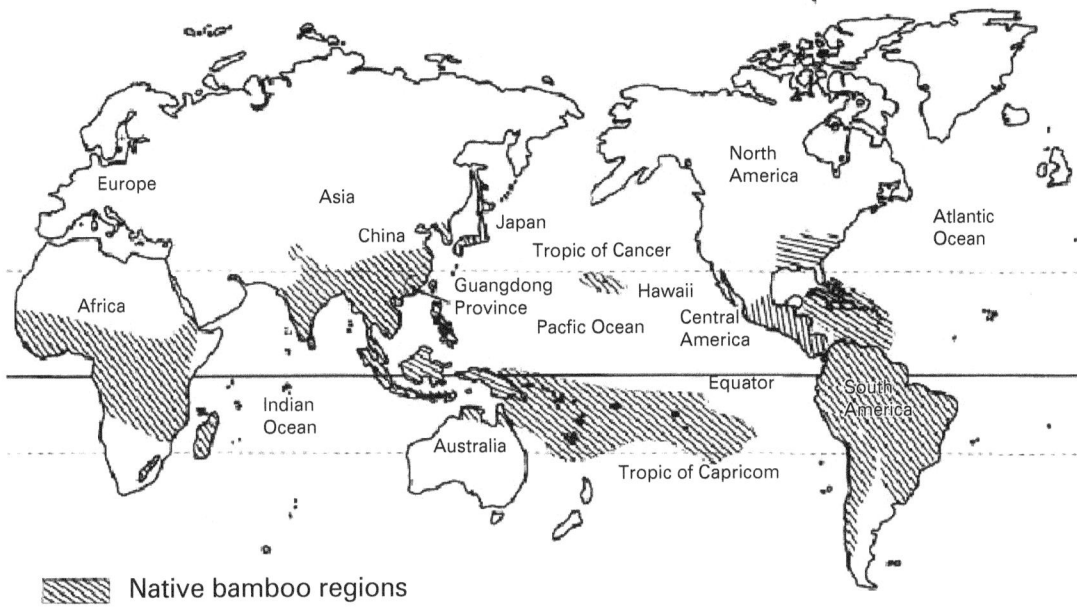

Source: El Bassam and Jakob, 1996

Figure 10.13 Native bamboo regions

Source: www.cepolina.com/freephoto/a/a.asp?N=bamboo.forest2&S=bamboo&C=All

Figure 10.14 The large leaf canopy of *Phyllostachys pubescens* allows it to absorb 95 per cent of available sunlight

Source: www.comp.nus.edu.sg/~wuyongzh/my-shoot

Figure 10.15 Bamboo sea in China

Bamboos of the same species, which in the same habitat would look the same, can take on different characteristics when grown in different climates. The majority of the bamboo species seldom flower, and even then some do not fruit. The flowering cycle can be anywhere from 15 to 120 years, with the average falling at about 30 years. Many bamboos or bamboo populations die once they have flowered.

Ecological requirements

Although bamboos are able to adapt quite well to varying environments, most species need warm and humid climates. An annual mean temperature between 20 and 30°C is preferred. Several species have the cold resistance necessary to be able to live in higher latitudes and elevations, where temperatures are as low as −30°C, but it is recommended that the temperature not fall below −15°C and the elevation remain under 800m.

In its original habitats bamboo usually needs precipitation levels of 300mm per month during the growing season and 1000–2000mm annually. There are drought-resistant strains, such as *Dendrocalamus strictus* in India which can survive on a minimum of 750–1000mm annual precipitation.

Bamboos can help to bioremediate saline soils and also help to prevent soil erosion. Bamboo has proved to grow well in all of Europe, but has shown a preference for areas similar to its native habitats. Italy, Portugal,

Spain and southern France are areas where bamboo has thrived. Bamboos prefer light, well-drained sandy loam to loamy-clayey soils, with abundant humus and nutrients. The soil should not dry quickly. Soils too high in salt content should be avoided. The optimal soil pH range is between 5 and 6.5. It is also recommended that the bamboo crop be protected from the wind as much as possible.

Propagation

Two categories for methods of propagating bamboo presently exist: conventional propagation and in vitro propagation. Conventional propagation has two traditional forms: seed and vegetative. Seed, though less expensive and easier to transport and handle, also introduces negative factors: in general, seeds have low viability, do not store well and are of questionable availability. Seed propagation usually involves the seed and the seedling first being reared in a laboratory or nursery, then being transplanted into the field. It takes the seedling approximately 2–3 months to reach a height of 5–10cm, at which time it is ready for transplantation. Sowing the seed directly into the field is confronted by the yearly necessities of weeding and thinning of the crop.

The main forms of vegetative propagation involve the planting of offsets, culm cuttings or branch cuttings. Propagation by offsets consists of cutting 30–50cm lengths of 1- or 2-year-old rhizomes with nodes and buds present, then burying them in the ground on the assumption that within a few weeks new shoots will appear. This method is not suitable for large plantations because of the limited quantity of rhizomes per clump of bamboo culms.

Propagation by culm cuttings involves planting 0.5–1.0m lengths of culm with nodes, 7–15cm deep into a rooting medium. Because of the large diversity of rooting capabilities across the different bamboo species it is difficult to generalize this method for all types. Culm cuttings from thick-walled bamboos have displayed a success rate of 45–56 per cent.

Propagation by the planting of branch cuttings involves the direct planting of branch lengths into the ground in the field. This method has not proved very successful; it takes several months for the branch to produce roots and one to three years for it to form rhizomes. During this time it is very susceptible to negative weather conditions.

Figure 10.16 Arrow bamboo (*Pseudosasa japonica*)

Vegetative propagation has the advantage that the characteristics of the new plant are known to be those of the parent plant. But vegetative propagation also has disadvantages in that these methods can only be performed at certain times of the year, the success rates are low, and the amount of work and costs involved make the approach infeasible for large plantations.

The other category, in vitro propagation, can be divided into two forms: somatic embryogenesis and micropropagation. Somatic embryogenesis involves plantlet production using somatic embryos, which are non-gametic fusion-formed cells that have the shoot and root pole. Somatic embryo cells can be obtained from the callus of either seeds or young inflorescences. Germination of the cells leads to plant formation. The work is performed in a laboratory on a tissue culture medium until mature plantlets reaching heights of 8–10mm are obtained, at which time the plantlets are transferred to pots containing soil, sand and manure. During this acclimation period the plants will spend 2–3 weeks in an acclimation chamber, then 2 months in a glasshouse or PVC greenhouse. After the 2 months the plants are transferred to bags, where they are kept until they are approximately 8 months old. These are then handed over to the foresters who plant them when they are about 1 year old.

Micropropagation begins with the selection of the starting material. This is usually from seedlings or mature culms. It is currently recommended that the axillary buds from young non-flowering stems be used. The buds have the ability in tissue culture to grow into complete plantlets. Experiments using *D. strictus* have resulted in over 10,000 plantlets from a single seedling in a year. After 6–7 weeks in media culture the nodal explants each produced 8–10 shoots. The shoots were then grown in flasks until rooting occurred. Upon reaching a height of 50–60mm the plantlets were transferred into pots with soil and sand. After acclimation, the plants could eventually be transferred into the field. Micropropagated plants have shown changes such as earlier flowering, earlier culm production and increased growth rates, as compared to those from conventional propagation methods.

In vitro propagation offers advantages in that a large number of plants can be propagated, and in a short period of time. The disadvantages of this form of propagation are its high cost; its labour-intensive character; the fact that characteristics of the starting material are often not known; and that at present it is not possible to link qualities of in vitro growth with desirable qualities of mature plants, so the production of inferior plants is often the result.

Crop management

The period of new growth is from approximately April/May until the end of August. Planting density tests of *Dendrocalamus strictus* in India concluded that the highest biomass production of 27t/ha, during an 18- month period, was seen at a density of 10,000 plants/ha.

Because little work has been performed with growing bamboo as a crop in Europe, there is very little information covering diseases and pests that attack bamboo. Even in other countries that depend on bamboo, the subject of diseases and pests is not extensively researched. In its native habitat bamboo is subject to disease attack on the roots, rhizomes, culms, foliage, flowers and seeds. The difference between the climate of Europe and that of bamboo's native climate means that time and research will be necessary to determine the effects of diseases and pests on bamboo. In addition, it is likely that bamboo will be affected by new diseases and pests native to Europe.

Bamboo is subject to a number of common diseases in its native habitats: damping-off caused by *Rhizoctonia solani*, rhizome bud rot caused by *Pythium middletonii* and *Fusarium* spp., rhizome decay caused by *Merulis similis,* root infections, basal culm decay caused by *Fusarium moniliforme* and *Fusarium* spp., bamboo blight, culm rot suspected to be caused by *Fusarium equiseti* and *Fusarium* spp., culm necrosis, culm canker caused by *Hypoxylon rubiginosum,* culm stains, inflorescence and seed infections, foliage infections, and rusts. Many of the fungi that attack bamboo are native to tropical areas, where bamboo is made susceptible to them because of excessive humidity and lack of sunlight. Allowing bamboo clumps to become run-down makes them more prone to diseases. Excessive humidity, lack of sunlight and allowing clumps to become run-down, by tropical climate standards, are three circumstances that are unlikely to be experienced in Europe.

Bamboo is attacked by several common pests in its native habitats. The larvae from *Estigmina chinensis* tunnel throughout the internodes, causing shortened and crooked internodes. The weevil *Cyrtotrachelus longipes* attacks the developing tip of new culms. The plant louse *Asterolecanium bambusae* may attack the buds. Insects such as *Atrachea vulgaris* and *Chlorophorus annularis* bore into bamboo to lay their eggs. Over-mature and stored culms are vulnerable to the beetle *Dinoderus minutus,* whose larvae feed on the culm's parenchyma tissue. Aphids, locusts and termites have been shown to attack bamboos, along with other animals that are known to eat bamboo shoots or chew on the rhizomes, such as goats, deer, porcupines, rats and squirrels.

Because of the proposed use of bamboo in Europe as biomass for energy production, bamboo plantations in Europe will have an advantage over those of bamboo's native habitats. Many of the diseases and pests that attack bamboo cause changes such as discoloration and crooked growth, which negatively affect the quality of the bamboo in terms of its usage for food, construction materials, furniture, decoration, etc. When bamboo is grown for biomass production the appearance and durability are not important factors.

In India, research was conducted to determine the influence of planting densities and felling intensities, among other factors, on the performance of the bamboo *Dendrocalamus strictus*. The cultivation took place on marginal land with red silt loam soil. The mean annual precipitation was 827mm and temperatures ranged from 13 to 36°C. Planting densities are related to the purpose for which the bamboo is intended. This is demonstrated by the

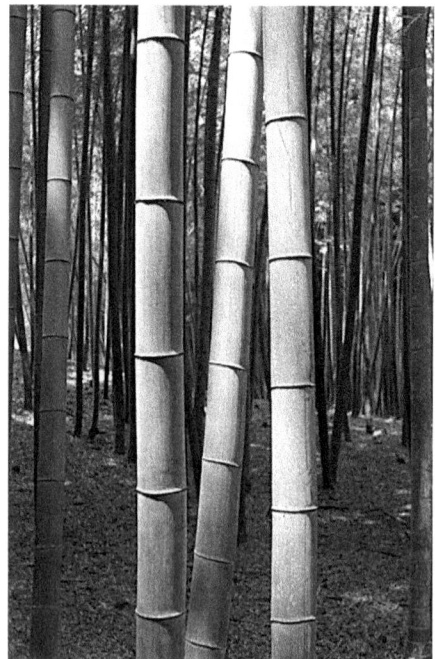

Source: Paul Vlaar 2004 http://en.wikipedia.org/wiki/File: BambooKyoto.jpg

Figure 10.17 Bamboo forest in Kyoto, Japan

Table 10.9 Suggested fertilizer levels for bamboo
resulting from tests in India

Fertilizer	Amount (kg/ha)
Nitrogen (N)	100
Potassium (as K$_2$O)	50
Phosphorus (as P$_2$O$_5$)	50

observations that high-density planting leads to the production of a greater biomass yield and more plants but with smaller culms, while as the planting density decreases so also does the number of culms, but the size of the individual culms increases.

Fertilizer application tests in India resulted in recommendations for fertilizer amounts for the area in which it was tested. These are shown in Table 10.9. These fertilizer amounts resulted in a threefold increase in biomass production over non-fertilized bamboo. Doubling the amounts showed a fourfold increase in biomass production over non-fertilized. Fertilization has been shown to lead to an increased number and weight of rhizomes. A nitrogen-rich, fast-release compound fertilizer should be used in the spring, about a month before sprouting, but avoided in the autumn. Studies in China have concluded that for every 1000kg of bamboo vegetable matter produced, there needs to be in the soil 2.7kg of nitrogen, 3.6kg of potassium and 0.36kg of phosphorus.

Once established, the bamboo crop can be maintained for many years. As it is a perennial plant species, a crop rotation is not valid for bamboo.

Production

Until now, bamboo has had a very wide range of uses, but only a little is used as firewood. Bamboo's principle uses for processed products mean that a market and market value do not yet exist for using it as an energy and fuel source.

The yield is greatly affected by the species and the growth conditions. The value of bamboo as a fuel source is connected to its dry matter content.

The culm evaluation (Table 10.10) shows differences in culm diameters, culm wall thickness and internode length. There are several bamboo species for different fields of application by indigenous people.

Bamboo was introduced into Europe about 200 years ago, mainly for horticultural uses. Since then a wide range of varieties and bamboo types has been developed for European conditions. Of great importance in northern European countries is bamboo's ability to tolerate temperatures as low as –25°C. Due to its potential fast growth rate and excellent chemical and physical characteristics, it is suggested that this plant species can be developed as a new alternative crop for the production of raw materials for industry and energy use.

Processing and utilization

Bamboo is currently harvested manually with knives, which is the most economical, though labour-intensive, way of bamboo harvesting. Mechanized harvesting methods need to be developed, but this would also depend on the extent of annual harvesting and the species of bamboo to be harvested. In addition, the many differences between the geography, climate, labour, technology, etc. of Europe and the countries of bamboo's natural habitat will need to be taken into consideration.

It was previously believed that clear felling was harmful to bamboo stands, but tests in India have shown that clear felling of the stand led to vigorous

Table 10.10 Culm evaluation of different bamboo species

Species	Height (m)	Culm diameter (cm)	Internode length (cm)	Wall thickness (cm)	Culm rating
B. pallida	14–17	3–5	30–60	0.2–0.6	2
B. tulda	12–14	2–4	45–50	0.4–0.7	1
D. asper	14–16	4–7	Up to 100	0.4–0.7	4
D. membranaceus	12–18	3.6–4	30–50	0.3–0.6	3
G. apus	16–18	4–5	40–50	0.1–0.2	5

Table 10.11 Annual biomass yield of air-dried bamboos as found in their native habitats

Bamboo species	Native country	Yield (t/ha)
Bambusa tuldo	India	3
Dendrocolomus strictus	India	3.5
Meloconno baccifero	India	4
Meloconno bombusoides	Bangladesh	10–13
Phyllostochys bombusoides	Japan	10–14
Phyllostochys reticulata	Japan	5–10
Phyllostochys edulis	Japan	5–12
Phyllostochys edulis	Taiwan	8–11
Phyllostachys mokinoi	Taiwan	6–8
Phyllostachys pubescens	China	5–10
Thyrsostochys siamensis	Thailand	1.5–2.5

regrowth of the clumps. Removal of all of the culms can lead to an increase in the number and size of culms, which means an increase in the amount of biomass per year, while providing more room for the growth of new shoots.

In its native habitat, bamboo is transported manually, by animals, carts or trucks, or floated down river. This is usually to a paper mill or a place for storage or further processing. Storage usually amounts to piling the bamboo culms in hills where, while waiting to be used, they are often attacked by disease and pests, which leads to deterioration of the culm's quality. Processing produces a large variety of durable, long-lasting products.

The analysis of fuel properties is listed in Table 10.12. Summed to the fuel rating, the potential of the different bamboo species are described. *B. tulda, B. pallida and D. membranaceus* show the greatest potential as a biofuel. However, *D. membranaceus* competes with other utilization purposes and the fuel properties are less

good than *B. pallida, B. tulda* and *G. apus*. Compared with other species, *B. tulda* and *B. pallida* have less use at the local level, but show great culm and fuel properties. Concerning the utilization pressure in different application fields, *B. tulda* and *B. pallida* have the greatest potential (Dannenmann, 2007).

Bamboo is considered a possible candidate for bioenergy production because of its high biomass yield and its high heating value of about 4600cal/g (wood has a heating value of 4700–4900cal/g). Bamboo is used to make furniture, building materials, paper, instruments, weapons and decorative ornaments, along with many other products. It is also used as a source of food and fibre. Although, to a small extent, bamboo is burned for heat production, this does not approach the scale of using bamboo as a substantial energy or biofuel producer.

The use of bamboo as an energy and biofuel source has been closely investigated by El Bassam et al (2001). The energy content of bamboo stems (17.1MJ/kg DM) and the cellulose content of over 40 per cent emphasize the suitability of bamboo as a raw material for energy, pulp and paper production. Measurements of plant height, number of tillers as well as chemical analysis of the plant material were conducted in order to establish the most important parameters for bamboo as energy feedstocks: energy contents, proportion of cellulose and lignin, nitrogen, potassium and phosphorus percentages.

Bamboo stems (90 per cent) mixed with some leaves (10 per cent), were flash-pyrolysed at 487°C in a continuously operating fluidized bed reactor. About 70 per cent of the feed material was converted into a pyrolysis liquid (bio-oil) which can be used as a fuel or for chemical feedstocks. Table 10.13 shows the characteristics of the bamboo material. The whole plant material was separated into leaves and stems and only the stem fraction was subjected to flash pyrolysis.

Table 10.12 Potential of different bamboo species for use as a biofuel

	B. pallida	B. tulda	D. asper	D. membranaceus	G.apus
Use rating	4	3	2	1	5
Culm rating	2	1	4	3	5
Fuel rating	2	1	5	4	3
Potential as a biofuel	8	5	11	8	13

Table 10.13 Specifications of the feedstock

Feedstock	Bamboo stems (*Bambusaceae* L.)
Particle size	2–4 [mm]
Glow residue (850°C@4h)	1.03 [wt.%]
Moisture content	7.1 [wt.%]

Table 10.14 Experimental conditions

Name of experiment		TP 29
Temperature	[°C]	488
Vapour residence time	[s]	1.36
Particle size feed	[mm]	3
Particle size sand bed	[mm]	0.55
Run time	[h]	1.5
Gas flow (cold)	[m³/h]	5.23
Gas flow (hot)	[m³/h]	12.93
Throughput	[g/h]	1961

A flow scheme of the flash pyrolysis plant is given in Figure 10.18.

The main experimental parameters of the thermal decomposition of bamboo are presented in Table 10.14.

Results and discussion

Field trials established in 1990 with 17 genotypes of bamboo have been monitored for their morphological characteristics. Table 10.15 indicates some tillers or stems were more than 3m long and the number of the tillers was up to 286 (El Bassam et al, 2001).

The bamboo produced approximately 7t DM/ha/yr and thus, it did not reach the yield potential of conventional crops. Both stem heights and the shoot number are an important potential for high biomass production. The average dry matter content was 83 per cent. The proportion of stems was 70–80 per cent, the side tillers 8–15 per cent and the leaves 12–15 per cent of the total biomass. The analysis of the stems and the leaves (Table 10.16) indicates that leaves contain much higher nitrogen, phosphorus and potassium than the stems.

The energy content of bamboo stem (17.1MJ/kg DM) and the cellulose content of over 40 per cent

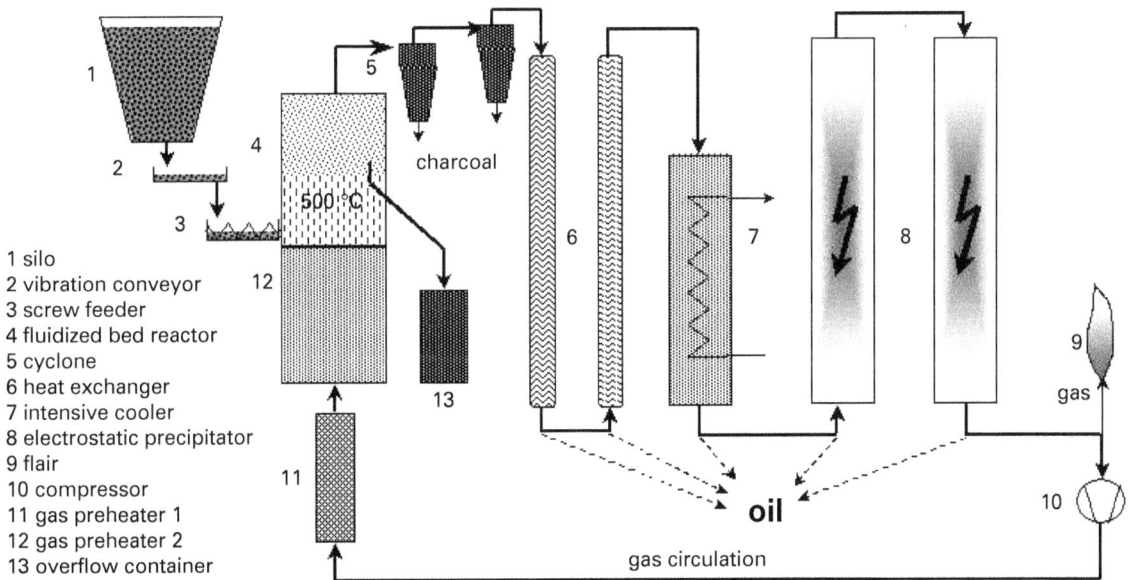

1 silo
2 vibration conveyor
3 screw feeder
4 fluidized bed reactor
5 cyclone
6 heat exchanger
7 intensive cooler
8 electrostatic precipitator
9 flair
10 compressor
11 gas preheater 1
12 gas preheater 2
13 overflow container

Figure 10.18 Scheme of the flash pyrolysis plant for biomass

Table 10.15 Some morphological data of the 17 genotypes after 5 years

Bamboo genotype	Plant height (cm)	Longest tiller (cm)	Number of tillers (n)
Fargesia murielae	100	130	144
Fargesia nitida 'Eisenach'	140	189	87
Fargesia nitida 'Nymphenburg'	170	223	256
Phyllostachys aureosulcata 'Spectabilis'	128	163	19
Phyllostachys humilis	115	185	10
Phyllostachys nigra	175	280	143
Phyllostachys praecox	70	135	6
Phyllostachys viridiglaucens	85	118	81
Phyllostachys vivax	104	117	22
Phyllostachys 'Zwijnenburg' (?)	178	217	286
Pseudosasa japonica	40	73	60
Sasa disticha (syn. *Pleiblastus distichus*)	167	227	127
Sasa keguma	215	325	79
Sasa kurilensis	30	30	9
Sasa palmata	48	60	6
Sasa pumila (syn. *Pleiblastus pumilus*)	205	273	130
Semiarundinaria kagamiana	53	68	18
Average	119	165	87

Table 10.16 N, P, K and Mg contents of the stems and the leaves of bamboo (percentage)

	N	P	K	Mg
Leaves	1.54	0.13	0.76	0.073
Stems	0.47	0.08	0.54	0.036

Table 10.17 Laboratory analysis of bamboo lignin, cellulose and energy contents

Lignin (%)		Cellulose (%)		Energy (MJ/kg DM)
Stems	Leaves	Stems	Leaves	Stems
10.7	8.7	40.9	22.0	17.1

emphasize the suitability of bamboo as a raw material for energy, pulp and paper production (Table 10.17; El Bassam et al, 2001).

Pyrolysis

Flash pyrolysis of plant material generally gives three fractions of products:

1.   a liquid, or bio-oil;
2.   a gas; and
3.   a charcoal fraction.

Figure 10.19 shows the distribution of these fractions obtained after the pyrolysis of bamboo stems.

Bio-oils are complex mixtures of degradation products of cellulose, hemicellulose and lignin, having a wide range of molecular weight distribution and polarities. Various analytical methods are necessary to give a chemical description of a bio-oil. A typical scheme of analytically accessible fractions is presented in Figure 10.20.

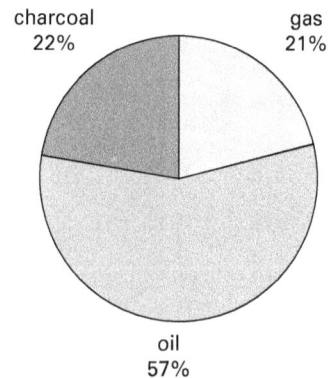

Figure 10.19 Yield (per cent weight) of pyrolysis products of bamboo (oil percentage is corrected by the wood moisture 7.37 per cent weight, water content of oil 18.69 per cent weight, reaction water 11.32 per cent weight)

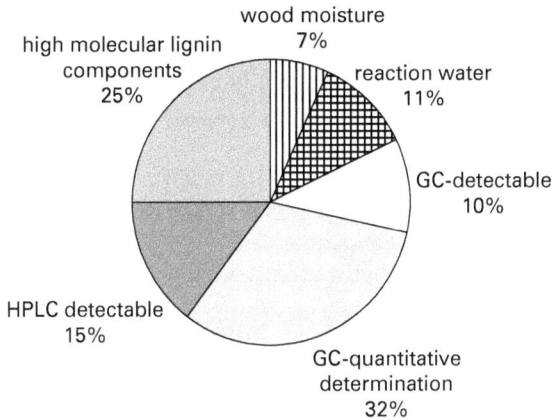

wood moisture 7%

high molecular lignin components 25%

reaction water 11%

GC-detectable 10%

HPLC detectable 15%

GC-quantitative determination 32%

HPLC: high performance liquid chromatography, GC: gas chromatography

Figure 10.20 Analytical accessibility of the bio-oil from bamboo

The volatile compounds of the bamboo bio-oils were separated by gas chromatography, quantified with a flame ionization detector (FID) and identified by a mass selective detector. The main compounds are presented in Table 10.18.

As mentioned above, bio-oils can be used to substitute for liquid fuels. In contrast to fossil fuels, bio-oils have unique characteristics due to their completely different composition. As can be seen in Table 10.18, all compounds have at least one oxygen functional group making the oils water miscible to a certain extent (El Bassam et al, 2001).

Table 10.19 gives an overview of the main differences between bio-oils and mineral oils. Despite their relatively low heating value, successful trials have been made to burn bio-oil in burners, diesel engines and gas turbines (Kaltschmitt and Bridgewater, 1997). In parallel to the activities in using bio-oils for energy, their potential use as a chemical feedstock has been investigated (Radlein and Piskorz, 1997).

Bamboo has had a very wide range of uses, but only a limited amount is used as firewood. The value of bamboo as a fuel source is connected to its dry matter content. Bamboo's uses for processed products mean that a market and market value do not exist for using it as an energy and fuel source. Although, to a small extent, bamboo is burned for heat production, this does not approach the scale of using bamboo as a substantial energy or fuel producer. Bamboo is considered a possible candidate for bioenergy production because of its high biomass yield and its

Table 10.18 Volatile compounds in pyrolysis-oil (corrected by wood moisture), based on dry oil

Volatile compounds	wt.%
Carbohydrates (main components)	
Acetic acid	24.18
Hydroxypropanone	9.95
Hydroxyacetaldehyde	5.76
Levoglucosan	3.49
2-Furaldehyde, 2-furfural	2.49
(5H)-Furan-2-one	0.96
2-Hydroxy-1-methyl-1-cyclopentene-3-one	0.70
Dihydro-methyl-furanone	0.69
alpha-Angelicalactone	0.51
gamma-Butyrolactone	0.43
2-Furfuryl alcohol	0.33
4-Hydroxy-5,6-dihydro-(2H)-pyran-2-one	0.30
3-Methyl-2-cyclopentene-1-one oil	0.14
2,5-Dimethoxy-tetrahydrofuran (cis) oil	0.10
Σ	50.04
Decomposition products of lignin	
Guajacyl components	
Isoeugenol (trans)	1.07
Guaiacol	0.71
4-Methyl guaiacol	0.51
Vanillin	0.33
Isoeugenol (cis)	0.26
Guaiacyl acetone	0.13
Coniferylaldehyde	0.13
Acetoguaiacone	0.12
4-Propyl guaiacol	0.07
4-Ethyl guaiacol	0.07
Eugenol	0.06
Σ	3.46
Syringyl components	
Syringol	0.65
Acetosyringone	0.36
4-Vinyl syringol	0.35
4-Methyl syringol	0.31
4-Propenyl syringol (trans)	0.31
Syringaldehyde	0.24
4-Allyl- and 4-Propyl syringol	0.19
4-Propenyl syringol (cis)	0.16
4-Ethyl syringol	0.10
Syringyl acetone	0.08
Sinapaldehyde	0.04
Isomer of Sinapyl alcohol	0.02
Propiosyringone	0.01
Σ	2.83
Other Lignin components	
3-Methoxy catechol	3.87
1,2 Ethanediol	1.04
Phenol	0.66
4-Hydroxy benzaldehyde	0.26
Meta-cresol	0.22
4-Vinyl phenol	0.17
2,4- and 2,5-Dimethyl phenol	0.14
4-Methylanisol	0.12

Table 10.18 Volatile compounds in pyrolysis-oil (corrected by wood moisture), based on dry oil (*cont'd*)

Volatile compounds	wt.%
Resorcin	0.12
Ortho-cresol	0.11
3- and 4-Ethyl phenol	0.10
3-Methyl catechol	0.08
Acetophenone	0.05
Benzylalcohol	0.03
2,6-Dimethyl phenol	0.02
Σ	6.99
Σ, Total	63.33

Table 10.19 Typical properties of bio-oil and mineral oil

	Bio-oil	Mineral oil
pH	2.6	–
Water content, wt.%	20	0.03
Viscosity (50°C), cSt	30	6
Density (15°C), g/cm³	1.24	0.89
Heating capacity (Hu), MJ/kg	17	40
Ashes, wt.%	0.03	0.01

high heating value of about 4600cal/g, whereas wood has a heating value of 4700–4900cal/g (Chen et al, 1995). The use of bamboo as an energy and biofuel source is currently being investigated. Beside briquetting, charcoal and production of powder to be used as biofuel, gasification tends to be a solution for the future, by reducing exhaust gases and their adverse influence on the environment, such as NO$_x$.

Selection of adapted bamboo genotypes and further breeding, as well as the development of bamboo crop management from the time of planting to the harvest, including harvesting systems, are factors that can improve the yield of bamboo, which is the basic requirement for the introduction of bamboo as an energy source.

BANANA (*Musa × paradisiaca* L.)

Description

Banana plants are of the family Musaceae. They are cultivated primarily for their fruit, and to a lesser extent for the production of fibre and as ornamental plants. As banana plants are normally tall and fairly sturdy they are often mistaken for trees, but their main or upright stem is actually a pseudostem (literally 'fake stem'). For some species this pseudostem can reach a height of up to 2–8m, with leaves of up to 3.5m in length. Each pseudostem can produce a bunch of yellow, green or even red bananas before dying and being replaced by another pseudostem.

Bananas are cultivated in the tropics and in many of the subtropical regions. The wild forms originated in Southeast Asia. They were brought to Africa in the first millennium BC. In America they were first introduced after 1500 AD. The banana is a herbaceous perennial with an underground rhizome, from whose sprouts the new fruit-bearing shoots are formed. The sprouts are also used for propagation. The sheaths of the 4–6m long and 1m wide leaves form a pseudostem. The inflorescence develops 7–9 months after the appearance of the sprout. After the fruiting shoot has grown through the hollow pseudostem, it bends and grows more or less vertically downwards. In the axils of the red-violet bracts, groups of flowers are formed, first females, then hermaphrodites, then males. The fruits develop from female flowers without pollination. The hermaphrodite flowers do not form fruits and can be absent (Rehm and Espig, 1991). The fruits are 10–25cm long with a green, yellow or brown colour, and with seeds or seedless (FAO, 1996a). The cultivated forms are diploid, triploid or tetraploid, based on the wild species *M. acuminata* Colla (genome A) and *M. balbisiana* Colla (genome B). So their genomes (each with many cultivars) are: 2n: AA or AB, 3n: AAA, AAB, ABB or RBB, 4n: AAAA or ARBB.

Ecological requirements

Bananas require a uniformly warm climate with an annual average temperature above 20°C, a large amount of sunshine and evenly distributed precipitation of about 2500mm/yr. The banana prefers more than 60 per cent humidity. Depending on the cultivar it may be killed by frost and damaged if the temperature falls below 7–12°C for more than several hours. The dwarf varieties cultivated in the subtropics endure temperatures as low as 0°C (Rehm and Espig, 1991). The species *Musa japonica*, originating from the Japanese Ryukyu archipelago, can be grown as an ornamental shrub even in central Europe. The stem and the leaves of this species

Source: Ofeky 2006 http://commons.wikimedia.org/wiki/File:Musa_acuminata5.JPG

Figure 10.21 Banana field

may be killed by the frost, but the plant will regenerate in the following spring if the frost does not reach the roots (van der Palen, 1995). The ground should be loose, well drained and rich in organic material in the top 20cm. Fertile volcanic or alluvial soils are preferred. The optimum pH value is 5–7, but both higher and lower values are tolerated. For high yields the uptake of nutrients is high, 1 tonne of bananas contains 2kg N, 0.3kg P, 5kg K, 0.4kg Ca and 0.5kg Mg.

Dwarf bananas also thrive in the subtropics and withstand temperatures as low as 0°C, though this brings growth to a standstill. Dwarf bananas are planted close together and develop very quickly, so they can produce yields similar to the tropical cultivars if they are sufficiently irrigated and receive high levels of mineral fertilizer. The banana is very susceptible to wind and hail damage (FAO, 1996a).

Propagation

Propagation of the banana is carried out vegetatively with suckers. Preferred are suckers with leaves that are not unfurled. These 'sword suckers' as well as 'maiden suckers' (young sprouts with unfurled leaves) or older sprouts with leaves are used for propagation, but from the older sprouts the leaves should be removed and the terminal growing point cut out, with the aim that a new sprout develops from an axillary bud. The sets should be treated against nematodes with hot water (56°C for 5 minutes) and with pesticides against pests and diseases. Propagation by way of meristem culture has also been used commercially for several years.

The in vitro production of meristem cultures provided a great advance in the rapid vegetative multiplication of Musa. Taken from the central growing point and lateral shoots, the tissue is placed on a growth medium and incubated. From 15 to 20 shoots are produced from each meristem in about 30 days. The shoots are transferred into fresh tubes of medium and take about 50 days to develop sufficient roots for transfer to soil.

The somaclonal variation resulting from tissue culture is a problem with respect to plant storage, but it offers a means to generate mutants for resistance screening to major diseases and pests (Stover and Simmonds, 1987). Resistance to Panama and sigatoka diseases is an aim of breeding. Furthermore, fruit characteristics and tillering capability are important.

Crop management

Planting distances are 3m or more for tropical cultivars and 2m for 'Dwarf Cavendish'. Only enough side shoots are allowed to grow and to build a fruit bunch every 4–6 months; excess side shoots are cut off, or killed with chemicals. From the appearance of the tip of the new daughter plant to the ripening of the fruit takes 12–14 months. The cultures last for 15–20 years before they need renewing (Franke, 1985). The most important feeding roots lie close to the surface, so the working of the soil for weed control should be kept to a minimum to avoid damaging the roots. A 20cm thick mulch layer on the ground restricts most weeds. For the preparation of this layer, grass, banana remnants or other organic materials are used. This will also contribute to the supply of potassium. Fertilization with nitrogen compounds is always necessary.

The leaf spot diseases yellow sigatoka (*Mycosphaerella musicola* Leach) and black sigatoka (*M. fijiensis*) are controlled by regular spraying with fungicides. Where the banana wilt (Panama disease) caused by *Fusarium oxysporum* Schlecht var. *cubense* is extensively present, sensitive cultivars cannot be grown. The selection of healthy planting material and disinfection of the sets, together with orchard sanitation and the removal of infected plants, are the most important preventive measures against other fungal, bacterial and insect diseases.

Production

Yields can vary considerably. Peak yields can reach 50t/ha per year. The dwarf forms produce similar

Source: Jean_Pol Grandmont 2006 http://commons.wikimedia.org/wiki/File:Musa_JPG01.jpg

Figure 10.22 Banana fruit cluster

yields per hectare because they are planted closer together than the tall cultivars and develop very quickly. The yield of plantains is similar to the yield of bananas: in well-kept plantations yields are 38–50t/ha, though with less care this can fall to 15–20t/ha (Rehm and Espig, 1991).

In a study, surface irrigation consumed 2500mm of water and gave a banana yield of 28t/ha, whereas drip irrigation once in 2 days at 24 litres/plant consumed 1081mm of water and gave a banana yield of 31t/ha.

The production of bananas under cover is well established and common practice in some desert-like regions (such as Tenerife). This system allows very high productivity and reduced water and nutrient inputs.

Figure 10.23 Kooky fruitmobile

Processing and utilization

To harvest the bananas the trunk is broken down and the fruit bunches are cut off. At the packing stage they are washed and treated with disinfectants. For export they are harvested green-ripe and shipped in cold storage chambers.

Bananas are mainly used for nutritional purposes, either for fresh consumption or dried or processed to banana flour. The carbohydrate content in the edible part of fresh bananas is about 20 per cent. During the ripening period the starch is converted into sugar, though in the cultivars known as plantains this process does not take place. Starch is produced from unripe bananas and from plantains (Franke, 1985).

Because of its sugar and starch content, the banana is also a promising energy plant. Unsaleable or surplus bananas could be used to produce energy. Sugar and starch can be used as source of ethanol, with utilization as a biofuel, solvent or raw material for the chemicals industry. The fermentation of sugars is a well-known process for producing alcohol. The sugar can be directly fermented to form ethanol. The starch has to be broken down into sugars before fermentation, using acids or hydrolytic enzymes. Another source of energy is certainly required, but the main cost is the raw material (Wright, 1996). The development of continuous fermentation techniques and the use of better varieties of yeast have improved efficiency and lowered the costs (FAO, 1995b).

The trunks are cut down to harvest the bananas, so a great quantity of organic material remains as a by-product of the harvest. It should be possible to use the harvested remains of trunks and leaves, which are currently only waste, for energy purposes – for example, after anaerobic fermentation to obtain biogas. The residue can be used as a fertilizer, as are the complete trunks today. In 1993, Costa Rica alone generated 1,750,346t of waste tree stems and leaves, 317,592t of wasted substandard fruits, and 143,528t of raceme stem waste (Hernandez and Witter, 1996). The methanogenic bacteria of the anaerobic fermentation can convert about 90 per cent of the feedstock energy content into a biogas that is a readily usable energy source. The biogas contains approximately 55 per cent methane. The sludge produced in this process is non-toxic and odourless, and has lost relatively little of its nitrogen and other nutrient content (FAO, 1995b).

Australian engineers have created an electricity generator that is fuelled by decomposing bananas and converts them to methane. They hope to build a full-size fruit-fuelled power station (BBC News, 2004). Growcom, an Australian horticulture/biofuel organization, has started the pre-construction process of a commercial biomethane plant. The plant will produce biomethane from banana waste to provide fuel to vehicles that run on natural gas (Walsh, 2008).

Internet resources

http://news.bbc.co.uk/go/pr/fr/-/2/hi/science/nature/3604666.stm, accessed 27 August 2004
www.nextautos.com/plantsmanufacturing/banana-waste-to-produce-fuel-in-australia

BLACK LOCUST (*Robinia pseudoacacia* L.)

Description

Robinia is a genus of about 10 species native to eastern North America and Mexico. The genus *Robinia* is named for Jean Robin (1550–1629) and his son Vespasian Robin (1579–1662), herbalists to kings of France and first to cultivate locust in Europe (Holzarten Lexikon).

Robinia pseudoacacia L. is endemic to North America between the latitudes 43°N and 35°N. Its natural range is divided into two zones. Its main zone of distribution is the Appalachian Mountains of central Pennsylvania to northern Alabama and Georgia. The second zone covers the Ozark Plateau of southern Missouri and parts of northern and western Arkansas and eastern Oklahoma.

There are many species within the genus *Robinia* used for floriculture, but only *R. pseudoacacia* L. plays

Figure 10.24 Black locust

an important role concerning forestry and wood production. *R. pseudoacacia* L. is also known as black locust, white locust, false locust, common locust, false acacia and white acacia.

Black locust is a multi-purpose tree with interesting characteristics. It is a pioneer plant with very rapid juvenile growth. Nonetheless, the wood is very dense, high in decay resistance, has excellent colour and makes a good pulp. This combination of high growth rate and high wood density is not so often seen with other trees (Hanover et al, 1991). In addition, black locust is a leguminous plant that can fix atmospheric nitrogen. This gives it the ability to tolerate low-fertility site conditions. Locust uses the C_4 photosynthetic pathway for utilizing solar energy. The leaves are high in nitrogen and can be used as animal feed. A very aromatic honey can be produced from the flowers.

R. pseudoacacia L. reaches a height of 15–35m. The stalk is characterized by a very high tendency towards warping. The wood is very heavy, hard, extremely durable and elastic. The fresh wood is green-yellow. Black locust is characterized by a very high net photosynthetic rate, high light saturation, low stomatal diffusion resistance, long leaf retention time and a high transpiration rate (Hanover et al, 1991).

The inflorescence of black locust is raceme-like when developed. Normally, time of flowering is May to June, after leaves have appeared. Pollination is by insects only. Black locust flowers at an early age of about six years. Seed maturity is at the end of October. The fruit is a red-brown pod containing 6 to 10 seeds. The seeds are kidney-shaped, brown-grey to black, and start to fall off in February. The seeds are characterized by a very high skin hardness.

The root develops a turnip-shaped taproot during the juvenile growth. Black locust produces strong lateral roots with increasing age. The whole root system is characterized by intensive, wide-branched and dense fibrous upper roots. Roots near the soil surface can reach a length of 20m and more (Göhre, 1952). There is a rapid and extensive growth of strong roots in combination with the development of intensive fine roots during the first years. Thus, black locust is able to bind loose soils in a short time. Black locust is a so-called pioneer plant and has a very high root-suckering capacity. Its root cuttings show vigorous sprouting.

Black locust is cultivated in many countries. Keresztesi (1983) lists some of the European countries with the largest areas of black locust stands as of 1978: Hungary (275,000ha), Romania (191,000ha), the then Soviet Union (144,000ha), France (100,000ha), Bulgaria (73,000ha) and the then Yugoslavia (50,000ha). Other European countries with black locust stands are the Czech Republic and Slovak Republic (28,000ha together) and Spain (3000ha), as well as Austria, Belgium, Greece and Ireland. Outside Europe, South Korea has cultivated about 1,017,000ha of black locust forests on infertile soil for firewood, erosion control and feedstocks. The present area of black locust growth is estimated to be about 3.25 million ha worldwide (Hanover et al, 1991). This is the third largest area in the world within the group of fast-growing trees, after eucalyptus and poplar.

Ecological requirements

The climate of origin of black locust has a mean annual precipitation between 1000 and 1500mm. The maximum temperatures in July are 30–38°C, and the minimum temperatures in January lie between −10°C and −25°C.

Black locust is adaptable to a wide range of sites because it does not make specific demands on its environment. It has shown to be resistant to environmental stresses such as drought, high and low temperatures, and air pollutants. It grows in a wide range of soils, but not in very dry or heavy soils. Black locust prefers sites with loose structural soils, especially silty and sandy loams. Soil aeration and water regime are the most important soil characteristics for good black locust growth. Also of importance is the soil's rootable depth. Compacted and waterlogged subsoils are not suitable for black locust cultivation. Black locust has the ability to flourish rapidly on disturbed sites, and has therefore been widely planted in erosion control programmes and for soil improvement in the United States (Hanover et al, 1991).

Black locust is a leguminous plant whose roots are infected with nitrogen-fixing nodule bacteria. The nitrogen-fixing capacity of symbiotic bacteria is difficult to determine because it depends not only on the effective combination of the amount and phyla of nodule bacteria but also on the photosynthesis of infected plants and soil conditions.

Experiments using black locust for reclamation on a spoil mound site in a lignite mining region showed that nitrogen fertilization reduced the growth of black locust trees. Potassium and phosphorus application improved nitrogen uptake, which resulted in increasing nitrogen content of the leaves. In addition, potassium positively affected the nodule growth of nitrogen-fixing bacteria (Hoffmann, 1960; Heinsdorf, 1987). Studies from Reinsvold and Pope (1987) have shown that the addition of phosphorus increased nodule number and nodular dry weight. With the addition of nitrogen, however, nodulation was reduced with the lowest and with the highest levels of added nitrogen. Pope and Anderson (1982) found an increased nodular biomass under trees fertilized with phosphorus and potassium in comparison to non-fertilized trees.

Zimmermann et al (1982) reported that black locust plants treated with nitrogen at 50kg/ha had significantly higher fresh nodule weight than the non-fertilized control plants. According to Reinsvold and Pope (1987), a minimal level of mineral N is needed, at least initially, to support nodulation.

Nitrogen fertilization in two applications during the growing season at a rate of 113kg/ha increased biomass production after one year by more than 30 per cent, but only by 12 per cent by the end of the third year (Bongarten et al, 1992). The authors explain these results by stating that the nitrogen fixation consumes more energy than the assimilation of nitrate, but as trees increase in size and nitrogen demand there are not enough developed nitrogen-fixing bacteria, and soil nitrogen sources are not adequate for the increased biomass as the fertilization level is held constant.

It is known that the black locust positively affects the nitrogen content of the soil. In experiments with sterilized cultures, nitrogen output from roots or nodules of black locust could not be found. Otherwise, black locust positively influenced the growth of mixed forest trees. In black locust stands where the litters of the trees were removed this positive effect could not be proved. These results are in agreement with results from Keresztesi (1988). He found that old black locust forest stands (16–20 years) enriched the upper 50cm of soil with 590kg/ha of nitrogen, but in young stands (5–10 years) this could not be observed. The conclusion to be drawn from this is that nitrogen-fixing bacteria do not directly influence the nitrogen content of the soil. Nonetheless, they increase the nitrogen content of the plant material. It is litter decomposition that causes total nitrogen enrichment of the soil.

Hoffmann (1960), reported that 35kg/ha of nitrogen are fixed in a two-year-old black locust stand (28,000 plant/ha) from which nearly 15kg/ha in the form of litter returns to the soil. The remainder is fixed in the stem, roots and bark. Nitrogen levels after three years were 116kg/ha fixed with 30kg/ha returning to the soil as litter and after four years 305kg/ha fixed with 200kg/ha returning to the soil as litter.

The soil pH value can influence not only the distribution of *Rhizobium* bacteria but also the availability of the nutritive material. Wide distribution of the soil's calcium carbonate places certain restrictions on soil acidity. Therefore, available phosphorus and pH value are the main factors influencing nitrogen fixation (Deng and Liu, 1991).

Propagation

Propagation of black locust is possible through root cuttings, green-wood cuttings, seedlings or micropropagation. Propagation using root cuttings and green-wood cuttings provides a guaranteed quality, but is more expensive than seed propagation. The optimal time for obtaining root cuttings seems to be spring, though it is possible to obtain root cuttings throughout the year (Göhre, 1952). Root pieces are cut with a length of about 8–10cm and diameter of 2 to 5cm and are planted vertically 1–2cm under the surface with the 'root side' downwards. They should not be put into the soil below a depth of 5cm. Plant spacing in the row should be 5–8cm, with 80–100cm between the rows. The best period for planting is April, and after 3 weeks the root cuttings should begin to sprout. Propagation using green-wood cuttings is also possible. After cutting, the pieces are prepared with a phytohormone and put into a special soil substrate for root development under controlled conditions in a greenhouse (Müller, 1990).

Propagation by seed is not a problem because of black locust's frequent and abundant seed production and the high survival rate of the seedlings. The germination rate is between 40 per cent and 60 per cent, but can be increased by special methods like scouring. The seedlings are produced in a nursery and planted into the field in spring or autumn.

Micropropagation of black locust is still at the research stage, though the first methods of in vitro shoot propagation have been established. The success of regeneration processes depends on the genotype and age of the basic tree. Plants produced with the help of

in vitro propagation can be used as basic plants for propagation by cuttings (Ewald et al, 1992).

Crop management

For the production of fast-growing stands, one-year-old seedlings are preferred. When choosing a location for black locust stands it is important to remember that *Robinia* is shade intolerant. Both in its young and mature forms, black locust cannot survive in shady conditions.

The most popular plant spacing is 2.4m × 0.7–1.0m. This spacing requires 4000 seedlings per ha. Other studies show that plant spacings from 1.0m × 1.0m (10,000 plants/ha) and 0.50m × 0.25m are necessary for high biomass yield from short rotation forestry (Baldelli,

Source: MaxFrear http://en.wikipedia.org/wiki/File:Spartium_junceum_ginesta.jpg

Figure 10.25 Black locust provenance/progeny test plantation in Michigan containing over 400 half-sib families was established in 1985. The first trees bloomed in 1987, in three years from seed.

1992). It is important to control weeds in the first season after planting to prevent the weeds from overgrowing the young trees. The density and planting design should be chosen to fit weed control and harvest options. One of the aims should be to try to achieve a closed canopy as soon as possible to maximize production and minimize weed competition. A stand that is too dense will be self-thinned. If the intention is to establish a black locust plantation, stems and roots have to be cut to ensure tree development. The results of different experiments concerning plant density have shown that plant spacing does not affect biomass production. It seems that black locust stand density does not depend on the initial plant spacing because the plantation will adjust its density through self-thinning.

Phosphorus and potassium should be added if soil analyses have shown these to be present in extremely low quantities. In most cases, nitrogen fertilization is not recommended. It is probably better, for economic and environmental reasons, to take advantage of the plant's nitrogen-fixing ability.

Black locust stands have been observed to be attacked by numerous diseases and pests throughout Europe. The diseases and pests vary greatly with climate and European region.

The cultivation of black locust stands, especially on plantations, is designed to encompass numerous harvests carried out over many years. Because of this, there is no need to develop a crop rotation. In addition, because of the tree's excellent coppicing ability, a stand established for energy production can be successfully regenerated from sprouts several times over without the need for replanting crops.

Production

Trees grow to 40–100ft (12–30m) in height. Trees grow upright in forests, but develop an open growth form in more open areas. Leaves are pinnately compound with 7–21 small, round leaflets per leaf. Leaflets are 4cm long. A pair of long, stipular spines is found at the base of most leaves. Flowering occurs in the spring, when flowers develop in 20cm-long clusters. The showy, fragrant, white to yellow flowers give way to a smooth, thin seed pod that is 5–10cm in length. The bark of black locust is light brown, rough, and becomes very furrowed with age (forestryimages.org).

In comparison to other wood species, black locust produces the highest biomass yield as a result of its early

growth and high density. In field experiments in Austria, annual dry matter production between 5 and 10t/ha was reached by three- and four-year rotations of black locust stands from 10,000 trees/ha (Müller, 1990). The four-year rotation reached an increase of yearly dry matter yield that was some 1.4 times higher than that for the three-year rotation. The moisture content of the wood

Source: Bofaurus stellaris 2007 http://commons.wikimedia.org/wiki/
File:Robinia-pseudoacacia-12-V-2007-6032.jpg

Figure 10.26 Stand of black locust

ranged between 30 and 38 per cent. All moisture contents were below the maximum moisture content for maintenance of combustion.

Examinations of different black locust stands have shown that a nine-year-old black locust stand produced the highest annual woody biomass yield of 3.43t/ha compared with stands aged from two to ten years. The annual biomass production ranged from 1.1 to 2.99t/ha within two- to six-year rotations. After six years the annual woody biomass production increased to over 3t/ha with a small loss in the tenth year. On a five-year rotation, black locust can produce 57.7GJ/ha per year (Stringer and Carpenter, 1986). After two years 43.5GJ/ha per year can be reached, with a maximum from a nine-year-old stand at 76.8GJ/ha per year.

One three-year rotation produced woody biomass that ranged from about 3 to 8t/ha per year. The maximum of about 24t/ha after three years was reached by the best black locust families with irrigation and nitrogen fertilization (Bongarten et al, 1992). All the yield data are only examples and not comparable because of the very different growing conditions of the stands with regard to planting density, soil properties, fertilization, climate, etc.

Processing and utilization

Results indicate that the current annual increment of black locust peaks at seven to eight years. But by this time, the stems may be too thick to harvest using available techniques. The grower should examine the economic options before deciding to harvest, taking into account the ease and possibility of harvesting. On short rotation plantations, the small stem diameters necessary for mechanized harvesting require a shorter rotation period of between three and five years. Black locust's yearly average of dry matter yield lies between 12 and 15m³/ha. In fact, after coppicing, the tree sprouts from the main stumps and also from the roots, which are diffused in various directions.

A study of its morphological characteristics has shown differences between black locust and other fast-growing trees such as willow and poplar. Black locust has thorns, which make it difficult to handle manually, and it is thus preferable to chip it in the field. Black locust can sprout from the roots, so after the third or fourth cut regrowth will also occur between the rows. In addition, black locust has harder wood than other

fast-growing trees. Therefore the cutting apparatus must be more durable and powerful than that used normally. There are different machines for direct harvesting (Baldelli, 1992).

Black locust was used primarily as fuel and for agricultural purposes. A common use has been for fence posts. The hardwood contains flavonoids which make it extremely decay resistant. The good durability and strength of black locust mean that the timber industry has uses for this species, especially in mining and hydraulic applications. Other products made from black locust are sawn logs, sawn wood, pit props, parquet, railway sleepers, round wood framing and interior fittings for agricultural buildings. Black locust may also serve as a source of pulp for the paper industry. The final felling of black locust takes place after 30–40 years and produces high-quality sawn wood.

Although Black locust has many uses, its highest value can be in biomass for energy feedstock. Many characteristics of black locust show interesting qualities for biomass production in energy plantations. The wood has excellent properties for timber production and for energy use. The heating value of the wood and bark is high and it burns well even when it is green. In Austria, heating values of about 17.82 for wood and 20.08MJ/kg for bark were measured (Müller, 1990). Experiments have shown that the heating value of the bark is higher than that of the wood: Stringer and Carpenter (1986) reported average heating values of 20.81MJ/kg for bark compared with 19.4MJ/kg for wood. In addition, the moisture content of bark (34.7 per cent) was lower than that of the wood (55.5 per cent). The leaf tissue had a moisture content of 60.6 per cent. The total moisture content of whole black locust trees ranged from 46.1 per cent for two-year-old trees to 39.1 per cent for nine-year-old trees.

The tree's nitrogen-fixing ability produces high nitrogen levels in the plant material, an advantage that makes black locust suitable for fodder production. But the nitrogen content of plant material for energy use plays an important role in combustion. Demand for low NO_x emissions requires a low nitrogen content in the plant material, so the nitrogen content of the combustion material must be as low as possible. Analysis of black locust shows that nitrogen content, especially of the leaves, is very high compared with

Table 10.20 Stem diameter variation in wood properties

Diameter class (cm)	Specific gravity[a]	Caloric content (cal/g)[b]	Moisture content (%)[b,c]	Heartwood content (%)[a]
		Mean ± SE		
0.1–2.5	0.549	4641 ± 52	41.1 ± 16.5	absent
2.6–5.0	0.588	4644 ± 58	38.0 ± 16.1	3.4
5.1–7.5	0.644	4637 ± 34	33.2 ± 9.7	28.2
7.6–10.0	0.658	4665 ± 42	26.7 ± 6.4	38.0
Mean	0.609		33.1	

Notes: Includes main stem and branch material from 2- to 10-year-old black locust trees.
[c]% wet-weight basis.
Source: [a]Stringer (1981); [b]Stringer and Carpenter (1986); [c](Hanover et al, 1991).

other trees. Different authors reported nitrogen contents of leaves between 2.3 and 5.5 per cent.

Detailed analysis of 2- to 10-year-old black locust biomass components provided an accurate energy content and energy yield determination as well as an index defining a relative ceiling on the delivered cost for black locust biomass in comparison with other fuels such as coal and forest residues. Total energy content of individual biomass components ranged from 20.97 × 10⁶J/kg for leaf tissue to 19.23 × 10⁶J/kg for current years' growth. Annual total energy yields ranged from 33.75 to 76.79 × 10⁹J/ha/yr for 3- and 9-year-old stands, respectively. Nine-year-old stands also exhibited the highest annual net whole-tree and woody biomass (whole-tree less foliage) energy yields of 32.71 and 30.73 × 10⁹J/ha/yr. The net annual energy yields were consistently greater for whole-tree biomass compared with woody biomass due to the foliage included in the whole-tree biomass. The relative cost indices maintained a high degree of variability between comparison fuels. Bituminous coal and forest residues were lowest with an overall mean cost index for woody biomass of US$13.28/Mt and US$13.72/Mt, respectively. Woody biomass maintained a greater relative index than whole-tree biomass over all age classes due to its inherently higher conversion efficiency (Stringer and Carpenter, 1986).

There are at least six ways in which black locust can be used as a fibre crop or to generate large amounts of biomass at relatively low energy inputs. These include pulp for paper, and leaves and young stems for fodder, and for solid, liquid or gaseous fuels.

Black locust is suitable for many purposes. As one of the most adaptable and rapidly growing trees available for temperate climates, it is valued for erosion control and reforestation on difficult sites. Vast new forests of fast-growing species may be needed to slow the accumulation of CO_2 in the atmosphere. Energy from woody biomass will be more important in the future when fossil fuel use is reduced, either to preserve the global ecosystem or simply due to rising costs (Barrett et al, 1990).

Internet resources

www.forestryimages.org
Holzarten Lexikon, www.holz.de/holzartenlexikon/de_zeigeholzart.cfm?HolzartenID=531

BROOM (*Genista*) (*Spartium junceum* L.)

Contributed by C. Baldelli

Description

The genus *Genista* (Leguminosae) includes a variety of characteristic plants, from creeping ground covers and massing shrubs to one small tree. With a single exception, all species bear bright golden yellow flowers in late spring or summer (*Genista monosperma* or *Retama monosperma* has white flowers). Among the many varieties (70–90 species of *Genista*) it is important to list *Spartium junceum* (Spanish broom, or ginestra), *Genista anglica, G. germanica, G. pilosa, G. sagittalis, and G. tinctoria. Genista aetnensis* (of Mount Etna, Sicily, Italy) is a small tree.

Genista are typically Mediterranean species but they are also present in some North American areas (west coast, southeast, Pacific northwest and mid-Atlantic) and Mexico. In South America, *Genista* is present in Brazil. There are also many occurrences in northern Europe and in Asia (the natural range of *G. tinctoria* even reaches into western Siberia).

Spartium janceum, Spanish broom, has several advantages, including its suitability for several marginal Mediterranean lands (from the south European Mediterranean to North Africa), its favourable industrial conversion, and its presence in the family of Leguminosae as a nitrogen fixer. Broom produces good quality biomass and precious fibre. The botanical name of the species (*Spartium*) comes from the Greek word 'sparton', which means rope.

The leaves of broom are sparse, small and deciduous, but the plant's green, reed-like twigs suggest the appearance of evergreens in winter. The plant is in fact defined by its thin, bright green, graceful and useful, nodding reed-like branches, which change very little during the season when the plant is not flowering. They contain an interesting natural fibre, which will be described below.

Broom has a very strong root apparatus, which explains the plant's application to soil protection. The influence of broom in erosion control is great. It helps to cut down the runoff and wash down of soil particles, which contain significant amounts of nutrients, and so has a direct influence on the productivity of the soil. Moreover, thanks to its strong, twined root system and light above-ground weight, broom encourages an efficient soil consolidation. Historically, broom has been used for this purpose because of its ability to block landslides on steeply sloping soils. The most significant demonstration of this property is seen in broom's large-scale employment along roadside slopes.

Broom is an old fibre crop. The Greeks and Romans used its fibre for textiles and ropes. During the Middle Ages, clothes, carpets and furniture coverings were made from broom. A major development of this use began in Italy in the 1930s, when government policy led Italian scientists and developers to increase study into this crop and its availability for fibre production. A significant amount of research work was carried out.

Source: Hans Hillewaert 2006 http://commons.wikimedia.org/wiki/File:Spartium_junceum_ginesta.jpg

Figure 10.27 Ginestra

The requirements of environmental protection, rural restructuring and new local jobs have led to a re-evaluation of earlier experiences of broom cropping, but utilizing the technological innovations now available. In particular, as will be discussed below, SRF (short rotation forestry) applied to broom could offer an interesting way to relaunch broom cropping and utilization.

Broom's suitability is mainly demonstrated by its good productivity in terms of biomass; it accepts very closely timed cutting cycles and is capable of fast regrowth after cutting. (The need for short cycles is the result of harvest restraints: mechanical harvest can be carried out only if the trunk diameter is below a certain size). Other favourable characteristics are: good physical-chemical qualities for industrial conversion; large-scale availability; no legal harvesting restraints (in Europe it is very difficult to cut and harvest biomass trees); good biomass production in the order of 6 to 10t/ha/year of dry matter; and very low cultivation costs.

Besides these characteristics, this species belongs to the legume family and is a rhizobial nitrogen fixer. Thus nitrogen fertilization can be avoided, preventing the pollution of aquifers, which is a problem today. In addition, it does not require chemical pest or weed control, an environmentally friendly feature that is in marked contrast to other fibre crops such as cotton, hemp, flax, kenaf, miscanthus, etc.

Ecological requirements

Broom is at its best in full sun, but tolerates some shade. Soil quality is also important. Like most other legumes, broom thrives in calcareous soils with low fertility, but performs adequately over a wide range of soil acidity (pH 5–8). Just as broom is a good choice for dry, infertile soils, so its continued health depends on the avoidance of excessive water and moisture, soggy soils, and heavy fertilization. Broom requires very free soil drainage, but is otherwise highly adaptable to many pedoclimatic situations, and to poor and even stony soils.

Propagation

Broom is not difficult to propagate. It can be planted using cuttings, seed or seedlings. Softwood cuttings taken in early summer root easily; semi-hardwood cuttings taken later in the summer and hardwood cuttings taken in the autumn have also been successful. These cuttings are then planted in October or February.

Direct seed propagation of broom is similar to that of many other legumes with hard seed coats. A half-hour soak in concentrated sulphuric acid is sufficient to make the seed coats permeable to water. Boiling water treatment may also work well, though some species or seed lots may not require either. A rapid mechanical treatment with a special knife mill is also possible.

Nursery and seedling transplantation can use various spacings. The most positive results have been seen with 1m × 1m spacing (a density of 10,000 plants/ha).

Crop management

Propagation by seeds is the easiest and cheapest way to cultivate broom. The best timing for the sowing is late autumn (November) or late winter (February). If a strong weed infestation is foreseen, late sowing is advisable. Sowing can be performed with the normal mechanical grass sowing machines with suitable regulation of parameters (density, diameter of seeds). The procedure includes good soil surface preparation with removal of stones and production of a fine tilth. Seeds have to be prepared before sowing, as mentioned above.

Seedlings are easy to prepare in the open air as well as in a greenhouse. Cuttings require more care, and it is better to cut pieces in late autumn and preserve them under refrigeration. Hormonal application before planting is preferable. Planting of cuttings or seedlings requires deep ploughing to ensure good contact between roots and deep soil, and surface preparation and stone removal are less important. Before planting, it is necessary to plough the rows deeply at the selected spacing (1m × 1m is a good average). All planting operations can be mechanized using existing planting machines. The best time for planting is from late January through to the end of March, when there are no problems with weeds.

If the soil is very stony, with large stones present, it is easier and quicker to plant into holes. The seedlings or cuttings are placed in holes 50–100mm deep and covered with soil to close the holes. These operations must be performed by hand. Planting in holes is less mechanized and slows the initial growth of young plants. This technique is favourable if the spring rains

Source: Carsten Niehaus, http://commons.wikipedia.org/wiki/File: Spartium_junceum_a.jpg

Figure 10.28 Ginestra

are light, because the thin ground cover over the holes retains more water than deep soil, and this benefits the roots. Obviously, seedlings and cuttings start to grow a year sooner than seeds and are more uniformly organized in the row.

Short rotation forestry (SRF) technology is the core of the proposed cultivation technique. The cropping methods follow the main aims of mechanizing the entire production cycle and minimizing agricultural problems, so it will be possible to reduce the cost of the final product (wood chips). To achieve this objective it is necessary to work with a very high plantation density (from 10,000 to 20,000 plants/ha) and with very short cropping cycles (3 years). Spacing in broom plantations is from 1m × 1m to 0.25m × 0.25m (10,000–80,000 plants/ha).

With regard to the environmental impact of broom cropping, a pre-summer harvest will help to overcome fire problems. Harvesting during the hottest and thus highest fire hazard period means that vegetable material will no longer remain in the fields and therefore there is nothing to catch fire. The coppiced plants sprout after 20–25 days, and in the following spring the new root suckers are already about 1m high, favouring the subsequent cycle. This procedure has already been verified.

Broom is a typical early succession species. Many people consider it a weed, and thus a dangerous plant. If the aim is to maximize biomass production, an easily established species is preferable, given the low amount of chemical input necessary.

Few diseases have been reported on broom plants, and none is a very frequent problem. The only true problem is vermene spot associated with plants in old age. The cutting cycle of every 3–5 years avoids spot problems.

Following are the important major characteristics of broom grown using SRF methods:

- no fertilizer is needed (broom is a nitrogen fixer);
- there is no weed control (the species is self-weeding);
- there is no agricultural work after planting, only harvesting every three years; and
- harvesting is possible in any season.

No crop rotation is recommended, because of the extended life of the crop.

Production

To offer some values, it can be considered that the optimum cycle is 2–3 years, in the context of vegetative cycles of over 10 years, with productions in the order of 6–10t/year of dry matter (40% maximum moisture).

Yield trials have shown that broom, 3 years from planting seedlings or 3 years from coppicing, at a spacing of 1m × 1m (10,000 plants/ha), reaches an average height of 3m and produces about 9t/ha of dry matter per year after the third year (moisture 7 per cent).

The efficiency of broom is very high, considering that 3 years after planting, most fast-growing forestry plants produce only an insignificant quantity of dry matter. Other trials performed on an old, wild broom plantation, with an a natural density of 8000 plants/ha, gave an average dry matter yield of 8t/ha per year.

The conclusion is that the yield of broom has the advantage of short harvesting cycles once roots are established. In this case, the yield can reach 10t/ha per year of dry matter, which is the target foreseen by forestry technicians for SRF applications.

Processing and utilization

Broom should be harvested in 2- to 3-year cycles. The optimum coppicing cycle is 3 years from planting, when the trunks have a diameter of about 5cm. In this condition, it is appropriate to call this kind of cultivation 'wood grass'.

As already mentioned, mechanical harvesting can be carried out only if the diameter of the trunks is below a certain value: the small diameter of the trunks of young plants permits mechanized harvesting by a special combine harvesting machine (under construction) that will overcome this main bottleneck of all forestry production.

Broom alone or mixed with other raw materials can produce, through special processes, several marketable final products such as biofuels, paper pulps, textiles, composite materials and agrocompost.

The state of the art of broom exploitation has created innovative paths, from cropping to final treatment, for the cultivation of extensive broom plantations on marginal lands suitable for the production of industrial raw materials; ecological, smokeless solid fuels; and vegetable fibre suitable for biocomposites and/or paper pulps. Another option is the utilization of broom biomass for the production of agrocomposts suitable for upgrading soil quality.

Broom production and its industrial conversion seem to be conducive to sustainable development in such areas as environmental protection, economic yield and new job creation. The plantation's long life, its several coppicing cycles, and the absence of chemical inputs contribute to a good potential from the point of view of conservation and biodiversity. This potential will be maximized if degraded land is used and if some parts of natural woodlands are included among the new plantations.

Table 10.21 summarizes the results of a chemical-physical analysis of dry broom chips. These data support the great interest in this species as a possible biofuel. Further positive attributes can be seen in the results obtained from a pilot pyrolysis plant (Table 10.22). The analyses confirm broom's good qualities as a biofuel, and in particular the virtual absence of sulphur.

There is also interest in the possibility of converting broom twigs into a fibre for making clothes and textiles. New techniques (mechanical fractionating, heating, etc.) are presently being studied.

Composite materials using vegetable rather than fossil-derived fibres as reinforcement constitute the most promising application of broom fibre at present, and some research and industrial laboratories are already working on this matter with broom and other fibre crops. In particular, the car and aircraft industries are very interested in developing new materials suitable for external and internal bodywork that utilize vegetable fibres.

Table 10.21 Chemical and physical characteristics of broom

Characteristic	Value
Ash (%)	4.63
Volatile substances (%)	71.05
Fixed carbon (%)	24.32
High heating value (kcal/kg)	3.906
Fibre length (mm)	8.50
Fibre diameter (μm)	30.20

Table 10.22 Broom pyrolysis performance element analysis

Characteristic	Charcoal	Bio-oil
C (%wt)	80.96	61.90
H (%wt)	1.70	6.00
N (%wt)	1.45	1.05
S (%wt)	0.01	0.03
Ash (%wt)	7.34	1.50
Moisture (%wt)	0.34	14.60
O by diff.	8.20	14.92
Char content (%wt)		9.20
Viscosity at 50°C (cP)		200
Viscosity at 70°C (cP)		55
HHV (kcal/kg)	7060	6290
LHV (kcal/kg)	6956	5980
Specific gravity (15/4°C)		1.195

BUFFALO GOURD (*Cucurbita foetidissima* Kunth ex H. B. K.)

Description

The buffalo gourd is a vigorous perennial. It grows wild on wastelands in the deserts of Mexico and south-western USA. It forms large, dahlia-like tubers reaching up to 5m deep to obtain and store water. The plant is covered with a wax coating to avoid water loss. The fruits are spherical, yellow and hard shelled, and contain pulp and flat white seeds. The fruits have a diameter of about 8cm; the seeds are 12mm long and 7mm wide. Wild gourds, belonging to the family Cucurbitae, are a potential source of oil and protein. Several species are highly drought tolerant, particularly the buffalo gourd.

On barren land the buffalo gourd may match the performance of traditional protein and oil sources like peanuts and sunflower, which require more water. The buffalo gourd is not yet commercially cultivated anywhere. Great yield differences occur between individual plants. The size of the fruit varies. Some plants are essentially barren, whereas others are prolific. The size of the fruit varies. Some plants have a preponderance of male, others of female flowers (NAS, 1975).

Ecological requirements

Buffalo gourd requires long periods of warm, dry weather. The optimum temperature is 20–30°C. The

minimum; the optimum is 400–600mm. The soil should be well drained with a pH of 6–7 (FAO, 1996a).

Propagation

The plants can be propagated vegetatively from nodal roots. The long running vines are stapled to the soil and watered (NAS, 1975).

Crop management

The roots are reported to live for more than 40 years, but the herbaceous vines are annual (FAO, 1996a). The vines are highly resistant to cucumber beetle and squash bug (NAS, 1975).

Production

Yield estimation: each fruit contains about 12g seed. On the basis of 60 fruits per plant, 1ha of buffalo gourd can produce 2.5t of seed. The seed contains 30–65 per cent protein and 34 per cent oil (NAS, 1975). The crude oil can be extracted by pressing or by a solvent process. Linoleic acid is present in the oil at about 50–60 per cent. The remaining seed meal contains about 45 per cent protein and 45 per cent fibre, and may be used as animal fodder. The vines grow along the ground and because of their protein content of 10–13 per cent and digestibility may have forage value (Johnson and Hinman, 1980). The plant forms a big taproot, which can weigh 30kg (70 per cent moisture) after only two growing seasons. The root contains starch. The root, leaves and fruits contain bitter-tasting glycosides. However, the starch can be separated from them by soaking in a dilute salt solution (NAS, 1975).

Processing and utilization

The fruit can be mechanically harvested. The flesh dries completely in arid areas, so threshing is possible (NAS, 1975).

Buffalo gourd oil is a seed oil, extracted from the seeds of the *Cucurbita foetidissima*, which is native to southwest North America. As the Latin name of the plant indicates, the vine has a foul smell. The seeds of

Source: Kurt Schaefer www.opsu.edu

Figure 10.29 Buffalo gourd *Cucurbita foetidissima*

vines are frost sensitive and may be killed by frost, though the roots may survive winter temperatures as low as −25°C. An annual rainfall of 250mm is the

the buffalo gourd are rich in oil and protein, and were used by American Indians to make soap. The oil's fatty acid composition is dominated by linoleic acid (up to 64.5 per cent) and oleic acid (17.1 per cent) (davesgarden.com).

The Cucurbitaceae family include gourds, melons, pumpkins and squashes. They are characterized by their fleshy fruits. The seeds of many members of the group have been noted for their oil-bearing properties. However, little information has been found that describes existing methods of oil-extraction. In most cases the raw materials are grown primarily as a food, and the oil bearing seeds are used to make supplementary foodstuffs, thereby making dietary use of the oil. In such cases it would seem unlikely that it would be economically viable to extract oil from these raw materials, unless a large processing plant is involved that processes seeds in quantity. It has been suggested, however, that certain types of Cucurbitaceae, and in particular those that grow wild in arid areas, could be a potential source of oil. It is reported that the Arid Land Agricultural Development Institute, Lebanon, has carried out considerable research in this field.

The New Mexico Solar Energy Institute at New Mexico State University (NMSU) has conducted a two-year investigation into the technical and economic feasibility of using the buffalo gourd plant as an energy feedstock in eastern New Mexico. The New Mexico buffalo gourd project conducted field planting trials to determine optimum planting density, fertilizer levels, and irrigation regime. Starchy roots produced by the field plantings were evaluated as an ethanol feedstock at both laboratory and pilot scale. These studies indicate that buffalo gourd is well suited for root production in eastern New Mexico. Current cultivars of buffalo gourd can be most efficiently produced under dry land farming conditions with little, if any, supplemental fertilizer. Traditional plant breeding techniques can be profitably employed on the buffalo gourd to breed a size and shape of root more easily harvested by existing farm machinery. Because of its sensitivity to root rot, buffalo gourd must be grown in well-drained soils. Finally, buffalo gourd has been shown to be an excellent feedstock for ethanol production provided necessary pre-fermentation processing (chopping of roots) is performed correctly.

Internet resources

www.davesgarden.com.
http://attra.ncat.org/attra-pub/PDF/biodiesel_sustainable.pdf

CARDOON (*Cynara cardunculus* L.)

Partially contributed by Professor J. Fernández González, Universidad Politécnica de Madrid, Spain

Description

Cardoon is originally from the Mediterranean area, where it was known by the ancient Egyptians, Greeks and Romans. Nowadays it can be found spontaneously growing in the coastal countries of the Mediterranean, both in the continental zone and in the isles. It is also found in south Portugal, and on the Canary Islands and the Azores. Naturalized cultivars are also found in California, Mexico and in the southern countries of South America (Argentina, Chile and Uruguay), where it is known as 'Cardo de Castilla' (Castilian thistle). This species is also naturalized in Australia.

The leaves of the basal rosette are petiolate, very large (more than 50cm × 35cm), subcoriaceous and bright green. They are usually deeply divided. Segments are ovate to linear-lanceolate, with rigid, 15–35mm yellow spines at the apex and clustered at the base. The intensity of the spiny character changes among the different varieties. The leaves on the stem are alternate and sessile. The plant can reach a height of more than 2m. The flowers are grouped in large globose capitula (up to 8cm in diameter). Involucral bracts are ovate to elliptical, gradually or abruptly narrowed into an erecto-patent spine (10–50mm × 2–6mm), which can be either glaucescent or purplish. The corolla can be blue, lilac or whitish in colour. The achenes (6–8mm × 3–4mm) are shiny and brown-spotted. The pappus can measure 25–40mm. The chromosome number is 2n = 34.

According to Wiklund (1992), the species *Cynara cardunculus* may be divided into two subspecies (ssp. *flavescens* and *cardunculus*) related to the geographical distribution. The ssp. *flavescens* is found in Macronesia, Portugal and the northwest Mediterranean region, whereas ssp. *cardunculus* has mainly a central and

Source: Eugene Zelenko, http://upload.wikimedia.org/wikipedia/commons/8/8e/Cynara-cardunculus.jpg

Figure 10.30 Wild cardoon vegetation

northeast Mediterranean distribution. Naturalized cultivars in America and Australia are very similar to ssp. *flavescens*.

Ecological requirements

Cardoon is a species that belongs to the Asteraceae family (Compositae), which also includes artichoke, sunflower, safflower and Jerusalem artichoke. It is a perennial plant that during its natural cycle sprouts in autumn, passes the winter in a rosette form and in spring develops a floral scape that dries in the summer while the remnant roots stay alive. Beginning in the autumn, the buds in the upper part of the roots develop a new rosette in order to continue the cycle for several years. Thanks to its deep root system, it is able to extract water and nutrients from very deep soil zones and as a result, in non-watering conditions, using the rainwater accumulated during autumn, winter and spring, total biomass production can rise to 20–30t/ha/year dry matter, with 2–3t of seeds rich in oil (25 per cent) and protein (20 per cent).

Cardoon is traditionally cultivated in some areas as a horticultural plant but its cultivation cycle is completely artificial in the Mediterranean area, where it is sown at the end of spring and stays in a vegetative state during the summer.

Consequently, it needs watering during this period. After a bleaching period, which usually takes about one month, it is harvested at the beginning of the winter. The enlarged petioles of the basal leaves are the commercial product.

Cardoon is a characteristic species of the Mediterranean climate. It is quite sensitive to frost in the seedling state. Winter frost may have a significant effect on the rosette's leaves both in the first and successive years. It can cause tears in the leaves; these leaves will die, but the plant remains alive, recovering from harm as soon as the period of freezing is over.

For good development of the plants, rainfall during autumn, winter and spring months should be about 400mm or more. With lower levels of precipitation, biomass production decreases substantially.

This species requires light, deep and limy soils, with the capacity of retaining winter and spring water in the subsoil (1–3m).

Crop management

Autumn sowing should be performed as soon as conditions allow (soil is moist after rainfall) in order to let the plant develop the cold-hardier rosette before the first frost (1–2 months, depending on growth speed). Note that the crop can tolerate temperatures below $-5°C$ once the seedlings have four leaves. Production is low during the first year but beginning with the second year production rises, reaching a steady level during the second year, depending mainly on weather conditions. Spring sowing is recommended for those areas where the first autumn frosts are very early. In this case, sowing can take place as soon as there is no more risk of frost.

Soil preparation is analogous to land preparation for cereal sowing. Before sowing cardoon, it is recommended that an adequate basal dressing is applied, depending on soil fertility. During subsequent years, use restoration fertilization to replace the nutrients exported with the harvest. Because cardoon is a great biomass producer, it is a crop that consumes a considerable amount of nutrients. It is estimated that a

Source: Hans B., http://commons.wikimedia.org/wiki/File:Cynara-cardunculus11.jpg

Figure 10.31 Cardoon

20t/ha harvest of the aerial parts extracts 277kg/ha of N, 56kg/ha of P and 352kg/ha of K from the soil. These figures and the soil fertility can be used to calculate the required fertilizer dose.

Rows should be sown approximately 1m apart, though this distance may vary according to the desired density. A pneumatic sowing drill can be used. Optimal final density might be established with approximately 10,000 plants/ha. This can be increased up to 15,000 plants/ha if the ground is fresh/moist and does not lack water, or decreased to 7500 plants/ha if the winter water reserve is too low; 3 to 4kg of seeds are required for 1 hectare.

Weeds can be controlled with herbicides (trifluralin, alachlor, linuron, etc.) or by passing with

the cultivator twice until the rosettes have covered the ground. This task is very important during the first year of crop establishment – that is, during sprouting and development of the seedlings – because at first a large portion of the ground is still free of the crop. As the rosettes keep growing, they cover the soil, making it harder for weeds to develop.

In the second year of cultivation, the quick regrowth and the development of a larger rosette of basal leave in early autumn gives weeds little opportunity to establish themselves, so it can be said that weeds will be no trouble from the second year onwards.

For obvious environmental reasons, mechanical weed control is preferable. Where this is not possible a cheap, effective herbicide is a mixture of alachlor and linuron (4 litres of 48 per cent alachlor and 1.4kg linuron per ha) dissolved in 300–400L/ha of water.

Among the main pests that might attack cardoon are aphids (*Aphs* spp.), the stem borer (*Gortyna xantenes* Germ.), the leaf borer (*Apion carduorum* Kirby) and the leaf miner (*Sphaeroderma rubidum* Graells), as well as cutworms *(Agrostis segetum* and *Spodoptera litoralis)* and several flies *(Agromyza* spp., *Terellia* spp.) and moths (*Pyrameis cardui* L.). They can be treated with either specific or broad-spectrum insecticides. Organophosphates work well in most of the cases, but biological control methods are available.

Among the main fungal diseases are downy mildew, powdery mildew and botrytis blight rot. Against the downy mildews, treatments based on copper or Zineb, Maneb or Captan are recommended. Powdery mildew and botrytis blight rot are successfully controlled with sulphur or Benomyl-based treatments. Fungal diseases can also be controlled through no irrigation or decreased irrigation of the crop.

Production

Aerial biomass production of cardoon depends mostly on water availability during the active growing period (spring). For a 450mm average rainfall, distributed in accordance with the Mediterranean climate pattern, the average yield of harvestable biomass is estimated at approximately 20t/ha dry matter. At harvesting time, the moisture content of the biomass is rather low (10 to 15 per cent), because harvesting takes place when the aerial biomass is dry. The average distribution of this biomass among the different plant parts, according to data obtained at the Universidad Politècnica de Madrid over several years, is presented in Tables 10.23 and 10.24.

In Mediterranean regions the capitulum (flowering heads) are often harvested separately from leaves stems and branches with a combine harvester.

Table 10.23 Distribution of biomass among cardoon plant parts

Plant part	Weight (%)
Basal leaves	21.0
Stem leaves	12.1
Stems and branches	21.9
Capitula	45.0
Total	100.0

Table 10.24 The main components of the capitulum as proportions of the total biomass

Capitulum parts	Weight (%)
Receptacle	9.5
Bracts	13.2
Pappi	9.1
Seeds	13.2
Total	45.0

Processing and utilization

Aerial biomass is harvested in summer (from July to September), as soon as it is dry, and always before seed dissemination. Two scenarios are worth considering: harvesting seeds and other biomass separately, or together.

A combine harvester can be used if seeds are to be harvested. The remaining biomass is then swathed and baled. The whole biomass can be harvested with a self-propelled baler, otherwise two operations will be

Source: Google image, www.biol.uni.wroc.pl/obuwr/archiwum/2/lista. html

Figure 10.32 Thistle of *Cynara cardunculus* L.

Table 10.25 Heating values of the different fractions, and of each one of them relative to cardoon total biomass

Fraction	%	kcal/kg GHP	fraction LHP	kcal/kg GHP	total biomass LHP
Basal leaves	21.0	2655	2449	558	514
Stalk leaves	12.1	4096	3809	496	460
Stems and branches	21.9	4204	3914	921	857
Capitula (45%)					
Receptacle	9.5	3605	3333	342	316
Bracts	13.2	4181	3878	551	512
Pappi	9.1	4353	4043	396	368
Seeds	13.2	5576	5208	736	687
Total	100.0			4000	3714

needed: cutting the biomass with a swathe mower followed by baling. If the mower is not able to make rows, swathing should be performed before baling.

The dry aerial biomass of cardoon can be used as raw material for fuel in large-scale combustion plants, either for electricity production or for heating applications. The heating value of the different components of *Cynara* biomass are presented in Table 10.25. The average values can vary between 3714 and 4000kcal/kg of dry biomass.

These values refer to 1kg dry matter, either for each fraction (kg fraction) or for the total biomass (kg total biomass). GHP = gross heat power, LHP = low heat power.

The seeds, having a 25 per cent oil content, appear to be an interesting source of oil; they represent a high percentage of the total harvested dry biomass (13.2 per cent), which is about 2640kg/ha. Linoleic acid is the main component (59.0 per cent), followed by oleic (26.7 per cent) and palmitic (10.7 per cent) acids.

Silymarin is present, which is an important characteristic from a nutritional point of view because it can act as a regenerator for hepatic cells. Oil is easily extracted by cool pressing (at 20–25°C), which is really useful for dietetic applications since cool pressing does not appreciably alter the oil's components.

Table 10.26 lists the main characteristics of cardoon oil with regard to its possible utilization as a fuel. The most significant characteristics are its high cetane number and the low pour point, which would be advantages for its use as an engine fuel, either unmixed in indirect injection diesels or mixed with gasoil for use in normal diesel engines.

Several laboratories have studied the possibility of using cardoon biomass for paper pulp production,

Table 10.26 Characteristics of cardoon's oil

Characteristic	Value
Density (g/mL)	0.916
Viscosity (mm²/s at 20°C)	95
Melting point (°C):	−21
Heat power (MJ/kg):	32.99
Cetane number	51
Flash point (°C):	350
Iodine value	125
Saponification value	194

among them Instituto Papelero de España (Spain), Ordinariat für Holztechnologie at Hamburg University (Germany), Departamento Florestal del Instituto Superior de Agronomia de Lisboa (Portugal) and l'Institut National Polytechnique de Tolouse (France). Although investigations are still being carried out to optimize the different processes, the outlook for this use of cardoon biomass is promising.

Table 10.27 shows the cellulose, hemicellulose and lignin content of the different parts of cardoon, excluding the leaves and the seeds.

The seeds comprise approximately 13 per cent of the total biomass of the whole plant and are rich in valuable oils. Commercial production of mature seed in cooler climates is probably not a viable option for cardoon growers, however, all of the plant can be considered as biomass and can be harvested in late summer with a swathe mower and baler before the new autumn growth appears. Moisture content at harvest is very low at between 10 and 15 per cent and yields in Mediterranean regions are around 20–30t/ha/yr dry matter. This material can be combusted to produce electricity and heat and has a heating value of 16.7MJ/kg (www.walesbiomass.org/grass-cardoon.htm).

Figure 10.33 Seeds of *Cynara cardunculus* L.

Table 10.27 Fibre content of the different parts of cardoon biomass (percentage dry matter basis)

Part of the plant	Cellulose	Hemicellulose	Lignin
Thin stems	46.4	24.1	7.5
Thick stems	49.3	21.5	13.2
Average stems	47.8	22.8	10.3
Branches	41.0	21.3	5.9
Receptacle	23.6	15.9	7.1
Bracts	38.5	23.8	6.6
Pappi	59.7	26.5	2.6

Internet resources

www.walesbiomass.org/grass-cardoon.htm

CASSAVA (*Manihot esculenta* Crantz)

Partially contributed by H. K. Were and Professor Stephen Gaya Agong, Maseno University, Kenya

Description

Tropical crops such as cassava have a significant advantage over biofuel crops grown in temperate climates: they convert sunlight more efficiently into biomass and yield far more of it. The result is that the energy balance of biofuels made from such crops is considerably stronger. Or in other words, to produce one unit of energy in the form of a liquid fuel, tropical crops require far less land and resources than crops grown in temperate regions.

Cassava (*Manihot esculenta* Crantz) originated and was domesticated in South America in about 4000–2000 BC, though only recently has it been distributed worldwide. The Portuguese began importing cassava into the Gulf of Guinea in Africa in the sixteenth century. In the eighteenth century, they introduced it to the east coast of Africa and the Indian Ocean islands of Madagascar, Reunion and Zanzibar. Portuguese ships probably carried cassava to India and Sri Lanka after the middle of the eighteenth century. Cassava was little accepted at first, but since the nineteenth century it has extended rapidly across Africa and is now grown in 39 countries.

Cassava is a dicotyledonous plant belonging to the botanical family Euphorbiaceae, and, like most other members of that family, the cassava plant contains latifers and produces latex. The normal cassava plant has a chromosome number 2n = 36. Polyploids are not common. Numerous cassava cultivars exist in each locality where the crop is grown. The cultivars have been distinguished on the basis of morphology (for example, by leaf shape and size, plant heights, petiole colour, etc.), tuber shape, earliness of maturity, yield, and the cyanogenic glucoside content of the roots. This last characteristic has been used to place cassava cultivars into two groups: the bitter varieties, in which the cyanogenic glucoside is distributed throughout the tuber and is at a high level, and the sweet varieties, in which the glucoside is confined mainly to the peel and is at a low level. The flesh of the sweet varieties is therefore relatively free of the glucoside, though it still contains some (Purseglove, 1968). The photosynthesis pathway is C_3 II (FAO, 1996a).

Ecological requirements

The crop is cultivated throughout the year up to a maximum altitude of 2000m. It does well in a warm moist climate where mean temperatures range from 25–29°C. It performs poorly under cold climates and at temperatures below 10°C, where growth of the plant is arrested. It tolerates drought, but grows best where annual rainfall reaches 1000–2000mm. When moisture

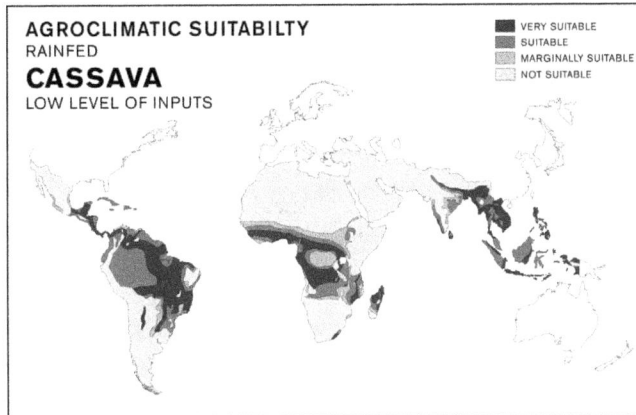

Source: www.fao.org/docrep/T8300E/t8300e0q.jpg

Figure 10.34 Agroclimatic suitability for rainfed cassava

availability is low, the plant ceases growth and sheds some of its older leaves, thereby reducing its transpiring surface. When moisture is again amply available the plant quickly resumes growth and produces new leaves, making cassava a valuable crop in places where, and at times when, the rainfall is low or uncertain or both. It is only during the first few weeks after planting that the cassava plant is unable to tolerate drought to an appreciable extent.

The best soil for cassava cultivation is a light, sandy loam soil of medium fertility and good drainage. On clay or poorly drained soils, root growth is poor, hence the tuber to shoot ratio is considerably reduced and root rot is increased readily. Gravelly or stony soils tend to hinder root penetration so these soils and saline soils are unsuitable. Cassava can grow and yield reasonably well on soils of low fertility where production of most other crops would be uneconomical. Under conditions of very high fertility, cassava tends to produce excessive vegetation at the expense of tuber formation. Continuous light delays tuberization and lowers yields (Mogilaer et al, 1967). For this reason, cassava is most productive when daylight duration is up to 12 hours a day and at latitudes between 30°S and 30°N.

Cassava tends to grow slowly and to yield poorly when grown at altitudes above 1000m; most of its present-day cultivation is at lower altitudes. It does reasonably well in windy regions.

Propagation

In agricultural production, cassava is propagated almost exclusively from stem cuttings. Ripe wood with 4–6 eyes is used and the planting depth is 5–15cm. Deeper planting may be useful in dry regions, but the deeper development of the tubers makes the harvest more difficult. In nature and in the process of plant breeding, propagation by seed is quite common. The main aims of breeding are the development of types that are resistant to the mosaic virus and bacterial blight, and to further the root form and quality (such as starch content, low fibre and linamarin content). The procedure is based mainly on selection and combination of suitable local strains, but crossing with other *Manihot* species is also possible. Experiments for better storage quality are being carried out in several countries (Rehm and Espig, 1991).

Spontaneous sexual and asexual polyploids occur in cassava. Some polyploidal breeding strategies use these processes. Some of the polyploids are very vigorous. Several tetraploids have performed as well as improved varieties, and some triploids out-yielded, by over 200 per cent, the best improved varieties, indicating that triploids are more promising than tetraploids (Hahn et al, 1994). Since the introduction of improved varieties of cassava, developed from crosses between Southeast Asian and Latin American germplasm, yields have increased by 20–40 per cent. The roots of the new varieties also have higher starch contents (CIAT, 1997).

In each inflorescence, the female flower opens first; the male flowers do not open until about a week later. Cross-pollination is the rule. Insects are the main pollinating agents. After pollination the cassava ovary develops into a young fruit that requires 3–5 months to mature.

Crop management

In traditional agriculture, seedbed preparation for cassava planting may take various forms, but the most common are planting on unploughed land and planting on mounds. When planting is done on unploughed land, no tillage is done beyond that required to insert the stem cuttings into the soil. The soil is opened up with machete, hoe or dibble and the cutting is inserted vertically, horizontally or at an angle. Horizontal planting seems to yield best results on unploughed land (Takyi, 1974). In Kenya and Uganda, cassava is not planted on mounds but on unploughed land or well-prepared land, because it is always accompanied by intercrops such as maize, beans, millets and sesame, among others.

In improved agriculture, land is first ploughed and then harrowed; thereafter cassava may be planted on the flat, on ridges or in furrows. For plantings on the flat, the cuttings are inserted directly in the earth after it has been harrowed. On ridges or furrows, the earth is ridged or furrowed after it has been harrowed. Flat plantings of cassava produce greater yields of tuber than ridge or furrow plantings (CIAT, 1976). However, flat planting is unsuitable on heavy soils because the tubers tend to rot.

The age of the stem cutting has a profound effect on the tuber yield. Generally, cuttings taken from the older, more mature, parts of the stem give a better yield than those taken from the younger portions (IITA, 1974). Therefore, stem cuttings for planting should be as mature as possible. The number of nodes on a cutting is very important because they serve as origins of shoots, and of roots if they are submerged. A cutting with one or two nodes runs the risk of failing to establish the stand if the buds or sprouts from those nodes are destroyed. For this reason, it is generally recommended that cuttings used for planting should have at least three nodes (Krochmal, 1969).

Cassava is normally planted in rows 80–100cm apart with 80–100cm between the plants, depending on cultivars and the growing conditions. Cultivars with a spreading habit should be spaced further apart than those with an upright habit. In areas of high soil fertility and high rainfall the plants tend to be luxuriant in foliage growth, so plants should be spaced further apart.

Cassava is drought tolerant except in the first few weeks after planting. It is therefore important that it should be planted at a time when there is ample moisture, and when the likelihood of further moisture supply is good. In cooler subtropical or high-altitude regions, planting should be done early in the warm season to allow for ample growth before the dormant cold season. In traditional agriculture where intercropping is practised, planting is often delayed until the later part of the rainy season when the other intercrops are nearly ready for harvest.

If cultivated on slopes, intercropping with tree crops (such as *Leucaena* or *Eucalyptus)* can reduce runoff and soil loss. Erosion can be 70–80 per cent less than with monocrops of cassava, because of the better canopy coverage of the soil surface. The disadvantage of intercropping trees with cassava is that the yields fall after the first year, because the trees are very efficient in removing nutrients from the soil (Ghosh et al, 1996)

Weed competition is most detrimental to cassava during the early stages until 2–3 months, when it has formed a closed canopy.

About 30 diseases of cassava are described. In many regions cassava is not normally affected by diseases or pests. However, in others it may be attacked by virus diseases – mosaic, brown streak and leaf curl of tobacco, resident in many parts of Africa – and bacterial diseases – *Phytomonas manihotis* (Brazil), *Bacterium cassava* (Africa) and *Bacterium solanacearum* (Indonesia).

When the soil is fertile and growth is normal, cassava removes large amounts of nutrients from the soil. Cassava has a high requirement for potassium (CIAT, 1976), and a lack of its adequate supply limits cassava yields on many soils. When the potassium level in the soil is low, the response of the crop to nitrogen or phosphorus is poor.

Excessive nitrogen may be disadvantageous, because it results in annual luxuriant shoot growth at the expense of the tuber. Also, a high level of nitrogen fertilization produces a high level of cyanogenic glucosides in the tuber (de Brujin, 1971). Tuber glucoside level is greater if nitrogen is applied to the soil than if it is applied as a foliar spray.

Because of its high uptake of nutrients cassava is used in a shifting cultivation system, preferably as the last crop before the leaves fall off the bush. In permanent systems it is good if a green manure plant can be integrated in the crop rotation system.

Production

Globally, the benefits of cassava genome sequencing could materialize as new higher-yielding or more pest- and disease-resistant cultivars. Of particular interest is coaxing more protein from cassava to better supplement the dietary needs of more than 600 million people in Asia, Africa and Latin America who rely on the crop as a main source of calories. On the industrial front, ratcheting up cassava's starch production under a wider range of conditions could set the stage for developing countries to use the crop for making fuel ethanol. Indeed, cassava can maintain high productivity under conditions that cause other crops to fail, including corn, whose starch costs more.

Cassava is the third largest source of carbohydrates for human food in the world, at an estimated annual yield of 136 million tonnes. Africa is the largest centre of production with 57 million tonnes grown on 7.5 million hectares in 1985. On a world basis, average yields are estimated to be about 5–10t/ha. On research experimental stations, yields of 80t/ha have been noted.

Processing and utilization

Young tubers are known to contain much less starch than older tubers, so harvesting must be delayed until an appreciable amount of starch has accumulated in the tubers. However, as the tuber becomes older, it tends to become more lignified and fibrous and the starch content as a percentage of the dry weight of the tuber tends to decrease or remain constant.

Cassava plots are rarely harvested in one pass, either in traditional or improved agriculture, because cassava deteriorates rapidly and it can only be kept in good condition for one or two days after harvesting. The exact time of harvesting depends on the cultivar, but ranges from 10 to 30 months. Cassava may be harvested manually (using hand tools) or by machines. The stem is cut a few centimetres from the ground. Then the ground is loosened using a hoe or a machete, and a pull of the remaining stub of stem is enough to lift out the tubers.

Sometimes cassava is replanted in the field as it is harvested: a piece of stem from the plant being harvested is planted in the same hole or very close by. Mechanical harvesting of cassava has met with some

Source: news.mongabay.com

Figure 10.35 Cassava harvest

success in Brazil and Mexico and is being vigorously developed in other parts of world.

Cassava plays a major role in alleviating food crises in Africa because of its efficient production of food energy, year-round availability, tolerance to extreme stress conditions, and suitability for present farming and food systems in Africa (Hahn and Keyser, 1985; Hahn et al, 1987). Traditionally, cassava roots are processed by various methods into numerous products and utilized in various ways according to local customs and preferences.

Constraints on cassava processing range from environmental to agronomic factors. During the rainy season, sunshine and ambient temperatures are relatively low for processing cassava, particularly in the humid lowland areas where cassava is mainly grown and utilized. In other localities, particularly in savannah zones, the water that is essential for processing is not easily available. In the dry season, the soil is so hard that harvesting and peeling cassava tubers for processing are difficult and result in more losses. Furthermore, cassava root shape varies among cultivars. Roots with irregular shapes are difficult to harvest and peel by hand, resulting in great losses of usable root material. Varietal differences in dry matter content, and in starch content and quality, influence the output and quality of the processed products.

Figure 10.36 Drying cassava chips

Thailand exploits the industrial prospects of cassava on a large scale. About 52 per cent of the country's cassava now goes to starch production. About one-third of this is further processed into various modified starches. Quantities of dried cassava chips and pellets are exported as fodder (CIAT, 1997). When dried chips are produced, 98 per cent of the linamarin is destroyed during the drying process (Rehm and Espig, 1991). Sun drying of peeled cassava roots is practised in many parts of Africa. Peeled roots are split and cut into smaller pieces to standardize or shorten the drying process. Shortening is necessary to reduce the chance of microbial spoilage in case of humid or rainy weather, or because of food shortage, or to comply with financial and market demands. This method reduces the cyanogenic glucoside levels from 400eq/kg dry weight to 56eq/kg (van der Grift et al, 1995).

Fermentation consists of two distinct methods: aerobic and anaerobic. For aerobic fermentation, the peeled and sliced cassava roots are first surface dried for 1–2 hours and then heaped together, covered with straw or leaves and left to ferment in air for 3–4 days until the pieces become mouldy. They are then sun dried after the mould has been scraped off. The processed and dried pieces (called 'mokopa' in Uganda) are then milled into flour, which is prepared into a 'fufu' called 'kowan' in Uganda. The growth of mould on the root pieces increases the protein content of the final products by three to eight times (Amey, 1987; Sauti et al, 1987).

In anaerobic fermentation, grated cassava for processing into 'gari' is placed in sacks and pressed with stones or a jack between wooden platforms. Whole roots or pieces of peeled roots for processing into frfuti are placed into water for 3–5 days. During the first stage of gari production, the bacterium *Corybenacteria manihot* attacks the starch of the roots, leading to the production of various organic acids.

In the second stage, the acidic condition stimulates the growth of the mould *Geotrichum sandida*, which proliferates rapidly causing further acidification and the production of a series of esters and aldehydes that are responsible for the taste and aroma of gari (Odunfa, 1985). The optimum temperature for gari processing is 35°C increasing to 45°C. The roots are peeled and boiled before they are eaten, but this does not effectively remove HCN. With processed leaves, the cyanide in pounded leaves ('pond' or 'sakasaka') remains high at 8.6mg/100g, though 95.8 per cent of the total cyanide in leaves is removed through further processing into soup.

The dried root pieces are fermented, dried and then milled into flour by pounding in a mortar or using hammer mills. The dried cassava roots (both fermented and unfermented) are often mixed in a ratio of two to three parts cassava with one part of sorghum, millet and/or maize and milled into a composite flour. This increases food protein, and enhances palatability by improving consistency.

Tissues may be disintegrated in the presence of excess moisture during grating or fermenting in water permitting the rapid hydrolysis of glucosides, effectively reducing both free and residual cyanide in the products.

During or after fermentation of roots for gari production, the grated pulp is put into jute or polypropylene sacks, and then pressed by stones or jacked between wood platforms. In this way much of the cyanide is lost together with the liquid, through a dewatering process.

Gari is very popular in Nigeria but less so in Cameroon, Benin, Togo, Ghana, Liberia and Sierra Leone. In Brazil, this method is used for the production of 'farinha de madioca'. Fresh roots are peeled and grated, mainly by women and children using a simple traditional grater though the work is done by men if a power driven grater is used.

Besides the described uses, cassava is an energy crop with a high potential, because cassava is one of the richest fermentable substances for the production of alcohol. The fresh roots contain 30 per cent starch and 5 per cent sugars; the dried roots contain about 80 per cent fermentable substances.

For processing in factories, only the outermost cork layer is removed. The starch in the skin layer is also used. The roots are crushed into a thin pulp. Saccharification is carried out by addition of sulphuric acid to the pulp in pressure cookers. Later, the pH value is adjusted with bases and then yeast fermentation is allowed. The alcohol is separated by distillation (Grace, 1977).

In Brazil, the cultivation of cassava has been intensified in recent years to produce alcohol for use in the bioethanol fuel project (Diercke, 1981). One tonne of cassava root can be processed to make 180 litres of ethanol (Wright, 1996). The potential yield of alcohol

is 8000L/ha, applying adequate agrotechnology. In 1978, the first Brazilian cassava alcohol distillery began production with a capacity of 60,000L per day. The alcohol production process used in this plant is very simple and is suitable even for farmers, though it is not yet sufficiently developed (Kišgeci et al, 1989). Especially where water availability is limited (not enough for the cultivation of sugar cane), cassava is a preferred feedstock for ethanol production.

The agricultural wastes produced in large amounts during harvest should not be thrown away but utilized instead. It is not only specially grown crops that are suitable for energy production; agricultural wastes should be considered as an energy source too. Energy from the cut-off stems and leaves can be obtained by using anaerobic fermentation to produce a methane-rich biogas. Methanogenic bacteria convert about 90 per cent of the feedstock energy into biogas. The odourless, non-toxic residue can be used as fertilizer (FAO, 1995b). Anaerobic digestion of starchy by-products has several advantages, including the use of the biogas during the transformation cycle (drying), pollution control, and the agronomic use of residues. Cassava by-product mechanization has been evaluated at pilot and industrial scales. The stalk can also be used for fermentation (Kišgeci et al, 1989).

Tables 10.28, 10.29 and 10.30 show, respectively, the composition of the cassava tubers, edible portion, and leaves (Onwueme, 1982).

Table 10.28 Composition of the cassava tuber

Tissue	Composition (%)
Peel	10–20
Cork layer	0.5–2.0
Edible portion	80–90

Table 10.29 Composition of the edible portion

Component	Composition (%)
Water	62
Carbohydrate	35
Protein	1–2
Fat	0.3
Fibre	1–2
Ash	1

Table 10.30 Composition of the leaves 100g (edible part)

Component	Amount
Water	80g
Carbohydrate	7g
Protein	6g
Fat	1g
Calcium	0.2g
Iron	0.3g
Vitamins:	
B1 (thiamin)	0.2mg
B2 (riboflavin)	0.3mg
C	200mg
A	1000 IU
Niacin	1.5mg
Calories	50

Efforts to sequence the genome of cassava, a staple food for millions of people worldwide, could yield the genetic keys to unlocking new traits for improved yield, more protein and even novel industrial applications – like putting fuel in the gas tank (Suszkiw, 2006).

An assessment of net energy and supply potentials was performed to evaluate cassava utilization for fuel ethanol in Thailand. In 2006, the Thai government gave approval for 12 cassava ethanol plants with the total output of 3.4 million litres per day to be constructed in the following two years. The cassava fuel ethanol (CFE) system involves three main segments: cassava cultivation including processing, ethanol conversion, and transportation. All materials, fuels and human labour inputs to each segment were traced back to the primary energy expense level. The positive net energy value and net renewable energy value, 8.80MJ/L and 9.15MJ/L, respectively, found for the CFE system in Thailand proved that it is energy efficient. Without co-product energy credits, CFE in Thailand is even more efficient than CFE in China and corn ethanol in the United States. Regarding supply potentials, about 35 per cent of the national cassava production would be used to feed approved CFE factories. A shift of cassava to ethanol fuel rather than its current use for chip/pellet products could be a probable solution. Using scaled-up data from this existing pilot plant, Nguyen et al (2007) from the Thonburi University of Technology calculate the net energy value (NEV) of cassava-based ethanol as 10.22 megajoules per litre (MJ/L), an overall positive yield. The most optimistic assessment for corn shows an

NEV of around 4.51MJ/L, meaning cassava is more than twice as efficient. NEV is a measure of the energy content of ethanol minus the net energy used in the production process.

Internet resources

http://news.mongabay.com/bioenergy/2007/04/first-full-energy-balance-study-reveals.html

CASTOR OIL PLANT (*Ricinus communis* L.)

Description

The castor oil plant is native to East Africa, though its present-day dispersal is in all warm regions of the world (Gliese and Eitner, 1995). Castor grows wild or in cultivation. Originally it was a perennial plant, a tree growing over 10m high. In the temperate zones castor grows only 1–2m high and dies off at the first frost. The agricultural cultivars are usually grown annually, particularly the dwarf cultivars (60–120cm high), which are best suited for dry areas and for mechanized cultivation. In perennial cultivation the plants are kept 2–3m high and are harvested for 2–4 years (Franke, 1985).

The shoot formation is sympodial (an inflorescence at the end of the shoot and branching under each inflorescence). The leaves are very large, have a long stalk and are formed like a hand. With most types the female flowers are in the upper part of the inflorescence, and the male in the lower. The capsule is composed of three carpels and splits off in wild forms, but it remains closed in recently bred cultivars so the whole crop can be harvested. Short day length causes the increased formation of male flowers. The seeds are egg shaped or oval, and of different sizes and colours. The seed coats cover the white, oil-containing kernels. The ratio between seed coat and kernel can vary. The seeds contain 42–56 per cent oil. The seed coats form 20–40 per cent of the weight. Other components of the seeds, mainly concentrated in the seed coats are ricin, a protein (3 per cent of the seed), and ricinin, an alkaloid (0.03–0.15 per cent). The ricinin can be easily removed because of its solubility in water. The ricin produces the toxic effect of the press cakes (Lennerts, 1984). The castor oil plant uses the photosynthesis pathway C_3 II (FAO, 1996a).

Ecological requirements

The best conditions for growth are in tropical summer rainfall areas. Castor grows well from the wet tropics to the subtropical dry regions. The optimum temperature is 20–25°C. The warmer the climate of the location, the higher the content of oil in the seeds, but at temperatures over 38°C seed setting is poor. Frost will kill the plant. Dwarf cultivars can be grown in summer rainfall areas with only 500mm annual precipitation, but 750–1000mm is optimal. It can be grown between the latitudes 40°S and 52°N and up to 1500–3000m elevation, but it is best grown between 300 and 1800m. Because of its deep-reaching root system the castor oil plant is quite drought tolerant. Short day conditions delay the flowering, but otherwise castor has little sensitivity to day length. The soil should be a deep sandy loam. The castor oil plant has a high demand for moisture, warmth, fertility and calcium content of the soil. The range of the pH value is 5–8, with an optimal pH of 6.

The nutrient removal for 1 tonne of seed (with capsules) is about: 30kg N, 12kg K, 4kg Ca and 3kg Mg. Unbalanced nitrogen fertilization encourages the growth of foliage at the expense of seed formation. Castor does not tolerate salt. The best performance is achieved with a relative humidity of 30–60 per cent; ideally, humidity should be higher during vegetative growth and lower during the period of maturity and harvest. Excessive humidity can lead to increased insect problems.

The castor-oil plant is easy to grow and is resistant to drought, which makes it an ideal crop for the extensive semi-arid region of northeast Brazil.

Propagation

The castor oil plant is propagated by seeds. The aims of breeding are cultivars with non-opening capsules, which allow mechanized harvesting and minimize the loss of seed. Their height and capsule-shattering habit mean that the wild forms are not ideal for commercial production. Improved cultivars vary in height between 1 and 4m, depending upon environment, and behave as annuals.

Crop management

The seed is sown 5cm deep in rows with 90cm between the rows and 50cm within rows. The plants develop quite slowly; weed control is important in the

Source: Karlheinz Knoch www.Botanik-Fotos.che

Figure 10.37 *Ricinus communis*, flowers and fruit

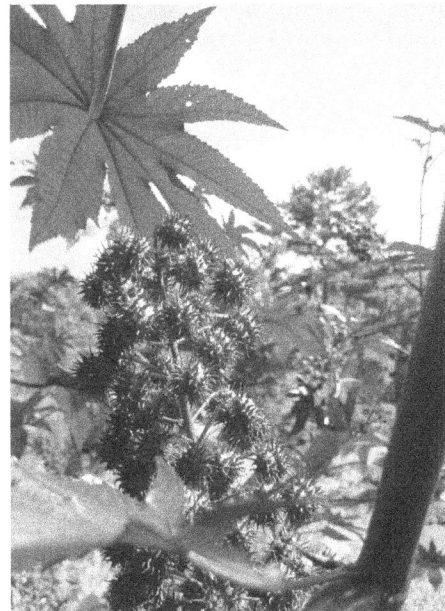

Source: http://upload.wikimedia.org/wikipedia/en/0/07/ Castor_plant.jpg

Figure 10.38 Castor bean plant with palmate-lobed leaf and cluster of spiny fruits

first weeks. Good preparation of the soil and eradication of persistent weeds before sowing are a substantial help. In the main areas of cultivation that have low air humidity, diseases do not play a major role. Harvesting can take place 4–6 months after sowing the seed. The yield can be increased by nitrogen fertilization at up to 80kg/ha, though tillage alone leads to an increase of the yields, even in the absence of nitrogen fertilization.

Where water is the limiting factor for productivity, improving the productivity of the soil should involve the reduction of runoff, using ridges and furrows, and increasing the water intake rate through in situ moisture conservation (Uma et al, 1990). For castor, the bed furrow system has proved to be more beneficial than the ridge furrow system: it was observed that initial plant vigour was greater in the bed furrow system than for ridges. This could be the result of better root penetration and establishment in the 20cm high loose beds (Venkateswarlu et al, 1986).

Insect pests can be controlled by sanitary measures and by crop rotation. Because of its deep root system and resistance to nematodes *(Meloidogyne* spp. and *Striga* spp.), castor is an excellent crop for growing in rotation with tobacco, cotton, maize, millet, etc. (Rehm and Espig, 1991).

The plant has also been bred to mature at a shorter height. Whereas the castor-oil plant traditionally reaches 3m in height, making mechanized harvest difficult, there are now varieties that grow to just 1.7m.

Production

Production results from Brazil indicate that the average seed production of castor is about 1500kg/ha with an oil content of 47 per cent. Total castor oil production amounts to 90,000m^3 per year.

Oil yields peak at 1000L/ha of high-quality industrial oil (Gliese and Eitner, 1995). The global average yield of seed is 0.6t/ha/year. Under more favourable conditions 1–1.5t/ha/year are possible. In the USA, seed yields of 3t/ha/year have been achieved under irrigation (Rehm and Espig, 1991).

Source: R. Yepez, http://rambiocom.files.wordpress.com/2008/05/dsc07353.jpg

Figure 10.39 Castor oil plant

Table 10.31 Seed and oil yields of various oil plants in Brazil

	Castor	Sunflower	Soya	Oil Palm	Cotton
Seed Production kg/ha	1500	1500	3000	20,000	3000
Percentage of Vegetable Oil	47%	42%	18%	20%	15%
Production in 2005 in Brazil in m³/yr	90,000	23,000	5,600,000	151,000	315,000

Source: Cohen (2007)

Recent studies and genetic improvements have increased the oil content of the castor bean from 24 to 48 per cent. In comparison, soya is just 17 per cent oil. Castor oil is the best substance for producing biodiesel because it is the only one that is soluble in alcohol, and does not require heat and the consequent energy requirement of other vegetable oils in transforming them into fuel.

Processing and utilization

Indehiscent cultivars are harvested by picking the capsules by hand or mechanically. Later, the capsules are threshed to separate the seeds from the fruit walls. With cultivars that burst open it is necessary to harvest the infructescence before full ripeness and then let them dry, to avoid the loss of the seeds which are ejected if the capsules burst open (Lennerts, 1984).

The aim of cultivating castor is to produce oil for several purposes. Nowadays, the oil is almost all used industrially. The seeds are pressed or extracted with solvent to obtain castor oil. Before pressing or extraction, the seeds are usually shelled. Tables 10.32 and 10.33 show, respectively, the glyceride contents of castor oil, and some additional properties of castor oil. Thousands of years ago, oil from the seed of the castor plant was used as a fuel for wick lamps (Bilbro 2007).

Most castor oil is used at present for technical purposes, with much smaller quantities going to medicinal and energy uses. Castor oil has some properties that make it an attractive material for chemical and technical purposes. The oil retains its viscosity at high temperatures and it is non-drying, and its adhesion is also strong. Therefore it is used as a lubricant. It does not corrode rubber, so it is used as an

Source: Slinger http://de.wikipedia.org/w/index.php?title= Datei: Castor_beans.jpg

Figure 10.40 Oil Castor beans

Table 10.32 The glycerides making up castor oil

Glyceride	Amount (%)
Ricinoleic acid	80–85
Oleic acid	7
Linoleic acid	3
Palmitic acid	2
Stearic acid	1

Source: Diercke (1981)

Table 10.33 Some properties of castor oil

Property	Value
Density (g/mL) (highest density of all vegetable oils)	0.961–0.963
Melting point (°C)	−18 to −10
Iodine number	82–90
Saponification number	176–190

Source: Römpp (1974)

additive in the production of rubber and in hydraulic systems (such as brake fluids). In the plastics industry it is used as a plasticizer. It is used in cosmetic products, inks and in the textile industry as a solvent for certain colours. Polymerized castor oil gives permanent adhesion and plasticity to paint material. Because of a double bond on the ninth carbon atom and a hydroxyl group at the twelfth, ricinoleic acid is more reactive than other fatty acids. Dehydration gives a fatty acid with two conjugated double bonds, which can be polymerized. At 275°C in an alkaline medium the oil is split, forming sebacic acid and octanol. Vacuum distillation results in undecenylic acid and heptaldehyde.

The press cake is poisonous because of its ricin content and is therefore only used as fertilizer. The shells and stems are used as fuel or for mulching (Rehm and Espig, 1991). If castor is grown annually, a large amount of agricultural waste is generated. Advanced energy production techniques such as gasification should be considered for the shells and stems.

In July 2008 Global Energy Ethiopia (GEE) completed sowing 5000 hectares of Chinese hybrid castor seeds for its alternative energy project in Ethiopia. The project entailed planting and harvesting castor for the production of non-edible oil for the biodiesel industry and for other uses.

GEE expected to harvest the produce between August 2008 and September 2008. The expected yield of the harvest from 28,000 tonnes of seeds was approximately 12,000 tonnes of castor oil. At that time, the commodity price for castor ranged from US$700 to US$1100 per tonne.

GEE had done the necessary groundwork for an agricultural cooperative in the regions of Waletia and Goma Gofa, in Ethiopia. This included signed agreements with over 25,000 families, to farm castor on approximately 7500 hectares of their land (Reuters, 2008).

The castor oil seed is 47–49 per cent oil. Biodiesel obtained from castor oil has a lower cost compared to the ones obtained from other oils, as due its solvability in alcohol trans-esterification occurs without heating. The use of biodiesel will allow a reduction on the consumption of petroleum-derived fuels, minimizing the harmful effects on the environment. This work wants to provide a thermoanalytical and physical-chemistry characterization of castor oil and biodiesel. Biodiesel was obtained with methyl alcohol and characterized through several techniques. Gas chromatography indicated methyl ester content of 97.7 per cent. The volatilization of biodiesel starts and finishes under temperatures inferior to the beginning and final volatilization temperatures of castor oil. Biodiesel data are very close to the volatilization temperatures of conventional diesel. (Conceição et al, 2007)

Castor oil has quite a few characteristics that could make it a suitable candidate for biodiesel, but one disadvantage is its viscosity. Castor oil in its straight vegetable oil form is about 100 times as viscous as diesel fuel, and while transesterification does reduce the viscosity significantly, research is still under way on whether the final viscosity for castor oil biodiesel is within acceptable limits for use in diesel engines.

Internet resources

http://waynesword.palomar.edu/plmar99.htm#castor
www.reuters.com/article/pressRelease/idUS103560+27-Mar-2008+BW20080327

COCONUT PALM (*Cocos nucifera* L.)

Description

The first palms appeared some 85 million years ago in a wide range of environments, but the 2800 species that exist today are primarily found in the intertropical zone. Palms are perennial plants, of every imaginable size. They are not trees: they have a stem rather than a trunk. Since time immemorial, palms have provided people with foodstuffs, personal hygiene products, medicinal preparations, building materials, etc. Coconut palms are tropical plants that bear fruit all year round. They are found on all sorts of soils, even very poor ones (coastal sands, peat, etc.) that are unsuitable for other crops.

There are tall, dwarf and hybrid coconut varieties. They begin to bear after four to ten years, grow up to

Figure 10.41 Coconut palm farm, La Digue, Seychelles

12m tall in the case of dwarf varieties and 30m for tall, and live for up to a hundred years.

The coconut palm grows up to 25m high. The trunk is slender, unbranched and leafless, and from a height of 1.5m to the top has a diameter of 25–35cm. The crown is built of 2.5–5.5m long, 1–1.7m wide leaves. About 10–15 leaves are formed new every year, and the same number are shed. The coconut palm originated in the south Pacific region, and is distributed from the latitudes 15°N to 15°S (Rehm and Espig, 1991) at elevations below 750m (Gliese and Eitner, 1995), even at the equator. The fruits can float for weeks in salt water without losing fertility (demonstrably, as far as 4500km), so the coconut palm is found on the coast everywhere in the tropics (Franke, 1985). It is an evergreen tree.

The inflorescences can develop in every leaf axil. Male and female flowers are formed in the same inflorescence, the female on the lower part of the panicle branches. The male flowers open 10–20 days before the female, so cross-pollination is the rule. Pollination is possible by insects and by wind. It takes 12–14 months from the time of pollination to the ripeness of the fruits. Dwarf forms open their male and female flowers at the same time; for them, self-pollination is the rule (Franke et al, 1988). Most have smaller fruits that ripen 3 months earlier than larger types.

The fruits are ovoid and 15–25cm long. The outer covering is a thick fibrous husk. A bony endocarp is lined with albumen and has three eyes at the apical end. Flowering begins at an age of 5–7 years (dwarf varieties 3.5–4.5 years). The palm reaches full bearing after 10–12 years, maturity at about 15 years. It will live up to 60–100 years in the wild state and 50–70 years under cultivation. Flowers and fruits are formed during the whole year. There is a 360–365 day yield cycle. The inflorescence is initiated 16 months before the spathe opens, and the nut takes about 1 year to mature from time of pollination. After 250 days, the fruit has reached the final size and the nutshell is formed; now the fruit flesh begins to develop. The photosynthesis pathway is C_3 (FAO, 1996a).

Ecological requirements

The coconut palm requires a mean annual temperature in the vicinity of 26°C and low diurnal variations of temperature; this is the cause of the absence of the coconut palm in regions with higher variations, even in the equatorial zones. The lowest temperatures that are tolerated for longer periods are for the palm 10°C, for the leaves 15°C, and for the flowers 25°C. It will withstand a small amount of frost. The palm needs a large amount of sunshine (Rehm and Espig, 1991).

If sufficient groundwater is available, the coconut palm grows even in dry regions. Where it depends on rainfall, a precipitation of 1250–2000mm is regarded as necessary for optimal growth (Gliese and Eitner, 1995). The palm needs well-drained and well-aerated soil, but otherwise it is not very demanding on the soil. Groundwater in dry areas should be available at a depth of 1.0–2.5m. The tree tolerates up to 1 per cent salt in the groundwater. An air humidity of 84–97 per cent is required for good production; 63 per cent is the absolute minimum. The monthly mean should not fall below 60 per cent.

Propagation

Whole fruits are put horizontally into a seedbed, the narrowest end downwards and covered with earth so that the top edge is above the surface. First roots are formed inside the fibrous hull. After 4–5 months, more robust roots are developed that penetrate into the soil. The major aims of breeding are earlier yielding, and resistance against 'lethal yellowing'. Very few cultivars are truly tolerant to lethal yellowing, a disease that is linked to phytoplasma-like organisms. Several hybrids look promising, but the most tolerant material will have to be tested on a larger scale (Mariau et al, 1996). To attain these goals, tall palms are crossed with dwarf types. The F1 hybrids between selected dwarf and tall growing parents have better productivity than the original strains. They are already cultivated in substantial numbers.

Propagation by callus culture is possible but has not yet reached the stage of practical application. After 4–6 months the seedlings can be transferred to their final position. The development of the dwarf types is quicker.

Methods are being developed to collect and transport samples of coconut cultivars and wild forms as embryo culture (Ashburner et al, 1996). Investigations are under way to achieve vegetative propagation by in vitro culture techniques (Verdeil et al, 1993).

Crop management

Tall and short varieties require a spacing of 9m and 6–7m, respectively (Gliese and Eitner, 1995). One hundred fruits contain: 10kg N, 2kg P, 10kg K; the use of a potassium fertilizer is unnecessary if the fibrous husks are returned to the soil. The coconut palm needs an adequate supply of Cl; chlorosis is prevented in areas far from the sea by giving each plant 2kg of cooking salt per year. If there is enough rainfall, cropping between plants is possible. High yields result if legumes are planted as ground covering plants (Rehm and Espig, 1991). Especially where the soil is impoverished after decades of coconut palm monoculture, the planting of legumes between the palms helps to regenerate the soil, though soil fertility and the organic and mineral content have to be increased to support the crops.

According to the Philippine Coconut Authority website, there are four systems of planting coconuts:

- Square – palms are set at fixed equal distance at the corner of each square, the distance between palms in each row and the distance between adjacent rows being the same. The most common distances are 10m by 10m at 100 trees per hectare.
- Rectangular – rows are set at right angles to one another but the distance between the palms in the row is closer than those between the rows. This system provides for a slightly lower number of palms in a stand but allows for more room for growing intercrops.
- Triangular – palms are set at fixed distance at the corners of an equilateral triangle. About 15 per cent more palms can be accommodated per unit area under this system. This system is used for monoculture when distances between coconuts are less than 10m.
- Quincunx – this system is used for replanting old coconut plantations where the old palms will be removed as soon as the new seedlings are established. Seedlings are planted in the centre of each square of old palms. This system is applicable only in square plantings.

Holes should be dug with a size of at least 50cm by 50cm, starting as early as 2 months before planting to allow for weathering of the soil on the sides and bottom

Figure 10.42 Layout of a coconut plantation

of the holes to promote early root contact (Philippine Coconut Authority website).

Experiments have shown that suitable plant species are *Acacia mangium* and *A. auricmliformis* (Dupuy and Kanga, 1991).

The greatest dangers to fully grown palms are lethal yellowing in Jamaica, Cuba and Haiti; Kerala disease in India; and cadang-cadang in the Philippines. The causal factors of these diseases are not yet sufficiently clear. Lethal yellowing is a mycoplasm; Malayan dwarfs are resistant to this disease. Of the fungus diseases, bud rot *(Phytophthora palmivora)* is particularly dangerous in regions with high rainfall.

Insect pests are the rhinoceros beetle *(Oyctes rhinoceros* L.), the palm weevil *(Rhynochophorms* spp.), and other leaf-eating beetles and caterpillars. Sanitation (e.g. pesticides, fumigation) of the plantation is the most important action to be taken to control pests. Insecticides are only necessary for severe attacks. Rats and shrews have to be controlled, as do squirrels and bats that eat the growing fruit (Rehm and Espig, 1991).

Production

Each palm produces 30–50 nuts per year; that means 8000 nuts per hectare (Gliese and Eitner, 1995). Hybrids of dwarf and tall forms yield 200–600 smaller nuts per year. High yields are achieved up to the 40th year, after which production markedly decreases.

Asia is the main production zone, with 84 per cent of global output and relatively stable yields (5 tonnes of nuts/ha). Global copra oil production totals about 3 million tonnes a year. In Africa, the Caribbean and Oceania, copra is still the only source of income and trade for some smallholders. The main copra oil-producing countries are the Philippines, Indonesia and India. Most of the oil is consumed in producing countries. Exports account for less than half of the total output (1.3 million tonnes a year). The European

Source: KM Kutty 2009 http://en.wikipedia.org/wiki/File:Young_Coconut.jpg

Figure 10.43 Coconut fruits

Union is the leading importer, followed by the United States. A biofuel, palm oil methyl ester is set to grow in importance, like all renewable energies.

Processing and utilization

Harvesting for oil production takes place when the fruit is fully ripe. The fruits are cut off with a knife attached to a bamboo pole, or someone climbs up to harvest the fruits. In Indonesia, trained monkeys are used for harvesting. They manage to pick up to 1000 nuts per day. It is also possible to wait for the nuts to fall and then to collect them. The nuts can be stored for several weeks before processing.

All the parts of the coconut palm, from root to tip, have uses, which include the following (Philippine Coconut Authority website):

- The coconut palm's roots may be used to produce astringents and antidiarrhoea medicines, as well as beverages and dyestuffs.
- The coconut trunk produces durable wood as well as pulp for paper making.
- The bud of the coconut palm's inflorescence produces a sweet juice called 'tuba', which may be drunk as a fresh beverage, or may be fermented to produce either a sweet coconut toddy or a potent gin called 'lambanog' which is 80 to 90 per cent

proof. Tuba is also used for making vinegar, sugar, and a honey-like syrup called 'coco honey', and as a source of yeast for making bread. Cut, dried and varnished, coconut flowers may also be used in creating candy trays and decorative objects.

- Coconut fibre, or 'guinit', may be used in producing helmets, caps, wooden shoe straps, handbags, fans, picture and house decor like lamp shades and guinit flowers for the table.
- Coconut heart, or 'ubod', is used to make the 'Millionaire's Salad', so called because getting the heart of the palm kills the tree, making the ubod very costly. Cubed in fairly large bits, it may be added to Spanish rice, or in long strips, to Arroz a la Cubana. It may also be used in coco pickles, guinatan and lumpia.
- Coconut leaves produce good quality paper pulp, midrib brooms, hats and mats, fruit trays, waste baskets, fans, beautiful midrib decorations, lamp shades, placemats, bags and utility roof materials. They are also used to wrap delicacies such as the suman sa ibos, as well as steamed rice, such as the puso. They figure prominently every year on Palm Sunday, when they are used for making palaspas, the coconut-leaf decorations that, after having been blessed with holy water by the priests at church, are taken home and placed on altars and fastened on doors and windows for good luck and protection against evil.
- Coconut husks are a cheap source of firewood, and are also used as bunot, used to buff waxed floors. Fibres from coconut husks are used in making brushes, doormats, carpets, bags, ropes, yarn fishing nets, and mattresses, as well as for making pulp and paper. They can also be used as substitute for jute in making rice, copra, sugar, coffee, bags and sandbags. Coir dust and short fibres from coconut husks may be used in manufacturing wallboard, which is termite-proof due to the presence of creosote. No binding materials are needed as lignin is inherent in the coconut husk. Coir yarn, coir rope, bags, rugs, husk decor, husk polishes, mannequin wig, brush, coirflex, coco gas, lye insulator, insoflex, plastic materials and fishing nets are other products that can be obtained from coco husk.
- Coconut shells are used in creating household products and fashion accessories, such as shell

necklaces, shell bags, cigarette boxes, shell ladles, buttons, lamp shades, fruit and ash trays, guitars, placemats, coffee pots, cups and wind chimes, as well as briquetted charcoal. Charcoal made from coconut shells is also used in producing activated carbon, used in air purification systems such as cooker hoods, air conditioning, industrial gas purification systems, and industrial and gas masks. Whole coconut shells, cleaned and polished, have traditionally been used in Filipino culture as coin banks.

- Young coconut meat produces buko, often used for salads, halo-halo(crushed ice with sweetened fruit), sweets and pastries, such as the well-known buko pie of Laguna. Coconut meat also produces coco flour, desiccated coconut, coconut milk, coconut chips, candies, bukayo or sweetened shredded coconut meat, latik, copra and animal feeds. Coco chips, which are curved and wrinkled coconut meat, are crisply toasted and salted. Coconut flour can be used as a wheat extender in baking certain products without affecting their appearance or acceptability. Coconut milk is a good protein source. Whole coco milk contains about 22 per cent oil, which accounts for its laxative property.

- The makapuno, or 'sport fruit' of the coconut, is considered a delicacy and largely used for making preserves and ice-cream. However, other uses of makapuno have been found, such as facial, hand and hairdressing creams, shampoo, toothpaste, vitamin carrier in pills, salicylic acid ointments, sulphur ointments and even muscle pain relievers.

- Dried coconut meat, or copra, has a high oil content, as much as 64 per cent, and is used in making coconut oil. Virgin coconut oil (VCO), taken orally, is purported to retard aging; counteract heart, colon, pancreatic and liver cancer; and is easy to digest. Coconut oil is also used to make soap and shampoo due to its high saponification value in view of the molecular weight of most of the fatty acid glycerides it contains. Other products from coconut oil are lard, coco chemicals, crude oil, pomade, margarine, butter and cooking oil. Coconut oil is also used as an alternative fuel or biofuel, as it is used in coconut methyl ester (CME).

- Coconut water, the liquid endosperm inside the coconut fruit, can be used in making coconut water vinegar, coconut wine, and chewy, fibre-rich nata de coco, good as a dessert and as a laxative. Coconut water can also be used as a growth factor and as a substitute for intravenous fluid or dextrose. It also makes a good and economical thirst quencher and is used in coconut water therapy, or 'bukolysis', to cure renal disorders.

- Coconut fibreboard is a novel and innovative product made up of cement, coir, shredded wood, fronds and other lignocellulosic materials that are available in coconut farms which are otherwise considered as agricultural waste.

Source: Koehler's Meclicinal-Plants 1887 http://en.wikipedia.org/wiki/File:Koeh_187.jpg

Figure 10.44 Illustration of a coconut tree

At the processing stage, the fibrous husks are removed and the nuts are broken. The opened nuts are dried for a short period, and then the meat is removed and dried to get copra. Drying takes place in the sun or in drying ovens. Copra is an intermediate product that can be used for oil production or for export. Copra contains 65–70 per cent oil. Peak yield is 9t/ha/year of copra, which gives 6t/ha/year of oil (Gliese and Eitner, 1995). The copra has a dry matter content of 94 per cent and contains 66 per cent fat.

The copra is pressed to obtain the oil. The remaining press cake contains about 20 per cent protein and is used as fodder. The pressing of the undried fruit flesh has not yet replaced the production and pressing of copra. The oil contains very small amounts of unsaturated fatty acids and does not turn rancid. Tables 10.34 and 10.35 show the glyceride content of coconut oil, as well as other properties. The main part of the coconut oil is used for nutritional purposes. Other parts are utilized in the soap and stearin industry.

In the tropics coconut oil is used as fuel. The oil is a source of liquid fuel that can be used for the running of special engines. Transesterified oil is a substitute for diesel. The shells are used as solid fuel – for example to run the ovens in which the fruit flesh is dried into copra. Other parts are processed to charcoal (Diercke, 1981).

Biodiesel produced from coconut oil blended with normal diesel can improve fuel efficiency from 5 per cent to 25 per cent, particularly on older vehicles (Bioenergywiki).

Malaysia is working to develop new, cost-effective fuels from alternative sources, after prices of palm oil have lately risen to a record high level. The sharp rise in prices has neutralized the commodity's advantage as a cheap raw material for producing biofuels. Chow Mee

Table 10.34 Composition of coconut oil

Glyceride	Amount
Caprylic-, lauric- and myristic acids	50–60%
2 X lauric- and myristic acids	15–20%
(Mixed) glycerides of other fatty acids,	Small
like oleic acid and palmitic acid	amounts

Source: Diercke (1981)

Table 10.35 Some properties of coconut oil

Property	Value
Density (g/mL)	0.88–0.90
Melting point (°C)	20–23
Iodine number	7.5–9.5
Saponification number	255–260
Acid number	2.5–10

Source: Römpp (1974)

Chin, head of the Malaysian Palm Oil Board's (MPOB) Energy and Environment Unit, said that palm-based materials such as palm kernel cake, empty fruit bunches, palm fibre and palm shells are under consideration.

So far, palm-based fuels have mostly been made from methyl esters obtained from crude palm oil. Chow said MPOB has embarked on a research programme to try to develop a second generation of environment-friendly fuels. Other than crude palm oil and palm kernel oil, the raw materials from oil palm trees are in solid form. Chow added that if these could be successfully processed into gaseous and liquid fuels, it would open up new alternatives to crude palm oil. On an annual basis, the energy potential per hectare of all oil palm-based raw materials is equivalent to 45 barrels of petroleum-based crude oil (UCAP, 2009).

On many Pacific islands, coconut oil is used for generating power and fuelling cars. This is the case of Ouvea Island (New Caledonia) or Bougainville (the largest island in the Solomon Archipelago, Papua New Guinea). A lot of mini-refineries produce the plant oil that replaces diesel. This alternative fuel source has triggered interest from Europe and Iran.

Besides being much cheaper, the new energy is a much more sustainable alternative, as these islands are a heaven of coconut palm plantations and many locals refine coconut oil in backyard refineries. Coconut oil is also employed for cooking and making cosmetics like soap, for example. Engines that function on coconut oil act as generators, which feed a desalting centre delivering drinking water for 235 families on Ouvea. The system provides 165kW, competing with diesel engines on power and fuel consumption (Anitei, 2007).

The first biomass power plant in the Philippines is to be constructed in the Bondoc Peninsula area in Quezon province. The proposed power plant will use coconut husks, palm and dry leaves for fuel. The plant is expected to generate 10MW of electricity, which will be used for the additional power needs of some towns. With extensive coconut farms in the area, the biomass power plant is assured of a ready and abundant supply of raw materials (Philippine Information Agency, 2006).

Internet resources

Bioenergywiki, www.bioenergywiki.net/index. php/Coconut_oil
Philippine Coconut Authority website,
'Coconut: The tree of life', http://pca.da.gov.ph/tol.html
UCAP, www.ucap.org.ph/011708.htm
Wikipilipinas.org, http://en.wikipilipinas.org/index.php?title=Coconut#Plantation
http://en.wikipilipinas.org/index.php?title=Coconut#Plantation

COMMON REED (*Phragmites australis* (Cav.) Trin. ex Steud.; syn *P. communis* Trin.)

Description

Common reed is an upright perennial that ranges in height from 1.5m to 5m. Long, narrow leaves alternate on its tall stalks. Culms (flower-bearing stems) have smooth nodes and hollow internodes. Leaf blades are flat or rolled. Plants grow in dense, single-species or monocultural stands. Plume-like flower spikes 15–30cm long form at the tops of the plants. Flowers are tiny with lots of silky hairs. Large purple flower heads turn grey and fluffy in late summer as they go to seed. They remain on the plant throughout the winter. The plant spreads through the growth of rhizomes or by seed. Aerial stems rise from joints in the rhizomes and aerial shoots that are knocked over can take root and produce new shoots at the nodes. The prostrate stalk sends out runners that generate new plants. Stout root stalks, often exceeding 6m in length, interlock to form a dense network that can withstand fires, mowing and other forces that damage stalks and leaves. The underground network of rhizomes

has an expansion rate of about 1m per year, but in nutrient-rich areas can spread up to 10m. Plants can spread by wind-blown or bird-deposited seeds, by movement of the rhizomes, by maintenance equipment in highway ditches, or by the action of tidal ice.

Common reed is a wetland grass, most often found in temperate zones, subtropics and tropical highlands growing naturally in swamps, drains and on moist headlands. The stem is erect and partially branched and has a diameter of up to 20mm. The smooth, flat lanceolate leaf blades are light green to grey-green in colour, and vary in length from 15 to 45cm and width from 1 to 5cm. The inflorescence is a 20–50cm-long open panicle. The plant flowers between July and

Source: Bruce W. Hoagland www.biosurvey.ou.edu/wetland/emergent_grasslike_poa_herb_multiple_florets.html

Figure 10.45 Common reed stand, green

Figure 10.46 Common reed stand, mature

September. Common reed effectively prevents water erosion along drains and river banks (Tothill and Hacker, 1973; Kaltofen and Schrader, 1991).

Ecological requirements

Common reed is a cane-like perennial grass, forming dense stands. Stems are round and hollow with flat leaves along its length. Leaves are long (up to 60cm by 5cm wide) and gradually taper to a point. The seed head is at the end of the stem and is multi-branched. Silky hairs along the flowers axis give a silky appearance. Common reed can propagate from seeds or from its creeping rhizomes.

Submerged portions of all aquatic plants provide habitats for many micro- and macro-invertebrates. These invertebrates in turn are used as food by fish and other wildlife species (e.g. amphibians, reptiles, ducks, etc). After aquatic plants die, their decomposition by bacteria and fungi provides food (called 'detritus') for many aquatic invertebrates. Many species of birds utilize common reed seeds and use the plant's thick colonies for shelter (Texas AgriLife Extension).

Firm mineral clays are the optimum soils for common reed. The plant prefers damp soils rather than wet soils. The appropriate rainfall range for common reed crops is between 750 and 3000mm, with seasonally wet locations being suitable. The plant's most favoured temperature range for growth is 30–35°C with full sunlight (Skerman and Riveros, 1990). Common reed is also salt tolerant. Weeds should not be a significant problem to the crop, because common reed is often considered a dominant weed itself.

Propagation

There is considerable variability in glaucousness of leaves, shape and denseness of panicle and growth habit. The variegated form, or spire-reed, is sold as an ornamental grass. The Eurosiberian Center of Diversity reported that reed tolerates fire, frost, high pH, salt, weeds, and waterlogging. The chromosome number is 2n = 48, 36, 54.

Stands may be started by transplanting young plants or rooted stolons. Growth starts in February in southern locations, later further north. Foliage stays green until frost. New shoots grow from buds at nodes of old stems, stolons or rhizomes. Common reed can be propagated using seed or rhizomes.

Crop management

In the spring, the seed can be sown or the rhizomes planted. In its most favoured habitat, sunny wetlands, there is no need for maintenance of the crop. Because common reed is an invasive plant it usually crowds out other weeds.

The plant is perennial and is often considered undesirable, so there is no established crop rotation recommendation.

Production

Giant reed cannot withstand prolonged heavy grazing. Its upright growth makes it easy for livestock to remove all the leaves. For maximum production, no more than 50 per cent of current year's growth by weight should be grazed off during the growing season. Water control that lowers the water level but does not drain the area increases production. Grazing deferments of 60–90 days improve plant vigour. The straight hollow stems are cut in autumn and dried for arrowshafts, pipestems, loom rods, screens, roofing for houses and adobe huts, etc. Leaves are also gathered and used for weaving mats and other objects.

Although common reed has proved to have a fairly large variance in dry matter yield, ranging from 5t/ha up to 43t/ha, average yield is about 15t/ha (Hotz et al, 1989). Common reed crops in Sweden have yielded 7.5–10t/ha dry matter (Björndahl, 1983).

Among other grass plants, common reed (*Phragmites australis*) is considered to be a promising source of renewable energy in Finland and Estonia. First, naturally growing common reed is abundant in the coastal areas of these countries, and it produces a sufficient amount of dry biomass (3–15t/ha). Second, since common reed is a perennial fast-growing plant, supply of this source of energy is practically inexhaustible. Third, provision of reed is relatively cheap and, in addition to direct cutting of plants from the coastal areas, dry reed material could be obtained as waste from roof manufacture and insulation material made from reed.

Reed swamps in Europe produce 7.5–13.0Mt/ha/yr. According to the Phytomass files, annual productivity ranges from 40 to 63Mt/ha. Reeds are currently being harvested from Swedish lakes at a cost of around US$50/Mt (US$2.86/GJ gross thermal value), which rises to US$60/Mt after transportation and final processing (US$3.43/GJ gross thermal value). These costs are expected to diminish as machines and methods are optimized (Palz and Chartier, 1980).

Processing and utilization

Once the crop has established itself, harvesting can take place in the autumn. At this time the food reserves in the leaves and stems are relocated into the rhizomes and the plant has a higher dry matter content.

Traditionally, common reed has been used for grazing (but only in its very young stages), thatching, matting, primitive construction and as a solid fuel for heat production, with a heating value of 14–15MJ/kg (Hotz et al, 1989). More modern applications include the use of extracts from reed wastes and rhizomes to produce alcohol and pharmaceuticals. The high cellulose content of about 40–50 per cent and the long fibres make reed components useful for the paper and chemical industries (Cook, 1974; Skerman and Riveros, 1990).

Additionally, common reed has been planted for water purification and the protection of lake and river banks. Common reed, along with a few other plants, is currently being used as plant cover for wastewater treatment in the Czech Republic. This is part of a new constructed wetlands technology (Vymazal, 1995).

By compressing of dry reed biomass, it is possible to form compact briquettes (140–170kg/m^3) or pellets (500–700kg/m^3), which are more convenient to handle and to keep in store. Heat production at reed incineration is 13–15MJ/kg, and such reed briquettes and pellets could be used by small- and middle-sized farmers and also power plants.

To develop the optimal strategy of utilization of reed in the coastal areas and to exchange information and experience between interested parties, the project 'Reed strategy in Finland and Estonia' was initiated in March 2005. The project is implemented by Southwest Finland Regional Environment Centre in cooperation with Tallinn University of Technology, Turku University of Applied Sciences, South-East Finland Regional Forest Center and Kotka–Hamina Region of Finland Ltd. The project represents an interdisciplinary approach focusing on different themes, such as bioenergy, water protection, landscape, construction and building. The increased cutting of reed from the coastal areas would have, moreover, a positive impact on water protection and for maintenance of landscape and biodiversity.

Internet resources

Texas AgriLife Extension, http://aquaplant.tamu.edu/index.htm
Union of the Baltic Cities (UBC.net), www.ubc.net/bulletin/bulletin1_06/p43.html
University of Maine Cooperative Extension, www.umext.maine.edu/onlinepubs/htmpubs/2532.htm

CORDGRASS (*Spartina* spp.)

Partially contributed by Dr Dudley G. Christian, IACR, Rothamsted, UK

Description

Spartina pectinata is commonly known as prairie cordgrass, marsh grass, slough grass or ripgut. The last name describes the mature leaves that are long and flat with sawtooth edges. The seed head – with long, comb-like, one-sided spikes – is the plant's most distinctive feature. The sod-forming grass favours low, wet areas such as road ditches and tolerates alkaline soils, although it also grows well on the same type of soils where corn and switchgrass flourish.

Source: Ben Kimball, NH Natural Heritage Bureau www.nhdfl.org/about-forests-and-lands/bureaus/natural-heritage-bureau/photo-index/high-salt-marsh.aspx

Figure 10.47 Salt meadow cordgrass

There are 16 species of *Spartina*. All are temperate halophytes, rhizomatous perennials and major components of salt marsh ecosystems (Figure 10.48) Distribution is mainly along the European and American seaboards of the Atlantic ocean. *S. pectinata* is also found in fresh water marshland and dryland habitats in the eastern USA from the east coast to the great plains (Hitchcock, 1935).

A classification and distribution of *Spartina* was made by Mobberley (1956), and a good summary of the distribution of species can be found in Long and Woolhouse (1979). Many species form monotypic communities dominating a habitat; roots and rhizomes help to collect and stabilize the mud. This characteristic has made it important as a marshland reclamation plant. *S. anglica* was introduced into China for this purpose (Chung, 1993).

Cordgrasses have the C_4 photosynthetic pathway. The benefits of better light conversion efficiency compared to C_3 plants make C_4 plants more productive (Monteith, 1978). Recent studies of *S. cynosuroides* and *Miscanthus × giganteus* have shown that the superior light conversion efficiency of C_4 photosynthesis also functions under cool (temperate) conditions (Beale and Long, 1995). As a forage plant, *Spartina* has little economic value. Shoots emerge late in spring and palatability is lost early in the growing season.

Production

Three productive species of cordgrass have been investigated as potential biomass crops: *S. anglica, S. pectinata* and *S. cynosuroides*. *Spartina anglica* Hubbard, a fertile amphidiploid, is thought to have evolved about 1870 in Southampton Water, southern England, by a natural crossing between *S.alterniflora* Lois and *S.maritima* Fern. The hybrid *S. × townsendii* H. & J. Groves evolved from *S. anglica* Hubbard (Gray et al, 1990).

The productivity of primary stands is variable and depends on location and season (Long et al, 1990). The time of harvesting can affect long-term production. In one study, yield fell from 16t/ha/year to 8t/ha/year over a three-year period following summer harvests which resulted in the loss of some plants. The crop did not respond to fertilizer applications (Scott et al, 1990). The decline in yield may be prevented by including a rest year after two years (Callaghan et al, 1984).

Figure 10.48 Cordgrass (*Spardina pectinata*), UK

Dryland production of *S. anglica* has not been successful (Ranwell, 1972). Two *Spartina* species more suited to dry land, *S. pectinata* and *S. cynosuroides,* were successfully established in southern England and Ireland in the 1980s (Jones et al, 1987; Jones and Gravett, 1988). Selected sites were on productive soils in England and on marginal soils in Ireland. The experiments were established with plants produced by ramets from a single clone. These species produce fertile seed but seed set can be arable with both species, and special methods of seed storage are required (Jones et al, 1987). *S. pectinata* did not grow as well as the other species on the marginal land sites and all crops had low yields in the first season.

The yield of both species has been studied for seven years at two sites in eastern England (Potter et al, 1995). Yield was low in the first year and both species produced their heaviest yields in the following year, 20t/ha and 15t/ha on a clay soil and on a lighter, less fertile soil, respectively. Although yields were less in subsequent years they showed no evidence of decline with time. Averaged over all years, *S. pectinata* yielded 12t/ha and 14t/ha for non-fertilized and fertilized treatments, respectively. *S. cynomroides* yielded less. The yield of both species was significantly greater on the heavier soil than on the lighter soil. Fertilization improved yield by a small but non-significant amount.

In another study, observation plots of *S. pectinata* were grown on a fertile peaty loam over fen clay in eastern England and on a low-grade silty clay over chalk at another site approximately 100km further south. Both crops received 60kg/ha of nitrogen in the spring. Yields at the peaty loam site were 4.9t/ha and 14.0t/ha in successive years and 2.8t/ha and 7.8t/ha at the silty clay site. Yield improvement between years was approximately the same at each site, suggesting that the yield reflected differences in soil quality (Bullard et al, 1995). In northern Germany, *S. pectinata* achieved a yield of 12.8t/ha/year (El Bassam and Dambroth, 1991). Although some of the studies show that fertilizer is of little importance in the long term, nutrients may need to be applied to compensate for harvest losses.

In experimental field trials at South Dakota State University in 2001–2004, even unimproved prairie cordgrass genotypes produced nearly 10t/ha, roughly twice as much as the best switchgrass varieties. In south-eastern England, average yields in excess of 10 t/ha were achieved over seven years of harvesting (Boe and Lee, 2007).

Seven populations of cordgrass and 'Cave-In-Rock', 'Summer', and 'Sunburst' switchgrass were harvested in October in 2001–2004 (Boe and Lee, 2007). Mean biomass production across years ranged from 5.1 to 7.9t/ha among cordgrass populations. Yields of cordgrass (6.0t/ha) were similar to Cave-In-Rock (6.8t/ha) for the first two years. However, production in the fourth year was greater for cordgrass (6.8t/ha) than Cave-In-Rock (2.0t/ha). Two cordgrass populations produced more biomass (9.3t/ha) than Summer and Sunburst (4.8t/ha) in the fourth year. Leaf comprised 70 per cent of the biomass of cordgrass, and differences occurred among phytomers for leaf and internode traits. Cellulose and hemicellulose concentrations were similar for cordgrass and switchgrass, but cordgrass had higher levels of total N and ash. Narrow-sense heritability estimates for biomass production in Summer and Sunburst switchgrass were 0.6. Biomass production of native warm-season grasses intended for biofuel purposes in the northern Great Plains may be enhanced by selecting from among populations of cordgrass and among families within cultivars of switchgrass.

Processing and utilization

Spartina has been planted to reclaim estuarine areas to supply feed for livestock, and to prevent erosion. Various members of the genus (especially *Spartina alterniflora* and its derivatives, *Spartina anglica* and *Spartina × townsendii*) have spread outside of their native boundaries and become invasive. *Spartina* species are used as food plants by the larvae of some Lepidoptera species including Aaron's skipper (which feeds exclusively on smooth cordgrass) and the engrailed (Wikipedia.org).

Harvesting of cordgrass occurring naturally in coastal salt marshes is not considered worthwhile, though it is technically feasible using reed harvesters (Callaghan et al, 1984). With dryland planting harvesting takes place after stems have died in the winter. Dieback may not occur in mild climates. Dead stems could be cut and collected using standard grass harvesting or forage harvesting machinery, though no references to using these methods have been found. The moisture content of collected material will depend upon season and location.

Figure 10.49 Mature cordgrass (*Spardina pectinata*)

Cordgrass can be considered a potential crop for producing a solid feedstock which could be converted into a wide variety of biofuels: biogas, ethanol, synthetic gas through fermentation, combustion and gasification.

Internet resources

Britannica.com, www.britannica.com/EBchecked/topic/137327/cordgrass
Wikipedia.org, http://en.wikipedia.org/wiki/Cordgrass

COTTON (*Gossypium* spp.)

Description

The origins of cotton were in Africa (as forms of *Gossypium herbaceum*). Four cotton species of the genus *Gossypium* are now under cultivation: *G. herbaceum* L., *G. arboreum* L., *G. birsutum* L., and *G. barbadense* L.

Cotton species are potentially perennial, even though they are normally grown for only one year in modern agriculture. The plants form a strong taproot, even at the seedling stage. The roots can reach a depth of 3m. The shoot system is dimorphic and the fruiting branches are sympodial. Each flower is surrounded by three deeply divided bracts and the fibres themselves are single celled hairs. The most useful ones are long hairs (lint), which are more than 20mm long in modern cultivars. The fruit (bolls) grow very quickly after pollination. When the seeds ripen, the hairs die and their wall collapses. The wall of hairs is composed of many layers of cellulose fibres.

Ecological requirements

The optimal temperature for germination is about 25°C. For further development, 27°C is optimum. At temperatures over 40°C, and with strong insolation, the bolls will be damaged and fall off. Cotton is extremely sensitive to frost, and its cultivation is only possible where 200 frost free days can be relied on. The highest yields are achieved in dry areas under irrigation. Ripening should occur in a rainless period, because rainfall after the opening of the bolls damages the

Source: Dan Mott, ncsu.edu (http://ipm.ncsu.edu/cotton/InsectCorner/photos/cotton.htm)

Figure 10.50 Cotton field at flowering stage

quality of the fibres. The plants are drought tolerant to some extent because of their long root system.

Cotton needs deep soil with sufficient drainage. The pH should be between 6 and 8. The plant is relatively salt tolerant at low salinity levels.

Crop management

The nutrient uptake ability is strong and the nutrient requirements are moderate. Too much nitrogen fertilization encourages vegetative growth and extends the vegetation period. Sufficient potassium is important for attaining food fibre quality and for disease resistance. The calcium requirement is decidedly high. Cotton withstands relatively high boron concentrations in the soil, but adequate boron fertilization (by spraying) is required when its deficiency appears.

Good preparation of the land is especially important before sowing so that the seedlings (which germinate epigeally) can penetrate easily. Before sowing, the fuzz must be removed from the seed mechanically or chemically, because the seeds otherwise cling together. The row spacing lies between 50 and 120cm, the spacing within the row between 20 and 60cm, depending upon local factors. For mechanical harvesting, the spacings are reduced to 15–20cm between the rows and 8–10cm within the rows (*G. birsutum* recommended for this practice). The seeds should be sown no deeper than 5cm. Cotton can be sown on level soil, in furrows, or on ridges (in poorly drained soils).

The most economical method of weed control has been the application of a strip of soil herbicides over the row of seeds, and later the flaming of weeds between the rows.

Genetically modified (GM) cotton is widely used throughout the world. Globally, GM cotton was planted on an area of 67,000km² in 2002. This is 20 per cent of the worldwide total area planted in cotton. The US cotton crop was 73 per cent GM in 2003 (blackherbals.com). Genetically modified cotton was developed to reduce the heavy reliance on pesticides.

Source: Dan Mott, ncsu.edu (http://ipm.ncsu.edu/cotton/InsectCorner/photos/cotton.htm)

Figure 10.51 Cotton stand shortly before harvesting

Production

GM cotton acreage in India continues to grow at a rapid rate increasing from 50,000 hectares in 2002 to 3.8 million hectares in 2006. The total cotton area in India is about 9.0 million hectares (the largest in the world, about 25 per cent of world cotton area) so GM cotton is now grown on 42 per cent of the cotton area. This makes India the country with the largest area of GM cotton in the world, surpassing China (3.5 million hectares in 2006).

The yield of cotton (cotton seed) can reach 4t/ha/year under optimal conditions, but in practice it is seldom over 2.5t/ha/year. With primitive cultivars, the yield of fibres (ginning out-turn) is 20–25 per cent; good upland cultivars nowadays yield at least 35 per cent and the best cultivars more than 40 per cent.

The surge in Bt cotton cultivation as well as favourable rainfall and weather conditions are likely to help Gujarat increase its average cotton yield in 2008–2009. The cotton yield in Gujarat was expected to rise to 797.95kg per hectare, which may well be higher than the record world average yield of 787kg per hectare in 2007–2008.

As per the data available from the International Cotton Advisory Committee (ICAC), world cotton yield reached a record level of 787kg per hectare in 2007–2008. The world yield was projected to fall by 1 per cent to 779kg per hectare. Globally, cotton production to decline in 2008–2009 following the fall in output in the US, China and the few other countries. 'Considering the present scenario, Gujarat may surpass average world cotton yield for the first time', said Kishor Shah, president, Central Gujarat Cotton Dealers' Association (Damor, 2008).

Processing and utilization

The seeds that are left over after ginning are also a valuable product because they provide linters (5%), cottonseed oil (24%), oil cake (33%) and hulls (34%). The cake and meal contain 23–44 per cent protein and are valuable animal fodder. The linters have high-cellulose fibres, coarse threads used as a material for cushions and in paper making. Cotton seed hulls can be used as solid fuel, roughage for livestock, as bedding and fertilizer. This waste can be successfully converted to clean hot gas. The dried cotton stems can be used as fuel.

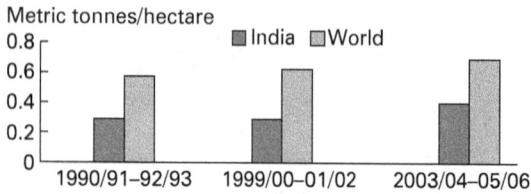

Source: USDA Production, Supply, and Distribution database.

Figure 10.52 Growth in Indian cotton yields

Cotton seed oil is a valuable by-product. The cottonseeds contain 16–24 per cent oil. The oil is light yellow and half drying. It has a high linoleic acid content and congeals between +4^{0}C and – 6°C. After refining (which also destroys toxic gossypol) it is used for nutrition, energy or industrial purposes.

If a large amount of cotton by-products are produced they can be used as energy sources. For example, stems, seed hulls and linters are cellulose-rich materials, which after pyrolysis and gasification could be prime material for the industrial and energy sectors. Likewise, cotton seed oil could be transformed into biodiesel by the process of transesterification.

The production of charcoal, pellets and briquettes from cotton hulls and stems is now being developed in several countries in Africa and Asia.

There has been a great deal of attention given to the 'energy balance' of agricultural production, especially in relation to biofuel crops such as soya beans for biodiesel, and corn for ethanol. The energy input considers obvious factors such as the amount of fuel used by agricultural equipment, but also includes the energy associated with the manufacture of inputs into the system such as fertilizers and crop protection products. Applying these methods to the cotton production system, to produce a typical cotton crop in the United States requires approximately 4900MJ of energy, or in terms of the energy content in gallons of diesel fuel, 36 gallons.

The full energy of cottonseed has to be adjusted to account for its being processed into fuel. But, considering that potential for cottonseed energy is already twice the energy input, it is conservative to say that cotton can be energy self-sufficient. The oil alone from the seed can generate almost 20 gallons of biodiesel per acre, and that does not take into account the energy content represented by other parts of the plant. (Business-standard, 2008).

The technological fundamentals of briquetting cotton stalks as a biofuel has been investigated by Saeidy (2004). The results indicate that the briquetting of cotton stalks represents an effective mean to upgrade the residues to fuel.

Internet resources

Business-standard, www.business standard.com/india/storypage.php?autono=336512
www.blackherbals.com/GM_crops_in_the_US.htm

CUPHEA (*Cuphea* spp.)

Description

The *Cuphea* species are perennial or annual herbs. *Cuphea* is a genus with 250 undomesticated species, native to Mexico, Central and South America, and the Caribbean area. One species, *Cuphea viscosissima*, is native to the USA. Several species are adapted to temperate agriculture and have seed oils rich in capric and lauric acids (Knapp, 1990). First attempts to domesticate some species as a new oilseed crop began only a few years ago.

Cuphea is a plant of open, often disturbed, mesophytic areas, roadsides, pastures and rocky road cuttings. It is 50cm to 2m tall, erect or sprawling. The leaves are opposite or whorled and finely scabrous, especially on the underside. Stem and leaves are covered with sticky hairs. In North America it flowers primarily from July to December. The seed capsules are thin walled, the capsule and the persistent floral tube each

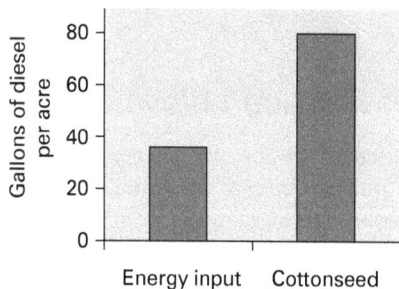

Figure 10.53 Total energy input (excluding solar energy capture by the plant) to produce an acre of cotton compared to the total energy content of cottonseed on the same acre, expressed as the equivalent content of a gallon of diesel fuel

rupturing along a dorsal line. The placenta and attached seeds emerge through the ruptures. The seeds are bilaterally compressed, and ovoid to orbicular.

Ecological requirements

The vegetative growth of *Cuphea* is aided by warm to hot weather with sufficient moisture, but seed yield decreases under hot and dry conditions. The *C. viscosissma* phenotypes would probably grow well in the mid-west and northwest USA, southern Africa, south-eastern South America, eastern and northern Australia, and in many parts of Europe (Mediterranean area).

Propagation

Cuphea shows characteristic features of a wild plant; domestication began only a few years ago. Wild populations of *Cuphea* are characterized by seed shattering and seed dormancy, and these problems are not yet overcome. In collected wild populations of *Cuphea* no differences in fruit morphology and seed retention were observed. All were seed shattering (Knapp, 1990).

The aim of breeding is the generation of non-shattering, auto fertile cultivars that do not show seed

Source: Melburnian http://commons.wikimedia.org/wiki/File:Cuphea_hyssopifolia.jpg

Figure 10.55 *Cuphea* flower

dormancy. These varieties may be the base for further improvement, like adaptation to colder climates, increased yields and improved shape. Inducing mutations is complicated because some species are polyploid. The chromosome numbers are: × = 8, n = 6–54. Often, mutations are not visible because in the polyploid species the effect of the mutation is masked by the non-mutated genes. The discovery of non-shattering phenotypes within the *C. viscosissima* × *C. lanceolata* silenoides population VL-119 were reported. This is the only known non-shattering phenotype in the genus. A fully non-dormant inbred line of *C. lanceolata* has been developed and is the first to have been reported.

> ARS researchers have produced oils by the barrelful from cuphea and other alternative crops, which may represent new domestic sources of industrial products ranging from soap to biofuels for cars, trucks and – in the case of Cuphea – even jet fuel. (Sciencedaily.com, 2008)

This brightens the outlook for the further domestication of *Cuphea*; these are the first steps of the development of auto fertile, non-dormant, non-shattering *C. viscosissima* × *C. lanceolata* silenoides (Knapp, 1993). The *C. viscosissima* × *C. lanceolata* hybrids are fertile. The development of auto fertile,

Source: Jack Scheper www.floridata.com/ref/C/images/cuph_hy2.jpg

Figure 10.54 *Cuphea* could be cultivated as a summer annual plant, and has the advantage of being a widely adapted warm season species

non-dormant and non-shattering germplasm has created a basis for the development of profitable *Cuphea* cultivars.

Insect pollination is required in the allogamus species such as *C. laminuligera, C. lanceolata* and *C. leptopoda*. A suitable pollinator for commercial plantings has not been found. Experimental plantings are mainly pollinated by bumblebees. The long floral tubes of the allogamus species prevent honeybees from gaining access to the nectar (Knapp, 1990). Insect pollinated allogamus species of *Cuphea* probably cannot be produced commercially. Even if honeybees or some other domesticated pollinators effectively pollinate these species, their commercial use is impractical and prohibitively expensive. Until the problem of pollination is solved, it is more promising to work with autogamus species like *C. lutea, C. viscosissima, C. tolucana, C. wrightii* and *C. carthagenensis* (Knapp, 1990).

Production

Cuphea is an oilseed crop in the spotlight of late as a potential producer of biochemicals and biofuels. While Cuphea apparently takes 2–3 weeks to germinate, it can produce seeds just six weeks after that, and to the tune of from 450 to 1350lb per acre. Other experiments showed that the mean value for harvest index was 2.8 per cent and seed yield 400kg/ha averaged across treatments and environments. Mean seed oil content was 297g/kg (Berti et al, 2007).

Economical seed yields cannot be obtained without seed retention. The crop could then be harvested after a killing frost and seed yield could accumulate to the end of the growing season. The problem is the capturing of the seed, not the seed yield (Knapp, 1990).

Most Cuphea species are characterized by sticky or glandular hairs covering their leaves, stems and flowers. This is unpleasant, but it is no barrier to the commercialization of *Cuphea*. Sticky non-shattering *Cuphea* can be harvested like other summer annuals. The hairs may be a defence against insect pests, and it has been no prior aim of breeding to remove them. The wild-type crop architecture poses no problem for the harvest: the plant grows upright, and very strongly upright if planted densely. It might be useful to develop determinate flowering species, but indeterminate flowering does not cause a problem for the harvest or seed production of *Cuphea* (Knapp, 1993).

Figure 10.56 *Cuphea* fruit and seeds

Experimental seed yields of *C. lutea* from swathing on tarpaulins and subsequent threshing ranged from 400 to 1200kg/ha at different harvest dates in 1987 in Medford, Oregon. The harvesting methods are under research and are not standardized or optimized. The yield data are useful for comparing species, but not for predicting yields under commercial cultivation. The single harvest seed yield represents about 10–40 per cent of the yield potential of equivalent populations with seed retention (Knapp, 1990). The seed and oil yields seem to be sufficient for *Cuphea* to compete as an oilseed crop (Knapp, 1993).

Processing and utilization

Some species are rich in specific single fatty acids – for example, *C. painteri* with 73 per cent caprylic acid, *C. carthagenensis* with 81 per cent lauric acid, and *C. koehneana* with >95 per cent capric acid. Some examples of the fatty acid composition are provided in Table 10.36, where they are compared with coconut, one main source of medium-chain fatty acids (MCFAs). No other temperate oilseed crop supplies these lipids.

As a source of lauric acid, *Cuphea* species have more to offer than coconut oil, because the concentration of lauric acid in the oil is potentially much greater. The isolation of single fatty acids should be easily accomplished, and tailor-made fatty acid compositions should be possible. Oil and protein values were determined on only a few occasions because only a few seeds have been collected from many of the species. In the tested species the oil content of the seeds varied from 16 to 42 per cent. The protein content (N × 6.25) ranged from 15 to 24 per cent.

The triglyceride analysis of *C. lanceolata* showed that the combination of fatty acids in the ester is not the result of random distribution but the specific combination of certain fatty acids (Kleiman, 1990).

Cuphea has an enormous potential as a source of renewable, safe and economical energy. The range of fatty acids in this genus is unique. Manipulated biosynthesis pathways can produce medium-chain acids similar to those from coconut. It can produce unique short-chain acids with low viscosity that can be used directly as a diesel fuel without chemical processing. Uneconomical and environmentally unfriendly transesterification is not necessary.

The *Cuphea* species are a source of MCFAs, which are useful raw materials. Caprylic acid, capric acid and myristic acid are potentially useful for industrial and nutritional purposes. Lauric acid is an important raw material for detergent products (Table 10.36).

Geller (1998) reported that:

- Normal vegetable oils have a high viscosity, which leads to injector coking and eventual engine failure.
- The reduced viscosity of *Cuphea* oil makes it a candidate for a fuel *without* transesterification.
- This is not biodiesel but a straight vegetable oil (SVO) fuel.
- The plant is indigenous to North America, and can be grown in many US regions including Georgia.
- It produces predominantly short-chained fatty acids C8:0 and C10:0.
- *Cuphea* oil and mixtures with #2 Diesel performed well in engine durability tests.
- Performance exceeded vegetable oils and petroleum #2 Diesel.
- Results suggest short-chain triglycerides may provide adequate diesel fuel substitute.

Internet resources

Sciencedaily.com (2008) www.sciencedaily.com/releases/2008/10/081013195803.htm

Table **10.36** Fatty acid composition of some *Cuphea* seed oils

Species	Caprylic acid (8:0)	Capric acid (10:0)	Lauric acid (12:0)	Myristic acid (14:0)	Others
C. painteri	73.0	20.4	0.2	0.3	6.1
C. hookeriana	65.1	23.7	0.1	0.2	10.9
C. koehneana	0.2	95.3	1.0	0.3	3.2
C. lanceolata		87.5	2.1	1.4	9.0
C. viscosissima	9.1	75.5	3.0	1.3	11.1
C. carthagenensis		5.3	81.4	4.7	8.6
C. larninuligera		17.1	62.6	9.5	10.8
C. wrightii		29.4	53.9	5.1	11.6
C. lutea	0.4	29.4	37.7	11.1	21.4
C. epilobiifolia		0.3	19.6	67.9	12.2
C. stigulosa	0.9	18.3	13.8	45.2	21.8
Coconut	8	7	48	18	19

Source: Kleiman (1990)

DATE PALM (*Phoenix dactylifera*)

Description

Few plant species have developed into an agricultural crop so closely connected with human life as has the date palm. One could go as far as to say that, had the date palm not existed, the expansion of the human race into the hot and barren parts of the 'Old World' would have been much more restricted. The date palm not only provided a concentrated energy food, which could be easily stored and carried along on long journeys across the deserts, it also created a more amenable habitat for the people to live in by providing shade and protection from the desert winds. In addition, the date palm also yielded a variety of products for use in agricultural production and for domestic utensils, and practically all parts of the palm had a useful purpose. But if the palm had an impact on human life, the influence was reciprocal, because through a long process of learning and experience, date palm cultivation was gradually adapted to man's needs. If left undisturbed, in its wild state, the date palm would, favourable growth conditions permitting, expand in an impenetrable forest of highly competitive clusters of an approximate equal number of male and female palms with relatively few reaching appreciable height or fruit-producing capacity. Examples of such uninhibited growth can still be found in some of the more remote areas of the Sahara (FAO, 1993a).

The date palm (*Phoenix dactylifera*) is a palm in the genus *Phoenix*, extensively cultivated for its edible fruit. Due to its long history of cultivation for fruit, its exact native distribution is unknown, but probably originated somewhere in the desert oases of northern Africa, and perhaps also southwest Asia. It is a medium-sized tree, 15–25m tall, often clumped with several trunks from a single root system, but often growing singly as well. The leaves are pinnate, 3–5m long, with spines on the petiole and about 150 leaflets; the leaflets are 30cm long and 2cm broad. The full span of the crown ranges from 6 to 10m. Dates have been a staple food of the Middle East for thousands of years. They are believed to have originated around the Persian Gulf, and have been cultivated since ancient times from Mesopotamia to prehistoric Egypt, possibly as early as

4000 BCE. There is archaeological evidence of date cultivation in eastern Arabia in 6000 BCE. In later times, Arabs spread dates around South and Southeast Asia, northern Africa, and Spain. Dates were introduced into Mexico and California by the Spaniards by 1765, around Mission San Ignacio.

The fruit is a drupe known as a date. They are oval-cylindrical, 3–7cm long, and 2–3cm in diameter, and when unripe, range from bright red to bright yellow in colour, depending on variety. Dates contain a single seed about 2–2.5cm long and 6–8mm thick. Three main cultivar groups of date exist; soft (e.g. 'Barhee', 'Halawy', 'Khadrawy', 'Medjool'), semi-dry (e.g. 'Dayri', 'Deglet Noor', 'Zahidi'), and dry (e.g. 'Thoory'). The type of fruit depends on the glucose, fructose and sucrose content.

The date palm is dioecious, having separate male and female plants. They can be easily grown from seed, but only 50 per cent of seedlings will be female and hence fruit bearing, and dates from seedling plants are often smaller and of poorer quality. Most commercial plantations thus use cuttings of heavily cropping cultivars, mainly 'Medjool' as this cultivar produces particularly high yields of large, sweet fruit. Plants grown from cuttings will fruit 2–3 years earlier than seedling plants.

Dates are naturally wind pollinated but in both traditional oasis horticulture and in the modern commercial orchards they are entirely pollinated manually. Natural pollination occurs with about an equal number of male and female plants. However, with assistance, one male can pollinate up to 100 females. Since the males are of value only as pollinators, this allows the growers to use their resources for many more fruit-producing female plants. Some growers do not even maintain any male plants as male flowers become available at local markets at pollination time. Manual pollination is done by skilled labourers on ladders, or in some areas such as Iraq they climb the tree using a special climbing tool that wraps around the tree trunk and the climber's back to keep him attached to the trunk while climbing. Less often the pollen may be blown onto the female flowers by wind machine. Parthenocarpic cultivars are available but the seedless fruit is smaller and of lower quality.

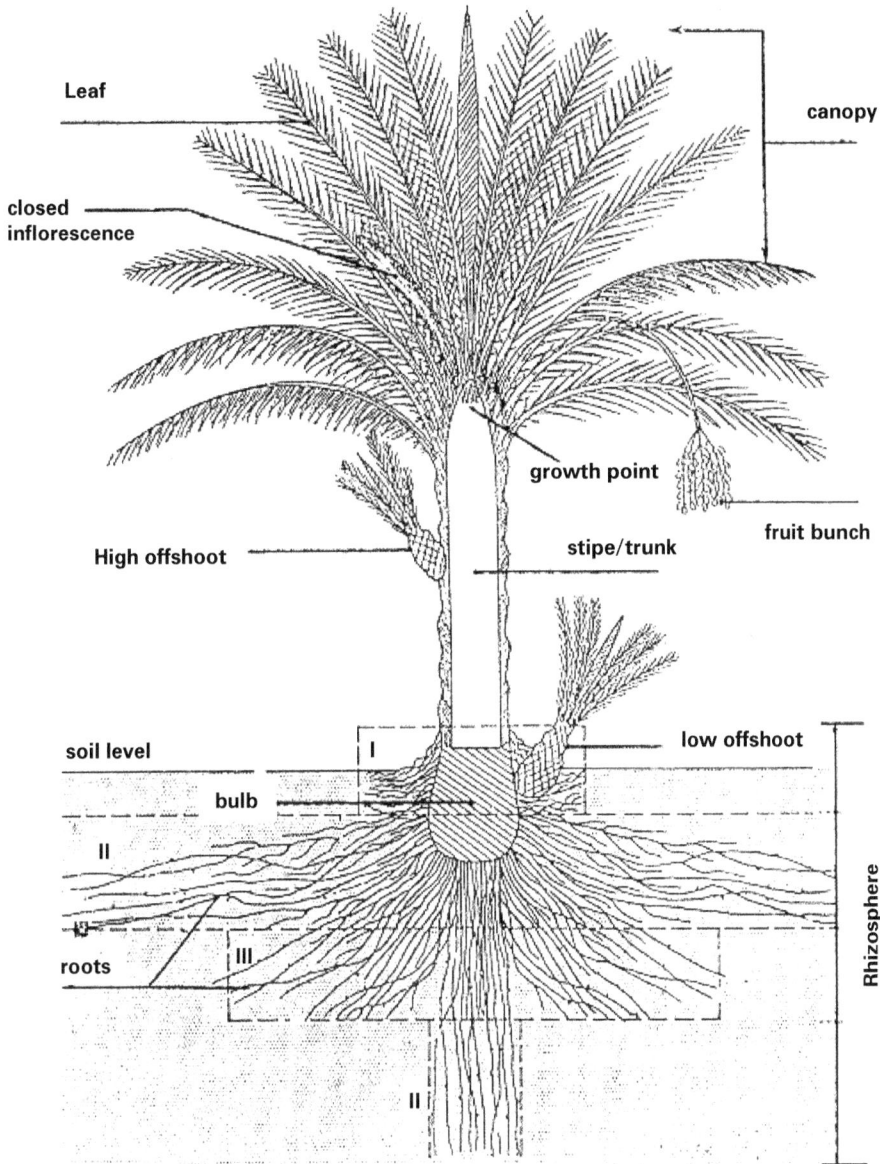

Source: Munier (1973) and Oihabi (1991)

Figure 10.57 Diagrammatic construction of a date palm with its root system

Dates ripen in four stages, which are known throughout the world by their Arabic names *kimri* (unripe), *khalal* (full-size, crunchy), *rutab* (ripe, soft), *tamr* (ripe, sun-dried). A 100g portion of fresh dates is a premium source of vitamin C and supplies 230kcal (960kJ) of energy. Since dates contain relatively little water, they do not become much more concentrated upon drying, although the vitamin C is lost in the process.

Dates are an important traditional crop in Iraq, Arabia, and North Africa west to Morocco and are

Source: Oasis Ltd

Figure 10.58 Date palm with fruit

mentioned in many places in the Quran. In Islamic countries, dates and yogurt or milk are a traditional first meal when the sun sets during Ramadan. Dates (especially Medjool and Deglet Noor) are also cultivated in southern California and Arizona in the United States. Iraq used to be a major producer of dates but in recent years the amount of dates produced and exported has decreased; however, it is still the world's major producer of dates with 87 per cent of the world's exports (wikipedia.org). Date palms can take 4 to 7 years after planting before they will bear fruit, and produce viable yields for commercial harvest between 7 to 10 years. Mature date palms can produce 80–120kg (176–264lb) of dates per harvest season, although they do not all ripen at the same time so several harvests are required. In order to get fruit of marketable quality, the bunches of dates must be thinned and bagged or covered before ripening so that the remaining fruits grow larger and are protected from weather and pests such as birds. Dry or soft dates are eaten without further preparation, or may be pitted and stuffed with fillings such as almonds, walnuts, candied orange and lemon peel, tahini, marzipan or cream cheese. Pitted dates are also referred to as stoned dates. Dates can be chopped and used in a range of sweet and savoury dishes, from tajines (tagines) in Morocco to puddings, ka'ak (types of Arab cookies) and other dessert items. Dates are also processed into cubes, paste called 'ajwa', spread, date syrup or 'honey'

called 'dibs' or 'rub' in Libya, powder (date sugar), vinegar or alcohol. Recent innovations include chocolate-covered dates and products such as sparkling date juice, used in some Islamic countries as a non-alcoholic version of champagne, for special occasions and religious times such as Ramadan. Dates can also be dehydrated, ground and mixed with grain to form a nutritious stockfeed. Dried dates are fed to camels, horses and dogs in the Sahara. In northern Nigeria, dates and peppers added to the native beer are believed to make it less intoxicating.

Young date leaves are cooked and eaten as a vegetable, as is the terminal bud or heart, though its removal kills the palm. The finely ground seeds are mixed with flour to make bread in times of scarcity. The flowers of the date palm are also edible. Traditionally the female flowers are the most available for sale and weigh 300–400g. The flower buds are used in salad or ground with dried fish to make a condiment for bread.

In India, North Africa, Ghana and Côte d'Ivoire, date palms are tapped for the sweet sap which is converted into palm sugar (known as jaggery or gur), molasses or alcoholic beverages. In North Africa the sap obtained from tapping palm trees is known as lāgbī. If left for a sufficient period of time (typically hours, depending on the temperature) lāgbī easily becomes an alcoholic drink. Special skill is required when tapping the palm tree so that it does not die.

Date seeds are soaked and ground up for animal feed. Their oil is suitable for use in soap and cosmetics. They can also be processed chemically as a source of oxalic acid. The seeds are also burned to make charcoal for silversmiths, and can be strung in necklaces. Date seeds are also ground and used in the manner of coffee beans, or as an additive to coffee.

Stripped fruit clusters are used as brooms. In Pakistan, a viscous, thick syrup made from the ripe fruits is used as a coating for leather bags and pipes to prevent leaking.

Date palm sap is used to make palm syrup and numerous edible products derived from the syrup. Date palm leaves are used for Palm Sunday in Christian religion. In North Africa, they are commonly used for making huts. Mature leaves are also made into mats, screens, baskets and fans. Processed leaves can be used for insulating board. Dried leaf petioles are a source of cellulose pulp, used for walking sticks, brooms, fishing floats and fuel. Leaf sheaths are prized for their scent, and fibre from them is also used for rope, coarse cloth,

Table 10.37 Nutritional value per 100g (3.5oz) of dried dates (edible parts)

Energy	280kcal 1180kJ
Carbohydrates	75g
- Sugars	63g
- Dietary fibre	8g
Fat	0.4g
Protein	2.5g
Water	21g
Vitamin C	0.4mg (1%)
Manganese	0.262mg

Source: Wikipedia.org

and large hats. The leaves are also used as a lulav in the Jewish holiday of Sukkot.

Date palm wood is used for posts and rafters for huts; it is lighter than coconut and not very durable. It is also used for construction such as bridges and aqueducts, and parts of dhows. Leftover wood is burned for fuel.

Where craft traditions still thrive, such as in Oman, the palm tree is the most versatile of all indigenous plants, and virtually every part of the tree is utilized to make functional items ranging from rope and baskets to beehives, fishing boats, and traditional dwellings (Wikipedia.org).

A pragmatic, well-thought-out strategy

Contributed by Brahim Zitouni, President of OASIS Ltd, Dubai

Some economic theories are founded on agricultural concepts, others on industrial ones, while still others are based on the capacity of intelligence to devise ever more ambitious development models. Bitter experience tells how, all too often, these economic models settle ideological scores at the expense of people.

For our part, OASIS has preferred to take the recognized wealth of the Arab territory and the know-how of the Arab people as the basis for proposing, with the aid of new technologies, a form of sustainable development that is in keeping with Arab countries', real potential. In this sense, we have taken a practical, pragmatic approach to the OASIS (Organization of the Agriculture in the Sahara by Integration and Substitution) project, far from pointless, sterile theoretical debates, because most of the Arabic world has already lost enough time.

Date groves are an inexhaustible source of wealth scarcely recognized by man owing to unfavourable historical circumstances in the Arab world. The date palm has nourished men in some of the harshest regions on the planet. In ancient times, it was without question a vector of prosperity at the service of man. It might justifiably expect some sort of recognition from the modern world for all the services it has rendered. Yet, owing to unfavourable historical circumstances, the date palm, having prospered in essentially Arab-Muslim geographical eras, has not enjoyed the fate it deserves. The decline of Science in the Arab-Muslim era, after Baghdad had been, ten centuries earlier, the world's most dazzling scientific centre, has been the major determining factor behind the failure to rationally exploit date groves.

As nature would have it, where the date palm flourishes, so too are fossil fuels found in abundance. What a strange fate that two quite different resources, yet both of which are economically profitable, should thus be found in the same territories; the date palm on the surface, hydrocarbons deep in the bowels of the Earth. And what if man, through science, were to combine these two sources of prosperous human activity into one, so that the qualities of one were used to offset the harmful consequences of the other? By opening up date palm cultivation to energy-sector continuity solutions, we intend to use modern man's scientific knowledge to settle the ancient debt owed by humanity to the date palm.

Source: http://en.wikipedia.org/wiki/File:Phoenix_dactylifera1.jpg

Figure 10.59 Date palm orchard

The date palm challenge

On the initiative of a small group of civil servants at the Brazilian ministries of Energy and Agriculture, a bold national ethanol programme, ProAlcool, was devised in 1973 and launched in 1975 in Brazil. In a country with such a drastic energy shortage, these intelligent individuals took their cue from Brazil's exceptional richness in biomass, a true gift from God, to devise and begin production of a sugar-cane-based agricultural fuel which today ranks the country as the number-one producer of 'green gold' on the planet. The Brazilians thus demonstrated how intelligence and the mobilization of natural resources in an appropriate setting can generate inexhaustible wealth.

In the past, intelligent populations that were lucky enough to have date palms reaped the benefits of oasis cultivation in regions where it was difficult to combat excessive sunshine. The agricultural surplus of the fertile gardens of Lower Mesopotamia (25,000km², 5000 BC) led to the founding of the first significant towns. What is known today about the role of the date palm in the growth of Mesopotamian civilization? Is it no more than coincidence that the cradle of urban humanity prospered, first and foremost, under the shade of date groves, even before the rise of the civilization of the Nile Valley?

> Having tamed the waters, Lower Mesopotamia became the 'Garden of Eden', to which men flocked in ever-increasing numbers, with its abundance of cereals, fruit trees, sesame and, *wonder of wonders, the date palm*. (in *Les Mémoires de la Méditerranée* by Fernand Braudel)

The date palm is highly adaptable. It is a species that thrives in warm conditions, showing activity from a temperature of 7°C. In addition, it withstands high soil salinity.

Dates can be classified in three major categories according to their characteristics:

- dried dates, with a maximum 15 per cent moisture content;
- semi-dried or semi-soft dates, with between 15 and 25 per cent moisture content; and
- soft dates, with over 25 per cent moisture content.

All dates are sweet, while their tastes and aromas depend on the proportions of sucrose, fructose and glucose they contain. High-fructose dates have the strongest aroma, high-sucrose dates are similar to sugar and high-glucose dates have a stronger flavour on the palate. But all cultivars show sugar content above 60 per cent. In 2006 the world production stood at around 6,551,000 tonnes of dates, of which 25 per cent could be considered as very low quality and eventually a waste product rich in sugars, which opens the path to all the fermentations processes – among them bioethanol production from dates of low quality. This over-abundance of supply will be accentuated over the coming years as a result of date production subsidization policies to support the oasis populations in each country, in addition to the development of the domestic consumption market, which for a long time was depressed due to the urbanization of populations in the date grove countries.

The natural increase in date supply to the world market will take the form of processed dates. Demand for this type of product is beginning to emerge in connection with minor products, such as date syrup or date jams, for example.

The significant contribution of the Baghdad Date-Palm Research Centre

In 1974, the Date-Palm Research Centre was set up in Baghdad. It was dissolved in September 1980 to be transformed into the Agriculture and Water Resources Research Centre, under the control of Dr Samir El Chaker. The Date-Palm Research Centre became a department of the new Centre, developing innovative ideas on date processing under the enlightened leadership of Dr Hassan Chabana, the department's head until 1985, then Dr Alaa El Baker, head until 1987 when the Centre was closed by the Iraqi authorities. New date-processing ideas, innovative at the time, were tested in a pilot FAO project directed by Dr Namrod Benjamin. A close partnership was forged with another Iraqi research centre – the Biology Research Centre – whose Industrial Microbiology Department was headed at the time by Dr Raad El Bassam, an expert in date fermentation. Baghdad's Date-Palm Research Centre played a significant role in the modernization of date-growing practices, paving the way for the processing of dates into various products. As a result of war the Centre was closed down, but these Iraqi elites have shown that it was possible to consider the dates sector as central to an agricultural model to be implemented, even if they did not think at that time of the ethanol date palm option.

In the 1970s there was a lack of reflection on the 'Economic Model' and a lack of boldness. While the political authorities in Iraq may have seen the potential in date processing, they were far too quick to declare the industrial plants not financially viable. By isolating the processing industry from the rest of the date sector, reflection on the 'Economic Model' resulting from the date groves became a secondary consideration.

The link with the issues of the new carbon economy was yet to be established, but the basic tools for growth already existed. However, like any fledgling industrial sector, processed-date products clash with 'equivalent' products on the market or which fulfil the same functions. Since the sector is at the start-up stage, it has problems pricing itself competitively compared to the liquid sugars for example.

For the processing industry to avoid these start-up difficulties, a powerful growth engine is needed, which requires very large quantities of dates and offers strategic characteristics with the capacity to spark the interest of the institutional community. Once coupled to this essential growth engine, the sector, through crop intensification, will be in a position to keep pace with market sectors to which it was previously hard pushed to gain access due to the dominant position of 'established actors'.

Validity of the ethanol sector based on the proportion of sugar per tonne: the date palm compared to sugar beet and sugar cane

If we compare the sugar content per tonne of crop on a dry basis, dates (65% sugar content on dry basis) compared to sugar cane (13%) and beet (18%), dates are by far in an advantageous position. Meanwhile, palm trees, sugar beet and sugar cane have different crop densities because we are comparing a primary crop with secondary crops. As temporary cultivation crops, sugar cane (90t/ha) and sugar beet (60t/ha) are by far the crops which offer the highest possible density. The date palm is a permanent cultivation plant and offers poor density (100–150 date palms per hectare, 22.5 tonnes of dates per hectare).

Of course, the key figure remains the proportion of sugar expressed per hectare for each crop. If we

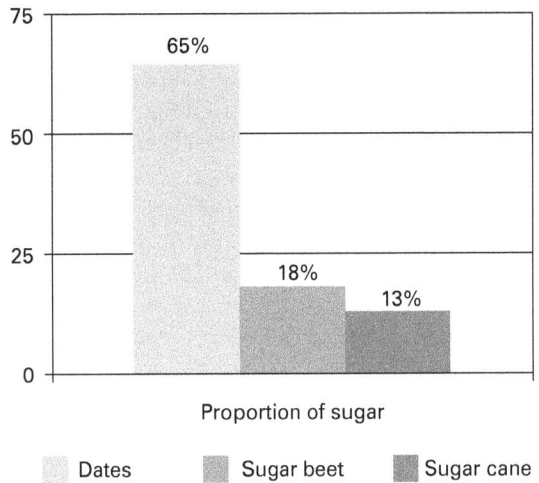

Source: UNCTAD – Market information on basic commodities; sugar and Sugar Trading Manual)

Figure 10.60 Sugar content per tonne of crop on a dry basis

compare the proportion of sugar/tonne per hectare dates come out on top (14.6 per cent). Sugar beet comes in last place (11.8 per cent) in terms of sugar yield per hectare whereas sugar cane can yield per hectare 11.7 per cent of sugar.

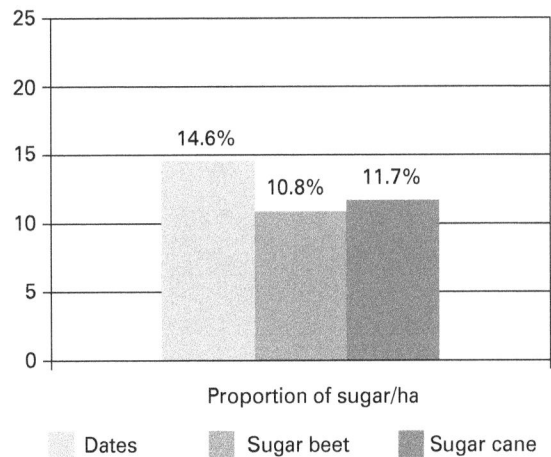

Source: Philip Digges and Dr James Fry, LMC International – costs of production for sugar beet and sugar cane figures; Dr Hassan Chabana, palm trees cultivation specialist – dates figures

Figure 10.61 Proportion of sugar/tonne per hectare

For these theoretical reasons, the bioethanol industry will have a very solid foundation in the Arabic world, essentially due to the high proportion of sugar per hectare which can be obtained from date palms.

If we compare the transformation into bioethanol of biomass from sugar beet, sugar cane, maize and dates: one tonne of sugar beet produces 116 litres of ethanol, one tonne of sugar cane 60 litres; one tonne of maize 375 litres and one tonne of dates produces 280 litres. Productivity per hectare puts sugar beet at 7000 litres/ha, sugar cane at 5500 litres/ha and maize at 3000 litres/ha whereas one hectare of date palms will produce around 6300 litres of ethanol.

The handicap of being a primary crop does not affect the date palm's performance. These basic economic conditions are favourable to the birth of a new agro-energy date sector.

Most countries with strong date-growing potential also have highly developed oil industries. These countries have managed to create the necessary labour force to maintain the hydrocarbon production apparatus. The date-based agro-energy-processing sector has all the more reason to concentrate on fuel bioethanol since it will find national executives trained in economic patriotism who are sensitive to the energy issues facing them. A number of these executives will be tempted to draw on their experience gained in the refineries and downstream hydrocarbon industry to offer to the fledgling date sector.

By positioning itself with a view to integrating with the industrial oil sector through the production of bioethanol from the date surplus, the sector will be seen as complementary to an industrial apparatus that is already in place and under control. Technological synergies, such as the marketing of bio-ETBE, for example, could amount to major points of convergence of interest upon which the budding sector has every reason to capitalize.

The existence of a dense liquid-energy distribution network is also an exceptional advantage for the date sector, serving as a captive outlet.

Logistical issues in connection with the sector's integration with the downstream hydrocarbon sector are also of paramount importance, and the sector's competitiveness in bioethanol in particular may be dependent upon them. Integrating the biorefineries near the hydrocarbons refinery will be of a critical importance from the point of view of its economical efficiency.

Furthermore, high-level political will is required on the part of the state, such that the development of the sector is not seen merely as the preservation of an element of national cultural identity, indeed of civilization as a whole, but rather as a sector that brings with it exceptional advantages and developments of benefit to the nation. It is in the interests of a state with considerable agricultural potential to stabilize its rural populations so that they rally to the issues relating to the country's food security. The date sector is naturally centred around this theme because the calorific value of dates is quite simply unique (3.150 calories per kg).

For its part, the energy sector must make a point of showing that it sees renewable energies as a priority, something which in practice is not always the case, since the hydrocarbon sector tends to follow its own dynamic to the exclusion of everything it may consider, quite wrongly, in the long run to be a dangerous competitor to its future, when in fact it is entirely complementary.

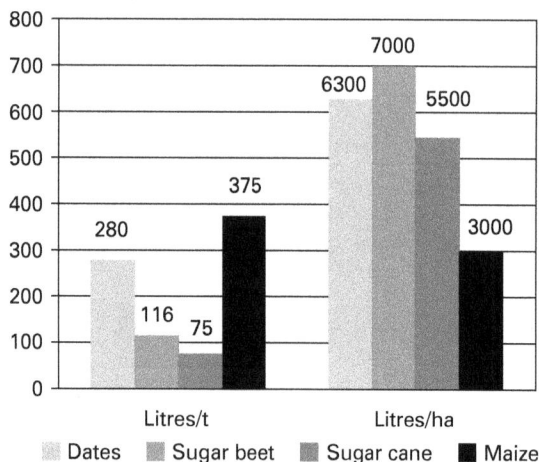

Source: Dr Christopher Berg – World Fuel Ethanol – Analysis and Outlook

Figure 10.62 Comparison between sugar beet, sugar cane, maize and dates in litres/t and litres/ha

The sector offers unique comparative advantages

Whether viewed from the point of view of energy or agriculture, the date sector offers unique comparative advantages. By producing fuel energy, it aids the downstream energy sector to export more fossil fuels, helping it to cater better for global demand in this respect. The date-based bioethanol sector thus meets the growing, pressing demand of developed economies.

By producing processed products for human consumption (natural liquid sugars, yeasts, vinegar, proteins, fibres etc.), the date sector offers unique products which farming in the dominant countries is unable to provide because it lacks the agricultural raw material that is the basis of all of these products.

Due to its specific characteristics (date groves are unique) and their geographic positioning (mainly in hydrocarbon-producing countries, in semi-arid areas unchallenged by other crops), the sector will not face competition from agribusiness in developed countries or from other local agricultural producers.

Moreover, its economic development model will be easily accepted and encouraged, because it tends to cater better for the liquid-energy needs of world economies and maintain oasis populations sustainability.

One of the harshest criticisms of the fuel-bioethanol industry is that the move to produce bioethanol from biomass means stripping the world of important resources for feeding its human population, at a time when we are experiencing a demographic explosion the like of which has never been seen in the history of mankind.

This observation is fundamentally sound, but should be tempered in the case of the date sector. First of all, because, in the case of the date palm, unlike other biomasses, the products associated with fuel-bioethanol production are for human consumption. Furthermore, in terms of value, date-based bioethanol production accounts for 15 per cent of the turnover of the processing sector, while its main associated products – date coffee and date fibres, both for human consumption – could amount to more than 85 per cent of processed-product value.

Through the rational exploitation of date groves, adopting silviculture methods, it will be possible to recycle 4 per cent of all date groves, particularly the ones which are not yielding any more, as animal feed each year. This will mean that areas used to grow grain for animal feed are free to turn their attention to human consumption. This is only possible in the case of date palm cultivation, since it is a prolific permanent crop, unlike the other bioethanol biomasses, which are intensive secondary crops low in biomass.

There are a great many oases in the Sahara, although the exact number is not known. A land register of the Arabic countries' oases and its date-growing potential needs to be drawn up, so that the expansion of the sector can be better thought through.

This is all the more justified since, in its areas of expansion, the date palm does not encroach upon any other crop. Giving absolute priority in the south to date palms in no way impinges on the development of other crops in the north of the Arabic countries. As a permanent primary crop, the date palm possesses its own means for growth entirely independently of other arable crops, yet without standing in the way of their expansion.

Source: www.oasisnakhoil.com

Figure 10.63 Date palm – fuelling the future

Given its nature as a permanent crop that develops effective protection against sunlight, the date palm aids the better exploitation of an agricultural area hitherto considered of little use due to its exposure to sunlight. Furthermore, the oasis system is built around this principle, and the possibility of extending the oasis model beyond its traditional boundaries warrants consideration, for deployment with emphasis on the intensification of the date groves and on complementarity with other co-planted agricultural crops.

Nakhoil, the ethanol from the dates, would in a sense become the Arab's primary agricultural crop and the corner stone of the breeding sector, playing a fundamental role as an agent of growth for the entire Arab agricultural sector, through the adoption of an oasis model in constant expansion in the desert and semi-arid areas of these countries. In consequence, the increased area under date palm cultivation will no longer cater solely for a need to create employment and economic activity by stimulating date supply; it will also open up new strategic horizons. In a sense the date palm elegantly solves the equation of the bioethanol industry and demonstrates in a brilliant way how bioethanol feeds the world.

An exceptional opportunity for the hydrocarbon sector to integrate with the national economy

For a country that has plentiful natural resources such as oil, the rise in the price of resources could be a real benefit, but it could also push up currency value, produce wealth outflow and stifle the growth of industries not dependent on Energy. This phenomenon is known as 'The Dutch syndrome'.

One means of avoiding 'The Dutch syndrome' is to invest massively in basic infrastructure and/or give priority to investments derived from high energy prices in sectors that favour import substitution or, better still, sectors with high export potential other than hydrocarbons. This is precisely what the Economic Model for the Date Sector offers through its multi-sectorial agricultural and industrial facets. Whereas today bioethanol is produced from the dates, tomorrow it could be produced from date palms, which are plants (rather than trees). Today's expansion of the date groves, thanks to Nakhoil derived from dates, will

mean access to significant quantities of biomass tomorrow. Date groves will then be in position to supply dates for other processing purposes, for example the production of 'sugars'.

Moreover, the expansion of a primary crop like the date palm carries fewer dangers in specialization than the expansion of intensive secondary crops such as maize, sugar cane or sugar beet.

The bioethanol-fuel processing sector could constitute the main underlying dynamic ingredient of the OASIS programme, providing a number of conditions were met, particularly the downstream hydrocarbon sector's involvement in the date sector.

In this regard, the hydrocarbon sector will not be seen any more as an industry only integrated into the rest of the economy of the nation financially, but will be understood as the true growth engine driving all the rest of the agro-industrial sectors.

Each of the following areas has a surprisingly high degree of potential for substitution from the dates groves sector: hydrocarbons, petrochemicals, wood, packaging, paper, cereals, animal feed, food-processing (coffee, sugar, vegetable proteins), cosmetics and pharmaceuticals. The miracle of the date palm – this 'wonder of wonders' – will be repeated, so long as existing market technologies in different sectors are applied and adapted intelligently to dates and date palms.

By providing the date sector with the means for growth in the form of Nakhoil, the surpluses produced by date growers will quite naturally go on investments in co-planted crops, following a principle of import substitution, focusing initially on high added-value agricultural production.

All of these co-planted crops will require food-processing industries. The densification of industrialization in the date grove regions will occur harmoniously on the basis of the date sector's comparative advantages once it has settled into a steady rhythm.

But the date sector's modus operandi of growth by substitution does not only concern imports. The date sector will enable grain farming to turn its attention towards growing cereals for human consumption.

In a hydrocarbon market tending towards growth, Brazilian bioethanol fuel from sugar cane emerges with the USA as a major actor in renewable energies in the strategic liquid fuels sector. By comparaison, the Arab bioethanol, the Nakhoil, the true growth agent of the date groves will allow food security to be reached in a semi-arid environment thanks to palm tree intensification by introducing in vitro propagation of prolific cultivars.

When the price of wheat was low, was it not linked to the even lower cost of energy, equivalent to 5 per cent of global GDP, whereas the contribution made by energy is more like 50 per cent? If, in this climate of deflated energy costs, the Arab world is still able to pay for its food bill, tomorrow, as energy prices are steadily driven up by the carbon economy, what will happen to wheat and rice prices on the international markets? These are the challenges for which biotechnology must partly provide answers, particularly those derived from date palm cultivation.

As a matter of fact the up-coming Arab model has got an asymmetric configuration if we compare it to the Brazilian model. Actually it goes from an excess of hydrocarbons to food security in a semi-desert environment thanks to Nakhoil, the Arabic bioethanol. The Brazilian model to contemplate was to the contrary steering from agricultural abundance to energy security. Both countries are using the bioethanol as a steering drive to reach what is perceived a strategic goal.

An exceptional environmental model in keeping with the carbon economy

Date palm cultivation not only offers the advantage of preparing us in its own way for the post-oil era, it also falls within the scope of the concerns expressed in the Kyoto Protocol.

Solar power, windmills and date- and palm-based bioethanol will be major assets for the Arabic countries. The use of 5 per cent of bioethanol in fuels reduces carbon monoxide in the atmosphere by 30 per cent.

It is obvious that the transformation of dates into Nakhoil fits naturally in the scheme of development of renewable energy for the years to come but the palm trees are also an essential element of the carbon economy. The date palm's 'biological zero' is 7°C. This means quite simply that the date palm continues its photosynthesis practically all year round. Unlike other trees or plants that lose their leaves during the autumn and winter months, the date palm remains green

throughout the year. Date palms are 'carbon sinks' that act as an effective, economical means of combating rising atmospheric carbon dioxide levels in semi-arid areas.

The date palm thrives in warm conditions, and amounts to a kind of insurance policy for the continuity of plant production. If current climate change leads to an increase in temperature of 1–2°C, this will have few consequences for biomasses. On the other hand, if the temperature were to rise by as much as 4–5°C, it would lead to significant breakdowns in the world agricultural system, with the exception of date groves, which would cope with such a change.

An agro-energy policy based on palm trees would have the tremendous benefit of combating desertification as an associated goal, by the gradual planting of a 'sustainable, productive green belt' of date palms along the main axes of the transportation system.

The date sector will gradually emerge as a sector of strategic interest for energy-producing countries that do not as yet see the need to prepare for the post-oil era. The US president's 2005 State of the Union Address ought to have incited people to reflexion, since he announced profound changes in American energy policy.

Date palm cultivation not only offers the advantage of preparing us in its own way for the post-oil era, it also falls within the scope of the concerns expressed in the Kyoto Protocol. The development of palm forests, the increasing use of bioethanol in vehicles to combat the greenhouse effect, the covering of semi-arid land with vegetation, and its capacity to push back the desert, all amount to major points in favour of the date sector.

Setting up the date sector also benefits from a favourable overlap with the international agenda. The UN is likely to proclaim 2010–2020 the decade for combating desertification. The international agenda and that of the development of the date sector coincide harmoniously.

A new economy – that of carbon – is gradually coming into being. It is already at work in Europe, preparations are being made in many US states, and it is beginning to be taken into account by planners in India and China. In this broader context, the date sector will reveal its unsuspected potential and will, to some extent, offset CO_2 emissions from the energy sectors.

Revive a plant that is as old as humanity

The challenge for the Arabic world is to develop its territory so that the semi-desert areas constitute a buffer zone that is needed to restabilize national development. It is not merely a matter of good governance, in the sense that the investment pumped into the semi-desert area will be far less costly than the health, social and political costs that will be borne by towns and cities in the event of an uncontrolled rural exodus.

It is also a matter of satisfying the country's deep aspirations, which demonstrate an attachment to 'their lands' that are still very much alive, unlike in the Northern industrialized nations, due to a different background to the social development history.

By rallying all efforts around its production, Nakhoil, the bioethanol from dates, develops the territory and creates varied agricultural and industrial forms of economic activity. It fixes populations and gives pride and dignity to the populations of the country's southern regions, who all too often feel 'useless' as they are not integrated with the rest of the national economy.

The national and Arabic dimension of the OASIS project are obvious. The underlying principles of OASIS are in keeping with sustainable development, i.e. globalization and the new carbon economy that are being set in place. Industry as we know it will adapt or else perish under the blows of the new order.

Through its cultural and civilizational resonance, the OASIS project speaks to the entire Arab People, indeed all Muslims, because of the status of the date as the preferred fruit of the prophet and beyond to all people that have deserts in common. Through its environmental aspects, it speaks to all peoples across the planet, and as a result places the Arab states as valid representatives and partners of choice in the dialogue of civilizations and the preservation of the world's ecological balance. The Arab states will appropriate globalization in order to better incorporate it into their historic development thanks to the palm tree, which at last has access to the rank it deserves: to be a major crop coming from the depths of the birth of humanity.

An Omani entrepreneur plans to start producing biofuel and marketing the same by 2010 through biofuel stations across the country. The biofuel refinery, to be set up in Sohar, will have a capacity of 4.8 million tonnes within four years. In the first two years the capacity will be 900,000 million tonnes annually, utilizing 10 million of the region's ubiquitous date palms as a feedstock for ethanol. Ethanol, used as biofuel, is produced by fermentation of glucose, to be derived from date palms in Oman, by yeast. The biofuel project is expected to create more jobs for Omanis, employing over 3500 Omanis in the first five years (Vaidya, 2007).

Internet resources

OASIS Ltd, www.oasisnakhoil.com/dataimg/English%20Brochure.pdf
Electronic Journal of Biotechnology, (2002) vol 5, no 3, www.ejbiotechnology.info/content/vol5/issue3/index.html
http://archive.gulfnews.com/articles/07/06/23/10134279.html
Wikipedia.org, http://en.wikipedia.org/wiki/Phoenix_dactylifera

EUCALYPTUS (*Eucalyptus spp.*)

Partially contributed by C. D. Dalianis

Description

Eucalyptus is a fast-growing tropical tree species used as a biomass source for bioenergy, and for pulp and paper manufacturing. Major research efforts are under way to map the tree's genome with the aim to improve it as an energy crop. Eucalyptus is on the agenda of the US Department of Energy's Joint Genome Institute (DOE JGI), with an international team working on increasing biomass production and the carbon sequestration capacities of the species.

Eucalyptus species are native to Australia. During the last two centuries, eucalyptus were spread from Australia into many tropical and subtropical regions of the world. The genus *Eucalyptus*, with more than 550 species, belongs to the Myrtaceae. The name of the genus was derived from the Greek words *eu* and *kalyptos*, meaning well covered in relation to the operculum of the flower bud (Brooker and Kleinig, 1991). A few eucalyptus species have proven to be excellent commercial crops: *Eucalyptus grandis* and *Eucalyptus tereticornis* in tropical regions and *Eucalyptus globulus* and *Eucalyptus camaldulenis* in temperate climates. The last is also extensively grown in tropical regions. *E. grandis* is a dominant species in Brazil, *E. tereticornis* in India, *E. globulus* in Spain and Portugal, and *E. camaldulensis* in Morocco and Spain. Some other important species are *E. saligna*, *E. urophylla*, *E. citriodora*, *E. viminalis* and *E. deglupta*. *E. citriodora* is being planted on a substantial scale in southern China (Eldridge et al, 1994). Although several eucalyptus species are adapted to certain Mediterranean countries, the two most important species for the Mediterranean climate are *E. camaldulensis* and *E. globulus*.

Ecological requirements

Many eucalyptus species are highly adaptive and grow rapidly in a wide range of climatic conditions. Sensitivity to low temperatures is the most important environmental factor limiting the latitudinal and altitudinal range over which eucalyptus can be planted. Many eucalyptus species can be grown successfully not only in the regions where they occur naturally, but in most parts of the tropics, subtropics and warm temperate zones. About 100 eucalyptus species have been planted outside Australia with various degrees of success.

The amount and distribution of rainfall affects eucalyptus species adaptation. Some species thrive in areas with summer rainfall, others in winter rainfall areas. As a general rule, the transfer of species from a native habitat of winter rainfall areas to summer rainfall areas is unsuccessful (Hillis, 1990).

The genus as a whole grows on old and well-leached soils, with a fairly low pH. Although many eucalyptus species can grow on soils of low nutrient status, especially on those deficient in nitrogen and phosphorus, eucalyptus respond well when nutrients are not limiting (Turnbull and Pryor, 1984; Hillis, 1990; Lapeyrie, 1990). Soil drainage and acidity may affect the adaptation of certain eucalyptus species. Some species have been used to assist control of waterlogging, water and wind erosion; others have shown smog tolerance, and others have proved invaluable for planting in arid regions (Hillis, 1990).

Eucalyptus camaldulensis is the most widespread in Australia as well as in several Mediterranean countries (Figure 10.64). Because of its wide distribution, several provenances have been developed that differ in

Figure 10.64 Short rotation and dense populations of *E. camadulemis* 4 months after harvesting (in front) and *E. globulus* 16 months old (in the background).

Source: Martin253 wikipedia.org

Figure 10.65 Eucalyptus plantation in Galicia in Northwest Spain

performance and adaptation. As a general rule, *E. camaldulensis* is able to produce acceptable yields on relatively poor soils with a prolonged dry season, exhibits some frost resistance, tolerates periodic waterlogging and some soil salinity, and becomes chlorotic on highly calcareous soils. It is a drought-resistant species and grows in areas receiving 200mm rainfall per annum, though growth is better where the annual rainfall exceeds 400mm (Turnbull and Pryor, 1984).

E. camaldulensis is a vigorous coppice and has several uses. In many countries it is used for fuelwood, charcoal, poles, posts, shelterbelts and hardboard production. In Argentina it is used for charcoal for industry. In Morocco and to a lesser extent in Spain and Portugal it is used for pulp production, though pulp quality is inferior to that of *E. globulus*. It is considered as a drought-tolerant species suitable for afforestation in arid or semi-arid regions.

Eucalyptus globulus Labill. ssp. *globulus* is widely spread in certain Mediterranean countries. Adult plants attain a height of up to 55m. The juvenile leaves are opposite, sessile, stem clasping, and glaucous to bluish. This subspecies is widely cultivated in the Iberian Peninsula and other parts of the world. It is well adapted to mild, temperate climates and high elevations in cool tropical regions. Ideal conditions in Europe are along the north-western coasts of Spain and Portugal, where the mean annual precipitation is above 900mm, the dry season is not severe and the minimum temperature above −7°C. It is considered very sensitive to moisture stress. In south-western Spain with 465mm annual precipitation and about four months dry season, it grows on deep soils with available soil moisture. On drier and shallower soils *E. camaldulensis* is superior.

Eucalyptus globulus is well adapted to a wide range of soil types. Although the best development is on deep, sandy clay soils, good growth is also attained on clay loams and clay soils, provided they are well drained. Insufficient depth, poor drainage and salinity are limiting factors. It is considered as one of the best eucalyptus species for paper making. It is also highly regarded for high and heavy construction, poles, piles, railway sleepers and as fuel.

Crop management

Most of the eucalyptus plantations established outside Australia are managed as coppice crops. Coppice shoots develop from dormant buds situated in the live bark or from lignotubers, buds found near the junction of root and stem in many eucalyptus species (Jacobs, 1979). Since eucalyptus regenerates itself after each harvest operation, biomass accumulation can be generated over several growth cycles with only one initial cost of site preparation and planting. A number of eucalyptus coppice regrow readily, sometimes up to six or seven times (Hillis, 1990). The number of coppice stems per stool is influenced by the initial planting density, genotype, stump height and climatic conditions immediately before and after harvest. Total volume growth of the coppice may be greater than initial seedling growth (Dippon et al, 1985).

Throughout the EU, several plants are considered as short rotation coppice plants exclusively to provide biomass for energy. In the southern EU, candidates for such a purpose are black locust (*Robinia pseudoacacia* L.) and eucalyptus. Both plants are still in the experimental stage. In Greece, both *E. camaldulensis* and *E. globulus* are being tested under various environmental and soil conditions, very short rotation cycles and dense populations. On fertile fields with a plant density of 20,000 plants/ha (1m × 0.5m) and appropriate irrigation and fertilization management, dry matter yields of two-year rotating *E. camaldulensis* of up to 32t/ha/year have been obtained. After each cutting a large number of sprouts (15–25) develop from each stump. However, very quickly a few of them, usually one to three per stump, dominate; the rest either die or remain thin and stunted. In contrast, on marginal and abandoned fields after each cutting the many sprouts are competing with each other and the plantation has a bushy appearance. It is reported (Borough et al, 1984) from Australia that the mean annual increment for above ground woody biomass of densely planted *E. globulus* (0.3m × 0.3m and 0.6m × 0.6m, and with 140g per tree of 3:3:2 NPK fertilization) varied from 11.0 to 14.3t/ha/year on a three-year rotation cycle.

Eucalyptus have often been accused of causing nutrient depletion. However, many scientific papers (e.g. Bara, 1970; Bara et al, 1985) have proved their positive effects on the soil and their relatively low levels of nutrient removal. An extensive literature review demonstrated that afforestation with eucalyptus improved soil fertility in the long term in several areas of the world (Philliphis, 1956; Karschon, 1961; Ricardo and Maderia, 1985). Annual litterfall could be

Source: http://en.wikipedia.org/wiki/File:Sherbrooke_forest_Victoria_220rs.jpg

Figure 10.66 Three-year-old eucalyptus forest

an important proportion of the organic material and nutrients being cycled through eucalyptus woodlands (McIvor, 2001). Miguel (1988) states that as a general rule nutrient removal by eucalyptus plantations in northern Spain is relatively light and can usually be compensated by nutrient inputs from rainfall and rock weathering. In southern Spain some nitrogen and/or phosphorus fertilization is needed.

Taking into account the large volume of biomass produced, eucalyptus are less nutrient depleting than many other agricultural crops. This is partly because the largest proportion of the biomass produced consists of lignocelluloses. The nitrogenous compounds are either absent in wood or limited in quantity, and any nitrate movement downwards, which pollutes underground waters, has an increased chance of being trapped by the rich root system of thick eucalyptus plantations. However, long-term sustainability and site fertility still need to be key concerns in any eucalyptus plantation scheme.

Eucalyptus plantations have been accused of absorbing more water from the soil than any other tree species. The uptake of soil water depends mainly on the architecture of the root system and the depth of root penetration (Lima, 1993). Most eucalyptus plantations develop superficial root systems (Reis et al, 1995), though some eucalyptus roots can grow to 30m in depth and extract water from 6 to 15m depth (Peck and Williamson, 1987). However, eucalyptus seem to behave as any other tree plantation or natural forest cover with respect to water dynamics and water balance of the watersheds (Couto and Betters, 1995).

Effects of eucalyptus plantations on soil water balance should also be considered from two other points of view. First, many eucalyptus plantations in certain Mediterranean regions are grown on hilly sites with sandy to sandy loam soils that are vulnerable to erosion and unsuitable for many other crops. Eucalyptus coverage of such hilly sites, beyond providing an income to the farmers and employment to local people, offers valuable protection against water and wind erosion. Second, since eucalyptus are heavy biomass producers compared to slow-growing plants (for example pine trees) it is reasonable to absorb larger amounts of soil water in absolute numbers, though water use efficiency, as previously mentioned, is more or less the same. This increased amount of soil water absorption may have negative effects only if it is affecting water tables, springs or wells in lowlands; otherwise there is no reason to worry because eucalyptus are depleting the soil moisture to the benefit of producing more biomass.

Production

Productivity in Brazil is among the world's highest. Dry matter yields of up to 50t/ha/year *Eucalyptus grandis* have been reported (Rossillo-Calle, 1987). The Brazilian average productivity for eucalyptus is estimated to be the vicinity of 25–30m³/ha/year. In the coastal area, average increases have been obtained in the order of 50 to 60m³/ha/year. For a long time, *E. grandis* and *E. saligna* were planted almost exclusively. The area per plant is estimated at around 6m². The volumetric conversion rate is 1.8–2.0m³ of wood per 1m³ of charcoal (wood's moisture content between 25 to 30 per cent) (Mangales and Rezende, 1989).

The crop yields below are from the University of Florida's research test plot area in Orlando (~6 acres) for cottonwoods and *Eucalyptus* species (*amplifolia* and *grandis*) at planting densities of ~1000 trees per acre (traditional planting of one row of trees per bed). The highest yields have occurred where weed control (either mulching, composting or a combination) was performed (Table 10.38).

Processing and utilization

Eucalyptus have many uses which have made them economically important trees, and have become a cash crop in poor areas such as Timbuktu, Africa and the Peruvian Andes, despite concerns that the trees are invasive in some countries like South Africa. Perhaps the Karri and the Yellow box varieties are the best known. Due to their fast growth the foremost benefit of these trees is the wood. They can be chopped off at the root and grow back again. They provide many desirable characteristics for use as ornament, timber, firewood and pulpwood. Fast growth also makes eucalyptus suitable as windbreaks, and when planted they reduce erosion (Wikipedia.org).

Heating values of 18.94MJ/kg for wood and 16.46MJ/kg for bark were reported. However, much higher heating values have been reported (Rockwood et al, 1985) for various eucalyptus species, ranging from a low value of 19.7 up to 21.0MJ/kg dry fuel for some heavier species such as *E. paniculata* (Jacobs, 1979).

Biomass from short rotation eucalyptus plantations, beyond stems and branches, includes leaves. Therefore, the raw material for combustion is quite different from the raw material obtained from willows. A full understanding of the physical and chemical properties of eucalyptus biomass is necessary to properly design systems for utilizing short rotation eucalyptus biomass for energy.

In a two-year rotation cycle, high nitrogen concentrations were found in both *E. camaldulensis* and *E. globulus*, probably the result of previous over-fertilization of the field. Among the biomass components, the highest nitrogen concentrations were found in leaves. In contrast, the sulphur content was very small in all plant components. Consequently, harmful SO_x emissions from the combustion of eucalyptus fuel will be practically insignificant, while the nitrogen levels indicate the need for NO_x removal equipment in medium- to large-scale eucalyptus-fuelled energy systems.

Biomass components of both *E. camaldulensis* and *E. globulus* have high heating values (Table 10.39). Ash

Table 10.38 Crop yields in green tonnes per acre per year with mulching and composting

Tree species:	Year 1 yields	Year 2 yields	Two-year average
Cottonwood	7	30	19
E. amplifolia	7	20	14
E. grandis	12	63	38

Source: www.treepower.org/yields/main.html

content is also critical for the type of energy system. In both species, ash content is higher in leaves, lower in stems and intermediate in branches. In the case of *E. globulus,* the ash content of branches was almost double that of stems, and the ash content of leaves was almost double that of branches. The high ash content of leaves requires automatic ash removal equipment in combustion systems.

The fast growth rates of eucalyptus under appropriate conditions, and the good tree form characteristics (tall and slender trees with little branching) combined with adequate wood properties, make some eucalyptus species suitable for pulp production (Pereira, 1992). The use of eucalyptus wood as a source of pulp for paper manufacture has increased markedly in recent decades and this trend is likely to continue. There has been a rapid increase in the world production of chemical eucalyptus pulp, from about 40,000t/year in the early 1960s to nearly 6 million t/year in 1985 (Eldridge et al, 1994). Presently, over 3 million tonnes of eucalyptus pulp, mostly from Brazil, Spain, Portugal, South Africa and Morocco, are sold annually on the market (Borralho and Coterril, 1992).

Other uses include the less well-known utilization of eucalyptus for the production of minor forest products such as floral nectar for honey, bark for tannin, and rutin and leaf oils for pharmaceutical and industrial purposes (Boland et al, 1991).

Among the many uses of eucalyptus wood, probably more is used for fuel than for any other purpose.

A few eucalyptus species are increasingly being planted in some global regions to provide wood for fuel and/or charcoal production for industrial and home use (Hillis, 1990). The high density of eucalyptus wood makes it a good fuel. Eucalyptus charcoal is of major economic importance for the iron and steel industry of Brazil and for domestic use in many other countries (Eldridge et al, 1994). Significant amounts of eucalyptus residues and wastes, such as bark and black liquors produced during the pulp process, are used by the pulp mills to meet their own energy needs.

A team of Taiwanese, from the Taiwan Forestry Research Institute (TFRI), and US scientists has succeeded in developing eucalyptus trees capable of ingesting up to three times more carbon dioxide than normal strains, indicating a new path to reducing greenhouse gases and global warming. The new trees also have properties that make them more suitable for the production of cellulosic ethanol. In this sense, they can be seen as part of third-generation biofuels. This generation is based on crops modified in such a way that they allow the application of a particular bioconversion technology. Analyses show that there is a very large potential for the production of sustainable biomass from eucalyptus in Central Africa and South America. The gene modification project aims not only

Table 10.39 Fuel analysis of short rotation eucalyptus species

Elementary analysis (% of dry matter)	E. camaldulensis			E. globulus		
	Stems	Branches	Leaves	Stems	Branches	Leaves
C	44.54	44.00	47.76	44.40	43.18	45.80
H	5.47	5.62	5.92	5.40	5.53	5.86
O	45.58	44.41	37.84	46.59	45.60	37.81
N	1.31	1.22	2.65	1.36	1.19	2.05
S	0.05	0.10	0.10	0.10	0.10	0.11
Residues	3.05	4.65	5.73	2.15	4.40	8.37
Proximate analysis (% of dry matter)						
Volatiles	75.17	77.15	77.29	78.33	75.51	75.50
Fixed carbon	21.85	18.24	17.15	19.67	20.08	16.26
Ast2.98	4.61	5.59	2.00	4.41	8.24	
Heating value (MJ/kg dry matter)						
Gross heating value	18.97	18.45	19.62	19.16	18.61	19.23
Net heating value	17.86	17.31	18.41	18.07	17.48	18.03

to create eucalyptus with a higher than normal CO_2 absorptive capacity, but also to cause them to produce less lignin and more cellulose (Biopact, 2007e).

Eucalyptus is an interesting crop for the production of solid biofuels as well (woody biomass), that can be co-fired with coal or used in dedicated biomass power plants. Estimates show that there is enormous potential for the establishment of eucalyptus plantations in the tropics. A European project analysing the production of 'green steel' based on utilizing biomass from the tropics indicated that some 46 million hectares of land are available in Central Africa alone. In Brazil, another 46 million hectares are suitable. The land in question can sustain eucalyptus plantations without any major negative environmental footprint.

Internet resources

Wikipedia.org, http://en.wikipedia.org/wiki/Eucalyptus #Photo_gallery
Wikipedia.org, http://en.wikipedia.org/wiki/Eucalyptus# Cultivation.2C_uses.2C_and_environmental_cost-benefits

GIANT KNOTWEED (*Polygonum sachalinensis* F. Schmidt)

Description

Polygonum sachalinensis (also known as *Fallopia sachalinensis*) is a member of the buckwheat family. Giant knotweed is the biggest of the three invasive knotweeds. The stems are smooth, hollow and light green, resembling the canes of bamboo, and sparingly branched. The leaves are 15–40cm long, with a deeply heart-shaped base and a blunt leaf tip (Kingcounty.gov). Knotweed is a C_3 perennial shrub that occurs in many varieties. Although knotweed has its origins in east Asia, it is often found growing naturally in meadows around rivers or on the banks of streams and creeks. It was originally brought to Europe as a fodder crop from east Asia around the middle of the nineteenth century. The plant grows to a height of 2–5m, depending on the variety (Kowalewski and Herger, 1992; Janiak, 1994). It has a strong hollow stalk with a diameter of up to 5cm

(Janiak, 1994). The large ovate leaves are approximately 17cm long and 12cm wide. The leaves have been found to contain a compound that can be extracted and used as a prophylactic treatment on plants to induce resistance against powdery mildew (Herger et al, 1988).

Flowering of the plant takes place from September to October. The underground biomass consists of a rhizome network, with finer roots developing from the rhizomes.

Ecological requirements

Results from three different test sites in Germany have shown that the site with the most soil moisture and highest nitrogen content produced the highest plant growth (Kowalewski and Herger, 1992). For large yields it is important that the crop receives sufficient irrigation and fertilization, especially if it is to be grown in an area with low levels of precipitation and low soil nutrient content. The giant knotweed has an advantage in that the strong root system often enables the plant to obtain sufficient nutrients and water from the soil even during dry years.

The plant is able to grow well through a wide pH range (4–8), but a pH value of about 7 is recommended.

Propagation

Knotweed propagation is most commonly performed using seedlings or rhizomes (Herger et al, 1990). Seed taken from the plant is stored for about 14 days in the refrigerator at 1°C to alleviate seed dormancy (Bradford, 2005), then sown into pots containing well-fertilized, nitrogen-rich soil. The pots are stored at 20°C during the day (16 hours) and at 10°C during the night (8 hours) until the seedlings reach a height of about 5cm. At this time the seedlings are then transferred to larger pots and stored in a greenhouse under long day conditions. By the beginning of May they should have reached a height of about 60cm and are ready for planting into the field.

Plant propagation using rhizomes has not been as successful as with seedlings. Test plots using rhizomes have not shown uniform growth of the plants throughout the plots. This is believed to be a potentially greater problem if giant knotweed is ever planted on a large (farm) scale.

Source: MdE, http://en.wikipedia.org/wiki/File:Fallopia-japonica_MdE_2.jpg

Figure 10.67 Japanese knotweed with flower buds

Crop management

Seedlings are planted at the beginning of May at densities between 1 plant per $75cm^2$ and 1 plant per $1m^2$ (Herger et al, 1990). Greater densities result in the plants shading each other, thus causing the plant's lower leaves to die. Along with planting, a layer of mulch should also be placed on the field to help prevent weeds. Weed removal will be necessary primarily in the first year; once the crop reaches maturity it usually crowds out weeds. Any removal of weeds in the first year should be done by hand because heavy machinery may damage the developing rhizome bed.

Tests in New Zealand using varying levels of inorganic fertilizers to grow seedlings showed that increasing fertilizer levels, especially nitrogen, led to increased seedling height and a greater dry matter production from the seedlings. The highest inorganic fertilizer levels of 250kg/ha P, 50kg/ha K and 200kg/ha N led to a plant height of about 75cm and a dry matter yield of about 3.4t/ha. When only 100kg/ha N was used the seedling height was about 65cm and the yield about 2.8t/ha (Herger et al, 1990).

Field observations have shown that knotweed is not significantly affected by disease, but knotweed seedlings and younger plants have shown to be attacked by rabbits and the black bean aphid *(Aphis fabae)* (Herger et al, 1990).

Because knotweed is a perennial plant there is no need to establish a crop rotation. Its rhizomes mean that care should be taken to remove all residues of the knotweed crop before a different crop is planted, otherwise knotweed may appear in the new crop.

Production

Giant knotweed is a hardy, herbaceous, rhizomatous perennial. Plants are dioecious, with some plants functionally female (male-sterile) and some male-fertile. Leaves are dock-like, ovate-cordate, glabrous except for scattered, long, wavy trichomes on the paler undersides; tips somewhat acuminate. The stems are stout, hollow, up to 5 metres in height; often reddish-brown, little branched. The flowers are a greenish-white; male-fertile examples with exserted anthers, male-sterile flowers with well-developed stigmas. The panicles are dense, up to 10cm long, and fruits are 4–5mm long.

Knotweed is a very high-yielding crop. In Germany, harvests of the entire plant in August and September have provided dry matter yields of 20 to 30t/ha, while harvests in the winter, after the leaves have fallen off, have produced dry matter yields of 8 to 10t/ha. Above-ground yields on the island of Sakhalin in Russia have been as high as 27.2t/ha, with 10.5t/ha of below-ground matter (Morozov, 1979). In northern Japan, above-ground yields of 11.9t/ha have been seen (Iwaki et al, 1964). Other investigation indicate that the average yield amounts to 12t/ha dry matter (Salter, 2006).

Processing and utilization

As mentioned above, harvesting in the autumn produces the largest dry matter yield because leaves are also harvested, whereas harvesting in the winter produces a smaller yield but with a lower plant moisture content. If heavy machinery is used during the harvest, care should be taken to avoid damaging the underground rhizome system of the crop. This is most important during the first year, at which time it is best to harvest the crop by hand to prevent excessive rhizome damage. Starting with the second year, machinery such as a corn harvester can be used for the harvest. In addition, there is also the possibility of harvesting the crop more than once per year, in which case the successive harvests should be at least four weeks apart (Herger et al, 1990).

For storage or further processing, the moisture content should be in the 8–12 per cent range. If the crop has a low enough moisture content at the time of harvest then no additional drying is necessary, but if not, the harvested material must be quickly dried and not left in a damp field for any long period of time, otherwise the plant will begin to ferment. The crop can be dried in a facility similar to those used for drying tobacco or grains using circulating fresh air. Once dried the crop can be pressed into bales and transported (Herger et al, 1990).

Fresh or dried leaves can be processed to obtain aqueous or ethanolic extracts, which tests have shown to be an effective prophylactic powdery mildew inhibitor on plants (Kowalewski and Herger, 1992). If

Source: Al Schneider, www.swcoloradowildflowers.com

Figure 10.68 *Bistorta bistortoides*

the leaves are to be harvested for the extraction of this compound, it is advisable to wait until at least the middle of June. Research has shown that the highest amount of the active agent in the leaves is present from about this time (Kowalewski and Herger, 1992).

The unbleached plants as harvested can be processed into paper, insulation or a solid biofuel. A lighter coloured paper can be obtained through bleaching (Janiak, 1994). To be used as a solid biofuel the plant needs to have a moisture content of approximately 12 per cent or less. The compacted plant material can then be burned to produce heat or for the production of electricity. It is possible to combine the plant material to burn with other biofuels, but more research needs to be done to determine the most beneficial combinations. Table 10.40 lists the fuel attributes of giant knotweed dry matter obtained during tests conducted by Vetter and Wurl (1994).

Research has shown that knotweed is able to absorb and store high amounts of heavy metals such as lead and cadmium. Selective breeding and cultivation of plants is under way to enhance this trait. It is believed that these plants could save and restore soil that is not usable as a result of toxic metal contamination. On a contaminated field, the plants would be harvested two or three times a year; then the rhizome and root system would be removed after two years (Haase, 1988).

Dry plants (stems) are of a woody nature and they could be converted to a wide variety of biofuels. The plant can grow without additional fertilization or other inputs.

Table 10.40 Fuel attributes of giant knotweed dry matter

Characteristic (%)	Value
Lignin content	18.9
Ash content	6.3
Volatiles content	75.9
Carbon	47.7
Hydrogen	6.6
Silicon dioxide	9.0
Chlorine	0.22
Nitrogen	0.54
Potassium	0.75
Sulphur	0.17
Heating value (MJ/kg)	17.2

Source: Vetter and Wurl (1994)

Internet resources

www.paflora.org
www.invasivespecies.gov
http://wiki.bugwood.org/Polygonum_cuspidatum
www.paflora.org/Polygonum%20spp.pdf
www.kingcounty.gov/environment/animalsAndPlants/
noxious-weeds/weed-identification/invasive-knotweeds/
giant-knotweed.aspx/invasive-knotweeds/giant-
knotweed-aspx

GIANT REED (*Arundo donax* L.)

Partially contributed by C. D. Dalianis

Description

Giant reed grows wild in southern European regions (Greece, Italy, Spain, southern France and Portugal) and other Mediterranean countries. It also grows wild in other parts of the world (China, southern USA etc.) (Xi, 2000). Although giant reed is a warm climate plant, certain genotypes are adapted to cooler climates and can be grown successfully as far north as the United Kingdom and Germany.

Arundo donax has been cultivated throughout Asia, southern Europe, northern Africa, and the Middle East for thousands of years. Ancient Egyptians wrapped their dead in the leaves. The canes contain silica, perhaps the reason for their durability, and have been used to make fishing rods, walking sticks and paper.

Giant reed has several attractive characteristics that make it the champion of biomass crops. Certain natural, unimproved populations give dry matter biomass yields of up to 40t/ha. This means that giant reed presents a good starting point in terms of yields, being one of the most productive among the biomass crops currently cultivated in Europe, and that it has a good chance, through selection and genetic improvement, of becoming the leading biomass crop in certain European regions.

Giant reed is also an environment-friendly plant:

- Its robust root system and ground cover, and its living stems during the winter offer valuable protection against soil erosion on slopes and erosion-vulnerable soils in southern European countries.

- It is a very aggressive plant, suppressing any other vegetation under its canopy.
- During the summer it is green and succulent, and has the ability to remain undamaged if an accidental fire (very frequent in southern EU conditions) sweeps across a giant reed plantation.
- It is an extremely pest (disease, insect, weed)-resistant crop, not requiring any of the chemical inputs (pesticides) that under certain conditions pollute the environment.

Giant reed is also considered to be one of the most cost-effective energy crops, because it is perennial and its annual inputs, after establishment, are very low. Only harvesting costs will occur and, depending on site and climate, irrigation and/or fertilization costs. Giant reed is also a lodging-resistant plant. All these attributes make giant reed a very attractive and promising candidate species for biomass production in European agriculture.

Giant reed, also known as Provence reed or Indian grass, is a grass that belongs to the *Arundo* genus of the Gramineae family. The *Arundo* genus consists of two reed-like, perennial species, *Arundo donax* L. and *Arundo plinii* Turra. They have coarse, knotted roots, cauline, flat leaves and large, loose, plumose panicles. Their spikelets are laterally compressed with few, usually bisexual, florets. The glumes are nearly equal, as long as the florets, with three to seven nerves. The lemmas have three to five nerves, with long, soft hairs on the proximal of the back. The rachilla is glabrous.

The two species are quite similar in appearance. However, there are certain morphological and growth differences. In *Arundo donax* L. the spikelets are at least 12mm long with three or four florets. The hairs of the lower lemma are almost the same length as the glumes and the lower lemma is two-pointed. In *Arundo plinii* Turra the spikelets are not more than 8mm long with one or two florets. The hairs of the lower lemma are shorter than the glumes and the lower lemma is entire at the apex. *Arundo plinii* Turra differs also from *Arundo donax* L. in that the plants are shorter, usually less than 2m, with stems that are always slender and with leaves that are rigid and that stick out stiffly from the stem at a right angle or less with tips sharply pointed. In *Arundo donax* L. the chromosome number is 2n = 110, 112, while in *Arundo plinii* Turra the number is 2n = 72.

Figure 10.69 Giant reed (*Arundo donax*) plantation in Greece

Giant reed is probably the largest grass species in the cool temperate regions (Figure 10.69), only exceeded in size by some of the bamboos. It is a vigorously rhizomatous perennial species with a stout, knotty rootstock. Rhizomes are long, woody, swollen in places, covered in coriaceous, scale-like sheaths. The stems are stout, up to 3.5cm in diameter and up to 10m tall. The leaves are regularly alternate on the stems and the leaf blades are up to 5cm wide and up to 3.3m long. They are almost smooth, green, and scabrous at the margin. The leaf sheaths are smooth, glabrous, covering the nodes. The largest leaves and most vigorous stems are produced on plants that are cut to ground level at the end of each season.

The inflorescence is a highly branched panicle up to 60cm long, erect or somewhat drooping. Its colour is initially reddish, later turning white. In cool regions the stems will not achieve flowering size.

In the warm Mediterranean regions, the above-ground giant reed parts remain viable during the winter months. If plants are not cut, in the following spring new shoots emerge at the upper part of the stem from buds located at stem nodes. After cutting a giant reed plantation, usually in autumn or winter, new growth starts early next spring. New shoots emerge from buds located on the rhizomes and they develop very rapidly. Later in the season, in June–July, peak growth rates up to 7cm per day have been observed.

In fertile fields, new shoots continue to emerge until early August under a huge, well-developed canopy. These late shoots develop at a faster rate and attain the same height as the early ones, though the leaves are smaller and the stem diameter is much larger – as much as twice as large.

There are three off-type giant reeds that are used for ornamental purposes:

1 *Arundo donax* 'microphylla', in which the leaves are even more glaucous and broader than the basic type, up to 9cm wide.
2 *Arundo donax* 'variegata', known in the USA as *Arundo donax* 'versicolor', in which all plant parts are usually smaller and the plants very much more frost sensitive than the basic type, and the leaves are white striped, usually with broad white bands at the margins.
3 *Arundo donax* 'variegata superba', a name used to distinguish a superior variegated form in which the leaves are much broader than in *Arundo donax* 'variegata' – the leaf blades are normally as much as 6.5cm wide, about 30cm long and borne on stems that attain heights of up to 1m, and the internodes are shorter and the leaves grow much closer on the stems.

Because giant reed is wild growing and entirely unknown as a cultivar or crop, the 'state-of-the-art' production knowledge is missing. This means that neither selection of wild grown genotypes nor genetic improvement has been attempted so far, and the most appropriate cultural techniques for maximizing biomass yields are unknown.

Ecological requirements

In its wild state, giant reed is usually found along river banks and creeks and on generally moist soils, where it exhibits its best growth. However, it is also found in relatively dry and infertile soils, at field borders, on field ridges or on roadsides, where it grows successfully. Giant reed can be grown on almost any soil type from very light soils to very moist and compact soils. When there is an underground water table it has the ability to absorb water from the table.

Propagation

In nature, giant reed populations spread outwards through their rhizomes' growth. Where farmers have planted giant reed on their field borders to serve as windbreaks, the plant creates problems by spreading into the fields, reducing the available cropland. In such cases, the unwanted rhizomes need to be eradicated every few years so that giant reed growth remains limited to the borders.

Giant reed is a seedless plant. For agricultural purposes it could be propagated either by rhizomes or stem cuttings. Rhizome propagation is implemented early in the spring before the new shoots start emerging in the mother plantation. Propagation by stem cuttings is implemented later in the season when the soil warms up and promotes mobilization of the node buds to develop new shoots.

Giant reed rhizomes are irregular in shape and variable in size and bud bearing. Rhizomes range from 1cm up to 10cm in diameter. Their abundant reserves promote vigorous new growth. Several buds are mobilized and up to 10 stems per rhizome may emerge by the end of the first growing period. However, propagation by rhizomes is labour intensive and very expensive. After collection, rhizomes have to be cut into pieces and sorted according to their bud bearing capacity.

It is much cheaper to use stem cuttings or whole stems. Stem cuttings consist of one node with sections of adjacent internodes. Stem cuttings could be either planted directly in the field or planted in plastic bags for transplanting into the field after they sprout. In the field, stem cuttings are covered to a depth of 4 to 8cm, depending upon the soil temperature and soil moisture.

Whole stems could be used instead of stem cuttings. Stems are laid down into soil furrows at a depth of 6 to 8cm and covered by soil.

However, propagation by stems or stem cuttings has not always proved successful in experiments in northern China (Xi, 2000).

Crop management

Giant reed has no special soil preparation requirements. A simple ploughing and/or disc harrowing is considered sufficient. Natural populations are usually very dense: more than 50 stems per m^2 is quite common. When establishing giant reed plantations with rhizomes, care should be taken that each piece has at least one bud in order to avoid gaps in the field. Distances of 70cm between rows and 50cm within rows result in a relatively thick plantation with an average stem number of up to 10 per m² at the end of the transplanting growing period. In the subsequent two years stem density increases and results in a thick stand.

Giant reed plantations established with stem cuttings are much thinner at the end of the growing period. Emerging plants have to develop rhizomes in order to overwinter in case the above-ground part is destroyed by low winter temperatures. Plant survival of between 70 and 82 per cent has been reported (Jodice et al, 1995a).

The biomass yields at the end of the establishment year are much lower – less than one-third of the biomass obtained by rhizome planting. Because stem cuttings are much cheaper it is advisable to plant them closer within the rows.

Although giant reed can be grown without irrigation under semi-arid southern European conditions, its response to irrigation is significant. However, the effect of irrigation rates on fresh and dry matter biomass yields is insignificant. It was reported (Dalianis et al, 1995a) that fresh and dry matter biomass yields of giant reed, averaged over three years for autumn harvests, were respectively 59.8t/ha and 32.6t/ha for the high irrigation rate (700mm/year) and 55.4t/ha and 29.6t/ha for the low irrigation rate (300mm/year).

Giant reed is a perennial crop that lasts for several decades and is also a high biomass-yielding crop, so before establishing a new plantation it is necessary to incorporate sufficient phosphorus into the soil by ploughing – more than 200kg/ha – especially in phosphorus-deficient fields. Most fields in semi-arid Mediterranean regions are rich in potassium, so potassium fertilization is not required.

Annual applications of nitrogen at up to 100kg/ha, especially in nitrogen-poor soils, are recommended. Applications should be implemented before the new sprouts start emerging early in spring. However, Dalianis et al (1995a) reported that high nitrogen rates (240kg/ha) have no significant effect on biomass yields compared to low rates (60kg/ha). This leads to the conclusion that the application of reduced nitrogen rates is justified, at least during the initial growing periods.

Figure 10.70 *Arundo donax* vegetation

Giant reed is one of the most pest-resistant plants. So far, no diseases have been reported or observed. Occasionally during the early growth stages of the new sprouts, while they are still in a succulent condition, they may be attacked by *Sesamia* spp. and die. However, very soon new sprouts appear from the rhizome buds and replace the damaged ones.

Giant reed develops a huge canopy that suppresses any weed growth. Even during the establishment year there is no need for herbicide applications if rhizomes are used as planting material. However, if establishment is implemented by stem cuttings, pre-planting herbicide applications help the establishment and early growth of the giant reed plantation.

Production

There are only a few references to giant reed biomass yields in the world literature. In an old study, in southern France, giant reed produced 20–25t/ha dry matter (Toblez, 1940), while in northern Italy a survey made of an almost fully grown stand of giant reed determined 35t/ha, taking into account only air-dried stems (Matzke, 1988). In recent studies (Dalianis et al, 1995a; Jodice et al, 1995a, 1995b; Morgana and Sardo, 1995), giant reed's high biomass yield potential has been confirmed.

It should be stressed that these high yields are obtained from unimproved wild populations with almost no crop management. This indicates the great biomass potential of this biomass plant for the future.

Processing and utilization

The stem material is both flexible and strong enough to be used as a reed for woodwind instruments such as the oboe, bassoon, clarinet and saxophone. It is also often used for the chanter and drone reeds of many different forms of bagpipes. Giant reed has been used to make flutes for over 5000 years. The pan pipes consist of ten or more reed pipes. Its stiff stems are also used as support for climbing plants or for vines. Further uses are walking sticks and fishing poles. Since *Arundo* species grow rapidly, their use has been suggested for biomass for energy and a source of cellulose for paper; at least one North American paper mill was considering planting it for a source of pulp fibre (Samoa Pacific, on

Humboldt Bay, CA, in 2002), but abandoned the plan by early 2003.

Giant reed can be harvested each year or every second year, depending on its use. For example, for constructing sun protection shelters it is usually harvested every second year because the durability of the shelters is increased. For energy or pulp production giant reed is harvested each year and new growth starts in the spring.

In southern European regions the giant reed could be harvested either in the autumn or in the late winter. However, it should be noted that a significant reduction in biomass yield is observed between the autumn and late winter harvests. The dry matter yield reductions are the result of losses of the leaves and many of the tops, especially if hard winters are accompanied by strong winds. Dry matter losses of up to 30 per cent were reported by Dalianis et al (1995a).

In semi-arid Mediterranean climates, the moisture content of the autumn-harvested plants ranges between 36 and 49 per cent, and weather conditions are suitable for natural drying in the field after cutting. These results indicate that not only is delaying harvesting time until after November useless, but there is a danger, depending on the prevailing weather, of significant biomass losses (Dalianis et al, 1995a). However, the autumn harvest, especially in fertile fields and warm regions, may result in an early sprouting next spring. If a late winter frost occurs these early sprouts are usually destroyed, but are quickly replaced with new ones that emerge from buds at the rhizomes.

A significant advantage of giant reed is its good storability compared to many other biomass crops. It can be stored outdoors without any shelter protection with minor losses. Storage losses occur mainly in the leaf fraction (blades and sheaths), which represents a small percentage, about 10 to 15 per cent, of the total biomass production. Stems can be stored with almost no losses.

Giant reed has not so far been exploited on a commercial basis throughout the world. In certain windy areas it is used as a windbreak to protect other crops, while its stems are used for the construction of sun protection shelters or to make baskets. Its rot-resistant stems are used as supporting poles for

Source: Chris Wagner (CalPhotos)

Figure 10.71 *Arundo donax* dry vegetation

climbing vegetables and ornamental plants. Giant reed stems, being tough and hollow, are also used as a source of reeds for musical instruments.

Initial results indicate that giant reed is a promising alternative to conventional non-wood fibre options and is a useful biofilter, C sequestration and biofuel crop.

Recently, giant reed has been considered as a new source of raw material for energy, pulp and/or the construction of woody building materials. Heating values of 3600kcal/kg were determined for giant reed (Dalianis et al, 1994). Based on these values and the dry matter yields obtained so far, the estimated energy potential is up to 11.8toe/ha/year.

Only a few references are available concerning the possibility of exploiting giant reed for pulp production (Arnoux, 1974; Faix et al, 1989). The cellulose and hemicellulose content of its stems are about 45 per cent

on a dry matter basis, while its lignin content is about 25 per cent. Because giant reed is a pithless plant (in contrast to miscanthus, for example) it is considered to be a very suitable non-wood plant for pulp production since no depithing is required.

Giant reed produces an average of 25 tonnes of high-quality fibre per acre twice annually. One of its most significant uses will be to produce chips for the manufacture of high-grade biofuel pellets or dried chips. Highly significant also is the importance of a crop with a growing cycle of over 20 to 25 years without annual replanting, and the ability to exclude many costly fertilizers and weed killers that are also an environmental concern, which will return agriculture to a more profitable basis than many crops.

Giant reed is an ideal biofuel (8000BTU/lb) that produces methanol from gas diffusion as a by-product in manufacturing cellulose. The option to gasify this product is to produce a valuable energy product. It is possible to utilize new high-efficiency gasification systems to convert giant reed into a multitude of different energy sources, such as syngas, standard steam turbine electrical generation, ethanol and biodiesel.

Florida-based Biomass Investment Group is embarking on a project using *Arundo donax* as an energy crop that will be grown on 20,000 acres (8000ha). The biomass will be converted into bio-oil, a heavy fuel oil, via a fast-pyrolysis process. This carbon-neutral oil will then be used in a power plant that will provide electricity to some 80,000 Floridian households.

Internet resources

http://thefraserdomain.typepad.com/energy/2006/12/
arundo_donax_fo.html
www.cal-ipc.org/ip/management/wwh/pdf/19646.pdf
http://en.wikipedia.org/wiki/Arundo_donax

GROUNDNUT (*Arachis hypogaea* (L.) Merr.)

Partially contributed by Professor Stephen Gaya Agong, Maseno University, Kenya

Description

The groundnut (*Arachis hypogaea* (L.) Merr.), also known as peanut, earthnut, goober, goober pea, pindas, jack nut, pinder, manila nut and monkey nut is a crop of global economic significance, not only in the geographically widespread areas of its production but also in the even wider areas of its processing and consumption (Smart, 1994). The crop belongs to the genus *Arachis,* which belongs to the family Papilionacea, the pea and bean family, and consists of some 22 described and 40 or more undescribed species (Gregory and Gregory, 1976). All the species are native to tropical and subtropical South America.

There is phytogeographical evidence that cultivated *Arachis* progenitors evolved in Eurasia (Leppick, 1971). After much collection and study, Krapovickas (1968) concluded that the cultivated peanut originated in Bolivia, along the eastern slope of the Andes. Krapovickas based this conclusion on linguistic,

cytological, genetic, morphological, chemical and geographical information. It is thus difficult to ascertain precisely the origin of the various species of the cultivated *Arachis,* simply because of the antiquity and generality of their worldwide distribution (Candolle, 1967).

The peanut, or groundnut (*Arachis hypogaea*), is a species in the legume Fabaceae native to South America, Mexico and Central America. It is an annual herbaceous plant growing to 30–50cm tall. The leaves are opposite, pinnate with four leaflets (two opposite pairs; no terminal leaflet), each leaflet 1–7cm long and 1–3cm broad. The flowers are a typical pea flower in shape, 2–4cm across, yellow with reddish veining. After pollination, the fruit

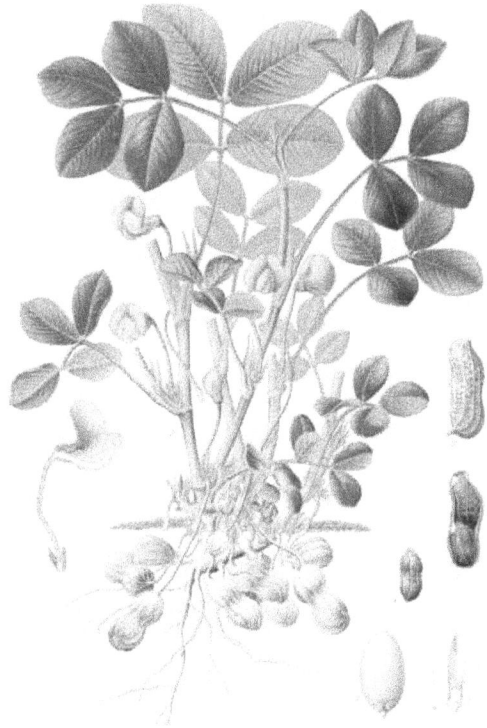

Source: http://en.wikipedia.org/wiki/Arachis_hypogaea

Figure 10.72 Peanut (*Arachis hypogaea*)

develops into a legume 3–7cm long, containing one to four seeds, which forces its way underground to mature.

The plant's name derives from a combination of the morphemes *pea* and *nut*, causing some confusion as to the nature of the fruit. In the botanical sense the fruit of the peanut plant is a woody, indehiscent legume and not a nut. The word pea describes the edible seeds of many other legumes in the Fabaceae family, and in that sense, a peanut is a kind of pea. Although a peanut is not a nut, in the culinary arts peanuts are utilized similarly to nuts.

Different varieties of groundnuts are used for different food products. The largest – red-skinned Virginia kernels – are used for cocktail and salted nuts. The medium-sized runner and the small Spanish varieties are best for peanut butter, oil and candies. The Valencia variety, with long shells containing three to four kernels each, are in demand for roasting in the shell.

New breeding technologies have produced a range of improved varieties, adapted for particular end uses or to specific growing conditions. Groundnuts with good shelling yields, high oil or protein content, or particular kernel shapes and sizes have been developed for specific end uses. High-yield varieties have been developed for specific locations, with characteristics such as early maturation; resistance to drought, diseases and pests; suitability for mechanized harvesting; adaptation to particular types of soils or farming requirements.

Production

Global groundnut production in 2008 was 34,856,007 tonnes. China leads in production of peanuts with about 37.5 per cent of overall worlds production followed by India (about 19 per cent) and Nigeria (11 per cent) (wikipedia.org/Peanut).

Yields in producing countries vary significantly, depending on climate, soil, farming systems and seed varieties. The spectrum is wide: over 2t/ha in the United States (International Trade Centre, 2001).

Groundnuts constitute more than half of all tropical legume seed production (Ashley, 1984). Asia and Africa have the largest areas under groundnut cultivation and the greatest production out of a world total of more than 25 million tonnes in shell (shelling percentage of about 70 per cent) as reported by FAO (1994). It is estimated that 67 per cent of the world's groundnut production is grown in the semi-arid tropics (SAT), almost entirely by small-scale farmers. The average yield of 780kg/ha of dried pods compares unfavourably with the 2900kg/ha grown in countries with developed agricultural practices, and yields of over 3000kg/ha are not uncommon on research farms in the SAT (McDonald, 1984; Smart, 1994). This indicates good potential for improving farmers' yields.

A comparison of peanut and other crops clearly demonstrates that this crop is very rich in energy (Table 10.41). The quantity of oil production on a crop basis shows that peanut is a key crop that would double as a cash crop as well as food crop. Groundnut is one of the food crops that can be utilized as a source for biofuel.

Studies performed by Anders et al (1992) can be used to confirm the proportions of the groundnut biomass. This biomass may be subdivided into above-ground and below-ground matter. Both fresh and dry yields are compared in Table 10.42.

From these data it can be estimated with respect to above ground matter that the fresh fodder (2.62 t/ha) makes up approximately 52 per cent of the total fresh biomass and the dry fodder (1.88 t/ha) makes up approximately 46 per cent of the total dry biomass.

A study of groundnut sowing dates conducted by Mayeux (1992) further divides up the above ground biomass into stems and leaves in the distribution of biomass per plant (Table 10.43).

Processing and utilization

Groundnut is an important oilseed crop and grain legume with high oil and protein contents. It is estimated that groundnuts contain 40–50 per cent oil, which consists mainly of the glycerol esters of oleic acid (50 per cent) and linoleic acid (25 per cent). As the nuts contain 30 per cent protein, they are a useful addition to a meatless diet (Vickery, 1976). Norden (1980) also reported a similar high percentage of oil (42 to 52 per cent) and protein (25 to 32 per cent) in groundnut seed. Furthermore, the development to vegetable oil crops in most oil-deficient developing countries would help reduce import requirements. Groundnut is also used directly for food and may be processed into oil and animal cake, or used for

Source: DOB (2008)

Figure 10.73 Mynahs predating on caterpillars in groundnut

confectionery products (Musangi and Soneji, 1967). As an edible legume, peanut is popular for its horticultural importance and used largely as peanut butter, salted

Table 10.41 Oil production for typical oleaginous plant crops

Crop	Oil yield (kg/ha/year)[a]
Soya bean	374
Safflower	653
Sunflower	801
Peanut	887
Rape seed	999
Castor bean	1188
Babassu	1541
Coconut	2260
Oil palm	7061

[a]Based on total annual volume production worldwide.

Table 10.42 Comparison of fresh and dry matter yields of groundnut (t/ha)

Tissue	Above ground Stems + leaves	Below ground Pods Roots[a]
Fresh	2.62	2.41
Dry	1.08	1.28

[a] Harvested but not determined.

table nuts or roasted in the shell. As an oil crop the main by-products are protein-rich press cakes, which can be used as fodder as well as an ingredient of fish meal (Nyina-wamwiza et al, 2007).

As a renewable energy source, groundnut was the first vegetable oil used in internal combustion engines, dating as far back as Rudolf Diesel (1858–1913) and his experiments with peanut oil, and Fujio Nagao, who achieved operation with pie oil in 1948 (Nagao et al, 1948). When vegetable oils are used as fuel for diesel engines, there is usually a decrease in the smoke and particulate matter released into the atmosphere. The thermal efficiency is much better than when diesel oil is used. Renewable fuels will continue to be used to help where farm policies have failed, as a source of energy for local consumption in tropical countries suffering from a lack of foreign fuels, or to recycle discarded oils that are difficult to dispose of.

The world community is rapidly growing and the demand for food will continue to escalate. Groundnut's suitability for the industrial production of high-quality oil products cannot be underrated. Improving groundnut production in the semi-arid tropics would enhance proportionally the energy yield from shells, vegetative components and nuts. The plant tissues may readily be used for human and livestock nutrition, as in the case of nuts. The shoots may readily be used for feeding livestock and as a source for cheap fuel. The shell is also considered a source of cheap fuel, whereas the groundnut residues would always go into improving the soil nutrients and structure. Groundnut oil can be considered as a potential biofuel source in semi-arid regions.

In October 2006, mobile phone companies, Ericsson, GSM Association and South Africa's MTN issued a press release announcing that they had teamed up to establish biofuels as an alternative source of energy for wireless networks in the developing world. The three mobile giants have already set up a pioneering project in Nigeria to demonstrate the potential of biofuels to replace diesel as a source of power for mobile base stations located beyond the reach of the energy grid. Groundnuts, along with pumpkin seeds, jatropha and palm oil will be used in the initial pilot tests (Mayet, 2007).

Table 10.43 Distribution of dry mass in stems, leaves and pods with respect to sowing date

Sowing date	Total dry mass (g)	Distribution of dry mass (%)		
		Stems	Leaves	Pods
Early	98.1	35.7	29.9	30.3
Mid	67.7	29.6	8.1	57.7
Late	23.2	29.7	18.1	47.0
Very late	14.7	38.8	40.8	13.6

Source: Mayeux (1992)

Source: DOB (2008)

Figure 10.74 Experimental nursery beds – groundnut in India

Internet resources

www.dbtbiopesticides.nic.in/event/EventDetails.asp?EventId=251
http://en.wikipedia.org/wiki/Peanut

HEMP (*Cannabis sativa* L.)

Description

The hemp plant is not only one of the oldest cultivated plants, it is also one of the most versatile, valuable and controversial plants known to man. The industrial hemp plant has a long history, which has proven its innate worth and its stalks and seeds can serve as raw material for an exciting array of many diverse products: paper, construction materials, in the car industry, and for fuels. The plant's Latin name actually means 'useful hemp', and it definitely measures up to its name.

Hemp has its origins in central Asia. It is an annual, short-day plant that is wind pollinated. It has a high cellulose and lignin content in its stems, and a high fat and protein content in its seeds. Although some varieties can reach heights of up to 5m, the average height is about 2.5m. The leaves are finger like with serrated edges. Both the stalk and the leaves are covered with glandular hairs. Its root system is dominated by a taproot. The plants vegetative period is about 100 days, with the main growth period in June and July, followed by flowering in August.

Hemp belongs to the oldest group of plants used by humans. Approximately 260,000ha of hemp are cultivated throughout the world, including about 55,000ha in Europe. The seeds contain an oil that is approximately 35 per cent fat and 46–70 per cent linoleic acid. The stalk contains a very strong and durable fibre. The amount of this bast fibre ranges from about 8 to 32 per cent depending on the hemp variety; 15 per cent fibre content is considered the cut-off between the high-fibre and low-fibre varieties (van Soest et al, 1993). The cellulose content of the stems averages about 57–60 per cent, while the lignin content averages about 8–10 per cent.

Because of the presence of delta-9-tetrahydrocannabinol (THC) the cultivation of hemp is completely prohibited in Germany. Hemp usually contains 8–12 per cent THC (Long, 1995), which is produced by the glandular hairs mainly in the leafy parts of the inflorescences of female and hermaphrodite plants. For THC to be used as a drug it needs to be present at over 2 per cent. Because of this, EU regulation permits the cultivation of hemp varieties with THC contents of 0.3 per cent or less. The following French varieties meet this criterion: 'Fedora 19', 'Felina 34', 'Fedrina 74', 'Ferimon', 'Fibrimon 24', 'Fibrimon 56' and 'Futura' (Hoppner and Menge-Hartmann, 1994).

Industrial hemp is a distinct variety of the plant species *Cannabis sativa* L. Hemp is a member the mulberry family. This annual herbaceous plant has a slender stem, ranging.from 0.5 to 2cm in diameter. The innermost layer is the pith, surrounded by woody material known as hurd. Outside of this layer is the growing tissue which develops into hurd on the inside and into the bast fibres on the outside. The stem is more or less branched, depending on how densely the crop is planted. When sown thickly the stems do not branch. The leaves are of a palmate type and each leaf has seven to eleven leaflets with serrated edges. The strong taproot penetrates deeply into the soil. However, if the soil conditions are unfavourable, the main root remains short, and lateral roots become more developed.

Ecological requirements

Hemp exhibits optimal growth in a moderate climate at 13–22°C, and is sensitive to frost during germination. Although the crop thrives on most soils, it prefers deep, humus-rich, calcareous soils with a good water and nitrogen supply. Sand, heavy clay and excessively wet soils are less suitable, as are compacted soils. A pH value of 7 is recommended, though slightly basic soils are also appropriate. The soil should be well prepared to a fine tilth and well supplied with water, but not excessively wet (Hoppner and Menge-Hartmann, 1994). A precipitation level of 700mm is necessary for a substantial yield.

Propagation

Hemp is propagated using seed. Because of the many varieties, care should be taken when choosing the variety that it is appropriate for the climate and soil where it will be planted, and for its intended use. The varieties are classified according to their fibre, oil and THC contents, along with the time of ripening. One of the main breeding objectives currently under way is to increase the bast fibre content of the plant while still maintaining a low THC content (van Soest et al, 1993).

Crop management

Sowing should be performed between the middle of April and the end of May. Approximately 35–50kg/ha of seed should be planted, and at a depth of 2–4cm. The distance between rows should be 10–20cm. These conditions will lead to a crop density of about 200–350 plants/m². A higher crop density, about 300 plants/m², is used if the crop is grown for the production of long fibres (Hoppner and Menge-Hartmann, 1994). The higher density leads to a reduced production of undesired leaves. The seed has a germination period of 4–6 days.

Source: Erik Fenderson, http://commons.wikimedia.org/wiki/File:Male_hemp_flowers-2.jpg

Figure 10.75 Male hemp plants

Figure 10.76 Hemp plantation

Because of the plant's rapid growth, weeds have not been found to be much of a problem. Herbicides may not even be necessary. Farms in England report that after drilling the seed there is no need for spraying or top dressing: the crop is left alone until it is time to check its progress (Long, 1995).

Several diseases and pests have been known to attack hemp crops. The detrimental presence *of Botrytis, Fusarium* and *Sclerotinia* has been observed, along with damage to germinating plants caused by snails and wireworms. The plant fibres may suffer damage from *Spherella cannabis* and *Phoma herbarum.*

Table 10.44 shows the fertilizer levels for hemp as recommended by the Federal Agricultural Research Centre in Germany.

Because of its deep and extensive root system, hemp makes a good preceding crop in a crop rotation. In England a crop rotation of wheat and hemp has proved to be appropriate (Long, 1995). Hemp can also follow itself in a crop rotation.

Production

Seeding rate research on hemp in The Netherlands led to stem dry matter yields in 1988 and 1989 of 11.9t/ha and 13.6t/ha respectively. The tests concluded that final harvest was not affected by seeding rate (Meijer et al, 1995). Under optimal

Table 10.44 Recommended fertilizer amounts for hemp cultivation

Fertilizer	Amount (kg/ha)
Nitrogen (N)	60–100[a]
Phosphorus (P$_2$O$_5$)	70–100
Potassium (K$_2$O)	150–180

[a]Divided into two or three applications.

circumstances a fibre yield of between 2 and 5t/ha could be obtained. Research in Germany has produced similar results, showing whole plant dry matter yields of 8–16t/ha and fibre yields of 2–4t/ha (Hoppner and Menge-Hartmann, 1994).

A typical hemp harvest consists of 52 per cent usable refuse, 31 per cent fibre (40 per cent long and 60 per cent short), 9 per cent non-usable refuse and 8 per cent seeds (Karus and Leson, 1995).

The subsidy for hemp cultivation in the European Union in 1995 was €774.74/ha. This subsidy is recalculated yearly.

Hemp grain yields over the years have varied from 100 to 1100 pounds per acre. New growers should expect 300–400lb/acre grain yield (clean, dry basis), while with experience the potential of current varieties could be in excess of 600lb/acre. In crops grown and managed solely as fibre crops, average yields of 3 tonnes per acre are expected with a range of 1 to 6 tonnes per acre. Moisture content at time of harvest will be about 40 per cent and will need to be reduced to 12 to 14 per cent for baling and storage.

Processing and utilization

Harvesting takes place in September, depending on the variety planted and its intended use. Research is under way to develop an appropriate hemp harvester. Although success has been achieved using a row-independent maize harvester, what has been deemed necessary is a harvester that not only cuts the stalks but also removes the fibres. Research to develop an appropriate harvester is currently in progress.

In England, success was achieved with hemp grown for fibre production by cutting the crop in August and turning it twice in a six week period before it was baled

(Long, 1995). While the crop is in windrows, rain helps to loosen the fibres.

The many uses of hemp are related to the four main parts of the plant: bast fibres, leaves, seeds and processable remains. The bast fibres can be used by the textile industry for making products ranging from clothing to carpet. The fibre can be used by the paper industry to produce a wide range of paper products of varying degrees of quality. The fibres can also be processed to make building materials. The leaves can be used for animal bedding, mulches and composts. The seeds can be used as bird feed and as a cereal ingredient or, after the oil has been extracted, they can be used for animal food and high protein meal. The oil can be processed into products such as fuel, paints, machine oils, soap, shampoo and cosmetics, and is also processed by the food industry – because of the high fat and linoleic acid contents – into edible oil, margarine and food additives (Parker et al, 2003). The processable remains can be used to make a range of paper products and construction materials, or, like the leaves, can go to make animal bedding, mulches and composts.

The entire plant can also be used as a solid fuel when compacted (Karus and Leson, 1995). It can be converted to other biofuels by a variety of conversion technologies. The oil from hemp can be used as biodiesel. Hemp's high total biomass means that productivity can be as high as 18t/ha. Hemp can be considered as a potentially important energy crop species. Special oil hemp varieties can produce up to 600L/ha of oil. Biofuels such as biodiesel and alcohol fuel can be made from the oils in hemp seeds and stalks, and the fermentation of the plant as a whole, respectively. The energy from hemp may be high, based on acreage or weight, but can be low based on the volume of the lightweight harvested hemp. It does, however, produce more energy per acre per year than corn, sugar, flax, or any other crop currently grown for ethanol or biodiesel.

Henry Ford grew industrial hemp on his estate after 1937, possibly to prove the cheapness of methanol production at Iron Mountain. He made plastic cars with wheat straw, hemp and sisal (*Popular Mechanics*, 1941). In 1892, Rudolph Diesel invented the diesel engine, which he intended to fuel 'by a variety of fuels, especially vegetable and seed oils'.

Hemp fabrics have added beneficial qualities of being stronger, more insulative, more absorbent and more durable than cotton and they don't stretch out of shape. Natural hemp fibre 'breathes' and is biodegradable. It is remarkable that hemp will produce 1500lb of fibre per acre, whereas cotton will produce only 500lb per acre, and it is estimated that half of all agricultural chemicals used in the US are employed in the growing of cotton.

Energy and fuel can be derived from hemp stalks through pyrolysis, the technique of applying high heat to biomass with little or no air. Reduced emissions from coal-fired power plants and automobiles can be accomplished by converting biomass to fuel, utilizing pyrolysis technology. The process can produce from lignocellulosic material (like the stalks of hemp), charcoal, gasoline, ethanol, non-condensable gases, acetic acid, acetone, methane and methanol. Process adjustments can be made to favour charcoal, pyrolytic oil, gas or methanol, with 95.5 per cent fuel-to-feed ratios. Around 68 per cent of the energy of the raw biomass will be contained in the charcoal and fuel oils – renewable energy generated here at home, instead of overpaying for foreign petroleum (Hemphasis.net).

Source: D-Kuru, http://commons.wikimedia.org/wiki/File:Hemp_bunch-dried_out_-seeds_close_up_%CE%940063.JPG

Figure 10.77 Hemp seeds

Internet resources

Hemphasis.net, www.hemphasis.net/Fuel-Energy/fuel.htm

www.gov.mb.ca/agriculture/crops/hemp/bko05s00.html

www.oilgae.com/energy/sou/ae/re/be/bd/po/hem/hem.html

Wikipedia.org (http://commons.wikimedia.org/wiki/File:Hemp_plants_bcig.jpg)

JATROPHA (PHYSIC NUT) (*Jatropha curcas* L.)

Description

Jatropha is a genus of approximately 175 succulent plants, shrubs and trees (some are deciduous, like *Jatropha curcas* L.), from the family Euphorbiaceae. The name is derived from (Greek *iatros* = physician and *trophe* = nutrition), hence the common name physic nut. Jatropha is native to Central America and has become naturalized in many tropical and subtropical areas, including India, Africa and North America. Originating in the Caribbean, jatropha was spread as a valuable hedge plant to Africa and Asia by Portuguese traders. The mature small trees bear separate male and female flowers, and do not grow very tall. As with many members of the family Euphorbiaceae, jatropha contains compounds that are highly toxic.

> The hardy *Jatropha* is resistant to drought and pests, and produces seeds containing up to 40 per cent oil. When the seeds are crushed and processed, the resulting oil can be used in a standard diesel engine, while the residue can also be processed into biomass to power electricity plants. (Wikipedia.org, 2009)

Jatropha curcas L. is a perennial crop that grows in regions around the equator. It grows as a bush of up to 6m in height and can live up to 50 years. After about three year the first seeds can be harvested. Jatropha can grow in areas that are unsuitable for other plants, because they are too dry or too arid, or because they have been left by humans because of soil depletion after excessive agriculture. The plant requires little water, fertilizer or pesticides. Many parts of jatropha plants have been used historically by local cultures. The oil from the seeds has applications as a medicine, a lubricant or as a fuel. At present it is mainly used as a feedstock for soap production. Jatropha soap (Sabuni ya Mmbono) is said to have healing properties for those who suffer from acne (Diligent-tanzania, 2009).

As the jatropha plant and seeds are toxic, they are not eaten by goats or other animals. This means that a hedge of jatropha plants keeps cattle out of fields where food crops are grown. Jatropha plants can also provide shade in harsh conditions, allowing more delicate crops to be grown in between. Good results have been reported using this method of intercropping. Growing, harvesting and processing of jatropha provides local employment. For this reason Diligent Tanzania Ltd is working with local farmers and presses the seeds to oil in Tanzania. Another advantage is the useful application of the remaining press cake in local communities.

Jatropha curcus is a drought-resistant perennial, growing well in marginal/poor soil. It is easy to establish, grows relatively quickly and lives, producing seeds, for 50 years. Jatropha produces seeds with an oil content of 37 per cent. The annual nut yield ranges from 0.5 to 12 tonnes.

Jatropha is presumed to be native to South America, but nowadays it is common worldwide in the tropics and parts of the subtropics (Münch and Kiefer, 1989; Francis et al, 2005). The 6–15cm-long leaves are lobed three or five times and are sometimes heart shaped; young leaves often are reddish.

Male and female flowers are produced on the same plant but at separate sites on the plant. Flowering probably does not depend on the day length, but on the climatic conditions and the variety of the plant. Short day length causes an increased formation of male flowers. In regions with dry seasons, flowering is associated with the rainy season (flowering at the beginning and the end of the rainy season). In the continuous wet tropics, Jatropha flowers during the whole year.

The plant is well adapted to drought (stem and root succulence, shedding of leaves) (Heller, 1992). Pollination is by insects (Münch and Kiefer, 1989).

The fruit is a capsule containing three seeds. Ripe fruits change colour from green to yellow and open a little. The castor-like seeds are black, 1.5–2cm long and 1cm wide, and contain 33 per cent oil. The embryo is embedded between two oil-containing halves. The number of chromosomes is 2n = 22 (Heller, 1992).

The genus *Jatropha* contains 160 species, most of which can be crossed with *Jatropha curcas*. This opens up opportunities for the breeding of hybrids – for example, some hybrids show earlier flowering or more strength than *Jatropha curcas* and the other ancestor. Many varieties of jatropha are known, with yellow, pink or green flowers, red or green leaves, poisonous or non-poisonous seed, different number of chromosomes, and different composition of the oil. This is an interesting starting point for breeding (Münch and Kiefer, 1989).

The current distribution shows that introduction has been most successful in the drier regions of the tropics with annual rainfall of 300–1000mm. It occurs mainly at lower altitudes (0–500m) in areas with average annual temperatures well above 20°C, but can grow at higher altitudes and tolerates slight frost. It grows on well-drained soils with good aeration and is well adapted to marginal soils with low nutrient content.

Jatropha includes the following species (Wikipedia.org):

- *Jatropha cuneata*, limberbush, whose stems are used for basket making by the Seri people in Sonora, Mexico, who call it haat. The stems are roasted, split and soaked through an elaborate process. The reddish colour dye that is often used is made from the root of another plant species, *Krameria grayi*.
- *Jatropha curcas*, also known as physic nut, piñoncillo and Habb-El-Melúk, is used to produce the non-edible jatropha oil, for making candles and soap, and as a feedstock for producing biodiesel. Prior to pressing, the seed can be shelled with the universal nut sheller which reduces the arduous task of removing the seeds from the shell by hand. Once the seeds have been pressed, the remaining cake can be used as feed in digesters and gasifiers to produce biogas for cooking and in

engines, or be used for fertilizing, and sometimes even as animal fodder. The whole seed (with oil) can also be used in digesters to produce biogas. Large plantings and nurseries have been undertaken in India by many research institutions, and by women's self-help groups who use a system of microcredit to ease poverty among semi-literate Indian women.
- *Jatropha gossypifolia*, also called bellyache bush: its fruits and foliage are toxic to humans and animals. It is a major weed in Australia.
- *Jatropha integerrima* Jacq., or spicy jatropha: ornamental in the tropics, continuously crimson, flowers almost all year.
- *Jatropha multifida* L., or coral plant: bright red flowers, like red coral, characterized by strongly incised leaves.
- *Jatropha podagrica*, or buddha belly plant or bottleplant shrub, was used to tan leather and produce a red dye in Mexico and the south-western United States. It is also used as a house plant.

Ecological requirements

Jatropha grows well in arid regions (Lutz, 1992). The soil requirements of jatropha are modest; it can grow well on oligotrophic soil. Well-drained soils are preferred. It will not grow on saturated soils. Jatropha enters symbiosis with root fungi (mycorrhizae) as a defence against phosphate deficiency (Münch and Kiefer, 1989). It requires very little water and can survive long periods of drought by shedding most of its leaves in order to minimize the loss of water. Jatropha grows in regions with wide temperature and rainfall ranges. Jatropha has been reported growing between 480 and 2380mm annual rainfall, but for high yields the plant needs 625 to 750mm precipitation (Jones and Miller, 1991).

Drought for several years causes damage. At first, fewer fruits are developed; later, fruit development stops and in the end the plant dies. Jatropha prefers warm temperatures, but manages with lower temperatures, and even slight frost if the plant is tough. Seeds developing under dry climatic conditions have an elevated oil content.

Propagation

There are generative and vegetative methods of propagation, depending on the local situation. The seeds should be selected with attention to size and weight. They are drilled 2–3cm deep. Sowing takes place prior to or at the beginning of the rainy season, or in the perennial humid tropics at any time of the year.

Cuttings should be set well before the beginning of the rainy season – that is, before the plants begin to sprout. The cuttings, with a diameter of 3–4cm and a length of 0.4–0.5m or 1–1.5m, are taken from the base of the branch and planted at a depth of 20cm. The planting distances depend on the soil quality, moisture and the intended use, and can vary widely from 20cm to about 2.5m. Seeds can be planted in situ or in nursery beds with later replanting, or wild seedlings can be transplanted. The cuttings too can be planted directly into place or in nursery beds. Pre-cultivated plants and direct-planted cuttings show higher rates of survival than directly seeded plants. In vitro propagation of jatropha is possible. Regenerated shoots could be rooted and, after hardening, transferred to soil (Sujatha and Mukta, 1996).

Crop management

The plants are cultivated as hedges to prevent erosion while enhancing soil productivity (Rehm and Espig, 1991). One goal is to stimulate the production of jatropha nuts for oil extraction. Only the seedling needs careful nurturing. Young plants are fragile in the rainy season, and in the dry season they have to be protected from termites; 30 per cent of them have to be replanted (Lutz, 1992). Cultivation methods are well known and, in part, long established in many countries. The useful life of jatropha is 50–60 years. Jatropha is very tolerant to pests, the most common of which include scale and wax lice (Coccina). The jatropha shows a vigorous root development and good infiltration, and hence improves the soil's water absorbing capacity. It is suitable for soil stabilization and to hinder erosion on slopes and on steep inclines. jatropha is also a pioneer plant that can be used to cultivate wasteland. Hedges are planted to protect the fields from cattle and other animals and to stop the deterioration of the ecosystem, to stop the loss of agricultural land. Goats do not eat the leaves and twigs

Source: www.schulemachtzukunft2007-014.de/mediapool/52/524467/data/Biokraftstoffe_Teil_3_Biodiesel_-_Langfassung.pdf

Figure 10.78 Indian women working in jatropha plantation

Source: Sami Globe Inc.

Figure 10.79 Jatropha nursery

Source: PierLuigi Susani, www.biofuelsdigest.com/blog2/2008/06/
20/biofuels-digest-special-outlook-report-on-india/

Figure 10.80 Jatropha plantation

Spaceloma manihoticola, a cassava disease, infests jatropha too (Münch and Kiefer, 1989). Fungal diseases of the fully grown trees occur, but are not significant. There are no problems with weeds, because Jatropha is competitive. All in all, jatropha shows high resistance or tolerance to diseases and pests.

Self-incompatibility does not seem to occur. Protandry and protogyny could not be observed in Cape Verde (Heller, 1992).

Production

It appears very difficult to estimate unequivocally the yield of a plant that is able to grow in very different conditions.

Yield is a function of water, nutrients, heat, the age of the plant and other factors. Many different methods of establishment, farming and harvesting are possible. Yield can be enhanced with the right balance of cost, yield, labour and finally cost per tonne.

Seed production ranges from about 2 tonnes per hectare per year to over 12.5t/ha/year, after five years of growth. Although not clearly specified, this range in production may be attributable to low and high rainfall areas (Jatrophabiodiesel.org).

Table 10.45 Seed yield of jatropha with and without irrigation

Dry	Yield (Mt/ha)		
	Low	Normal	High
Year 1	0.10	0.25	0.40
Year 2	0.50	1.00	1.50
Year 3	0.75	1.25	1.75
Year 4	0.90	1.75	2.25
Year, 5	1.10	2.00	2.75
Irrigated	Yield (Mt/ha)		
	Low	Normal	High
Year 1	0.75	1.25	2.50
Year 2	1.00	1.50	3.00
Year 3	4.25	5.00	5.00
Year 4	5.25	6.25	8.00
Year 5	5.25	8.00	12.50

Source: Jatrophabiodiesel.org

because of their content of bitter and poisonous constituents.

Jatropha belongs to the Euphorbiaceae family, like other important cultivated plants such as cassava and castor. Some pests are common to these, and may be important for plant protection if jatropha is planted on a larger scale. In Brazil the pathogen

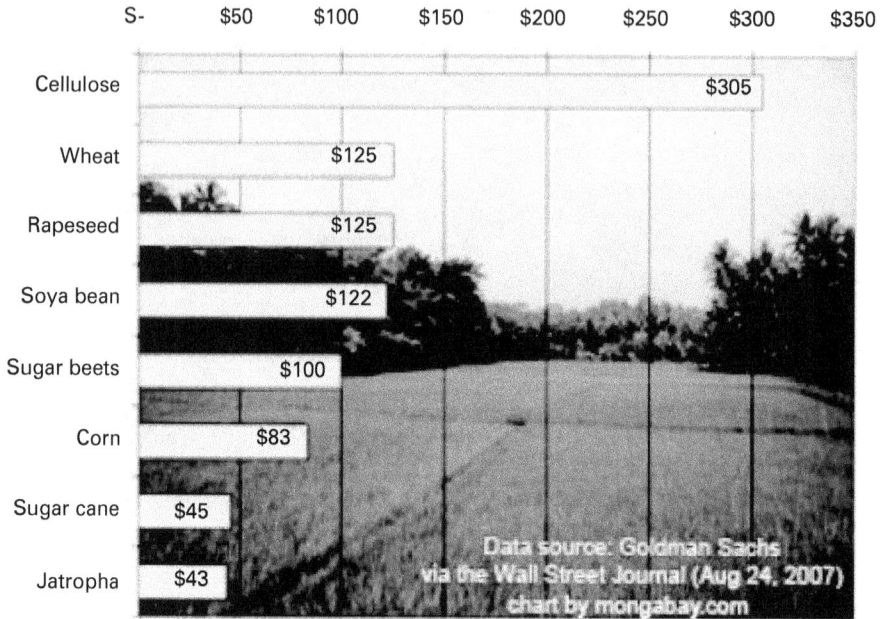

Source: Goldman Sachs via *The Wall Street Journal*. Graphic by Rhett A. Butler

Figure 10.81 Chart showing relative costs, in dollars per barrel of fuel, of biofuels derived from various bioenergy feedstocks

Once big enough to be planted out, jatropha can bear first fruits within a year (Lutz, 1992). The yield from hedges is 1kg of nuts per metre, which gives 0.21kg of oil. Three harvests per year are possible, depending on the rainfall; in semi-arid areas one harvest is more likely. It was reported that under very favourable conditions Jatropha can yield 8t/ha of seed. Under less favourable conditions 0.2–2t/ha can be harvested (Cape Verde) (Münch and Kiefer, 1989). Yields from Jatropha hedges can reach 200L/km of purified oil. Extension work from Udaipur in Rajasthan indicated an average seed yield of respectively 0.125, 0.250, 0.550 and 0.750t/ha in rainfed 2-, 4-, 6- and 8-year-old plantations, and 0.30, 0.60, 0.90, 1.05t/ha under irrigation (Henning, 1995). Maximum seed and oil yield in Rajasthan are 2t/ha/year of seed and 688 kg/ha/year of oil at 75 per cent field capacity and 1.9t/ha/year of seed and 632kg/ha/year of oil at 40 per cent field capacity;

1.26t/ha/year of seed and 404kg/ha/year of oil can be achieved under rainfed conditions. These results indicate that jatropha does not need excessive irrigation for optimum yield (Jones and Miller, 1991).

In comparison to other oil-producing species, jatropha has a low productivity. If appropriate conditions are provided, economic production may become feasible – for instance, in remote areas. In addition, if jatropha is grown in hedges, as windbreaks or to stop erosion, the oil is a welcome by-product.

The Indian government recently expressed support for bringing marginal lands under cultivation of jatropha to produce biodiesel (Francis et al, 2005).

The more oil that is produced, the more press cake is available as fertilizer to improve the soil and stabilize the field crop yield without expensive chemicals. The CO_2 balance by using the oil in engines is neutral, which helps to combat the greenhouse effect.

Processing and utilization

The ripe fruits are harvested using a picking device; fallen fruits are collected. The dry pericarp is removed. The kernels can be stored. The efficiency of the harvest is a crucial point. It is estimated that a single person may harvest up to 5kg/hour.

The oil of jatropha is bitter and poisonous (Lutz, 1992). For the extraction of the oil, the nuts are milled and pressed. Therefore the nuts must be clean, dry, properly stored and preheated for pressing. Heating increases the yield as much as 25 per cent, but the seed must not be overheated. Typically, the nuts will be spread out on a black sheet of plastic, and solar heat will suffice for good drying.

The purified oil is used as a fuel for stationary engines – for example, for generating electricity or driving water pumps and grist mills. The oil is also in demand for the production of soap. Jatropha oil and its constituents (phorbol esters) have insecticidal effects. Simple methods of extracting the active substances have been developed. The press cakes make a valuable organic fertilizer.

The phorbol esters were found to have molluscicidal effects on the water snails that carry bilharzia and on snails like *Lymnaeaamicularia rubigniosa,* an intermediate host of the river fluke. The effect on water snails is very pronounced. The pure oil is toxic, even more than the extracted esters (0.01 per cent of the esters have an insecticidal effect). For use as an insecticide, the oil is mixed with water in the ratio 1:2 and used at 40L/ha. The esters can be extracted from the oil with methanol, which is distilled off in vacuum to obtain the pure esters.

The energy content in comparison with diesel is about 3 per cent lower, related to the volume. Before use it is only necessary to filter the cold pressed oil. The oil can be stored in vessels, if possible airtight, to avoid polymerization (Henning and Mitzlaff, 1995).

The oil does not need to be processed before use in special diesel engines (Elsbett engines); careful filtering is enough. Diesel engines can run with transesterified jatropha oil. Attempts to produce an ester of the oil have been undertaken, but production on a large scale would necessitate major investments in equipment and personnel. A 2000t methyl ester pilot plant was to be built in Nicaragua in 1996.

Source: Diligent Energy Systems 2009

Figure 10.82 Jatropha fruits and seeds

The by-products can be used for energy production. The shells may be transformed into biogas in a fermenter (Foidl et al, 1996). The press cake is used as organic fertilizer, and contains more nitrogen and organic matter than that provided by conventional mineral fertilizer. Turned under, it mineralizes more quickly than cattle dung. It contains more nitrogen and phosphorus than poultry and cattle dung. After drying, it can be used as a solid fuel, which makes energetical use of the remaining oil in the press cakes.

The ratio between fuel and soap production depends on the price of diesel fuel. In countries like Mali, where diesel fuel is expensive, it is economic to use the jatropha oil as fuel. Tables 10.46 and 10.47 show, respectively, the glyceride composition of the oil, and other physical properties.

Table 10.46 The glycerides of Jatropha oil's fatty acids

Glyceride	Number of C atoms and double bonds	Content (%)
Myristic acid	(C 1510)	1.4
Palmitic acid	(C 1610)	10–17
Stearic acid	(C 1810)	5–10
Oleic acid	(C 1811)	36–64
Linoleic acid	(C 1812)	18–45

Source: Münch and Kiefer (1989)

Table 10.47 Some physical properties of Jatropha oil

Property	Value
Density (g/cm³)	0.92
Flash point (°C)	340
Melting point (°C)	–5
Kinematic viscosity (10⁻⁶m²/s)	75.7
Cetane number	23,151
Iodine number	103
Heat of combustion (MJ/kg)	39.628

Source: Münch and Kiefer (1989).

Source: R. K. Henning www.Jatropha.org

Figure 10.83 Oilseed mill

The human toxicity of the oil is caused by curcin, a toxic albumin. In some investigations it was discovered that jatropha oil has a carcinogenic effect after contact with the skin (Heller, 1992), but other investigations gave no indications of mutagenic or carcinogenic effects (Henning and Mitzlaff, 1995).

The oil's insecticidal effects resulting from the 2 per cent phorbol esters content mean that it is a source of insecticides that can be locally produced and applied.

The phorbol esters are toxic, even at a concentration of 250ppm.

Currently the oil from *Jatropha curcas* seeds is used for making biodiesel fuel in the Philippines. Likewise, jatropha oil is being promoted as an easily grown biofuel crop in hundreds of projects throughout India and other developing countries. The railway line between Mumbai and Delhi is planted with jatropha and the train itself runs on 15–20 per cent biodiesel. In Africa, cultivation of jatropha is being promoted and it is grown successfully in countries such as Mali. In the Gran Chaco of Paraguay, where a native variety (*Jatropha matacensis*) grows, studies have shown the suitability of jatropha cultivation and agro-producers are starting to consider planting in the region.

China plans to plant 13 million hectares – an area the size of England – with jatropha trees for biodiesel, according to the State Forestry Administration (SFA). Jatropha is currently grown on around 2 million hectares across the country. The 13-million-hectare forest, mostly spread over southern China, is expected eventually to produce nearly 6 million tonnes (1.8 billion gallons US) of biodiesel every year.

Jatropha may be a more economic biofuel than corn-based ethanol, reported *The Wall Street Journal* on Friday [in August 2007], citing research from Goldman Sachs. Analysis of the bioenergy market suggests that jatropha, which can be grown in variable conditions with little water or fertilizer, could be used to produce a barrel of fuel for around US$43, less than the cost of sugar cane-based ethanol (US$45 per barrel) or corn-based ethanol (US$83 per barrel) currently favoured in the United States.

Further, because jatropha isn't edible and grows on land unsuitable for foods crops, its expansion doesn't compete with traditional food production. (www.jatropha.de).

Estimates of jatropha seed yield vary widely, due to a lack of research data, the genetic diversity of the crop, the range of environments in which it is grown, and jatropha's perennial life cycle. Seed yields under cultivation can range from 1500 to 2000kg/ha, corresponding to extractable oil yields of 540 to 680L/ha (58 to 73 US gallons per acre).

Jatropha can also be intercropped with other cash crops such as coffee, sugar, fruits and vegetables.

On 30 December 2008 Air New Zealand successfully completed a test flight from Auckland using a 50:50 mixture of Jatropha oil and Jet A1 in one of the four Rolls-Royce RB211 engines of a 747 jumbo jet. The two-hour test flight could mark another promising step for the airline industry to find cheaper and more environmentally friendly alternatives to fossil fuel. Air New Zealand announced plans to use the new fuel for 10 per cent of its needs by 2013. Jatropha oil is significantly cheaper than crude oil, costing an estimated US$43 a barrel or about one-third of the 4 June 2008 closing price of US$122.30 for a barrel of crude oil. However, the falling cost of oil has changed the dynamic, with crude oil selling for US$43.71 a barrel, as of 9 December 2008.

Internet resources

Diligent Tanzania Ltd, www.diligent-tanzania.com/ index. php?id=3
Jatrophabiodiesel.org (2009) www.jatrophabiodiesel. org/ jatrophaPlantation.php
Reuk, www.reuk.co.uk/Jatropha-for-Biodiesel-Figures.htm
Sami Globe Inc., http://samiglobe.com/jatorpha.html
Wikipedia.org, http://en.wikipedia.org/wiki/Jatropha
www.jatropha.de/news/Jatropha-may-be-a-more-economic-biofuel-than%20corn%85.pdf

JOJOBA (*Simmondsia chinensis* (Link) Schneid.)

Description

Jojoba (*Simmondsia chinensis*), is a shrub native to the Sonoran and Mojave deserts of Arizona, California and Mexico. It is the sole species of the family Simmondsiaceae, and sometimes placed in the box family, Buxaceae. It is also known as goat nut, deer nut, pignut, wild hazel, quinine nut, coffeeberry, and gray box bush.

Jojoba is a dioecious shrub. It is cultivated mainly in Arizona, northern Mexico, Argentina and Israel; there is some small-scale planting in Australia, Chile and India (Benzioni, 1997). The plant is evergreen, and the leaves are covered with wax to prevent excessive loss of water. The trunk is branched directly over the ground. Various types are observed: some grow like shrubs, others grow up to a height of 5m. Some types bear fruits regularly, others do not. The roots reach depths of as much as 10m (Huber, 1984). The fruits normally contain one seed, but up to three are possible. The seeds contain 47–62 per cent wax, which becomes liquid at 7°C (Rehm and Espig, 1991).

Source: AP Photo/NZ Herald; Estcourt (2008)

Figure 10.84 Test Pilot Captain Keith Pattie, right, Air New Zealand's Chief Pilot Captain David Morgan, left, pose with the company's CEO, Rob Fyfe before their test of a biofuel mixture in the left-hand engine of Boeing 747 in Auckland, New Zealand, Tuesday, 30 December 2008

Ecological requirements

Jojoba tolerates drought. It thrives under soil and moisture conditions that are not suitable for most other agricultural crops. The shrub has been known to survive as long as a year with no rainfall at all. Jojoba requires water during winter and spring months to set its flowers and seeds. Its summer requirements are low (NAS, 1975). An annual rainfall of 250mm is enough for the plant to survive, but 370–450mm allows a quicker and luxuriant growth; 750mm precipitation or irrigation is adequate; 2000mm is acceptable as maximum (Huber, 1984). The plant needs a cool period with a mean temperature of about 15°C for about 2 months to break the dormancy of the flower buds. At constantly high temperatures it never flowers. The optimum temperatures are 21–36°C, but it tolerates extreme desert temperatures (NAS, 1975). Seedlings may tolerate frost of −5°C, older plants −9°C. In addition, it seems that irrigated plants are less tolerant to frost than non-irrigated plants.

The pH value of the soil can range from 5 to 8, the optimum pH value is 6–7. Jojoba is not very demanding on soil fertility and is to a certain degree tolerant to salt. It needs a well-drained, deep and light, sandy soil, and will not tolerate undrained wet soils. The plant is adapted to intense sunshine and the changes associated with high day temperatures and low temperatures at night. At first the young plants grow rather slowly, because they are endeavouring to develop long roots to ensure a good supply of water.

Propagation

Propagation is possible by seed or with cuttings. The seed may be planted at a depth of 2–5cm in wet, well-drained soil. A seeded plantation of jojoba has genetic heterogeneity, and low average yields. Half of the seedlings are males that should be rooted out, because only 8–10 per cent males are sufficient for pollination. All new plantations are from vegetative propagated plants originating from cuttings from selected clones. Rooted cuttings are produced in a greenhouse and after hardening planted out (Benzioni, 1997). Breeding is in progress for high-yielding monoecious and early-bearing cultivars and for types adapted to local climates. In the meantime, vegetative propagation of highly productive genotypes is the most promising approach. Different varieties in growth and regularly nut-producing varieties are selected. The aim of breeding is to obtain cultivars that are suited to mechanized picking and that are resistant to diseases. Jojoba can also be propagated by tissue culture (Huber, 1984).

Crop management

Jojoba thrives best in areas of low humidity and is drought hardy. Wind affects the shape and habit of growth and the plant often only reaches a height of 0.6–0.9m because of the harshness of the environment. Jojoba can be grown at altitudes from sea level to 1500m. In plantations the ratio between male and female plants is 1 to 5–10.

Plantation cultivation is possible with or without irrigation, but in regions with an annual rainfall of less than 500mm intensive cultivation is not feasible. The low yields per hectare are compensated by using a bigger area. The decision depends on the price of land and irrigation water. In plantations the jojoba is planted in hedges with rows 3m apart. The spacings between the plants range from 23cm to 75cm or more (Johnson and Hinman, 1980).

Jojoba is said not to be very susceptible to diseases and pests. In Mexico the drilling insect *Saperda candida* damages the wild jojoba shrubs. Some fungus diseases (*Phytophthora parasitica*, *Phytium aphanidermatum*, *Verticillium dahliaae* and *Macophomina phaseolina*) can be dangerous to the plant (Huber, 1984). The most serious problems in the first two years of a plantation are weeds. Phytopathological problems in the nursery are caused by *Fusarium solani* and *F. oxysporum* (Benzioni, 1997).

Production

The first yields are obtained after 4–5 years, and full yields after about 10 years. Jojoba bears for 100 to 200 years. The fruits mature 90–180 days after fertilization. Jojoba has a high yielding potential: up to 4.5t/ha of seed has been reported (Rehm and Espig, 1991), though for 7–10-year-old plants a yield of 3t/ha is a realistic estimation. The oil achieves a

Source: Stan Shebs 2005 http://commons.wikimedia.org/wiki/File:Simmondsia_chinensis_form.jpg

Figure 10.85 Jojoba foliage

Source: www.jojoba.cl/en/home/gallery/galjojo.html

Figure 10.86 Jojoba fruit

high price – about 2–4 times higher than that of other vegetative oils. The world's demand is estimated at 20,000–200,000t. The seed yields can vary from 0.5–15kg per tree. The average yield obtained from planted stands is 2.25t/ha, or about 1.5kg per plant.

Jojoba's high value has led to its cultivation in many localities, largely with disappointing results due mainly to unsuitable climatic conditions. Optimum temperatures for vegetative growth are 27–33°C with at least one month at 15–20°C to break the dormancy of the flower buds – at constantly high temperatures jojoba never flowers. Frosts of below −3°C damage the crop, and if frosts occur during flowering the flower buds are damaged and the crop may be lost. Although jojoba is very tolerant of drought, 750mm of rain or irrigation are necessary for a good yield. The plant is susceptible to waterlogging and low pH. The product which is traded is jojoba oil. Potential world production of this product is currently around 3500 tonnes per year, the major production areas being the USA, Mexico, Costa Rica, Australia, Brazil and Paraguay. Commercial plantations also exist in Argentina, Egypt, Israel and Peru. The total area covered by the crop throughout the world in 2002 was around 8500ha. In 1999–2001 Israel produced one-third of the world production of jojoba, average yields are around 3.5 tonnes/ha (potential yield is 4.5 tonnes/ha), this is relatively high compared with production potential in Argentina and the USA. World demand in 2002 was estimated at between 64,000 and 200,000 tonnes/year. Because there appear to be few natural products that would directly compete with jojoba oil the future of the product appears to be bright (Ienica).

Farming

Jojoba has the distinction of being the first native plant since corn to be successfully domesticated. The methods used by jojoba farmers in the past have been varied, as there were no real records of the performance of cultivated plants in existence. Subsequent research, however, has led to a greater understanding of the classic farming requirements for jojoba.

Native populations occur between the latitudes of 23° and 34° north, but a general rule is that jojoba will grow well wherever avocados do well and the days of full sun are greatest. Temperatures are critical but only in the low range. Temperatures in the high 20s

Source: www.jojoba.cl

Figure 10.87 Jojoba plantation with drip irrigation

(Fahrenheit) will freeze the buds and new growth in mature populations, and in juvenile plantations (three years old and younger) a very large number of plants will be killed. In general the older the plant the less it will be damaged on a permanent basis by low temperatures. Jojoba handles heat very well.

Jojoba grows best in sandy or decomposed granite or rocky soils, and slowest in heavy clay soils such as adobe. Even if the fertility of the soil is marginal, jojoba is still able to produce well without the use of fertilizers. However, jojoba plants kept in containers seem to do better with some fertilization.

Irrigation systems are necessary when establishing jojoba plantations whether by planting seeds or seedlings. The plants seem to do well on their own after two years of intensive watering in early winter and spring when the jojoba plant maximally utilizes water for growth. The timing of this watering period is a bonus for the jojoba farmer as jojoba's water requirements will not conflict with the watering requirements for traditional crops. Under ideal conditions of soil, water and sun, the taproot will grow an inch a day; within two years the roots should reach the level of the aquifer thus enabling sufficient growth for seed production without supplemental watering.

In the wild, plants will produce a crop solely utilizing groundwater, and they also do so in plantations if an underlying aquifer is available to the roots. If it is possible (and economically viable) watering should be continued every winter and spring as this will keep layers of water moving downward, thereby causing the root systems to develop at greater and greater depths each year. In this case, if the aquifer should drop because of over-drawing, the plants will still have water each year for good seed production.

Seedlings can be expected to flower after three years' growth. The plants are wind pollinated; there are no known insect pollinators other then accidentals. The flowers form in the winter and after pollination grow until they are mature seed in July. The seed skin will dry, shrink and split, and the slightest breeze will send hundreds of the seeds to the soil below. Seed oil content may vary from 45 to 65 per cent. The properties of the oil are constant regardless of geographical origin of the seed.

Rodents collect the seeds but like humans have no enzymes to digest them; so they waste energy in eating them. The largest native plants have been observed in areas around 500m elevation, with rocky sandy soil, with 380–460mm of rain a year and where abundant rainwater drains into low-lying local depressions where the plants grow. The smallest native plants have been found at about 1400m elevation even where precipitation was the same as that in areas with large plants. Jojoba tolerates salinity very well whether in the substrate (soil) or in water. It has been observed doing well in brackish water along the coast of California. It is grown successfully in Israel and is irrigated with water from the Dead Sea.

In order to maximize production, it would seem advantageous to plant rooted cuttings from sexed plants which are known high producers or known to have seeds with high oil content. Rooting the cuttings takes some technical know-how but it would be worthwhile to have a plantation with 90 per cent to 95 per cent female plants, leaving the 5 per cent or 10 per cent males to produce adequate pollen for all the female flowers.

Processing and utilization

The high costs of collecting the small fruits may be a limiting factor. The harvest of the nuts can be mechanized. One principle is to use 'over the row' machines like those used for mechanized grape harvesting. The fruits are shaken off and fall into collectors. As the nuts do not all ripen at the same time, the process has to be repeated. Another option is to wait for the ripe fruits to fall off and later collect them from the ground with a kind of vacuum collector (Huber, 1984).

Jojoba oil has special properties that distinguish it from other vegetable oils. It is mainly used for technical purposes. The oil is obtained by pressing the seeds or by extraction with solvents. In the raw state it has a light yellow colour. Strictly speaking, jojoba oil is a wax: it is not a triglyceride like other vegetable oils, but is composed mainly of straight chain monoesters of C-20 and C-22 fatty acids and alcohols. The ester molecule has two double bonds, one at each side of the ester bond. Table 10.48 shows the make-up of the oil with respect to wax esters. Other esters occur in amounts lower than 1 per cent, and small amounts of free acids and alcohols

Source: Steve Hurst, USDA-NRCS plants database

Figure 10.88 Jojoba seeds and husks

Table 10.48 The main wax ester composition of jojoba oil

Wax ester	Amount (%)[a]
C-40	30.56
C-42	49.50
C-44	8.12
C-38	6.23

Note: [a]Other esters occur in amounts lower than 1%.

Table 10.49 Chemical and physical properties of jojoba oil

Property	Value
Freezing point	7–10.6°C
Melting point	3.8–7°C
Boiling point (757mm, N)	398°C
Smoke point[a]	195°C
Flash point[a]	295°C
Fire point	338°C
Heat of fusion	21cal/g
Refractive index (25°C)	1.4650
Specific gravity	0.863
Total esters	52%
Average molecular weight of wax esters	606
Hardness (Brinell number) of hydrogenated oil	1.9

[a]Determined according to Cc9A-48 of the American Oil Chemists' Society.

Table 10.50 Viscosity of jojoba oil

Temperature (°C)	Viscosity	
	Pure oil	Sulphurized oil (9.88% sulphur)
37.8	127	3518
99	48	491

Source: Johnson and Hinman (1980)

can be found. (Additional chemical and physical properties of jojoba oil can be found in Table 10.49). The almost complete absence of glycerine means that jojoba oil differs radically from all other known oils (Ayerza, 1990).

The oil is stable against oxidation and does not become rancid, and is a high-quality lubricant that withstands high pressures and temperatures. Jojoba oil is a high-quality substitute for sperm oil. Because of its chemical structure (double bonds) and natural purity it is processed to waxes, creams, polishes and cosmetics. It can be used as a source of long-chain alcohols.

The oil is not damaged by repeated heating, and the viscosity does not change after repeated temperature changes (NAS, 1975). Table 10.50 shows the viscosity of pure and sulphurized jojoba oil. The hydrogenated oil is a hard crystalline wax, which is comparable to

carnauba wax (Franke, 1985). Because of its manifold practical uses in chemistry and industry and its high price, jojoba oil is not yet used for energy purposes.

Human contact dermatitis can be caused by jojoba or jojoba oil (Berardino et al, 2006).

The defatted meal has about 30 per cent protein. However, the seeds also contain about 11 per cent antinutritional compounds like simmondsin,

simmondsin 2'-ferulate, 5–desmethylsimmondsin and didesmethylsimmondsin. Several methods are under development to eliminate these materials in order to use the meal as animal feed (Kleiman, 1990). The glycosides are found in leaves, stems and roots (Benzioni, 1997).

An experimental investigation has been carried out to examine for the first time the performance and combustion noise of an indirect injection diesel engine running with new fuel derived from pure jojoba oil, jojoba methyl ester, and its blends with gas oil. A Ricardo E6 compression swirl diesel engine was fully instrumented for the measurement of combustion pressure and its rise rate and other operating parameters. Test parameters included the percentage of jojoba methyl ester in the blend, engine speed, load, injection timing and engine compression ratio. Results showed that the new fuel derived from jojoba is generally comparable to and a good replacement for gas oil in a diesel engine at most engine operating conditions, in terms of performance parameters and combustion noise produced (Huzayyin et al, 2004).

Jojoba is grown for the liquid wax (commonly called jojoba oil) in its seeds. This oil is unusual in that it is an extremely long (C36–C46) straight-chain wax ester and not a triglyceride, making jojoba and its derivative esters more similar to human sebum and whale oil than to traditional vegetable oils. Jojoba oil is easily refined to be odourless, colourless and oxidatively stable, and is often used in cosmetics as a moisturizer and as a carrier oil for speciality fragrances. It also has potential use as a biodiesel fuel for cars and trucks, as well as a biodegradable lubricant. Because sperm whales are endangered, plantations of jojoba have been established in a number of desert and semi-desert areas, predominantly in Argentina, Israel, Mexico, Palestinian Authority, Peru and the USA. It is currently the Sonoran Desert's second most economically valuable native plant (overshadowed only by the Washington palms used in horticulture). Selective breeding is developing plants that produce more beans with higher wax content, as well as other characteristics that will facilitate harvesting (Phillips and Wentworth Comus, 2000).

Internet resources

http://news.mongabay.com/2007/0824-biofuels.html
www.jojoba.cl/en/home/gallery/galjojo.html?start=1
http://cat.inist.fr/?aModele=afficheN&cpsidt=14641173
http://en.wikipedia.org/wiki/Jojoba

www.armchair.com/warp/jojoba1.html
Ienica, www.ienica.net/crops/Jojoba.pdf

KALLAR GRASS (*Leptochloa fusca*)

Contributed by Professor Dr Kauser Abdulla Malik, Forman Christian College, Lahore, Pakistan

Salinity and sodicity are major problems in the arid and semi-arid regions of the world. They do not only affect agriculture production but also lead to environmental degradation and deterioration of the natural resource base. The situation is further compounded by uncontrolled increase in population especially in South Asia.

Salt-affected soils are defined as formations under the dominant influence of different salts in their solid or liquid phase that will then have a decisive influence on the physico-chemical properties which negatively influence the soil fertility. The causes of salinization differ from region to region, ranging from naturally occurring salt-affected soils due to parent rock material to man-made salinization due to artificial irrigation without appropriate drainage.

Historically salinity contributed to the downfall of several ancient civilizations. Old civilizations in Mesopotamia, China and Pre-Columbian America collapsed due to salinization of irrigated lands. A similar situation continues to exist in many parts of the world especially in arid and semi-arid regions. According to the FAO and UNESCO as much as half of all irrigated lands of the world are more or less under the influence of secondary salinization, alkalinization and waterlogging. This has resulted in large tracts of salt-affected wastelands.

Similarly, availability of fresh water for agriculture is becoming scarce in many areas of arid and semi-arid regions. However, there are abundant sources of brackish waters available which could be used for economic utilization of salt-affected soils by introducing crops which could tolerate both salinity of soil and water. Based on the concept of living with salinity, various technologies have been developed which constitute biosaline agriculture technology (BAT) which is defined as 'the profitable and integrated use of genetic resources (plants, animals, fish, insects and micro-organisms) and improved agricultural practices to obtain better use from saline land and saline irrigation water on a sustained basis'.

The most important source of biomass on such soils is the introduction of plants which can tolerate salt-affected soils and brackish underground water. Kallar grass happens to be one such grass which has been used as a primary colonizer of highly saline sodic soils and can be irrigated with brackish underground water (Sandhu and Malik, 1975). This grass, earlier identified as *Diplachne fusca* (L.) Beauv., belongs to the Gramineae (Poaceae) family and has been reported in many of the floras pertaining to Pakistan and northern India (Blatter et al, 1929; Ahmad, 1954, 1977, 1980; Chaudhry, 1969) and has generally been referred as grass of salt marshes. Its name has been changed to *Leptochloa fusca* (L.) Kunth (Booth, 1983). This grass has also been referred to in the vernacular as 'Kallar mar ghas' (salt-eradicating grass) and Australian grass in Pakistan (Hussain and Hussain, 1970). In India, it has the common name Karnal grass. However, the name Kallar grass is now commonly used in the literature. Its distribution in the tropics of the Old World in places like Egypt, India, Sri Lanka, tropical Africa, Asia and Australia has been reported by Blatter et al (1929).

Cultivation of Kallar grass on a large scale

In order to demonstrate large-scale cultivation of this grass, the BioSaline Research Station (BSRS) was established on 65 hactares of highly saline sodic soil near Lahore. The prime objective of establishing BSRS, Lahore, was large-scale cultivation of Kallar grass.

There is a great success story on Kallar grass from this station. The grass was grown on some 40ha of highly salt-affected land using hazardous tube-well water. No particular field preparation was done before cultivation of the grass. The fields were levelled, ploughed once, irrigated and Kallar grass was sown in the beginning of summer by spreading the root stubble and stem cuttings as propagation material and the farmer simply walking on them to partially bury them in soil. After a few weeks' time, a nice tuft of grass developed and at maturity the grass reached to a height of 1–1.5m. The grass provided nearly 40–50 tonnes of biomass per hectare per year in 4 to 6 cuttings.

Agronomic studies of Kallar grass

Some agronomic studies conducted on this grass showed that the maximum amount of biomass from this grass was obtained when it was harvested at intervals of 4 months and was irrigated twice weekly. A biomass yield of 50t/ha/year was achievable under optimum management conditions. The grass did not respond to N fertilization and it was shown that it could fix atmospheric N through a system of associative symbiosis. Similarly the grass also did not respond to P fertilization and it was found that the soils at BSRS, Lahore had a very high level of available P. Kallar grass is capable of maintaining P uptake under salinity stress. The P concentrations in shoots of plants grown in saline and normal soil were similar or even higher in the case of saline soil at later growth stages (Mahmood, 1995). However, application of 5ppm of Cu significantly increased the growth of Kallar grass while the

Source: K. A. Malik

Figure 10.89 Kallar grass planting

Source: K. A. Malik

Figure 10.90 Kallar grass being harvested

application of Zn had no effect. The favourable effect of application of Cu disappeared when it was applied in combination with 10ppm Zn.

Ecology of Kallar grass fields

After a few years' growth, the grass slowly lost its vigour and was overtaken by other weeds. A survey of the vegetation of fields, where Kallar grass was grown for different time periods, indicated a successive invasion by weed species. This process of weed establishment was associated with improved soil conditions with Kallar grass growth and indicated the degree of soil amelioration (Mahmood et al, 1989). An investigation on the possible causes of the decline in growth of grass indicated interference by weeds against Kallar grass that appeared on relatively improved soil. The invading species, *Suaeda fruticosa*, *Kochia indica*, *Sporobolus arabicus* and *Cynodon dactylon*, showed strong competition against Kallar grass; however, the degree of reduction in Kallar grass growth varied with competing species, soil salinity and miosture levels (Mahmood, 1997; Mahmood et al, 1993). A number of weed species also showed allelopathic effects on Kallar grass that further accentuated the interference effects (Mahmood et al, 1999). Autotoxicity of Kallar grass was also observed. All these mechanisms were the possible cause of decline in biomass yield of older Kallar grass stands.

Improvement of saline soil

The cultivation of Kallar grass is very beneficial in improving salt-affected soil and leads to redistribution of salts in the soil profile. The salt accumulation decreases in the top 100cm of the soil but the salts leached with water accumulate in the deeper layers. The extensive root system of Kallar grass help leaching of salts from the top layer and the decrease in soil pH throughout the profile might have been due to root exudates or due to products of decomposition of grass litter (Malik et al, 1986). The physical properties such as soil structure, porosity, infiltration etc. also improve. The downward movement of water in saline-sodic soil where Kallar grass had been growing for three years was studied by using a neutron moisture probe. Water penetration and thus moisture content in the cropped soil profile was significantly higher than in unplanted soil. Properties of saline sodic soil irrigated with poor quality groundwater were

monitored for five years to evaluate the effects of Kallar grass growth on chemical, physical and minerological characteristics of soil (Akhter, 2001). Soil salinity, sodicity and pH decreased exponentially by growing Kallar grass due to leaching of salts from the surface (0–20cm) to lower depths (>100 cm). The reduction in soluble cations (Na^+, K^+, Ca^{2+} and Mg^{2+}) and anions (Cl^-, SO_4^{2-}, HCO_3^-) occurred due to leaching from upper to lower depths of soil (Akhter et al, 2003). A significant decline in soil pH was attributed to release of CO_2 by grass roots and solublization of $CaCO_3$. Both soil salinity measured as electrical conductivity (EC) and soil pH were significantly correlated with most of the chemical characteristics Na^+, Ca^{2+}, Mg^{2+}, K^+, Cl^-, HCO_3^-, and sodium adsorption ratio. Significant correlations were found between soluble cations (Na^+, Ca^{2+} and K^+), soluble anions (Cl^-, SO_4^{2-} and HCO_3^-) and sodium adsorption ratio. In contrast, the results showed negative correlations between soil organic matter content and all chemical properties (Akhter et al, 2003).

Detailed studies on its ecology, physiology and agronomy have been documented by Malik et al (1986). Kallar grass has been shown to have the C_4 photosynthetic pathway (Zafar and Malik, 1984). It has also been reported that this grass supports nitrogen fixation associated with roots (Malik et al, 1981; Zafar et al, 1986). Salt tolerance studies carried out in gravel culture showed a 50 per cent reduction in biomass at the salinity level of 22dS/m which is nearly half the salinity of sea water.

Energy

Nearly all the biomass has a major component of cellulose which is a polymer of sugars and is amenable to enzymatic hydrolysis, thus converting into sugars which can then be converted to ethanol or methane gas through fermentation. Presently ethanol is being produced from starch and sugars obtained from corn, soya beans and sugar cane. These are all products of agriculture and impact the food security of the world. The recent international food crisis has been attributed to the fact that large tracts of arable land were used for producing ethanol/energy rather that meeting the food demands of the world population.

It is in this scenario that production of ligno-cellulosic biomass from non-arable saline wastelands becomes attractive. Moreover, significant progress through biotechnology has been made in the area of enzymology whereby cellulolytic enzymes with high expression are possible. The salt-tolerant plants grown on saline lands could then be used as energy crops. An evaluation of the calorific value of biomass produced on saline land showed that these values were comparable with the biomass produced on normal soils. Energy values differed with species; grasses and shrubs had overall higher mean values (5220kcal/kg) than mean value (4794kcal/kg) for tree species (Table 10.51). However, the tree species capable of good growth on saline lands, *Eucalyptus camaldulensis*, *Acacia ampliceps*, *Prosopis juliflora* and *Leucaena leucocephala*, had comparable energy values. These observations suggest the potential of biomass produced on saline wastelands for developing renewable energy resources and biopower plants.

The vast areas of salt-affected wastelands available can be utilized for the production of energy crops by growing salt-tolerant plants for the economic production of fuel and other products. Because of its C_4 system of photosynthesis, a high tolerance to salinity

Table 10.51 Calorific values of salt-tolerant species growing on saline land at BSRS-II, Pacca Anna near Faisalabad (Pakistan)

Plant species	Energy (kcal/kg)
Grasses	
Leptochloa fusca (Kallar grass)	4753
Sporobolus arabicus	5308
Cynodon dactylon	5685
Desmostachya bipinnata	5351
Imperata cylindrica	5009
Shrubs	
Kochia indica	5107
Suaeda fruticosa	4815
Atriplex lentiformis	4331
Capparis aphylla	5922
Calatropis procera	5922
Trees	
Eucalyptus camaldulensis	4890
Acacia ampliceps	5332
Prosopis juliflora	5018
Acacia nilotica	5040
Leucaena leucocephala	5024
Tamarix aphylla	4196
Parkinsonia articulata	5611
Pongamia glabra	2992
Azadirachta indica (Neem)	5234
Dalbergia sissoo	4606

and sodicity, and associative nitrogen fixation, *Leptochloa fusca* (Kallar grass) can produce 40t/ha/year and could be a cheap way of capturing solar energy in developing countries. *Atriplex* sp. and *Panicum maximum* can also be utilized for biotechnical conversion for production of gaseous or liquid fuels.

Fermentative ethanol production from Kallar grass and fibrous wastes can give 222 million gallons and 37.8 million gallons of ethanol per year, respectively (Rajoka et al, 1996). Micro-organisms such as genetically modified *S. cerevisiae* could be tailored to produce a variety of enzymes as co-products with ethanol (Rajoka et al, 1996). These enzymes could include those required for large markets such as the detergent, textile, food and wood pulp industries and biocatalyst-based industries for novel chemicals.

Kallar grass is also commonly used as a forage crop. The fodder value of Kallar grass (*Leptochloa fusca*) is recognized by local farmers in Pakistan who depend on it to feed their buffaloes; the stocking capacity of *L. fusca* estimated from forage production data is reported to be 1–2 cattle/ha/yr (Kernick, 1986). *L. fusca* excretes salts through specialized glands and is therefore reasonably palatable to farm animals (Kumar, 1996). Kallar grass is being cultivated on large areas in Punjab and Sindh provinces, the worst affected regions by salinity in Pakistan. In the Jhang (Punjab) area, Kallar grass is the main fodder crop and spreads over vast areas covering 30 to 50 per cent of the land at private farms. The farmers are raising cattle mainly for milk and meat production. Livestock raising on Kallar grass is an economically feasible activity; an estimated budget of a farmer in the area is presented in Table 10.52. It is evident that saline land and saline groundwater could be managed and used profitably by growing salt-tolerant plants like Kallar grass.

The work at Nuclear Institute for Agriculture and Biology (NIAB), Faisalabad, Pakistan, indicated that this grass could also be used for energy production. Estimates have shown that direct conversion of Kallar grass to methane would yield nearly six times more energy than when it is consumed by livestock (Table 10.53). Further, a look at the composition of different parts of Kallar grass (Table 10.54) indicated that it is a suitable substrate for microbial degradation. For this purpose, many types of fungi and bacteria, collected from saline environments, have been screened for their cellulolytic ability and salt tolerance, and some *Cellulomonas* spp. have been found to be very efficient (Malik et al, 1986).

Table 10.52 Farm budget of a private farmer raising buffaloes on Kallar grass

Area under grass	100 acres	41.66 hectare
Number of buffaloes	100	
Purchase price of animals (average per head)	Rs.2500/-	US$43.85
Sale price of animals after 9–12 months (average per head)	Rs.4500/-	US$78.94
Profit per head	Rs.2000/-	US$35.08
Gross income	Rs.200,000/-	US$3508.77
Charges on labour, water rates and land revenue, etc.	Rs.60,000/-	US$1052.63
Total profit	Rs.140,000/-	US$2456.14
Net income from each acre (0.41 hectare) under grass	Rs.1400/-	US$24.56

Source: Malik et al (1986); Sandhu and Qureshi (1986)

Note: The estimates were prepared based on the prices during the 1980s; the cost of inputs will be higher but the net income may be up to four fold at present market rates.

Table 10.53 Total energy obtained from Kallar grass when used for livestock or for production of methanol

Utilization of Kallar grass for fuel production	
Green matter production	40/t/ha
Dry matter	16.8/t/ha
Total digestible nutrient (TDN) (proteins, cellulose, hemicellulose, lignin).	7.88/t/ha
Potential for methane production (0.18m^3/kg dry matter)	3024m^3/ha/year
Sludge (0.72kg/kg dry matter)	12t
Nitrogen in the sludge	240kg
Total energy yield from Kallar grass-derived fuel	15 × 10^{6}kcal/ha
Utilization of Kallar grass for food production	
Total buffaloes	3
Total milk	4320L
Calorific value of milk	2.86 × 10^{3}kcal/ha
Calorific value of meat	0.28 × 10^{5}kcal/ha
Calorific value of biogas from dung	2.64 × 10^{6}kcal/ha
Total energy yield from buffaloes	2.9 × 10^{6}kcal/ha

Table 10.54 Composition of different parts of Kallar grass plants grown under high salinity (EC = 22.3dS/m)

Parameter	Leaves	Stems	Roots
Dry matter (%)	36.3	38.1	53.3
Lignin (%)	19.6	16.5	18.5
Hemicellulose (%)	4.9	5.8	4.7
Cellulose (%)	18.7	22.9	16.12
Nitrogen (%)	1.05	0.63	0.35
Ash (%)	17.7	8.5	27.3
Na$^+$ (%)	0.39	0.39	0.66
K$^+$ (%)	0.94	0.78	1.55
Ca^{2+} + Mg^{2+} (%)	3.33	2.74	5.90

KENAF (*Hibiscus cannabinus* L.)

Description

Kenaf is a short day, annual, herbaceous plant possessing high-quality cellulose. It is a member of the Malvaceae family along with cotton and okra, and is endemic to Africa. In Switzerland the plant grows to a height of between 1.5m and 3m (Ott et al, 1995). The rough-haired stems, most having small spines or thorns, consist of a central pith with short wood fibres and a bark with long fibres. Lower-quality paper can be made from the short wood fibres, while high-quality paper can be made from the long fibres (Manzanares et al, 1995). At medium height the plant possesses heart shaped to deeply lobed leaves (Rehm and Espig, 1991).

Ecological requirements

Kenaf is able to adapt to a wide range of soil types and climates. The limits for cultivation are roughly 45°N to 30°S, which consists mainly of the warm temperate zone to the equator, with a maximum elevation of approximately 500m. To avoid the reduced growth associated with flowering and fruit formation, it is recommended that the day length be greater than 12.5 hours during the growing season (Rehm and Espig, 1991).

Kenaf is very cold sensitive. The optimal temperature range is from 15 to 27°C. The optimal pH ranges from 6 to 7. Kenaf grows well on light to medium-weight quickly warming soils, with sandy soils also showing very good growth. Precipitation requirements are 500–700mm rainfall. Very wet soils are not suitable and kenaf cannot tolerate waterlogging. Fields with high weed levels should be avoided (Rehm and Espig, 1991). A test plot on sandy soil in Sardinia showed that kenaf needs frequent irrigation and more attention than sorghum (Bombelli et al, 1995). Kenaf is one of the fibres allied to jute and shows similar characteristics. Other names include bimli, ambary, ambari hemp, deccan hemp and bimlipatum jute.

Propagation

Kenaf is most commonly propagated by seed. It is an annual or biennial herbaceous plant (rarely a short-lived perennial) growing to 1.5–3.5m tall with a woody base. The stems are 1–2cm in diameter, often but not always branched. The leaves are 10–15cm long, variable in shape, with leaves near the base of the stems being deeply lobed with three to seven lobes, while leaves near the top of the stem are shallowly lobed or unlobed lanceolate. The flowers are 8–15cm in diameter, white, yellow or purple; when white or yellow, the centre is still dark purple. The fruit is a capsule 2cm in diameter, containing several seeds.

Crop management

Sowing takes place in the middle of May, though later sowing dates may benefit from the higher ground temperature. Test plots in Spain resulted in the recommendation of crop sowing in June. This later date provides warmer temperatures that lead to quicker seed germination and seedling growth (Manzanares et al, 1995).

The ground temperature should be at least 15°C, with warmer temperatures resulting in an increase in growth rate. The dry ground should be ploughed to a depth of 20cm, then left for 10–15 days if possible (Ott et al, 1995). Seeds are sown at 45–50 seeds per m², which amounts to 13–15kg/ha of seed, and a depth of 2–3cm. A spacing of 45–50cm between rows is recommended. A 70 per cent success rate results in a crop density of 30–35 plants per m². The duration of germination is 4 to 7 days.

Seeds should be disinfected against anthracnose (*Colletotrichum hibisci* Poll.) and fungal diseases including *Rhizoctonia solani* Kühn and *Macrophomina phaseolina* Tassi.

In the early stages a pre- or post-emergence herbicide can be used, or a single hoeing after germination may prove sufficient for combating weeds. If a more persistent weed problem is present, then hoeing twice may be necessary. This would be performed after the kenaf is at least 15cm high and the weeds are in the germinating leaf to two leaf stage (Ott et al, 1995). Because of kenaf's fast growth, weeds are not much of a problem once the plant is established.

Nematodes *(Meloidogyne* spp.) are significant kenaf pests and *Sclerotinia* is one of the few diseases affecting kenaf. *Botrytis* fungi may occur, but do not need combating because they do not affect fibre quality. During the vegetation period, which can be between 70 and 140 days, violet discoloration and leaf necrosis may occur after a cold period.

(a)

(b)

Figure 10.91 Kenaf plantations, two different
varieties at different places, (a) Asia; (b) Largo, FL

(a)

(b)

Figure 10.92 Kenaf: (a) at 6 months; (b) at seed stage

The following herbicides are acceptable for use
according to conditions in Switzerland. Herbicides
containing trifluralin (Trifluralin PSO) are acceptable
for pre-sowing, herbicides containing alachlor (Lasso,
Micro-Tech), dimethenamid (Wing), pendimethalin
(Wing, Stomp) or metolachlor (Dual 960) are
acceptable for pre-emergence and herbicides containing
cycloxydim (Focus Ultra), haloxyfop-(R)-methyl ester
(Gallant 535), or quizalofopethyl (Targa Super) are
acceptable for post-emergence (Ott et al, 1995).

Table 10.55 Recommend fertilizer rates for kenaf

Fertilizer	Amount (kg/ha)
Nitrogen (N) at time of sowing	20–30
Nitrogen (N) at approx. 20cm plant height (June)	50–60
Potassium (as K_2O)	120
Phosphorus (as P_2O_5)	60

Source: Ott et al (1995)

Adequate fertilization is needed for large yields. Before ploughing, an initial soil fertilization using dung or liquid manure is possible. The recommended levels for crop fertilization as used in Germany and Switzerland are shown in Table 10.55.

Kenaf has proved to be very resistant to wilting and if the plant has been broken or blown over it will continue to root and grow.

Kenaf can be placed into a normal crop rotation. Because the harvest is in winter, summer crops with sowing dates starting on or after the middle of March can follow kenaf, but avoid preceding kenaf with crops that carry *Sclerotinia* spp. A cultivation break of 3 years is recommended for kenaf (Ott et al, 1995).

Production

Although a strong market for kenaf has yet to be established, price and cost estimations have been made. For example, with an estimate of 5t/ha dry matter at a value of at least €150/€740/ha per dry tonne, this gives. After the subtraction of estimated costs of €640 and the addition of a set-aside land subsidy, this leaves approximately €485/ha (H.W., 1995).

Test plots in northern Italy, using 10 May sowing and 10 October harvesting dates, have provided the following cultivar fresh biomass yields: 'Khon Kaen-60' 164t/ha, 'Kenalf' 149t/ha, 'Tainung-1' 140t/ha, and 'Everglades-41' 137t/ha. These yields correspond to dry matter yields of approximately 30t/ha (Petrini and Belletti, 1991).

A study in Spain was conducted to determine the appropriate cultivar under Spanish climatic conditions (Manzanares et al, 1995). From the three cultivars tested – 'El Salvador' (long cycle), 'Everglades-71' (intermediate cycle) and 'PI-343 129' (short cycle) – 'Everglades-71' proved to be most appropriate for the test plot's latitude and climate. The mean dry matter

(a)

(b)

Source: (a) Kenaf Green Industries, www.kenafibers.com/pics.html; (b) David Nance, USDA

Figure 10.93 (a) Kenaf grows in the desert with irrigation; and (b) kenaf harvesting machine

stem yields for 'Everglades-71' for 1992 and 1993 were 8.9t/ha and 8.03t/ha respectively.

The cultivated kenaf area is estimated to about 200,000ha and the main producers are Thailand, China, India and Mexico.

In southern Europe a production of 20t/ha dry stem has been reported. Other research works have reported up to 26t/ha dry matter yields for the period from 1993–2000.

Yields in research plots have varied widely, from 2.5 tonnes/acre at Rosemount, Minnesota, to 15 tonnes/acre at College Station, Texas. One-acre blocks at Fort Gibson, Oklahoma yielded nearly 10 tonnes/acre in 1988. Commercial-scale production in Texas has produced dry-weight yields of 7.5 tonnes/acre under irrigation and 6.0 tonnes/acre on dry land. (www.hort.purdue.edu/ newcrop/afcm/kenaf.html)

Field research evaluation was conducted to examine the yield components of five kenaf cultivars. A two-year study was conducted at Lane, OK, on a Bernow fine sandy loam, 0 to 3 per cent slope, (fine-loamy, siliceous, thermic Glossic Paleudalf). Kenaf cultivars Tainung #2, Everglades 71, Everglades 41, Cuba 108 and Guatemala 51 were planted on 8 May 1989 and 23 May 1990. In each year plots were arranged in a randomized complete block design with four replications. Plots were hand harvested on 24 October 1989 and 1990. Data collected included stalk and leaf yield, stalk and leaf percentage, bark and core percentage, bark to core ratio, stalk diameter, plant height and plant population. Everglades 41 had the greatest percentage of stalk by weight (78.5 per cent). Cuba 108 had the least percentage core material (62 per cent), and the greatest bark to core ratio (0.61), while Tainung #2 had the greatest percentage core material (69 per cent) and greatest plant height (299cm). Tainung #2 also had greater stalk yields (13.8t/ha) than either Guatemala 51 (11.4t/ha) or Cuba 108 (10.8t/ha) (Weber, 1993).

Processing and utilization

After the plant has died due to frost and achieved its highest dry matter percentage (60–70 per cent), it is ready to be harvested. Kenaf can be harvested with a row-independent maize harvester. Only the stems are harvested; the leaves will have died and fallen because of the frost. The harvest takes place in the winter between December and February. At this time the fibre and wood are still easily separable. Fresh stems consist of 5–6 per cent fibre, while the dry weight is 18–22 per cent fibre. Under average conditions a fibre yield of 1–2t/ha can be expected, while under favourable conditions yield reaches 3–3.5t/ha (Rehm and Espig, 1991) or more (Stop Global Warming, Inc.). After being harvested, the kenaf is dried for 12 hours in a gas dryer to prepare it for transport.

The US Department of Agriculture identified kenaf as the best non-wood paper alternative for its rapid growth, high yield and paper-making characteristics (Stop Global Warming, Inc.). The entire plant can be used to produce a pulp yield of 20t/ha for the paper industry. The fibre is mechanically or manually removed and used for weaving such items as sacks. After the fibre has been removed the remaining stems can be used as firewood or by the paper industry. The paper industry uses kenaf to make mulch paper for gardening and agriculture, for corrugated board, and to improve the tensile strength of recycled paper (Rehm and Espig, 1991). Kenaf as a high-yielding plant species is a potential energy crop when used as a whole crop. Also, the residues from different industrial processes can be utilized as energy sources.

Kenaf has a long history of cultivation for its fibre in India, Bangladesh, Thailand, parts of Africa, and to a small extent in southeast Europe. The stems produce two types of fibre, a coarser fibre in the outer layer (bast fibre), and a finer fibre in the core. It matures in 100 to 1000 days. About 9000 cultivars are produced. It has been grown for over 4000 years in Africa where its leaves are consumed in human and animal diets, the bast fibre is used for cordage, and the woody core of the stalks burned for fuel. This crop was not introduced into southern Europe until the early 1900s. Today, principal farming areas are throughout China, India, and it is also grown in many other countries including the following: Mackay, Australia in trial stages; Seed farms in Texas, USA and Tamaulipas, Mexico; North Carolina, USA and Senegal, to name a few.

The main uses of kenaf fibre have been rope, twine, coarse cloth (similar to that made from jute) and paper. In California, Texas and Louisiana, 3200 acres (13km²) of kenaf were grown in 1992, most of which was used for animal bedding and feed (Wikipedia).

Emerging uses of kenaf fibre include engineered wood, insulation and clothing-grade cloth (Panasonic has set up a plant in Malaysia to manufacture kenaf fibre boards and export them to Japan), oil absorbent (based on a patent issued to H. and C. Willett), soil-less potting mixes, animal bedding, packing material, organic filler for blending with plastics for injection moulding (using the technology developed and patented by Fibre Packaging International, Inc.), as an additive for drilling muds, and various types of mats, such as seeded grass mats for instant lawns and mouldable mats for manufactured parts and containers.

Non-food uses of Kenaf could be summarized as follows:

• Kenaf seeds yield a vegetable oil that is edible and high in omega antioxidants. The kenaf oil is also used for cosmetics, industrial lubricants and as biofuel.
• Core uses (paper pulp, oil/chemical absorbents, insulation panels, horticultural mixes, bedding

materials for animals, etc.). Apart from the industrial uses the core material can be used for thermochemical process (combustion, gasification and pyrolysis).

- Bark uses (paper pulp production).
- The whole plant has high protein and good digestibility and may be pelletized.

Kenaf absorbs CO_2, storing the carbon and oxygen releasing more than any other plant or tree – twice as much as the rain forest and three to eight times as much as trees. Because kenaf grows so fast, its unique characteristic is that it stores an abundance of carbon in the cells, which makes kenaf the best absorber of carbon dioxide (Stop Global Warming, Inc.).

Internet resources

Loftus, Bill, Stop Global Warming, Inc., www.stop-global-warming.org/
www.hort.purdue.edu/newcrop/afcm/kenaf.html
www.wanabo.com/the-raw-feed/kenaf
www.wiu.edu/AltCrops/kenaf.htm
Wikipedia, http://en.wikipedia.org/wiki/Kenaf//de.wikipedia.org/wiki/Kenaf
www.stop-global-warming.org/Kenaf_pics.html

KUDZU (*Pueraria lobata*)

Description

Kudzu is native to Japan and China. Kudzu is a vine that when left uncontrolled will eventually grow over almost any fixed object in its proximity including other vegetation. Kudzu, over a period of several years will kill trees by blocking the sunlight and for this and other reasons many would like to find ways to get rid of it. The flowers which bloom in late summer have a very pleasant fragrance and the shapes and forms created by kudzu vines growing over trees and bushes can be pleasing to the eye during the summer months. Kudzu vines will cover buildings and parked vehicles over a period of years if no attempt is made to control its growth.

Kudzu *Pueraria lobata* (syn. *P. montana, P. thunbergiana*), (sometimes known as foot a night vine,

(a)

(b)

Source: (a) Pollinator 2004 and (b) US Government, http://en.wikipedia.org/wiki/File:Kudzu4903.JPG

Figure 10.94 Kudzu vegetation growing on (a) shrubs; and (b) trees

mile a minute vine, Gat Gun, Ge Gan and The vine that ate the South) is one of about 20 species in the genus *Pueraria* in the pea family Fabaceae, subfamily Faboideae. It is native to southern Japan and southeast China in eastern Asia. The name comes from the Japanese word for this plant, *kuzu*. The other species of *Pueraria* occur in Southeast Asia, further south.

Kudzu is a climbing, woody or semi-woody, perennial vine capable of reaching heights of 20–30m in trees, but also scrambles extensively over lower vegetation. The leaves are deciduous, alternate and compound, with a petiole 10–20cm long and three broad leaflets 14–18cm long and 10cm broad. The leaflets may be entire or deeply two or three lobed, and are pubescent underneath with hairy margins. The flowers are borne in panicles 10–25cm long with about 30–80 individual blooms at nodes on the stems.

Each flower is about 1–1.5cm long, purple and highly fragrant. The flowers are copious nectar producers and are visited by many species of insects, including bees, butterflies and moths. Flowering occurs in late summer and is followed by production of brown, hairy, flattened seed pods in October and November, each of which contains three to ten hard seeds. Seeds, however, are only produced on plants that are draped over vegetation, fences, and other objects. Only one or two viable seeds are produced in a cluster of seed pods.

Establishment

Once established, these plants grow rapidly, extending as much as 20m per season at a rate of about 30cm per day. This vigorous vine may extend 10–30m in length, with basal stems 1–10cm in diameter. Kudzu roots are fleshy, with massive taproots 10–20cm or more in diameter, reaching depths of up to 3.7m in older patches, and weighing as much as 180kg. As many as 30 stems may grow from a single root crown.

Kudzu grows well under a wide range of conditions and in most soil types. Preferred habitats are forest edges, abandoned fields, roadsides and disturbed areas where sunlight is abundant. Kudzu grows best where winters do not drop below −15°C, average summer temperatures are regularly above 27°C, and annual rainfall is 1000mm or more. This fast-growing plant does not do as well in less temperate areas.

Biomass productivity and utilization as biofuel feedstock

There is little information related to the biomass productivity of kudzu. Enzenwa et al (1996) found that the dry matter yields of ruzi grass–tropical kudzu in the open and under oil palms were 5.9 and 2.7t/ha respectively.

Kudzu is receiving serious consideration as a biofuel. It has even been suggested that kudzu may become a valuable asset for the production of cellulosic ethanol.

As concerns rise over corn ethanol creating competition between food and fuels, ethanol made from kudzo, one of the country's most invasive plants, could be part of the solution, according to Rowan Sage of the University of Toronto and colleagues at the US Department of Agriculture.

His team collected samples of kudzu from different locations in the southern US at different times of year, and measured the amount of carbohydrate present in leaves, vines and roots. The roots were found to contain the highest concentration of carbohydrate up to 68 per cent carbohydrate by dry weight, compared to a few per cent in leaves and vines.

The researchers estimate that kudzu could produce 2.2 to 5.3 tonnes of carbohydrate per acre in much of the southern US, or about 270 gallons per acre of ethanol, which is comparable to the yield for corn of 210 to 320 gallons per acre. They recently published their findings in *Biomass and Bioenergy* (Sage et al, 2009).

Crucial to implementation of the plan work would be determining whether kudzu could be economically harvested, especially the roots, which can be thick and grow more than 2m deep. Even if equipment could harvest the roots, a large fraction of kudzu vines blanket steep hillsides and would be difficult to access. The team estimated that about one-third of kudzu plants would be harvestable. If so, they calculate that kudzu could offer about 8 per cent of the 2006 US bioethanol supply (Discovery News Channel).

Recently, tremendous effort has been put forth to identify plants with potential to be used as biofuels. Kudzu (*Pueraria lobata* (Willd.) Ohwi), while native to the Orient, has proliferated as an invasive weed throughout the southern US. It is currently at or near the top of invasive species lists for virtually every southern state. Kudzu, as a member of the Fabaceae family, is a natural nitrogen fixer and, thus, grows rapidly across the landscape with no inputs (e.g. fertilizers). Given its perennial growth habit, its rapid growth rate, and the fact that kudzu has high starch content (particularly its root system), its potential as a biofuel could be tremendous. However, to date, this potential has gone unstudied. We propose to initiate an investigation into this potential by quantifying above- and below-ground kudzu biomass production, and associated starch and nutrient content. This initial work will lead to more in-depth studies of potential kudzu production systems, harvesting techniques, and cost-benefit analyses. (USDA Agricultural Research Service)

Internet resources

www.nps.gov/plants/alien/fact/pumo1.htm
Discovery News Channel http://dsc.discovery.com/news/2008/06/16/kudzu-biofuel-ethanol.html
www.jjanthony.com/kudzu/
USDA Agricultural Research Service, www.ars.usda.gov/research/publications/publications.htm?SEQ_NO_115=202385

LEUCAENA (horse tamarind) (*Leucaena leucocephala* (Lam.) de Wit)

Description

Leucaena is a genus of about 24 species of leguminous trees and shrubs, distributed from Texas in the United States to Peru. It belongs to subfamily Mimosoideae of the legume family Fabaceae. Some species (namely *Leucaena leucocephala*) have edible fruits (as unripe) and seeds. The seeds of *Leucaena esculenta*, called guaje in Mexico, are eaten with salt, but high levels of mimosine in other species may leads to loss of hair and infertility.

Leucaena originates in Mexico and Central America, though its origin is obscured by wide distribution by humans. Today, leucaena is grown in many parts of the tropics. It grows naturally as a weed in many regions (Brewbaker, 1995). It is an arborescent deciduous small tree, up to 20m high, with a 10–25cm trunk diameter. The alternating leaves are evergreen, 10–25cm long and bipinnate with 3–10 pairs of pinnae. The numerous flowers are white, sitting axillary on long stalks, occurring in dense global heads. The fruit pods are 10–15cm long, 1.6–2.5cm wide and thin. They become dark brown and hard, and contain 15–30 seeds. Flowers and fruits are formed nearly throughout the year. The plant develops a long, strong taproot.

Leucaena is grown for many purposes. It can be grown as a protein-rich fodder plant, for land reclamation, erosion control, water conversation, reforestation or soil improvement (Duke, 1983). Leucaena is one of the few woody tropical legumes that is highly digestible and relatively non-toxic. It is planted for forage in many sites. All phytophagous animals favour leucaena. It is a supplement to grass pastures. The foliage carries a mild toxin, the amino acid mimosine, which causes depilation in non-ruminant animals. The use of the wood includes fuelwood, lumber, pulpwood, craftwood and charcoal. Improved varieties allow the harvesting of lumber and paper pulp on a large plantation basis, with clear 30cm boles on 14m-high trees in 8 years. Leucaena coppices well and withstands almost any type or frequency of pruning or coppicing (Brewbaker, 1995). Seedlings less than one year old will produce viable seeds (Duke, 1983).

Today, the species used commercially outside the New World is almost exclusively *L. leucocephala* (Lam.) de Wit (2n = 104). However, hybrids of this and other species are grown to an increasing extent: notably, *L. pallida* Britton and Rox. (2n = 104) and *L. diversifolia* (Schlecht) Renth. (2n = 104) are used for hybridization. These three polyploid species intercross with ease; the last two provide cold tolerance, and *L. pallida* provides resistance to psyllid insect damage. The common cultivar is a self-pollinating *L. leucocephala* with virtually no genetic diversity outside the centre of its origin. New cultivars of arboreal type out-yield the common cultivars and are less weedy. Less known species with some commercial interest include *L. collinsii*, *L. diversifolia* (2n), *L. diversifolia* (4n), *L, esculenta*, *L. pallida* and *L. pulverulenta*. All *Leucaena* species are woody perennials of the New World, ranging from Texas south to Ecuador, in dry and mesic secondary forests (Brewbaker, 1995).

The leaves are used for mulching around other crops. They are said to increase the yields. The plant can fix nitrogen at 500kg/ha/year. Leucaena wood is hard and heavy; the specific gravity is $0.7g/cm^3$ (Duke, 1983).

Ecological requirements

The tolerated climatic conditions range from the warm temperate dry to moist, through tropical very dry to wet forest life zones. Leucaena requires long, warm wet growing seasons. It does best under full sun. In the wild it is mostly found below 500m, though in Indonesia and Java it is grown at altitudes of 1000m and above. The growth rate is slower at high altitudes (Duke, 1983). Interspecific hybrids greatly extend this range. An annual precipitation of 650–1500mm is typical, but leucaena can be found on drier or wetter sites depending on drainage or competing vegetation (Brewbaker, 1995).

The plant accepts a wide range of soils, but does best where the soil is a deep, fertile, moist and alkaline clay (Duke, 1983). Leucaena does not like poor drainage and waterlogging or acid soil that is low in calcium and/or high in aluminium and manganese (Middleton, 1995). The low tolerance to aluminium and the high calcium demand contribute to the poor growth on soils with a pH below 5. The performance is excellent on coralline or other calcareous sites up to pH 8. It can be found on alkaline soils, but the tolerance to salinity is low. Growth is poor when the mean annual temperature is below 20°C (Brewbaker, 1995). Low temperatures cause a decline in growth. Leucaena is susceptible to frost, which will kill the top growth. Regrowth occurs quickly from aerial stems (slight frost) or from the base (heavy frost). Cold temperatures depress the germination of the seeds. Young seedlings will be killed by heavy frost (Middleton, 1995). The plant is known for drought tolerance and endures a wide range of heat and desiccation. The success of leucaena in the tropics may be the result of their relative promiscuity (permissiveness) allowing nodulation with diverse indigenous rhizobial types (Bala et al, 2003).

Propagation and crop management

Plant breeders are crossing *Leucaena leucocephala* with *L. diversifoia* to develop psyllid-resistant and more cold-tolerant varieties. However, their feed quality is lower than that of *L. leucocephala*, and weight gain of steers may be better on leucaena affected by psyllids than on the hybrids.

Leucaena can be propagated by seed or with cuttings. Seedlings are usually transplanted from the nursery at an age of 3–4 months. They may also be transplanted bare-rooted, rolled in mud, if the soil moisture is adequate. Typical growth results in canopy closure in 4–6 months (Brewbaker, 1995). Some seedlings less than one year old will produce viable seed. The seeds are covered with a hard, waxy seed coat, so must be scarified before planting to ensure rapid and even germination. The seeds are either immersed in water at 80°C for 3–5 minutes, or in boiling water for 4–5 seconds (Middleton, 1995).

Inoculation with rhizobia is recommended for many soils, though rhizobia are almost universal in tropical soils. For forage, seeds should be sown 2.5–7.5cm deep. Planting should take place at the onset of the wet season. The crop soon produces a dense stand (Duke, 1983). The space between the rows depends on the annual rainfall. In areas with little precipitation, the spacings are increased.

Planting leucaena may also help to manage weed in the long-term fallow period for maize–cassava cultivation (Akobundu et al, 1999). Brucchid, which predate the seeds of leucaena, can be used to regulate the invasiveness of leucaena where it occurs as a problem (Raghu et al, 2005).

Leucaena is best suited to deep well-drained fertile soils of neutral to high pH; its deep root allows it to produce new leaf after shallow-rooted grasses have run out of moisture. Its leaf is killed by frost, but its height protects it from ground frosts and it shoots again with the onset of warm conditions.

Seed can be pelleted with lime for more acid soils. In dry regions, leucaena is planted as a crop 2–4cm deep in a fully cultivated seedbed in rows 5–9m apart. The rows are kept weed-free with herbicide and the inter-row area is kept weed-free through cultivation for the first year, as the legume seedlings are very susceptible to competition (Tropical Grasslands).

Leucaena can be cultivated as separately spaced trees, or as closely placed bushes. The bush growth pattern follows when plants are placed close together and are harvested frequently. The best harvest conditions are found when plants are separated by about 5cm; this configuration, with alleys between rows, leads to convenient harvesting with a silage harvester (Appropedia.org).

Production

Yield depends on climatic conditions. It takes several years before a plant is ready for full yield harvests (Prine et al, 1997). On 3–8-year-old trees the annual wood increment varies from 24 to 100m³/ha, averaging 30–40m³/ha (Duke, 1983). Oven-dry stem wood produced by leucaena harvested annually averaged 31t/ha/year over four growing seasons for 12 selected accessions. The average total biomass yields of 10 leucaena accessions in Gainesville, Florida, planted in 1979 and harvested January 1994 after 4 seasons growth from 1990 to 1993, was 76.9t/ha, or 19.2t/ha/year. The yields ranged from 43.1 to 102.2t/ha (Prine et al, 1997). Rahmani et al (1997) report an annual dry matter yield of 35t/ha.

Reported wood yields of leucaena are based primarily on experimental plantings and small plantations, and range from 7 to 77m³/ha/yr. In general, yields below 15m³/ha/yr are considered poor, indicating lack of adaptability, soil fertility or management. Yields above 30m³/ha/yr are very good, indicating good sites and management.

Unusually low yields have been reported (less than 1m³/ha/yr) in regions of Taiwan with low soil pH and cool temperatures. Exceptionally high yields exceeding 130m³/ha/yr) have been reported, but should be taken with great caution until experimental and mensurational techniques have been verified.

Yield increments of 90 sample plots in 20 Philippine dendrothermal plantations (average age = 2.5 years) ranged from 0.25 to 47.0t/ha/yr. This range is a reflection of a variety of sites, ages and management practices. Yield increments of the best 10 plantations ranged from 13 to 32t/ha/yr.

As these observations make clear, leucaena plantations must be established on good sites to attain high growth rates without soil amendments. Fertilizer applications improve leucaena yields on poor quality sites, just as they do for major farm crops, and must be considered long-term investments necessary to obtain maximum yields on poor quality sites. In many cases,

Source: William M. Ciesla, Forest Health Management International, Bugwood.org, www.invasive.org/browse/detail.cfm?imgnum=3948088

Figure 10.95 Leucaena fuelwood planting near Morogoro

yields will be greater for coppice regrowth, where weed competition no longer exists and existing root systems are well established.

Processing and utilization

Fuelwood and lumber harvests optimize at 3–8-year rotations when growth is not severely limited in some way. Trees coppice rapidly to 8m in one year (Brewbaker, 1995). In regions with frost, leucaena will be harvested annually or every second year. New growth will coppice from stumps if the plant is cut or frozen. If frost kills the top growth, the dead stems will stand for one season while the next growth occurs and both seasons' growth can be harvested at the same time during the following winter (Prine et al, 1997). The wood can serve as a biofuel for heat and electricity production using different conversion technologies.

Leucaena has the highest-quality feed of any tropical legume, and the potential to produce the highest weight gains. Steers can gain 300kg of live weight in a year with adequate leuceana, and irrigated leuceana has produced over 1000kg of live weight gain per hectare per year.

The plant can be used for such purposes as livestock fodder, green manure, firewood crops and soil conservation. Its seeds can be used as beads (jumbie beans). The crop can be raised as a charcoal source and for energy (1 million barrels of oil per annum from 120km²). Extract from its seeds has anthelmintic medicinal qualities in Sumatra, Indonesia.

Leucaena, when used as an anaerobic fermentation feedstock, undergoes, on a dry basis, 60 per cent conversion to methane-rich gas and carbon dioxide. The unconverted solids form the core of the anaerobic compost – the major ingredient of an organic fertilizer. The methane-rich gas product fuels an inexpensive, reliable supply of electricity and/or thermal/steam energy.

In many countries in developing areas that do not have indigenous energy sources but are prime areas for leucaena agro-energy development, methane-rich gas produced by anaerobic fermentation of leucaena would economically and efficiently supply reliable electric power systems.

This reasonable-cost, electric generation coupled with an ample labour force will encourage the establishment of new manufacturing facilities with low facility operations and maintenance, as well as significantly foster the use of vacant agricultural land, further creating job opportunities in agriculturally based economies.

Additionally, if leucaena is augmented with other available crop residues, the anaerobic digestion process then produces further quantities of valuable methane, carbon dioxide and organic fertilizer. Anaerobic fermentation eliminates the adverse environmental impact of burning and crop harvest residue decay, thereby reducing greenhouse effects due to methane released to atmosphere upon decomposition of tilled crops.

Internet resources

www.hort.purdue.edu/newcrop/CropFactSheets/leucaena.html
www.hort.purdue.edu/newcrop/duke-energy/leucaena-leucocephala.html
http://leaky.rock.lap.csiro.au/facts/leucext2-txt.html

Source: Tony Pernas, USDI National Park Service, Bugwood.org, www.invasive.org/browse/detail.cfm?imgnum=5280034

Figure 10.96 *Leucaena retusa* tree

Source: Prine et al (1997)

Figure 10.97 One-year-old leucaena logs, USA

Tropical Grasslands, www.tropicalgrasslands.asn.au/
pastures/leucaena.htm
www.horticopia.com/hortpix/html/pc3308.htm
Appropedia.org, www.appropedia.org/Leucaena_Wood_
production_and_use_9

LUPIN (*Lupinus* spp.)

Description

Lupin, often spelled lupine in North America, is the
common name for members of the genus *Lupinus* in
the legume family (Fabaceae). The genus comprises
between 200 and 600 species, with major centres of
diversity in South America and western North America
(subgenus Platycarpos) and in the (Mediterranean
region and Africa subgenus Lupinus).

Lupin originated from the Mediterranean region
and South America. Although the genus *Lupinus* is
quite large, with a large genetic variability, only a few
species are found to be agriculturally important. The
domesticated species considered here are the 'Old
World' species *L. albus* (white lupin), *L. luteus* (yellow
lupin) and *L. angustifolius* (blue lupin), which are of
Mediterranean origin, and the 'New World' species *L.
mutabilis* (pearl lupin), which is of South American
origin (Hondelmann, 1984). One of the largest
differences between the Old and New World species is
seed size. Old World plants have larger seeds, with a
diameter greater than 5mm and a weight of more than
60mg. New World plants have smaller seeds, with a
diameter less than 5mm and a weight of less than 60mg.

L. albus grows to a height of up to 1.2m. The stalk
and leaf stems are slightly silky. The stems have five to
nine egg-shaped leaves with a ciliated edge and a tip
that comes to a fine point. The racemes are 5–30 cm
long and almost stemless. The corollas are white with a
blue-violet tint. The plant's pods contain three to six
seeds each. The seeds are smooth and white with a
salmon/pink tint or with dark brown speckles.

L. angustifolius grows to a height of up to 1.5m. The
stalk is slightly silky. The stems have five to nine linear
spatula-shaped leaves. The racemes are 0.5–2.0cm long
and are almost sessile. The corollas are light to dark blue
with a violet tint. The plant's pods contain four to seven
seeds each. The seeds are smooth and variably coloured.

L. luteus grows to a height of up to 0.8m. The stalk
is hairy. The leaves are linear to ovate and pointed. The
racemes are 0.5–2.5cm long. The corollas are bright
yellow and sweet smelling. The plant's pods contain
four to six seeds each. The seeds are smooth and white
or white with brown to black speckles.

L. mutabilis grows to a height of up to 2m. The
stalk is more or less smooth to adjacently haired. The
stems have five to nine leaves which are long, ovate and
pointed with a ciliated edge. The racemes are 5–30cm
long. The corollas are blue and/or pink and white with
distinct yellow flecks. The plant's pods contain two to
six seeds each. The seeds are smooth and black, brown-
black, red-brown or white.

Elaboration on these descriptions of the plants is
found in Hondelmann (1996). All four of the *Lupinus*
species mentioned here are annual plants. Lupins are
able to fix nitrogen from the air and to dissolve
phosphorus and other minerals in the soil.

The value of lupins as a food source is seen in the
high protein contents of their seeds, with averages
ranging from 31 per cent for *L. angustifolius* up to 44
per cent for *L. mutabilis*. *L. albus* and *L. mutabilis* are
considered rich in fat/oil due to their 10 to 20 per cent
dry matter lipid contents (Hondelmann, 1996).

World production of lupin legumes is estimated at
1.5 million tonnes, coming mainly from Australia and
the Commonwealth of Independent States (Rehm and
Espig, 1991).

In 1872, a disease was first described that occurred
in Germany with animals that were fed the alkaloid-
containing seeds from lupins as food. The disease,
referred to as 'lupinose', was detected by the presence of
one of two effects, mainly in sheep. There was either
irreversible central nervous system damage or liver
inflammation caused by mycotoxicosis from *Phomopsis
leptostromiformis*. Both effects were often deadly. This
alkaloid poisoning was linked to the species *L. luteus*
and *L. angustifolius* (Hondelmann, 1996). The first
species that were low in alkaloid content began to
appear at the beginning of the 1930s.

Ecological requirements

L. luteus has the smallest number of ecological
demands. It prefers a slightly acidic to acidic sandy soil

or sandy loam soil, has a low water demand and can tolerate light frost.

L. angustifolius, on the other hand, is sensitive to drought and requires a high humidity, but has been shown to tolerate light to medium frost. It prefers a weakly acidic to neutral sandy loam or loamy sand soil.

L. albus prefers a slightly acidic loamy sand to loam soil. The plant is lime sensitive and has been shown to have the best growth of the four lupins on acidic soils. It is also slightly frost tolerant except during seed ripening. It is proved to behave as a long-day plant (Keeve et al, 1999).

Both *L. mutabilis* and *L. albus* have their highest seed yields in climates with warm summers. An advantage of lupin cultivation is that the plant can be grown on marginal land. In general, the recommended soil types are sandy to sandy loam soils which are in the slightly acidic to acidic pH range. The climate should be warm/mild and very sunny. Summers should not be too wet (Sator, 1979; Bramm and Bätz, 1988; Plarre, 1989).

Propagation

Lupins are most commonly propagated by seed. Here it is important that the seeds are of high quality and that the appropriate seed variety is selected according to the proposed use of the crop. It should also be ensured that the seeds are free of contaminants.

Crop management

To help maximize the yield and because lupins use nitrogen-fixing bacteria to obtain nitrogen it may be necessary to inoculate the lupin seeds with a strain of nitrogen-fixing bacteria if the bacteria isn't already present in the soil. Although the exact strain of bacteria depends on which type of lupin is being planted, it has been found that lupins are most commonly the host plants for *Bradyrhizobium* spp. and *Rhizobium loti* (Rehm and Espig, 1991). Before inoculation the seeds should be disinfected.

When preparing the field, although ploughing and harrowing are normal, it may be sufficient just to loosen the earth. Seed sowing/drilling in subtropical Mediterranean regions takes place in the autumn,

(a)

(b)

Source: (a) Europe House, www.europehouse.com; (b) Borealis55, http://commons.wikimedia.org/wiki/Category:Lupinus

Figure 10.98 (a) Lupins, Cutler ME; (b) lupin field in Russia

from the end of August to November, with the beginning of the rainy season. Seed sowing/drilling in temperate regions takes place in the spring. The suggested crop density for *L. luteus* and *L. angustifolius* is 80 to 90 plants per m² and for *L. albus* 50 to 60 plants per m². The distance between rows should be 18 to 25cm with a seed planting depth of 2 to 4cm (Bramm and Bätz, 1988).

Table 10.56 Recommended fertilizer rates for lupins

Fertilizer	Amount(kg/ha)
Phosphorus (P$_2$O$_5$)	60–90
Potassium (K$_2$O)	90–120
Magnesium (MgO)	25–35

Table 10.57 Lupin seed yield with respect to species

Species	Yield (t/ha)
L. albus	0.5–4
L. angustifolius	0.5–4
L. luteus	0.8–2.6
L. mutabilis	0.3–2

Source: Bellido (1984)

In Germany, the use of organic fertilizers and nitrogen fertilizers is not recommended. Suggested fertilizers and amounts are shown in Table 10.56.

Keeping weeds under control is very important, especially during the early phases of plant growth. Because of this a herbicide and mechanical weeding may be necessary. Pre-emergence herbicides have shown to be appropriate (Bellido, 1984).

The diseases and pests which attack lupin crops usually vary from region to region. Lupins in general have been found to be attacked by fungi and rusts such as *Fusarium* spp. and *Uromyces lupini*. *L. mutabilis* is often afflicted with anthracnose caused by *Colletotrichum gloeosporioides*. Viruses also play a role with lupin crops, in particular cucumber mosaic virus, which can cause significant damage to *L. angustifolius* crops.

Pests that are known to attack lupin crops include, among others, insect larvae from butterflies and flies, lice, root parasites, snails and rodents.

It is recommended that lupins be cultivated after cereal crops and before crops such as potatoes, turnips and cabbage. Legume crops should not be planted directly before or after a lupin crop. A lupin crop cannot follow another lupin crop in the crop rotation; in addition a cultivation break of 4 years for lupins should be adhered to (Bramm and Bätz, 1988).

Because of the presence of nitrogen-fixing bacteria from the lupins, the following crops will benefit from a significant amount of nitrogen remaining in the soil (Hondelmann, 1996).

Production

Lupins have a wide range of seed yields that are affected by the location and species of cultivation along with the form of crop management. This wide yield range is shown in Table 10.57.

Since the beginning of the 1990s, average annual Australian seed yields have been in the 1.0–1.2t/ha range (Nelson, 1993). Germany has had *L. luteus* seed yields ranging from 1.71t/ha to 1.93t/ha, depending on sowing density (Sator, 1979).

Observed yields from experiments in Australia were 2000kg/ha in 1995 and 1700kg/ha in 1996 and simulated yields were 2370kg/ha in 1995 and 1740kg/ha in 1996. Measured evapotranspiration in the period from 2 July to 15 September 1996 was 178mm and the simulated value was 147mm (Farre et al, 2003).

Processing and utilization

Harvesting in Mediterranean regions takes place in June and July with dead ripeness of the crop. The entire plant is harvested using a harvester (Hondelmann, 1996).

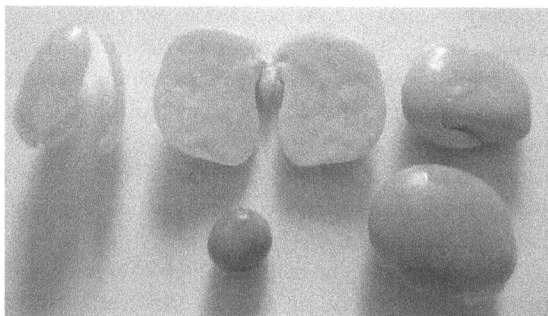

Source: Vi Cult, http://commons.wikimedia.org/wiki/File:Lupin_jaune,-peau,-lupin_sans_sa_peau,-petit_pois,-lupin_intact_(avec_peau).JPG

Figure 10.99 Lupin seeds

The above-ground portion of the plant can be used as fodder or hay and for green manuring. The seeds can also be directly used as animal feed and for human food (Hondelmann, 1996). Sweet lupins such as *L. luteus* and *L. angustifiolius*, which have no or low alkaloid contents, are planted if the crop is to be used as feed or food (Hoffmann et al, 1985). Lupin kernel meal can also be processed and used as fish dietary protein resources (Glencross and Hawkins, 2004).

The similarities in the amounts of raw nutrients of lupins and soya bean as fodder can be seen in Table 10.58. Table 10.59 shows the raw protein and fat contents of seed dry matter from lupin crops. The characteristic that makes *L. albus* and *L. luteus* of interest is the percentage of fatty acids with respect to the total fat content of the seed oil.

Research has shown that grain yields can achieve 2.5–6.0t/ha with an oil content from 8–20 per cent. This can be used as a biofuel source. The total biomass

Table 10.58 Nutrient contents of lupins and soya bean

Raw material (%)	L. albus	L. angustifolius	L. luteus	Soya bean
Dry matter	14.1	15.0	13.4	16.2
Raw protein (N × 6.25)	3.4	2.7	2.2	2.9
Raw fibre	3.5	3.7	4.4	4.4
Raw fat	0.3	0.5	0.4	0.6
Raw ash	1.0	2.0	1.2	1.8
N-free extract materials	5.9	6.1	5.2	6.5

Source: Kellner and Becker (1966)

Table 10.59 Raw protein and fat contents of lupin seed dry matter

Dry material	L. albus	L. angustifolius	L. luteus	L. mutabilis
Raw protein (%)	34–45	28–40	36–49	32–49
Raw fat (%)	10–16	5–9	4–9	13–23

Source: Plarre (1989)

Table 10.60 Fatty acid contents of *L. albus* and *L. luteus*

Fatty acid	L. albus	L. luteus
Oleic acid	52.6–60.6	23.8–39.1
Linoleic acid	16.5–23.4	45.0–48.8
Linolenic acid	2.5–8.1	0.9–7.6

Source: Sator (1979)

can reach 5 to 15t/ha. The energy balance is very adequate because lupin is a legume crop and does not need substantial nitrogen fertilization, and it is a very convenient crop for improving the physical characteristics of the soil.

MEADOW FOXTAIL (*Alopecurus pratensis* L.)

Description

The meadow foxtail (*Alopecurus pratensis*), also known as the field meadow foxtail, is a perennial belonging to the family Poaceae. It is native to Europe and Asia.

This plant is common on grasslands, especially on neutral soils. It grows to a height of about 110cm. The stem is erect or geniculate at the base. The leaves are about 5mm wide. Meadow foxtail has a cylindrical inflorescence, with glumes about 5–10mm wide and spikelets about 4–6mm long. It flowers from late spring to early summer and often again from late summer to autumn.

This species is widely cultivated for pasture and hay, and has become naturalized in many areas outside of its native range, including Australia and North America.

Meadow foxtail is a loose-growing perennial grass most commonly found on damp to moderately wet meadows and other similar locations. The flat, bare leaf blades are grooved and mid-green to dark green, with a glossy underside. The inflorescence is a 5–6cm long and 8mm wide panicle with four to six or more spikelets on each lateral branch (Siebert, 1975; Kaltofen and Schrader, 1991).

Source: (a) Richard Old, XID Services, Inc., Bugwood.org; (b) Rasbak, http://commons.wikimedia.org/wiki/File:Alopecurus_pratensis_Grote_vossenstaart_in_berm.jpg; (c) Richard Old, XID Services, Inc., Bugwood.org

Figure 10.100 Meadow foxtail plants

Ecological requirements

Meadow foxtail has a relatively high demand for nutrients and water. Pastures with an insufficient water supply are not very suitable for foxtail. Most appropriate are heavy to medium heavy, humus and nutrient-rich soils with good aeration. The plant can tolerate occasional flooding, snow cover, cold and late frosts (Kaltofen and Schrader, 1991).

Crop management

Meadow foxtail is propagated using seed. The seed is sown around the middle of May. A nitrogen fertilizer rate of three times 30kg/ha of liquid manure (NH_4-N) or three times 30kg/ha mineral fertilizer (N) can be applied.

A recommendation for a crop rotation including meadow foxtail has not been established, partly because the plant is perennial.

Production

Most of the seed production and use of this grass is in Oregon. However, it can be grown on any site on which creeping foxtail is adapted. Because of the light, fluffy seed and the indeterminate seed ripening, seed production is quite difficult. Seed should be harvested when seeds begin to shatter from the tips of the seed heads. Strippers can be run through the field every two to three days as the seed continues to mature. Combining from a windrow is the most common method of harvest. The air flow on the combine must be minimized by covering the openings on the fan housing or inactivating the fan. Seed must be run through a hammermill or debearder prior to clean-ing. Seed production is between 200 and 400 pounds per acre depending upon seed recovery success. Seed production responds significantly to nitrogen fertilization (Animal & Range Sciences, Extension Service).

Meadow foxtail dry matter yields consist of approximately 56 per cent leaves, 37 per cent stems and 7 per cent inflorescences (Siebert, 1975). Mediavilla et al (1993, 1994, 1995) provide the dry matter yields from meadow foxtail field tests that were conducted using liquid manure fertilizer at two locations in Switzerland: Reckenholz (R) and Anwil (A) (Table 10.63).

Source: http://wahlens.se/bildsidor/2008_05/stora/alopegurus_pratensis_ugg_080515STOR.jpg

Figure 10.101 Meadow foxtail

The crop was harvested three times in 1993 and twice in each of 1994 and 1995. At Reckenholz between 70 and 90kg/ha/year NH$_4$-N liquid manure was used, while Anwil used up to 180kg/ha/year NH$_4$-N. Generally speaking, the biomass yield of meadow foxtail in Europe ranges from 6 to 13t dry matter/ha/year (European Biomass Industry Association, 2007).

Processing and utilization

The crop is harvested by mowing. Success has been achieved with harvests of one, two and three times per year. The crop is then dried in the field to prepare it for storage or for economic utilization.

Meadow foxtail is among the grasses that are seen as suitable crops for the production of biomass for use as a biofuel. This involves burning the biomass for the production of heat (directly) and electricity (indirectly). The primary utilization of the crop is as fodder, especially in the early growth stages in the springtime. If it is cut later, the crop will have a higher lignocellulose content. This will negatively influence the quality of the crop as fodder, but improve the possibility of its utilization as an energy crop (pelleting,

Table 10.61 Quantity of *Alopecurus pratensis* seeds produced, yield and proportion of *A. pratensis* (*A. prat.*) in the sward in the different cutting regimes

Cutting regimes	Seeds (kg/ha/yr) Mean yr 1 to 5	Yield (t DM/ha/yr) yr 1	yr 5	A. prat. (%) yr 5
1st cut at early shooting (ESh)	16 b	11.1 a	11.3 b	50 b
1st cut at shooting (Sh)	10 b	11.1 a	10.8 ab	45 b
1st cut at early heading (EHe)	1 a	12.8 b	10.2 a	26 a
1st cut at seed maturity (SMa)	31 c	10.3 a	11.4 b	67 c

In a column, the means followed by a common letter are not significantly different at the 5% level by LSD (Huguenin-Elie et al, 2008)

Table 10.62 Overview on perennial grasses tested as energy crops including meadow foxtail in Europe and the reported yields

Common English name	Latin name	Photosynthetic pathway	Yields reported [t dry matter/ha/year]
Miscanthus	*Miscanthus* spp.	C$_4$	5–44
Switchgrass	*Panicum virgatum* L.	C$_4$	5–24
Giant reed	*Arundo donax* L.	C$_3$	3–37
Reed canarygrass	*Phalaris arundinacea* L.	C$_3$	7–13
Meadow foxtail	*Alopecurus pratensis* L.	C$_3$	6–13
Big bluestem	*Andropogon gerardii* Vitman	C$_4$	8–15
Cypergras, galingale	*Cyperus longus* L.	C$_4$	4–19
Cocksfoot grass	*Dactylis glomerata* L.	C$_3$	8–10
Tall fescue	*Festuca arundinacea* Schreb.	C$_3$	8–14
Ryegrass	*Lolium* spp.	C$_3$	9–12
Napier grass	*Pennisetum purpureum* Schum	C$_4$	27
Timothy	*Phleum pratense* L.	C$_3$	9–18
Common reed	*Phragmites communis* Trin.	C$_3$	9–13
Sugar cane	*Saccharum officinarum* L.	C$_4$	27
Giant cordgrass	*Spartina cynosuroides* L.	C$_4$	5–20
Prairie cordgrass	*Spartina pectinata* Bosc.	C$_4$	4–18

Source: Lewandowski et al (2002)

Table 10.63 Three years of meadow foxtail yields (t/ha) from field tests at two locations in Switzerland

	1993 (R)	(A)	1994 (R)	(A)	1995 (R)	(A)
Dry matter yield	10.3	13.3	7.1	9.5	8.7	7.5

gasification, fermentation for biogas and direct combustion for heat and electricity generation).

Meadow foxtail biomass (dry matter content 26.1 per cent) contains 17.2g crude protein, 4.6g fat, 21.5g crude fibre, 10.7g ash and almost equivalent to coal in energy content.

Internet resources

www.eubia.org/193.0.html
Animal & Range Sciences, Extension Service, www. animalrangeextension.montana.edu/Articles/Forage/ Species/Grasses/Meadow-foxtail.htm
Old, Richard, XID Services, Inc., Bugwood.org, www.invasive.org/browse/detail.cfm?imgnum=5225058

MISCANTHUS (*Miscanthus* spp.)

Partially contributed by Dr Qingguo Xi, Agricultural Institute of Dongying, Shandong, China

Description

Miscanthus is a genus of woody, perennial, tufted or creeping rhizomatous grasses that originated in East Asia. It is high in lignin and lignocellulose fibre and uses the C_4 photosynthetic pathway. Among all the 23 species (Liu, 1989; Xi, 2000) in this genus there are 3 species, *M. sacchariflorus*, *M. sinensis* and *M.* × *giganteus*, that are of great significance for energy purposes. Another species, *M. floridulus*, is a robust evergreen plant, which is important in warm regions and is considered to be a breeding material of energy plants.

Miscanthus sacchariflorus (Maxim.) Benth (*M. lutarioriparius* Liu ex S. L. Chen & Renvoize; *Triarrhena Lutarioriparia* L. Liu)

M. sacchariflorus is a perennial plant growing in temperate regions, with long creeping rhizomes covered with scale-like sheath blades. Culms are erect, 2–7.5m tall, 0.8–3.5cm in diameter near the base; culms are hollow or partly pith filled towards the apex; with or without wax powder on the surface, and with or without branches; usually making an extensive colony. The leaf

blade is 90–100cm long and 1.5–4cm wide with scabrous margins and somewhat glaucous underneath. The ligules consist of a fringe of minute hairs (cilia); auricles present or not; leaf sheath with or without hairs. The panicle consists of numerous, sub-digitally arranged racemes on a short axis; racemes are 20–35cm long, somewhat pendulous. The axis of the panicle is shorter than the racemes, i.e. shorter than half of the whole panicle. The spikelets are paired, one short and one long pedicellate, both alike, 4–6mm long and bearded at the base with white hairs, which are far longer than the spikelet. The glumes are sub-equal and faintly 3-nerved. The lower glume is as long as the spikelet and sparsely hairy on the edge, while the upper glume is shorter. The upper lemma is awnless; the lower lemma is shorter and

Source: Xi

Figure 10.102 *Miscanthus sacchariflorus*, China

nerveless. The anthers are 2mm long. The seeds are about 2mm long and dark brown.

The plants in this species are sharply diverse in heights and could be sorted as different forms. The taller forms have higher yields and are interesting for energy production. The tallest form is found in Middle-East China, named as *Triarrhena Lutarioriparia* firstly (Liu, 1989) and later renamed as *M. Lutarioriparius* (Chen and Renvoize, 2005). Yet the name *Miscanthus sacchariflorus* remains for general use up to now.

The plant in the tallest form of *M. sacchariflorus* is easily branching, especially when the top of the main stem is destroyed e.g. by insect or by wind. The normal height of the plants is 4–6m with a stem diameter of about 20mm. In contrast to *M. × giganteus* the tall

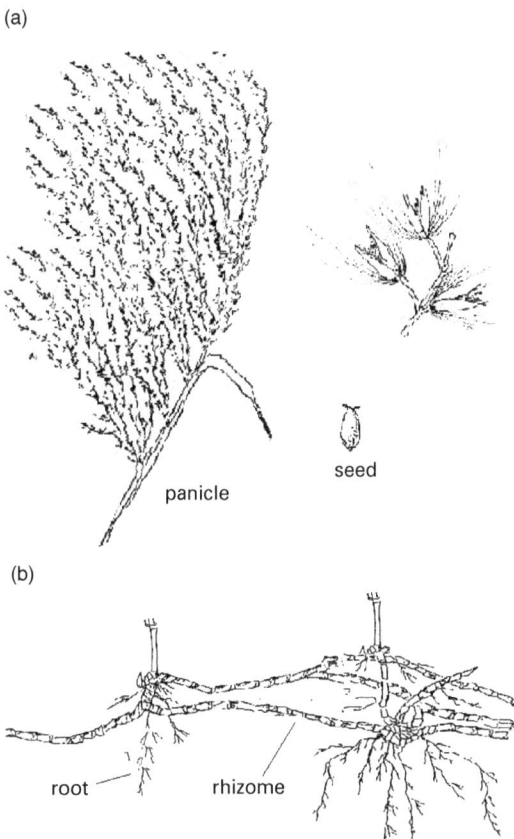

(a)

panicle

seed

(b)

root   rhizome

Source: Xi

Figure 10.103 (a) Inflorescence; and (b) rhizomes of *Miscanthus sacchariflorus* (M. × *giganteus* has similar inflorescence without seed)

plant of *M. sacchariflorus* has less leaves. This is because the older leaves on the lower part of the plant die off along with the plant growth. The strong and creeping rhizomes are distributed in the soil about 5–20cm under the ground surface. Roots can reach a depth of 1–2m or more. The flowering time is in September to October. The plants are self-infertile and the seeds ripen in November to December, when growing in populations. Though in warm areas flowering can occur in the first year of growth, more vigorous growth happens in the second year and thereafter.

Miscanthus sinensis Anderss

M. sinensis is a perennial grass growing on hillsides and field margins in temperate regions, with short rhizomes. Culms are densely tufted, erect or half-erect 0.5–3.3m tall and 3–7mm in diameter near the base. Leaf blades are 20–70cm long and 0.6–1.2cm wide with very rough margins; the midrib is white. The ligules are conspicuous, short ciliate on upper margin. The panicle has 10–25 racemes which are 10–30cm long, slightly nodding; the central axis shorter than the racemes, i.e. shorter than half of the whole panicle. The spikelets are paired, one short and one long pedicellate; both alike, 4–7mm long, bearded at the base with white hairs, which are as long as the spikelet. The glumes are equal and as long as the spikelet; the upper 3-nerved and the lower 1–nerved. The upper lemma is deeply bifid, awned between the teeth; the awn is 8–10mm long, exserted and geniculate. The lower lemma is membranous, hyalin. The anthers are 2mm long. The seeds are purplish brown or dark brown. The flowering and seed-ripening time is from July to December.

This species is distributed widely in East Asia. It is easily distinguished from *M. sacchariflorus* with their awns on spikelets, but in the southern regions it is sometimes confused with *M. floridulus*, because there are frequently transition forms between *M. sinensis* and *M. floridulus*, perhaps arising from natural crossing.

Miscanthus floridulus (Lab.) Warb. ex Schum. et Laut.

M. floridulus is a robust perennial plant growing in warm temperate, subtropical and tropical regions, with thick and short rhizomes. The culms are pith filled, 2.0–4.1m tall, 8–10 mm in diameter near the base,

Source: Xi

Figure 10.104 *Miscanthus sinensis*, China

densely tufted in large clumps. The surface of the culms, especially at the base of the leaf blade, is covered with wax powder. The leaf blade is 30–100cm long and 1.5–3.5cm wide, with very rough margins and a white midrib, glabrous except for the basal tomentose portion. The ligules are tall, with short fimbriate on the upper margins. The radical leaves usually remain green through the winter. The panicle is 30–50cm long and has numerous racemes which are 10–20cm long, evidently shorter than the central axis, i.e. the central axis is relatively thick and longer than two-thirds of the whole panicle. The spikelets are paired, one short and one long pedicellate, 3–4mm long and bearded at the base with white hairs, which somewhat longer than the spikelet. The glumes are alike, as long as the spikelet, glabrous or short hairy on back. The upper lemma is deeply bifid, awned from the sinus the lawn is

8–10mm long. The lower lemma is hyalin. The anthers are 2mm long. The seeds are brown. The flowering and seed-ripening time is from May to October.

M. floridulus is an evergreen plant and is distributed widely towards the south, often dominant in its habitats. It is more robust than *M. sinensis*.

Miscanthus × giganteus Greef et Deu. (*M. sinensis* 'Giganteus')

M. × giganteus is a triploid perennial plant, with thick and stout rhizomes. The culms are 2.5–3.5m long; the nodes without hairs. Root promodia and aerial branches at lower nodes sometimes observed. The leaves are cauline; the leaf blades linear, >50 × 3.0 cm; the ligule truncate with hairs (2–3mm). The inflorescences are about 30cm long; the rachis of the panicle 15cm long. The spikelets are 4–6mm long; longest pedicel 4mm long. The three anthers are 2mm long. The callus hairs are about twice as long as the spikelet. The first glume is 5–6mm long, 1.5–2mm wide, faintly 2-nerved, with hairs on the back, 3–6mm long. The second glume is 4–5mm long, 1.5–2mm wide, 3-nerved, with hairs on back; hairs shorter than on the first glume. The upper lemma is not awned, 3–3.5mm long, <1mm wide, nerveless, filiate haired at the edge. The lower lemma is 3–3.5mm long, 3-nerved, sometimes the lower (fertile) lemma is longer than 3.5mm with an additional soft awn, which is more than 1mm long. Palea is present, 1mm long. Lodicule is present. No seed is produced. The flowering time is between September and November.

M. × giganteus is a natural hybrid between *M. sacchariflorus* and *M. sinensis* and it was originally introduced to Europe as an ornamental garden grass (Greef and Deuter, 1993). This hybrid is no longer found in nature in its origin Japan but has been re-introduced into East Asia recently (Xi, 2008). The erect stems are slim yet vigorous and are not usually branched. The solid pith stems are approximately 10mm in diameter and can reach a height of a little over 2m the first year and up to 4m each following year in Europe (El Bassam, 1994b). The lower height in the first year is a result of a great amount of the plant's energy being used to establish its extensive root and rhizome system. The rhizomes make up a highly branched storage system. The roots usually penetrate down well over one metre into the soil. Although much of the underground growth takes place in the first year, the crop usually does not reach maturity until after 2–3 years.

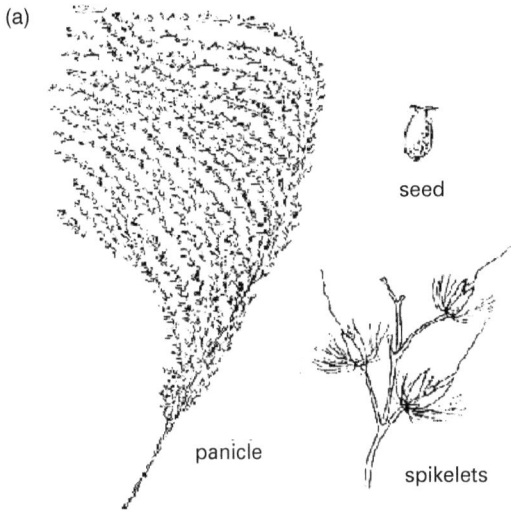

(a)

seed

panicle

spikelets

(b)

root

rhizome

Figure 10.105 (a) Inflorescence; and (b) rhizomes of
Miscanthus sinensis (M. × *giganteus* has similar
rhizomes)

Breeding

The genus *Miscanthus* has a genome of n=19. Most
miscanthus gynotypes are diploids (*M. sinensis*; *M.
sacchariflorus* from China) or tetraploids (*M. sacchariflorus*
from Japan). Other heteroploids or aneuploids are also
possible (Liu, 1989). *M. × giganteus* has is triploid and is
considered to be a hybrid of *M. sacchariflorus* and

Figure 10.106 *Miscanthus floridulus*, China

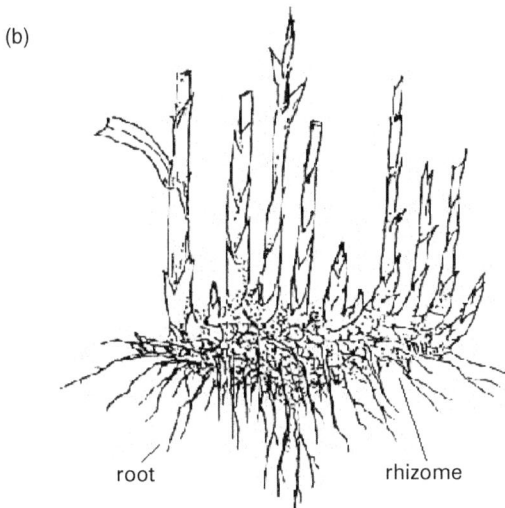

Figure 10.107 *Miscanthus × giganteus*, Germany

M. sinensis from Japan (Greef and Deuter, 1993). Miscanthus breeders are now endeavouring to get new hybrids similar to *M. × giganteus* by crossing. Such hybrids should have advantages of both *M. sacchariflorus* and *M. sinensis* or other pairings of species. Miscanthus species are self-incompatible but cross easily among each other; the problem is mainly the meeting of different flowering times of different species or eco-types. In nature miscanthus genotypes are normally fertilized only when they grow in a population and have the same or overlapped flowering times.

Antimitotic treatments with colchicine or oryzalin were tested in *M. sinensis* to develop an efficient chromosome doubling method for *M. sinensis* to enable the production of triploids and so avoid seed dispersal to the environment (Petersen et al, 2002).

It was reported in China that a new high-yield tetraploid of *M. sacchariflorus* (*Triarrhena lutarioriparia*) had been created by colchicine treatments (Untitled, 1994).

Ecological requirements

Overall, miscanthus is a group of highly resistant plants against disadvantageous ecological factors. It has evolved in regions of the world that have large temperature fluctuations between summer and winter. While some miscanthus species grow well in habitats under environmental stresses in Taiwan (Chou et al, 2001), others are distributed far north to Siberia. This evolution has led to the development of characteristics that make the plant resistant to heat, frost, drought and flood, though their biomass productivities may vary in different conditions, correlated with different species and genotypes.

Temperature, overwintering and frost tolerance

The overwintering ability of the plants depends on the cold tolerance of their rhizomes. An artificial freezing test with rhizomes removed from the field in January 1998 showed that the lethal temperature at which 50 per cent were killed (LT_{50}) for *M. × giganteus* and *M. sacchariflorus* genotypes was −3.4°C. However, LT_{50} in one of the *M. sinensis* hybrid genotypes tested was −6.5°C and this genotype had the highest survival rates in the field in Sweden and Denmark. Moisture contents correlated with frost hardiness (LT_{50}) in most cases (Clifton-Brown

and Lewandowski, 2000). In soil condition of rich humus in Germany *M. sacchariflorus* can overwinter under lowest air temperature to −21°C without problem (Xi, 2000). Some miscanthus genotype has proved able to do well in the Great Lakes region of Canada, where for several weeks during the winter the mean temperature can be as low as −40°C (Green, 1991).

With the presence of a thick bud around the growing point, and the latent buds on the rhizomes still being under the soil, miscanthus is able to tolerate light spring frosts (Rutherford and Heath, 1992). At temperatures below −5°C the developing shoots and leaves are destroyed.

The first occurrence of frost represents the end of the growing season for miscanthus (Bunting, 1978). It is around this time that crop senescence is accelerated, nutrients are sequestered into the rhizomes and the plant begins to dry out.

The growing season temperature has a large effect on crop yield for miscanthus. This is because miscanthus uses the C_4 photosynthetic pathway, which is more efficient at high temperatures and light intensities. Although the crop prefers warmer climates it has been shown that miscanthus can be grown with favourable results throughout Europe. No optimum growing temperature or range has been determined, yet experiments have shown strong interaction between environment and genotypes. Trials show that in northern regions of Europe, *M. sinensis* hybrids have high yields of up to 25t/ha, but non-hybrid genotypes of *M. sinensis* have higher combustion qualities. In mid- and south Europe, *M. × giganteus* (yielding up to 38t/ha) or specific high-yielding *M. sinensis* hybrids (yielding up to 41t/ha) are more suitable for biofuel production (Clifton-Brown et al, 2001; Lewandowski et al, 2003).

Wind and lodging

Wind also plays a role in the success of miscanthus crops. Before the extended stems can lignify there is the chance of strong winds leading to lodging and/or damage to the leaves. Sheltering the crop from wind also reduces crop cooling, but it can also be responsible for hindering plant growth (Bunting, 1978).

Water

Generally speaking, *M. sacchariflorus* prefers more humid condition and is more tolerant to water flooding than *M. sinensis*. *M. × giganteus* is situated between *M. sacchariflorus* and *M. sinensis* because of its hybrid property. *M. floridulus* has less cold tolerance.

In Japan it was observed that *M. sacchariflorus* produced more dry matter in wet areas where the water is 50mm from the surface than in dry areas where water is 150mm from the surface. Although the wet plot crops produced more shoots, they also showed a lower number of rhizomes compared to the dry plots. The lower number of rhizomes was attributed to flooding. In the long run, the insufficiency of rhizomes will probably lead to the crop's inability to maintain itself (Yamasaki, 1981).

Different genotypes may have different water use efficiency. It is found that Stay-green genotypes, against rapid- and slow-senescing genotypes, appear to make the most effective use of available water and are likely to be important for further breeding of miscanthus (Clifton-Brown et al, 2002).

Soil

Miscanthus does not make many demands on the soil, a fact demonstrated by its ability to grow on many types of arable land. Sands and sandy loams consisting of up to 10 per cent of clay have shown to be the preferred soil types in Denmark (Knoblauch et al, 1991). Success on sands and very stony soil is dependent on sufficient rainfall. A good yield has also been experienced on well-drained soils with high humus.

In order for the crop to establish itself in April and May it is necessary that the soil be sufficiently aerated and have a fine tilth, thus making soils of greater than 25 per cent clay probably unsuitable. The soil must also be able to support farm machinery during the harvest. This makes wet, peaty soils inappropriate for miscanthus crops. Soil with 70 to 95mm of available water per 500mm depth of soil is recommended for miscanthus (El Bassam et al, 1994b).

Miscanthus is a deep-rooting crop with roots going down more than 1.0m. This leads to the crop's preference for deeper soils such as on plains and in valleys. Results in Japan have also shown that *M. sinensis* grows better on the deep humus of a concave slope (Numata, 1975).

Soil texture, colour and pH can also affect miscanthus growth rate. Maize has displayed an increased growth rate as result of light textured, dark soil's ability to warm up rapidly (Bunting, 1978). A darker soil colour

and lighter texture is also believed to assist in a faster miscanthus growth rate. In China *M. sacchariflorus* grows mostly in dark, acid soils of lowlands in nature, but it also develops well in yellowish loess soils with pH 7–8 (Xi, 2000). Data from Denmark and the UK suggest that the optimum pH range is between 5.5 and 7.5 for growing miscanthus (Knoblauch et al, 1991). This excludes soils that are very acidic or very chalky as potential supporters of optimum miscanthus growth.

Saline soils are not preferred by miscanthus plants (Xi, 2000), though some genotypes of *M. sacchariflorus* can tolerate salt content up to 1 per cent (w/w), but the plant growth and biomass yield are decreased.

Symbiotic bacteria

Anaerobic nitrogen-fixing consortia consisting of N_2-fixing clostridia and diverse non-diazotrophic bacteria were previously isolated from various gramineous plants. A group of clostridia were found to exist exclusively in *M. sinensis* (Miyamoto et al, 2004).

When the grass *M. sinensis* was inoculated with an anaerobic nitrogen-fixing consortium (*Clostridium* sp. and *Enterobacter* sp., a functional community of bacterial endophytes), it was more tolerant to salinity stress than uninoculated plants. So it is concluded that some endophytic bacteria enhance the salinity tolerance of the host plant (Ye et al, 2005).

Propagation

Seed

Although seed may prove to be the cheapest form of plant propagation, it has the limitations of seed heterozygosity and availability. In addition, some species do not produce viable seeds. *M. × giganteus* has no seeds because of its triploid property. For other species the seed's longevity has been tested. The results show that chilling and long-term storage do not seem to have negative effects on the viability of fertile seeds, but moisture decreases the viability. The germination ability of the seeds stored in ambient conditions for 6 months was reduced drastically. No germination was observed after storing in ambient conditions for periods of 12 months or more. The germination ability of seeds stored in a refrigerator for up to 24 months was not affected. It was concluded that miscanthus seeds might lose their germination ability 6 months after being dispersed by the wind under natural conditions in Taiwan (Hsu, 2000). In cool regions miscanthus seeds may survive longer. Seeds of *M. sacchariflorus* germinate well up to 24 months after ripening.

Studies testing the effects of gibberellic acid, pH and temperature on the germination of seeds revealed no substantial preferences. But the results suggested that pH extremes should be avoided and that the temperature range 20–30°C may be best (Aso, 1977). Increasing water stress has been proved to decrease the growth rate of seedlings (Hsu, 1988).

The sowing depth must be very shallow because of the small size of the miscanthus seeds. Maintaining humidity is important for the germination and the survival of young seedlings. In normal conditions (20–30°C) seeds germinate 7–14 days after sowing.

Division

Propagation by division is a slow process, but it is simple and quickly leads to a large plant. *M. Sinensis* 'Gracillimus', *M. sinensis* 'Variegatus' and *M. sinensis* 'zebrinus' do not produce viable seeds, so plant division is used for their propagation. Stock plants are grown, from which 5–7.5cm-long sections are cut in the spring. The sections are given a covering of slow release fertilizer and are ready for sale after three to six months (Gibson, 1986). Similar methods can be used for *M. × giganteus*.

Stem cuttings

Mature, non-flowering stems are used to obtain tip and basal cuttings with two or three nodes. The cuttings are then planted. The use of stem cuttings has gained success with many perennial grass species. It is especially easy with *M. sacchariflorus* (Xi, 2000), but little success has been achieved with other *Miscanthus* species (Corley, 1989).

Rhizome cuttings

Rhizomes form an intertwined matrix 10–15cm below ground. This form of propagation has the advantage that more potential plants can be produced from a single plant, but it takes longer to achieve a full-grown plant. It is presently recommended that in November, stock plants two to three years old be used for dividing up the rhizomes

into lengths of 8–10cm. Shorter rhizome lengths can be cut, but they are more susceptible to winter kill. The cuttings can be stored between −1°C and 1°C. Tests have had success with cuttings planted in the beginning of May at a depth of 3–6cm and a density of 10,000 rhizomes per hectare (Nielsen, 1987). Data from the UK point to *M. sacchariflorus* as having the best success rate for propagation from rhizome cuttings, though plantations from rhizome cuttings in Denmark have suggested a 65–75 per cent success rate. With respect to *M. × giganteus*, Huisman and Kortleve (1994) explain that farmers can grow and harvest rhizomes using a rotary cultivator to chop up the above-ground plant, then a flower bulb harvester to harvest the rhizomes. Pot-grown transplants, 30–35cm high, possessing at least one shoot and noticeable robust roots, have repeatedly provided 100 per cent establishment (Knoblauch et al, 1991).

Micropropagation and tissue culture

This method of propagation has the advantage that a large number of plantlets can be produced in a short period of time. In addition, this method can be easily converted into large-scale production (Rutherford and Heath, 1992). At Piccoplant Laboratories in Germany, *M. sinensis* has been used to develop a micropropagation system in which meristem tissue is removed from selected plants, then grown in culture. Field tests have indicated that the micropropagated plants are stronger than other propagated plants. Piccoplant has suggested that they could produce 2 million plantlets per year. In the USA, immature inflorescences have displayed positive enough results to suggest that they could be used for tissue cultures (Gawel et al, 1987).

Pre-culture and transplanting

For various propagation methods the plants are normally pre-cultivated in protected conditions for 2–3 months, when the plantlets reach a height of about 30cm, and then transplanted onto the field. To make the work easier plug cells could be used before transplanting.

An experiment was conducted to determine the optimum plug cell size and medium composition for germination and seedling growth of *M. sacchariflorus* for mass propagation. The results show that seed

Source: Xi

Figure 10.108 *M. sacchariflorus*, propagated with rhizomes and cultivated in North China

germination percentage was highest in a medium of peat moss + perlite at 1:1 ratio (v/v) in 200 plug cell trays. After germination the seedlings grew well and formed branches (Lee et al, 2005).

Crop management

Transplanting

Crop management studies have concentrated on *M. × giganteus* in Denmark (Knoblauch et al, 1991) and Germany (Sutor et al, 1991). It is recommended that young plants and stored rhizomes be planted when the planting depth temperature of the soil is 10°C or higher. This corresponds to approximately late April to early June. The important factors are sufficient soil moisture, a fine tilth, and the avoidance of young plant destruction from frost.

Rhizome cuttings have been successfully planted using machines designed for planting seed potatoes, while forestry and vegetable transplanting machinery have demonstrated positive results when planting young miscanthus plants.

The results of experiments in Denmark recommend that for transplanting, young plants should have a height of at least 30–35cm. The plants should also have vigorous roots and a minimum of one shoot. For planting rhizomes, it is recommended that lengths of at least 10cm be used for rhizome cuttings, though

under favourable conditions lengths as small as 5cm have been successful. Here a planting depth of 5–7cm has been recommended. For planting pot-grown plants it is suggested that the tops of the pots be covered with 2cm of soil.

Germany and Denmark have developed slightly different recommendations concerning plant spacing of *M. × giganteus*. The suggested spacing from Germany is 0.7–1.0m between plants and 0.8–1.0m between rows. Tests using spacing of one, two and three plants per m² have also shown that higher crop density results in higher dry matter yield. One exception was found, in that *M. × giganteus* provided the highest second-year yield at a spacing of two plants per m² (Kolb et al, 1990). The suggested spacing from Denmark is 0.75m between two rows and 1.75m between these groupings of two rows, with 0.8–1.0 plants per m².

In the UK, planting of miscanthus takes place in March or April using hand-planted micropropagated plantlets or rhizome pieces. Although new shoots are quick to emerge after planting, the period until May or June is marked by slow growth. With the more favourable temperatures of summer come rapid growth and the development of stalks of over 2m in the first year and 4m in the following years (Heath et al, 1994).

Weed, disease and pest control

Weed control is an important factor, especially during the establishment and first two years of the crop. It is recommended that before planting the field is completely cleared of all perennial weeds. Because miscanthus is a perennial crop that can live for 15 years or more, a thorough pre-planting weeding is important, because once the crop is in place weed clearing will be greatly hindered. The predominant problem weeds are wild oat (*Avena* spp.), creeping thistle (*Cirsium arvense*) and couch grass (*Elymus repens*), in Germany and Denmark. A number of herbicides have proved successful, including atrazine, fluroxypyr, mecoprop, propyzamide and some sulfonyl ureas.

Diseases that attack or are likely candidates for attacking miscanthus in its native habitats can be grouped into Uredinales (rusts), Ustilaginales (smuts), Sphaeropsidales, Clavicipitales, Hyphomycetes, Peronosporaceae (downy mildews) and Pythiaceae. Miscanthus has been shown to be attacked by one virus disease, miscanthus streak virus (Rutherford and Heath, 1992). In the areas where miscanthus has been relocated it has not yet shown much sign of being attacked by pathogens. Miscanthus has already proven itself to be quite resistant against diseases, but the more miscanthus is bred and grown throughout Europe and the Mediterranean region, the greater the chance of it being attacked by diseases already common to these areas.

Many pests on maize, sorghum and rice are also dangerous for miscanthus in its native vegetation. The most harmful insects on *M. sacchariflorus* in China are stem borers (*Chilo* spp., *Sesamia* sp., etc.). Application of chemical pesticides did not solve the problem but made it severe, because it killed the natural enemies of the borers. Biological control is especially suggested for miscanthus stem borer control in its natural vegetation.

Fertilization

Little research has yet been completed on the nutrient requirements of miscanthus, but some calculations have been made in order to determine the amounts of nutrients that would need to be replaced in order to maintain a proper soil balance (Table 10.64).

Because of the low nutrient requirements of miscanthus it is believed that the soil and atmosphere will be able to supply/return much of the nutrients, but that the addition of nitrogen, phosphorus and potassium may at times be necessary. It may also be necessary to add magnesium on sandy soils with less than 3–4mg/L magnesium concentration. It seems that 50kg N, 21kg P_2O_5, and 45kg K_2O/ha are sufficient to support adequate yields.

Table 10.64 Nutrients needed to maintain a proper soil balance for miscanthus

Nutrient	Amount (kg/ha)
Nitrogen (N)	50
Potassium (K_2O)	45
Phosphorus (P_2O_5)	21
Sulphur	25
Magnesium	13
Calcium	25

Source: Rutherford and Heath (1992)

The best time for nutrient application is in the spring, before the new growth season but after the previous harvest. Applicators for liquid or granular fertilizers are believed to be adequate.

Irrigation

Water plays a large role in crop yield. During the crop's growing season it is estimated that 600mm of precipitation would be necessary to produce 20t/ha dry matter yields. Greater yields become possible with increased irrigation. To provide for adequate establishment and first-year growth of young plants, irrigation has been deemed necessary in Germany and Denmark (El Bassam et al, 1994b).

Production

Yield

There was no significant difference in the yield between plants established from rhizome cuttings or by micropropagation (Clifton-Brown et al, 2007), but the yields of miscanthus are found to vary considerably according to site and climate, with the highest yields being recorded in southern European sites when water was not a limiting factor. Yields of over 24t/ha dry matter were recorded in Portugal, Greece and Italy in the third year (all with irrigation) in comparison with third-year yields of 11t/ha in Ireland, 16t/ha in Britain and 18.3t/ha in northern Germany. It has been reported that in Portugal a hybrid of *M. sinensis* produced 40.9t/ha dry matter after the third year of growth (Clifton-Brown et al, 2001). In China, yields up to 40t/ha and more (air-dry matter, equivalent to dry matter of about 28–36t/ha) could be achieved with *M. sacchariflorus* (*M. lutarioriparius*) in their natural vegetation on lower lands. There the plants are harvested for the paper-making industry. The yields in Ireland and Britain compare very favourably with the annual dry matter production from short rotation coppice, which is grown as an energy crop in these countries.

Harvest

Good biomass combustion quality depends on minimizing moisture, ash, K, chloride, N and S. Delaying the harvest by three to four months improves the combustion quality by reducing ash, K, chloride,

N and moisture (Lewandowski et al, 2003). Although the delayed harvest also decreased mean biomass yields, harvesting of miscanthus should be performed in February/March, roughly late winter/early spring. This is the time at which the highest dry matter content, up to 80 per cent, can be achieved and much of the nutrients are already sequestered in the rhizomes. Machinery for harvesting must be used carefully so as to limit compaction and damage to the underlying rhizome bed. Currently, mechanical harvesters are being studied to determine what form of machinery will work best. One promising option is a modern sugar cane harvester that does not remove the leaves as conventional harvesters do.

Processing and utilization

For energy use

At present, modifications to machinery are necessary for the handling of the miscanthus crop. Several general handling methods are presently seen as possible: pelleting, briquetting, baling and wafering. For the production of pellets and briquettes the crop is cut, then the biomass is immediately processed in the field or taken to a processing facility where it is highly compressed into pellets or briquettes. Baling consists of cutting and windrow drying, then using a large baler to produce large dense crop bales. Conventional baling produces a bale compressed to $120kg/m^3$ while in Germany a new technique can produce a density of up to $450kg/m^3$. Wafering takes the windrowed miscanthus and compresses it into a wafer. Wafering has the advantage of creating a denser final product than conventional baling, but at a significantly higher operational cost.

Drying and storage are possible, but have been little researched. It is believed that after harvest the product can be covered and left outside or moved to a storage facility, or the excess heat from combustion of already dry crop matter can be used for crop drying, leading to long-term storage. The important factor with storage is the moisture content. Jørgensen and Kjeldsen (1992) explain that big bales can be stored at 25 per cent moisture content, but that denser bales must have moisture content below 18 per cent.

To minimize costs it is best to process and utilize the crop as much as possible on the farm where it is grown. The main goal for the utilization of miscanthus is whole plant compaction to produce a solid fuel, with gasification as a large-scale option. Compaction can involve in-field or stationary pelleting or briquetting of the crop, or further compaction of bales. Compaction makes the product both easy to transport and ready for combustion for heat or electricity generation. These are possible directly on the farm, thus cutting costs. The solid fuel could also be transported to larger user facilities. The compact biofuel can be burned in individual homes, communities, or industries to produce heat, or regionally and nationally to generate electricity.

The energy density of miscanthus is 18.2MJ/kg. Miscanthus can be considered as one of the most important future energy crops due to its high yield potential under different climatic conditions, and can be converted to wide varieties of energy feedstocks for heat, electricity generation through combustion, gasification and liquefaction. Miscanthus biomass contains extremely low ash and nitrogen contents (El Bassam, 2008b).

For other purposes

Miscanthus plants have good fibre length and quality that make it possible to use them for different purposes, e.g. for paper-making, for plank and packing block production and for thatching (Table 10.65)(Xi, 2000).

In China large areas of natural miscanthus vegetation (*M. sacchariflorus*/*M. lutarioriprius*) have been harvested by hand using a sickle up to now. This is partly because the wetland conditions are not easy for machinery work. The mass products are left in the field for air-drying and then transported by ship to paper-making factories. Miscanthus mainly has been used as paper-making material in China up to now.

It was reported that *M. sinensis* (Chinese reed) could be used for edible fungi cultivation. This method could easily be expanded to other giant grass species (Xi, 2000).

An experiment was conducted to compare physico-chemical and physico-technical properties of *M. sinensis* to other materials like pine and coconut fibres and recycled polyurethane mats. The result indicated

Table 10.65 Chemical composition and fibre properties of miscanthus culms

Species	Cellulose (%)	Hemicellulose (%)	Lignin (%)	Fibre length (mm)	Ash (%)
M. sacchariflorus (*M. lutarioriparius*)	46–53	21.21–24.91	18.62–20.04	1.6–3.11	2.16
M. sinensis	42.02**				3.46
M. floridulus	39.91				5.36
*M. × giganteus**	68–70***		21–24	1.46	3

*cut-pieces of whole plant. **crude fibre content, sample on 29 September. ***holocellulose content.
Source: Xi (2000)

that miscanthus fibres could be an environmentally sound organic substrate for soil-less culture (Li et al, 2002).

Impact on environment

Carbon emission mitigation

Biomass crops mitigate carbon emissions by both fossil fuel substitution and sequestration of carbon in the soil (Clifton-Brown et al, 2007). In gross figures, 1t biomass consumption could mitigate 0.5t carbon emissions by using fossil fuels. Policies to promote the utilization of biomass crops require yield estimates that can be scaled up to regional, national and continental areas. The only way in which this information can be reliably provided is through the use of productivity models (Clifton-brown et al, 2004).

A model (MISCANMOD) coupled with a geographic information system (GIS) environment is used to estimate the contribution that miscanthus could make to projected national electricity consumption, and thus to the purpose of carbon mitigation. If miscanthus was grown on 10 per cent of suitable land area in the European Union (EU15), 231TWh/yr of electricity could be generated, which is 9 per cent of the gross electricity production in 2000. The total carbon mitigation could be 76Mt C/yr, which is about 9 per cent of the EU total C emissions for the 1990 Kyoto Protocol baseline levels (Clifton-brown et al, 2004). Using the same scenario, if miscanthus was grown on 10 per cent of the arable land in set-aside in the European Union (EU25), 282TWh/yr of electricity could be generated. This, corresponds to about 91Mt C/yr of carbon mitigation, would meet 39 per cent of the EU-25 target of 723TWh/yr of electricity from renewable energy sources (RES) by 2010 (Stampfl et al, 2007).

(a)

(b)

Source: Xi.

Figure 10.109 *Miscanthus sacchariflorus:*
(a) harvesting; and (b) shipping in South China

Miscanthus crops add soil organic matter (SOM) by pre-harvest losses and harvesting residues. It was reported that the potential supply to SOM could be

3.1t/ha carbon annually accumulated by the litter, and 9.1t/ha carbon accumulated by rhizomes and roots in the long term (Beuch et al, 2000).

Established miscanthus stands are able to produce about 8.2t/ha organic substance, which is comparable with farmyard-manure (FYM) in terms of SOM impact. This kind of calculation showed higher values for miscanthus than for the agricultural crops investigated to date. An SOM increase of about 0.5 per cent on sandy soils and 0.2 per cent on silt soil was determined after 6–8 years of cropping *Miscanthus × giganteus* (Beuch et al, 2000).

Miscanthus-derived soil organic carbon sequestration detected by a change in ^{13}C signal was 8.9 ± 2.4t C/ha over 15 years. It is estimated that total carbon mitigation by this crop over 15 years ranged from 5.2 to 7.2t C/ha/yr depending on the harvest time (Clifton-Brown et al, 2007).

Nitrate leaching mitigation and water purification

Experiments show that miscanthus, once established, can lead to low levels of nitrate leaching and improved groundwater quality compared with growing arable crops (Christian and Riche, 1998).

Investigation of matter production and water-purification efficiency, with grasses including *M. sinensis*, was conducted by floating culture on natural water and provided a positive result (Hirose et al, 2003).

NEEM TREE (*Azadirachta indica* A. Juss.; syn. *Antelaea azadirachta* (L.) Adelbert)

Description

The neem tree *Azadirachta indica* A. Juss. (Meliaceae), is a tropical evergreen related to mahogany. Native to east India and Burma, it grows in much of Southeast Asia and West Africa; a few trees have recently been planted in the Caribbean and several Central American countries, including Mexico. The people of India have long revered the neem tree; for centuries, millions have cleaned their teeth with neem twigs, treated skin disorders with neem-leaf juice, taken neem tea as a tonic, and used neem leaves to repel insects. Trees can reach up

Source: www.hort.purdue.edu

Figure 10.110 Neem tree nursery with 216 trees in a 7 × 7m square planting pattern in Yaqui Valley, Sonora, Mexico

to 30m tall and a trunk diameter of 30–80cm with limbs 15–40cm in diameter. The shiny dark green pinnately compound leaves are up to 30cm long. Each leaf has 10–12 serrated leaflets that are 7cm long by 2.5cm wide. It can be found at altitudes up to 1500m and occurs naturally within the latitudinal range 10–25°N (Gliese and Eitner, 1995, Muñoz-Valenzuela et al, 2007).

Ecological requirements

The optimum temperature for growing the neem tree is 26–40°C, but a temperature range of 14–46°C is acceptable. Frost will kill the tree. An annual rainfall between 450–1200mm is adequate. The best pH value of the soil is 5.5–7. The neem tree needs bright light and short-day conditions. It accepts a wide range of soil texture, but the soil should be well drained and deep. The tree can be successfully grown in soils with low fertility, but it will not grow on seasonally waterlogged soils (FAO, 1996a).

The Neem has adapted to a wide range of climates. It thrives well in hot weather, where the maximum shade temperature is as high as 49°C and tolerates cold down to 0°C on altitudes up to 1500m. Today, the Neem is well established in at least 30 countries worldwide, in Asia, Africa and Central and South America. Some small-scale plantations are also reportedly successful in the United States of America (www.svlele.com).

Propagation

Modern techniques used for propagation include in vitro culture of leaf discs: 12–15 plantlets can be obtained from a single leaf disc within 6 months (Ramesh and Padhya, 1990). It is also possible to culture the embryos after dissection from the fruits. Embryos of 3–5mm were ready for transplanting within 45 days (Thiagarajan and Murali, 1994).

Crop management

The neem tree is planted as a windbreak and for shade and shelter. The growing period is 150–210 days per year. The tree can reach a height of 4–7m after 3 years and 5–11m after 8 years. In West Africa it is grown on a fuelwood rotation of 8 years. It withstands a dry season of 5–7 months. The tree is drought hardy, it regenerates rapidly and is shade tolerant in youth. The plantations need control because the neem may aggressively invade neighbouring areas and become a weed. It will not grow on seasonally waterlogged soils or where the dry season water table lies below 18m. Apart from the deep root system the tree also has lateral roots that may extent up to 15m away from the trunk (FAO, 1996a).

Production

The yield of neem oil can reach 600L/ha. The optimum wood yield of 8-year-old trees is 169 m³/ha. The annual wood production is 5–18m³/ha.

Source: Emmanuel D'Silva, www.gms-eoc.org/CEP/Comp2/docs/
Thailand/BCI_Emmanuel_DSilva_biofuels.pdf

Figure 10.111 Fully developed neem trees

The neem tree plantations located in southern Sonora, Mexico, which consists of 216 trees in a square planting pattern of 7 × 7m, show significant contrast in morphology and oil content. In 11 selected trees, tree height varied from 3.5 to 5.0m, weight of 10 fruits from 3.5 to 7.75g, kernel percentage from 21.1 to 31.9 per cent, and oil content from 15.4 to 24.5 per cent.

In seed from India, maximum oil content was obtained in seed from Hizar. Seed oil content in most of the provenances was not consistently correlated with seed morphology. Average fatty acid content was oleic acid 45.6 per cent, linoleic acid 16.8 per cent, palmitic acid 17.21 per cent, stearic acid 15.2 per cent and linolenic acid 1.3 per cent (Muñoz-Valenzuela et al, 2007) (www.hort.purdue.edu).

Satish Lele indicates that fully grown neem trees yield between 10 and 100 tonnes of dried biomass per hectare, depending on rainfall, site characteristics, spacing, ecotype or genotype. Leaves comprise about 50 per cent of the biomass; fruits and wood constitute one-quarter each. Improved management of neem stands can yield harvests of about 12.5m³ (40 tonnes) of high-quality solid wood per hectare (www.svlele.com).

Processing and utilization

The fruits are one-seeded drupes 1 to 2cm long with a woody endocarp, greenish yellow when ripe. The seeds are ellipsoid; cotyledons thick, fleshy and oily. The fruits ripen from June to August and possess a germination capacity of 70–90 per cent for a very short period. The kernel yields an average of 25–30 per cent of oil. The oil is yellow in colour and bitter in taste. It

Source: http://parisaramahiti.kar.nic.in

Figure 10.112 Flowers, fruits and seeds of
a neem tree

is said to have medicinal properties and is used for skin diseases. Neem oil contains limonoid, mahmoodin, protolimonoid, naheedin, tetranortriterpenoids, azadirone, epoxyazadirone, nimbin, gedunin, azadiradione, deacetylnimbin, 17-hydroxy azadiradione, nimbocinol and 17-epinimbocinol. It is extensively used in the soap industry. It is also used by the poorer classes as an illuminant (http://parisaramahiti.kar.nic.in).

The neem tree can be used as an energy plant in two ways. It can be grown as a source of fuelwood, or its oil can be used as fuel. The seeds contain about 45 per cent oil, used primarily in the manufacture of soap, insecticides and insect repellents and for various medicinal purposes (Rehm and Espig, 1991). 2 per cent neem-based formulations have been developed for personal protection against biting midges (Blackwell et al, 2004) Neem tree oil is opaque, bitter and unpalatable, though it can be transformed into an edible oil containing 50 per cent oleic acid and 15 per cent linoleic acid (Gliese and Eitner, 1995). The neem tree is not only used as an oil-producing plant, but also as a pesticide-producing plant. The seeds and leaves contain the insect repellent azadirachtin. The oil, bark extracts, leaves and fruits have medicinal properties. The oil can be used in soaps, disinfectants or as a lubricant.

Research is being undertaken by organizations such as the UN to see if neem can be used as a biofuel, particularly to manufacture biodiesel. Neem and jatropha are being grown on a commercial basis for their possible and effective use in manufacturing biodiesel. A blend of up to 20 per cent of biofuel in diesel will not require much modification to automobile engines, also this biofuel will be cheaper than the regular fuel. According to certain pilot projects undertaken, neem-based biofuel was used in generators and it has proven to be successful. Biofuel containing neem oil is being used in developing countries for rural electrification (www.neem-products.com).

Indian renewable energy company Bhoruka Power Corporation Ltd, has received a grant of US$100,000 from the US government to conduct a detailed feasibility report for a biodiesel project in the State of Karnataka. The study envisages use of neem or pongamia non-edible oilseeds for the production of biodiesel as well as power (www.oilgae.com).

Internet resources

www.svlele.com/neem.htm
http://parisaramahiti.kar.nic.in/PDF/publications/Bio Fuels.pdf
www.neem-products.com/neem-biofuel.html
www.oilgae.com/energy/sou/ae/re/be/bd/po/nee/nee.html
www.hort.purdue.edu/newcrop/ncnu07/pdfs/munoz126-128.pdf

OIL PALM (Elaeis guineensis Jacq.)

Partially contributed by Professor Dr Thamer Ahmed Mohammed, Universiti Putra Malaysia, Serdang, Malaysia

Description

The oil palm is indigenous to the humid tropics of West Africa. It occurs wild along the banks of rivers in the transition zone between the rain forest and the savannah. Today, the most common oil palm is *Elaeis guineensis,* which makes up the largest part of the stocks in the plantations. Only a little is known about the origin of the oil palm, which is found in Africa and America. The African oil palm is native to tropical Africa, from Sierra Leone in the west to the Democratic Republic of Congo in the east. It was domesticated in its native range, probably in Nigeria, and moved throughout tropical Africa by humans who practised shifting agriculture at least 5000 years ago. It is assumed either that the oil palm was brought from its habitat in Africa to America by (slave) traders, or that the oil palm existed already in the territory when Africa and South America were connected by a land bridge (Lennerts, 1984).

Palm was introduced to Java by the Dutch in 1848 and Malaysia (then the British colony of Malaya) in 1910 by Scotsman William Sime and English banker Henry Darby. The first plantations were mostly established and operated by British plantation owners, such as Sime and Darby. From the 1960s a major oil palm plantation scheme was introduced by the government with the main aim of eradicating poverty. Settlers were each allocated 10 acres of land (about 4 hectares) planted either with oil palm or rubber, and given 20 years to pay off the debt for the land. The large plantation companies remained listed in London until the Malaysian government engineered their 'Malaysianization' throughout the 1960s and 1970s (wikipedia.org).

In Asia, the first plantations were established on Sumatra in 1911, and in 1917 in Malaysia. Oil palm plantations were established in tropical America and West Africa about this time, and in 2003, palm oil production equalled that of soya bean, which had been the number one oil crop for many years.

The oil palm is an unbranched evergreen tree, reaching a height of 18–30m and having a stout trunk with diameter of 22–75cm and covered with leaf bases. Optimal plant density is 145 trees/hectare with a distance of 10m between trees.

The tree sprouts adventitious roots, growing from the bottom 1m of the trunk. A few deeply penetrating roots anchor the tree, but most of the roots grow horizontally in the top 1m of the soil, and as far away as 20m from the trunk. The roots lignify in the early states of growth; the intake of nutrients takes place through the tips of the roots. This explains the importance of the soil quality for the growth of the oil palm. The stem terminates in a crown of 70–100 leaves at the very top. The leaves are up to 7m long. The photosynthetic pathway is C$_3$ II (FAO, 1996a).

The palm is monoecious, and male and female flowers do not appear on the plant at the same time. The African oil palm (*E. guineensis)* and the American oil palm (*Elaeis oleifera)* can be crossed to obtain fertile hybrids. Each leaf axil produces a blossom that develops into an infructescence weighing 15–25kg, of which 60–65 per cent is fruit. An infructescence contains 1000–4000 oval fruits measuring 3–5cm in length.

The largest proportion of oil is in the fruit pulp (mesocarp), and about an eighth is in the endosperm. Fruit flesh and kernel deliver different types of oil. As Table 10.66 shows, three types of palms are distinguished, depending on the thickness of the nutshell of the palm kernels (endocarp).

Each tree changes periodically between male and female inflorescences. In the first years of a newly founded plantation, it may be necessary to pollinate the female flowers artificially if the palms do not produce enough male flowers. The pulp of the fruit contains an average of 56 per cent oil. The ratio between the pulp and the kernel depends on the climatic conditions and on the variety of the tree.

The oil content of the whole fruit can vary widely, depending on the cultivar and the growth conditions. The dura, grown in dry regions, has an oil content of 14 per cent, the wild and semi-wild dura fruit 20–27 per cent, and the tenera 36 per cent (Gliese and Eitner, 1995).

The inflorescences begin to develop 33 to 34 months before flowering; the sex is determined 24 months before flowering. The development of the fruit takes 5–9 months from pollination to ripening (Rehm and Espig, 1991). One month before the final ripening, delicate oil drops and carotenes begin to form in the fruit. As fruits ripen, they change from black (or green in *virescens* types) to orange, but have varying degrees of black cheek colour depending on light exposure and cultivar. However, fruit abscission is the best index of bunch ripeness.

Ecological requirements

The oil palm requires mean temperatures between 24 and 28°C. The cultivation zone is restricted to the humid tropics between the latitudes 10°N and 10°S with an altitude of under 500m. The yields are low in cool conditions. The annual precipitation should be in the 1500–3000mm range. Dry periods lasting not longer than 3 months are tolerated (Rehm and Espig, 1991). The oil palm needs a large amount of sunshine – 5–6 hours per day are optimal. The soil should be deep and well drained. Brief and occasional flooding does not cause any damage, but undrained soil is incompatible.

The best pH range is 5.5–7, though the palm thrives even at pH 4 if provided with the correct fertilizers. The majority of nutrient-absorbing roots lie 0–30cm deep, in a circle with a radius of 5m. The nutrient intake is high: a 1t fruit bunch contains 6kg N, 1kg P, 8kg K and 0.6kg Mg; B and Cl are important trace elements (Rehm and Espig, 1991). In climates like that of Malaysia and the Congo where no distinct dry periods occur, harvesting is possible through the whole year. In areas like Cameroon, Nigeria and Sierra Leone, with dry and rainy periods, the main flowering takes place in the former and the harvest in the latter (Lennerts, 1984). Fruit bunches are harvested using chisels or hooked knives attached to long poles.

Propagation

The natural germination is slow and irregular. In modern cultivation the seeds are treated in a special way. First the seeds are kept in dry heat for 40 days at a temperature of 38–40°C. Then follows a water absorption period lasting 7 days. The moist kernels are then kept at 28–30°C. The germination begins after 10 days and is complete after 30–40 days. The germinated seeds are transplanted into nursery beds or containers. After 4 or 5 leaves have sprouted, they are transferred into a growing field or larger containers. They can be planted out into the plantation at an age of 10–12 months. At first the trunk does not grow in height, but it grows in width until the diameter of the normal trunk is reached. Thereafter the palm begins to grow in height. High temperatures caused by solar radiation are good for the development of the seedlings.

The aim of breeding is the selection of high-yielding strains and the production of hybrid seeds, such as tenera (= dura × pisifera). Apart from the major aims of a thin endocarp and a thick mesocarp, breeders are striving to achieve earlier fruiting and slower stem elongation, higher fruit weight in the fruit bunch and resistance against *Fusarium* and dry basal rot (Rehm and Espig, 1991). Propagation by tissue culture technique is practised on a large scale and can speed up the breeding of improved palm cultivars, but there are problems because of somaclonal variations. The crossing of *E. guineensis* and *E. oleifera* has not yet resulted in commercially usable cultivars (Rehm and Espig, 1991).

An oil palm seed company in Australia has propagated new fast-growing varieties which will produce up to 15,000kg of crude oil per hectare per year. This is more than double the industry average. The seeds will produce a crop within their second year if properly cultivated (www.palmplantations.com.au/oil-palm-seeds.htm).

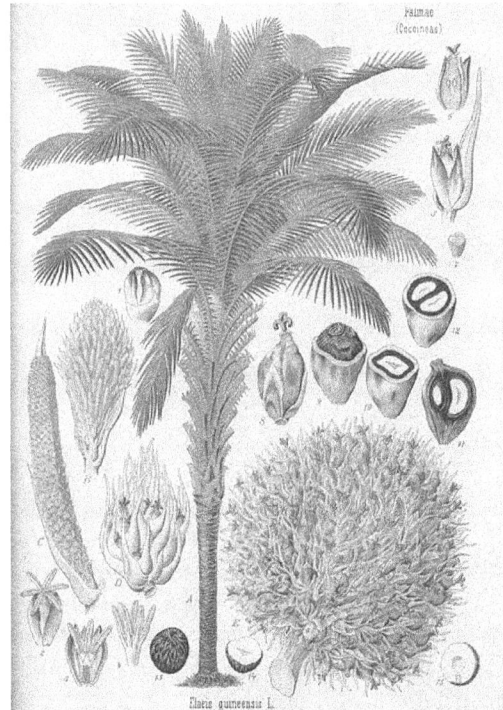

Table 10.66 Palm type based on nutshell thickness

Palm type	Nutshell thickness	Nutshell as % of total fruit weight
Dura	2–8mm	35–55
Tenera	0.5–3mm	1–32
Pisifera	No shell	

Source: Rehm and Espig (1991)

Source: http://en.wikipedia.org/wiki/File:Koeh-056.jpg

Figure 10.113 Various parts of the palm oil type *E. guineensis*

Source: Jim Crutchfield, USDA, www.pecad.fas.usda.gov/highlights/
2007/12/Indonesia_palmoil/

Figure 10.114 Oil palm nursery

Source: A. Mohammed

Figure 10.116 Harvesting of palm oil fruits

Source: Norpalm AS, www.norpalm.no/gallery-A.html

Figure 10.115 Oil palm plantation

Source: A. Mohammed

Figure 10.117 Fruit collection by a tractor

Crop management

If there are not enough male inflorescences, artificial pollination is possible. The pollen can be stored for a month without special treatment. Hybrid seeds of tenera varieties are produced in specialized breeding stations (Rehm and Espig, 1991). The distance between the palms in the plantation is 8–9m (130–150 palms per hectare) (Figure 10.119).

Between the plants, legumes can be sown as a ground cover and to provide the oil palm with nitrogen. In the first years intercropping is possible. Most diseases and pests can be avoided by:

- raising healthy plants (treatment of the nuts with TMTD and streptomycin, weeding out weak plants in the nursery, applying pesticides in the nursery beds);
- choosing an appropriate site for the plantation; and
- breeding cultivars resistant against the main diseases.

The main diseases of the oil palm are spear and bud rot (*Fusarium oxysporum* Schlecht. *F. elaeidis* Toovey, and

Figure 10.118 An 8-year-old oil palm tree in a seed nursery produced 535kg of fruit

other micro-organisms), leaf spot disease (*Cercospora elaeidis* Stey.), and dry basal rot *(Ceratocystis paradoxa)*. The cause of tip rot or fatal yellowing has not yet been discovered. It is suspected that it is a physiological disorder, perhaps associated with micronutrient deficiency (Lieberei et al, 1996). With good care of the plants, disease control measures in the stand are not usually necessary. Insect pests like *Oyctes* spp. and *Rhynochophorus* spp. are held within acceptable limits by sanitary measures. The use of insecticides is only occasionally necessary (Rehm and Espig, 1991). After 30 years, replanting is necessary because the palm is then so tall that harvesting becomes too difficult. Since

the import of *Elaeidobius* beetles from West Africa, where they improve pollination, yields in Malaysia have been increased. The ratio between kernel and fruit flesh was increased from 0.2:1, to 0.27:1, without decreasing the oil content of the flesh (Schuhmacher, 1991).

Production

The palms begin to bear fruit after 4–5 years, while still in the rosette stage. The yield potential has increased because of breeding and selecting high-yielding strains. Unimproved African palms yield about 0.6t/ha, and bred varieties produce 6t/ha. Annual yields in Malaysia are 7t/ha of mesocarp oil and 1.25t/ha of kernel oil (Rehm and Espig, 1991).

Global oil palm production is about twice that of any other fruit crop. Oil palm is produced in 42 countries worldwide on about 10.93 million hectares. Based on the above yield per hectare, oil palm is more than fourfold that of any other oil crop, which has contributed to the vast expansion of the industry over the last few decades. Table 10.68 shows the Top 10 Countries in oil palm production. The biggest producer in oil palm is Malaysia which produces 44 per cent of the total world production.

Processing and utilization

The palm fruit is the source of both palm oil (extracted from palm fruit) and palm kernel oil (extracted from the fruit seeds). Palm oil itself is reddish because it contains a high amount of beta-carotene. It is used as

Table 10.67 A comparison between SIRIM (Malaysian Government Palm Oil Seed Certifying Authority) and commercially available 'Super Yield' performance

	SIRIM minimum standard	'Super Yield'
Minimum dura yield	160kg	250kg
Oil to bunch	16%	18%
Oil yield / palm	25.6kg	45kg
Minimum progeny yield	160kg	250kg
Oil to bunch	24%	27%
Progeny oil yield / palm	38.4kg	67.5kg
Minimum kernel yield	3%	5%
Oil yield/ha (136 palms)	5.22 tonnes	9.18 tonnes
Oil yield/ha (156 palms)		10.5 tonnes

Figure 10.119 On average, a hectare of oil palm field accommodates 136–160 trees

Table 10.68 Oil palm production

Top 10 Countries (% of world production)	
1. Malaysia (44%)	6. Côte d'Ivoire (1%)
2. Indonesia (36%)	7. Ecuador (1%)
3. Nigeria (6%)	8. Cameroon (1%)
4. Thailand (3%)	9. Congo (1%)
5. Colombia (2%)	10. Ghana (1%)

Source: www.uga.edu/fruit/oilpalm.html

cooking oil, to make margarine and is a component of many processed foods. Boiling it for a few minutes destroys the carotenoids and the oil becomes colourless. Palm oil is one of the few vegetable oils relatively high in saturated fats (like coconut oil) and thus semi-solid at room temperature. Palm is also used in biodiesel production, as either a simply processed palm oil mixed with petrodiesel, or processed through transesterification to create a palm oil methyl ester blend which meets the international EN 14214 specification, with glycerine as a by-product. The actual process used varies between countries and the requirements of any export markets. Second-generation biofuel production processes are also being trialled in relatively small quantities (http://en.wikipedia.org/wiki/Palm_oil).

The fruits are harvested when they change colour (black to orange), or after the fall of individual fruits. The fruit bunches are cut off or knocked down. The fruits must be processed within 24 hours after harvesting because some fruits are always damaged and the oil is cracked by lipase into fatty acids and glycerine, reducing the value of the oil (Rehm and Espig, 1991). The red raw palm oil has a high carotene content (Gliese and Eitner, 1995). The oil obtained from the fruit flesh differs significantly from the kernel oil (Table 10.69).

The fruit bunches are treated in an autoclave to destroy the lipases and to facilitate threshing. The separated fruits are treated in a digester and stirred to a pulp at a temperature of 95–100°C for 20–75 minutes. Then the fruit flesh is pressed. The palm nuts are separated from the pulp, cleansed and dried. After the

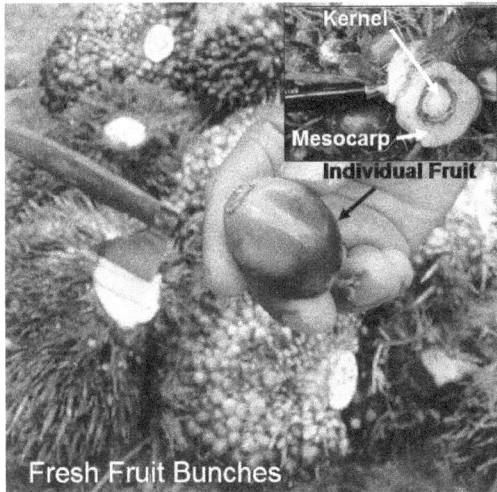

Source: Jim Crutchfield US Department of Agriculture, pecad.fas.usda.gov

Figure 10.120 Oil palm fruits and seeds

shells are broken and the shells and kernels separated, the kernel oil is extracted by pressing or solvent extraction.

In Australia, oil palm seed has been bred from the highest-yielding oil palms in the world. The seeds will produce a crop within their second year if properly cultivated. Superior genetics have been bred over the past 60 years from the best trees selected for their early maturity, high fruit and oil yields, disease resistance and long productive life (www.palmplantations.com.au/).

As a summary, oil extraction is not a simple process, but a complex one which is carried out by large mills located in industrial urban areas processing up to 60 tonnes of fruit per hour, or by small-scale mills which are located in rural villages that produce only about 1 tonne of oil in an 8-hour shift. Oil extraction from fruit follows the same basic steps in either case:

1 steam sterilization of bunches (inactivates lipase enzymes and kills micro-organisms that produce free fatty acids, reducing oil quality);
2 stripping fruit from bunches;
3 crushing, digestion, and heating of the fruit;
4 oil extraction from macerated fruit (hydraulic pressing);
5 palm oil clarification;
6 separating fibre from the endocarp;
7 drying, grading, and cracking of the endocarp;
8 separating the endocarp from the kernel;
9 kernel drying and packing.

The product of step 5 is termed crude palm oil, which must be refined to remove pigments, free fatty acids, and phospholipids, and to deodorize it. The final product, termed 'refined, bleached, deodorized' palm oil is produced.

The oil palm is a rich source of vegetable oil. It can also be used as fish feed. The ever-expanding oil palm cultivation in Malaysia and other tropical countries offers the possibility of an increased and constant

Table 10.69 Composition and properties of palm oil (mesocarp oil) and kernel oil

Composition of palm oil (mesocarp oil)		Composition of kernel oil	
Component	Amount (%)	Component	Amount (%)
Palmitic acid	35–40	Palmitic acid	7
Oleic acid	34	Oleic acid	18.5
Myristic acid	1.5	Myristic acid	12
Stearic acid	5	Stearic acid	4
Linoleic acid	10	Capric acid	6
		Lauric acid	55
Property			
Melting point	27–42°C		23–26°C (fresh) 27–28°C (old)
Iodine number	53.6–57.9		10–18[a]
Saponification number	200–205		245–255[b]

Source: Diercke (1981).
[a]Lennerts, 1984; [b]Römpp, 1974

Figure 10.121 Delivering the fruit to the factory

availability of palm oil products for aquafeed formulation (Ng, 2002).

Palm oil can be used as liquid fuel and in industry as a lubricant or for soap and candle production. Palm oil can be used as a vegetable substitute for diesel, and not only in agricultural machinery: in Malaysia processed (transesterified) palm kernel oil is used to run buses, and plants have been built to produce the transesterified oil on an industrial scale (Lutz, 1992).

The press cakes from the fruit have a high fibre content. After drying they can be used as solid fuel. The ashes are used as fertilizer. The press cakes from the kernel have a high protein content and are used as fodder (Franke, 1985). The fruit bunches can be used as a high-value energy source; they are dried to a water content below 50 per cent and chopped for use as a fuel. The ashes contain 41 per cent K_2O, 3.7 per cent P_2O_5, 5.8 per cent MgO and 4.9 per cent CaO, and are used as fertilizer. Increasing quantities of fruit bunches are used for mulching to provide the soil with organic matter.

The fibres and nutshells are suitable for energy provision because of their low water content. They are usually burned at the oil mill to provide it with heat and power through turbine-powered generators.

During processing in the oil mill, for each tonne of palm oil 2.5t of sewage is produced. One tonne of sewage can produce $28m^3$ of biogas. After the concentration of hydrogen sulphide has been reduced to below 1000ppm, the gas is suitable to operate special gas units that can drive generators and supply 1.8kWh per m^3 of biogas. Conventional treatment of the sewage in sewage treatment plants releases climate-affecting gases into the atmosphere during the process. It is possible to get 110kWh from the by-products of 1t of fruit bunches. If the energy needed for the production of palm oil, palm kernel oil, sewage treatment and threshed fruit bunches is subtracted from this amount of energy, the oil mills could provide 370kWh from each tonne of palm oil. The use of by-products could reduce the emission of CO_2 equivalents during the production of palm oil, especially by avoiding methane emission from the sewage. Using the by-products as an energy source is

Source: A. Mohammed

Figure 10.122 The extracted oil before processing

economically feasible and has no negative impacts on the environment (Germer and Sauerborn, 1997).

Suporn (1987) found that using 100 per cent refined palm oil in a Kubota diesel engine model KND 5B resulted in the best power output and lowest emissions, while using 70 per cent refined palm oil blended with 25 per cent diesel resulted in the best specific fuel consumption. In Thailand, Gumpon and Teerawat (2003) successfully used the palm oil as fuel for agricultural diesel engines.

A train runs daily from Hatyai District (South Thailand) to Sugaikolok district (Thailand–Malaysia border), a distance of 214km, using 50 per cent biodiesel (palm oil) as shown in Figure 10.123.

From 2007, all diesel sold in Malaysia had to contain 5 per cent palm oil. Malaysia is emerging as one of the leading biofuel producers with 91 plants approved and a handful now in operation, all based on palm oil. The main aim is to meet the regional demand, though exports to Europe are also planned, with China currently the main importer of Malaysian products for biodiesel.

Malaysia opened its first biodiesel plant in the State of Pahang, which has an annual capacity of 100,000 tonnes and also produces by-products in the form of 4000t of palm fatty acid distillate and 12,000t of pharmaceutical grade glycerine. With the cooperation of Finland, Malaysia planned to produce 800,000t of biodiesel per year from Malaysian palm oil in a new Singapore refinery by 2010, which would make it the largest biofuel plant in the world, and 170,000t from its first second-generation plant in Finland from 2007–2008, which can refine fuel from a variety of sources (Wikipedia.org).

Source: *Songklanakarin Journal of Science and Technology*

Figure 10.123 Train fuelled by palm oil in
Thailand

Using waste biomass to produce energy can reduce the use of fossil fuels, reduce greenhouse gas emissions and reduce pollution and waste management problems. A publication by the European Union (EEA, 2006) highlighted the potential for waste-derived bioenergy to contribute to the reduction of global warming. The report concluded that 19 million tonnes of oil equivalent is available from biomass by 2020, 46 per cent from bio-wastes: municipal solid waste (MSW), agricultural residues, farm waste and other biodegradable waste streams. In Malaysia, the waste from palm oil processing is estimated to be 5.5t/ha, while the amount is 20 million tonnes.

In Malaysia, the waste from palm oil fruit (biomass) is used to generate electricity. A large plant, located in the State of Sabah, uses the palm husks, known as empty fruit bunches, and processes them to generate electricity. The plant saved as much as 75 per cent in

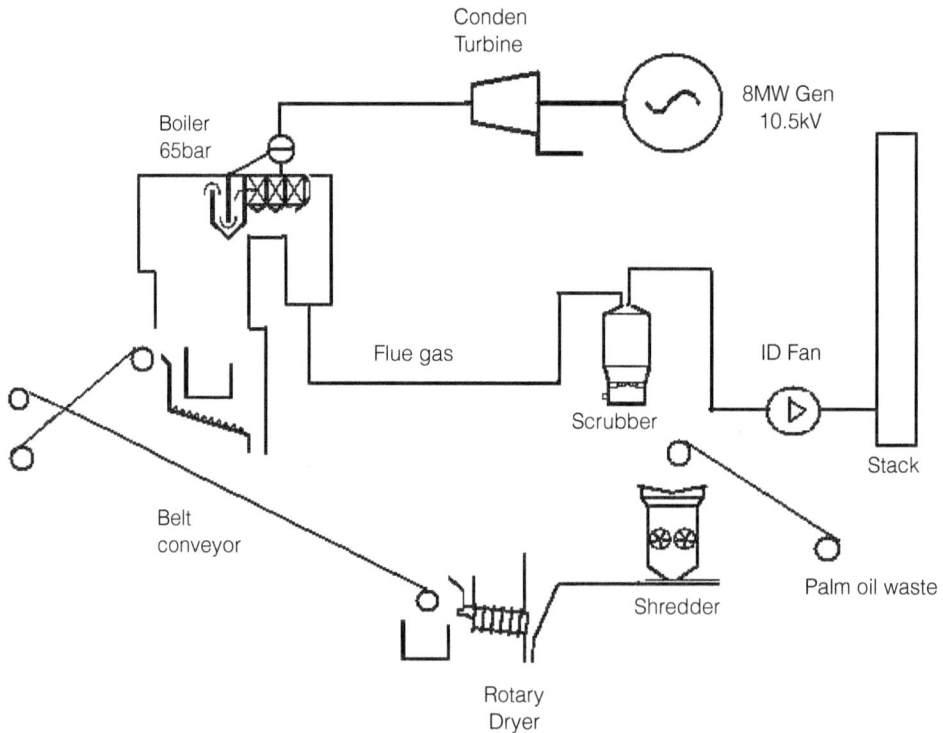

Source: Li and Tay (2006)

Figure 10.124 Schematic diagram of plant using palm oil waste to generate electricity in Malaysia

electricity generation cost compared with the diesel-fired plants. At the plant, 30 fruit bunches can be processed per hour to generate 7.5MW of electricity. Palm oil-processing factories and households on the plantation have so far utilized the electricity generated from the oil palm biomass. Figure 10.124 shows the schematic diagram of the plant to generate electricity from the palm oil waste.

Other scientists and companies are going beyond merely using the oil from oil palm trees, and are proposing to convert the entire biomass harvested from a palm plantation into renewable electricity, cellulosic ethanol, biogas, biohydrogen and bioplastic. Thus, by using both the biomass from the plantation as well as the processing residues from palm oil production (fibres, kernel shells and palm oil mill effluent), bioenergy from palm plantations can have an effect on reducing greenhouse gas emissions. Examples of these production techniques have been registered as projects under the Kyoto Protocol's Clean Development Mechanism.

By using all the biomass residues from palm oil processing for renewable energy, fuels and biodegradable products, both the energy balance and the greenhouse gas emissions balance for biodiesel from palm oil is improved. For each tonne of crude palm oil (CPO) produced from fresh fruit bunches, the following residues, which can all be used for the manufacture of biofuels, bioenergy and bioproducts, become available: around 6t of waste palm fronds, 1t of palm trunks, 5t of empty fruit bunches (EFB), 1t of press fibre (from the mesocarp of the fruit), 0.5t of palm kernel endocarp, 250kg of palm kernel press cake and 100t of palm oil mill effluent (POME). In short, a palm plantation has the potential to yield a very large amount of biomass that can be used for the production of renewable products. However, regardless of these new innovations, first-generation biodiesel production from palm oil is still in demand globally and will continue to increase. Palm oil is also a primary substitute for rapeseed oil in Europe, which too is experiencing high levels of demand for biodiesel purposes. Palm oil producers are investing heavily in the refineries needed for biodiesel. In Malaysia companies have been merging, buying others out and forming alliances in order to obtain the economies of scale needed to handle the high costs caused by

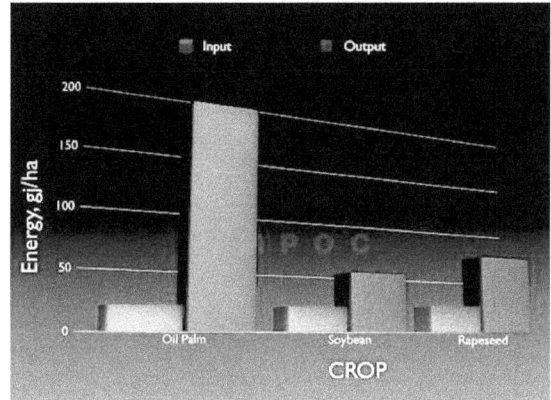

Source: www.mpoc.org.my/Palm_Oil_Fact_Slides.aspx

Figure 10.125 Oil palm is the most efficient crop, producing the highest energy output:input ratio. In absolute terms, oil palm also requires the lowest inputs of pesticides, fertilizers and fuel for unit production of oil

increased feedstock prices. New refineries are being built across Asia and Europe (en.wikipedia.org/wiki/Palm_oil).

The Univanich Palm Oil Public Company Ltd is focused on the growing market in Thailand, where a population of 62 million consumers and an expanding economy is increasing the demand for palm oil. However, Univanich also operates an export facility at the nearby Krabi port of Leam Phong. From this deep water port Univanich's exports in 2003 exceeded 50,000t of CPO and palm kernel oil (PKO), making the company Thailand's leading exporter of palm oil.

The Company's newest factory in Lamthap district of Krabi Province was completed in October 2004. Consulting engineers from Malaysia were engaged to supply the latest in new crushing mill technologies. In 2001 Univanich's Topi factory, situated in Thailand's Krabi Province, was renovated and expanded to its present capacity of 60t of fresh fruit bunches per hour (FFB/hour) and kernel crushing of 5t palm kernel per hour (PK/hour).

Palm shells share the same characteristics as coconut shells. They both have a highly complex pore

Source: http://univanich.com/c4.htm

Figure 10.126 Principal overseas customers are in India and Malaysia but shipments have also been sent to China and Vietnam

structure and fibre matrix, making them the most suitable raw materials for the production of premium activated carbon. No other type of activated carbon has such high iodine values and superior hardness.

As a raw material for fuel briquettes, palm shells offer the same calorific characteristics as coconut shells. Palm shell charcoal is characterized by high energy content that is released slowly during combustion, owing to its complex fibre matrix structure. Its smaller shell size makes it easier to carbonize for mass production, and the resulting palm shell charcoal can be pressed into a heat-efficient biofuel briquette (http://envirocarbon. com.my/7-PSC1.0.htm).

The Malaysian palm oil industry has grown tremendously over the last four decades, and since then it has maintained its position as the world's leading country in the production of palm oil. Nevertheless, the industry has also generated vast quantities of palm biomass, mainly from milling and crushing palm kernel. The types and amounts of these biomass generated in year 2005 are tabulated in Table 10.70.

Biomass briquettes are mostly used for cooking, heating, barbequing and camping in the regions such as the USA, the EU, Australia, Japan, Korea and Taiwan. In the developing countries, biomass briquettes are mainly for household usage only. On a larger commercial scale, it can be used as fuel in producing steam, district heating and electricity generation.

Source: univanich.com

Figure 10.127 New factory at Lamthap

Nasrin et al (2008) found from their investigations that, overall, converting palm biomass into briquettes increased its energy content and reduced its moisture

Source: EnviroCarbon

Figure 10.128 Palm shells from mills

(a)

(b)

Source: EnviroCarbon

Figure 10.129 Palm shells (a) and; (b) shell charcoal

content about 5 per cent and 38 per cent respectively compared to its raw materials. Palm biomass briquettes could become an important renewable energy fuel source in the future.

Reducing dependency on fossil fuels is but one part of the biodiesel advantage. Its positive impact on the environment and global warming could well be its larger contribution. Palm-based biodiesel reduces 'greenhouse' carbon dioxide concentrations in the atmosphere, not only through cleaner emissions, but also through a significant role in the carbon cycle. Oil palms draw carbon dioxide from the atmosphere as they grow at an average of 29.3t/ha of plantation, and hence carbon dioxide from biodiesel emissions is returned to the palm trees reducing accumulation in the air. In turn, 21.3t of oxygen is released. Studies show that full biodiesel use would reduce net carbon dioxide by over 78 per cent compared to petroleum diesel, and up to 16 per cent with the use of blends comprising 20 per cent biodiesel. Further research will aim at identifying the breadth of the positive impact Malaysian oil palm planting makes in the carbon cycle through the enormous carbon sink that is created (www.malaysiapalmoil.org/product/biofuel.asp).

Malaysia grows about 500 million oil palms. When combined with some 64 per cent of forest, the green tree cover is 83 per cent. That makes Malaysia one of the largest carbon sink nations on Earth.

Oil palm is a highly sustainable and energy-efficient crop to produce better and provides a biofuel source than other vegetable oils such as corn, soya and rapeseed. It ensures sustainable production for Malaysia's biofuel programme (malaysiapalmoil.org/product/biofuel).

Table 10.70 Palm biomass generated in 2005

Biomass	Quantity, million tonnes	Moisture content, %	Calorific value, kJ/kg	Main uses
Fibre	9.66	37.00	19,068	Fuel
Shell	5.20	12.00	20,108	Fuel
Empty fruit bunches	17.08	67.00	18,838	Mulch
Palm kernel	2.11	3.00	18,900	Animal feed

Source: Nasrin et al (2008)

Internet resources

http://findarticles.com/p/articles/mi_7109/is_/ai_n28
458738
http://en.wikipedia.org/wiki/Palm_oil
www.khunheng.com/product.html
http://en.wikipedia.org/wiki/Biomass
http://en.wikipedia.org/wiki/Biofuel
www.malaysiapalmoil.org/product/biofuel.asp
http://envirocarbon.com.my/7-PSC1.0.htm
http://univanich.com/c4.htm
www.palmplantations.com.au/
www.dipbot.unict.it/palms/descr05.html
www.pecad.fas.usda.gov/highlights/2007/12/Indonesia
_palmoil/
www.sabah.gov.my

OLIVE TREE (*Olea europaea* L.)

Description

The olive is one of the plants most mentioned in literature. In Homer's Odyssey, Odysseus crawls under two olive shoots growing from a single stock. The Roman poet, Horace describes his own diet as very simple: 'As for me, olives, endives, and smooth mallows provide sustenance.' Lord Monboddo comments in 1779 that the olive is one of the foods preferred by the ancients and one of the most perfect foods.

The leafy branches of the olive tree – the olive leaf as a symbol of abundance, glory and peace – were used to crown the victors of games and war. As emblems of benediction and purification, they were also ritually offered to deities and powerful figures: some were found in Tutankhamen's tomb.

The olive (*Olea europaea*) is a species of small tree in the family Oleaceae, native to the coastal areas of the eastern Mediterranean region, from Lebanon, Syria and the maritime parts of Turkey and northern Iran at the south end of the Caspian Sea. Its fruit, the olive, is of major agricultural importance in the Mediterranean region as the source of olive oil.

The olive tree is an evergreen tree or shrub native to the Mediterranean, Asia and parts of Africa. It rarely exceeds 8–15m in height. The silvery green leaves are oblong in shape, measuring 4–10cm long and 1–3cm wide. The trunk is typically gnarled and twisted.

The small white flowers have four-cleft calyx and corolla, two stamens and bifid stigma, and are generally borne on the last year's wood, in racemes growing from the leaf axils.

The fruit is a small drupe 1–2.5cm long. Wild plants have thinner flesh and are smaller than orchard cultivars. Olives are harvested at the green stage or left to ripen to a rich purple colour (black olive). Canned black olives may contain chemicals to turn them black artificially (http://en.wikipedia.org/ wiki/Olive).

The leaves have a grey-green colour on their upper surface and are silver-white beneath. They are dropped after two years.

The tree can live for several hundred years. The buds arise from the leaf axils. The tree blooms in early June with yellow-white flowers. The olives ripen between October and December (Gliese and Eitner, 1995). At first they are green, then later they become

Source: Wikipedia.org/wiki/Olive

Figure 10.130 Nineteenth-century illustration of the olive

dark green to black. The oil is formed in the flesh during the change of colour. The tree requires 210–300 days from flowering to harvest. The roots reach up to 12m away from the trunk and grow to 6m deep (Rehm and Espig, 1991). Its vitality allows the tree to live for several hundred years and form new branches even if the trunk has become hollow (Franke, 1985).

The fruit flesh (mesocarp) contains 23–60 per cent oil (40–60% from oil cultivars), and the seed (endosperm and embryo) contains 12–15 per cent oil. Oil from fruit flesh and seed both belong to the oleic acid group (Rehm and Espig, 1991).

Ecological requirements

Olive trees can grow in nutrient-poor, but well-drained soils. They need full sun for fruit production and slight winter chill for the fruits to set. Olive trees should not be planted in areas where temperature falls below $-5°C$ because they do not tolerate very low temperatures and get seriously damaged by winter and spring frosts. A safe criterion for choosing an area is the presence of undamaged olive trees for at least 20 years in the vicinity. Olive trees are also damaged by hot and dry air, particularly during flowering and fruit setting. Also, in areas with low air circulation and high humidity, diseases such as leaf spot appear more easily. Another criterion for the selection of the planting area is the availability of manpower, especially during the harvesting period, as well as the presence of processing units nearby. The decision must also take into account the annual rainfall. Thus, in low rainfall areas (200–300mm), olive yield is only satisfactory in soils with good water-retaining capacity, unless irrigation is applied. In high rainfall areas (400–600mm) olive yield is good on condition that adequate drainage is provided. In fields with steep slopes, contour cultivation on terraces must be employed. In this case, specialized tractors (caterpillar or crawler tractors) and other vehicles should be used to minimize the danger of overturn (www.biomatnet. org/publications).

Olive trees show a marked preference for calcareous soils, flourishing best on limestone slopes and crags, and coastal climate conditions. They

Source: (a) www.olives101.com; (b) Arturo Reina Sánchez, http://commons.wikimedia.org/wiki/File:Olivar_en_Granada.jpg

Figure 10.131 Overview of olive plantations in (a) Pakistan; and (b) Spain

tolerate drought well, thanks to their sturdy and extensive root system. Olive trees can be exceptionally long-lived, up to several centuries, and can remain productive for as long, provided they are pruned correctly and regularly.

A temperature of 12–15°C is sufficient for blooming and fructification, while the olives require 18–22°C to develop well. The tree tolerates temperatures as high as 40°C. An annual rainfall of 200mm is adequate (Gliese and Eitner, 1995). Where the annual rainfall is 400–600mm 100 trees per hectare can be planted, but dry regions can support only 17–20 trees per hectare. The olive's extended root system means that the tree can take in water from a large volume of soil, so it is rather drought tolerant (Rehm and Espig, 1991). The trees may survive a temperature of −10°C, but it is better not to plant olive trees where the temperature regularly falls below –5°C in the dormancy period. Late spring or early autumn frosts are undesirable (FAO, 1996a).

Propagation

The olive tree is reproduced vegetatively by grafting on seedlings, cuttings or wood of the trunk (Rehm and Espig, 1991).

The olive can be propagated in various ways, but cuttings or layers are generally preferred; the tree roots easily in favourable soil and produces suckers from the stump when cut down. However, yields from trees grown from suckers or seeds are poor; it must be budded or grafted onto other specimens to do well (Lewington and Parker, 2002). Branches cut into lengths of about 1m and planted deeply in manured ground will soon vegetate; shorter pieces laid horizontally in shallow trenches and covered with a few centimetres of soil rapidly throw up sucker-like shoots. In Greece, the cultivated tree is commonly grafted on the wild form. In Italy, embryonic buds, which form small swellings on the stems, are carefully excised and planted beneath the surface, where they grow readily, their buds soon forming a vigorous shoot.

Occasionally the larger boughs are marched, and young trees thus soon obtained. The olive is also sometimes raised from seed, the oily pericarp being first softened by slight rotting, or soaking in hot water or in an alkaline solution, to facilitate germination.

Where the olive is carefully cultivated, as in Languedoc and Provence, the trees are regularly pruned. The pruning preserves the flower-bearing shoots of the preceding year, while keeping the tree low enough to facilitate gathering of the fruit. The spaces between the trees are regularly fertilized. The crop from old trees can be enormous, but they seldom bear well two years in succession, and in many instances a large harvest only occurs every sixth or seventh season (Wikipedia.org).

Breeding aims are to obtain varieties with special properties like larger fruits, early ripening and different oil yields (Parlati et al, 1994).

Crop management

The olive tree planting scheme is determined by the cultivation system (intensive or non-intensive). For intensive cultivation, in areas with fertile soil and sufficient rainfall or irrigation, trees are planted densely. A planting density of 200–300 trees/ha is not unusual, depending on variety. Often trees are planted very densely (400–500 trees/ha), but as they grow, half of them are removed, particularly those planted in the intermediate rows. In areas with less fertile soils and low rainfall, planting density is reduced.

Before planting, some cultivation tasks must be carried out, such as uprooting (other trees and bushes), levelling the soil, construction of terraces, etc. If the field has been uprooted, it is advisable to cultivate grains or legumes for a period of 1–2 years, in order to remove all remaining roots from previous crops and minimize the incidence of root decay in the new trees. Deep ploughing, with or without the use of herbicides, may also be necessary to destroy weeds.

The field is then ploughed to facilitate the growth of the root system of the new trees. With the last ploughing there is an addition of phosphate and potash fertilizers that will be used by the trees during the first years of growth. Before adding any fertilizer, it is strongly recommended to perform soil analysis of samples taken from different spots and depths in the field (30, 60, 90 cm) (Biomatnet.org).

The olive fruit fly *Dacus oleae* Gmel., scale insects, the olive moth *Prays oleae* and other pests are difficult to control (Rehm and Espig, 1991). New plantings of olive trees for oil production are economically sound only in dry regions or where the soil is too stony for other forms of cultivation. The tree can be grown up to an altitude of 1220m and is well adapted to dryness. Most, but not all, cultivars require a chilling period through the last two months in the winter with temperatures between 1.5 and 15.5°C to stimulate the initiation of the flowers. Late spring or early autumn frosts are undesirable, as are hot dry winds or cool wet weather during flowering.

Production

The tree starts bearing at an age of 8–10 years. The maximum yields are obtained between 60 and 100 years of age. An average tree produces 60–65kg fruit per year. The oil yield is rather low: in the Mediterranean region the average is 400kg/ha/year. On the best sites of California, yields of 12.5t/ha of table olives have been achieved; 14–16L of oil can be obtained from 100kg of fruits (Gliese and Eitner, 1995). The bearing alternates: a full yield can be expected only in every second year. There are high labour costs during harvesting (Rehm and Espig, 1991).

Source: Giancarlo Dessì, http://commons.wikimedia.org/wiki/File:
Olea_europaea-g4.jpg

Figure 10.132 The olive road, Italy

The trial was carried out on a mature Manzanillo olive orchard in a very dry region of California. The trees were planted at a spacing of 15 × 30 ft (93 trees/acre). Eight different rates of irrigation were applied to a number of plots within the orchard over three years (1990–1992). There were a total of six plots under each rate of applied irrigation (6 plots × 8 different irrigation rates = 48 plots total). The applied annual irrigation ranged from just 9 inches to 40.6 inches . Each plot was assessed for a number of variables, the most relevant of which was the fruit yield in pounds/acre.

Trees which were only 'supplementary' irrigated with 9 inches (on top of the 4 inches natural rainfall) yielded an average of 113.8lb/tree (10,580lb/acre), whereas, trees fully irrigated with up to 40.5 inches yielded an average of 210.5lb/tree (19,580lb/acre). Average yields per tree with their corresponding irrigation levels were as follows: 9 inches – 114lb, 13.3in – 118lb, 16.7in – 125lb, 23.6in – 147lb, 28.7in – 167lb, 33in – 188lb, 37.2in – 209lb, 40.5in – 211lb. All of these figures were averaged over the two years 1991 and 1992 to take into account the effects of alternate bearing. NB. Alternate bearing is reduced when irrigation is applied to an orchard. It should also be noted that irrigation increased the actual dollar value of the fruit due to its healthier weight and appearance.

Although this Californian paper gives an average of up to 211lb/tree in a mature fully irrigated Manzanillo orchard and an Australian Mildura trial averaged 205lb/tree under similar conditions, Santa Cruz Olive Tree Nursery believes that a more conservative value would be an estimated mature tree yield in the same conditions at 154lb/tree. (www.santacruzolive.com)

Processing and utilization

Harvesting takes place from December to February. The oil olives are stripped off using machines, shaken down (or knocked down) from the tree, and then collected from the ground. The fruits are harvested before they are fully mature, at a time when the oil content of the flesh has reached its maximum at 50 per cent (Gliese and Eitner, 1995). The oil content does not increase after this, but the proclivity of the oil to become rancid does. Manually picked olives provide the best oil. Damaged olives may go mouldy, and therefore the oil quality decreases (Franke, 1985).

The main part of the olive oil is used for nutrition; only a small part is used as a source of energy. To obtain the oil, the whole olives are crushed into a paste and then pressed (Rehm and Espig, 1991). First (cold) and second (warm) pressings give a high-quality edible oil. A third (hot) pressing gives oil for fuel and industrial purposes (lubricant, soap production). The press cake contains 8–10 per cent oil. By treatment with hexane all but about 2 per cent of the remaining oil can be extracted (Gliese and Eitner, 1995). The oil is yellow to greenish-yellow and non-drying. The extraction remains are used as fuel or as fertilizer.

Olive oil contains 99.6 per cent fat and 0.2 per cent each of water and carbohydrates. Table 10.71 shows the acid composition of the oil, and Table 10.72 some of olive oil's properties.

The production of fermentable sugars from olive tree biomass was studied by dilute acid pretreatment and further saccharification of the pretreated solid residues. Pretreatment was performed at 0.2, 0.6, 1.0 and 1.4 per cent (w/w) sulphuric acid concentrations while temperature was in the range 170–210°C. Attention is paid to sugar recovery both in the liquid fraction issued from pretreatment (prehydrolysate) and that in the water-insoluble solid (WIS). As a maximum, 83 per cent of hemicellulosic sugars in the raw material were recovered in the prehydrolysate obtained at 170°C, 1 per cent sulphuric acid concentration, but the enzyme accessibility of the corresponding pretreated solid was not very high. In turn, the maximum enzymatic hydrolysis yield (76.5 per cent) was attained from a pretreated solid (at 210°C, 1.4 per cent acid concentration) in which cellulose solubilization was detected; moreover, sugar recovery in the prehydrolysate was the poorest one among all the experiments performed. To take account of fermentable sugars generated by pretreatment and the glucose released by enzymatic hydrolysis, an overall sugar yield was calculated. The maximum value (36.3g sugar/100g raw material) was obtained when pretreating olive tree biomass at 180°C and 1 per cent sulphuric acid concentration, representing 75 per cent of all sugars in the raw material. Dilute acid pretreatment improves results compared to water pretreatment. (Cara et al, 2008)

In Greece, where the olive tree is cultivated, during the processing of the olives the olive kernel wood is produced. It has a high heating value and is a very good fuel for heat or power generation. Today, it is used extensively on Crete for heat production; in the future, it has very good prospects for being used in power generation and/or heat and power cogeneration (Vourdoubas, 2005).

Since a large proportion of power on Crete is generated today from wind energy, it is likely that, in the future, biomass will also contribute to the generation of olive kernel wood, which is mainly used for heating purposes today on Crete for greenhouses and various small-sized industries. Its heating value is 3500–4000kcal/kg (with a moisture content of 12 per cent) and its price is approximately €0.05/kg; thus, it is a very attractive option as a fuel in comparison to oil. Olive kernel wood, however, has not yet found applications in power generation or cogeneration of heat and power in Greece. Because it can be easily burned and the combustion technology is widely known, it could be used as a solid fuel for power generation in the future. Presently, it is used in houses and in greenhouses for the production of hot water. Also, it is used in various industries for drying purposes and/or for the production of hot water. Greece exports small quantities to other European countries each year, where olive kernel wood is used

Table 10.71 Acid composition of olive oil

Acid component	Amount (%)
Oleic acid	66.3
Linoleic acid	12.3
Palmitic acid	8.9
Eicosenoic acid	4.9
Palmitoleic acid	4.7
Arachidic acid	0.3
Myristic acid	0.2

Source: Diercke (1981)

Table 10.72 Some properties of olive oil

Property	Value
Density (g/mL)	0.91–0.92
Melting point	6°C
Cloud point	10°C
Iodine number	75–94
Saponification number	185–203
Acid number	0.3–1

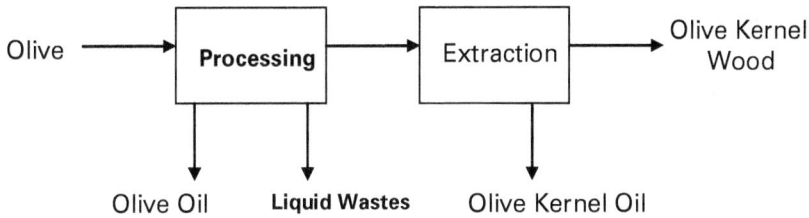

Source: Vourdoubas (2005)

Figure 10.133 Production scheme of olive oil, olive kernel oil and olive kernel wood from olives

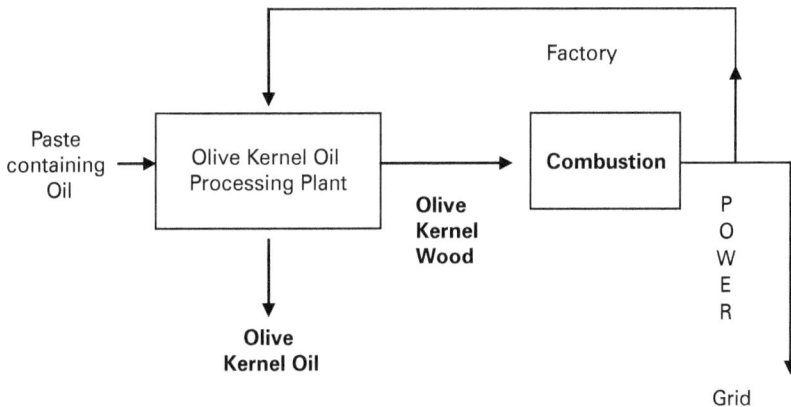

Figure 10.134 Power generation scheme from olive kernel wood

for fuel. The required machinery for olive kernel wood combustion is the boiler (including the burner), which is quite simple to use and not expensive. These boilers are reliable and made locally. Table 10.73 shows the characteristics of a power plant on Crete fuelled with olive kernel wood.

In a laboratory study of fixed-bed gasification of olive kernals and olive tree cuttings, it was found that gas from olive tree cuttings at 950°C and with an air equivalence ratio of 0.42 had a higher LHV (9.41MJ/Nm³) in comparison to olive kernels (8.60MJ/Nm³). Olive kernels produced more char with a higher content of fixed carbon (16.39 per cent w/w) than olive tree cuttings; thus, they might be considered an attractive source for carbonaceous material production (Skoulou et al, 2007).

Endesa, one of Spain's biggest utilities plans to build two olive oil-fired power stations at a cost of almost

Table 10.73 Characteristics of a power plant using olive kernel wood as a fuel on Crete

Nominal power	3MW
Operating hours	8000 hours/year
Energy generation	24,000 MWh/year
Plant efficiency	20%
Olive kernel wood consumption	27,180t/year
Heating value of the olive kernel wood	3800kcal/kg
Capital cost of the plant	€4.8 million
Subsidies (40%)	€1.92 million
Cost of olive kernel wood/year	€1,359,000/year
Consumption of olive kernel wood in the plant	3.4t/hour
CO_2 emissions which are avoided	29,137t/year
Energy generation from the plant as % of total energy consumption on Crete	1.04%

Source: Vourdoubas (2005)

£25 million. using solid olive residue – known in the trade as orujillo – left after the oil has been pressed from the fruit. Spain is well placed as a provider of orujillo. It has almost 200 million olive trees, making it the world's largest producer of olive oil and consequently of fuel for the new plants. When the plant come into operation it will produce a total of 32MW. They will be the first in the world to use olive residues to generate electricity (www.guardian.co.uk).

Internet resources

http://en.wikipedia.org/wiki/Olive
www.biomatnet.org/publications/1859cu.pdf
www.olives101.com/page/2/
www.santacruzolive.com/planting.asp?page=irrigation
www.guardian.co.uk/business/2000/feb/17/ethicalliving

PERENNIAL RYEGRASS
(*Lolium perenne L.*)

Description

Perennial ryegrass (*Lolium perenne*) is a grass from the family Poaceae. It is native to Europe, Asia and northern Africa, but is widely cultivated and naturalized around the world.

Perennial ryegrass is a low-growing, tufted, hairless grass, lacking stolons or rhizomes. The leaves are dark green, smooth and glossy on the lower surface, with untoothed parallel sides and prominent parallel veins on the upper surface. The leaves are folded lengthwise in bud (unlike the rolled leaves of Italian ryegrass, *Lolium multiflorum*) with a strong central keel, giving a flattened appearance. The ligule is very short and truncate, often difficult to see, and small white auricles grip the stem at the base of the leaf blade. Leaf sheaths at the base are usually tinged pink and hairless. Stems grow up to 90cm.

Lolium perenne is a perennial plant whose greatest importance has been its utilization as a feed. The plant produces above-ground stolons and grows to heights ranging from 30 to 70cm or more. The growth of the mature crop varies between semi-erect and semi-prostrate. The leaf blades are small, with a width of no more than 5mm. They are dark green with a grooved top side and a keeled, smooth and glossy underside. The inflorescence is a two-rowed ear that grows to a length of up to 20cm. The plant flowers in May or June.

Perennial ryegrass is commonly infected with the endophytic fungus *Acremonium lolii*. In this mutualistic relationship the fungus produces alkaloids that have been shown to lead to disorders in livestock, but confer insect resistance upon the plant.

Ecological requirements

Temperate zones, subtropics and tropical highlands are regions with the most suitable climate for perennial ryegrass cultivation. The plant does not tolerate drought, wet soil and salt. Most important for the crop's growth are firm, nutrient-rich soil and a mild, humid climate. Depending on the variety, it is also sensitive to snow cover and late frosts, though it has been shown to have good regenerative capabilities (Villax, 1963).

Propagation

Perennial ryegrass is propagated using seed.

Crop management

The crop is sown at the end of summer or the beginning of autumn using about 25–35kg/ha of seed.

Source: Richard Old, XID Services, Inc., Bugwood.org,
www.invasive.org/browse/detail.cfm?imgnum=5328048

Figure 10.135 Perennial ryegrass

Table 10.74 Fertilizer rates for perennial ryegrass, as used in field tests in Germany

Fertilizer	Amount (kg/ha)
Nitrogen (N)	284–380
Phosphorus (P_2O_5)	93–160
Potassium (K_2O)	176–463
Magnesium (MgO)	49

Perennial ryegrass has a high demand for fertilizers, especially nitrogen. In Germany, ryegrass field tests at several locations have used fertilizer rates that varied as shown in Table 10.74. A large percentage of these quantities was supplied in the form of liquid manure. Fertilizer rates vary with soil composition.

Among the diseases most damaging to annual ryegrass crops are mould (*Fusarium nivale*) and rusts (*Puccinia* spp.). Little research has been undertaken to determine an appropriate crop rotation for perennial ryegrass. It is a perennial plant with several years of significant yields once the crop has reached maturity.

Production

Field tests at four locations in Germany where perennial ryegrass was harvested once a year at the onset of flowering provided yields averaging between 7.4 and 12.7t/ha of dry matter. The plant's dry matter content averaged around 32–33 per cent. The heating value averaged approximately 17MJ/kg.

Additional research in Germany that tested different varieties of perennial ryegrass resulted in late-ripening plants producing the highest fresh and dry matter yields. In 1994 the average fresh yield was 78.1t/ha and dry yield 13t/ha. The highest fresh matter yields were produced by the varieties 'Morenne', 'Phoenix t' and 'Sambin', and the highest dry matter yields were produced by the varieties 'Parcour' and 'Tivoli t'. Late-ripening plants produce a consistent yield throughout the entire summer.

Field trials in Lithuania indicated that the herbage yield of perennial ryegrass were between 39t/ha/yr and 64t/ha/yr and the dry matter yield lies between 9.6 and 14.6t/ha/yr, depending on the type of the varieties used in the investigations (images.katalogas.lt/maleidykla).

Processing and utilization

The perennial ryegrass crop is harvested by mowing. Research is under way into harvests varying from one to five times per year. If the crop is harvested five times per year, the first harvest takes place when the crop is approximately 40–60cm high. This corresponds to harvests taking place about every 3–4 weeks from around the end of May through until the middle of October. The crop should be carefully harvested and quickly dried when it is to be used as fodder. When the grass is grown simply for biomass production the feed quality of the harvest is not significant. It is important that the harvested material be sufficiently dried for storage and/or economic utilization as a biofuel.

Traditionally, perennial ryegrass has been an important feed crop; now it is being examined as a producer of biomass for conversion into biofuel. New technologies are under development for the utilization of biomass and biofuels to produce energy in the form of heat and electricity, and existing technologies have been improved.

Researchers at the Institute of Grassland and Environmental Research (IGER), Aberystwyth, believe grass could be a more viable alternative to traditional biofuel crops such as maize and oilseed rape. It would also be more sustainable, as land would not be taken out of food production. 'A London Transport bus, for instance, could run for a whole year on grass-derived ethanol grown on 11.2 hectares.' Perennial ryegrass enjoys a long growing season, has low nitrogen requirements and, because it is native to Wales, can grow on marginal land. The studies indicate a hectare of grassland would produce 14 tonnes of dry matter, from which 5000 litres of fuel could be produced. Grassland managers often take three or more grass crops off the land each year. In early summer grass contains up to 40 per cent soluble sugars, which could be easily fermented to bioethanol. The remaining cellulose element could be ensiled for animal feed or further broken down with enzymes to produce more fuel. All cuts could be used for bioethanol but peak harvest crop would be utilized. Farmers and contractors already have all the necessary harvesting equipment. And it would be quite sustainable as few, if any, additional inputs would be required (Forgrave, 2008).

Internet resources

http://images.katalogas.lt/maleidykla/bio32/B-62.pdf

PIGEONPEA (*Cajanus cajan* (L.) Millspaugh)

Partially contributed by A. K. Gupta

Description

Pigeonpea is a C_3 plant that belongs to the family Phaseoleae and subfamily Cajaninae under suborder Papilionaceae of the order Leguminosae.

The cultivation of the pigeonpea goes back at least 3000 years. The centre of origin is most likely Asia, from where it travelled to East Africa and by means of the slave trade to the American continent. Today pigeonpeas are widely cultivated in all tropical and semi-tropical regions of both the Old and the New World.

The pigeonpea plant is a woody, short-lived perennial shrub growing up to 4m high but is most often cultivated as an annual reaching up to 2m in height. Pods are usually short (5–6cm) and contain four to six seeds. The cultivated species show a wide range of diversity of various morphological characters and also of maturity periods. Peninsular India or eastern Africa is credited with its homeland but pigeonpea has already been acclimatized for multiple uses in more than 60 countries in the world. It is produced commercially in India, Myanmar, Kenya, Malawi, Uganda, the Dominican Republic, Haiti and Puerto Rico. In India, the crop is grown on more than 3.6 million hectares (FAI, 1993).

Pigeonpea is an ecologically highly desirable plant species because it is endowed with several unique features that give it many excellent qualities and make it an important crop for drier regions and wastelands of poor fertility. Its prolific root system allows the optimum moisture and nutrient utilization. Pigeonpea ameliorates poor soils through its deep, strong rooting system, leaf drop at maturity and addition of nitrogen (Nene and Sheila, 1990). Its characteristic root exudates give pigeonpea a high ability to take up phosphorus even from normally unavailable forms (Ae et al, 1990; Otani and Ae, 1996). Pigeonpea has tremendous potential for use in agroforestry systems.

Source: Taŭolunga, http://en.wikipedia.org/wiki/File:Cajanus_cajan.jpg

Figure 10.136 Pigeonpea

With the increasing shortage of energy, pigeonpea sticks are likely to be in demand for fuel, and many farmers will be tempted to grow more pigeonpea (Nene and Sheila, 1990). A number of benefits of growing pigeonpea, when weighed against its low input requirements, justify its expansion to non-traditional areas such as the hilly lands of tropical Asia and marginal lands in Australia and the Americas.

Ecological requirements

Pigeonpea is predominantly a crop of the tropics and subtropics, but a wide range of maturity groups enables the crop to adapt to diverse agroclimatic areas and cropping systems. The range of maturity varies from 90 to 300 days, depending mainly on genotype and time of planting. Its new dwarf, short duration cultivars and hybrids can be cultivated up to 45° latitude on both sides of the equator. Although it is highly drought and heat resistant (Gooding, 1962) the crop performs better when annual precipitation exceeds 500mm. The plant also grows in subhumid ecologies where ripening can occur during the dry season (Rachie and Roberts, 1974).

The traditional landraces and cultivars are highly photosensitive (short-day) plants, whereas new lines have been shown to be highly insensitive. Regardless of plant type, pigeonpea as a sole crop is relatively inefficient because of its low initial growth rate (Willey et al, 1981), and is frequently mixed with other crops – mainly short-term, hot-weather cereals or other grain legumes to make efficient use of growth resources. Short-term and intermittent waterlogging severely impairs plant growth, so areas of high rainfall and/or impermeable soils are unsuitable for pigeonpea. Furthermore, the plant is very susceptible to frost in the winter.

Pigeonpea is often reported to be a crop well adapted to marginal conditions (Whiteman et al, 1985) and is grown on a wide range of soils types from gravel to heavy clay loam. It can grow in the pH range 5–8. In highly acidic soils, crop growth is adversely affected by aluminium toxicity or calcium deficiency. Farmers in India often grow pigeonpea on poor soils where no other crop can be used so profitably. However, well-drained, deep fertile loam with neutral pH is ideal.

Crop management

Pigeonpea is propagated using seed. The optimum temperature range for germination is wide (19–43°C),

with the most rapid seedling growth occurring between 29 and 36°C (de Jabrun et al, 1981). Traditionally, the crop is sown in India at the beginning of the monsoon or rainy season, which is from the middle of June to the end of July. In eastern Africa, pigeonpea is sown in October/November at the onset of short rains. Pigeonpea does not require special land preparation. Ploughing to a depth of 15cm is sufficient to obtain a good crop. The optimum depth of seeding is 4–5cm. The seed shape is very suitable for machine sowing. The inoculation of seeds with rhizobium culture ensures good performance.

Once established, pigeonpea roots are capable of penetrating hard pan layers. Kampen (1982) found the broad bed and furrow system of sowing to be more useful. Dry sowing is usually preferred as it is difficult to work on vertisols when they are wet. Ratoon cropping – a multiple harvest system in which stubbles of the first sown crop are allowed to regenerate for subsequent production – is also becoming popular using some short duration (100–140 days) varieties in parts of India where the winter is mild. This system minimizes the cost of cultivation, avoids the risks associated with sowing a second crop in rainfed conditions, and provides high returns (Ali, 1990).

Historically, pigeonpea has been grown in mixtures without a distinct row arrangement, but the modern method is to sow in rows for inter-row cultivation and mechanical harvesting. A plant population of 60, 80, and 80–100 thousand plants per hectare has been recommended for summer-sown long-, medium- and short-duration varieties respectively; winter planting as practised in the north-eastern hill regions of India requires a higher plant population of 250–300 thousand plants per hectare (Chandra et al, 1983). The most popular spacings are 60cm × 18–20cm.

To produce 1t of pigeonpea grain requires about 50kg N, 5kg P and 22kg K (Kanwar and Rego, 1983). Under intense management systems, more nutrients will be required to produce higher biomass. Rao (1974) estimated that to produce 2t of grain and 6t of stalks from 1ha, the short duration variety removed 132, 20 and 53kg of N, P and K respectively. It may be stated that 20kg N, 60kg P and 5–6t of farmyard manure per hectare will be necessary to obtain a good yield.

With regard to water management, more than 90 per cent of the pigeonpea growing area is rainfed (Singh and Das, 1985), but both biomass and seed yield may be enhanced in pigeonpeas of all maturity groups by the application of irrigation (Venkataratnam and Sheldrake, 1985). Flowering and pod-setting stages are most crucial with respect to water stress. Pigeonpea, being widespread and slow growing, suffers heavily from weed infestations; these can be tackled biologically by inserting a vigorously establishing intercrop or chemically by the pre-emergence incorporation of Basalin at 1kg per hectare. Pigeonpea when intercropped with a cereal ensures higher biomass production and a measure of income stability.

Production

Pigeonpea is grown in a multitude of contrasting production systems aimed at obtaining higher grain yield of better quality. The composition of seed is presented in Table 10.75.

Favourable growing conditions can result in a seed yield of up to 2.5t/ha; a high yield of 5t/ha of dry seeds was reported from India (Anon, 1967). Akinola and Whiteman (1972) recorded the highest yield of 7.6t/ha of dry seed from experimental plots in Queensland, Australia. The production potential in India, especially in the farmer's field, is severely reduced by a moisture deficit during critical growth stages, the use of traditional low-yielding varieties, or by poor management including the lack of any plant protection measures (Chandra et al, 1983). A moderately well-managed crop routinely yields about 2t/ha of grain and 7–8t/ha of stalks. Figures given in Table 10.76 project the average breakdown of dry matter in new genotypes of pigeonpea.

Production of biomass can be further enhanced by introducing a compatible intercrop. When pigeonpea is to be planted for energy, seed quality is not important. In India, growers have recently been encouraged to

advance the sowing of short-duration pigeonpea to April in order to produce more stalks (up to 15–17t/ha) and up to 50 per cent more seed yield compared to conventional June sowings. Besides this, the succeeding wheat crop is reported to produce a 25 per cent higher yield than that following the June-sown crops (Panwar and Yadav, 1981). However, the April sowing is feasible only where there are ample irrigation facilities.

Pigeonpea is normally grown as an annual shrub, but is a perennial which may grow for several years and develop into small trees. It gives additional yield after the first harvest if sufficient moisture is available, and it has great flexibility in a wide range of cropping systems. The crop has a wide range of maturity (80 to 250 days) and time to maturity is greatly affected by temperature and photoperiod. Thus, there exist maturity types of pigeonpea for many different cropping systems. Pigeonpea is a superb intercrop for planting with cereals and other crops. However, short-duration types have been developed in Australia and India that mature in less than 100 days with a yield potential of over 5000kg/ha and can be grown as a sole crop in multiple cropping systems. Pigeonpea stems are used as fuelwood in the energy-short villages of several African countries. Stems are also used for fencing crop fields (Phatak et al, 1999).

The improved production technologies gave higher yields and recorded a mean grain yield of 1.61t/ha which was 204 per cent higher than that obtained with the farmers' practice yields of 0.53t/ha. In addition to increased grain yields, improved technology also resulted in higher stalk yield of 2.93t/ha compared to 1.10t/ha of farmers' practice. The increased grain and stalk yields with improved production practice were mainly because of increased total dry matter, increased pod weight, higher shelling

Table 10.75 General composition of pigeonpea seed (percentage)

Crude protein	Starch	Soluble sugars	Fat	Crude fibre	Ash	Gross energy (MJ/kg)
21–25	50–60	3–5	1–2	1–5	3–4	16–18

Table 10.76 Partitioning of dry matter (per cent weight) in pigeonpea

Roots	Stem sticks	Branches and leaves	Husk	Seed
1.0 ± 0.4	38.6 ± 3.7	29.0 ± 4.0	7.2 ± 1.8	14.2 ± 1.7

Source: Gupta and Rai (unpublished)

Source: icrisat.org

Figure 10.137 Pigeonpea grown as a vegetable crop is a profitable venture in dry lands

percentage, higher 100-grain weight and harvest index. Total dry matter was 5.26t/ha, pod weight 2.33t/ha. (Ramakrishna et al, 2005)

Processing and utilization

Pigeonpea has traditionally been a valuable pulse crop. It can simultaneously satisfy the need for the '4 Fs' (food, fodder, fibre and fuel) and thus ensures a self-sustaining system.

Its *dal* (a thick soup cooked from dry decorticated split cotyledons) is a rich source of protein for vegetarians in India; the tender green pods serve as a vegetable; crushed dry seed forms an excellent feed for cattle, pigs and poultry; green or dry leaves provide palatable fodder; and the woody stem serves as fuel and raw material for making huts, brooms and baskets. Recent studies in Bangladesh (Akhtaruzzaman et al, 1986) and in Pakistan (Shah, 1997) indicate the possibility of using pigeonpea to produce paper pulp. This is a crop that can be used in many diverse ways. Presently, little is known about the efficient utilization of the entire plant as a biofuel. Pigeonpea can be regarded as a potential source of bioenergy because of its high biomass yield in comparison to the amount of off-farm inputs used.

> Pigeonpea stalk is a widely available biomass species in India. In this article the potential use of pigeonpea stalk as a fuel source through thermochemical conversion methods such as combustion, gasification, and pyrolysis has been investigated through experimentation using a thermogravimetric analyzer and pilot-plant-scale equipment. It has been proposed that pigeonpea stalks can be effectively utilized in two ways. The first is to pyrolyze the material to produce value-added products such as char, tar, and fuel gas. The second alternative is to partially pyrolyze the material to remove tar-forming volatiles, followed by gasification of reactive char to generate producer gas. (Katyal and Iyer, 2000)

Pigeonpea sticks are an important household fuel in many areas. Productivity more than makes up for comparatively poor fuel characteristics (low specific gravity and high moisture content). Stick yields of 7–10t/ha/yr dry matter are routinely reported for medium and early duration lines, and yields of 30t/ha/yr from irrigated, early duration varieties have been reported in India. Perennial varieties can produce 10t/ha/yr of dry material over a 2–3 year period on good sites. Sticks also produce thatch and basket materials.

Source: Rastaseed.com

Figure 10.138 Pigeonpea grains

Internet resources

www.icrisat.org/PigeonPea/PigeonPea.htm
www.rastaseed.com/?p=225
www.winrock.org/fnrm/factnet/factpub/FACTSH/
C_cajanbckup.html

POPLAR (*Populus* spp.)

Description

Different *Populus* species (family Saliaceae) exist in temperate climate zones. *Populus alba* is grown mainly in southern and central Europe, and *P. tremula* L. in Europe and Asia (Franke, 1985), while *P. tremuloides* Mic. is grown mainly in North America, Canada and in parts of Alaska (Schütt et al, 1994). The hybrid *Populus × canadensis* (Canadian poplar) is cultivated in Europe (Franke, 1985). *Populus* spp. are pioneer trees with a capacity for fast growth. In Europe, cultivated poplars grow to heights of between 30 and 35m, and the hybrid *Populus × canadensis* is able to reach 50m. Poplar wood is long grained, relatively soft, and easy to divide. Poplar is being used as a short rotation coppice (SRC) crop, and is cultivated on biomass plantations for use for direct combustion, making briquettes or gasification.

Populus is a genus of 25–35 species of deciduous flowering plants in the family Salicaceae, native to most of the northern hemisphere. English names variously applied to different species include poplar, aspen and cottonwood.

The genus has a large genetic diversity, and can grow to anywhere between 15 and 50m tall, with trunks of up to 2.5m diameter.

Ecological requirements

Like willow (*Salix* spp.), poplar grows successfully on a wide range of soil types – on sandy soils as well as on loamy-clay soils with a pH range of 6.0 to 8.0 and an optimum pH of 6.5 (FAO, 1996a). Irrigation is beneficial to yield if the yearly rainfall is lower than 600mm (Parfitt and Royle, 1996a). Poplar grows well on fine sandy-loamy soils with organic matter and a good supply of water. It is more dependent on water than other agricultural crops, so extremely dry land

Source: Acftu.org.cn

Figure 10.139 Poplar trees in Dunhuang, Gansu Province

should be avoided (Johansson et al, 1992). The tree's consumption of water is as much as 4.8mm per day. The minimum growing temperature of the different *Populus* species has a range between 5°C and 10°C and the maximum growing temperature between 30°C and 40°C; the optimal growing temperature range is between 15°C and 25°C. Poplar species cannot survive if the temperature reaches −30°C or below.

Propagation

Many poplars are grown as ornamental trees, with numerous cultivars used. They have the advantage of growing very big, very fast. Almost all poplars take root readily from cuttings or where broken branches lie on the ground.

Many poplar clones could be used as short rotation crops and are planted as plantlets or cuttings. Poplar cuttings incorporating primary and undeveloped secondary buds should be planted as vertically as possible with no more than 2.5 to 3cm remaining above ground (Parfitt and Royle, 1996). For ease of planting, successful rooting and subsequent management, the site should be subsoiled if necessary, deep ploughed (25–30cm) in autumn and power harrowed to produce a level and uncompacted tilth. Planting is practicable in the spring. About 12,000–18,000 of 20cm-long and more than 0.8cm-thick cuttings can be planted per hectare (Johansson et al, 1992).

The *Populus* cuttings are placed in double rows with a row spacing of 75cm and 125cm between the double rows. The plant spacing in the row is 55cm (Johansson et al, 1992).

First-year growing shoots on each cutting are cut back to the ground level in the first winter and can be used for making cuttings to establish new plantations. The regrowth of cut-back poplar induces the generation of a large number of new shoots.

Crop management

Uncontrolled weeds in the first season will reduce growth by up to 50 per cent and dry matter yield by up to 20 per cent of that from poplars in weed-free conditions (Parfitt et al, 1991). Therefore, land for growing poplar must always be cleaned of perennial weed rhizomes before planting, and weed control is necessary during the first few years. Poplar plantations are able to suppress weeds from the second growing year on.

No fertilizer is applied in the planting year because of the risk that the weeds benefit more than the poplars. An average application of 60–80kg N, 10–20kg P and 35–70kg K per hectare per year is suitable in the following years, depending on soil fertility. The fertilizer regime should be adapted to the soil type and the natural mineralization, and to the provision of nitrogen via rainwater.

Poplar can be attacked by numerous leaf-consuming, stem-sucking and wood-boring insects (Hunter et al, 1988). In most cases, none of these pests represents a serious problem. A poplar plantation is used by a great number of different insects, birds and animals. The insects' natural enemies and the high number of shoots and leaves on 1–3-year parts of the rootstocks should ensure that normal levels of insect attack are not seriously damaging. Diseases are more serious, and numerous pathogens are described, especially different species of rust. Different clones have varying susceptibilities to fungi and insects and can be selected for better resistance. Generally, there is little reason to introduce inputs for insect and disease control. Considerable damage may be caused by mice, rabbits, deer and farm animals during the establishing phase of poplar. Established cultivars may be at less risk.

Source: Puyallup.wsu.edu

Figure 10.140 Hybrid poplar plantations

Production

Populus plantations grown as SRC can be harvested for the first time 3–4 years after cut-back. This holds true for all subsequent harvest periods. The best harvest time is in winter with a dry matter percentage of 50 per cent. Current expectations are that there are eight or more cycles of harvests possible during the life of a poplar plantation. Annual yields up to 12–15ODT/ha/year and more (ODT = oven dried tonne) are obtained in Germany from the second and third ratoon on (El Bassam, 1997). Increased yields are expected when the full benefits from improved materials have been obtained and when agricultural practices have been optimized.

To define the produced biomass of individual clones of poplar, the increment elements were measured after cycles of one and two years. Average dry biomass yield reached 21 and 12t/ha/yr. Based on califoric values of oven dry wood bark of each clone, average energy potential researched popolar clones was estimated up to 395GJ/ha/yr (Klasnja et al, 2006).

Processing and utilization

There is interest in using poplar as an energy crop for biomass or biofuel, in energy forestry systems, particularly in light of its high energy out–energy in ratio, large carbon mitigation potential and fast growth.

In the United Kingdom, poplar (as with fellow energy crop willow) is typically grown for two to five years (with single or multiple stems), then harvested and burned – the yield of some varieties can be as high as 12 oven dry tonnes every year.

Harvesting of stems takes place in winter when the leaves have fallen off. The two available harvesting methods – direct chipping and whole shoot harvesting – need different techniques. Direct chipping is possible if chips can be combusted at the harvest moisture content (50 per cent) in large heating plants. Chips with a high moisture content will quickly deteriorate through microbial activity and must be combusted directly or ventilated for storage. Drier chips are needed in smaller plants and stationary ovens on farms. In this case, shoots are harvested whole to dry in piles during the summer. This material reaches a dry matter content of about 70 per cent.

Dry wood chips for energy production need no further processing and can be burned or gasified (Parfitt and Royle, 1996). *Populus* chips can be utilized in two ways. Direct combustion in special automated boilers will give low-grade space heating of 50 to 400kWh. With high-pressure steam techniques, steam

Source: Puyallup.wsu.edu/poplar/

Figure 10.141 (a) Boise Cascade chipper operation; and (b) Boise Cascade chip pile

turbines can be powered with up to 10MW installed capacity (Graef, 1997). Another possibility is the gasification (pyrolysis) of wood chips to drive an engine to power a generator and produce heat (a system known as combined heat and power, or CHP). A growth rate of 15ODT/ha/year is equivalent to the 7000L oil needs of two one-family homes.

Swedish forest industries are supposed to face a new threatening shortage of wood. Therefore, demonstration of new ideas of producing wood for pulp and fuels (heat, ethanol, hydrogen gas, dimethylether, electricity) in the most southern part of Sweden was very welcome.

One example of new ideas is that some farmers planted hybrid poplars on abandoned farming land at Sångletorp (33ha), Johannesholm (15ha), Kadesjö (11ha) and Näsbyholm (2ha) in the southernmost part of Sweden in 1991. Some of these plantations were harvested in 2004, but some are still growing. The harvested wood had been used for pulp and fuel. The results of the harvest and of the plantations that are still growing are analysed here from an economic, ecologic and energy point of view. (Christersson, 2008)

Pretreatment has been recognized as a key step in enzyme-based conversion processes of lignocellulose biomass to ethanol. The aim of this study is to evaluate two hydrothermal pretreatments (steam explosion and liquid hot water) to enhance ethanol production from poplar (*Populus nigra*) biomass by a simultaneous saccharification and fermentation (SSF) process. The best results were obtained in steam explosion pretreatment at 210°C and 4 min, taking into account cellulose recovery above 95 per cent, enzymatic hydrolysis yield of about 60 per cent, SSF yield of 60 per cent of theoretical, and 41 per cent xylose recovery in the liquid fraction. Large particles can be used for poplar biomass in both pretreatments, since no significant effect of particle size on enzymatic hydrolysis and SSF was obtained. (Negro et al, 2007)

In Colorado, ZeaChem raised $34 million in Series B financing from Globespan Capital Partners, PrairieGold Venture Partners, MDV-Mohr Davidow Ventures, Firelake Capital and Valero Energy, to develop cellulose-based green fuels and a chemicals biorefinery platform that converts poplar biomass into ethanol (Biofuelsdigest.com).

Internet resources

http://biofuelsdigest.com/blog2/2009/01/08/zeachem-raises-34-million-for-high-yield-cellulosic-ethanol-plant/
www.puyallup.wsu.edu/poplar/photos/gallery1.htm
www.puyallup.wsu.edu/poplar/photos/gallery2.htm
www.acftu.org.cn/template/10002/page.jsp?cur_page= 1&aid=370&cid=61
http://en.wikipedia.org/wiki/Poplar#Energy

RAPE (*Brassica napus L.*) and CANOLA (*Brassica napus L. and B. campestris L.*)

Description

Rapeseed (*Brassica napus*), also known as rape, oilseed rape, rapa, rapaseed and (in the case of one particular group of cultivars) canola, is a bright yellow flowering member of the family Brassicaceae (mustard or cabbage family). The name derives from the Latin for turnip, *rāpum* or *rāpa*, and is first recorded in English at the end of the fourteenth century. Older writers usually distinguished the turnip and rape by the adjectives 'round' and 'long(-rooted)' respectively. *Brassica napobrassica* may be considered a variety of *Brassica napus*. Some botanists include the closely related *Brassica campestris* within *B. napus* (wikipedia.org/wiki/ Rapeseed).

Canola is one of two cultivars of rapeseed or *Brassica campestris* (*Brassica napus* L. and *B. campestris* L.). Their seeds are used to produce edible oil that is fit for human consumption because it has lower levels of erucic acid than traditional rapeseed oils and to produce livestock feed because it has reduced levels of the toxin glucosin. Canola was originally naturally bred from rapeseed in Canada by Keith Downey and Baldur R. Stefansson in the early 1970s, but it has a very

Source: Ron Wiebe, www.canola-council.org/gallery/

Figure 10.142 Rapeseed field in Prince George, BC

Figure 10.143 Winter canola in early bloom April 2008

different nutritional profile in addition to much less erucic acid. The name 'canola' was derived from 'Canadian oil, low acid' in 1978. A product known as LEAR (low erucic acid rapeseed) derived from cross-breeding of multiple lines of *Brassica juncea* is also referred to as canola oil and is considered safe for consumption (wikipedia.org/wiki/Canola).

Rape is an annual C_3 plant which originated from the Mediterranean region. It is a member of the Cruciferae family and has both winter and spring forms. The plant germinates quickly, forming a deep-growing taproot and a rosette of blue-green leaves from which emerge 7–10 lateral shoots. On the ends of the branched stems grow the gold-yellow flowered racemes. Each plant has approximately 120 long slender seed pods; 40 to 60 of these are found on the main shoot. Each seed pod contains 18–20 seeds (2000–3000 seeds per plant) (Honermeier et al, 1993). The seeds are small, round and black.

Based on its seed oil, rape belongs to the erucic acid group of oil plants. Of the total fatty acid content, about 6 per cent is saturated fatty acids and 94 per cent is unsaturated fatty acids. Among the unsaturated fatty acids, 14 per cent is oleic, 45 per cent erucic, 14 per cent linoleic and 10 per cent linolenic (Rehm and Espig, 1991). Low erucic acid and acid free cultivars

(0–rape) and cultivars also low in glucosinolate have been developed in Europe and Canada because of health problems associated with the consumption of oils containing erucic acid. Cultivars with no erucic acid and low glucosinolate content have also been bred (00–rape).

Ecological requirements

Rape is an ecologically demanding plant. A deep sandy loam rich in humus and nutrients and with an optimal lime content is the most appropriate soil for rape cultivation, followed by humic loam, loam and clayey loam in decreasing order of suitability. Humic and loamy soils can be appropriate when a sufficient water supply is guaranteed in April, though in general the success of the crop is greatly influenced by a sufficient water supply during the vegetation period. Because of spring rape's weaker root development it is more sensitive to water deficit than winter rape. Marshy or waterlogged soils are unsuitable (Honermeier et al, 1993).

After sowing winter rape, the plant should have about 100 days with temperatures over 2°C to reach the 8–10 leaf stage and develop the taproots necessary for wintering. The winter can be moderately cold with a light snow cover. Temperatures below −15°C may be damaging. There should be very little spring frost, and the spring should be moderately warm. In late winter or early spring when the temperature remains constantly above 5°C the plant will begin leaf growth (Honermeier et al, 1993). Summer rape is planted, usually in northern latitudes, where the winters are too severe for winter rape to survive the hibernation period.

Propagation

Rape is propagated using seed. Variety selection is based on the desired glucosinolate and erucic acid in the end product. Other factors that influence seed selection are location, climate, yield potential, date of ripening, special tolerances and planting time (that is, winter and spring forms). The most commonly used forms fall into the categories '0–rape' (elimination of erucic acid) and '00–rape' (elimination of erucic acid and greatly reduced glucosinolate content). These characteristics are necessary if the crop's seeds are to be used by the food

industry (Gross, 1993). In Germany the best yields have been obtained from the winter rape varieties 'Lirajet', 'Wotan' and 'Silvia', with 'Falcon', 'Liberator' and 'Vivol' having reached middle-value yields in comparison. Spring rapeseeds with high yield and oil contents are 'Lisonne' and 'Evita' (Rottmann-Meyer et al, 1995).

Since 1991, virtually all rapeseed production in the European Union has shifted to 00–rape, with low contents of erucic acid and of glucosinolates. The production of rapeseed in the European Union is still 'conventional', that is it does not contain GMO.

Crop management

Tests comparing conventional tillage (CT) (25cm deep ploughing) and minimum tillage (MT) (10–15cm deep disc harrowing) were conducted over a three-year period on very sandy soil in Italy. The results showed that there was no significant difference in rapeseed grain or biomass yields between the two forms of tillage, but MT was found to lead to a progressive worsening of soil conditions for plant root growth. MT was found, however, to lead to an average reduction of 55 per cent for working time, fuel consumption, energy requirement and cost when compared with CT (Bonari et al, 1995).

Winter rape is sown from the middle of August to the beginning of September. As a guide, areas in northern Germany usually plant rape between 10 and 20 August, while in southern Germany planting takes place between 20 August and the beginning of September. The desired crop density is between 60 to 80 plants/m², which is achieved by using 3–4kg/ha of seed. The seed should be planted at a depth of 1.5–3.0cm, with a distance between rows of 13.5cm, as with wheat. The distance between rows can be increased to 20–28cm if the soil is in optimal condition. Spring rape is sown from the end of March to the beginning of May. Up to twice as much seed as with winter rape may be necessary to obtain the same desired crop density (Rottmann-Meyer et al, 1995).

Pre-emergence or post-emergence herbicides can be applied to the field before sowing. Problem weeds are: chickweed, field foxtail grass, couch grass, chamomile, blind nettle, pansy, annual wild oats and reappearing wheat, among others.

Fungicides may also be used. Among the most common fungi to attack rape crops are *Sclerotinia*

Table 10.77 Fertilizer levels for winter rape as recommended in Germany

Fertilizer	Amount (kg/ha)
Nitrogen (N)	0–50 (in the autumn)
	Up to 100 (at the beginning of year with start of vegetation)
	80–100 (4 weeks later)
Phosphorus (P$_2$O$_5$)	80–100
Potassium (K$_2$O)	180–220
Magnesium (MgO)	25–30

Table 10.78 Recommended fertilizer levels for spring rape

Fertilizer	Amount (kg/ha)
Nitrogen (N)	80–100 (with sowing)
	60–80 (at 6–8 leaf stage)
Phosphorus (P$_2$O$_5$)	80
Potassium (K$_2$O)	120
Magnesium (MgO)	40

Source: Rottmann-Meyer et al (1995)

sclerotiorum, Botrytis cinerea, Phoma lingam and *Alternaria brassicae.*

Insecticides can be used to combat the variety of pests that infest rape crops. These pests include fleas (*Psylloides chryocephala*), lice (*Brevicoryne brassicae*), snails (*Deroceras agreste*), pollen beetles (*Meligethes aeneus*) and weevils (*Ceutorhynchus* spp.).

Organic fertilization using liquid manure is possible and recommended. Nitrogen fertilizer should be used according to the development of the crop. Table 10.77 gives the fertilizer levels recommended in Germany for winter rape. Recommended levels for spring rape are shown in Table 10.78.

Crops coming before rape should be early harvested crops such as winter barley or peas. Because of the danger of nematodes infesting the crop, rape should not include sugar beet in its crop rotation. Nor should host plants of *Sclerotinia* be used in rape's crop rotation. A break in rape cultivation of 4–5 years should be adhered to.

Studies by Christen and Sieling (1995) in northern Germany demonstrated that the best yield was obtained when rape was cultivated following peas.

Table 10.79 The average annual rapeseed yields since 1998

Year	Yield (T/ha)
1998	2.9
1999	3.2
2000	2.9
2001	2.6
2002	3.4
Average	3

Table 10.80 Canadian canola yield – updated 4 December 2008, (tonnes/acre)

1986	0.572
2008	0.769

Source: Field Crop Reporting Series – Statistics Canada (canola-council.org)

Table 10.81 Worldwide rapeseed production (million tonnes)

1950s	3.5
1965	5.2
1975	8.8
1985	19.2
1995	34.2
2006	47.0

Table 10.82 Top rapeseed-producing countries (million tonnes)

China	12.2
Canada	9.1
India	6.0
Germany	5.3
France	4.1
United Kingdom	1.9
Poland	1.6
Australia	0.5
World Total	**47.0**

Source: soyatech.com/rapeseed facts

Rotations with cereal crops also had favourable yields. The lowest yields were obtained when rape was grown in monoculture. It was found that in general the yields of oilseed rape increased with the length of the rotation and the length of the break between two rape crops.

Production

One way of looking at the value of the biomass from a crop is to look at its coefficient of energy performance (energy balance). This is the ratio of total energy obtained from the crop to total energy used by the crop, so ratios greater than one mean that energy is generated. Rapeseed methyl ester (RME), which can be used as a substitute for diesel fuel, has an energy balance of 5.5 including straw and 2.9 not including straw; on set-aside land this becomes 6.1 including straw and 3.2 not including straw. In comparison, ethanol from wheat has energy balances of 3.6, 3.9 on set-aside; and ethanol from sugar beet has energy balances of 2.46, 2.53 on set-aside. Fossil fuel refining has an energy balance of 0.94, not including losses from extraction, transport, etc. (Poitrat, 1995). This shows that biofuels have a positive energy balance, while mineral fuels have a negative energy balance.

In a study of oil crops in the USA by Goering and Daugherty (1982), it was determined that the total energy input for a non-irrigated spring rape crop was 7624MJ/ha. This includes the entire range of costs from seed price to oil recovery costs. The energy output was determined to be 20,066MJ/ha, based on a seed yield of 1233kg/ha and an oil content of 41 per cent. The energy output/input ratio was 4.18. This was the second highest ratio of nine different crops tested under conditions of non-irrigation. Soya bean had the highest output/input ratio at 4.56.

Table 10.79 shows the average annual rapeseed yields since 1998, the average yield over this time is 3t/ha. This value was used to calculate how much biodiesel could be produced from rapeseed (biodiesel-expo.co.uk).

The potential yield of finished biodiesel from rapeseed is 1 hectare of rapeseed = 1322L or 1 acre of rapeseed = 535L.

According to the BMVEL (German Federal Ministry of Food, Agriculture and Consumer Protection) estimate, the average winter rapeseed yield decreased by 8.4 per cent from 4.13Mt/ha to 3.78Mt/ha. From 2003 to 2004 the average yield had increased by 40 per cent.

Worldwide production of rapeseed/canola rose to 47 million tonnes in 2006, of which the total EU-25 production accounts for 16 million tonnes.

Processing and utilization

The winter rape crop is harvested in July, after a vegetation period of approximately 330 days. The seed pods and upper shoots are grey-brown; the seeds are black-brown to black, hard (when fully ripe), and rattle in the seed pods when moved (Honermeier et al, 1993). At this time the moisture content has fallen to under 20 per cent. Drying may be necessary to achieve the desired 9 per cent moisture content that makes the seed storable and ready for further processing. The straw can be immediately chopped to be used for fodder, or dried for storage or further processing. Winter rape has a yearly average seed yield of about 3t/ha and a dry matter straw yield of 10–12t/ha (Honermeier et al, 1993).

The spring rape crop is ready for harvesting when the plants take on a brown colour and the seeds have become loose in their pods. Spring rape

Source: www.bioenergiedorf.info/index.html

Figure 10.144 Rapeseed field for biodiesel production

has a seed yield of 1.5–2.5t/ha (Rottmann-Meyer et al, 1995).

Graef et al (1994) provide a look at the components that are processed from a rape crop. A whole-plant rape crop of 8650kg/ha consists of 5470kg/ha straw and 3180kg/ha seed. The seed can be processed into 1848kg/ha rapeseed meal and 1332kg/ha rapeseed oil. After it is refined, 1279kg/ha of the oil remains. This is mixed with 139kg/ha methanol to produce 1285kg/ha (1460L/ha) of rapeseed oil methyl ester and 133kg/ha of glycerol. The 1460L/ha rapeseed oil methyl ester has an energy content of 47.8GJ/ha.

Raw rapeseed is approximately 40 per cent oil and 60 per cent meal. The seed is processed by first pressing the seed to separate the oil from the meal. The oil can be further processed for technical uses. Among its many uses are hydraulic oils, chain saw oils, lubricants, biodiesel (RME), biocomponents for heating oil and diesel fuel, and constituents in cosmetics, ointments, plaster, and for leather processing. If the oil has no or low erucic acid content then it can be used by the food industry for such products as cooking oil, margarine and mayonnaise, or as a feed ingredient for cattle, pigs and birds. The meal undergoes an extraction process to produce the remaining oil; the residual meal product can be used as fodder (Gross, 1993; Sipilä, 1995).

The straw from the rape crop can be used in boilers or flash pyrolysis power plants to obtain heat and bio-oil. The bio-oil can be used by existing oil boilers or diesel generators to produce power and heat (Sipilä, 1995).

Rapeseed oil has also become the primary feedstock for biodiesel in Europe (estimates for 2006: more than 4.0 million tonnes of rapeseed oil went into biodiesel). Processing of rapeseed for oil production provides rapeseed animal meal as a by-product. The by-product is a high-protein animal feed. The feed is mostly employed for cattle feeding, but also for pigs and poultry (though less valuable for these). The meal (from 00–rape) has a very low content of the glucosinolates responsible for metabolism disruption in cattle and pigs.

Germany is a leading country in producing biodiesel. Several plants, some of which are of huge capacities, have been constructed in the country.

Biodiesel production capacity is growing rapidly, with an average annual growth rate from 2002–2006 of

over 40 per cent. For the year 2006, the latest for which actual production figures could be obtained, total world biodiesel production was about 5–6 million tonnes, with 4.9 million tonnes processed in Europe (of which 2.7 million tonnes was from Germany) and most of the rest from the USA. In 2007 production in Europe alone had risen to 5.7 million tonnes. The capacity for 2008 in Europe totalled 16 million tonnes. This compares with a total demand for diesel in the US and Europe of approximately 490 million tonnes (147 billion gallons). Total world production of vegetable oil for all purposes in 2005/2006 was about 110 million tonnes, with about 34 million tonnes each of palm oil and soya bean oil.

Some typical yields in litres of biodiesel per hectare:

- Algae: 2763L or more
- Hemp: 1535L
- Chinese tallow: 772L (970 GPa)
- Palm oil: 780–1490L
- Coconut: 353L
- Rapeseed: 157L
- Soya: 76–161L in Indiana (Soya is used in 80 per cent of USA biodiesel)
- Peanut: 138L
- Sunflower: 126L

The average US farm consumes fuel at the rate of 82 litres per hectare of land to produce one crop. However, average crops of rapeseed produce oil at an average rate of 1029L/ha, and high-yield rapeseed fields produce about 1356L/ha. The ratio of input to output in these cases is roughly 1:12.5 and 1:16.5. Photosynthesis is known to have an efficiency rate of about 3–6 per cent of total solar radiation and if the entire mass of a crop is utilized for energy production, the overall efficiency of this chain is currently about 1 per cent. While this may compare unfavourably to solar cells combined with an electric drive train, biodiesel is less costly to deploy (solar cells cost approximately US$1000 per square metre) and transport (electric vehicles require batteries which currently have a much lower energy density than liquid fuels).

Once considered a speciality crop in Canada, canola has become a major North American cash crop. Canada and the United States produce between 7 and 10 million tonnes of canola seed per year. Annual

Table 10.83 Selected properties of typical diesel and biodiesel fuels

Fuel property	Diesel	Biodiesel
Fuel standard	ASTM D975	ASTM D6751
Higher heating value, Btu/gal	~137,640	~127,042
Lower heating value, Btu/gal	~129,050	~118,170
Kinematic viscosity, @ 40°C	1.3–4.1	4.0–6.0
Specific gravity kg/L @ 60°F	0.85	0.88
Density, lb/gal @ 15°C	7.1	7.3
Water and sediment, vol%	0.05 max	0.05 max
Carbon, wt.%	87	77
Hydrogen, wt.%	13	12
Oxygen, by dif., wt.%	0	11
Sulphur, wt.%*	0.0015 max	0.0 to 0.0024
Boiling point, °C	180 to 340	315 to 350
Flash point, °C	60 to 80	100 to 170
Cloud point, °C	−35 to 5	−3 to 15
Pour point, °C	−35 to −15	−5 to 10
Cetane number	40–55	48–65

Source: McCormick (2009)

Source: Elsbett Technologie GmbH

Figure 10.145 Elsbett single-tank straight vegetable oil fuel system

Canadian exports total 3 to 4 million tonnes of the seed, 700,000 tonnes of canola oil and 1 million tonnes of canola meal. The United States is a net consumer of canola oil. The major customers of canola seed are Japan, Mexico, China and Pakistan, while the bulk of canola oil and meal goes to the United States, with smaller amounts shipped to Taiwan, Mexico, China and Europe. World production of rapeseed oil in the 2002–2003 season was about 14 million tonnes (wikipedia.org/wiki/Biodiesel).

Rapeseed oil is becoming a growing source of biodiesel fuel for automobiles. But the plant oil is also being used to drive many European power plants.

Biodiesel or pure rape oil can be used in almost any application and is already being used in heat and electricity generation worldwide. For example, the whole heat and power needs of the German Parliament building is provided by biodiesel.

Cars, trucks and trains can be found worldwide which are powered by pure plant oils, biodiesel in proportions from 5 to 100 per cent. The first car engine powered by pure rape oil was the Elsbett car engine.

A British Virgin Voyager, billed as the world's first biodiesel train, number 220007 *Thames Voyager*, was

Source: Chris McKenna 2006, http://en.wikipedia.org/wiki/File:390018_at_Crewe_railway_station.jpg

Figure 10.146 Virgin train powered by biodiesel

Source: http://commons.wikimedia.org/wiki/File:Bus_Articulat_Barcelona.JPG

Figure 10.147 A bus fuelled with biodiesel

Source: Flometrics www.flometrics.com/rockets/B100_test

Figure 10.148 Biodiesel may be an effective replacement for kerosene rocket fuel

converted to run on 80 per cent petrodiesel and only 20 per cent biodiesel, and it is claimed it will save 14 per cent on direct emissions.

Similarly, a train in eastern Washington will be running on a 25 per cent biodiesel 75 per cent petrodiesel blend during summer, purchasing fuel from a biodiesel producer situated along the railroad tracks. The train will be powered by biodiesel made in part from Washington-grown canola (wikipedia.org/wiki/ Biodiesel).

Future trips to space could be powered by vegetable oil. In a test firing, the California-based engineering firm Flometrics announced that commercially available biodiesel produced almost the same amount of thrust as conventional rocket fuel.

Two massive engines operate inside the plant at the Deutsche Bank 24 building on the edge of the Rhine River in Bonn. Built in 1997, the facility has a 600kW electrical and 700kW thermal capacity. It burns over 10,000 gallons of rapeseed oil methyl ester per week to supply the 165,000 square feet of office space with heat, warm water and power.

Facilities like this are substantially more efficient than traditional power plants because they retain the heat that is generated as opposed to releasing it into the atmosphere. The generator is also particularly eco-friendly because it uses refined rapeseed oil methyl ester as its primary fuel.

Rapeseed oil methyl ester emissions are neutralized when they are released into the environment, while other potentially harmful compounds are filtered through a single smokestack.

Internet resources

blogs.edmunds.com/.../FuelsTechnologies/Diesel/
www.dw-world.de/dw/article/0,1564,1222452,00.html
www.biodiesel-expo.co.uk/biodiesel_production.htm
http://en.wikipedia.org/wiki/Rapeseed
http://en.wikipedia.org/wiki/Canola
http://en.wikipedia.org/wiki/Biodiesel
www.soyatech.com/rapeseed_facts.htm
http://journeytoforever.org/biodiesel_svo.html
www.newscientist.com/article/dn16471-could-biodiesel-power-future-rockets.html

REED CANARYGRASS
(*Phalaris arundinacea* L.)

Description

Reed canarygrass is a perennial C_3 plant. This rhizomatous grass is most commonly found growing on creek and river banks, in ditches and on damp to wet meadows in subtropical and temperate zones throughout the northern hemisphere. It commonly forms extensive single-species stands along the margins of lakes and streams and in wet open areas, with a wide distribution in Europe, Asia, northern Africa and North America. The stems can reach 2.5m in height. The leaf blades are blue-green when fresh and straw-coloured when dry. The flowers are borne on the stem high above the leaves and are pinkish at full bloom (wikipedia.org).

It produces an expansive and well-developed underground rhizome and root system. The stem is generally stiff and upright, with branching often present in the lower part. The plant grows to a height of between 0.5 and 2.0m or more and produces a strong leaf shoot. The ungrooved leaf blades are from 8 to almost 20cm long and 0.8–2.0cm wide, with a matt underside. The inflorescence of canarygrass is a panicle between 10 and 20cm long. Canarygrass usually flowers in June and July (Siebert, 1975; Kaltofen and Schrader, 1991).

Ecological requirements

Success of the crop depends on a sufficient supply of nutrients and oxygen-rich, non-stagnant water. Reed canarygrass does not tolerate drought or salt. It is very tolerant to wet soil, but waterlogged soil is not suitable because of the deficiency of oxygen (Kaltofen and Schrader, 1991).

Reed canarygrass grows best under cool, moist conditions along lakeshores and rivers and is one of the best grass species for poorly drained soils. However, it is also found on upland sites, where it can survive temporary droughts.

Propagation

Reed canarygrass is propagated from seed, and vegetatively. For pure stands, sow 8 to 10lb/acre of pure

(a)

(b)

Source: (a) Jan Jackson, lcrwc.com, (b) Jamie Nielsen, University of Alaska Fairbanks, Cooperative Extension Service, Bugwood.org, www.invasive.org/browse/detail.cfm?imgnum=1196192

Figure 10.149 Reed canarygrass at different growth stages: (a) early; and (b) mature

live seed; for mixtures with legumes sow 6lb/acre of reed canarygrass. Sow at 0.5 to 1cm soil depth. Deeper sowing is advantageous on sandy soils and for summer sowing when surface soil moisture may be limiting. Seeds sown on the soil surface or greater than 1cm deep have little chance of developing into seedlings.

Reed canarygrass may require two weeks to germinate and emerge and its seedlings are not highly

Source: berr.gov.uk

Figure 10.150 Experimentation field with Miscanthus, reed canarygrass and switchgrass growing at Invergowrie

competitive. Therefore, careful attention needs to be given to establishment practices (Sheaffer et al, 1990).

Crop management

Canarygrass is most appropriately sown in damp to wet fields, though the extent of dampness depends on the soil's ability to support the necessary farm machinery.

In field trials conducted by Lechtenberg et al (1981), reed canarygrass was given 112, 224 or 336kg/ha of N and was harvested two, three or four times a year. Yields at 336kg/ha N averaged 11.5t/ha.

Plant recovery of N applied to canarygrass averaged 64, 67 and 55 per cent for the 112, 224 and 336kg/ha rates respectively, when the plant was harvested four times a year. Recoveries averaged 82, 67 and 63 per cent respectively when it was harvested only three times. The yield increase resulting from an application of 112 and 224kg/ha N was 30 per cent greater with two and three cuts than with four cuts a year. In a biomass production system where herbage quality is not important, an infrequent harvest system may yield more biomass per unit of applied N.

Some of the benefits as seen in Sweden for reed canarygrass cultivation (Olsson, 1993) are:

- a low investment level for establishment;
- the machinery and facilities already exist;
- there is no need for contractors; and
- flexible land use.

Canarygrass is known to be affected by rusts and mildew, along with other fungi. Because reed canarygrass is a perennial plant there is currently no recommended crop rotation. A single crop duration of 10 years or more may be possible.

Producers desiring highest quality should harvest at boot stage, while those desiring highest yield should harvest at heading in the first growth in the spring, or at the end of stem elongation for the summer regrowth. Reed canarygrass will normally flower only once when subject to repeated harvests during the year. The spring growth will ultimately terminate in the production of an infloresence and seed. Subsequent regrowths will result in stem elongation but no infloresence.

Table 10.84 Characteristics of perennial cool-season grasses

Grass	Heat/drought tolerance	Flooding tolerance	Winter hardiness	Frequent cutting tolerance	Seedling vigour	Sod-forming capacity
Reed canarygrass	E	E	E	E	F	E
Smooth bromegrass	E	F	E	P	E	E
Orchardgrass	G	P	F	E	E	P
Tall fescue	E	P	F	E	E	F
Timothy	P	P	E	P	F	P
Perennial ryegrass	P	P	P	E	E	P
Kentucky bluegrass	P	F	E	E	F	E

E = excellent, G = good, F = fair, P = poor

Reed canarygrass is better adapted to diverse uses and environmental conditions than most other commonly used perennial grasses (Table 10.84).

Reed canarygrass has superior persistence on poorly drained soils, yet its yield and persistence under moisture deficits is equal or superior to other commonly grown cool-season grasses. Reed canarygrass is very winter hardy.

Of the perennial grasses adapted to Minnesota, reed canarygrass is among the most persistent. It maintains yield under cutting strategies designed to produce both low- and high-quality forage.

Production

Tests in Sweden comparing the varieties 'Palaton', 'Venture', 'Vantage' and 'Muttenwitzer' resulted in dry matter yields (after winter losses) of about 4.8t/ha, 3.3t/ha, 5.2t/ha and 4.2t/ha respectively. In addition, it was shown that despite the higher yields from summer-harvested crops, the spring-harvested crops had lower production costs (Olsson, 1993).

Mediavilla et al (1993; 1994, 1995) obtained the dry matter yields (t/ha) from reed canarygrass field tests that were conducted using liquid manure for fertilizing at two locations in Switzerland, Reckenholz and Anwil. The yields (t/ha) for 1993, 1994 and 1995 were respectively 12.4, 11.0 and 13.5 at Reckenholz, and 19.3, 17.0 and 12.3 at Anwil. The crop was harvested three times in 1993 and twice in each of 1994 and 1995. At Reckenholz between 70 and 90kg/ha NH_4-N liquid manure per year was used, while at Anwil the rate was up to 180kg/ha NH_4-N. It is also notable that different climatic conditions during the growing period may significantly affect the biomass yields.

In Finland, the cultivation area of reed canarygrass used mainly for combustion has increased quite rapidly and is estimated to reach 70,000 hectares in 2010 (Pahkala et al, 1994). For combustion, reed canarygrass is mostly harvested in spring when the dry matter content is high (about 85–90 per cent), but for methane production it should be harvested as fresh biomass, enabling harvesting usually at least twice a year in boreal growing conditions.

The biomass total solids (TS) yield per hectare was 79 per cent higher when harvesting twice per plot at flowering stage compared to harvesting twice per plot at vegetative stage in 2005. In 2006, the biomass TS yield per hectare was the same in both plots. The yields of TS in both years were in the same range as earlier reported for reed canarygrass harvested at spring in the US (1.6–12.2t DM/ha) and in Europe (7–13t DM/ha) (Lewandowski et al, 2003).

In 2005, about 10 per cent higher methane production per tonne of TS was gained when harvested at vegetative stage (270m³ CH_4/t TS) compared to harvesting at flowering stage (250m³ CH_4/t TS).

However, the total methane production potential per hectare was 64 per cent higher when harvesting at flowering stage because of the higher TS yield compared to harvesting at vegetative stage. The difference between methane productions per tonne of TS was higher in 2006 when the samples at vegetative stage produced 310m³ CH_4/t TS and the samples at flowering stage 260m³ CH_4/t TS, which means that the plot harvested at vegetative stage yielded 20 per cent more methane per hectare than the plot harvested at flowering stage. The potential methane yields of reed canarygrass per hectare per year corresponded to energy yields of 20–36MWh/ha (Paavola et al, 2007).

Processing and utilization

Canarygrass is most commonly harvested once a year by mowing and baling. Drying consists either of swathe drying or barn drying, with storage in a barn or silo or under canvas. Drying is important for reducing transportation costs and for maintaining good fibre quality.

In Sweden, research on crop production using a delayed harvesting method has been performed (Olsson, 1994). This method involves harvesting the canarygrass crop after the snow has melted in the spring (May) instead of the usual summer harvesting. This leads to death of the above-ground plant and, along with field drying, moisture contents as low as 10–15 per cent can be achieved with delayed harvest crops. In addition, during wintering, the plant returns much of its nutrients to the soil, but maintains the important cellulosic matter content.

Their generally lower density means that spring-harvested crops can be baled using a high-density baler. Spring-harvested crops have been shown to have improved quality for processing into fuel and pulp raw material. A disadvantage of this method is that winter losses of 20–30 per cent in dry matter have been experienced.

Large-scale trials comparing a traditional harvest time in August and the delayed harvest time in the spring showed that for the August harvest of 10.4t of total crop yield, 35 per cent was lost during harvesting, resulting in 6.8t net yield. For the spring harvest of 10.4t of total crop yield, 29 per cent was lost during the winter and 18 per cent was lost during harvesting, resulting in 5.5t net yield (Olsson, 1993).

One of the most heavily researched and promising uses of reed canarygrass is as pulp for fine paper making. Qualities that are required of short fibres in printing grade papers are a large number of fibres per unit weight, and stiff, short fibres (low hemicellulose content, and low fibre width to cell wall thickness ratio). Canarygrass has short, narrow fibres (0.72mm mean length) and produces pulp with a greater number of fibres than most hardwoods. The delayed harvest crops have been shown to lead to approximately 20 per cent more pulp per tonne of grass compared to summer harvest crops. The grass fibres are easy to pulp and can be cooked and bleached using existing techniques.

Because of its low moisture content, spring canarygrass makes a good candidate for fuel upgrading to pellets, briquettes and powder (with qualities similar to wood powder). In addition, the delayed harvest crop shows a reduction in undesirable elements, making it more suitable for combustion as well as better for the environment. The heating value of the grass is about 4.7kWh/kg dry matter.

Table 10.85 Fuel characteristics of reed canarygrass

Parameter	Summer harvest	Delayed harvest
	(% of dry matter)	
Net heating heat value (MJ/kg)	17.9	17.6
Ash	6.4	5.6
Carbon	46	46
Hydrogen	5.7	5.5
Nitrogen	1.33	0.88
Sulphur	0.17	0.09
Chlorine	0.56	0.09
Volatile matter	71	74
Initial ash deformation (°C)	1074	1404
Potassium	1.23	0.27
Calcium	0.35	0.20
Magnesium	0.13	0.05
Phosphorus	0.17	0.11
Silica	1.2	1.85
Cadmium (mg/kg)	0.04	0.06

Table 10.85 provides the fuel characteristics of reed canarygrass as found in summer harvest and delayed harvest crops. The information was compiled from 1991 and 1992 Swedish trial sites (Olsson, 1994).

Reed canarygrass, along with a few other plants, are currently being used as plant cover for wastewater treatment in the Czech Republic. This is part of a new constructed wetlands technology (Vymazal, 1995).

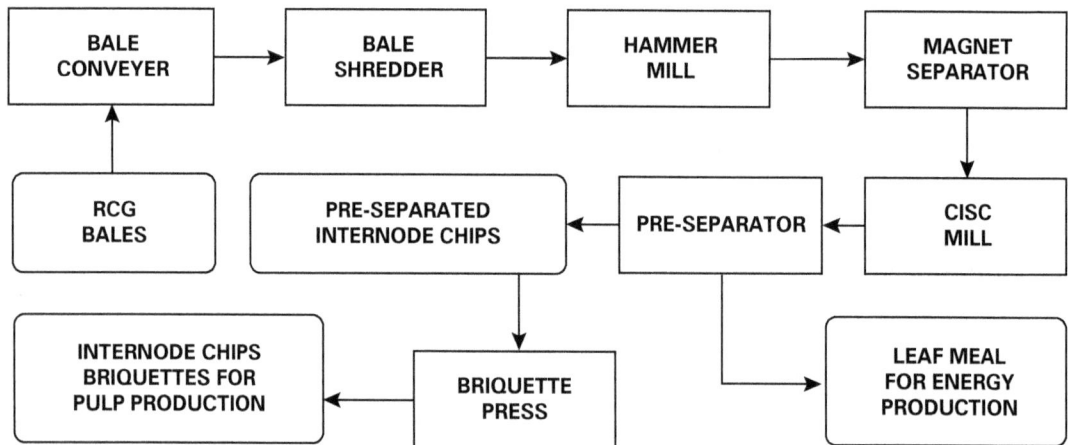

Source: Olsson (2005)

Figure 10.151 Scheme of the fractionation process, including briquetting of the chip fraction

Source: Ismo Myllylä, www.vapoviesti.fi/vapoview/index.php?id=136
8&articleId=86&type=9

Figure 10.152 Reed canarygrass was trialled last spring in a number of power plants. The photo shows the grass on its way to be crushed at the Rauhalahti power plant in Jyväskylä

Although canarygrass is presently part of breeding programmes to make it more advantageous for biofuel and pulp production, it is still a good fodder grass for hay and silage production, especially in floodplain meadows.

Reed canarygrass is presently used as pellets and briquettes as fuel feedstock for heat and power generation fuel in Scandinavia.

A R&D plant for fuel upgrading and local heating was constructed in November 1999. The plant produces the heat for the research buildings of the university in Röbäcksdalen Umeå and replaces the previous consumption of about 300m³ oil.

One of the users of reed canarygrass is Vapo's Lieksa power plant in eastern Finland, where it began to be combusted in combination with peat in March 2007. Jussi Tiihala, the power plant's manager, says that experiences have been positive.

The fluidized bed boiler at Lieksa was refurbished in summer 2006 and since March it has been combusting reed canarygrass mixed with peat. Experiences of use have been positive (Myllylä, 2007).

Reed canarygrass is considered as an ideal biofuel source for Michigan's Eastern Upper Peninsula (EUP). It grows luxuriously in fields that are too wet for other uses and it is already abundant across the EUP. Investigations showed that reed canarygrass could represent an economic potential for the EUP both in terms of reducing fuel costs and providing another source of income for the area's farmers.

Source: Ismo Myllylä, www.vapoviesti.fi/vapoview/index.php?
id=1368&articleId=86&type=9

Figure 10.153 The fluidized bed boiler at Lieksa refurbished 2006 combusting reed canarygrass mixed with peat

Internet resources

www.vapoviesti.fi/vapoview/index.php?id=1368&articl
eId=86&type=9
www.p2pays.org/ref/17/16274/ollson.pdf
www.cropgen.soton.ac.uk/publications/8%20Other/O
th_27_FCES2007%20paper_Paavola.pdf
http://en.wikipedia.org/wiki/Phalaris_arundinacea
www.lcrwc.com/projects/beaver-creek-project/
www.dnr.state.wi.us/invasives/fact/reed_canary.htm
www.berr.gov.uk/files/file34815.pdf

ROCKET (*Eruca sativa* (L.) Mill)

Partially contributed by A. K. Gupta

Description

Rocket is a collective name for many species within the Brassicaceae family whose leaves are characterized by a more or less pungent taste and are, therefore, used to flavour salads. Variation of taste and pungency is great, depending on the species, its genetic diversity and the environment. In the Mediterranean region three main rocket species can be found, along with several other taxa also occurring wild throughout the region. These three species used for human consumption are:

- *Eruca sativa* Miller: a diploid, annual species which flowers in spring and whose seeds are ready for collecting in late spring. It seems to prefer rather rich soils even though it can be found mixed with ruderal flora in very marginal areas. It is frequently cultivated, although domestication cannot be considered complete. A wild type, known as subspecies *vesicaria* (L.) Cav., is also rather well represented in the Mediterranean flora.

- *Diplotaxis tenuifolia* (L.) DC.: a diploid and perennial species, in the sense that the roots can survive winters and produce new sprouts in the next spring; it flowers from late spring to autumn and its seeds are generally ready for collecting in autumn. It seems to be very well adapted to harsh

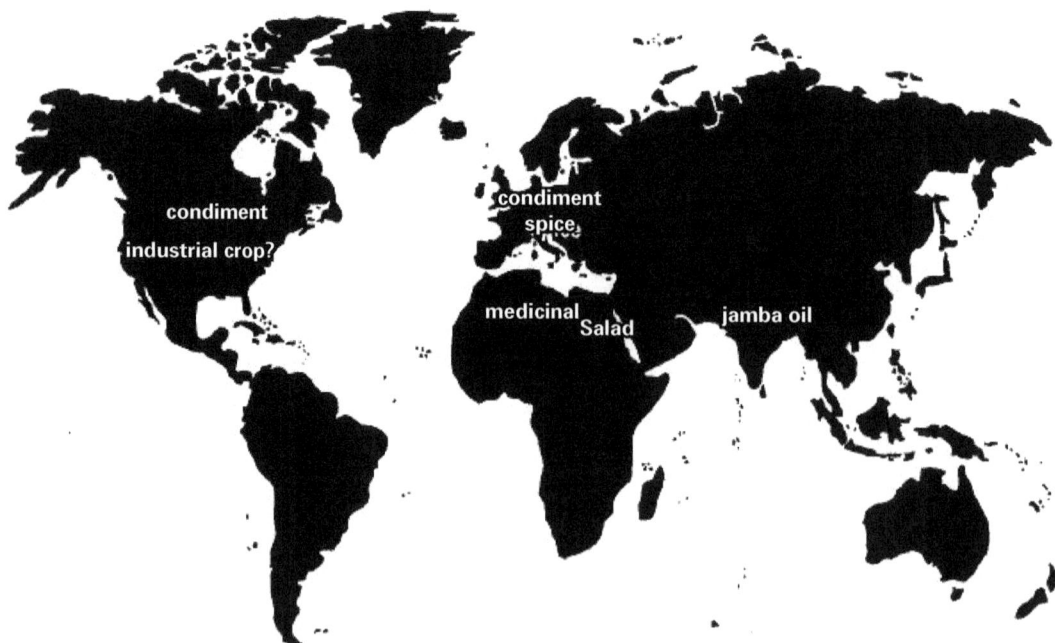

Source: Padulosi and Pignone (1997)

Figure 10.154 Uses of rocket throughout the world

and poor soils, and often it can compete well with other species in calcareous shallow soils. This species has succulent leaves and is much appreciated in cuisine. In some Italian areas *D. tenuifolia* is also cultivated, but it is mostly collected from the wild and sold in small bunches in local markets.

- *Diplotaxis muralis* (L.) DC.: polyploid and perennial, in the same sense as *D. tenuifolia*. It flowers from summer to autumn and its seeds are ready for collecting in autumn. It grows in similar habitats as *D. tenuifolia* and is also collected from the wild to be sold in the markets. It seems less adapted to cultivation because of its procumbent growth habit, which is the main character distinguishing it from *D. tenuifolia*.

The above-mentioned nomenclature follows the Flora of Italy (Pignatti, 1982).

Rocket is grown on a commercial scale in Portugal for export to the UK and other northern European countries as a fourth-generation salad product, i.e. prepared and sold in sealed bags after having been cleaned and mixed with other leafy vegetables (Silva Dias, 1997).

The oil of *E. sativa* is rich in erucic acid, an important industrial compound, and attempts to exploit the potential of this species as an industrial oil crop are also being made (Figure 10.154). With an average of 2.5 harvests for *Eruca sativa* and 1.3 for *Diplotaxis* spp., average production per hectare is around 16–18t/ha and 19–21t/ha respectively (Padulosi and Pignone, 1997).

High erucic acid (HEA) oil has special properties which include high smoke and flash points, oiliness and stability at high temperatures, the ability to remain fluid at low temperatures, and durability. HEA oil is used to produce erucamide which is used as a slip additive in polythene and polypropylene, to reduce surface friction and prevent adhesion between film surfaces. It is also used in printing inks, lubricants and has a range of other applications. Current world production of erucic acid is 25,000Mt/year. Consumption of HEA in the EU was 40,000t/year in 2000 and was predicted to increase to 55,000t/year by 2005. The largest current applications for HEA oil are in polymer additives and

Source: Richard Old, XID Services, Inc., Bugwood.org,
www.invasive.org/browse/detail.cfm?imgnum=5234057

Figure 10.155 Garden Rocket *Eruca sativa*

detergents. As the crude oil is biodegradable it provides an alternative to mineral oil in many industrial applications. It may have application as a lubricant for chainsaws.

Eruca matures from seed in 2–3 months; periods of very warm temperatures cause it to bolt rather quickly. It appreciates full sun, although shade should be provided from midday sun in summer. The crop also appreciates regular watering or adequate rainfall. It can tolerate temperatures down to $-4°C$.

High levels of nitrogen applied increase fresh yield by up to 50 per cent (Ahmed et al, 2002). In trials oilseed yields of 180–350kg/ha have been achieved, commercial yields are currently unknown although they are predicted to be similar (ienica.net/crops/eruca.pdf).

Rocket is a C_3 annual plant that grows to a height of 0.3–1m. It is a member of the rapeseed and mustard group of crops (n=11). The plant is native to the Mediterranean region but is distributed from India in the east, to the central USA in the west, and to Norway in the north. Rocket is known by different local names such as: Ölrauke or Senfkohl (Germany), roquette (France), ruchetta or ruccola (Italy), roqueta, oruga or jaramago (Spain), rocket, rocket-salad or hedge mustard (UK), taramira (in Hindi), etc.

The roots of the rocket plant are of spindle form with few branches (Schuster, 1992). The stem is branched and the leaves are compound, resembling those of spinach. The flavour of young, tender leaves is pungent and they used in southern Europe as a salad. However, rocket is commercially cultivated for its oil-bearing seeds, though the oil (about one-third of the seed weight) is non-edible because of the presence of erucic acid and it is used only for secondary purposes.

India, with more than 6 million hectares under oleiferous Brassicaceae, seems to be the largest producer of rocket. Here it is raised as a main crop on moisture-deficient and saline areas of the mustard belt in the north-western parts of the country. More frequently, rocket is chosen as a contingent crop in the event of failure of a winter crop, or because of delayed winter rains from a receding monsoon. These factors mean that there is no trend over time with respect to the area and production of rocket in India.

Ecological requirements

The ecological requirements of rocket are similar to those of rapeseed and mustard. It is a cool-season crop where the winter is mild. Nevertheless, the plant has some unique inherent features that allow it to survive under adverse soil and atmospheric conditions in which crop plants can hardly grow. These characteristics include an efficient and fast-penetrating root system that imparts drought hardiness, tolerance to salinity, a wide adaptability towards sowing time, tolerance of frost, and less susceptibility to biotic stresses (Kumar and Yadav, 1992). The crop thrives well on manured loamy soils with a pH range of 6–8 and having good provision for drainage.

'Rocket Improved', a spicy, improved cultivar, is less prone to bolting than the standard type, though it still grows best in cooler weather (Facciola, 1990).

Propagation

Rocket is propagated using seed. Because of its small seed size, rocket requires a seedbed of fine tilth. When the crop is to be sown in summer fallow, disc harrowing after every effective shower is the best method of conserving moisture. After the cessation of rains, harrowing should invariably be followed by planking to prepare a fine and compact seedbed. In rainfed areas, germination can be improved by presoaking the seed in water, or by using 'Jalshakti' starch polymer (Gupta and Agarwal, 1997). Because of its confinement in small and isolated pockets, rocket has been a neglected crop and, consequently, high-yielding seed is a great problem. A normal seed rate of 3kg/ha is adequate to obtain the desired plant population of 225–250 thousand plants per hectare, but a higher rate is usually adopted to compensate for the poor germination that is mainly due to suboptimal moisture conditions.

Seed is sown outdoors in spring in situ. Germination is usually very quick and free. In order to obtain a continuous supply of edible leaves, successional sowings can be made every few weeks until mid-August. A late summer/early autumn sowing can provide leaves in winter, though the plants might require some protection in very cold winters (Organ, 1960).

Seed treatment with a fungicide such as Bavistin, Thiram, etc. protects the crop from seed-borne diseases and in turn ensures uniform crop emergence and an optimum stand. Spacing of 30cm × 15cm is ideal for optimizing biomass yield. Seed should be sown to a depth of 3cm in open furrows with a ridge seeder. As with any other crop, the date of sowing (mainly governed by temperature) is an important factor in determining the growth and performance of rocket. For instance, around the first week of October has been shown to be the optimum sowing time in northwest India. Delayed sowing resulted in reduced yield both of biomass and oil primarily as a result of the shortening of the reproductive phase and an increased incidence of aphids. On the other hand, high temperature in early sowing led to improper canopy development and reduced branching of stems. The attack of seedling-killing insects such as *Bagarada hilaris* was also frequently observed at early sowing.

Crop management

Because rocket has been grown on wastelands, it has continued to rely on natural selection for survival. As a result, the prevailing landraces and cultivars do not respond to intense input management. The problem of soil-inhabiting insects can be tackled by applying methyl parathion at 25kg/ha. With regard to nutrient management, the available literature suggests that rocket is a moderate feeder. An application of 40kg/ha N, 20kg/ha P_2O_5, 15kg/ha K_2O and 90kg S will be enough. Further, pre-emergent application of Fluchloralin (1kg/ha) is effective against most of the weeds, whereas hand weeding requires 15 labourers per hectare.

Irrigation is not required if rain is well distributed. The pod formation stage is recognized to be most crucial with regard to water stress; ensuring adequate moisture at this stage significantly boosts the seed yield. Although the established crop is not prone to significant damage by insects, even here the probability of any economic injury can be prevented by spraying with a systemic insecticide. Rocket has been shown to be twice as frost tolerant as Indian mustard; however, a spray of 0.1 per cent sulphuric acid is reported in the literature to protect a crop against frost injury. Rust/blight and/or mildew can be controlled by spraying a 0.2 per cent solution of Dithane M-45.

Its deep root system makes rocket a good preceding crop in a crop rotation. In India, a crop rotation of cowpea or cluster bean/rocket has proved to be more profitable. In multiple cropping systems, rocket/summer green gram or a fodder crop/maize or cowpea can be an appropriate option. There are no recommendations or restrictions for possible crops to follow or precede rocket.

Production

A higher production level cannot be expected from the local varieties and the poor management they receive. Seed and straw yields of 1.5t/ha and 4t/ha respectively have been obtained in various locations in India. There is considerable scope for raising the productivity level of this crop through crop improvement and proper management.

Processing and utilization

Reported properties of rocket include its strong aphrodisiac effect, known since Roman times. Among other less intriguing medicinal properties, there is also a depurative effect and it is a good source of vitamin C and iron. In Egypt, particular ecotypes with large leaves are used as salad species instead of other more expensive and less adaptable species like lettuce. These large-leaved ecotypes are reported to lack a pungent taste.

In the Indian subcontinent, and in Pakistan in particular, special ecotypes of *E. sativa* are cultivated for seed production. The seeds are used to extract an oil often named 'jamba oil' which has many interesting uses such as for illumination or in the production of pickles (Padulosi and Pignone, 1997).

The crop is considered mature and ready to harvest when the siliquae have turned brown-yellow and the seeds inside the pods have become loose. In India, the crop matures in about 140–150 days. The crop is harvested manually using sickles. Harvesting needs 18 labourers per hectare. The harvested material is brought to an area for threshing to reduce the moisture content of the seed to about 9 per cent, which makes it storable and ready for processing. Currently, the seed is threshed out manually by beating the plants with a long stick or by hitting small bundles of the crop against a hard surface. Seed is sold in the market, and then oil is extracted and used for various purposes. The seed yields a semi-drying oil which is a substitute for rapeseed oil. It can also be used for lighting, burning with very little soot.

The aerial part of a mature plant consists of about 20 per cent seed and 80 per cent straw. The seed can be processed into approximately 33 per cent oil and 67 per cent oilcake. The energy value of rocket oil is not available, but is believed to be comparable with that of rapeseed. Oilcakes serve as a concentrated feed for dairy animals or are used as a manure.

Straw is used for bedding material in cattle sheds or burned to generate heat. The present method of disposing of straw cannot be called efficient, effective and environmentally sound. An alternative strategy is needed to make the best possible use of rocket biomass.

Internet resources

www.guenther-blaich.de/pflseite.php?par=Eruca+sativa&abs=pflti&fm=
www.ienica.net/crops/eruca.pdf

ROOT CHICORY (*Cichorium intybus* L.)

Description

The best known wild plant from the genus *Cichorium* (family Cichoriaceae) is *Chicory intybus* L. var. *intybus*. It is a biennial C_3 plant with strong cylindrical to cone-shaped roots. The first year is marked by the development of a flat adjacent leaf rosette and the roots. The inflorescence is developed in the second year and by the end of July the plant will produce blue flowers (Franke, 1985). One of the main values of chicory is its root content of inulin, a polyfructoside, which can be processed by the chemical, pharmaceutical and food industries for various uses. Extracts from roasted roots have also been used to improve the aroma of coffee.

Introduction of the fructan crops root chicory (*Cichorium intybus* L.) and Jerusalem artichoke (*Helianthus tuberosus* L.) into agricultural production systems is desirable to diversify crop rotation.

The traditional areas of chicory cultivation have been northern France and Belgium, though the total EU cultivated area was approximately 14,000ha in 1992.

Ecological requirements

Appropriate soils for chicory cultivation are sandy loams, loams and other soils that have less than a 30 per cent clay content. The soil should be well drained, but not too dry, and quick to warm up in the spring. The optimal pH value is between 6 on sandy soils and 7 on loamy soils. Not appropriate are soils that are excessively wet or acidic, have too many stones, or do not permit an even, level seedbed (Dhellemmes, 1987; Baert, 1991, 1993).

Chicory is most commonly produced in a mild, moist climate such as in north-western France and Belgium. There the crop can be harvested up until December.

The nutrient content of the roots depends, to a degree, on the variety of chicory planted. In general, 100g of root dry matter consists of 0.9–1.0 per cent of nitrogen, 1480—1986mg potassium, 34–42mg sodium, 212–237mg calcium, 88–94mg magnesium and 271–318mg phosphorus.

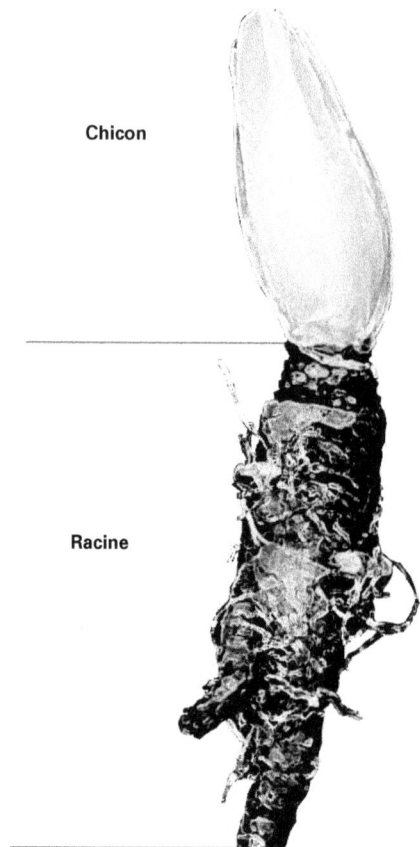

Chicon

Racine

Source: wikimedia.org

Figure 10.156 Belgian endive

Propagation

Chicory is most commonly propagated using seed. The plant has already undergone extensive breeding and is still the subject of breeding. Among the traits that are being improved are better emergence, adaptability to an earlier sowing, faster growth, higher sugar yield, salt tolerance in the germinating plants, resistance to disease and damage, etc. The plant may also be propagated by planting seedlings, which has the advantage of controlling undesired crossings. When choosing the variety to be planted it is important to take note of the potential yield and date of ripening, among other things.

Crop management

Following a spring ploughing of the field to a depth of about 30cm, the sowing of the crop by seed takes place between the end of April and the beginning of May. For smaller seeds it is important that the field is flat and level and has a soil temperature of at least 10°C. A machine for drilling sugar beet can be used. The seed can be either coated or uncoated.

Uncoated seed is sown at a depth of 0.5–1.0cm. Coated seed is sown at a depth of 1.0–1.5cm. The distance between rows should be 45cm and the distance between plants should be about 9cm. An appropriate crop density is 140,000 to 160,000 plants per hectare, which is achieved with approximately 250,000 seeds.

Chicory reacts very sensitively to poor irrigation and crop management during germination, or to an encrusted soil surface after a heavy rainstorm. If after 14 days no germinating plants can be seen, the crop should be resown (Dhellemmes, 1987; Baert, 1991).

Mineral fertilizers should not be used shortly before sowing because of possible erosion of the germ roots. The soil should be analysed to determine its composition and to help determine fertilizer levels, according to Bramm and Bätz (1988), but recommended fertilizer levels in Germany are as shown in Table 10.86.

Two-thirds of the nitrogen should be applied three weeks before sowing, and the remaining one-third at the six-leaf stage.

Source: Richard Old, XID Services, Inc., Bugwood.org
(www.invasive.org)

Figure 10.158 Root chicory foliage

Table 10.86 Recommended fertilizer levels in Germany for chicory crops

Fertilizer	Amount (kg/ha)
Nitrogen (N)	100–130
Phosphorus (P_2O_5)	60–130
Potassium (K_2O)	200–250
Magnesium (MgO)	40–60

Source: California Vegetable Specialties

Figure 10.157 The biennial life cycle of the chicory plant

Fields with serious weed problems should not be used for cultivating chicory. Because chicory is slow to establish itself, a pre-sowing and/or pre-emergence herbicide is recommended to help control weeds.

Chicory should not be cultivated after sugar beet, carrot, sunflower, bean, lettuce or endive (wild chicory) crops because of the danger of *Sclerotinia*. There are no recommendations or restrictions for possible crops to follow chicory. Chicory has a deep-growing and wide-spreading root system, so the soil is usually left in a very good condition after the crop.

Production

With yield variations depending on the climate, location and crop management, root yields of 32 to 50t/ha (Anon, 1993) have been achieved, and yields as high as 56t/ha with an inulin content of 18.2 per cent can be foreseen. This would produce an inulin-fructose yield of over 10t/ha (Baert, 1993). Root dry matter yields between 10.6 and 16.5t/ha have been observed. This resulted in inulin-fructose yields between 8 and 12.2t/ha (Meijer and Mathijssen, 1991).

Field experiments were conducted to compare the agronomic performance of root chicory (RC) and Jerusalem artichoke (JA) with sugar beet (*Beta vulgaris* L.; SB). One set of cultivars during 1995, 1996, and 1997 and an additional set in 1997 were grown on a sandy loam soil (Haplic Luvisol) at Braunschweig, Germany (Schittenhelm, 1999). Crops were cultivated with and without supplemental irrigation, complete and no weed control (1995 and 1996), and N fertilization rates of 0, 60 and 120kg/ha. Severe water stress caused significant but similar storage organ yield losses in all crops, whereas mild water stress mainly affected JA yields. Averaged across years and N levels, storage organ yield losses through weed competition under irrigation amounted to 70, 47 and 8 per cent in SB, RC and JA, respectively. Averaged across years, in the absence of water and weed stress, SB, RC and JA at their respective optimal N levels gave root and tuber dry matter yields of 14.8, 15.0 and 11.5t/ha, and sugar yields of 11.5, 11.2 and 8.1t/ha, respectively. Maximal SB and JA yields generally were achieved at the highest N rate, while RC peak yields were attained at 60kg N/ha. With the same amount of N taken up, RC in 1995 and SB in 1996 and 1997 produced the highest sugar yields. (Schittenhelm, 1999)

Processing and utilization

Chicory seeds are usually harvested in the first half of September of the second year. Seed moisture contents of 12–15 per cent make it appropriate to allow the seeds to dry naturally in an elevator. If the moisture content is more than 15 per cent then the seeds will need to be dried before storing. Although the seed yield largely depends on the year's weather, yields of 600 to 1200kg/ha can be expected (Delesalle and Dhellemmes, 1984).

Source: Chicory, S.A.

Figure 10.159 A chicory root harvest

Chicory roots are usually harvested in the autumn, around the middle of November depending on the time of ripening for the variety used. A modified six-row sugar beet rooter can be used to harvest the roots.

According to Frese et al (1991), chicory roots have a dry matter content of 22.2–26.8 per cent, a total sugar content of 17.4–21.9 per cent, a fructose content of 14.6–19.1 per cent and a glucose content of 2.5–3.5 per cent. Because the fructose fraction in syrups produced from inulin is up to 75 per cent, fructose syrup production from chicory is possible (Guiraud et al, 1983). The polyfructosides, which grow in chain length throughout the vegetation period, also have the potential to be used for ethanol production.

Internet resources

http://upload.wikimedia.org/wikipedia/commons/a/ab/
Intibum_Witloof_schema.jpg
www.humanflowerproject.com/index.php/weblog/com
ments/chicory_the_root_of_todays_coffee_break/
www.croptech.com.au/agrd.html

ROSIN WEED (cup plant) (*Silphium perfoliatum* L.)

Description

Rosin weed is a perennial wild C_3 plant originating in North America that grows to a height of 2 to 3.5m. The stalk is strong and relatively square with a diameter of 20–25mm. The plant has six to eight leaf pairs opposing each other on opposite sides of the stalk. The leaves are coarse with a length of over 30cm and a width of 25cm. The leaves towards the top of the stem grow around the stem to form a cup that captures and holds precipitation.

Flowering begins sometime in July, at which time the plant produces numerous yellow flowers, each having a diameter of 5–8cm. The flowers produce nectar and pollen that attract many insects to the plant. The seeds ripen at the end of September, with each flower producing 20 to 30 seeds. This amounts to approximately 400–500 seeds per plant. Rosin weed develops a very strong root system, which consists of one main root reaching a length of up to 100cm and numerous lateral roots that spread out near the surface of the soil.

Ecological requirements

Rosin weed has no special requirements from the soil. The plant thrives on soils with a high water level and marshy soil. Only soils with a high salt content are unsuitable. Rosin weed is also open to a wide range of climates. The plant has an excellent winter hardiness, being able to withstand temperatures below –30°C. The plant also has a distinct drought resistance. A sufficiently warm summer, such as in wine grape-growing areas, is recommended for rosin weed crops. Wet and cold years lead to a diminished harvest.

Propagation

Although rosin weed can be propagated using plant cuttings, the most common propagation method is by seed. The seed can be directly sown into the field, or it can be cultivated in pots first and then transplanted into the field at a height of 6–10cm.

(a)

(b)

Source: (a) Beverly Turner, Bugwood.org (www.invasive.org); (b) Van Buren County Community Center

Figure 10.160 Rosin weed (*Silphium integrifolium*)

Crop management

The field should be properly prepared to avoid weeds in the first year. The seed can be sown in the spring, after sufficient field preparation with ploughing and harrowing. Approximately 10kg/ha of seed is used, at a sowing depth of 1–1.5cm. The distance between plants is 25–50cm and between rows 50cm. The seeds germinate 3–4 weeks after being sown. Afterwards, the plants grow very slowly and the soil needs to be hoed two or three times to remove weeds. By the time the plants have reached a height of about 1m they are large enough to crowd out the weeds. However, a weed hoeing is recommended at the beginning of each additional year.

Diseases and harmful pests of rosin weed have not yet been encountered.

Rosin weed produces a large amount of above-ground and below-ground biomass, which is very positively influenced by application of fertilizer. Experience from plots in Germany has concluded that an organic fertilizer such as stable manure is recommended at the beginning of the growth cycle and then later liquid manure can be used. The additional inorganic fertilizer levels have also been advised (Table 10.87). Stable manure-fertilized crops such as potatoes, turnips and cabbage are good prior crops, but because rosin weed has a life of more than 10 years it may not be necessary to develop a crop rotation.

Production

No market or market value has been established for rosin weed. Because the plant has been used mainly as feed for animals, usually directly where it is grown, it has not been sufficiently established as a crop.

In Germany, yields have been recorded in four areas: fresh-cut yield in secondary mountains was approximately 70t/ha, on dry loess soil 100t/ha, on hilly land in Thuringen 130t/ha, and on the coast 140t/ha. These yields correspond to dry matter biomass approximations of 8, 12, 15 and 16t/ha respectively.

Processing and utilization

When used for animal feed the plant is harvested in September, at which time it has a dry matter content of

Table 10.87 Recommended fertilizer levels for rosin weed, as used in Germany

Fertilizer	Amount (kg/ha)
Nitrogen (N)	100–125
Phosphorus (P_2O_5)	110–115
Potassium (K_2O)	125

13.9 per cent. It is also possible, starting in the second year, to harvest the portion of the plant above 0.8–1.2m one to three times a year. This has been shown to lead to a higher total yield than a once a year harvest. Because planted fields have been small until now, the plant has been harvested manually with a sickle.

When used as feed, rosin weed is chopped into small pieces and fed directly to the animals. The plant as well as the techniques of harvesting, handling and storage need to be researched with a view to large-scale biomass production for energy production or conversion into a biofuel through gasification and hydrogen generation (El Bassam, 2002a).

Internet resources

www.vbco.org/government486655.asp

SAFFLOWER (*Carthamus tinctorius* L.)

Description

The safflower is probably native to an area bounded by the eastern Mediterranean and the Persian Gulf. Seeds have been found in Egyptian tombs over 4000 years old, and its use was recorded in China approximately 2200 years ago. After the breeding of higher-yielding cultivars it is now grown in other parts of the world too, for example in America or Australia (Diercke, 1981). Safflower is commercially cultivated to a large extent in India, Mexico and the USA (Schuster, 1989a).

Safflower (*Carthamus tinctorius* L.) is a highly branched, herbaceous, thistle-like annual, usually with many long sharp spines on the leaves. Plants are 30 to 150cm tall with globular flower heads (capitula) and commonly, brilliant yellow, orange or red flowers which bloom in July. Each branch will usually have from one to five flower heads containing 15 to 20 seeds per head. Safflower is very susceptible to frost injury from stem elongation to maturity (wikipedia.org/wiki/Safflower_oil).

The stalkless oval-lacerate dark green leaves are usually set with spines. Leaf size varies widely among varieties and on individual plants. Typically it ranges from 2.5 to 5cm in width and 10 to 15cm in length. The leaves are coated with wax. The plant develops a deep taproot, penetrating the soil 2–3m deep, that

Source: wikipedia.org/wiki/Safflower_oil

Figure 10.161 Safflower *Carthamus tinctorius*

enables it to thrive in dry climates (Dajue and Mündel, 1996).

A study by Dajue and Yunzhou (1993) found that the number of spines on the outer involucral bracts varies from none to many, the diameter of the capitula ranges from 9 to 38mm, the weight of 1000 seeds ranges from 25 to 105g, and the oil content in the seeds varies from 14.90 to 47.45 per cent. There is a negative correlation between oil content and 1000-seed weight, revealing that large seed is not very good for oil content. Lines that were used for the production of herbaceous medicine yielded the least oil (Dajue and Yunzhou, 1993). The photosynthesis pathway is C_3 II (FAO, 1996a). Originally, the plant was grown for pigments extracted from its petals, but now it is commonly grown as an oilseed (Rehm and Espig, 1991). The evaluation of more than 2000 accessions of the World Collection of Safflower Germplasm (WCSG) showed a considerable diversity. Safflower may be a good alternative because its deep roots take up water and nutrients that are out of reach of other crops.

Ecological requirements

The production is restricted to the region between the latitudes 20°S and 40°N. It is usually grown below 900m elevation. The climate, especially the temperature, plays an important role in safflower growth. Emerging plants need cool temperatures (15–20°C) for root growth and rosette development, and high temperatures (20–30°C) during stem growth, flowering and seed formation. The seedling in the rosette leafy state is frost resistant down to −7 to −14°C, depending on the variety. The mature plant is destroyed by a slight frost of −2°C.

The plant needs a light, medium-deep and well-drained soil. The best pH value is 6.5–7.5. Optimum precipitation ranges from 600 to 1000mm. Its deep root system makes it drought tolerant and it can survive with an annual rainfall of 300mm. The humidity should be medium to low. Bees or other insects are generally necessary for optimum fertilization and yields (FAO, 1996a). Irrigation may be necessary, depending on the precipitation levels. To help ensure a good yield it is important that the crop receives sufficient water at least during the flowering stage (Zaman and Maiti, 1990). A dry period is

Source: Jack Dykinga, USDA Agricultural Research Service, Bugwood.org, www.invasive.org/browse/detail.cfm? imgnum=1323093

Figure 10.162 Safflower field, *Carthamus tinctorius*

necessary for ripening (FAO, 1988). It does not survive standing in water in warm weather, even for a few hours. Excess rainfall, especially after flowering begins, causes leaf and head diseases, which reduce the yield or even cause the loss of the crop (Dajue and Mündel, 1996). The plant is salt tolerant, which is a very useful quality if it is grown under irrigation (Rehm and Espig, 1991). Because of its tolerance to drought, it can be grown in those areas where other oil plants fail (Diercke, 1981).

Safflower is generally considered to be a day-length-neutral or long-day plant. However, the origin of the variety is important: when summer crop varieties from temperate regions are sown during shortening days as a winter crop in tropical or subtropical regions, they show a delayed maturity (Dajue and Mündel, 1996).

Propagation

Safflower is propagated by seed. Seeding safflower into a firm, moist seedbed enhances emergence and stand and improves the vigour of the plant. Germination is followed by a slow-growing rosette stage, during which numerous leaves and the strong roots are formed. In this stage the plant is endangered by fast-growing weeds. After the rosette stage, the stem develops, elongates quickly and branches.

Varieties that are almost free of spines have been developed for hand harvesting of the floral parts and seeds (Dajue and Mündel, 1996). Selection for high oil content reduced the pericarp thickness. Breeding aims are the improvement of resistance against diseases, and tolerance of cold and saline-alkaline conditions.

A number of wild species can be used as good sources with resistance and tolerance to several diseases and pests. Drought hardiness and disease resistance have been incorporated into cultivated types by repeated back-crossing and selection. Resistance against the fungus *Alternaria carthami* together with resistance against the bacterial pathogen *Pseudomonas syringea* is incorporated into several varieties. Other lines are resistant against rust (*Puccinia carthami*), verticillium wilt (*Verticillium alboatrum* Reinke & Berth.), fusarium wilt and rot (*Fusarium oxysporum* Schlecht f. ssp. *carthami* Klis. & Hous.), rhizoctonia rot (*Rhizoctonia solani* Kuhn) and other diseases (Dajue and Mündel, 1996).

Crop management

The seed is sown 2–3cm deep, with planting distances of 10cm in the row and 30–60cm between the rows. The desired crop density is 40–50 plants per m^2, and up to 70 plants per m^2 on light soil (Rehm and Espig, 1991). Depending on the climatic conditions, safflower can be sown in autumn or in spring. The growing period is 200–245 days or 120–160 days, respectively. Organic fertilization is not recommended because of the non-calculable nitrogen release. Table 10.88 shows the inorganic fertilizer levels as recommended in parts of Germany.

No recommendations can be made for herbicide use, as this aspect of weed control is still being tested, but herbicides appropriate for sunflower and soya bean crops are believed to be appropriate for safflower too (Schuster, 1989b). Mechanical weed control is still commonly practised.

A number of diseases and pests are known to attack safflower crops. The most serious disease that affects safflower in humid climates during and after the flowering is botrytis (*Botrytis cinerea*). The use of botryticides at the proper time can lessen the effect of this disease. Further diseases are: flower and leaf rot (*Pseudomonas syringae*), leaf spot (*Cerospora carthami*),

Table 10.88 Fertilizer rates for safflower as recommended in Germany

Fertilizer	Amount (kg/ha)
Nitrogen (N)	50–70
Phosphorus (P_2O_5)	60–80
Potassium (K_2O)	150–210

root rot (*Phytophthora drechsleri*), rust (*Puccinia carthami*), seedling blight (*Phytophthora palmivora*) and wilt (*Sclerotinia sclerotiorum*). Pests include aphids (*Macrosiphum jaccae*), caterpillars (*Perigaea capensis*), fruit fly larvae (*Acanthiophilus helianthi*) and nematodes (FAO, 1988; Schuster, 1989a).

Although safflower does not have any specific crop rotation, it is suggested that it may be placed between two cereal crops in the rotation. Safflower improves the soil quality by leaving behind its taproot and root system. A five-year break in safflower cultivation should be adopted.

Production

Safflower is physiologically mature about one month after flowering and ready to harvest when most of the leaves have turned brown and only a tint of green remains on the bracts of the latest flowering heads. Safflower usually is directly harvested with a small-grain combine.

Worldwide, India is the largest producer of safflower for oil, but most of its production is consumed internally.

The yields of seed can reach 1.1–1.7t/ha/year (FAO, 1996a). With irrigation and good fertilization, yields of 2.8–4.5t/ha/year can be achieved, but the world average yield is 0.5t/ha/year (Rehm and Espig, 1991). The new cultivars have 36–48 per cent oil in the fruit. The oil contains 73–79 per cent linoleic acid, depending on the cultivar. The scrap cake portion amounts to 18–30 per cent.

In a study of oil crops in the USA by Goering and Daugherty (1982) it was determined that the total

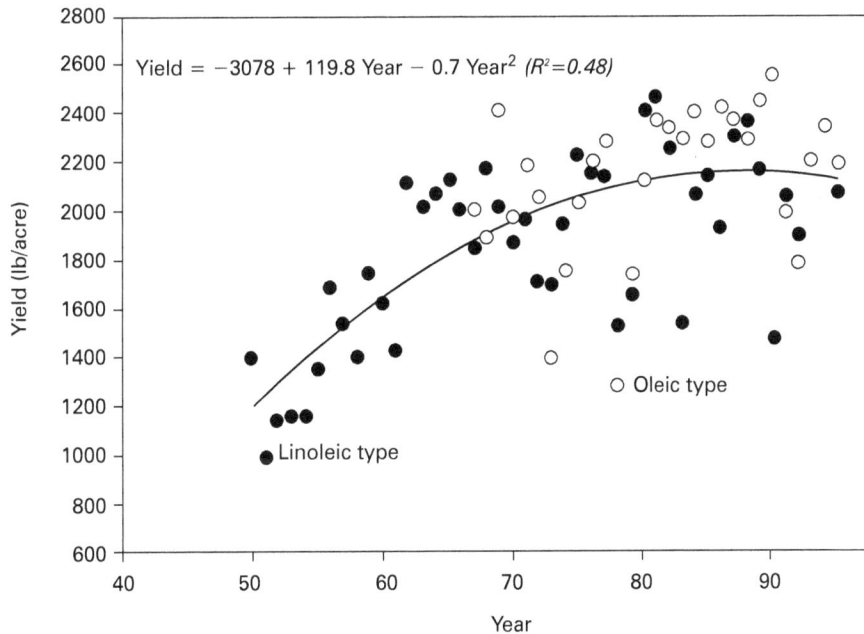

Yield = $-3078 + 119.8$ Year $- 0.7$ Year^2 ($R^2 = 0.48$)

Source: ucdavis.edu

Figure 10.163 Safflower yields in California, 1950 to 1995

energy input for non-irrigated safflower was 9313MJ/ha. This includes the range of costs from seed price to oil recovery costs. The energy output was determined to be 19,483MJ/ha. This was based on a seed yield of 1233kg/ha and an oil content of 40 per cent. On irrigated safflower, irrigation was determined to cost 48,180MJ/ha, and the total energy input increased to 63,062 MJ/ha. The energy output was 44,278MJ/ha, based on a seed yield of 2802kg/ha. The energy output/input ratio decreased from 3.39 without irrigation to 1.14 with irrigation. As a comparison, soya bean and rape had the highest non-irrigated output/input ratios at 4.56 and 4.18 respectively. Corn and sunflower had the highest irrigated ratios at 3.44 and 1.96 respectively.

Yields increased at approximately 120 pounds per acre (134.4 kg/ha) per year during the first two decades of production in California, but have since remained relatively stable. Yield = −3078 lb/acre + 119.8 lb/acre/yr − 0.7 year^2 (R^2 = 0.48).

The oil content of safflower cultivars used in California and Arizona has increased at approximately 0.2 per cent per year since approximately 1950. Linoleic oil per cent = 20.8 + 0.29 (year) (R^2 = 0.59); oleic oil per cent = 22.9 + 0.23 (year) (R^2 = 0.55).

Processing and utilization

The harvest takes place between the middle of August and the end of September. A more precise time can be determined when the stalks, leaves and flower tops dry out and turn brown. At this time the seeds should have a moisture content of 8–9 per cent, which makes them suitable for storage. If the seeds have a higher moisture content they need to be dried. Harvesting consists of cutting the crop at ground level, drying it in the field for several days and threshing it (FAO, 1988). The use of a combine harvester should be possible. In countries with low labour costs the seed can be harvested by hand; spineless cultivars make this easier. Tests in India showed that seed production from a ratooned crop is possible (Dajue and Mündel, 1996).

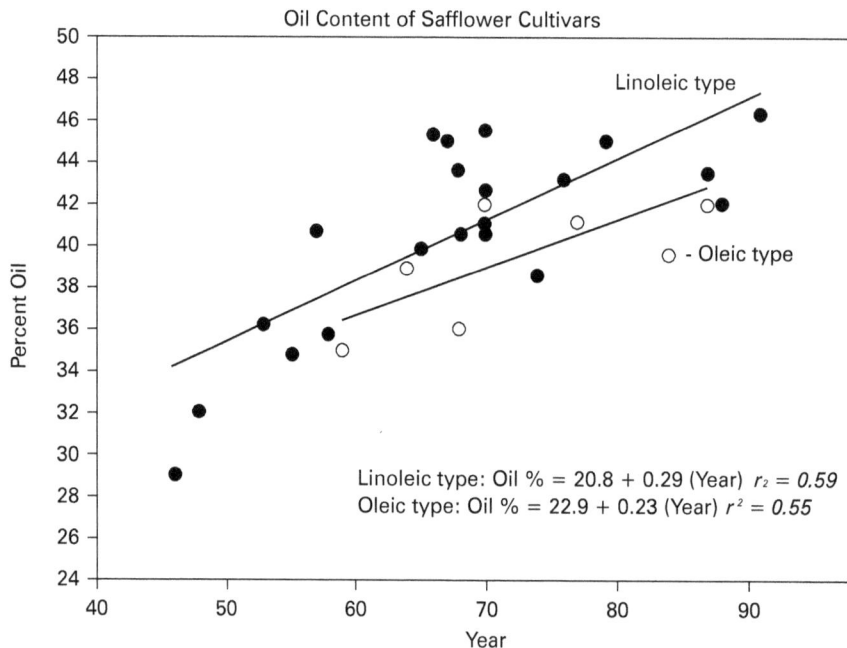

Source: ucdavis.edu

Figure 10.164 Oil content of selected safflower cultivars, 1940 to 1995

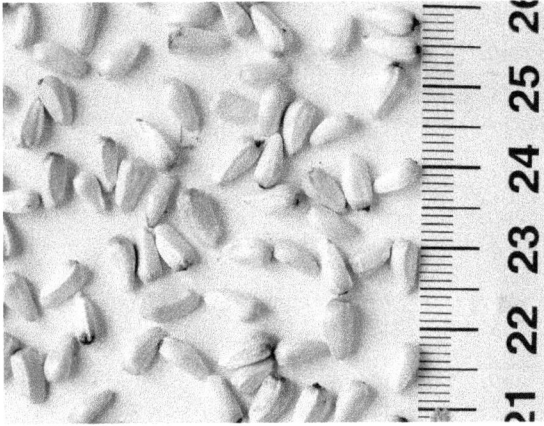

Source: Světlice barvířská, http://vfu-www.vfu.cz/vegetabilie/plodiny/czech/Saflor.jpg

Figure 10.165 Seeds of the safflower

Table 10.89 Agronomic performance of varieties, 2006 (average of eight site locations)

Variety	Seed yield (lb/acre)	Plant height (inches)	Test weight (lb/bu)	Oil (%)
Oleic types				
Montola 2003	1678	20	41.5	39.2
Montola 2000	1635	19	39.9	40.5
Mondak	1808	22	42.0	37.8
Montola 2004	1539	20	40.9	37.5
Linoleic				
Finch	1599	23	44.2	38.9
Nutrasaff	1393	24	37.5	48.0
Cardinal	1810	24	42.7	38.5
S–541	1814	24	40.9	43.8

Quadris fungicide applied at first flower.

Source: Berglund et al (2007)

The seeds are milled, and the oil is pressed out of the pulp or extracted. According to the procedure and the intended purpose of the press cake, the seeds are either shelled or left unshelled. The oil is edible, and because of its composition it is in demand in the food industry. It is used for technical purposes too. The oil is quick drying and does not become darker, so it is valued for paints and lacquers (Rehm and Espig, 1991). It is also used as a fuel for lighting and as a component in soap production. It could also be used as biodiesel, as a fuel additive or as a biodegradable chainsaw bar oil.

The hulls can also be used as fuel. They are a low-ash fuel that burns readily, but they are bulky. It is possible to process the hulls into a gas that can replace natural gas. Gasification systems that use agricultural wastes like rice hulls, cotton seed hulls, straw, etc. are available. These wastes can be successfully converted to clean hot gas (Bailey and Bailey, 1996). The hulls can be processed to charcoal, though volatile materials might also be produced. The pressed cakes are used as protein-rich fodder. The value of the fodder is higher if the hulls are removed before pressing.

In addition to these energy uses, safflower can be grazed by cattle or stored as hay or silage. The yields and the feed value are similar to alfalfa. Safflower straw is used in a similar way to cereal straw (Dajue and Mündel, 1996).

Safflower oils differ widely in composition, depending on the cultivar. Some safflower oils are free of stearic acid or nearly free of palmitic acid (Dajue and Yunzhou, 1993). Table 10.90 shows characteristics of safflower oil as well as of the linoleic and oleic acid portions. Table 10.91 shows the cultivar-dependent fatty acid contents of the oil.

High oleic safflower (*Carthamus tinctorius* L.) oil has promise as a pollutant-reducing diesel fuel additive to reduce smoke and particulate emissions. High oleic safflower oil as a diesel fuel additive would also reduce acid rain, the greenhouse effect, and surface pollution because safflower oil is virtually free of sulphur, totally

Table 10.90 Some properties of safflower oil

Property	Value
Density (g/mL)	0.972
Iodine number	140–150
Saponification number	188–194
Linoleic safflower oil	
Kinematic viscosity (mm²/s)	32.3
Heat of combustion (kJ/kg)	39,226
Oleic safflower oil	
Kinematic viscosity (mm²/s)	42.1
Heat of combustion (kJ/kg)	39,306

Source: Römpp (1974); Peterson (1985)

Table 10.91 The composition of safflower oil, depending on cultivar (per cent of the fatty acids)

Oil type	Palmitic acid (C–16/0)	Stearic acid (C–18/0)	Oleic acid (C–18/1)	Linoleic acid (C–18/2)
Very high linoleic	3–5	1–2	5–7	87–89
High linoleic	6–8	2–3	16–20	71–75
High oleic	5–6	1–2	75–80	14–18
Intermediate oleic	5–6	1–2	41–53	39–52
High stearic	5–6	4–11	13–15	69–72

Dajue and Mündel (1996)

lacks fossil carbon dioxide, and is biodegradable. High oleic safflower oil offers a promising technology for further research and development as a fuel extender and is adapted to Montana's growing conditions. Growing safflower in Montana and other northern Great Plains states is highly desirable as an alternative crop for inclusion in rotation with dryland wheat to break wheat disease and pest cycles. Montana dryland farms have the potential to produce more oil on a per acre basis from high oleic safflower than Iowa farms can produce from soya beans. Research in Montana is continuing to make high oleic safflower oil more economical for use as a biofuel by adding value to safflower meal through genetic breeding and improvement. (Bergman and Flynn, 2009)

Internet resources

www.ag.ndsu.edu/pubs/plantsci/crops/a870w.htm
www.sidney.ars.usda.gov/state/saffcon/abstracts/Safflo
werProducts/bergman.htm
www.safflower.jochinke.com.au/
http://en.wikipedia.org/wiki/Safflower_oil
www.ars.usda.gov/is/graphics/photos/apr98/k7908-
1.htm
http://agric.ucdavis.edu/crops/oilseed/safflower.htm

SAFOU (*Dacryodes edulis* H.J. Lam; syn. *Canarium edule* Hook.)

Description

The safou tree, *Dacryodes edulis* H.J. Lam, also reported under the name of *Pachylobus edulis* G. Don,

P. saphu Engl. and *Canarium saphu* Eng., *C. edulis* Hook or *C. mubago* Fichalo, belongs to the family Burseracea (Okafor, 1983). It is also improperly called bush butter tree (Youmbi et al, 1989), African pear (Omoti and Okiy, 1987), African plum and africado (Viet Meyer, 1990). The wood of its straight stem, which may reach 8 to 10m in height and up to 25m when growing in the dense forest, is of good quality. The tree may be found in the Gulf of Guinea, from Sierra Leone down to Angola, and eastwards up to Uganda. It is a promising non-conventional oil tree, both for human nutrition (pulp) and oil production (pulp and kernel) (Silou, 1996), that has been domesticated for many years in countries like Cameroon and the Congos.

The butterfruit tree is commonly 8–12m high when grown under cultivation in the open, but up to 45m in the forest and in old plantations. The trunk is generally cylindrical and straight. Although it can reach 1.5m in diameter, it is normally much smaller. The plant has compound leaves with 4–12 pairs of leaflets (odd-pinnate; with a single terminal leaflet). It is deciduous, losing its leaves in the dry season.

Although the species has male and female flowers on separate plants, there are hermaphrodite (male/female) trees as well. Male flowers are larger (8–25cm long) than female (5–15cm long). At least in some vigorous inflorescences, the terminal bud forms the flowering shoot for the following year. Only female flowers produce fruit of course. Although the amounts vary, female inflorescences tend to be very productive. Hermaphrodites are less productive.

The fruit is ellipsoid, globular, or conical; 4–15cm long, 3–6cm in diameter. It is rose pink to white when young, deepening to blue, purple or even black at maturity. The pericarp, which represents half the

Distribution of Safou

Source: International Centre for Underutilized Crops (2001)

Figure 10.166 Distribution of safou

Source: Honoré Tabuna

Figure 10.167 A safou tree

weight of the whole fruits, consists of a thin, waxy, and coloured epicarp and a pulpy mesocarp that is light pink, rose, light yellow, light green, or whitish in color. This pulp varies in flavour depending on the tree.

In popular literature this fruit is often called 'African pear' or 'bush mango', awkward terms that are botanically and culinarily misleading. A common English name in Central Africa is just 'plum', 'bush plum', or 'African plum', due to its shape and colour. It is called safou in Angola, Gabon, Cameroon and the Congos. It is also known in French as 'prunier' and the fruits called 'prunes'. Because they resemble avocado in composition and texture we suggest 'africado' for international marketing purposes.

The heavy heartwood's elastic quality makes it useful for axe-handles, mortars, pestles, and pillars for houses and buildings. It is also suitable for carpentry and fine woodworking. As noted, it is not unlike mahogany, and its woodworking qualities and interesting appearance make it suitable for veneers and for fine cabinetwork (NAS, 2008).

Environmental Requirements

Most of the world's oil-bearing plants are confined to narrow ecological areas (oilpalm and coconut, for instance, are restricted to hot and humid areas). Butterfruit, however, tolerates several. It thrives, for example, in all the ecological zones of Nigeria and Cameroon except the very dry northern provinces. Although it fits well into savannah zones, its fruit production is greatest in the humid forest zones. In general, performance is best in the shade and in good soil.

(a)

(b)

Source: (a) Ann Degrande, tropicallab.ugent; (b) Honoré Tabuna

Figure 10.168 (a) Safou fruits; and (b) Safou on sale at a local market in Cameroon

- Soil seems not to present a limitation. The species has been reported growing on oxisols, ultisols, loamy clay, sandy clay, humic ferralitic soils, deep loam rich in organic matter, andosols, and ferriginous (chalybeate).
- The plant tolerates rainfall from 600mm to 3000mm and more. By some accounts, low humidity at flowering time may frustrate fruiting.
- The plant is found at low-medium elevation, from sea level to 1500m.
- The minimum temperature is unknown. One contributor reports the minimum at his location as 9°C (in January). Possibly the plant requires 'low' night temperature for uniform flowering (22°C or 14°C have been suggested).
- High temperatures are not a limitation. The plant thrives where temperatures go above 40°C (NAS, 2008)

Crop management

With urbanization, the trade in safou, the local name of the fruit, has recently taken off locally and even among the countries of Central Africa (Tabuna, 1993). Production starts in June and ends in November in the northern hemisphere (Cameroon, Nigeria and northern parts of Congo and Congo ex Zaire) whereas in the southern hemisphere it starts in November and ends in April.

The tree could be planted on equatorial and moist tropical grasslands, such as the 2 million hectare Batteke Plateau in Congo (Cailhiez, 1990), instead of or together with eucalyptus trees, which today cannot be further developed because of insufficient demand for that quality of wood.

Propagation

There are male and female trees; the average ratio is 1:1 but the optimum ratio would be 5 male to 95 female. However, the seeds are viable only for a few days (Kengue and Nya Ngatchou, 1991). Vegetative propagation, long considered as difficult, has been solved using aerial layering (Mampouya, 1991; Silou, 1996). Layers start to bear fruits after only 18 months instead of after 5 to 6 years in the case of trees from seeds (Kengue and Tchio, 1994). This technique will now permit the creation of productive orchards.

Production

The fruits are ripe when they have darkened from pink to blue/black. Hand picking typically occurs in the morning, with men or boys climbing into the trees and using hooked poles to draw the fruits within reach. Mature ones are packed into bags (holding up to 10kg) that are usually hung on a branch. When filled, the bags are lowered on a rope to the ground.

Once picked, the fruits become quite perishable. In the oppressive heat and humidity of tropical lowlands, they do not keep much more than a week. Storage in a cool and airy spot, preferably a basket, helps. Deterioration occurs through moisture loss and the fruit consequently shrivels. Dampness is especially to be avoided; it causes the fruit to soften faster and turn mouldy. For this reason, butterfruit is not harvested on rainy days. Similarly, it cannot be packed densely or stored in airtight containers such as plastic bags (NAS, 2008).

The harvesting of fruits on layers in Congo is easier than on large trees from seeds, and is less damaging for the fruits. The fragile fruit may thus be conserved for a longer time (Silou and Avouampo, 1997). There is a great variability in fruits: they may be 6 to 8cm long and 3.5 to 4.5cm thick, and weigh between 50 and 65g. The pulp (mesocarp) is 0.2 to 1cm thick. Some promising trees with large seedless fruits, late maturation and excellent taste have been identified (Avouampo, 1996). But there are also acid fruits. The fruit, much appreciated in the human diet, contains, on a dry weight basis, 30 to 70 per cent lipids (average 50 per cent), 13 to 27 per cent proteins, carbohydrates and fibres in the pulp, and 10 per cent oil in the kernel.

Although this species has not yet benefited from any selection, tree breeding or fertilizer programmes, high fruit yields have been recorded: 30kg on 5-year-old trees, and from 100kg (in Congo) to 200kg (in Nigeria) on 10- to 15-year-old mature trees (Silou, 1996). Fruit yields of 10 to 15t/ha may thus be achieved.

Oil from safou pulp has the same composition as that extracted from the kernels. Unlike most other oil-producing species like palms, olives etc., the simultaneous extraction of oil from the kernel and the pulp is therefore possible. The main three fatty acids of this palmito-oleic oil are palmitic acid (C16:0; 35.6 to 58.4 per cent), oleic acid (C18:1; 16.9 to 35.5 per cent) and linoleic acid (C18:2; 3.9 to 31.5 per cent) – and these usually account for 95 per cent of the total amount of fatty acids. R, the ratio of unsaturated fatty acids to saturated fatty acids, is close to 1, as it is for palm oil, and thus safou oil falls between the fluid oils ($R = 4$) and vegetable butters ($R = 0.25$). Esterification with alcohol (either methanol or ethanol) would increase its fluidity, as is seen with palm oil.

Oil extraction has been performed in the laboratory and in preliminary experiments with a simple extrusion screw (Tchendji et al, 1987; Ndamba, 1989; Kiakouama and Silou 1990; Bezard et al, 1991; Silou et al, 1991, 1994; Silou, 1996; Silou and Avouampo, 1997). Oil cakes from the pulp are similar in composition to those from coconuts and oil palms: 13 to 16 per cent proteins, and 20 per cent of cellulose only slightly lignified and easily digestible. Oil cakes may be promoted as animal or fish food (Ndamba, 1989; Silou and Avouampo, 1997). A ten-year-old safou tree studied bears 2820 fruits gathered on 557 bunches with 1 to 31 fruits on bunch (Kinkéla et al, 2006).

Processing and utilization

At present, the safou tree has oil production potentials comparable to oil palms, one of the most promising species in the world for vegetable oil (Table 10.92). Oil yields, already 2–4t/ha, are similar to oil palm fields in the early stages. Now, with vegetative propagation being possible, yields could similarly be improved with a consistent breeding programme.

However, there are other reasons for promoting the species:

- It increases the diversity of candidates for highly productive vegetable oil species in the tropics.
- Oil can be extracted with a simple and low-cost extrusion screw, and requires much less time and effort than palm oil extraction on a village level.
- As a fruit tree, it enhances the security of food supplies.
- It fits into agroforestry systems (Hladick, 1993), and can be planted with other species – for instance, with coffee (Tiki Manga and Kengue, 1994);
- Oil esterification with locally produced ethanol could be performed either in relatively small rustic installations or in the more traditional facilities developed for rapeseed oil.

Table 10.92 Biofuels for transport: energy yield for alternative feedstock/conversion technologies

Production	Feedstock yield (dry t/ha/yr)	Transport fuel yield (GJ/ha/year)	Transport services yield (1000 vehicle-km/ ha/year)	Ratio between rapeseed oil and other biofuels
Rapeseed methyl ester, Netherlands	3.7t of rapeseed (1369 litres of oil)	47	21,000 (ICEV)[a]	1
Palm oil and safou oil, Africa	2–4 of oil (2000–4000 litres)	67 to 137	30,000 to 61,000	1.4 to 2.85
Hardnut (*Jatropha curcas*), Mali	1t of seeds (270 litres)	9.3	4200	0.2
Ethanol from maize, USA	7.2t	76	27,000 (ICEV)	1.3
Ethanol from wheat, Netherlands	6.6t of grain	72	26,000 (ICEV)	1.2
Ethanol from sugar beet, Netherlands	15.1t of beet	132	48,000 (ICEV)	2.3
Ethanol from sugar cane, Brazil	38.5t of cane	111	40,000 (ICEV)	1.9
Ethanol, enzymatic hydrolysis of wood (present technology)	15t of wood	122	44,000 (ICEV)	2.1
Ethanol, enzymatic hydrolysis of wood (improved technology)	15t of wood	179	64,000 (ICEV)	3
Methanol, thermochemical gasification of wood	115t of wood	177	64,000 (ICEV), 133,000 (FCV)[b]	3 (ICEV), 6.3 (FCV)
Hydrogen, thermochemical gasification of wood	15t of wood	213	84,000 (ICEV), 189,000 (FCV)	4 (ICEV), 9 (FCV)

[a] ICEV: internal combustion engine vehicle (standard).
[b] FCV: fuel cell vehicle.

Note: Consumption (in litres of gasoline equivalent per 100km) is assumed to be 6.30 for vegetable esterified oil, 7.97 for ethanol, 7.90 for methanol and 7.31 for hydrogen used in ICEVs, and 3.81 for methanol and 3.24 for hydrogen used in FCVs; 1 litre of gasoline equivalent = 0.0348GJ HHV.

Source: Second Assessment Report of IPCC (1995) chapter 19, p608 completed by A. Riedacker

The esterification of oil is a readily available technique. Diesel engines can use the resulting fuel without any problem, whereas gasification and methanol production from lignocellulosic material, and cars or trucks with internal combustion engines or methanol reforming devices and fuel cells, are still under development. More emphasis should therefore be laid on improving both the species and agronomic techniques related to this crop.

A small pilot project, with both oil production and extraction and alcohol production for esterification, incorporating a microdistillery and a small esterification facility, should be considered as a high priority by donors. Such a development would further increase incomes and would improve the potential for replacing fossil fuels by vegetable oil to reduce greenhouse gas emissions.

In Cameroon 11,000 tonnes of safou fruit (*Dacryodes edulis*), was commercialized in 1997, equivalent to a value of US$7.5 million.

The economic development of Safou fruit shows good prospects in some African countries. The fruit contains nearly 50 per cent oil of high nutritional value and classic oil extraction could provide new markets for its exploitation. Iodine value, peroxide value and the content of thiobarbituric reactive species were measured over a period of 2 months, at four temperatures from 4 to 50°C, to evaluate oxidation of free and maltodextrin-encapsulated safou oil. The present study showed the good stability of the native oil at low temperature and the good protection provided by a 6DE maltodextrin matrix against oxidation, at 50°C. Freeze-dried particles were more efficient against oil oxidation than spray-dried particles. (Dzondo-Gadet et al, 2005)

Internet resources

www.icuc-iwmi.org/files/News/Resources/Factsheets/
dacryodes.pdf
www.bioversityinternational.org/fileadmin/bioversity/d
ocuments/news_and_events/Rudebjer_et_al_2008_Bi
odiversity_in_Forestry_Education.pdf
www.new-ag.info/07/04/develop/dev2.php
www.tropicallab.ugent.be/ann.htm

SALICORNIA (*Salicornia bigelovii* Torr.)

Description

Salicornia bigelovii Torr. is an annual C_4 vascular plant originating in the Americas that is most commonly found in coastal estuaries and salt marshes. The plant has succulent, erect shoots with articulated and apparently leafless stems that are completely photosynthetically active and take on the appearance of green, jointed pencils. Seed spikes are found on the top third of the plant. Flowering takes place from July to November (Wiggins, 1980).

Salicornia bigelovii is a halophyte, or salt-tolerant plant. Halophytes are unique in that they expend energy to maintain a higher salt concentration in their vacuoles than is found in the soil. By having the higher salt concentration in the vacuoles of the leaves the flow of water into the plant is ensured. The vacuoles are able to achieve a concentration of more than 6 per cent salt, while pure seawater is only 3.2 per cent salt (Douglas, 1994).

Approximately 30 per cent of the seed's total weight is oil and about 30 per cent protein, in comparison to the soya bean which is 17–20 per cent oil. Salicornia's oil is about 72 per cent linoleic acid, which is a healthy polyunsaturated fat. For over a decade, salicornia has been 'selectively developed' to produce higher oilseed yields. Test plots in the United Arab Emirates, Egypt, Kuwait, Saudi Arabia and Mexico have all had success (Clark, 1994). It is estimated that there are approximately 130 million hectares worldwide that would be appropriate for salicornia cultivation. In addition, salicornia could help prevent erosion, lead to the return of nutrients to the soil, sequester carbon dioxide, remove salt and heavy metals from power plant

wastewater, and provide a new non-fossil biofuel (Clark, 1994; Douglas, 1994).

Nearly 60 species have been proposed for *Salicornia*. Some common species are:

- American, Virginia or woody glasswort, *Salicornia virginica*;
- common glasswort, *Salicornia europea*;
- slender glasswort, *Salicornia maritima*;
- dwarf glasswort, *Salicornia bigelovii*;
- perennial glasswort, *Salicornia perennis*;
- purple glasswort, *Salicornia ramosissima*; and
- umari keerai, *Salicornia brachiata*.

Source: wikipedia.org/wiki/Salicornia

The *Salicornia* species are small, usually less than 30cm tall, succulent herbs with a jointed horizontal main stem and erect lateral branches. The leaves are small and scale-like and as such the plant may appear leafless. Many species are green, but their foliage turns red in autumn. The hermaphrodite flowers are wind pollinated, and the fruit is small and succulent, and contains a single seed.

Salicornia species can generally tolerate immersion in salt water. *Salicornia* species are used as food plants by the larvae of some Lepidoptera species including the *Coleophora* case-bearers *C. atriplicis* and *C. salicorniae* (the latter feeds exclusively on *Salicornia* spp.) (wikipedia.org/wiki/Salicornia).

Ecological requirements

According to Glenn and Watson (1993), there are three types of areas that could be used for the development of large-scale farms for halophytes: coastal deserts (seawater irrigation), inland salt deserts (underground or surface water irrigation) and existing arid zone irrigation districts (brackish drainage water irrigation). Although sand or sandy soil will support the development of *S. bigelovii*, heavier soils with a larger water-holding capacity would be preferred in that this reduces the amount of irrigation, which would normally be in the form of flood irrigation.

Salicornia's incredible salt tolerance is seen in the fact that it can endure salt concentrations of up to

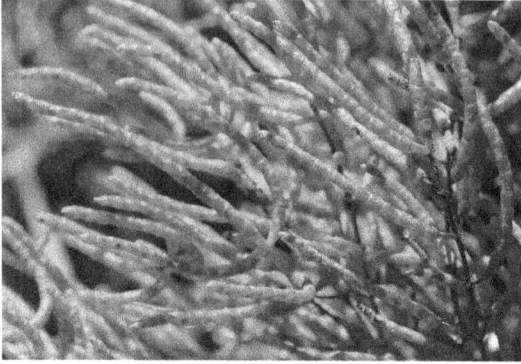

Figure 10.169 *Salicornia virginica*, Marshlands,
Near Rehoboth, Gratwicke

50,000ppm (5 per cent salt) without blighting.
Although the salt build up can be harmful, it can be
avoided by overwatering. This pushes the salt below
root level. On a test farm in Mexico this was
accomplished by estimating the crop's water needs,
then irrigating with an additional 25 per cent water
(Clark, 1994). From this it is estimated that S. *bigelovii*
needs yearly 1–3m³ of seawater per m² of soil (Douglas,
1994).

Research has been conducted that has shown that
euhalophytes, more extreme halophytes, have optimal
growth at salinities in the range of 100–200mol/m³.
Salinities above or below this range lead to decreases in
growth (Greenway and Munns, 1980; Munns et al,
1983).

The plant grows well in sand and sandy loam soils
with adequate drainage. The crop responds to increased
application of nitrogen and phosphorus (hindu.com).

Propagation

More than a decade of breeding and selection has led to
seed being the most prominent form of salicornia
propagation (Clark, 1994). Seeds can also be used for
the mass propagation of selected plants. This involves
the germination of seeds in a greenhouse followed
by the selection and growth of shoot tips in a tissue
culture medium. Research has shown that it is possible

Figure 10.170 *Salicornia europaea* L.

to produce 12 to 30 new shoots per culture every 8 weeks (Lee et al, 1992).

Crop management

S. *bigelovii* has its growing season from March to August. According to Clark (1994), new strategies that were planned to be implemented in the 1994–1995 season in Saudi Arabia included 'lowering seed density to produce fewer but bigger plants; applying phosphorus before planting to promote general crop growth; using "socks" or tubes to carry water from the irrigation sprinkler heads directly to the ground when plants begin to pollinate; and cutting off irrigation when the largest plants reach full size, instead of waiting for the entire crop to mature'.

S. *bigelovii* grows in its native habitats and on test farms without being significantly affected by disease or pest, though Stanghellini et al (1988, 1992) reported instances of S. *bigelovii* being attacked by *Metachroma* larvae and *Macrophomina phaseolina*. Test plots in Sonora, Mexico, showed stand losses varying from 0 to 35 per cent in 1988. It was determined that the small beetle larvae of the *Metachroma* genus were responsible for severing the plants' roots 3–5cm below the surface. Treatment of the soil with diazinon stopped plant damage by the larvae.

Macrophomina phaseolina, which is a soil-borne fungus of arid and semi-arid regions, has also been reported to attack S. *bigelovii* (Stanghellini et al, 1992). On test plots in Sonora, Mexico, it was found that the root pathogen M. *phaseolina* caused rotting of the plants' roots and led to stand mortality rates as high as 80 per cent in 1989 and 30 per cent in 1990.

The occurrence of the disease is attributed to the plant undergoing environmental and/or physiological stresses.

Because salicornia primarily grows in the wild and is only now starting to be cultivated there is no established crop rotation.

Production

The most significant cost of growing halophytes is the expense of irrigating/pumping with seawater (Douglas, 1994). The direct cost of raising the crop, which includes diesel fuel for irrigation and tilling, is

Source: Global Seawater Inc.

Figure 10.171 Salicornia plantation near Bahia Kino on the Gulf of California in Mexico

estimated at US$44–53 per tonne of fresh matter (Douglas, 1994).

Using planting dates in October and November 1993 and harvesting in September, several areas on a farm in Saudi Arabia have achieved their goals of 10t/ha of forage and 1t/ha of seed, while in a field 60km away, oilseed yields have reached as much as 3.5t/ha (Clark, 1994).

Source: hindu.com

Figure 10.172 Salicornia, oil-yielding plant for coastal belts

On survey plots in Mexico, *S. bigelovii* finished in the top five of halophytes for dry weight (DW) productivity in a one-year trial (Table 10.93).

At the same location in Mexico after six years of trials, *S. bigelovii* showed a mean biomass yield of 2.02kg/m²/year and a mean seed yield of 203g/m²/year (Glenn et al, 1991).

Seawater appears to contain sufficient quantities of other nutrients and micronutrients to eliminate the need for supplementation with other fertilizers. In seven months the crop reaches maturity, and in many locales a high yield of 18 tonnes of dry biomass has been harvested.

The field seed-yield is about 10–12 per cent of the total dry biomass. A 2000ha farm would yield a total biomass of 30,000 tonnes and a seed yield of 2500–3000 tonnes of seeds (hindu.com).

Processing and utilization

Salicornia is fit for human consumption as a vegetable and salad greens, and also represents a source of vegetable oil, fodder and meal. The meal is approximately 40 per cent protein and has had success as a poultry feed additive once the 'anti-feedant' saponin has been neutralized. The leftover straw, though high in salt and low in protein, may be used as feed (Clark, 1994).

Based on the calculations that halophytes have heat contents that fall in the range of lignite coal, research conducted by the Environmental Research Laboratory (ERL) at the University of Arizona has suggested that 'using a 2-to-1 ratio of coal to halophyte biomass in a 500MW power plant would produce about 10 per cent less heat per unit weight than coal alone but would result in 25 per cent less carbon being contributed to

Source: David Seth Michaels blogspot.com 2008

Figure 10.173 Salicornia field

the atmosphere (because of the recycling of carbon back to the biomass)'. The ERL also indicates that it is possible to use refined halophyte oil to blend with diesel fuel (Douglas, 1994).

There are experimental fields of Salicornia in Ras al-Zawr (Saudi Arabia), Eritrea (Northeast Africa) and Sonora (Northwest Mexico) aimed at the production of biodiesel. The company responsible for the Sonora trials (Global Seawater) claims that between 225 and 250 gallons of BQ-9000 biodiesel can be produced per hectare of salicornia, and is promoting a US$35 million scheme to create a 4900ha salicornia farm in Bahia de Kino (wikipedia.org/wiki/Salicornia).

Oil analysis of seeds of *Salicornia brachiata* was carried out in a study by Eganathan et al (2006). Hexane extraction yielded maximum oil content from seeds (22.4 per cent). High ester (538.32mg/g) and saponification (547.52mg/g) suggest a potential for industrial use of the oil (Eganathan et al, 2006).

On two Gulf of California plantations in the state of Sonora, Mexico, Phoenix-based Global Seawater Inc. is using coastal land and seawater to grow what it sees as an important biodiesel feedstock that will help solve the world's energy needs. At farms in Bahia Kino and Tastiota, Mexico, Global Seawater is growing salicornia. The company has used the oil as a feedstock to produce biodiesel which meets the BQ-9000 biodiesel accreditation standard. Between 225 and 250 gallons of biodiesel can be produce per hectare of salicornia (Christiansen, 2008).

Table 10.93 Productivity of five halophytes in Mexico

Plant	Productivity (DW g/m²/year)
Atriplex lentiformis	1794
Batis maritima	1738
Atriplex linearis	1723
Salicornia bigelovii	1539
Distichlis palmeri	1364

Source: Glenn and O'Leary (1985)

After taking into account environmental protection and other factors, Glenn et al (1991) estimated that 480,000 square miles of unused land around the world could be used to grow halophytes, and calculated that this could produce 1.5 billion barrels of oil equivalent per year, 35 per cent of the United States' liquid fuel needs.

Concerning which halophytes could be used for biofuel, McDermott (2008) indicated that *Salicornia bigelovii*, commonly known as dwarf glasswort yields 1.7 times the oil per acre from sunflowers, and that some others produce more than switchgrass.

Internet resources

www.biodieselmagazine.com/article.jsp?article_id=2600
www.treehugger.com/files/2008/12/biofuels-grown-in-saltwater-could-be-35-percent-united-states-fuels.php
www.hindu.com/seta/2003/09/05/stories/2003090500
300300.htm
http://en.wikipedia.org/wiki/Salicornia
www.guenther-blaich.de/pflseite.php?par=Salicornia+europaea&fm=pflfamla&abs=pfllst
www.usf.uni-osnabrueck.de/projects/expo2000/german/Pflanzen/pflanzesalicorniaeuropaea.html
http://dreamantilles.blogspot.com/2008/07/salicornia.html

Source: US Fish & Wildlife Service (treehugger.com/files/2008)

Figure 10.174 The chunky-looking green plants are *Salicornia bigelovii*, which have been proposed to be used as a biofuel crop

SHEA TREE (*Vitellaria paradoxa* Gaertn. f.; syn. *Butyrospermum paradoxum* Gaertn. f.; *Butyrospermum parkii* (G. Don) Kotschy)

Description

The shea tree (sheabutter tree) is indigenous to the Sahel and northern Guinea zones, where it grows wild or cultivated. It has local importance as a source of fat. The tree usually reaches a height of 7–15m, hut it may reach up to 25m with a trunk diameter of up to 2m and a dense crown. The bark is corky; the leaves are oblong and grow at the end of the twigs. In dry seasons the leaves are dropped.

The tree blossoms from December to March; later, green to yellow fruits are formed (Franke, 1985). The sheabutter fruits are ellipsoidal, 4–5cm long with fleshy pulp and usually with a single large seed. The fruits require 4–6 months to ripen (Gliese and Eitner, 1995). The soft pericarp is edible and very sweet. The seed contains 45–48 per cent fat, 10 per cent protein, and 25–60 per cent carbohydrates. The sheabutter tree is a main resource of fat in regions where neither the olive tree, cultivated in northern regions, nor the oil palm can be grown. The perennial, drought-resistant trees provide the inhabitants of the growth area with nutrients if the crops fail or during long dry seasons (ICRAF, 1997).

The shea tree grows naturally in the wild in 19 countries in the dry Savannah belt of West Africa from Senegal in the west to Sudan in the east, and onto the foothills of the Ethiopian highlands. According to legend among local people no one owns the shea tree, because it germinates and grows on its own.

When it passes the germination stage in about three to five years, it becomes fire resistant. Once it survives the first five years of its early stages of germination and growth, it grows slowly and takes about 30 years to reach maturity and can then live for up to 300 years. In the absence of any hazards, including tree felling, it can bear fruit for 200 years (sheabutterblog.blogspot.com).

Figure 10.175 Distribution of shea trees, 2009

Ecological requirements

The sheabutter tree is usually found at elevations from sea level to 1000m, but in Cameroon it reaches 1200m. It occurs in areas with a rainfall of 600–1000mm and a marked dry season of 6–8 months, or 900–1800mm and a shorter dry season of 4–5 months, but in both cases is subjected to annual burning (the tree is fire resistant). The absolute minimum of precipitation is 300mm. It needs a well-drained, deep soil, but its soil fertility requirements are low. The optimum temperature lies between 29 and 38°C. The plant does not withstand frost. Sheabutter occurs naturally within the latitudes 10°N and 20°N (FAO, 1996a).

The sheabutter tree, also known as 'the tree of life', grows very well on a wide range of soils, including highly degraded, arid, semi-arid and rocky soil. It has a thick waxy and deeply fissured bark that makes it drought and fire resistant (sheabutterblog. blogspot.com).

Production

The sheabutter tree occurs both wild and cultivated. The tree bears the first fruits after 10–15 years, but takes 20–25 years to reach full productivity. In a given year, roughly every third tree bears fruit. An average tree bears 15–20kg fresh fruits; 50kg fresh nuts yield

(a)

(b)

Source: (a) Marco Schmidt, http://commons.wikimedia.org/wiki/File:
Vitellaria_paradoxa_MS_6563.JPG; (b) Marco Schmidt, http://
commons.wikimedia.org/wiki/File:Vitellaria_paradoxa_MS4195.JPG

Figure 10.176 (a) Shea tree; and (b) seeds

20kg of dry kernels, from which approximately 4kg sheabutter can be made. The kernels contain 32–45 per cent fat (Gliese and Eitner, 1995).

Processing and utilization

The seeds are milled and pressed. An older method is to cook the mortared seeds with water and to skim off the oil. The compounds in the oil are glycerides of stearic acid (41 per cent) and oleic acid (48 per cent). The oil is not only used as edible oil (sheabutter), but also for technical purposes such as the traditional production of soap and candles (Franke, 1985).

In a traditional setting, the wood is being used for charcoal production, while sheabutter of poor quality is used as an illuminant or fuel in lamps or as candles (www.solutions-site).

Internet resources

www.solutions-site.org/artman/publish/article_10.shtml
http://sheabutterblog.blogspot.com/2007_12_01_archive.html
http://en.wikipedia.org/wiki/Shea_butter
www.savannatrading.com/BulkSheaButter.html

SORGHUM (*Sorghum bicolor* L. Moench)

Sorghum is a C_4 crop that belongs to the grass family. Most cultivated varieties of sorghum can be traced back to Africa, where they grow on savannah lands. The name 'sorghum' comes from Italian 'sorgo', in turn from Latin 'Syricum (granum)' meaning 'grain of Syria'.

Sorghum is a genus of numerous species of grasses, some of which are raised for grain and many of which are used as fodder plants, either cultivated or as part of pasture. *Sorghum* is in the subfamily Panicoideae and the tribe Andropogoneae (the tribe of big bluestem and sugar cane). Commercial *Sorghum* species are native to tropical and subtropical regions of Africa and Asia, with one species native to Mexico. Commercial sorghum refers to the cultivation and commercial exploitation of species of grasses within the genus *Sorghum*. These plants are utilized for grain, fibre and fodder (wikipedia.org).

Although rich finds of *S. bicolor* have been recovered from Qasr Ibrim in Egyptian Nubia, the wild examples have been dated to circa 800–600 BCE and the domesticated ones no earlier than CE 100. The earliest archaeological evidence comes from sites dated to the second millennium BC in India and Pakistan, where *S. bicolor* is not native. Sorghum is well adapted to growth in hot, arid or semi-arid areas. The many subspecies are divided into four groups – grain sorghums (such as milo), grass sorghums (such as Sudan grass) for pasture and hay, sweet sorghums (formerly called 'Guinea corn', used to produce sorghum syrups), and broom corn (for brooms and brushes). The name 'sweet sorghum' is used to identify varieties of *S. bicolor* that are sweet and juicy.

The FAO reports that 440,000 square kilometres were devoted worldwide to sorghum production in 2004.

Sorghum is being considered as a 4F crop: for food, fodder, fibre and fuel. This section considers the importance of grain, fibre and sweet sorghum as bioenergy feedstock and recent developments in this field in China, India, the USA, Europe and Japan.

Sorghum is a genus comprising numerous grass species, some of which are used for grain, fodder and forage (grain sorghum) and some of which are used for syrup production (sweet sorghum). Sorghum thus offers multiple pathways to ethanol (greencarcongress.com):

- Starch to ethanol from grain sorghum. About 15 per cent of the US grain sorghum crop currently goes into ethanol production with one bushel of grain sorghum producing the same amount of ethanol as one bushel of corn, according to the National Sorghum Producers.
- Sugar to ethanol from sweet sorghum.
- Cellulosic ethanol from:
  - residue/regrowth on grain sorghum;
  - forage sorghums;
  - bagasse from sweet sorghum; and
  - dedicated biomass sorghums.

In November 2005, the US Congress passed a Renewable Fuels Standard as part of the Energy Policy Act of 2005, with the goal of producing 30 billion litres of renewable fuel (ethanol) annually by 2012. Currently, 12 per cent of grain sorghum production in the US is used to make ethanol.

Texas A&M is interested in exploring the potential use of sorghum as a potential biofuel feedstock. Scientists at Texas A&M University's Agricultural Experiment Station (TAES) are breeding a drought-tolerant sorghum that may yield between 37 and 50 tonnes of dry biomass per hectare.

The research process adopted includes the following steps:

- Sorghum genetic resources will be screened for sources of improved yield and biomass composition

Table 10.94 Top sorghum producers, 2005

	United States	9.8Mt
	India	8.0Mt
	Nigeria	8.0Mt
	Mexico	6.3Mt
	Sudan	4.2Mt
	Argentina	2.9Mt
	China	2.6Mt
	Ethiopia	1.8Mt
	Australia	1.7Mt
	Brazil	1.5Mt
World Total		**58.6Mt**

Source: FAO

(sugar content, cellulose, hemicellulose, lignin) optimal for biofuels production.

- Sorghum germplasm, traits and genes that improve biomass yield, bioenergy composition and drought tolerance will be identified and introduced into cultivars and elite hybrids.
- Advanced material will be tested to identify cultivars that have optimal biomass-to-biofuels conversion properties and agronomic production parameters.
- Logistical approaches will be optimized for the harvest and transport of sorghum to facilities for biofuels and bioenergy production.
- High-yielding, drought-tolerant sorghum bioenergy cultivars and hybrids will be produced, specifically engineered to meet the needs of the US biofuels industry.
- Information and technology useful for improving corn, sugar cane, switchgrass, and other grass species for biofuels production will be generated.

Source: Biopact (2007a)

One Texas A&M initiative is to develop a high biomass sorghum for use in cellulosic ethanol production. The goal of the development is high biomass accumulation – about 20 tonnes/acre.

Source: Texas A&M, greencarcongress.com

Figure 10.177 A Texas A&M researcher working
with a high-biomass variant of sorghum

Chevron Corporation and the Texas A&M
Agriculture and Engineering BioEnergy Alliance (Texas
A&M BioEnergy Alliance) announced that they had
entered into a strategic research agreement to accelerate
the production and conversion of crops for
manufacturing ethanol and other biofuels from
cellulose (greencarcongress.com)

In 2008, Renergie, Inc. received US$1,500,483
(partial funding) in grant money from the Florida
Department of Environmental Protection's Renewable
Energy Technologies Grants Program to design and
build Florida's first sweet sorghum juice mechanical
harvesting system and ethanol plant capable of
producing fuel-grade ethanol solely from sweet
sorghum juice (freencarcongress.com)

The International Crops Research Institute for the
Semi-Arid Tropics (ICRISAT) research on ethanol for
biofuel from sweet sorghum (*Sorghum bicolor (L.)*
Moench) and biodiesel from pongamia and jatropha
crops, is not only ensuring energy, livelihood and food
security to the dryland farmers, but also reducing the
use of fossil fuel, which in turn can help in mitigating
climate change. These crops meet the main needs of the
dryland farmers – they do not require much water, can
withstand environmental stress and are not that
expensive to cultivate. After extracting the sugar, the
biomass residues of sweet sorghum can be used as a
solid biofuel in power (co)generation plants, or later as
a feedstock for next-generation biofuels. Alternatively,
it makes for a good animal feed. ICRISAT has
developed drought-tolerant sweet sorghum hybrids
that deliver grain, forage and sugar syrup all in one
plant. The hybrid produces potentially more ethanol
than sugar cane and needs much less water (Biopact,
2007b).

Sorghum Breeders at ICRISAT sum up the three
main advantages of the sweet sorghum hybrids
(Biopact, 2007c):

- **Food, feed, fibre, fuel**: the improved sorghum
  provides the dryland farmer with grain for food,
  fodder for livestock and an additional source of
  income through bioethanol, obtained from the
  sugar in the canes.
- **Water requirements**: sweet sorghum requires
  only one seventh of the water that is used by sugar
  cane.
- **Land use**: it has the advantage over other biofuel
  crops that it yields grain as well as ethanol. Rather
  than replacing land grown to food, the cultivation
  can stimulate increased yield of grain and stalk,
  and also fodder from bagasse, the by-product of
  the crushes canes. This allows for an integration of
  farming practises and the environmental benefits
  that come with this.

ICRISAT tested the new hybrid – called SSH 104 –
first in Andhra Pradesh, with such success that the
plant was immediately patented. The institute then
took it to the Philippines, where a vast region of land
was identified as suitable. Data from field trials, carried
out in collaboration with the Mariano Marcos State
University (MMSU) in Batac Ilocos Norte, indicate
following aspects (Biopact, 2007b):

- **Average yield**: in the MMSU study, the average
  yield was 110t/ha of sweet sorghum cane stalk
  for two cropping seasons in eight months (one
  main crop followed by one ratoon crop.) Ratoon
  is the outgrowth after the main stalk has been
  cut.

- **Sugar content:** the MMSU studies have shown that sugar cane has up to 14 per cent sugar content while sweet sorghum has 23 per cent.
- **Cropping season:** one hectare planted with sweet sorghum will yield 95–125t after a planting season of 100–115 days, compared to sugar cane's 65–90t/ha with a longer crop season of 300–330 days.
- **Water requirements:** sweet sorghum adapts well to drought and will not compete much for fresh water, needing only about 175m³ per crop, which is just one-quarter of sugar cane's 700m³ water need per crop.
- **Commercial viability:** the study estimates the net income for two cropping seasons with sweet sorghum to range from 65,000–72,000 pesos per hectare (€1000–1150/US$1300–1500), comparing favourably to sugar cane and most other commonly grown crops.
- **Ethanol potential:** at an extraction and processing rate similar to that of sugar cane and an average yield of 110t/ha, using first-generation bioconversion technologies, an ethanol yield of around 10,000L/ha can be expected.

ICRISAT has signed agreements with private companies in India and the Philippines to form a consortium to exploit sweet sorghum for ethanol. Further, ICRISAT are in the initial stages of exploring such consortia in Uganda, Nigeria, Mozambique and South Africa (Biopact, 2007c).

Li Dajue, from Beijing Botanical Garden, China, and one of the worldwide pioneers in sweet sorghum breeding, has developed superior cultivars with sugar, ethanol and biomass huge potentials (see section on sweet sorghum).

Scientists at the US Department of Energy (DOE) Joint Genome Institute (JGI) and several partner institutions have published the sequence and analysis of the complete genome of sorghum, a major food and fodder plant with high potential as a bioenergy crop. 'Sorghum will serve as a template genome to which the code of the other important biofuel feedstock grass genomes – switchgrass, miscanthus, and sugar cane – will be compared', said Andrew Paterson, the publication's first author and Director of the Plant Genome Mapping Laboratory, University of Georgia. The genome data will aid

Source: Mandy Gross, Robert M. Kerr Food & Agricultural Products Center, Oklahoma State University

Figure 10.178 Sorghum research field at ICRISAT

scientists in optimizing sorghum and other crops not only for food and fodder use, but also for biofuels production (Joint Genome Institute, 2009).

Fibre sorghum

Description

Fibre sorghum is a new hybrid between grain and broomcorn sorghums, making it a subspecies of *Sorghum bicolor* in the sub-tribe Sorghastrae of the tribe Andropogoneae (Jvanjukovic, 1981). Fibre sorghum's stalks are fibrous, not juicy. It is a annual C₄ plant species with high water efficiency, high biomass yield and one of the highest dry matter accumulation rates on a daily basis. The plant can grow to a height of 3.5–4m (Figure 10.179). Its habitat is similar to that of maize. Although sorghum is of tropical origin, the plant is well adapted in subtropical and temperate regions, especially the fibre sorghum hybrids, which are able to grow in northwest European countries. Temperature is the most critical growth factor of all climatic parameters (GEIE Eurosorgho, 1996).

Ecological requirements

Germination of fibre sorghum takes place between 8 and 10°C. The most productive growth stage is in the temperature range of 27–30°C and when annual precipitation is from 400–600mm. Sorghum's ability to tolerate drought is due to the high efficiency and growth of its root system, which is twice as high as with maize. This, and the closer spacing between the rows, reduces the danger of soil erosion.

Fibre sorghum has been successfully grown in temperate climates, such as in northern Germany. The new hybrids are very productive at higher latitudes (El Bassam, 1994a). It has a great ability to mobilize the natural nitrogen reserves and other minerals of the soil because of its well-developed root system (GEIE Eurosorgho, 1996). Soil requirements of sorghum are very modest. Sorghum tolerates salts and alkalis and can grow in a wide pH range between 5.0 and more than 8.0. Waterlogged soils and acidic soils should be avoided.

Crop management

Sowing of fibre sorghum should not be done before or after maize, as this could lead to crop rotation problems. Sorghum is propagated by seed. To prepare the seedbed a ploughing depth of 10–30cm is suggested. The seed is sown with a drilling machine to a depth of 2–4cm into the soil (El Bassam, 1994a). Pre-emergence herbicide application is recommended, as with maize. 'Safer' can be used as a seed dressing to minimize pre-emergence herbicide damage. The sowing date is dependent on the climate of the region, because sorghum is sensitive to frost. Therefore in northern parts of Europe, such as Germany, fibre sorghum varieties with a growth season shorter than six months should be chosen. The most suitable plant density ranges between 200,000 and 300,000 plants (20–70kg seed/ha) with a row spacing of 40–50cm (GEIE Eurosorgho, 1996). Because of the great ability of fibre sorghum to mobilize soil nitrogen, nitrogen fertilization does not need to exceed 80–100kg/ha and should be adjusted with regards to the soil contribution possibilities and the accessible yield. Recommended fertilization rates are 70kg/ha P_2O_5 and 140kg/ha K_2O (GEIE Eurosorgho, 1996).

Production

Total dry matter fibre sorghum yields (stem and leaves) of up to 40t/ha are possible. Dry matter contents range from 20 to 30 per cent (GEIE

Figure 10.179 Fibre sorghum, Germany

Eurosorgho, 1996). In northern parts of Germany, the highest total dry matter yield of 23t/ha was ascertained during a test of different fibre sorghum genotypes. The highest yield of stems was 14.5t/ha (El Bassam, 1994a). The analysis for lignin and cellulose of selected genotypes using the ADF (acid detergent–fibre) method showed 3–10 per cent lignin and 24–34 per cent cellulose.

Processing and utilization

Fibre sorghum is an annual plant which has to be harvested in autumn before winter begins. Because fibre sorghum varieties are sterile hybrids, no seed production is possible. The greatest interest in fibre sorghum is for biomass production. This involves using fibre sorghum as an energy plant or as pulp for the paper industry. The harvest and the raw material handling are decisive elements for using sorghum for the paper industry. Important for all harvesting systems is the prevention of breakdowns and jamming of the biomass, which is extremely dependent on the physiological stage of the harvested material and the weather conditions. In addition, protection against sand and earth during the harvest must be guaranteed. Chopping and baling, as well as ensiling, are possible harvesting and storing methods. For the utilization of fibre sorghum as a solid fuel, the biomass should be mowed and dried, and then afterwards baled with a high-density cubic baler or with a round baler. In humid regions, fibre sorghum should be ensiled and pressed mechanically before being used as a fuel feedstock for combustion. Fibre sorghum dry matter is suitable for producing pellets and briquette (El Bassam, 1994a). Fibre sorghum can also be used for producing the second generation of ethanol (El Bassam, 2007a).

Sweet sorghum (*Sorghum bicolor* L. Moench)

Contributed by Professor Dr Li Dajue, Institute of Botany, Chinese Academy of Sciences, Beijing, China and C. D. Dalianis

Description

Sweet sorghum originated in Africa and is grown for its sweet stems which are used mainly for syrup production (Figure 10.180).

It has a fibrous root system that branches profusely. Under favourable conditions, the above-ground nodes may produce strong adventitious roots that help to anchor the plant and thereby reduce lodging. The stems are similar to those of corn, being grooved and nearly oval. The top internode is not grooved. Sweet sorghum has sweet juicy pith in the stems. Crown buds give rise to tillers. Each tiller soon develops an independent root system, though it remains attached to the main stem.

Ecological requirements

Sweet sorghum, although native to the tropics, is also well adapted to temperate regions. Recently, sweet sorghum has been considered in China and the EU as a potential energy crop, mainly for ethanol production. It is expected to play a similar role in temperate climates to that which sugar cane plays in tropical regions, especially Brazil. In recent years there have been many studies in EU countries of sweet sorghum's adaptability, productivity and ethanol potential under various environmental and soil conditions as well as under various cultural practices.

Sweet sorghum is well adapted to the warm southern regions of the EU and moderately well adapted to several central EU regions with mild climates. Because sweet sorghum originated in the warm region of central Africa it is a cold-sensitive plant, so its adaptation in northern cooler climates is poor. Attempts to grow sweet sorghum in Ireland have failed as a result of problems with establishment and poor growth.

However, it should be mentioned that cool-tolerant sorghum genotypes are encountered at high altitudes in Africa and in temperate regions of America (Dogget, 1988). In the USA, cool-tolerant grain sorghum cultivars have been bred for extending grain sorghum cultivation in cool northern areas.

The minimum temperatures are 7–10°C for germination and 15°C for growth. The optimum temperature for growth is 27–30°C (Quinby et al, 1958). Sorghum withstands high temperatures better than many other grain crops.

Sorghum can be grown successfully on a wide range of soils, such as heavy clays, medium loams, calcareous soils and organic soils. It tolerates a pH range from 5.5 to 8.5, and also some degree of salinity, alkalinity and poor drainage (Hayward and Bernstein, 1958).

Sorghums are considered to be one of the most drought-resistant agricultural crops. They have the ability to remain dormant during drought and resume growth when favourable conditions reappear. For this reason sorghum is frequently called the 'camel' of field crops.

One of the important factors affecting its drought endurance is the effectiveness of its large fibrous root system, mainly the secondary roots, which can extend to a distance of up to 1m laterally and to a depth of 1.9m. The root system is about twice as active as that of corn in taking up water from the soil (Martin, 1941).

Sorghum, beyond its ability to absorb water more effectively than many other crops, has a better ability to regulate water loss to the atmosphere. With the onset of water stress, sorghum showed the lowest rate of decline in relative turgidity and the least diurnal depression as compared with other crops, because it reduced its transpiration rate to a much greater extent than did other crops (Slatyer, 1955). Sorghums also

Source: Dajue

Figure 10.180 The experimental field in Huairou District Beijing, China. The plant height of super hybrid sweet sorghums (SHSS) is about 5m

have a relatively impervious waxy epidermis that retards the desiccation of stalks and leaves. The leaf blades are covered with a waxy coating that reduces water loss.

Propagation

At present, sweet sorghum is being extensively bred to increase sugar content and biomass, and to improve its resistance to lodging, frost and disease. Sweet sorghum is propagated by seed, of which the cultivars 'Keller' and 'Korall' have had significant success in Europe. New highly productive sweet sorghum cultivars have been developed in China by Li Dajue. At present, there is no hybrid sweet sorghum cultivar used in

production. Some scientists use the A-line of grain sorghum to make hybrid sweet sorghum.

Even the plant height and grain yield had increased, but the stalk had less juice and the sugar content in the juice decreased as medulla versus rich juice where medulla is dominant, and non-sweet versus sweet where non-sweet is dominant. A male sterile line with high sugar content in the juice of the stalk was found by Beijing Green Energy Institute in a field in 2001. The line has been cloned using advanced biotechnology. When it was crossed with an excellent sweet sorghum mutant, the yield of hybrids increased more than 40–80 per cent. They were named super hybrid sweet sorghums (SHSS) (Figure 10.181).

Crop management

Sweet sorghum is sown with seed; 4.5–7.5kg/ha of seed is enough; and it is feasible to sow by machine. A warm seedbed free from weed growth is essential to good seed germination of sweet sorghum. Planting takes place in the spring and may start when soil temperature remains above 15°C. Seed germination occurs within 24 hours in warm and moist soil. Pneumatic sowing machines may be used for sweet sorghum. Recommended plant distances are 70cm between rows and 10–20cm within rows. The early growth of the plants is very slow, and places them at a disadvantage in competition with weeds.

Although sweet sorghum is one of the most drought-resistant crops, in southern EU regions it needs to be irrigated, otherwise the plants remain stunted and biomass yields are very low. Depending on irrigation rates, annual dry matter biomass yields of sweet sorghum in central Greece ranged from 22 to 33t/ha and in southern Italy from 18.5 to 47t/ha.

Compared with many other widely grown agricultural crops, sorghums have the highest water use efficiency. In a very old but classic work, Shantz and Piemeisel (1927) found that alfalfa requires 844kg of water for the production of 1kg of dry matter, oats 583kg, common wheat 557kg, barley 518kg and corn 349kg, but grain sorghum requires only 304kg of water.

Sweet sorghum, having very high water use efficiency, is very well suited for southern EU conditions. In central Greece, water use efficiencies, depending upon irrigation rates applied throughout the growing period by the drip method, ranged from 206L/kg of dry matter, produced under highly irrigated conditions (458mm), to 181L/kg of dry matter, produced when irrigation was restricted to only 157mm – 34 per cent of the highly irrigated rate (Dercas et al, 1996).

In a comparative study in southern Italy (Mastrorilli et al, 1996) the water use efficiency of sweet sorghum has been estimated at 192L/kg of dry matter produced, whereas the comparable figure for grain sorghum was 270L/kg, for sunflower 278L/kg and for soya bean 357L/kg.

Several nitrogen experiments have been carried out in various EU countries and sites. So far, no significant nitrogen responses have been observed in the various experiments (Dalianis et al, 1995b; Dercas et al, 1995; 1996; Cosentino, 1996), except in Portugal, where significant differences have been observed between the 75 and 100kg/ha nitrogen rates (Oliviera, 1996).

Beyond its low nitrogen requirements leading to less nitrogen fertilization inputs, sweet sorghum is characterized by a large and widespread root system. Therefore, even if some nitrate leaching occurs during the growing period there are increased chances for nitrates to be trapped during their downward movement by the plant's extensive root system compared to the poorer systems of many other agricultural crops.

The potassium requirements of sweet sorghum are very high. It is reported (Curt et al, 1996) that potassium uptake amounts to 617kg/ha for a dry matter yield of 25t/ha. The fact that all the above ground biomass is removed from the field – compared, for example, with a corn grain crop where only one-third of the potassium in the above ground part is removed with the grains – indicates that sweet sorghum is a heavily potassium-depleting crop.

There are certain indications that plant density affects biomass yields from sweet sorghum. With a stable between-row distance of 70cm and within-row distances of 5, 10, 15 and 20cm it was found (Dalianis et al, 1996) that the 15 and 20cm within-row distances were superior in terms of total fresh and dry matter stem yields. Significant also is the effect of plant density on sugar yields and on final stem diameter. Sugar yields

and stem diameter were higher with the 20cm within-row plant density.

The quantity of irrigation water needed by sweet sorghum is only one-half and one-third of that needed by maize and sugar cane respectively. In some particular areas with scarcity of water and poor quality soil (North of Brazil, some areas of Africa, China etc.) sweet Sorghum is an even more promising crop in comparison with sugar cane for bioethanol production (Source: a World Bank study on the bioethanol prospects and development opportunities in Sub-Saharan Africa).

Production

Sweet sorghum (*Sorghum bicolor* (L.) Moench) is also called sweet stalk, sweet bar and sugar sorghum. Beside producing 2250–7500kg/ha of grain, it also produces 60–150t/ha of stalk containing rich sugar. An excellent cultivar of sweet sorghum can match against sugar cane for sugar content in the juice. Because of its excellent characteristics, such as high yield, easy cultivation, wide adaptation, and resistance to drought, waterlogging and saline-alkali, it is suitable to be grown in China with the national conditions of a large population and a shortage of arable land. In addition, the sugar in the stalk and the starch in the grain are easily transformed into alcohol. Ethanol is a substitute for petroleum; therefore, it is obvious that sweet sorghum is a 'ground oil field' demanding prompt development.

Sweet sorghum grows fast and with a high yield. The plant height ranges from 250 to 580cm; a single stalk can weigh 3600g. Among the biomass energy systems, sweet sorghum is the principal competitor. In the Beijing area, the variety 'Wray' grows as high as 12 cm/day from 20 to 26 July on average. An experiment in the USA showed the ethanol yield of sweet sorghum was 6106L/ha while the yield of the sugar cane known as the best 'solar energy transformer' was only 4680L/ha. An experiment in Beijing Botanical Garden, Institute of Botany, Chinese Academy of Sciences also showed that sweet sorghum can be used to produce 6150L/ha of alcohol; the output of fresh biomass can reach 128t/ha. The result showed that the average yield of ethanol (anhydrous at 99.5 per cent concentration)

Figure 10.181 Panicles of newly developed
hybrids of sweet sorghum

Figure 10.182 A single stalk of SHSS.
The plant height is 576cm

of sweet sorghum is 5700–6500L/ha/year, while the yield of sugar cane is only 4000–5000L/ha/year in India. The results of experiments in F. R. Yugoslavia showed that it is possible to produce 4700L/ha of ethanol in rainfed condition and 8000L/ha under irrigation. The per unit area yield of sweet sorghum is much higher than that of other crops, its daily photosynthetic carbohydrate can be used to produce 48L/ha alcohol, while maize, wheat and grain sorghum would only produce 15L/ha, 3L/ha and 9L/ha, respectively.

Sweet sorghum is not always more short and slender than sugar cane. If we calculate, based on 2000g/stalk and 75,000 plant/ha, the yield of the stalk would be 150t/ha. Some plants are very tall; the highest plant is 576cm (Figure 10.182). Of course, it does not need to be so high, as there is a risk of wind logging. Generally speaking, around 450cm is acceptable. The utilization of SHSS would make a great contribution toward energy security.

Sweet sorghum adaptation and productivity have been studied around the world. Fresh biomass yields ranged from 22.4 to 44.8t/ha in Ohio (Lipinsky et al, 1978), 40 to 45t/ha in southern USA (Bryan et al, 1981), and up to 64t/ha in Alabama (Soileau and Brandford, 1985). It is reported from China (Nam and Ma, 1989) that for the varieties 'Wray' and 'Keller' fresh stem yields were around 49.5t/ha, with a simultaneous grain yield of 1.8t/ha for 'Wray' and 2.1t/ha for 'Keller'. Work in China (Dajue and Yonggang, 1996) reports fresh biomass yields, depending on variety, from 82.4 to 128.1t/ha with a simultaneous seed production of 1.43 to 6.67t/ha.

Under certain EU environmental conditions, sweet sorghum biomass yields are very high. From previous work, fresh biomass yields of sweet sorghum, depending upon site, weather conditions, soil fertility, varieties and cultural practices, were 80–100t/ha in Germany (Bludau, 1987), 92.6t/ha in Spain (Curt et al, 1994), and up to 129t/ha in Greece (Dalianis et al, 1995b). It should be noted that all these high biomass yields in the southern EU were obtained under irrigation.

Sweet sorghum biomass yields in the EU are significantly affected by the latitude (Figure 10.183). Thus, total above ground fresh biomass yields ranged from as low as 55t/ha in Belgium (Chapelle, 1996), through 120t/ha in Malaga in southern Spain (Curt et al, 1996) and to 127.3t/ha in northern Italy (Dolciotti et al, 1996), up to 141t/ha in southern Greece (Dalianis et al, 1995b).

Total dry matter yields were 12t/ha in Belgium (Chapelle, 1996), 40t/ha in southern Spain and northern Greece, and up to 45t/ha in southern Greece and southern Italy (Dalianis et al, 1995b; Cosentino, 1996; Curt et al, 1996).

Several varieties have been tested ('Keller', 'Wray', 'Mn 1500', 'Cowley', 'Dale', 'Sofra', 'Theis' and 'Korall') in various EU regions and under various environmental and soil conditions.

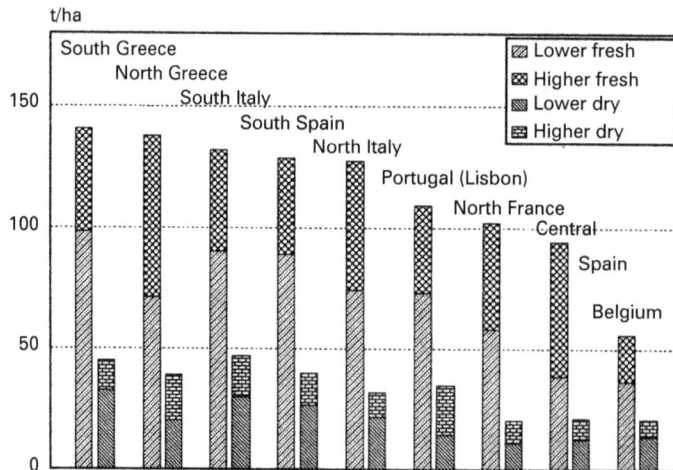

Figure 10.183 Total fresh and dry matter yields of sweet sorghum in various EU regions (ranges within regions are the result of variety, year and/or treatment effects)

On average, 'Keller' was proven the best, combining wide adaptation, excellent stability (from site to site and from year to year) and very high biomass and sugar yields. However, some other varieties were superior in certain EU regions. Thus, fresh biomass yields of 'Keller' were superior to those of 'Cowley' in Greece (Dalianis et al, 1995b); yields of 'Wray' were superior to those of 'Korall' in northern Italy (Petrini, 1994); and yields of 'Dale' and 'Theis' were superior to those of 'Wray', 'Keller' and 'Korall' in southern Spain (Curt et al, 1996), all grown under the same conditions at each site.

Processing and utilization

Historically, syrup production was the main use of the sweet sorghum grown in certain tropical and subtropical regions. Attempts to develop a sorghum sugar industry have not been successful because of certain difficulties and limitations that make the sugar more expensive than that from sugar cane or sugar beet.

Sorghum is gaining attention as a potential alternative crop for energy and industry because of its high yield in biomass and fermentable sugars. As soon as possible after harvesting, the stalks are transported to a processing facility where the stalks are cut and crushed for juice extraction to separate the sugar from the bagasse. Two extraction techniques are being investigated: in one the sorghum is crushed with a mechanical screw or roller press, and the other uses a crushing, diffusion and dewatering system. After extraction the sugar can be concentrated for storage or it can be sent to an energy conversion process. Bagasse can also be processed immediately or it can be dried for long-term storage.

Sweet sorghum can be converted into energy through two processes: biochemical and thermochemical. The biochemical process is used to convert the sugar into fuel ethanol. The sugar, which is two-thirds saccharose and one-third monosaccharides, can be processed into ethanol. The thermochemical process, which can consist of carbonization, combustion, gasification, or pyrolysis, is used to convert bagasse into bio-crude oil, gasoline, electricity, pulp for paper, compost and charcoal pellets.

The bagasse is mainly lignocellulose, has 55–62 per cent residual moisture, and contains two-thirds of the plant's energy. A cycle can be created when bagasse is used to produce the energy necessary for ethanol distillation.

The theoretical ethanol yield (in L/ha) can be calculated from the formula:

Total sugar content (per cent) in fresh matter (FM) × 6.5 (a conversion factor) × 0.85 (the process efficiency) × total biomass (in t/ha of FM).

Sugar percentages are around 9–14 per cent from fresh stem matter and 30–45 per cent from dry stem matter. Based on stem biomass yields and sugar percentages observed in various EU regions, sugar yields have been estimated Figure 10.184. In southern Spain, estimated sugar yields are as high as 13.8t/ha (Curt et al, 1996) and in Greece more than 12t/ha (Dalianis et al, 1995b).

The ethanol potential of sweet sorghum, under Greek conditions in fertile fields and under appropriate cultural management, has been estimated at 6750L/ha (Dalianis et al, 1995b); in China, the ethanol potential has been estimated at 6159L/ha (Dajue and Yonggang, 1996).

It should be mentioned that, beyond this high ethanol potential of sweet sorghum, a significant amount of lignocellulosic material is left as bagasse. The bagasse is sufficient to cover all the energy requirements of the chain from plant production to ethanol distillation; or it may be used for pulp production, or to ferment the cellulosic fraction, thereby increasing ethanol yields by at least 40 per cent. Up to 10,000L/ha of ethanol may be possible.

Among the crops for producing ethanol, the crop with highest yield is sweet sorghum. If we could grow 26.7 million ha/year of SHSS in the next 5–10 years in China, we would produce 2240 hundred million litres of alcohol. It is 67.9 per cent of the total yield in the world. It can be made into 11,200 hundred million litres of E_{20} gasohol and it is enough for consumption in China. Supposing the grain yield is 4.5t/ha and stalk yield is 90t/ha, the ethanol yield would be 8400L/ha. The income for the farmer would be US$1658/ha (grain US$123/t × 4.5 t + stalk US$17/t × 90 t − cost for production US$426/ha = US$1658/ha). The net income for 26.7 million ha will be US$442.7 hundred million.

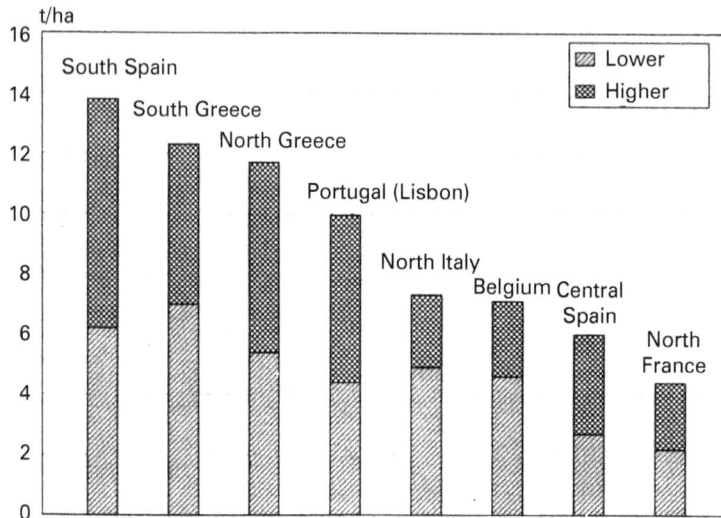

Figure 10.184 Sugar yields of sweet sorghum in various EU regions (ranges within regions are the result of variety, year and/or treatment).

The net income for the ethanol plant would be US$388 hundred million (if the price of ethanol is US$617/t and cost: raw material US$313/t, cost for processing US$86/t, the profit is US$217/t.) The total net income for peasants and ethanol plants for the 26.7 million ha of sweet sorghum would be US$831 hundred million.

Sweet sorghum potential in Japan

Contributed by Dr Youji Nitta, Ibaraki University College of Agriculture, Japan

A study was carried out to investigate the potential of sweet sorghum for biofuel production and bioremediation in the suburbs of the Tokyo Metropolitan Area.

Sweet sorghum (*Sorghum bicolor* Moench) produces a lot of sugars in its stem. Because sweet sorghum has a wide range of growing areas including the northern part of Japan, it has a great potential to produce a large amount of biofuel (ethanol) without having a bad influence on the food crop economy. In addition, great attention is focused on its great ability to absorb nutrients from the soil, including excess nitrogen and phosphorus which would have a harmful effect on the ecological environment. We investigated the sugar yield potentials of sweet sorghum to evaluate its biofuel productivity. The activity of bioremediation was also focused on in our experiment.

Three varieties were used in this experiment: FS501 (Kotobun sugar), FS902 (Big sugar sorghum) and KCS105 (Super sugar sorghum). Experiments were conducted in a field of Ibaraki University, Japan (36°02′10″N, 140°12′43″E), located about 50km northeast of Tokyo, which is in a temperate zone. Sowing was carried out on 1 June (I), 15 June (II) and 30 June (III) 2007. The spacings between and within rows were 80 and 15cm respectively. Slow release fertilizer (coated fertilizer) was applied at the rate of 9g/m^2 of N, P$_2$O$_5$ and K$_2$O. After that, quick-acting fertilizers were applied at the rate of 3g/m^2 of N, P$_2$O$_5$ and K$_2$O as basal dressing (Treatment A), and supplemental dressing at 45 days after sowing (Treatment B). Above-ground portions were harvested 102 to 153 days after sowing. The centre portion of the stem axis was pressed with pliers to remove some sugar juice, followed by brix measurement by refractometer. For estimation of bioethanol yield, we adopted six formulas cited in previous reports.

Among the three sowing dates, brix values and stem sugar contents of III (sowed 30 June 2007) was the

Source: Nitta

Figure 10.185 Sweet sorghum (var. FS902) growing in the field of Ibaraki University, Japan (36°02'10"N, 140 °12'43"E), in the suburbs of the Tokyo Metropolitan area, 3 October 2008

highest in all varieties and treatments. It was because the harvest dates were too early in I (sowed 1 June 2007) and II (sowed 15 June 2007), 102 days after sowing respectively to get enough sugar.

Figure 10.185 shows an example of sweet sorghum taken in the 2008 experiments. Plant height reached more than 4m about 4 months after sowing. There are many commercial varieties with different plant types in Japan.

Table 10.95 shows some growth characters and stem sugar content in III. In III, the fresh shoot weight ranged from 537 (FS501/B) to 929g/plant (FS902/B) which is equivalent to 44,750 to 77,416kg/ha. Plant height was lowest in FS501/B (292cm) and highest in FS902/B (480cm). Brix value and stem sugar content ranged from 14.0 to 16.2 per cent and from 47.0 to 95.9g/plant, respectively. As a result, sugar yields were estimated 3917 for FS501/B and 7992kg/ha for FS902/B, respectively.

For bioethanol yield estimation, we used four formulas: Lipinsky (1978), Pari and Ragno (1998), Soldatos and Chatzidaki (1999), and Mamma et al (1996). In formulas 3 and 4, 538 and 700g ethanol/kg sucrose, 0.052 and 0.084g ethanol/g total fresh weight were applied, respectively. Bioethanol yield was highest in FS902, followed by KCS195 and FS501. In FS902, the bioethanol yield ranged from 2952 to 6936L/ha using the six formulas. From 2810 to 5926L/ha or

from 1860 to 3960L/ha bioethanol yields were estimated for KCS105 and FS501, respectively. As a result, estimation of bioethanol yield was highest with 6936L/ha for FS902, and lowest 1860L/ha among three varieties and treatments.

Our research revealed that harvest should be done after securing enough days after heading. Moreover, plants should not only have large plant height; above-ground biomass is also an important practical factor for higher sugar production. As a result, ethanol production was estimated at more than 5t/ha. Recently, in the suburbs of Tokyo Metropolitan area, non-farming of idle land area has increased remarkably. But abundant nitrogen accumulation in the farm soil is also a serious problem. Sweet sorghum seems to have a great potential not only to act as a biofuel crop but also to prevent the leaching of nitrogen.

The importance of choosing an adequate variety of sorghum, fertilizer application and good management is also emphasized. We must also take into consideration weather conditions, including typhoon and cool temperatures, harvesting dates and plant of each variety. Biomass of above-ground portions largely differed among varieties and treatments. Therefore, the ability of nutrient absorption from soil is also different due to the ecological and physiological functions of each variety.

According to the media, Japanese car industries have realized that bioethanol fuel as a blended fuel with gasoline would be widely effective from the viewpoint of carbon dioxide reduction and consumption restraint of fossil fuel (oil). As an example, Honda E100 has equipped a car to run on E100 and Toyota has changed the materials of the fuel system of some some gasoline engines since June 2006, which allows them to use E10. In addition, Toyota has already sold the FFV (flex-fuel vehicle) fuelled by E20 (bioethanol mixture rate 20 per cent) car in Thailand since January, 2008. Furthermore, more bioethanol-ready cars have been developed in Brazil and the US by Toyota. However, if development and diffusion of bioethanol-ready cars is to move forward and if a high mixture percentage (E20 and above) is to be applied, car industries will need authorization by the Ministry of Land, Infrastructure and Transport in Japan.

Japan's five ministries (Ministry of Economy, Trade and Industry, Ministry of the Environment, Ministry of Agriculture, Forestry and Fisheries, Ministry of Land, Infrastructure and Transport, and

Table 10.95 Growth characteristics and stem sugar content in III

Variety/ treatment	Harvest (days after sowing)	Shoot				Plant height (cm)	Brix (%)	Stem sugar content	
		Fresh weight (g/plant)	Fresh weight (kg/ha)	Dry Weight (g/plant)	Water content (g/plant)			(g/plant)	(kg/ha)
FS501/ A	22 Nov (145)	558 d	46500	137 c	421 d	289 d	15.4 a	54.5 de	4542
FS501/ B	22 Nov (145)	537 d	44750	124 c	413 d	292 d	14.0 b	47.0 e	3917
FS902/ A	26 Nov (149)	852 ab	71000	288 a	565 b	486 a	15.9 a	85.1 b	7092
FS902/ B	26 Nov (149)	929 a	77416	294 a	635 a	480 a	16.2 a	95.9 a	7992
KCS105/ A	22 Nov (145)	811 bc	67583	325 a	487 c	352 b	15.9 a	62.0 cd	5167
KCS105/ B	22 Nov (145)	718 c	59833	220 b	497 c	322 c	15.3 a	63.4 c	5283
LSD 0.05	–	100	–	52	65	17	1.1	8.7	–

Note: Values including the same letter are not significantly different at 5% level according to Fisher's LSD test.

Table 10.96 Theoretical ethanol yield (L/ha) in III

Formula	Variety and treatment					
	KCS105		FS902		FS501	
	A	B	A	B	A	B
Formula 1	5925.99	5063.53	6241.97	6935.70	3959.92	3462.89
Formula 2	5470.95	4839.28	5748.19	6260.80	3760.04	3621.13
Formula 3 (538g ethanol/kg sucrose)	3555.58	3194.14	4071.30	4393.90	2454.54	2270.59
Formula 3 (700g ethanol/kg sucrose)	4625.55	4155.34	5296.47	5716.14	3193.18	2953.87
Formula 4 (0.052g ethanol/g total fresh weight)	2809.77	2485.36	2952.16	3215.43	1931.08	1859.74
Formula 4 (0.084g ethanol/g total fresh weight)	4504.82	3984.70	4733.10	5155.19	3096.04	2981.66

Formula 1: The ethanol production (L/ha) = total sugar content (%) × 6.5 (a conversion factor) × 0.85 (the process efficiency) × total fresh weight (t/ha) Lipinsky (1978)

Formula 2: The ethanol production (L/ha) = 0.081 × total fresh weight (kg/ha) Pari and Ragno (1998)

Formula 3: The ethanol production (L/ha) = 64.8 − 84.3 × total fresh weight (t/ha) × Percentage of stem in total fresh weight (%) Soldatos and Chatzidak (1999)

Formula 4: The ethanol production (g/ha) = 5.2 − 8.4 × (total fresh weight (g/ha) ÷ 100) Mamma et al (1996)

Ministry of the Cabinet Office) have advocated a 'Bioethanol Island Design' and conducted the proof of that cooperation in a Miyako-jima Island city. The design produces bioethanol from sugar cane and switches all gasoline (about 24 million litres a year) usage on the island to E3 (an automobile fuel produced by blending 3 per cent bioethanol into gasoline), and aims at 50,000 citizens.

Therefore, sweet sorghum may make a large contribution as a biofuel to satisfy the ethanol markets in Japan.

Internet resources

http://esciencenews.com/articles/2009/01/28/scientists.publish.complete.genetic.blueprint.key.biofuels.crop

http://en.wikipedia.org/wiki/Commercial_sorghum

http://news.mongabay.com/bioenergy/2007/05/us-scientists-develop-drought-tolerant.html

http://news.mongabay.com/bioenergy/2007/02/sweet-super-sorghum-yield-data-for.html

http://news.mongabay.com/bioenergy/2007/03/icrisat-launches-pro-poor-biofuels.html

www.greencarcongress.com/2007/06/texas_company_t.html

www.greencarcongress.com/2008/03/renergie-incre.html

SORREL (*Rumex acetosa* L.); (R. *patientia* L. × R. *tianshanious* L.)

Contributed by V. Petríková

Description

Common sorrel or garden sorrel (*Rumex acetosa*), often simply called sorrel and also known as spinach dock or narrow-leaved dock, is a perennial herb that is cultivated as a garden herb or leaf vegetable. Sorrel is a slender plant about 60cm high, with roots that run deep into the ground, as well as juicy stems and edible, oblong leaves. The lower leaves are 7 to 15cm in length, slightly arrow-shaped at the base, with very long petioles. The upper ones are sessile, and frequently become crimson. It has whorled spikes of reddish-green flowers, which bloom in summer, becoming purplish. The stamens and pistils are on different plants (dioecious); the ripe seeds are brown and shining. Common sorrel has been cultivated for centuries (Wikipedia.org).

Sorrel was improved in the 1970s in Ukraine as a hybrid of *Rumex patientia* × *Rumex tianshanious*. Sorrel is a perennial plant that can persist on the same stand for more than 10–15 years. This plant is similar in

(a)

(b)

Sources: (a) http://accipiter.hawk-conservancy.org/images/Sorrel.jpg;
(b) Richard Old, XID Services, Inc., Bugwood.org

Figure 10.186 Sorrel

features to the wild sorrel, but it is much larger. Sorrel grows as tall as 180–200cm and is a robust, rich flowering plant that creates well-ripened fruits. Sorrel is a photophilic plant that grows very quickly. It has a high content of proteins and vitamins: the protein content amounts to as much as 8t/ha (from the whole above-ground mass).

Grain yield reaches 7t/ha. In the most unfavourable conditions the plant provides 2t/ha of grain as a minimum. According to data from Ukraine, the growth of farm animals after feeding with sorrel is two to three times higher than with normal feed. Milk production and milk fat content increased.

Sorrel can also be used as a vegetable. Very early in the spring it develops the large brittle leaves that form the ground level rosette. The leaves are routinely used as the first high-quality green feedstuffs of spring. The data mentioned here are for sorrel of Ukrainian origin. For sorrel grown in central European conditions it will be necessary to verify and confirm all these data.

Ecological requirements

Sorrel is a modest plant that is resistant to cold and dry conditions, as well as being resistant to diseases. Sorrel is highly resistant to frost, and hibernates without problems at significantly decreased temperatures. The plant does not demand special treatment, and is covered abundantly with leaves every year. Sorrel is not particularly demanding with regard to nutrition, but it will benefit from rich organic manuring during the establishment of growth.

Propagation

Sorrel is propagated by seed. The seed should be treated by grinding off the small side wings.

Crop management

Sorrel is sown in April in rows 50cm apart at an average of 6kg/ha seeds. The recommended depth of sowing is approximately 1.5–2cm. Because sorrel is a perennial plant, growth can also be established during the summer and autumn if the humidity is suitable for the germination of the seeds.

Source: Nielsen et al (2006)

Figure 10.187 Sorrel field
(photographed 4 October 2006)

Its perennial nature means that there is no recommended crop rotation for sorrel. Sorrel can routinely be cultivated on arable soil that may be used to grow the usual agricultural crops after the sorrel has been removed.

Production

Sorrel will first produce seeds the second summer and the flower stalk will grow up to 2.6m.

Sorrel has been improved as a feeding crop. A large quantity of seed was imported into the Czech Republic in order to test sorrel as an energy plant for direct combustion at heating plants. Sorrel is a most suitable plant for this purpose. It reaches abundant yields of as much as 40t/ha of dry matter, and has the advantage of prolonged production from the same stand.

With only one annual harvest, sorrel offers on average almost 14–16t/ha/yr. The energy content of the above-ground mass is approximately 18MJ/kg, representing about 800GJ/ha (only the above-ground mass, excluding the seed and roots, is included in this assessment) (Ustjak et al, 1996).

Processing and utilization

In its second year, sorrel can already be harvested at least twice a year. Leaves for green fodder are harvested in the spring, and the fruits are harvested in the summer. If sorrel is utilized for energy purposes (for direct combustion), then the whole plant is harvested. The plants can be formed into large bales similar to straw bales.

Sorrel seems to be a very promising plant not only as fodder, but also as a source of renewable energy for direct combustion. It can be profitably used on set-aside soils in the interests of limiting food production. For these purposes, it would be very useful to undertake detailed research into the production of sorrel under different soil and climatic conditions and to verify the applicability of the Ukrainian data to other parts of the world, with regard to the plant's energy utilization. For energy purposes, sorrel can be regarded as in the same group as miscanthus or other recommended energy species, and may even outperform these plants; the energy content of above-ground biomass is approximately 18MJ/kg (800GJ/ha).

Internet resources

http://grimstad.hia.no/studium/energi/stralsundNielsen06.pdf
http://en.wikipedia.org/wiki/Sorrel

SOYA BEAN (*Glycine max* (L.) Merr.)

Description

The soybean (US) or soya bean (UK) (*Glycine max*) is a species of legume native to East Asia. The English word soy is derived from the Japanese pronunciation of (shōyu), the Japanese word for soya sauce; soya comes from the Dutch adaptation of the same word.

Soya bean is an annual C_3 plant originating from China. It is a cultural variety (a cultigen) with a very large number of cultivars. Beans are classed as pulses whereas soya beans are classed as oilseeds. It may grow prostrate, not higher than 20cm, or grow up to 2m high. The pods, stems and leaves are covered with fine brown or grey hairs. The leaves are trifoliolate, having three or four leaflets per leaf, and the leaflets are 6–15cm long and 2–7cm broad. The leaves fall before the seeds are mature. The small, inconspicuous, self-fertile flowers are borne in the axil of the leaf and are white, pink or purple. The fruit is a hairy pod that grows in clusters of three to five, each pod is 3–8cm long and usually contains two to four (rarely more) seeds 5–11mm in diameter. Soya contains significant amounts of all the essential amino acids for humans, and so is a good source of protein.

It grows to a height of up to 80cm with all plant parts being hairy. The plant produces a main root with smaller roots branching from it. The smaller roots possess numerous nodules inside of which develop the bacteria *Bradyrhizobium japonicum*. The bacteria is responsible for free nitrogen fixing. Above the cotyledonary nodes two single leaflets develop opposite each other. The plant's other leaves usually consist of three leaflets.

Soya bean is for the most part self-fertilizing. Numerous very small purple or white flowers are found in compact racemes in the leaf axils. These racemose inflorescences growing from the leaf axils are responsible for the production of the plant's fruit, which develops in the form of pods. Although one to five seeds per pod can be found, the usual number is two or three. The seeds are round to oval, and whitish yellow to brown or black-brown. Their composition is approximately 38–43 per cent protein, 18–25 per cent oil and 24 per cent carbohydrates.

Soya bean is a very important crop for the production of oil and protein. Among the largest

Figure 10.188 Soya bean pod

producers in the world are the USA, Brazil, China and Argentina. Originally, soya bean was considered a plant of subtropical areas, growing mainly in North America and East Asia in a band between 35 and 40°N from the equator, but it has since been widely cultivated in tropical zones. Since the 1900s soya beans have been cultivated in Europe.

Soya beans occur in various sizes, and in many hull or seed coat colors, including black, brown, blue, yellow, green and mottled. The hull of the mature bean is hard, water resistant, and protects the cotyledon and hypocotyl (or 'germ') from damage. If the seed coat is cracked, the seed will not germinate. The scar, visible on the seed coat, is called the hilum (colours include black, brown, buff, grey and yellow) and at one end of the hilum is the micropyle, or small opening in the seed coat which can allow the absorption of water for sprouting.

Ecological requirements

Soya bean plants are able to adapt to a wide range of environments, but this adaptability is possible because of the many different cultivars, each having its own characteristics and preferred conditions. Soya bean has similar ecological requirements to maize. The plant needs high temperatures in the summer and autumn to facilitate ripening. Loamy soil or loess and black soil with the ability to hold water are recommended. When enough water is present a light sandy soil can also be used. Heavy waterlogged soils are not appropriate. A constantly humid subtropical climate is most favourable for soya bean crops. Although 24–25°C is preferred for optimal growth, the 20–25°C range is still excellent for all stages of plant growth. Frost tolerance is better than that of maize.

A sensitivity to photoperiod means that a day length of 12 hours or less leads to premature flowering, which results in diminished yields. In the majority of cultivars, blooming is seen only when the day length is less than 14 hours.

The crop has an approximately 4–5-month growing period. The rainfall required for a good yield in warmer areas is 500–750mm, though less precipitation during ripening is preferred. The pH range of 6–6.5 has been shown to be the most desirable for soya bean crops, with certain cultivars performing better at slightly more acidic or basic levels. Approximately 15kg phosphorus and 50kg potassium are used from the soil for every tonne of seeds produced.

Rhizobium symbiosis is an important factor when it comes to crop growth and yield. Soya bean seeds need to be inoculated with *Bradyrhizobium japonicum*, and because many cultivars require a specific strain it is recommended that the seeds be inoculated with a composite of many strains.

Propagation

Soya bean propagation is performed using seed. Because of the wide variety of characteristics found in soya bean plants, it is necessary to choose seed of the plant that is most appropriate for the soil and climate in the area it will be grown. The use or end product of the crop also needs to be taken into consideration.

Soya beans are one of the 'biotech food' crops that have been genetically modified, and GM soya beans are being used in an increasing number of products. In 1995 Monsanto introduced Roundup Ready (RR) soya beans that have had a copy of a gene from the bacterium, *Agrobacterium* sp. strain CP4, inserted into its genome by means of a gene gun, that allows the transgenic plant to survive being sprayed by the non-selective herbicide, Roundup. Glyphosate, the active ingredient in Roundup, kills conventional soya beans. The bacterial gene is EPSP (5–enolpyruvyl shikimic acid-3-phosphate) synthase. Soya beans also have a version of this gene, but the soya bean version is sensitive to glyphosate, while the CP4 version is not. RR soya beans allow a farmer to spray widely the herbicide Roundup and so to reduce tillage or even to sow the seed directly into an unploughed field, known as no-till farming or conservation tillage. In 1997, about 8 per cent of all soya beans cultivated for the commercial market in the United States were genetically modified. In 2006, the figure was 89 per cent. The ubiquitous use of such types of GM soya beans in the Americas has caused problems with exports to some regions. GM crops require extensive certification before they can be legally imported into the European Union, where there is extensive supplier and consumer reluctance to use GM products for consumer or animal use (wikipedia.org).

Crop management

Sowing should take place in the middle of April or beginning of May. The ground temperature at a depth of 5cm should be at least 10°C. A sowing depth of 3–4cm and a distance between rows of 25–30cm is recommended. The desired crop density is 40–60 plants per m^2. This corresponds to about 70–90kg/ha of sown seed. Because the N_2 binding potential of soya bean's nodules is negatively affected by high soil temperatures, regions in southern Europe and the Mediterranean should use denser crop spacings, thus shielding the soil in hotter climates. Wider spacings usually lead to the production of more branches and seeds.

Because of soya bean's nitrogen-fixing bacteria, most soils do not need supplementary nitrogen fertilization, especially where the previous crop has been abundantly fertilized.

Weed control is essential during the first few weeks after sowing. This can be performed mechanically, or

(a)

(b)

Source: (a) Carl Dennis, Auburn University, Bugwood.org; (b) Pioneer Hi-Bred

Figure 10.189 (a) Soya bean field; and (b) yield testing and evaluation 2006

there are several herbicides that can be used preferentially before seed sowing or before germination. The time from planting to maturity is from 90–180 days, depending on climate and plant variety.

Drought or excessive dryness strongly reduces crop yield, especially before flower formation and during granulation. General irrigation may or may not be necessary depending upon the climate, but irrigation is necessary at germination, flowering and during the plant's seed development in semi-arid climates.

Disease damage to soya bean crops can be greatly reduced by proper cultivar and seed selection, and by maintaining sanitary practices and appropriate crop rotation. The most destructive diseases are, among others, frogeye leaf spot (*Cercospora sojina* Hara), stem canker (*Diaporthe phaseolorum*), *Pseudomonas glycinea* Coerper and soya bean mosaic virus.

Some of the more damaging pests are the soya bean moth (*Laspeyresia glycinivorella*), found in parts of Asia, the leaf eating beetle (*Plagiodera inclusa* Stål.), the Japanese beetle (*Popilliajaponica*) and the soya bean cyst *nematode* (*Heterodera glycines* Ichinohe). If these pests are present, insecticides may be necessary.

Soya bean can be used with a large variety of crop rotation plans. Usually, soya bean comes between two grain or cereal crops. Possible rotation combinations are: corn/soya bean/wheat, soya bean/wheat/sorghum, millet/winter wheat/soya bean, and corn/soya bean/cotton.

Production

In a study of oil crops in the USA by Goering and Daugherty (1982) it was determined that the total energy input for a non-irrigated soya bean crop was l2,044MJ/ha. This includes the entire range of costs from seed price to oil recovery costs. The energy output was determined to be 20,777MJ/ha, based on a seed yield of 2623kg/ha and an oil content of 20 per cent. The energy output/input ratio was 4.56. This was the highest ratio of the nine different crops tested under conditions of non-irrigation. Rape had the next highest output/input ratio at 4.18.

In November 2007, at Des Moines, Iowa, DuPont and its Pioneer Hi-Bred business congratulate Kip Cullers of Purdy, MO, for setting a new world record in soya bean production at 154 bushels per acre, 15 bushels higher than the record Cullers set in 2006.

Table 10.97 Top soya bean producers in 2006 (million tonnes)

	United States	87.7
	Brazil	52.4
	Argentina	40.4
	China	15.5
	India	8.3
	Paraguay	3.8
	Canada	3.5
	Bolivia	1.4
World Total		**221.5**

Source: FAO

Cullers achieved this with Pioneer® brand soya bean variety 94M80 in his winning entry for the Missouri Soya bean Association yield contest (pioneer.com).

At present, soya bean crops do not have a high enough yield to make affordable use of the vegetable oil as an energy source (diesel fuel substitute). Increased oil yields per hectare appear to be an achievable goal of research and plant breeding. Harvests in Germany resulted in 2.1t/ha of soya beans with an oil content of 18 per cent. This produced a 0.378t/ha oil yield.

Between 1930 and 2003, average corn yields jumped nearly sevenfold, from 20.5 bushels per acre to 142.2 bushels per acre. In that same period, average soya bean yields not quite tripled, from 13 bushels per acre to 33.4 bushels per acre. National soya bean yields have hovered around 40 bushels per acre for about a decade (Leer, 2004).

Processing and utilization

A combine harvester can be used for harvesting soya bean crops. The plants are harvested when the leaves turn yellow and the plant dies. By this time the seeds are fully ripe, but the pods have not yet ruptured. The water content of the seeds should be under 20 per cent at the time of harvest.

The harvested beans are prepared for storage by being cleaned and dried to a moisture content of

(a)

(b)

Source: (a) landgarten.at; (b) http://nfarms.com/2008Fall/

Figure 10.190 Soya bean: (a) plantation in Austria 2008; and (b) at maturity, ready to be harvested for seed

Source: pedigreedbypenner.com

Figure 10.191 Soya bean harvesting, USA

Source: wikipedia.org

Figure 10.192 Varieties of soya beans are used for many purposes

between 12 and 14 per cent, after which the beans can be stored in elevators or silos. This provides a continuous, year-round supply of beans for further processing.

Soya beans currently have a very wide range of food and non-food uses. The seeds can be used to produce edible products such as soya meal used for cooking and baking, soya milk, soya sauce, cheeses such as tofu, yogurt, and a product similar to meat called TVP (textured vegetable protein). The oil is used to make vegetable oil, margarine, shortening and salad dressings. The unripe seeds and the bean sprouts can be eaten as vegetables. The remaining green plant can be fed as forage to animals.

The seeds can also be used for non-food products. The lecithin in the seeds can be used as an emulsifier for pharmaceuticals, decorating materials, printing inks, pesticides, etc. The protein can be used for the production of synthetic fibres, glues, etc.

For purposes of bioenergy production the oil is the most important element. The oil is usually obtained from the seeds by solvent extraction, and consists of 3–11 per cent linoleic acid. Pure vegetable oil can be obtained through solvent extraction. With the additional step of transesterification, an esterified vegetable oil is obtained that can be used alone or mixed with diesel fuel to produce a biofuel.

In 1991, the market price of a litre of diesel oil in Italy was 327L untaxed and 1131L taxed, whereas the price of a litre of soya bean vegetable oil was 674L, or 830L with esterification. This shows that a subsidy is necessary if soya bean oil is to compete with diesel and be used as a biofuel.

A study showed that both corn grain ethanol and soya bean biodiesel produce more energy than is needed to grow the crops and convert them into biofuels. This finding refutes other studies claiming that these biofuels require more energy to produce than they provide. The amount of energy each returns differs greatly, however. Soya bean biodiesel returns 93 per cent more energy than is used to produce it, while corn grain ethanol currently provides only 25 per cent more energy. The study was be published in the 11 July *Proceedings of the National Academy of Sciences, 2006* (news.mongabay.com).

Source: http://lesterfeedandgrain.net/index.cfm?show=10&mid=16

Figure 10.193 Soya bean processing plant

An average US farm consumes fuel at the rate of 82L/ha of land to produce one crop. However, average crops of rapeseed produce oil at an average rate of 1029L/ha, and high-yield rapeseed fields produce about 1356L/ha. The ratio of input to output in these cases is roughly 1:12.5 and 1:16.5. Photosynthesis is known to have an efficiency rate of about 3–6 per cent of total solar radiation and if the entire mass of a crop is utilized for energy production, the overall efficiency of this chain is currently about 1 per cent. While this may compare unfavourably to solar cells combined with an electric drive train, biodiesel is less costly to deploy (solar cells cost approximately US$1000 per square metre) and transport (electric vehicles require batteries which currently have a much lower energy density than liquid fuels).

Feedstock yield efficiency per hectare affects the feasibility of ramping up production to the huge industrial levels required to power a significant percentage of national or world vehicles. Some typical yields in litres of biodiesel per hectare (wikipedia. org/wiki/Biodiesel):

- algae: 2763L
- hemp: 1535L
- chinese tallow: 772L (970GPa)
- palm oil: 780–1490L
- coconut: 353L

- rapeseed: 157L
- soya: 76–161L in Indiana (soya is used in 80 per cent of US biodiesel)
- peanut: 138L
- sunflower: 126L

The National Institute of Standards and Technology (NIST) now has a method to accelerate stability testing of biodiesel fuel made from soya beans. It has also identified additives that enhance stability at high temperatures (hindu.com).

Compared to normal diesel, soya biodiesel has the following advantages (soya.be):

- Soya biodiesel is better for the environment because it is made from renewable resources and has lower emissions compared to petroleum diesel. The use of biodiesel in a conventional diesel engine results in substantial reduction of unburned hydrocarbons, carbon monoxide and soot. The use of biodiesel does not increase the CO_2 level in the atmosphere, since growing soya beans consumes also CO_2. Biodiesel is also more biodegradable than conventional diesel. Studies at the University of Idaho have shown that biodiesel degraded by 95 per cent after 28 days compared to 40 per cent for diesel fuel.
- Lubrication tests have demonstrated that biodiesel is a better lubricant.

Internet resources

www.pioneer.com/web/site/portal/menuitem.205da85f935957934cfe78e1d10093a0/
www.hindu.com/seta/2009/01/22/stories/2009012250071400.htm
http://news.uns.purdue.edu/UNS/html4ever/2004/040715.Volenec.yields.html
http://news.mongabay.com/2006/0711-umn.html
http://en.wikipedia.org/wiki/Soyabean
http://en.wikipedia.org/wiki/Biodiesel
www.sdnotill.com/officers.htm
www.pioneer.com/CMRoot/Pioneer/media_room/media_kits/record_soyabean/blueprint/blueprint_research.pdf
www.landgarten.at/index.php?id=132&L=1
www.pedigreedbypenner.com/farm.htm

Source: soya.be

Figure 10.194 Biodiesel production plant

www.ecofriend.org/entry/electricity-from-soy-bean-oil-newest-in-alternative-fuel/
www.medicalrace.com/dictionary/Biodiesel
www.soya.be/soy-biodiesel.php

SUGAR BEET (*Beta vulgaris* L.) and Tropical Sugar beet

Description

B. vulgaris is the only species of agricultural importance in this small family which includes sugar beet, fodder beets and mangels. Several members of the family are common arable weeds. The wild forms of beet from which cultivated forms are thought to derive are seacoast plants of Europe and Asia and are very variable in habit and duration.

Sugar beet (*Beta vulgaris* L.), a member of the *Chenopodiaceae* family, is a plant whose root contains a high concentration of sucrose. Sugar beet is a hardy biennial plant that can be grown commercially in a wide variety of temperate climates. During its first growing season, it produces a large (1–2kg) storage root whose dry mass is 15–20 per cent sucrose by weight. If not harvested, during its second growing season, the plant uses the nutrients in this root to produce flowers and seeds. In commercial beet production, the root is harvested after the first growing season, when the root is at its maximum size (wikipedia.org).

Seed production and sugar production need to take place in different locations because frost resistance is poor, but plants need a cold shock to flower and produce seed. Syngenta AG has developed the so-called tropical sugar beet. It allows the plant to grow in tropical and subtropical regions (Syngenta Global).

Ecological requirements

Sugar beet is a hardy biennial vegetable that can be grown commercially in a wide variety of temperate climates.

It requires a deep well-drained stone-free soil that is not acid. A high standard of management of land is needed to provide a well-structured soil, free from compaction. Sowing date is quite crucial, early sowing gives better sugar yields due to increased water availability earlier in the season, but sowing too early leads to a high population of bolters. The seedling stage is a poor competitor with weeds and can be fatally damaged by millipedes, symphalids, spring tails and pigmy-mangel beetle. Beet cyst eelworm (*Heterodera schactii*) can be damaging and is only satisfactorily controlled by adequate rotation.

Production

In most temperate climates, beets are planted in the spring and harvested in the autumn. At the northern end of its range, growing seasons as short as 100 days can produce commercially viable sugar beet crops. In warmer climates, such as in California's Imperial Valley, sugar beets are a winter crop, being planted in the autumn and harvested in the spring. Beets are planted

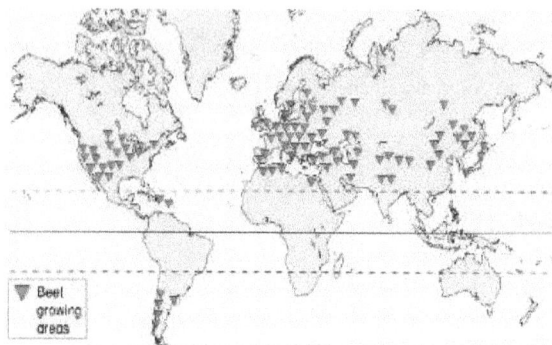

Source: britishsugar.co.uk

Figure 10.195 Map of the world showing the beet-growing areas

from a small seed; 1kg of beet seed comprises 100,000 seeds and will plant over a hectare of ground (1lb will plant about an acre) (solarnavigator.net).

Preparing the fields for sugar beet begins as early as the autumn when the soil is tested to see if phosphate, potassium and sodium (minerals) need to be applied before ploughing. Sometimes lime is spread on the fields to ensure that the soil is not too acidic.

Sowing generally takes place in late March and early April. Nitrogen fertilizer is applied at this time to help the crop grow, and specialist herbicides may be sprayed over the fields to stop weeds growing.

In the growing and harvesting of sugar beet, timing is critical. The harvesting period, known as the 'campaign' amongst farmers, takes place between September and Christmas when the amount of sugar in the beet is at its highest. A delay in harvesting can prove very costly to the farmer as sugar beet is easily damaged by frost. Harvesting is therefore completed as quickly as possible (britishsugar.co.uk)

Harvesters cut off the top leaves of the sugar beet. The tops are used as animal feed for cattle and sheep or are ploughed back into the land as a natural fertilizer. The root is then cleaned to remove any soil. As beet is a heavy and bulky crop, transport distances are kept as short as possible to reduce costs. The sugar factories have therefore been built in the beet-growing areas and are all located close to large

Different stages of growth

Source: Wilhelm Dürr (Südzucker AG Mannheim/Ochsenfurt)

Figure 10.197 Sugar beet field, Germany

because the transport distances involved are greater than in the cane industry. This is a direct result of sugar beet being a rotational crop which requires nearly four times the land area of the equivalent cane crop which is grown in monoculture. Because the beets have come from the ground they have to be thoroughly washed

Source: Südzucker AG Mannheim/Ochsenfurt

Figure 10.196 Sugar beet plants

towns, which can provide the workforce required. (solarnavigator.net)

The beets are harvested in the autumn and early winter by digging them out of the ground. They are usually transported to the factory by large trucks

Source: MSU Weeds.com Blog

Figure 10.198 Sugar beet field during harvesting

Table 10.98 Top ten sugar beet producers, 2005 (million tonnes)

	France	29
	Germany	25
	United States	25
	Russia	22
	Ukraine	16
	Turkey	14
	Italy	12
	Poland	11
	United Kingdom	8
	Spain	7
World Total		**242**

Source: FAO

Table 10.99 Area harvested: yield and production in some selected countries, 2008

Country	Area harvested ('000 ha)	Yield (t/ha)	Production ('000 MT)
Austria	51.7	56.2	2900
Bel-Lux	102.0	53.6	5470
Denmark	70.0	48.3	3384
Finland	34.6	26.6	920
France	459.5	74.3	34,154
Germany	504.0	56.5	28,487
Greece	49.2	67.0	3300
Ireland	33.0	40.3	1300
Italy	287.5	46.3	13,304
Netherlands	117.0	54.8	6416
Portugal	4.3	46.7	201
Spain	156.9	51.8	8128
Sweden	60.8	41.1	2500
UK	195	54.0	10,527

Source: solarnavigator.net

Table 10.100 Comparison between sugar beet and sugar cane

Character	Sugar Beet	Sugar cane
Duration (months)	6–7	10–12
Brix reading	23–24%	18–20%
Pol %	20–22%	13–16%
Sugar recovery	15–16%	11–12%
Average sugar recovery in factory	10–12%	8–10%
Yield (t/ha)	60–80	100
Water requirement	120cm	200cm

Source: tnau.ac.in

over the field. A modern harvester is typically able to cover six rows at the same time. The beet is left in piles at the side of the field and then conveyed into a trailer for delivery to the factory. The conveyor removes more soil – a farmer would be penalized at the factory for excess soil in his load.

If beet is to be left for later delivery, it is formed into 'clamps'. Straw bales are used to shield the beet from the weather. Provided the clamp is well built with the right amount of ventilation, the beet does not significantly deteriorate. Beet that is frozen and then defrosts produces complex carbohydrates that cause severe production problems in the factory. In the UK, loads may be hand examined at the factory gate before being accepted.

In the US, the fall harvest begins with the first hard frost, which arrests photosynthesis and the further growth of the root. Depending on the local climate, it may be carried out in a few weeks or be prolonged throughout the winter months. The harvest and processing of the beet is referred to as 'the campaign,' reflecting the organization required to deliver crop at a steady rate to processing factories that run 24 hours a day for the duration of the harvest and processing (in the UK the campaign lasts about 5 months) (www.solarnavigator.net/solar_cola/sugar_beet.htm).

Processing and utilization

After harvesting the beet are hauled to the factory. Delivery in the UK is by haulier or, for local farmers, by tractor and trailer. Railways and boats were once used in the UK, but no longer (some beet is still carried by rail in the Republic of Ireland).

and separated from any remaining beet leaves, stones and other trash material before processing.

Harvesting is now entirely mechanical. The sugar beet harvester chops the leaf and crown (which is high in non-sugar impurities) from the root, lifts the root, and removes excess soil from the root in a single pass

Figure 10.199 Sugar beet factory

Each load entering is weighed, and sampled before tipping onto the reception area, typically a 'flat pad' of concrete, where it is moved into large heaps. The beet sample is checked for

- soil tare – the amount of non-beet delivered;
- crown tare – the amount of low-sugar beet delivered;
- sugar content ('pol') – amount of sucrose in the crop; and
- nitrogen content – for recommending future fertilizer use to the farmer.

From these the actual sugar content of the load is calculated and the grower's payment determined.

The beet is moved from the heaps into a central channel or gulley where it is washed towards the processing plant.

The processing starts by washing and slicing the beets into thin chips called cossettes. This process increases the surface area of the beet to make it easier to extract the sugar. The extraction takes place in a diffuser where the beet is kept in contact with hot water for about an hour. A typical diffuser weighs several hundred tonnes when full of beet and extraction water.

The diffuser is a large horizontal or vertical agitated tank in which the beet slices slowly work their way from one end to the other and the water is moved in the opposite direction. This is called counter-current flow, and the water becomes a stronger and stronger sugar solution usually called juice. It also collects a lot of other chemicals from the flesh of the sugar beet.

The used cossettes, or pulp, exit the diffuser at about 95 per cent moisture but low sucrose content. Using screw presses, the wet pulp is then pressed down to 75 per cent moisture. This recovers additional sucrose in the liquid pressed out of the pulp, and reduces the energy needed to dry the pulp. The pressed pulp is dried and sold as animal feed, while the liquid pressed out of the pulp is combined with the raw juice or more often introduced into the diffuser at the appropiate point in the countercurrent process. The raw juice must now be cleaned before it can be used for sugar production. The raw juice contains many impurities that must be removed before crystallization. (solarnavigator.net)

Ethanol is a high-octane fuel that is most commonly used as a gasoline additive and or extender. Ever since the late 1070s, methyl tertiary butyl ether (MTBE) has been used to replace lead, and is the primary gasoline additive in the United States. Sugar beet is high in sugar that can rapidly be converted to 18 per cent alcohol. The yield of ethanol from one tonne of sugar beets is approximately 24.8 gallons (Ethanol-Still-Plans.com).

According to a 2003 survey, around 61 per cent of world ethanol production is being produced from sugar crops, be it sugar beet, sugar cane or molasses, while the remainder is being produced from grains, and here maize or corn is the dominating feedstock. Feedstocks crucially determine the profitability of fuel ethanol production.

There are various ways to look at the issue. In Figures 10.200–10.205, the theoretical per hectare ethanol yields, and cross-costs of the three major feedstocks currently in use are plotted.

Figure 10.203 shows that the industrial alcohol market is the smallest of the three. Moreover, it shows a rather modest rate of growth which is similar to the increase in gross domestic product. Demand for distilled spirits in most developed countries is stagnating or even declining, due to increased health awareness. This is unlikely to change in the future (Berg, 2004).

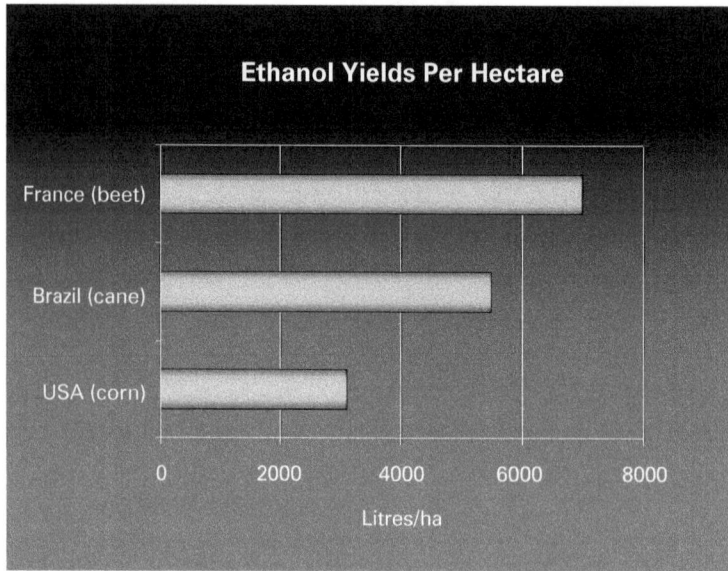

Figure 10.200 Ethanol yield per hectare

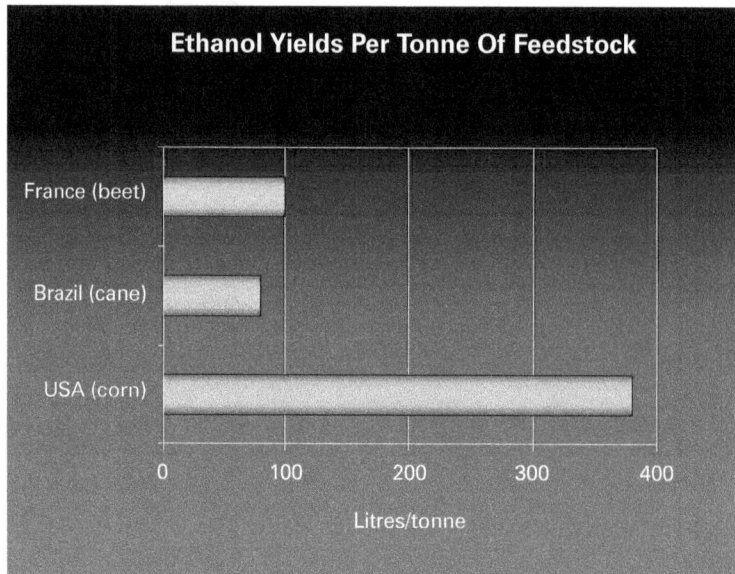

Figure 10.201 Ethanol yield per tonne of feedstock

The first sugar beet-fuelled ethanol production facility opened in Norfolk, eastern England in 2007. The opening of the plant operated by British Sugar was attended by Britain's Minister for Sustainable Food and Farming and Animal Health, Lord Rooker. The plant has a capacity of 18.5 million gallons a year from locally grown sugar beets. The British government was planning to implement a new regulation in spring 2008 that would require 5 per cent of all retail fuels to be from renewable sources and this plant would be a step toward achieving this.

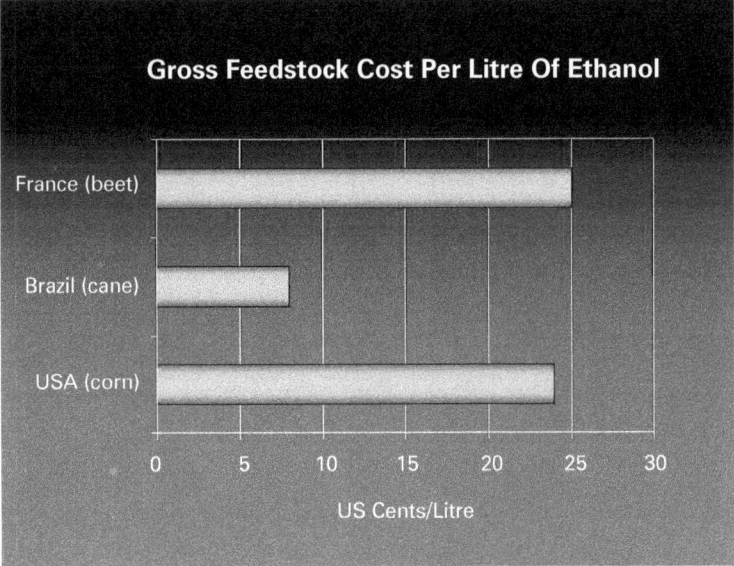

Figure 10.202 Gross feedstock cost per litre ethanol

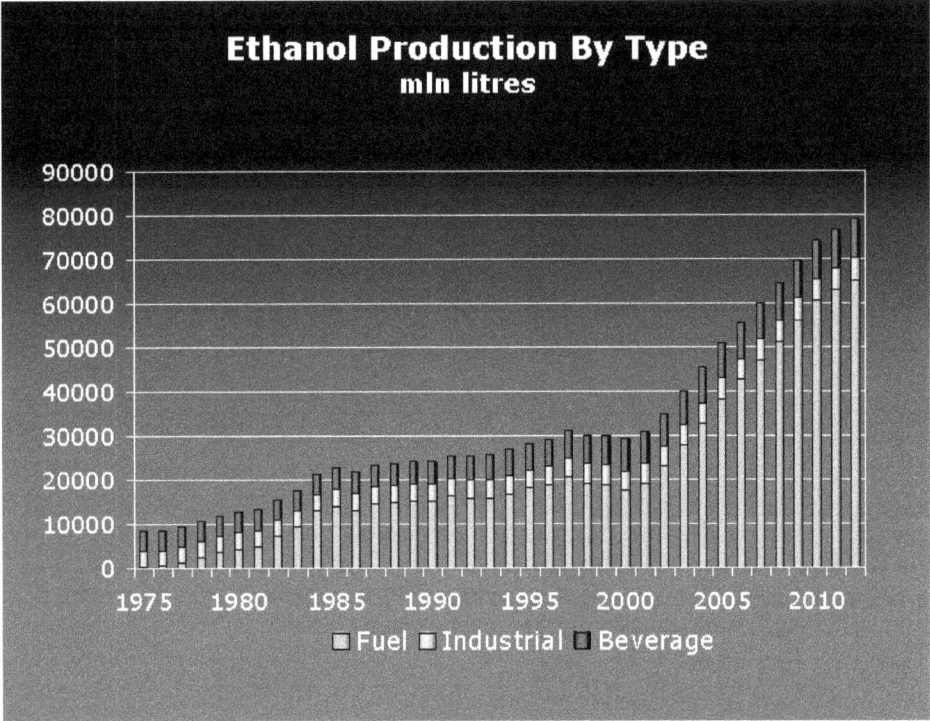

Figure 10.203 Ethanol production by type

The plant was the first of its kind in the UK. It produces bioethanol from locally grown sugar beet and has an annual production capacity of 70 million litres; made from 110,000 tonnes of sugar that is surplus to the needs of the food market (Abuelsamid, 2007).

Tiger Ethanol International has invested, together with the local government, in sugar beet production in China's Fujian province, as a way to reduce its reliance on corn. The 'flex factory' model can switch between producing fuel and sugar, depending on the market situation. Local farmers who will grow the new biofuel feedstock should see an increase of 60 per cent in their farm incomes. The company expects sugar beet to result in increased ethanol production as compared to corn, while adding an additional source of income from refined sugar.

By the end of the implementation cycle of the plan, sugar beet will be growing on 121,000ha of land to yield an estimated 1.5 million tonnes. There will be a capacity to produce 100,000 tonnes of ethanol and 100,000 tonnes of sugar (marketwire.com/press-release 2008).

By-product

Because of its sugar and raw protein content, molasses is used as a feed for beef cattle, dairy cows, horses, pigs and poultry. In some countries, molasses is mixed with grass, maize or grain as its rapidly fermentable sugar speeds up the silage-making process.

Molasses is also used as a source of carbohydrates for fermentation purposes (alcoholic fermentation, production of baking yeast).

The 'Carbokalk' is pressed to produce a solid matter content of about 70 per cent. The material obtained in this way, which has the same moisture content as soil, can be stored at the edge of a field without being covered. Because of its highly reactive surface, 'Carbokalk' is a very rapidly acting calcium fertilizer and is suitable for all kinds of crops.

Pellets are slices of sugar beet that have been enriched with molasses, dried and pressed. They are mainly used to feed ruminants, but are also given to pigs and horses. Pellets have a high energy content, which is comparable to the energy content of barley. The feed value per 1000g of pellets is as follows: net energy lactation: 6.82MJ NEL and releasable energy = 10.84MJ ME (freepatentsonline.com).

Tropical sugar beet

Tropical sugar beet can be grown in relatively dry areas as it requires substantially less water than sugar cane. The beets are also faster growing, allowing farmers to grow a second crop on their land in the same period as sugar cane crops take to mature. This increases farmers' productivity and income, bringing significant benefits to the agricultural sector of developing markets. Tropical sugar beet delivers similar output yields to sugar cane and can be used both for processing sugar for food and conversion to bioethanol. An alternative to cane, it supports biodiversity when used in areas with extensive sugar cane monocultures.

It took Syngenta 11 years to develop tropicalized sugar beet. In 2007, the beet was successfully introduced in India. In the State of Maharashtra, for example, Syngenta helped a cooperative of more than 12,000 smallholder farmers to build and operate a bioethanol production plant that runs on Syngenta tropical beet. In Colombia, the building has started of two beet-to-ethanol plants, which are expected to start processing tropical sugar beet in 2009.

Syngenta is currently conducting adaptation trials in many other tropical countries, such as China, Australia, Thailand, Vietnam, Kenya, South Africa, Ethiopia, Brazil, Peru and Mexico, and in the USA.

Tropical sugar beet, as a new but adapted crop which could meet the following requirements:

- high yielding (30–40Mt/acre);
- tolerant to high temperature;
- less water requirement and drought tolerant;
- improves soil conditions and excellent in performing on saline and alkaline soils;
- rotation with most other crops;
- water saving;
- easy for cultivation and harvesting; and
- an industrial crop ready to be processed when the factory needs it.

The first production data for tropical sugar beet are promising: the variety Posada in India, planting date 15 November 2003, harvesting date 23 April 2004 achieved a tuber yield of 72.73t/ha, a sucrose content of 15.50 per cent and a leafy matter content of 9.88 per cent.

Figure 10.204 Tropical sugar beet in India

Figure 10.205 Tropical beet on its way to be processed

The new sugar beet can be grown in relatively dry areas with substantially less water than typically required by sugar cane. It is faster growing and can be harvested after five months allowing farmers to grow a second crop on the same land, thus increasing agricultural output and raising farmer income. Syngenta is engaged in two tropical sugar beet projects:

- Sugar for food: at Ambad near Jalna, Maharashtra, Samarth Cooperative Sugar Mill has commissioned a pilot plant for processing tropical beet in cooperation with the Vasantdada Sugar Institute. First harvests delivered the expected high yield of top-quality sugar;

- Sugar for fuel: at Kalas, near Pune, Syngenta cooperates with over 12,000 farmers linked to Harneshwar Agro Products, Power and Yeast Ltd, which built and operates a bioethanol production plant processing Syngenta tropical beet. The faster growth of tropical beets increases annual ethanol output over sugar cane.

Syngenta has received the 2008 World Business and Development Award (WBDA) for the development and successful introduction of a new sugar beet that can be grown under tropical climate conditions and brings significant advantages to farmers, the environment, the sugar and ethanol industries and the economy (Syngenta Global 2008).

Source: Syngenta

Figure 10.206 Harneshwar Agro Products Power and
Yeast bioethanol plant

Internet resources

www.distill.com/World-Fuel-Ethanol-A&O-
2004.html
Ethanol-Still-Plans.com, www.ethanol-still-plans.com/
sugar_beet_ethanol.htm
http://en.wikipedia.org/wiki/Sugar_beet
www.marketwire.com/press-release/Tiger-Ethanol-
International-Inc-OTC-Bulletin-Board-TGEI.OB-
811217.html
www.solarnavigator.net/solar_cola/sugar_beet.htm
Syngenta Global
www.syngenta.com/en/media/mediareleases/en_08092
5.html
www.britishsugar.co.uk/RVE759fc523b5b2437b82f26
22dc81f3c2d,,.aspx
www.suedzucker.de/en/Presse/Pressefotos/
http://msuweeds.com/blog/?cat=9
www.syngenta.com/en/media/newstopics.28.08.2007.
html
www.tnau.ac.in/tech/swc/sugarbeet.pdf
www.freepatentsonline.com/4012535.html

SUGAR CANE (*Saccharum officinarum* L.)

Description

Sugar cane was originally from tropical South Asia and
Southeast Asia. Different species probably originated in

different locations with *S. barberi* originating in India
and *S. edule* and *S. officinarum* coming from New
Guinea. The thick stalk stores energy as sucrose in the
sap. From this juice, sugar is extracted by evaporating
the water. Crystallized sugar was reported 5000 years
ago in India. Around the eighth century AD, Arabs
introduced sugar to the Mediterranean, Mesopotamia,
Egypt, North Africa and Spain. By the tenth century
there was no village in Mesopotamia that didn't grow
sugar cane. It was among the early crops brought to the
Americas by Spaniards. Brazil is currently the biggest
sugar cane-producing country (wikipedia.org/wiki/
Sugar cane).

Sugar cane is a perennial C_4 plant that has many
positive qualities, including high sugar content, low fibre
content, good juice purity, good ecological adaptability,
and disease resistance. It is also a convenient raw material
for the production of alcohol and cellulose. Although
many cultivars barely flower, the presence of short days
can lead to undesired flowering (Rehm and Espig,
1991). Because of sugar cane's excellent capacity for
energy binding with respect to surface area and time, it
is considered to be one of the most important economic
plants in the world (Husz, 1989).

Sugar cane is a genus of 6 to 37 species (depending
on taxonomic interpretation) of tall perennial grasses
(family Poaceae, tribe Andropogoneae), native to warm
temperate to tropical regions of the Old World.
They have stout, jointed, fibrous stalks that are
rich in sugar and measure 2 to 6m tall. All of the sugar
cane species interbreed, and the major commercial
cultivars are complex hybrids (wikipedia.org/wiki/
Sugar cane).

Ecological requirements

Subtropical and tropical areas have proven to be the
most appropriate for growing sugar cane. The areas fall
into the latitudes between roughly 37°N and 31°S.
Because of significant variations in the conditions of
different growing areas, special cultivars are usually
grown at breeding stations in the same climatic area as
that in which the plant will be grown (Rehm and Espig,
1991).

Although sugar cane can be and is cultivated on a
greatly varying range of soils, those preferred for sugar
cane cultivation are heavy soils with a high nutrient

Source: britishsugar.co.uk

Figure 10.207 Map of the world showing the cane-producing areas

content and a high water-holding capacity, though extended periods of waterlogging are not tolerated. It is recommended that the water table be at or more than 1m below the soil surface (Alvim and Kozlowski, 1977).

Soil pH values ranging from 5.5 to 8.5 have been shown to have no negative effects on sugar cane production (Husz, 1989).

With the production of such a large amount of biomass there comes a high nutrient requirement. According to Husz (1972), 1t of fresh cane holds 0.5–1.2kg N, 0.2–0.3kg P, 1.0–2.5kg K, 0.3–0.6kg Ca and 0.2–0.4kg Mg. Here the leaf and top nutrient contents are not included because they are much higher. The presence of an endomycorrhiza to facilitate potassium uptake and free-living nitrogen-fixing bacteria in the soil are positive conditions.

The most favourable temperature for subtropical and tropical sugar cane cultivation is 25–26°C. Cold tolerance is an attribute of subtropical cultivars, but in the case of tropical cultivars plant growth is hindered at temperatures as low as 21°C, while temperatures around 13°C cause a halt in growth and temperatures below 5°C lead to chlorosis. Sugar cane needs a large quantity of rainfall – an average of around 1500–1800mm of rain under most conditions. In hot and dry conditions, as much as 2500mm or more may be needed (Rehm and Espig, 1991).

Propagation

Sugar cane is propagated from cuttings, rather than from seeds; although certain types still produce seeds, modern methods of stem cuttings have become the most common method of reproduction. Each cutting must contain at least one bud, and the cuttings are usually planted by hand. Once planted, a stand of cane can be harvested several times; after each harvest, the cane sends up new stalks, called ratoons. Usually, each successive harvest gives a smaller yield, and eventually

(a)

(b)

(c)

Source: (a, b) Hannes Grobe, wikimedia.org; (c) Rufino Uribe, http://commons.wikimedia.org/wiki/File:Cut_sugarcane.jpg

Figure 10.208 Sugar cane plantations and stalks

the declining yields justify replanting. Depending on agricultural practice, two to ten harvests may be possible between plantings (wikipedia.org/wiki/ Sugar cane).

Sugar cane is most commonly propagated using stem cuttings. This form of vegetative propagation is best performed using cuttings that have two nodes and are taken from unripened canes that are 8 or 9 months old. For best results the cuttings should come from stems that have been grown in the area where the cuttings will be planted. Roughly 5 to 6ha can be planted from the cuttings obtained from 1ha of sugar cane. A pre-planting fungicide and insecticide treatment is recommended. To destroy viruses and insects an additional hot water treatment of 1½ hours at 52°C can be performed (Rehm and Espig, 1991).

Crop management

Sugar cane is cultivated in almost all the world only for some months of the year, in a period called 'safra', the Portuguese word for harvest. The only places in the world where there is no 'safra', and therefore sugar cane is cultivated and produced year round are Peru and Colombia in South America (Wikipedia.org).

A row spacing of 1.2–1.5m is recommended, but the cultivar and machinery used also need to be taken into consideration. The ground is ploughed to a depth of 30–40cm and the stem cuttings are placed end to end in the furrows, either mechanically or by hand. The furrows are then closed so that the cuttings are 3–5cm beneath the soil surface, or deeper if the soil is sandy or the climate is dry. The root system takes several weeks to form (Rehm and Espig, 1991).

About 15,000 to 20,000 stem cuttings will be needed for each hectare to be planted, corresponding to 25 to 35 stem cuttings per 10m (Husz, 1989).

Until the leaf canopy can close, it is suggested that a herbicide be used on the plants; but the area between the rows should be weeded mechanically.

According to Husz (1989), the best time to apply fertilizer is at the time of planting, along with an initial watering, and then again before the third month.

Because of the greatly varying degree of nutrient requirements among the different cultivars, it is

Source: nayagarh.nic.in

Figure 10.209 Pit method planting. Demonstration plot on sugar cane cultivation under the pit method of planting in the village of Aswasthapada, Nayagarh

recommended that leaf or internode samples be analysed to determine the most appropriate fertilizer dosages for a particular cultivar (Husz, 1972). The sugar cane nutrient requirements listed earlier can be used as a guideline to help determine fertilizer requirements. Through resistance breeding, many of the diseases that affected sugar cane such as downy mildew, smut, rust, leaf blight and mosaic disease are no longer major problems. With the planting of resistant cultivars, disinfection of planting materials and good sanitary practices, many fungus and bacterial disease problems can be avoided.

Rehm and Espig (1991) draw attention to the eye-spot diseases (*Drechslera sacchari* (Butl.) Subram. et Jain.), pineapple disease (*Ceratocystis paradoxa* (Dade) Mor.) and gumming disease (*Xanthomonas vasculomm* (Cobb.) Dows). Also mentioned is the use of crop rotation and special disinfection steps to combat soil-borne diseases of the *Fusarium*, *Pythium* and *Rhizoctonia* species, among others.

Problem pests such as aphids, leafhoppers, mealybugs and stem borers (*Diatraea* spp.) can be properly controlled through the use of appropriately timed insecticides or other insects such as the Cuban fly, *Lixophaga diatraea* (Long and Hensley, 1972).

Sugar cane is harvested mostly by hand or sometimes mechanically. Hand harvesting accounts for more than half of the world's production, and is especially dominant in the developing world. When harvested by hand, the field is first set on fire. The fire spreads rapidly, burning away dry dead leaves, and killing any venomous snakes hiding in the crop, but leaving the water-rich stalks and roots unharmed. With cane knives or machetes, harvesters then cut the

(a)

(b)

Source: Krestavitis, http://en.wikipedia.org/wiki/File:Sugar cane_
drink.JPG

Figure 10.210 (a) Extracting juice from sugar cane;
and (b) a facão, or sugar machete, can be as sharp
as a razor blade

Source: Shinobu Miwa, www.yomiuri.co.jp/dy/columns/0003/
seasons025.htm

Figure 10.211 Harvesting tropical sugar cane

standing cane just above the ground. A skilled harvester can cut 500kg of sugar cane in an hour (wikipedia.org).

Most machine-cut cane is chopped into short lengths but is otherwise handled in a similar way to hand-cut cane. Machines can only be used where land conditions are suitable and the topography is relatively flat. In addition the capital cost of machines and the loss of jobs caused makes this solution unsuitable for many sugar estates (sucrose.com/Icane).

Sugar cane is harvested by chopping down the stems but leaving the roots so that it re-grows in time for the next crop. Harvest times tend to be during the dry season and the length of the harvest ranges from as little as 2.5 months up to 11 months. The cane is taken to the factory: often by truck or rail wagon but sometimes on a cart pulled by a bullock or a donkey (sucrose.com/Icane).

With mechanical harvesting, a sugar cane combine (or chopper harvester), a harvesting machine originally developed in Australia, is used. The Austoft 7000 series was the original design for the modern harvester and has now been copied by other companies including Cameco and John Deere. The machine cuts the cane at the base of the stalk, separates the cane from its leaves, and deposits the cane into a haul-out transporter while blowing the thrash back onto the field. Such machines can harvest 100 tonnes of cane each hour, but cane harvested using these machines must be transported to the processing plant rapidly;

Source: Scott Bauer, USDA Agricultural Research service, Bugwood.org, http://www.forestimages.org/browse/detail.cfm?imgnum=1322055

Figure 10.212 Harvesting sugar cane in south Florida

(a)

(b)

Source: (a) Commons/Wikipedia.org; (b) egypt-travel-guide.co.uk.

Figure 10.213 (a) Sugar cane on a bullock cart, Maharashtra, India, January 2007; and (b) narrow gauge railway near Luxor loaded with sugar cane, 2006

once cut, sugar cane begins to lose its sugar content, and damage inflicted on the cane during mechanical harvesting accelerates this decay (wikipedia.org/ Sugar cane).

Harvesting, using sugar cane harvesters, is most strategically performed after a dry period lasting approximately 2 months. This dry period results in the halt of plant growth and an increase in sugar content (Rehm and Espig, 1991).

The harvesting of one year's growth of the crop should be performed between 10 and 13 months after planting, while the harvesting of crops grown for two years should be performed between 19 and 24 months after planting.

The harvested cane should be sent, most commonly by truck, to the processing facility within 30 hours of harvest to prevent quality deterioration (Husz, 1989).

Production

About 195 countries grow sugar cane, producing 1324.6 million tonnes a year (more than six times the amount of sugar beet produced). As of the year 2005, the world's largest producer of sugar cane by far is Brazil followed by India (wikipedia.org/Sugar cane).

According to Husz (1989), the amount and quality of sugar cane harvested are affected by the correlation of growing periods with the time of the year. Table 10.102 compares 12-month growth cycles using planting dates in January and May with a theoretical maximum harvest.

Processing and utilization

Uses of sugar cane include the production of sugar, Falernum, molasses, rum, soda, cachaça (the national spirit of Brazil) and ethanol for fuel. The bagasse that remains after sugar cane crushing may be burned to provide both heat – used in the mill – and electricity, typically sold to the consumer electricity grid. It may also, because of its high cellulose content, be used as

Source: http://picasaweb.google.com/Vaufi1947/SouthAfrica
2006Bado#5096626721776203746

Figure 10.214 Sugar cane transportation
today

Table 10.101 Top ten sugar cane producers, 2008

Country	Production (Tonnes)	Footnote
Brazil	514,079,729	
India	355,520,000	
People's public of China	106,316,000	
Thailand	64,365,682	
Pakistan	54,752,000	P
Mexico	50,680,000	
Colombia	40,000,000	F
Australia	36,000,000	
United States	27,750,600	
Philippines	25,300,000	F
World	1,557,664,978	A

P = official figure, F = FAO estimate, A = Aggregate (may include
official, semi-official or estimates).

Source: FAO

Table 10.102 Characteristic and theoretical
maximum yield of sugar cane based on planting
month

Characteristic	Month of planting		Theoretical maximum
	January	May	
Dry matter (t/ha)	34	27.5	50
Fresh matter (t/ha)	136	110	200
Fresh leaves (t/ha)	46	49.5	40–60
Cane stalk	90	60.5	140–160
Saccharose (%)	14	8.5	–
Saccharose (t/ha)	12.6	5.14	–
White sugar (t/ha)	10	3.45	14–16

raw material for paper, cardboard, and eating utensils branded as 'environmentally friendly' as it is made from a by-product of sugar production.

Fibres from Bengal cane (*Saccharum munja* or *Saccharum bengalense*) are also used to make mats, screens or baskets etc. in West Bengal. This fibre is also used in Upanayanam, a rite-of-passage ritual in India and therefore is also significant religiously (wikipedia.org/wiki/Sugar cane).

74 tonnes of raw sugar cane are produced annually per hectare in Brazil. The cane delivered to the processing plant is called burned and cropped (b&c) and represents 77 per cent of the mass of the raw cane. The reason for this reduction is that the stalks are separated from the leaves (which are burned and the ashes left in the field as fertilizer) and from the roots that remain in the ground to sprout for the next crop. Average cane production is, therefore, 58 tonnes of b&c per hectare per year.

Each tonne of b&c yields 740kg of juice (135kg of sucrose and 605kg of water) and 260kg of moist bagasse (130kg of dry bagasse). Since the higher heating value of sucrose is 16.5MJ/kg, and that of the bagasse is 19.2MJ/kg, the total heating value of a tonne of b&c is 4.7GJ of which 2.2GJ comes from the sucrose and 2.5GJ from the bagasse.

Per hectare per year, the biomass produced corresponds to 0.27TJ. This is equivalent to 0.86W/m^2. Assuming an average insolation of 225W/m^2 the photosynthetic efficiency of sugar cane is 0.38 per cent.

Table 10.103 Planting method and planting date effects on cane yield, theoretical recoverable sugar (TRS), and sugar yield of plant-cane crop, harvested in 2001 and 2002

Planting date	Cane yield (tonnes/acre)		Total recoverable sugar (TRS) (lb/tonne)		Sugar yield (tonnes/acre)	
	billet	whole	billet	whole	billet	whole
			2001			
Aug 2000	40.2Ab[x]	43.1Aa	270Aa	274Aa	5.5Ab	5.9Aa
Sept 2000	36.1Ba	36.0Ba	270Aa	264Aa	4.9Ba	4.8Ba
Oct 2000	29.5Cb	34.0Ba	254Bb	262Aa	3.7Cb	4.4Ba
			2002			
Aug 2001	37.5Aa	35.8Aa	246Aa	246Aa	4.6Aa	4.4Aa
Sept 2001	34.7ABa	30.4Ba	246Aa	248Aa	4.3Ba	3.8Ba
Oct 2001	32.8Ba	30.0Bb	242Aa	240Aa	4.0Ba	3.6Bb

Note: [x] Means within a column and year followed by the same upper case letter or within a row and a particular measurement (cane yield, TRS, etc.) followed by the same lower case letter are not statistically different using Fisher's protected LSD at alpha = 0.10.

Source: Viator et al (2005)

Table 10.104 Planting method and planting date effects on cane yield, theoretical recoverable sugar (TRS), and sugar yield of first-ratoon crop, harvested in 2002 and 2003

Planting date	Cane yield (tonnes/acre)		TRS (lb/tonne)		Sugar yield (tonnes/acre)	
	billet	whole	billet	whole	billet	whole
			2002			
Aug 2000	30.7Aa[a]	32.9Aa	240Aa	242Aa	3.7Aa	4.0Aa
Sept 2000	28.9Aa	32.3Aa	236Aa	240Aa	3.4Aa	3.9Aa
Oct 2000	31.9Aa	32.4Aa	236Aa	238Aa	3.8Aa	3.8Aa
			2003			
Aug 2001	41.5Aa	40.4Aa	258Aa	256Aa	5.3Aa	5.2Aa
Sept 2001	38.2Ba	35.8Bb	260Aa	258Aa	4.9Ba	4.6Bb
Oct 2001	40.0ABa	39.7Aa	256Aa	260Aa	5.1Ba	5.2Aa

Note: [a] Means within a column and year followed by the same upper case letter or within a row and a particular measurement (cane yield, TRS, etc.) followed by the same lower case letter are not statistically different using Fisher's protected LSD at alpha = 0.10.

Source: Viator et al (2005)

The 135kg of sucrose found in 1 tonne of b&c is transformed into 70L of ethanol with a combustion energy of 1.7GJ. The practical sucrose–ethanol conversion efficiency is, therefore, 76 per cent (compare with the theoretical 97 per cent).

One hectare of sugar cane yields 4000L of ethanol per year (without any additional energy input because the bagasse produced exceeds the amount needed to distil, the final product). This, however, does not include the energy used in tilling, transportation, and so on. Thus, the solar energy-to-ethanol conversion efficiency is 0.13 (da Rosa, 2005).

World production of biofuels rose some 20 per cent to an estimated 54 billion litres in 2007. These gains meant biofuels accounted for 1.5 per cent of the global supply of liquid fuels, up 0.25 per cent from the previous year. Global production of fuel ethanol, derived primarily from sugar or starch crops, increased 18 per cent to 46 billion litres in 2007, marking the sixth consecutive year of double-digit growth. Production of biodiesel – made from feedstock such as soya, rape and mustard seed, and palm and waste vegetable oils – rose an estimated 33 per cent, to 8 billion litres.

The United States, which produces ethanol primarily from corn, and Brazil, which primarily uses sugar cane, account for 95 per cent of the world's ethanol production. Brazil increased its ethanol production by 21 per cent in 2007, to 19 billion litres. The United States remained the world's leading producer, boosting output 33 per cent to 24.5 billion litres in 2007, and now accounts for a little more than half of the world's ethanol production (Monfort, 2008).

The sugar cane is processed by diffusion or pressing to remove the saccharose-containing juice from the stalk. The remaining bagasse can be burned to produce heat and electricity or additionally processed to produce building materials, cellulose and paper, among other products. The saccharose juice is purified through crystallization and centrifuging, at which time the 96 per cent pure product is referred to as raw sugar. Additional washing and centrifuging leave the 99.8 per cent pure product, white sugar. Molasses, which is removed during these purifying steps, can be processed to produce alcohol. One tonne of sugar cane can produce approximately 100kg of sugar, 30kg of bagasse and 40kg of molasses (which can produce 10L of alcohol). If processed for alcohol production alone this 1t of sugar cane would produce about 70L alcohol (Husz, 1989).

Less than three years after a major government initiative, more than 70 per cent of the automobiles sold in Brazil, expected to reach 1.1 million in 2006, had flex-fuel engines, which entered the market generally without price increases. The rate at which this technology was adopted is remarkable.

Also, in the past, the residue left when cane stalks are compressed to squeeze out juice was discarded. Brazilian sugar mills now use that residue to generate the electricity to process cane into ethanol, and use other by-products to fertilize the fields where cane is planted. Some mills produce so much electricity that they sell their excess to the national grid.

Such energy generation is increasingly practical in a variety of agricultural, lumber and urban waste

Figure 10.215 Typical sugar and ethanol plant in Brazil

Source: Cohen (2007)

Source: Cohen (2007)

Figure 10.216 A car powered by ethanol, 1931 in Brazil

Source: David Monniaux, http://commons.wikimedia.org/wiki/File: Bagasse_dsc08999.jpg

Figure 10.217 Bagasse, or residue of sugar cane, after sugar is extracted

management settings. There is sufficient benefit to warrant the establishment by some states of a Renewable Portfolio Standard in the absence of federal policy.

Brazilian sugar cane farmers have recognized the economic benefit of such waste to energy. In addition to operating biomass-fuelled, steam-generated electric power plants to provide electricity for sugar production, heat that is generated from combustion of the biomass can be utilized in the fermentation process. Such use of biomass improves the environmental impact of ethanol as an alternative transportation fuel.

In September 2008 Paris-based Areva was awarded a contract to design and build a power plant fuelled by sugar cane bagasse by Brazil's largest private power generator, Tractebel Energia. The US$42 million plant with a 33MW capacity is being built at an existing ethanol plant in the state of São Paulo. Areva is already building two smaller biomass plants in Brazil running on saw dust, bark and waste by-products from woodworking and furniture factories. Tractebel Energia generates 8 per cent of Brazil's total power, mostly from renewable energy sources.

The world's largest biomass power plant (70MW installed capacity) is the Okeelanta sugar mill in Palm Beach county owned by Florida Crystals, which also runs on bagasse and other woody biomass, including hurricane debris and municipal waste (wikipedia.com).

In 2008 Pakistan's first renewable energy project to use sugar cane-waste biogas created from the production of ethanol began supporting the national grid. The plant is powered by eight of GE Energy's ecomagination™-certified Jenbacher biogas engines.

The commercial start-up of sugar cane milling company Shakarganj Mills Ltd's new biogas power plant in Jhang, Pakistan came as the country was working to overcome its 3500MW energy shortage. The new plant generates enough power to support more than 50,000 homes.

The new biogas plant also provides a reliable, on-site source of power to help the mill and other industrial operations meet production requirements and remain competitive. The biogas used to fuel the Jenbacher gas engines is extracted from spent wash – a residue of the ethanol production operation that uses sugar cane molasses as a raw material.

As a renewable energy project, the plant is eligible for carbon credits because it enhances energy efficiency at the mill and displaces the national grid's energy generated from fossil fuels. By using biogas instead of fossil fuels for power generation, the plant is expected to produce approximately 20,000 tonnes of certified emissions reductions (CERs) annually under the Kyoto Protocol. The expected income from these CERs was instrumental in the customer's financial decision-making process. The project will be registered with the United Nations Framework Convention on Climate Change (UNFCCC) by Carbon Services Pakistan and First Climate AG as a successful demonstration project for the region, with 225 million litres of ethyl alcohol produced annually in Pakistan (Business Wire, 2008).

Internet resources

http://blogs.tampabay.com/energy/2008/09/new-power-plant.html
www.worldwatch.org/node/5777
www.plantmanagementnetwork.org/pub/cm/research/2005/sugar cane/
www.sucrose.com/lcane.html#skiltop
http://en.wikipedia.org/wiki/Sugarcane
www.oecd.org/dataoecd/9/46/38253356.pdf
www.britishsugar.co.uk
www.ibmalaysia.com
http://en.wikipedia.org/wiki/File:Sugar_cane_madeira_hg.jpg
www.transfairusa.org
http://nayagarh.nic.in/govtsection/agri/sugarcane.jpg
www.spiegel.de/wirtschaft/0,1518,602457,00.html
www.yomiuri.co.jp/dy/columns/0003/seasons025.htm
http://jcwinnie.biz/wordpress/?p=2796
commons.wikimedia.org/wiki/File:Sugar_cane.jpg
www.egypt-travel-guide.co.uk/Egypt-Boats-Trains.html

SUNFLOWER (*Helianthus annuus* L.)

Partially contributed by C. D. Dalianis

Description

The cultivated species of sunflower is a native of America that was taken to Spain from Central America before the middle of the sixteenth century. It was being grown by Indians for food and for hair oil. Improved

Figure 10.218 A sunflower farm near Mysore, India

varieties had been developed in Europe before 1600. These varieties were introduced into the USA and cultivation of sunflower started.

Sunflower originated from America where it was present when Europeans arrived. It was reported to be present in Arizona and New Mexico 3000 BC. The Spanish explorer Monardes brought the plant in Europe in 1569. Tsar Peter the Great brought the plant from Europe in Russia where its production reached the first rank in the world (about 16% in Russia and 9% in Ukraine in 1999).

In recent years, the world's cultivated area of sunflower has been steadily increasing. This is mainly the result of the breeding of dwarf high-yielding hybrids that also facilitate mechanization. Another reason is the emphasis given to polyunsaturated acids for human consumption. Since sunflower oil is rich in

polyunsaturated acids, its use for human consumption has greatly increased.

Sunflower belongs to the genus *Helianthus* of the Compositae family. The genus *Helianthus* was named using the Greek *helios* meaning sun, and *anthos,* flower. The inflorescence of the plants of this family are heads in which the fertile flowers are aggregated and bordered by rays, the corollas of sterile flowers.

The genus *Helianthus* includes 67 annual and perennial species. The cultivated sunflower is an annual plant with the scientific name of *Helianthus annuus*. It is an erect, unbranched, coarse annual, with a distinctive large, golden head, the seeds of which are often eaten and are commonly crushed for oil production.

The stem can grow as high as 3m and the flower head can reach 30cm in diameter with large seeds. The term 'sunflower' is also used to refer to all plants of the genus *Helianthus*, many of which are perennial plants. What is usually called the flower is actually a head (formally composite flower) of numerous florets (small flowers) crowded together. The outer florets are the sterile ray florets and can be yellow, maroon, orange, or other colours. The florets inside the circular head are called disc florets, which mature into what are traditionally called 'sunflower seeds', but are actually the fruit (an achene) of the plant. The inedible husk is the wall of the fruit and the true seed lies within the kernel. The florets within the sunflower's cluster are arranged in a spiral pattern. Typically each floret is oriented toward the next by approximately the golden angle, 137.5°, producing a pattern of interconnecting spirals where the number of left spirals and the number of right spirals are successive Fibonacci numbers. Typically, there are 34 spirals in one direction and 55 in the other; on a very large sunflower there could be 89 in one direction and 144 in the other (wikipedia.org/wiki/Sunflower).

Ecological requirements

Sunflower is a well-adapted crop under various climatic and soil conditions. With its well-developed root system it is one of the most drought-resistant crops and considered suitable for the southern semi-arid countries. However, oil yields are substantially reduced if plants are allowed to become stressed during the main growth period and at flowering. Under moisture stress conditions the number and the size of the leaves are reduced. One of the mechanisms employed by sunflower to resist moisture stress is wilting, since it has been shown in controlled experiments that in limp leaves water loss was reduced to a greater extent than photosynthesis.

A satisfactory crop can be produced, without irrigation, even in winter rainfall regions of approximately 300mm.

Sunflower is well adapted to warm southern European regions. Growth is satisfactory when temperatures do not fall below 10°C, but it can resist

Source: Bruce Fritz wikipedia.org/wiki/Sunflower

Figure 10.219 Sunflowers growing near Fargo, North Dakota

far lower temperatures. The young plants can withstand considerable freezing until they reach the four to six leaf stage, and the ripening seeds suffer little damage from slight frost. Sunflower requires ample sunlight and is considered insensitive to day length.

Sunflower grows on a variety of soils ranging from sandy soils to clay. In low-fertility soils its performance is better in comparison with other crops such as corn, potato and wheat. The best pH range is between 6.5 to 8. It is considered slightly susceptible to salt.

Crop management

The seedbed for sunflowers should be prepared as for corn. Minimum tillage is recommended. However, because of the pericarp of the seed a seedbed with high moisture content is required for satisfactory germination. The seed should be placed in moist soil; rapid drying out of the seedbed should be avoided.

Sunflower seeds are capable of germinating even at 5°C, but in practice temperatures over 10°C are required for satisfactory germination and even higher for satisfactory emergence. Generally, it is possible to sow sunflower fairly early in spring before other summer crops, such as corn, cotton or sorghum. Usually, early plantings lead to higher seed yields and higher oil content of seeds.

Optimum plant population is a key factor for obtaining high yields. This is mainly because each plant produces only one disc, so sunflower has a limited ability to adjust yields when the deviation from the optimum number of plants is large. When deviations are not large, thinner populations may produce larger discs, but they usually have more empty seeds in the centre of the disc and consequently oil yields are usually reduced.

In areas with more than 500mm annual rainfall, populations range from 35,000 to 60,000 plants per hectare. Under irrigation, the number of plants per hectare is 25–50 per cent higher, but plant lodging probabilities are also higher. An empirical method for determining the optimum population density is the disc diameter at maturity. If it is smaller than 12cm, plant density is too high, and if it is larger than 25cm, plant density is too low.

Planting is done with a corn planter equipped with special sunflower plates, or with a grain drill with most of the feed cups closed off. Distances of 75cm between rows are very common. Under irrigation, smaller between-row spacings of 50–70cm and within-row spacings of 15–30cm give higher yields. If possible, the rows should be sown with an east–west orientation because most of the discs remain orientated towards east and mechanical harvesting is thus easier. Seed rates range between 5 and 15kg/ha.

In the early growth stages sunflower is sensitive to weed competition. A pre-emergence harrowing, followed by cross-harrowing of the crop in the seedling stage, is recommended. Immediately after plant establishment, cultivation should be undertaken as required. Pre-sowing and/or pre-emergence herbicides are used for the control of certain weeds.

Large areas of sunflower throughout the world are grown without irrigation, because farmers prefer to use irrigation for other crops that are more dependent on irrigation or more profitable. In summer rainfall areas growth and production are satisfactory. In winter rainfall areas, as in most southern European regions, although satisfactory yields are obtained without irrigation in comparison with many other summer crops, sunflower's response to irrigation is very pronounced. In the same field, irrigated sunflower yields have been observed to increase up to 100 per cent or more in comparison to non-irrigated plants. However, water use efficiency depends on fertilization.

A significant yield reduction is observed under water stress conditions, mainly because the proportion of fertile seeds is reduced as a result of poor fertilization, and also because the abortion of flowers shortly after fertilization is greatly increased. As a rule, a favourable soil moisture regime must be maintained throughout the period from the differentiation of floral organs, when the inflorescence bud is about 15mm in diameter, up to harvest.

In non-irrigated fields sunflower's response to fertilizers is very limited, as is the case with many other non-irrigated spring-grown crops. In irrigated fields of low fertility, nitrogen fertilization response is positive. Nitrogen fertilization rates from 50 to 80kg/ha are recommended. Nitrogen application must be related to the availability of phosphate and potash, for if these are deficient nitrogen usually depresses yield and seed oil content.

Production

In southern European countries a total of 2.66 million hectares of sunflower are grown each year, mainly for oil production. Annual sunflower seed production is equal to 3.92 million tonnes with an average seed yield of 1473kg/ha. However, there are very large productivity differences among the various southern European countries. These are related to the soil type, but mainly to climatic conditions in each country as well as to the irrigated versus non-irrigated areas. If we assume an average 50 per cent oil content of the sunflower achenes, oil production ranges from as low as 280kg/ha in Portugal to 1349kg/ha in Italy, with an average overall yield of 738kg/ha.

Recently, in southern European regions, sunflower has been considered as an energy crop for biodiesel production, as is rapeseed in northern European regions.

However, sunflower biodiesel can be produced only under strong and continuous financial incentives (subsidies and/or tax exemptions). In Italy there are tax exemptions of 125,000t/year for biodiesel either from sunflower or from rapeseed. In France, tax exemptions exist for all liquid biofuels, whereas in Spain, Portugal and Greece there are no incentives for liquid biofuels.

The world production of sunflower seed was about 30 million tonnes in 1999, yields between 500 and 2600kg/ha are reported. The whole seed contains about 40 per cent oil and about 25 per cent protein suited for animal feeding. Neutral triacylglycerols constitute the

Source: Robert G. Hayes, http://home.earthlink.net/~rghayes/perlingerphotos.html

Figure 10.220 Sunflower harvester

major lipid class in sunflower seeds. Phospholipids and glycolipids constitute less than 4 per cent of the total lipids.

The composition in triacylglycerol species is characterized by the presence of LLL (33 per cent), OLL (25 per cent), LOP (11 per cent), and OLO (6 per cent) (cyberlipid.org).

The production of sunflowers for cooking oil, confectionary and birdseed has increased throughout the United States. The 2007 sunflower crop totalled 2.89 billion pounds, up 35 per cent from 2006, and was valued at US$607 million. Yield also increased to

Table 10.105 Acreage, production and yields of seed sunflower in southern EU countries and Germany

Country	Acreage (ha)	Production (tonnes)	Yields (kg/ha, seed)
France	991,000	2,143,000	2162
Spain	1,455,000	1,363,000	937
Italy	127,000	349,000	2748
Portugal	73,000	42,000	575
Greece	14,000	22,000	1571
Germany	65,000	160,000	2461
Total	2,725,000	4,076,000	1496

Table 10.106 Sunflower: highest yield achieved by research oil trials, 2006

Company	Hybrid	Moisture (%)	Oil (%)	Seed yield (kg/ha)
Pioneer	64H41	10.7	48.6	3730

Sunflower oil, about 13 per cent of the world oil production, in EEC, Russia and Argentina, 9Mt in 2004–2005.

Source: seeds2000.net

1468 pounds per acre, up 21 per cent from the previous year. (agmrc.org)

Figure 10.221 shows the yield per harvested acre in the USA in 2004 in different counties. They lies between less than 1000 and more than 1800 pounds per acre.

Processing and utilization

Sunflower heads should be fully mature before harvesting. It is estimated that during the last 14 days of maturation the dry matter of the seeds may increase by 50–100 per cent. However, seed maturation in a head is not uniform. Delaying harvesting until all the seeds are fully mature may cause significant losses due to shattering and losses from birds that eat the seeds. A practical indication of the appropriate stage for harvesting is when the backs of the sunflower heads are yellow and the outer bracts are beginning to turn brown. However, even at this stage a fair proportion of seeds contain up to 50 per cent moisture. This should be taken into account when considering safe storage because the maximum moisture content at storage should not exceed 9 per cent.

Sometimes a pre-harvest desiccant is applied in order to facilitate harvesting. Diquat, applied when the leaves are turning yellow, is very effective and the yield losses are negligible. Dwarf varieties are harvested with a cereal combine machine equipped with a special attachment on the header.

The head of the matured sunflower contains about 50 per cent of the dry matter of the whole plant. Nearly one-half the weight of dried heads is seed.

In oil-producing varieties the pericarp ranges from 22 to 28 per cent and the kernel from 72 to 78 per cent. Kernels are rich in oil. The average oil composition of the achenes ranges from 40 to 50 per cent, the protein

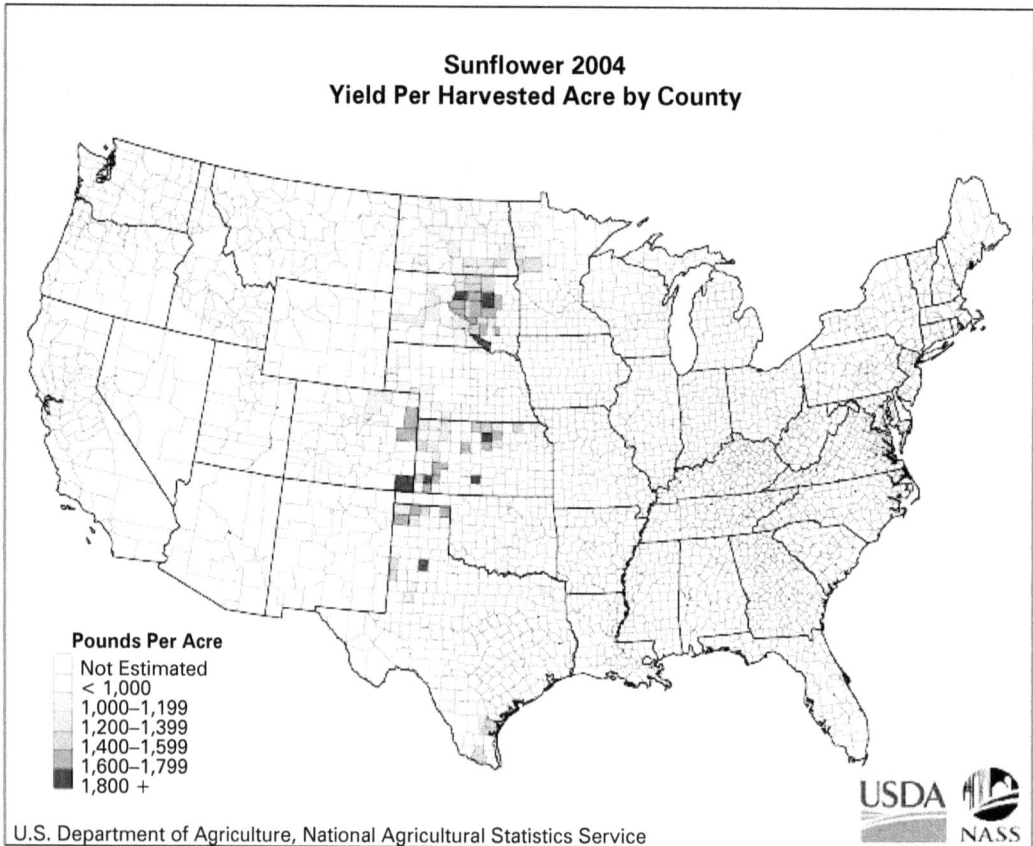

Sunflower 2004
Yield Per Harvested Acre by County

Pounds Per Acre
Not Estimated
< 1,000
1,000–1,199
1,200–1,399
1,400–1,599
1,600–1,799
1,800 +

U.S. Department of Agriculture, National Agricultural Statistics Service

USDA NASS

Source: unl.edu/nac/atlas/Map

Figure 10.221 Sunflower seed yield (pounds/acre) by county in the USA, 2004

(a)

(b)

Source: wikipedia.org/wiki/Sunflower

Figure 10.222 (a) Sunflower heads sold as snacks in China; and (b) sunflower seeds

content from 15 to 20 per cent, and the fibre content from 10 to 15 per cent. In non-oil- producing varieties the pericarp ranges from 45 to 50 per cent and the kernel from 50 to 55 per cent. These achenes have a lower oil content (30 to 35 per cent) and higher fibre content (20 to 25 per cent).

It is estimated that 90 per cent of the world's sunflower seed production is crushed for oil extraction. Sunflower oil is characterized by its high content of linoleic acid and medium content of oleic acid. Sunflower oil produced in cool, northern climates usually contains 70 per cent or more linoleic acid, whereas sunflower oil produced in southern warm regions contains 30 per cent linoleic acid. There is usually a negative relationship between linoleic and oleic acid contents. The nutritive

value of sunflower oil is equal to that of olive oil. Sunflower oil is mainly used for human consumption.

In growing sunflower for biodiesel production it should be taken into account that, beyond the seeds, large quantities of other sunflower plant parts could be exploited for energy purposes. A sunflower crop produces a significant amount of plant residues, mainly stalks, that remain in the field. On average, stalk residues are estimated as twice as much the seed production. These residues have a high gross heating value of about 17 to 18MJ/kg of dry matter.

The non-seed part of the heads is also an important biomass source that is almost equal to the seed production. In addition, if dehulling is practised prior to the oil-extraction process, a significant quantity of hulls is produced. In oil-producing hybrids about 25 per cent of the achene weight is the hull, which is composed of equal proportions of lignin, pentosans and cellulosic material, representing 82 to 86 per cent of the total weight. These materials also have a high heating value. The gross heating value is about 17MJ/kg of dry matter.

The long-term potential of sunflower oil for biodiesel is not encouraging for three main agronomic reasons. First, sunflower is a low-yielding crop, giving on average less than 1t/ha of oil. Second, as a worldwide cultivated crop of many years' standing it has been considerably improved, so future productivity improvements are expected to be small and difficult. Third, under the same soil and climatic conditions many other biomass crops give much higher yields of raw material that could be exploited for energy purposes. For example, irrigated corn may give, under southern European conditions, approximately 4000L/ha ethanol, whereas irrigated sunflower gives only 1000 to 1200kg/ha of biodiesel.

It is understood that from a strictly agronomic point of view in a future competitive market for liquid biofuels, sunflower biodiesel will be in a very disadvantageous position to compete with other liquid biofuels such as corn bioethanol. Sunflower is in a similar position with regard to other irrigated agricultural crops or wild-grown plants (giant reed, sweet and fibre sorghum), and when non-irrigated sunflower is compared with other non-irrigated crops (wheat, etc.). Of course, sunflower biodiesel has the great advantage that the oil extraction plants already exist in many rural areas, so there is usually no need for costly new investments.

It should be also taken into account that sunflower oil for biodiesel would have to compete with the strong market for human consumption that is steadily increasing both worldwide and Europe-wide. Its high quality and value for human consumption means that the bulk of present European sunflower oil production is directed towards this end.

Finally, it is well known that the GATT agreement imposes severe restrictions on the area in Europe that could be given to oil crops for biodiesel production. It is estimated that throughout Europe no more than a million hectares can be given to oil crops, rapeseed and sunflower for biodiesel production.

Sunflower oil, extracted from the seeds, is used for cooking, as a carrier oil and to produce margarine and biodiesel, as it is cheaper than olive oil. A range of sunflower varieties exist with differing fatty acid compositions; some 'high oleic' types contain a higher level of healthy monounsaturated fats in their oil than even olive oil. Sunflowers may also be used to extract toxic ingredients from soil, such as lead, arsenic and uranium. They were used to remove uranium, caesium-137, and strontium-90 from soil after the Chernobyl accident (wikipedia.org/wiki/Sunflower).

Sunflowers were chosen for research by the University of New Hampshire because of their ability to grow well in region and their higher oil yield, estimated at 130 gallons of biodiesel per acre of sunflowers. While a mobile biodiesel manufacturing station was built on a trailer, five hybrid varieties of sunflowers were planted on a four-acre plot at the university's Kingman Farm.

The first batch of sunflowers has been harvested, and a second crop has also been planted. Seed and oil yield measurements will be taken to determine the economic feasibility for New Hampshire growers. Biodiesel will be produced with the sunflower oil and then used to power farm equipment at the farm, and through further research it is hoped to identify a new feedstock that local farmers can use to fuel their energy needs without having to rely on something that is imported (biodieselmagazine.com).

Assuming a typical sunflower seed yield of 1800kg/ha, as obtained under normal conditions on fertile dry lands, and taking into account the energy value of the seed, the net energy value was estimated to be 36·87GJ/ha and the ratio of energy outputs to energy inputs approximately 4.5:1. The total energy input was calculated to be 10.49GJ/ha with fertilizers being the major inputs (Kallivroussis et al, 2002).

In 2008 a team in Africa calculated that 1ha of sunflowers could produce 600L of sunflower oil, a 100ha farm would potentially be able to produce

sufficient fuel from 10ha to plough and plant the whole 100ha of farmland (Cameron, 2008).

Internet resources

www.cyberlipid.org/glycer/glyc0051.htm
www.agmrc.org/commodities__products/grains__oilseeds/sunflowers.cfm
www.biodieselmagazine.com/article.jsp?article_id=1288
www.seeds2000.net/images/2006%20Sunflower%20Yield%20Book.pdf
http://en.wikipedia.org/wiki/Sunflower
www.unl.edu/nac/atlas/Map_Html/Protection_of_Crops-Etc/National/Sunflower_yield/Sunflower_yield.htm

SWEET POTATO (*Ipomoea batatus* (L.) Poir. ex Lam.)

Description

The sweet potato is a herbaceous vine with a creeping growth habit. Its habitat is the part of the Andes from northern South America to Mexico, where it was already in use in pre-Columbian times (Lötschert and Beese, 1984). The plant is hexaploid, 2n = 90. It is assumed that its ancestors were tetraploid and diploid species found in this region. Nowadays, it is grown everywhere in the tropics and subtropics (Rehm and Espig, 1991). The sweet potato can even be grown in some parts of the temperate zones with warm summers. The reason for its wide range is the rather short vegetation period. The sweet potato can be grown where the temperature during the growing period is above 18°C and temperatures below 10°C do not occur (Franke et al, 1988). All parts of the plant contain latex tubes with a neutral-tasting milky juice, which, in contrast to cassava, is not toxic (Franke, 1985).

Ecological requirements

The sweet potato grows well if there is an average daytime temperature of more than 18°C. At 10–12°C growth comes entirely to a standstill. The slightest frost kills all plant parts above the ground. The sweet potato is very sensitive to temperatures below 1°C for brief periods and below 3°C for longer periods. Sweet potato is a warm-season perennial crop that is cultivated as an annual and may be harvested after 80–360 days. A frost-free period of 110–170 days is necessary. However, the sweet potato's

Figure 10.223 Sweet potato plantation

Figure 10.224 Freshly dug sweet potato

need for warmth is less than most other root and tuber crops. In the tropics it grows up to 2500m.

Distinct differences between day and night temperatures encourage the formation of tuberous roots. Cultivars with short vegetation times grow well in the subtropics, up to 40° latitude. Long day conditions delay the flowering and reduce the formation of tuberous roots: a photoperiod of less than 11 hours induces flowering; less flowering occurs at 12 hours, and no flowering at all at 13 hours daylight.

Sweet potato cultivation is limited to regions with 750–1250mm rainfall. The plant survives long dry periods, but for high yields and for good tuber quality there must be a water supply (precipitation or irrigation)

that is evenly spread throughout the growing period. Dryness is desirable during the ripening period, for then the roots will be firm and keep well. In subtropical climates the roots contain more starch and less sugar than in tropical climates. High yields are only achieved in areas with large amounts of sunshine. Sweet potato needs a well-aerated and well-drained soil. Where there are heavy soils or there is a danger of waterlogging it is grown on ridges or mounds. Neither pH below 5 nor salt is acceptable (FAO, 1996a). Calcium deficiency of the soil makes planting impossible.

Propagation

Propagation is done vegetatively using stem cuttings of 30cm length, with the leaves stripped off before planting. The cuttings are placed into moist warm soil, and after 2–3 days they take root. The sweet potato cannot be planted in dry soils. Another way is to plant roots that were stored during the winter, but this is disadvantageous.

Because the plants are frequently self-sterile, cross-breeding is necessary as a rule. Parent plants are easy to bring to flowering at the same time in any particular climate if they are grafted onto an annual *Ipomoea* species. The in vitro technique is used in breeding improved cultivars – for example, to select cold-resistant cultivars (Irawati, 1994).

Crop management

The cultivation of the sweet potato can be completely mechanized. Planting distances are 90cm between and 60cm within the rows. Weeding by hoeing or with herbicides (linuron, diphenanid) is necessary in the first 6 weeks; thereafter the leaves suppress the weeds' growth.

The greatest damage is caused by several viral diseases; then less susceptible or new virus-free cultivars have to be planted. Losses by fungal diseases are generally slight because resistant cultivars are available in most regions. Infestation is possible by leaf-eating beetles whose larvae hollow the roots, and leaf-eating caterpillars can cause serious losses. Nematodes attack the roots and tubers. To prevent severe damage, insecticides, harvesting at the right time and crop rotation are useful measures.

The sweet potato has a positive influence on weed control; it can suppress perennial weeds like nutsedge

(*Cyperus rotundus* L.). It is self-compatible. Some tribes in New Guinea plant sweet potatoes in monoculture, but mostly the plant is grown in rotation with other field crops.

Production

According to the Food and Agriculture Organization (FAO) statistics world production in 2004 was 127Mt. The majority comes from China with a production of 105Mt from 49,000km². About half of the Chinese crop is used for livestock feed. The sweet potato should be harvested when the leaves begin to yellow because the starch content is highest at that point. Harvesting can be fully mechanized, as for potatoes. In small-scale farming the tubers can be dug up with a plough or manually with spade or hoe. For cultivars with a growing time of five months a successful yield is 20t/ha, and 40–50t/ha can be achieved with high- yielding cultivars. With early ripening cultivars, less improved local cultivars, or under bad cultivation conditions, the yields are much lower. The global average is a yield of 14.6t/ha.

Processing and utilization

The major portion of production is consumed fresh. Drying or starch extraction is possible. The sweet potato, like other starch- and sugar-producing plants, is a potential energy crop. The carbohydrate content of the edible part is 9–25 per cent starch and 0.5–5 per cent sugar.

The concentration of sugar becomes higher the nearer to the equator the plant is grown (Franke, 1985). If the sweet potato is grown as an energy source the starch of the roots is processed to alcohol. First, the roots are ground and the starch is cracked into sugars by treatment with acids or hydrolytic enzymes. After the conversion of starch into sugars, fermentation takes place to give ethanol. The development of continuous fermentation techniques and the use of improved varieties of yeast have improved efficiency and lowered costs (FAO, 1995b).

Industrial uses include the production of starch and industrial alcohol. Taiwanese companies are making alcohol fuel from sweet potato (wikipedia.org).

One tonne of sweet potatoes yields 125L of ethanol (Wright, 1996). Leaves, foliage, unsaleable tubers and remnants from processing are used as fodder. Wastes and by-products can be used to produce energy – the tubers for ethanol production, and leaves and foliage for biogas production.

In an investigation carried out by the USDA Agricultural Research Service, for sweet potatoes, carbohydrate production was 4.2 tonnes an acre in Alabama and 5.7 tonnes an acre in Maryland.

Source: (a) LSU AgCenter, www.lsuagcenter.com; (b) Mary Peet, NCSU, http://peet.hort.ncsu.edu/

Figure 10.225 (a) Sweet potato harvesting; and (b) workers sorting and packing at sweet potato factory

Table 10.107 Raw sweet potato values per 100g (3.5 oz)

Energy	90kcal/360kJ
Carbohydrates	12.7g
Starch	12.7g
Sugars	4.2g
Dietary fibre	3.0g
Fat	0.1g
Protein	1.6g

Source: nutritiondata.com

Carbohydrate production for cassava in Alabama was 4.4 tonnes an acre, compared to 1.2 tonnes an acre in Maryland. For corn, carbohydrate production was 1.5 tonnes an acre in Alabama and 2.5 tonnes an acre in Maryland.

The disadvantages to cassava and sweet potato are higher start-up costs, particularly because of increased labour at planting and harvesting times. If economical harvesting and processing techniques could be developed, the data suggests that sweet potato in Maryland and sweet potato and cassava in Alabama have greater potential than corn as ethanol sources. Another advantage for sweet potatoes and cassava is that they require much less fertilizer and pesticide than corn.

The additional research could help develop new biofuel sources without diverting field corn supplies from food and feed use to fuel (sciencedaily.com).

In a study in Africa, root yields ranged from 13.1 to 31.4t/ha for cassava, and from 10.2 to 14.0t/ha for sweet potato. The unavailability of acceptable improved varieties, high incidence of pests and diseases, and poor cultural practices are the main causes of relatively low yields (Moyo et al, 2006). The yield potential of sweet potato can be up to 30–40t/ha (Saleh and Hartojo, 2000).

In the highlands of Papua New Guinea, soils are rich and productive. Sweet potatoes are harvested sequentially, making it notoriously difficult to obtain reliable yield data. The team used single harvest yields to assess relative production from this 'one-off' farmer survey. In both gardens, old and new, single harvest yields of up to 30t/ha were observed with a median yield of 6t/ha. Due to the sequential harvesting these yields were considerably lower than the average total yields of 13–15t/ha for lowland and highland systems (Kirchhof, 2006).

The following characteristics make sweet potato almost perfect for ethanol production (welovenature.blogspot.com):

- It can be grown in tropical, subtropical and warm temperate regions and in any terrain.
- The plant can be grown in a variety of soils except for heavy clay types where the roots have little chance of development.
- It does not require much fertilizer and little maintenance. Additional fertilizers are not needed in fertile soil, since the plant will produce mostly leaf.
- Sweet potato has a short maturity period of 3.5 to 4 months.
- The tubers can be stored for three months after curing for seven days in any open space.

Source: Albert Cahalan http://upload.wikimedia.org/wikipedia/commons/3/38/5aday-sweet-potato.jpg

Figure 10.226 A sweet potato tuber

In 2005 the Canadian Sweet Potato Ethanol Alliance (CSPEA) unveiled plans to build a Can$100 million (about US$85 million) plant in Tillsonburg, Ontario, that will convert sweet potatoes into ethanol. The facility is expected to be operational within five years. CSPEA selected the Tillsonburg site because it is close to a major highway and a rail line and is situated on the border of three counties that will supply the sweet potato feedstock. The facility will also sell sweet potatoes to the food market, with approximately 40 per cent of the total crop reserved for ethanol production. The plant will generate its own electricity (London Free Press, 2005).

To satisfy domestic demand for vehicle fuel, control its dependence on foreign sources of oil and attempt to moderate fuel costs, China has embarked on a robust effort to increase fuel ethanol development. In 2008 the China International Project Consultancy Corp., under the direction of the National Development and Reform Commission (NDRC), completed an evaluation of five locations with respect to their suitability for non-grain ethanol projects. The report concluded that developing ethanol-refining capacity based on yam/cassava/sweet potato in those five provinces and cities – Hubei, Hebei, Jiangsu, Jiangxi and Chongqing – would be in keeping with the government's policy of substituting non-grain ethanol for grain ethanol and would not compete with food requirements. The first of the non-grain ethanol projects is being developed by the Chongqing-based World Petroleum Company using sweet potatoes as the feedstock; this project will produce 100,000 tonnes per year of ethanol. On 1 April 2008 Hong Kong Development, a publicly traded Hong Kong company, announced that it would purchase a 71 per cent interest in the World Petroleum Company for 163.4 million Yuan (US$23.4 million). In addition to the Chongqing World Petroleum Company, four other Chongqing companies are investing or plan to invest in ethanol projects. Chongqing is China's third largest producer of sweet potatoes with an output of approximately 20 million tonnes per year; two areas within Chongqing have emerged as centres of fuel ethanol refining: Changshou and Wanzhou. Against the background of these economic realities, the non-grain fuel ethanol industry is getting under way. With prices of yam/sweet potato/cassava feedstock rising, market forces would be expected to increase supply, allowing the non-grain fuel ethanol industry in China to flourish (Schwartz, 2008).

Internet resources

www.aciar.gov.au/project/SMCN/2005/043

www.allbusiness.com/operations/shipping/585121-1.html

http://knowledge.cta.int/en/content/view/full/2784

www.papuaweb.org/dlib/tema/ubi/psp-2003-saleh-hartojo.pdf

www.renewableenergyworld.com/rea/news/story?id=52450

www.sciencedaily.com/releases/2008/08/080825200752.htm

http://en.wikipedia.org/wiki/Sweet_potato#Non-culinary_uses

http://welovenature.blogspot.com/2007/04/ethanol-from-sweet-potato.html

SWITCHGRASS (*Panicum virgatum* L.)

Partially contributed by Dr Dudley G. Christian, IACR, Rothamsted, UK and H. W. Elbersen

Description

Switchgrass is a native of North America where it occurs naturally from 55°N latitude to deep into Mexico, mostly as a prairie grass. In North America it has long been used for soil conservation and as a fodder crop. Both in America and Europe it can be found as an ornamental plant. The grass is also found in South America and Africa where it is used as a forage crop. Switchgrass is a perennial C_4 grass propagated by seed that can be established at low cost and risk, and requires very low inputs, while giving high biomass yields even on marginal soils. Since the early 1990s the crop has been developed as a model herbaceous energy crop for ethanol and electricity production in the USA and in Canada and it is also being considered as a paper pulp production feedstock.

Switchgrass is a perennial grass with slowly spreading rhizomes and a clump form of growth. Culms (stems) are erect and 50–250cm tall, depending on variety and growing conditions. The stiff, tough stems are capable of withstanding moderate snowloads. The inflorescence is a panicle 15–50cm long. Seeds are hard, smooth and shiny. Seed weights are variable and range between 0.7 and 2.0g per 1000. Dormancy level may be high in newly harvested seed. Storing seed or other dormancy breaking techniques may be required before it can be sown.

The grass is highly polymorphic, outcrossing, and has a basic chromosome number of nine. Most cultivars are tetraploids or hexaploids. Two major ecotypes can be distinguished (Moser and Vogel, 1995). Lowland types are generally found in floodplains and are taller and coarser, while upland types are found in areas that do not flood, and have finer leaves and generally slower growth than lowland types (Sanderson et al, 1996). The geographical distribution of switchgrass is from North America to Central America, but it is also found in South America and Africa.

In the USA, switchgrass is an important warm-season native range grass. Warm-season grasses, which have the C_4 photosynthesis pathway, produce most of the biomass in hot summers when cool-season (C_3) grasses become semi-dormant. The more efficient C_4 carbon assimilation pathway potentially gives warm-season grasses much higher yields than cool-season grasses. Compared to C_3 grasses, C_4 grasses have higher base optimum temperatures; Hsu et al (1985a) found a minimum germination temperature of 10.3°C for switchgrass, while the optimum temperature for seedling growth was more than 30°C (Hsu et al, 1985b). Switchgrass has several characteristics that make it suitable as a biomass crop, including high potential productivity, low nutrient demand, efficient water use and good persistence (Jung et al, 1988;

Source: USDA Agricultural Research Service, Bugwood.org

Figure 10.227 Switchgrass trials

Hope and McElroy, 1990; Jung et al, 1990; Sladden et al, 1991).

Considerable information is available on the use and management of switchgrass as a forage grass. In recent years, switchgrass has been evaluated as a potential biomass feedstock for energy production in the USA. It has been selected as a model herbaceous crop for the Oak Ridge National Laboratory's Biofuel Feedstock Development Program (Parrish et al, 1993; McLaughlin et al, 1996; Sanderson et al, 1996). The suitability of this grass as a biomass crop in Europe has yet to be demonstrated. Therefore this discussion draws heavily on American publications and the reader should bear this in mind.

Ecological requirements

Switchgrass has considerable cold tolerance after winter hardening; switchgrass stands can be maintained up into Canada (Hope and McElroy, 1990). In a study at Rothamsted in southern England, seven varieties have successfully overwintered; however, the adaption of varieties to a range of European conditions has not yet been evaluated. The adaptability of switchgrass means there are cultivars available for a wide range of environmental conditions.

Switchgrass will tolerate acid conditions but establishment and growth will be better if soils are limed to correct the pH to neutrality (Jung et al, 1988). Seedbeds are normally prepared using traditional ploughing and secondary cultivation to produce a firm seedbed with a fine-textured surface. The seedbed should be weed free because switchgrass is not competitive during the establishment phase.

Crop management

Switchgrass is established by seed. Sowing is normally carried out when soil temperatures have reached at least 10–15°C. Because of dormancy problems in newly harvested seed, freshly harvested seed can be 95 per cent dormant; year-old seed often germinates better than fresh seed. The amount of pure live seed (PLS), not the weight of seed, is used to calculate the seed rate. PLS is calculated by multiplying the purity by the germination. Germination rates of above 75 per cent are desirable.

The seed can be sown using normal small seed drills or cereal drills fitted with fine seed rolls. Direct drilling (no-till planting) may be appropriate in some circumstances (Parrish et al, 1993). Seed depth should not be more than 1cm. The number of plants established can be up to 400 plants per m^2, depending upon weed control strategy after sowing (Vogel, 1987). In the experiment at Rothamsted a common seed rate of 10kg/ha was used to sow seven varieties of switchgrass. The number of plants established ranged from 189 to 301 per m^2 (Christian, 1994).

For optimum yields adequate fertility levels are required, though reasonable yields can be attained under low-fertility conditions (Moser and Vogel, 1995). If levels of phosphorus and potassium are low, these nutrients may be added to the seedbed to benefit seedling growth. Nitrogen should not be applied to the seedbed or during the early establishment phase because it may stimulate weed growth to the disadvantage of the slow-growing grass. There may also be a risk in the established crop that nitrogen carried over from the previous year and from early nitrogen application will stimulate weed growth at a time when switchgrass is still not competitive because of low temperatures in spring.

Weed competition is a major cause of seedling establishment failure (Moser and Vogel, 1995). It is important to reduce weeds prior to planting. Pre-emergence herbicides are often used to reduce weed competition when switchgrass is not competitive. Broad leaf weeds can be suppressed with herbicides until switchgrass seedlings are large enough. Grassy weeds can be mown to reduce light competition.

There are no herbicides approved for use in switchgrass in Europe, but simazine and atrazine will effectively control weeds even in young crops. Isoproturon has been used during the plant's dormant period in winter.

Production

In the UK, shoots from the rhizomes emerge in late spring, anthesis takes place during August or September depending on the cultivar, and senescence starts with falling temperatures in the autumn. Growth may stop sooner than for temperate grasses.

Biomass production in the year of planting is likely to be low but substantially higher in subsequent years (Table 10.108). The four years of evaluation at Rothamsted provided evidence of wide variation in the yield of different cultivars. The yield of the three best cultivars averaged 11t/ha in 1996.

The yield of individual cultivars has increased each year, except for 'Dacotah' which had the same average yield in 1995 and 1996. It is not known if maximum yields have been reached. It is noteworthy that yields have increased despite drought conditions in the summers of 1995 and 1996. In 1993/1994 the date of maturity varied from October to January, depending on the cultivar. No differences in maturity date were

Table 10.108 The yield (DM t/ha) of seven varieties of switchgrass between 1993 and 1996

Variety	1993		1994		1995		1996	
	N0	N60	N0	N60	N0	N60	N0	N60
Pathfinder	2.42	2.06	6.89	5.52	8.48	8.22	9.91	9.01
Sunburst	1.34	1.03	6.05	4.16	6.98	8.05	7.36	8.60
Cave-in-rock	1.66	1.66	6.92	6.06	8.11	8.03	10.88	10.11
Nebraska 28	1.55	1.51	5.34	5.25	8.59	7.30	10.56	9.27
Dacotah	1.11	1.35	4.44	4.10	5.28	5.70	5.40	5.37
Kanlow	2.15	1.64	6.77	4.34	7.10	5.34	12.48	10.78
Forestburgh	1.53	1.47	6.27	6.28	7.35	8.72	11.72	9.95
Mean	1.68	1.53	6.10	5.10	7.41	7.34	9.76	9.01
SED, Varieties XN	0.517		0.938		0.529		1.434	

Table 10.109 The amount of various minerals present in different varieties of switchgrass at harvest 1995/1996

Cultivar	Fertilizer (N at 0 or 60kg/ha)	Nitrogen (kg/ha)	(kg/t)	Phosphorus (g/ha)	(kg/t)	Potassium (kg/ha)	(kg/t)	Magnesium (kg/ha)	(kg/t)
Pathfinder	N0	43.9	5.2	3.5	0.4	11.8	1.4	7.3	0.9
	N60	43.8	5.3	3.2	0.4	8.2	1.0	5.4	0.7
Sunburst	N0	39.3	5.6	2.9	0.4	11.1	1.6	4.9	0.7
	N60	36.5	4.5	2.5	0.3	9.3	1.2	4.1	0.5
Cave-in-rock	N0	48.1	5.9	4.0	0.5	10.5	1.3	6.6	0.8
	N60	49.2	6.1	3.6	0.4	10.9	1.4	6.4	0.8
Nebraska 28	N0	42.5	4.9	2.8	0.3	9.0	1.0	4.2	0.5
	N60	38.2	5.2	3.4	0.5	9.3	1.3	4.4	0.6
Dacotah	N0	32.7	6.2	2.6	0.5	5.7	1.1	2.2	0.4
	N60	32.6	5.7	2.3	0.4	5.3	0.9	2.4	0.4
Kanlow	N0	39.3	5.5	3.6	0.5	11.1	1.6	10.4	1.5
	N60	44.1	8.2	3.6	0.7	10.1	1.9	9.5	1.8
Forestburgh	N0	35.6	4.8	3.0	0.4	8.8	1.2	4.0	0.5
	N60	40.5	4.6	2.8	0.3	8.4	1.0	4.0	0.5

Note: Mineral content is reported as total offtake (kg/ha) and kg/t of biomass

found in 1995, probably as a result of the drought conditions that occurred in the summer. Response to nitrogen has been inconsistent and frequently negative, possibly because in the first year there was adequate N supplied from soil sources whereas in the second season drought may have been influential. Work in the USA has shown 50kg/ha N to be about optimal for yield on marginal soils (Parrish et al, 1993).

Crops were sown in May 1993 and received no fertilizer (N0), or 60kg/ha/year (N60). Crops were harvested in winter when stems were dead. In the Rothamsted experiment the amount of the major nutrients removed at harvest was found to be similar for all cultivars except Dacotah.

Nutrient content at harvest is lower than in miscanthus grass which was of the same age, grown in the same field and harvested in the same winter. The miscanthus contained 8kg N, 0.6kg P and 8.7kg K per tonne of dry biomass. The ash content of switchgrass and miscanthus was 7.8 per cent and 8.4 per cent respectively. The low mineral content of switchgrass is a desirable characteristic for efficient combustion and low exhaust gas emissions.

The yields of switchgrass grown at four locations at different latitudes in the eastern USA are presented in Table 10.110. Although the Rothamsted site has a more northerly location, the yield of some cultivars compares favourably with yields obtained at the US sites. This would indicate that if switchgrass is grown in Europe heavy yields of biomass might be achieved, especially at more southerly locations. In northern Europe, switchgrass may have a lower yield than some alternative biomass crops such as miscanthus grass (Schwarz et al, 1994; Van der Werf et al, 1993) or energy coppice (Cannell, 1988). However, switchgrass has lower production costs compared to miscanthus because it can be established by seed and may require less fertilizer to replace minerals exported in biomass; and, in common with miscanthus, the annual harvest gives a quicker payback than short rotation coppice which has a 3–5 year harvest cycle.

Processing and utilization

There is no technical reason why the crop cannot be cut and harvested using traditional grass-harvesting machinery. The thin woody stems of switchgrass allow good dry-down in the winter. At Rothamsted, moisture content at harvest was down to 30 per cent; this may allow the baled crop to be stored for a short period before use without the need for drying. McLaughlin et al (1996) report moisture contents of 13 to 15 per cent at baling in the USA. No records of harvesting methods or storage in the European context have been found.

End uses of switchgrass being considered are ethanol production, combustion and thermal conversion. Yields of 280L of ethanol per tonne of dry switchgrass appear possible, which compares favourably to other species that have been examined (McLaughlin et al, 1996). The main attributes that determine usefulness for combustion are total energy content, moisture content, and fouling and corrosion characteristics. The energy content of switchgrass is about 18.4kJ/kg of dry weight, which is comparable to woody species but about 33 per cent less than that of coal (McLaughlin et al, 1996). Moisture content is variable and depends on time of harvest and storage methods, but compares favourably to woody species. McLaughlin et al (1996) reported the ash content of switchgrass to be variable (2.8–7.6 per cent), and generally lower than that of coal.

There may be some concern about the alkali and silica content of switchgrass, which could lower ash melting points and thus contribute to slagging in boilers. Particularly high potassium and phosphorus can be a problem. Ragland et al (1996) conducted experiments in which switchgrass was co-fired with pulverized coal. They found an ash content of 4.6 per cent for switchgrass, which was about half of that of coal. Also switchgrass had 40 per cent as much nitrogen and 5 per cent as much sulphur as coal. Co-firing switchgrass with coal can thus reduce air pollution emissions per unit of energy produced. The switchgrass ash contained 3.4 times more potassium and 50 times more phosphorus than coal ash.

Appropriate harvest management can reduce the mineral content of switchgrass and thus increase combustion quality. Sanderson and Wolf (1995) concluded that allowing switchgrass to reach maturity would minimize the concentration of inorganic elements in the feedstock. In Europe, combustion, co-firing and gasification are good options for switchgrass utilization in the short run, but other options for its use considered in the USA have so far not been evaluated.

With respect to switchgrass the following conclusions can be reached:

Table 10.110 The yield of switchgrass at four locations in the eastern USA (cutting regimes vary)

Region (location)	Nitrogen (kg/ha N)	Dry matter yield (t/ha)		Notes	Reference
Northeast (Pennsylvania)		1981–1984		Mean of five varieties cut at heading	Jung et al, 1990
	0	7.5	(6.9–7.8)		
	75	9.9	(9.1–12.3)		
East (Virginia)		1988–90		Various sites on marginal land. Single variety 'Cave-in-rock'. Reduced yield variation with November cut, wide yield variation with site and season.	Parrish et al, 1993
		Cut Sept.	Cut Nov.		
	0	8.4	8.1		
	50	10.7	9.1		
	100	10.6	9.3		
	Residual year, previously	1991		Cut after stems senesced. Heavier yield where crop previously cut in November.	
	0	5.8	8.2		
	50	8.2	9.8		
	100	9.8	10.2		
Upper southeast (Virginia, Tennessee, West Virginia, Kentucky, North Carolina)	100	One cut system		Seasonal yields from six varieties grown at eight locations. One cut in October or November or two cuts, at heading and in October or November.	Parrish et al, 1997
		13.3	(11.0–19.4)		
		Two cut system			
		15.7	(13.4–21.2)		
Southeast (Alabama)	84	1989		Mean of eight varieties. Yield is total of two cuts, anthesis and regrowth. Two varieties, 'Alamo' and 'Kanlow', gave outstanding yields in both years. Other varieties included 'Cave-in-rock' and 'Pathfinder'. Most had a higher yield in the second year.	Sladden et al, 1991
		10.4	(6.8–17.5)		
		1990			
		14.2	(6.8–34.6)		

- Evaluation of switchgrass as a biomass crop for European conditions is strictly limited at present.
- The range of cultivars available makes it likely that switchgrass could be grown in many regions of Europe and may suit drier areas.
- Switchgrass is established by seed and therefore has a low cost of establishment and potentially low costs of production. Establishment is the most critical part of switchgrass cultivation.
- Yields so far obtained in Europe have been comparable to yields obtained in similar climatic regions of the USA.

Switchgrass is often considered a good candidate for biofuel – especially ethanol fuel – production due to its hardiness in poor soil and climate conditions, rapid growth, and low fertilization and herbicide requirements.

Switchgrass has a large biomass output (the raw plant material used to make biofuel) of 6–10 tonnes per acre.

Switchgrass has the potential for enough biomass to produce up to 380 litres of ethanol per metric tonne harvested. This gives switchgrass the potential to produce 1000 gallons of ethanol per acre, compared to 665 gallons per acre of sugar cane and 400 gallons per acre of corn. Switchgrass is being used to heat small industrial and farm buildings in Germany and China through a process used to make a low-quality natural gas substitute. It can also be pressed into fuel pellets which are burned in pellet stoves used to heat homes, which typically burn corn or wood pellets.

In the spring of 2008, 1000 acres (400ha) of switchgrass will be planted near Guymon, Oklahoma, in the Oklahoma Panhandle, to study the feasibility of using the crop for biofuel. It will be the largest stand ever planted for such purposes. The project is being spearheaded by the Oklahoma Bioenergy Center (wikipedia.org/wiki/Switchgrass#Biofuel).

Switchgrass grown for biofuel production produced 540 per cent more energy than needed to grow, harvest and process it into cellulosic ethanol, according to estimates from a large on-farm study by researchers at the University of Nebraska-Lincoln. Switchgrass grown in this study yielded 93 per cent more biomass per acre and an estimated 93 per cent more net energy yield than previously estimated in a study done elsewhere.

Results from a five-year study involving fields on farms in three states highlight the potential of prairie grass as a biomass fuel source that yields significantly more energy than is consumed in production and conversion into cellulosic ethanol. The study involved switchgrass fields on farms in Nebraska, North Dakota and South Dakota. It is the largest study to date examining the net energy output, greenhouse gas emissions, biomass yields, agricultural inputs and estimated cellulosic ethanol production from switchgrass grown and managed for biomass fuel.

Switchgrass is not only energy efficient, but can be used in a renewable biofuel economy to reduce reliance on fossil fuels, reduce greenhouse gas emissions and enhance rural economies, The joint USDA-ARS and Institute of Agriculture and Natural Resources study also found greenhouse gas emissions from cellulosic ethanol made from switchgrass were 94 per cent lower than estimated greenhouse gas emissions from gasoline production. In a biorefinery, switchgrass biomass can be broken down into sugars including glucose and xylose that can be fermented into ethanol in a similar way to corn. Grain from corn and other annual cereal grains, such as sorghum, are now primary sources for ethanol production in the US.

In the future, perennial crops, such as switchgrass, as well as crop residues and forestry biomass could be developed as major cellulosic ethanol sources that could potentially displace 30 per cent of current US petroleum consumption. Technology to convert biomass into cellulosic ethanol is being developed and

Source: Chariton Valley Biomass Project, www.iowaswitchgrass.com/cofiring~harvesting.html

Figure 10.229 Switchgrass is harvested and baled just like hay – an advantage for farmers who would not have to buy new equipment if the grass becomes an economical way to make ethanol

Source: US Government, http://upload.wikimedia.org/wikipedia/commons/7/7a/Panicum_virgatum.jpg

Figure 10.228 Switchgrass trials

is now at the development stage where small commercial-scale biorefineries are beginning to be built with scale-up support from the US Department of Energy. Researchers point out in the study that plant biomass remaining after ethanol production could be used to provide the energy needed for the distilling process and other power requirements of the biorefinery. This results in a high net energy value for ethanol produced from switchgrass biomass. In contrast, corn grain ethanol biorefineries need to use natural gas or other sources of energy for the conversion process.

In this study, switchgrass managed as a bioenergy crop produced estimated ethanol yields per acre similar to those from corn grown in the same states and years based on state-wide average grain yields.

Six cellulosic biorefineries that are being co-funded by the US Department of Energy across the US should be completed over the next few years. These plants are expected to produce more than 130 million gallons of cellulosic ethanol per year, according to the US Department of Energy (sciencedaily.com).

Internet resources

http://en.wikipedia.org/wiki/Switchgrass#Biofuel
http://upload.wikimedia.org/wikipedia/commons/7/7a/
Panicum_virgatum.jpg
http://news.mongabay.com/2006/1205-switchgrass.
html
www.msnbc.msn.com/id/11518172/
www.sciencedaily.com/releases/2008/01/080109110629.
htm

TALL FESCUE (REED FESCUE) (*Festuca arundinacca* Schreb.)

Description

Tall fescue is a perennial cool-season grass commonly grown for forage, turf, or as a conservation grass. The plant grows to a height of 0.6–1.8m. The leaf blades are grooved and the underside is keeled. The inflorescence is a panicle. The main flowering period is from June to July (Siebert, 1975; Kaltofen and Schrader, 1991).

Most tall fescue is infected (mutualistic symbioses) by the fungal endophyte (*Acremonium coenophialum*). Although this fungus makes the plant toxic for grazing, it is ecologically necessary for the plant's success and has been proven to confer insect resistance on the plant.

Tall fescue plant material has mineral concentrations of approximately 9–10 per cent ash, 2–3 per cent silicon dioxide (SiO_2), 2.5 per cent nitrogen, 100mg/kg iron, 60–70mg/kg manganese and 5–6mg/kg copper (Pahkala et al, 1994).

Ecological requirements

Tall fescue can adapt to a wide range of soils and climates. It responds well to fertile soils, in which it develops open sods. Temperate zones, subtropics and tropical highlands are the areas with the most suitable climates for tall fescue. In mild climates, growth continues throughout winter. Sustained summer growth is usual. Tall fescue grows well even on poor soils. Although tall fescue is not tolerant of wet soils and salt, it has been shown to be tolerant to drought (Rehm and Espig, 1991).

Tall fescue is a deep-rooted, sod-forming grass best adapted to cool-season production. It is extremely well suited for use as a stockpile forage because it retains its quality and improves in palatability in the fall. It is well adapted to low-pH soils like those found in strip mine reclamation. It is more tolerant of animal and machinery traffic and of mismanagement than are other cool-season grasses (Hall, M., 1994).

Propagation

Tall fescue is propagated using seed.

Crop management

The amount of seed sown is approximately 20kg/ha. Fertilizer application tests on tall fescue by Moyer et al (1995) in the USA showed that subsurface (knife) application of fertilizer led to a 20 per cent increase in yield when compared with surface broadcast application. In addition, a yield increase of 69 per cent was observed when the nitrogen fertilizer level was increased from 13 to 168kg/ha.

Common weeds that infest tall fescue crops are, among others, crabgrass (*Digitaria* spp.) and white clover (*Trifolium repens* L.). Herbicide trials on tall fescue crops by Mueller-Warrant et al (1995), using pre-emergence herbicides in mid-October, post-emergence herbicides in early December, or both, showed that although herbicides may control weeds effectively, crop damage (yield reduction) does occur. The least damaging herbicide tested was pendimethalin applied pre-emergence with no post-emergence herbicide.

Among the most common fungi known to attack tall fescue plants are rhizoctonia blight, white blight, and rust (*Puccinia graminis* ssp. *graminicola*). Insect pests such as grass grub larvae are known frequently to feed on the roots of tall fescue.

Production

Field tests at four locations in Germany where tall fescue was harvested once a year at the onset of flowering provided yields averaging between 11.4 and 13.1t/ha dry matter. The plant's dry matter content averaged around 30 per cent at the time of harvest. The heating value averaged approximately 17MJ/kg (Feuerstein, 1995).

Processing and utilization

Tall fescue crops are harvested by mowing, starting in the first year they reach maturity. When grown for the production of biomass, the crop may be harvested several times each year beginning in early summer, usually with about 4 weeks between harvests. The harvested crop can be baled and allowed to dry for storage. It is important that the crop matter be sufficiently dried for its economic utilization as a biofuel.

In addition to its traditional uses for hay and pasture, tall fescue has also taken on the roles of turf or conservation grass and has been widely and successfully used to stabilize ditch banks and waterways.

One of the most heavily researched and promising uses of tall fescue is as pulp for making fine papers. The qualities that are required of short fibres in printing grade papers are: high number of fibres per unit weight, and stiff, short fibres (low hemicellulose content, and low fibre width to cell wall thickness ratio). Tall fescue has short, narrow fibres and produces pulp with a greater number of fibres than most hardwoods.

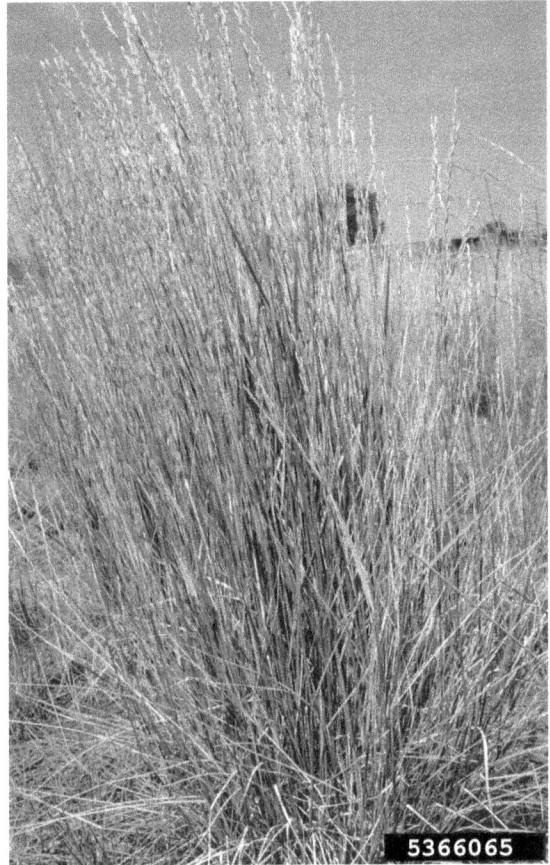

Source: Howard F. Schwartz, Colorado State University, Bugwood.org, http://invasive.org/browse/detail/cfm?imgnum=5366065

Figure 10.230 Tall fescue (*Festuca arundinacea*) growing as a roadside weed

Source: Jwarp, http://commons.wikimedia.org/wiki/File:Sodfarm.jpg

Figure 10.231 Harvesting tall fescue turfgrass sod in Annapolis, Maryland

Like many other grasses, tall fescue has the potential to be used as a biofuel. This involves combustion of the dry matter for the direct production of heat or combined with the production of electricity. This grass is also suitable to produce high-quality pellets.

Demonstration pilot cellulosic ethanol production facilities are planned in New York State, using perennial grasses such switchgrass, tall fescue and others (Mayton et al, 2008).

Internet resources

http://cropsoil.psu.edu/extension/facts/agfact28.pdf
www.stephenpastureseeds.com.au/content.asp?pid=39095

TALL GRASSES

Historically, the tall-grass prairie consisted of big and little bluestem, switchgrass and Indian grass. These species thrive in zones of 30–40 inch annual precipitation and reach 6–8 feet in height. Within this zone, cordgrass (*Stipa* spp.) and reed grass (*Phragmites* spp.) are dominant species in the wet lands (forages.oregonstate.edu).

The 'tall grasses' topic includes brief descriptions of various plant species that lack adequate information. The tall grasses have high biomass yields because of their linear crop growth rates over long periods of 140–196 days, and sometimes even longer. They can yield oven dry biomass of 20–45t/ha in colder subtropical or warmer temperate zones, and over 60t/ha annually in Florida. Grasses as bioenergy crops can be used as feedstocks for industrial processes or burned directly to produce energy. The juice of sugar and energy cane can be extracted for ethanol production. Cellulose and hemicellulose can be converted to sugars and then to alcohol, or used to make methane (Prine et al, 1997). Table 10.111 shows the cellulose contents of several tall grasses. Some of these crops have been already discussed in previous chapters. Others are described in this chapter.

Grasses are cosmopolitan plants that live from the equator to the arctic circle. Grasses have the power to grow again after they have been eaten or cut down almost to the roots. High-yielding grasses have been introduced into many parts of the world. First, the aim

Table 10.111 Cellulose composition of some tall grasses

Tall grass	Hemicellulose (%)	Cellulose (%)
Elephant grass	27.95	39.4
Erianthus	26.64	40.8
Energy cane	27.78	37.6
Sugar cane	23.31	31.8

was to use them as fodder plants. However, their high biomass yields make them interesting subjects for energy production (Burton, 1993). Desirable characteristics for energy feedstocks include efficient conversion of sunlight, efficient water use, capture of sunlight for as much of the growing season as possible, and low external inputs.

Perennial plants do not require annual establishment costs. In northern climates, the sunlight interception of perennial plants is more efficient, because annual plants spend much of the spring developing a canopy (Samson and Omielan, 1994). The plant should remain erect without lodging, and retain most of the biomass produced during the season in the standing plant (Prine and McConnel, 1996).

Grass tribes may be grouped into panicoid grasses and festacoid grasses. The panicoid group contains most of the African grasses and fixes carbon via the C_4 pathway. Their optimum temperature is 30–40°C and optimum light intensity is 50–60klux. Panicoid grasses can fix 30–50g/m²/day of dry matter whereas festacoid grasses fix up to 20g/m²/day of dry matter. However, panicoid grasses do not make use of this advantage if temperatures and light intensities are too low (Burton, 1993).

Napier grass, elephant grass, Uganda grass (*Pennisetum purpureum* Schumach.)

P. purpureum is a robust bunch grass that may reach a height of 6m, growing in dense clumps (Figure 10.232). The leaves are 30–90cm long, and up to 3cm broad (Duke, 1983). It grows on a wide range of well-drained soils and is drought tolerant.

The plant does not tolerate much frost. The above-ground biomass may be killed by frost, but it will thrive

Figure 10.232 Tall grass (elephant grass)

again from the rhizomes if the soil is not frozen (Burton, 1993). The plant requires a rich soil for best growth. The tolerated climate ranges from warm temperate dry to wet, through tropical dry to wet forest life zones. *P. purpureum* is reported to tolerate annual precipitation of 200–4000mm, annual temperatures of 13.6–27.3°C and pH of 4.5–8.2. In the African centre of diversity, elephant grass is reported to tolerate drought, fire, waterlogging, sewage sludge and monsoons.

The grass is usually propagated vegetatively. A good supply of nitrogen is required for high yields (Duke, 1983). The grass is established from stem cuttings or crown divisions. If cut once a year, *P. purpureum* can produce more dry matter per unit area than any other crop that can be grown in the deep south of the USA (Burton, 1993). It is one of the highest-yielding tropical forage grasses. The annual productivity ranges from 2 to 85t/ha. An investigation by Miyagi (1980) yielded 500t/ha wet matter (70t/ha DM), spacing the plants 50cm × 50cm.

Some reported dry matter yields were: 19t/ha/year in Australia, 66t/ha/year in Brazil, 58t/ha/year in Costa Rica, 85t/ha/year in El Salvador, 48t/ha/year in Kenya, 14t/ha/year in Malawi, 64t/ha/year in Pakistan, 84t/ha/year in Puerto Rico, 76t/ha/year in Thailand and 30t/ha/year in Uganda. The stems of elephant grass provide primarily lignocellulose with virtually no juice sugars. Experimental yields in Queensland, Australia, have attained 70t/ha/year DM,

of which 50t were stem. Expected farm yields might be 50–55t/ha/year DM.

Yields of other *Pennisetum* species: *P. americanum* is reported to yield 1–22t/ha/year, *P. clandestinam* 2–25t/ha/year, *P. pedicellatum* 3–8t/ha/year and *P. polystachyum* 3–10t/ha/year (Duke, 1983).

Weeping lovegrass (*Eragrostis curvula* (Schrad.) Nees)

E. curvula is a bunch grass with long, narrow drooping leaves, reaching a height of 0.5–1.5m. It grows best on sandy soils from southern Texas to northern Oklahoma (Burton, 1993).

Buffelgrass, African foxtail (*Cenchrus ciliaris* L.)

C. ciliaris is a drought-tolerant perennial bunch grass that produces forage for the livestock industry in Texas. The plant is 1–1.5m high (Burton, 1993). The leaves are green to bluish-green, 2.8–30cm long and 2.2–8.5mm broad. It thrives from sea level to an altitude of 2000m in dry sandy regions, with rainfall from 250 to 750mm, but it tolerates much higher rainfall. It grows on shallow soils of marginal fertility. The tolerated climate ranges from warm temperate thorn to moist through tropical desert to moist forest life zones. It is reported to tolerate an annual precipitation of 250–2670mm, an annual temperature of 12.5–27.8°C and pH of 5.5–8.2. The annual productivity ranges from 1 to 26t/ha/year, though some strains yield up to 37t/ha/year (Duke, 1983).

Pangola grass (*Digitaria decumbens* Stent)

D. decumbens is a sterile stoloniferous natural hybrid of unknown parentage. It lacks winter hardiness and is restricted to the southern two-thirds of Florida. It grows to a height of 0.4–0.8m (Burton, 1993).

Kleingrass, coloured Guinea grass (*Panicum coloratum* L.)

P. coloratum is a leafy bunch grass that grows well on sandy clay soils from Texas to Oklahoma. It is drought tolerant and grows earlier in the spring and later in the autumn than many warm-season grasses. Seed shattering is a serious problem that limits the plant's use. The plant grows to a height of 0.4–1.4m (Burton, 1993).

Guinea grass (*Panicum maximum* Jacq.)

Guinea grass grows to a height of 0.5–4.5m (Burton, 1993).

Bermuda grass (*Cynodon dactylon* (L.) Pers.)

Bermuda grass originated in Africa, but is now common in many parts of the tropics and temperate zones. It may become a weed. Improved cultivars out-yield the common and are suited to other climatic conditions.

Energy cane (*Saccharum officinarum* L.)

Energy cane and sugar cane are identical. 'Energy cane' means that the total biomass (leaves and stems) are harvested together and used as an energy feedstock. The crop can be harvested for several years. *S. officinarum* originated in the South Pacific islands and New Guinea. Nowadays it is found throughout the tropics and subtropics. Sugar cane is cultivated as far north as 36.7°N (Spain) and as far south as 31°S (South Africa). The tolerated climate ranges from warm temperate dry to moist through tropical very dry to wet forest life zones. *S. officinarum* is reported to tolerate an annual precipitation of 470–4290mm, an annual temperature of 16.0–29.9°C, and pH of 4.3–8.4.

Energy cane is propagated by stem cuttings. It is generally grown for many years without rotation or rest (Duke, 1983). Sugar cane is sensitive to cold, the tops being killed by frost. In the continental USA, where frost may occur, it is planted in late summer or early autumn and is harvested one year later. The growing period is 7–8 months. In the tropics it can be planted at any time of year, because there the plant does not have a rest period and grows continuously. The yields of sugar are higher under tropical conditions. The mature stems are 1.3–4m high, and have a diameter of 20–50mm (Magness et al, 1971).

Energy cane occurs gregariously, growing in sunny areas, on soils unsuitable for trees. It needs aeration at the roots and grows in sand, but not in loam. Lime is considered beneficial for the proper development of the sugar content. For good growth it requires a hot, humid climate, alternating with dry periods. It thrives best at low elevations on flat or slightly sloped land. Occasional flooding is tolerated (Duke, 1983).

At harvesting time the stems are cut and the leaves are removed. The cane should be processed without delay to avoid loss of sugar (Magness et al, 1971). The cane is cut as close to the ground as possible, because the root end is richest in sugar. Harvest takes place after 12–20 months. The canes become tough and are turning pale yellow when they are ready for cutting. The rhizomes will regenerate the crop for at least 3–4 years, and sometimes up to 8 or more years.

Reported dry matter yields are from 16 to 73t/ha/year. Yields achieved at some sites are: Hawaii 67.3t/ha/year and Brazil 54t/ha/year (Duke, 1983). During sugar manufacture the residual bagasse is a by-product. The bagasse can be used for energy production. The advantage of biomass energy derived from residues is that no additional land is demanded. Residues are an already available renewable resource (Swisher and Renner, 1996).

Table 10.112 displays the average yields of sugar cane, energy cane, erianthus and elephant grass

Table 10.112 Average yields of several cultivars of tall grasses on phosphatic clay soils

Tall grass cultivar	Annual average biomass (t/ha/year)	
	1987–1990	1992–1994
Sugar cane (US 78-1009)	49.7	32.3
Energy cane (US 59-6)	52.2	36.5
Erianthus (IK-7647)	48.8	17.9
Elephant grass (N-5l)	45.2	19.0

Source: Prine et al (1997).

Table 10.113 Promising warm-season grasses for biomass production in the USA and Canada

Dry prairie	Dry-mesic prairie	Mesic prairie	Wet-mesic prairie	Wet prairie
Sand bluestern, little bluestem, prairie sandreed (*Calamovilfa longifolia*)	Indian grass (*Sorghastrum nutans*), little bluestem	Big bluestem (*Andropogon geradii*), Indian grass (*Sorghastrum nutans*)	Big bluestem (*Andropogon geradii*), Indian grass (*Sorghastrum nutans*), switchgrass (*Panicum virgatum*), prairie cordgrass (*Spartina pectinata*)	Prairie cordgrass (*Spartina pectinata*)

cultivars on phosphatic clay soils in Polk County, central Florida. The yields were recorded from 1987 to 1990 and 1992 to 1994.

Elephant grass and energy cane are limited to the lower south of the USA (Prine and McConnel, 1996). If appropriate cultivars are chosen, the productivity of switchgrass is high across much of North America. In studies near the Canadian border, winter hardy upland ecotypes have yielded 9.2t/ha in northern North Dakota and 12.5t/ha in northern New York. Table 10.113 shows possible tall grass prairie species for the USA and Canada.

Some common tall grass prairie species like big bluestem or Indian grass have high levels of productivity. The seeding of all three of the major tall grass prairie species in mixtures rather than switchgrass monoculture may play an important role in reducing potential disease and insect problems.

Switchgrass is classified as a wet-mesic prairie species. Other species are better adapted than switchgrass to drier or wetter prairie conditions. For example, prairie sandreed has out-yielded switchgrass in the dry regions of the northern US Great Plains when 250–350mm of annual rainfall occurred. In the higher rainfall areas of the prairie region, eastern gamagrass (*Tripsacum dactyloides*) has also proven to be very productive. Prairie cordgrass has a more northern native range than switchgrass; it has a greater chill tolerance, which enables earlier canopy development. Nitrogen fertilization of native prairie cordgrass stands in Nova Scotia have produced yields of 7.7t/ha. Small plot biomass studies in England have yielded 8–23t/ha (Samson and Omielan, 1994). Scientists at Aberystwyth University are looking at ways of growing so-called

'energy grasses', such as 4m-high Asian elephant grass, in bulk. It is already burned in power stations to generate electricity, but there are plans to use it as a biofuel (news.bbc.co.uk/).

Studies by the Agrobiology Centre at the state Brazilian Agricultural Research Corporation (Embrapa) are finding that elephant grass has even greater potential. The popular eucalyptus tree, planted in Brazil to produce cellulose and charcoal, yields 7.5t/ha/yr of dry biomass, on average, and up to 20t/ha/yr in optimum conditions, while elephant grass yields 30–40t/ha/yr. A medium-sized electricity company, Sykue Bioenergia, has already commissioned a thermoelectric power plant that will be fuelled by elephant grass (IPS, 2008).

The thermal conversion of elephant grass (*Pennisetum Purpureum Schum*) to biogas, bio-oil and charcoal has been investigated under two heating rates of 10 and 50°C/minute. The energy required to pyrolyse elephant grass was evaluated using computer-aided thermal analysis techniques, while composition of the resultant biogas and bio-oil products were monitored with gas chromatographic and mass spectroscopic techniques. At 500°C, the biogas compounds consisted primarily of CO_2 and CO with small amounts of methane and higher hydrocarbon compounds. The heat of combustion of the biogas compounds was estimated to be 3.7–7.4 times higher than the heat required to pyrolyse elephant grass under both heating rates, which confirms that the pyrolysis process can be self-maintained. The faster heating rate was found to increase the amount of liquid products by 10 per cent, while charcoal yields remained almost the same at 30 per cent. The bio-oil mainly consisted of organic acids, phthalate esters, benzene compounds and amides. The amount of organic acids and benzene compounds were significantly reduced at 50°C/minute, while the yields of phthalate esters and naphthalene compounds increased. The difference in bio-oil composition with increased heating rate is believed to be associated with the reduction of the secondary reactions of pyrolysis, which are more pronounced under the lower heating rate (Strezov et al, 2008).

Research is needed to allow more efficient processing of lignocellulose from abundant plant biomass resources for production of fuel ethanol at lower costs. Potential dedicated feedstock species vary in degrees of recalcitrance to ethanol processing. The standard dilute acid hydrolysis pretreatment followed by simultaneous saccharification and fermentation (SSF) was performed on leaf and stem material from three grasses: giant reed (*Arundo donax* L.), Napier grass (*Pennisetum purpureum* Schumach.), and Bermuda grass (*Cynodon* spp). In a separate study, Napier grass, and Bermuda grass whole samples were pretreated with esterase and cellulose before fermentation. Conversion via SSF was greatest with two Bermuda grass cultivars (140 and 122mg/g of biomass) followed by leaves of two Napier grass genotypes (107 and 97mg/g) and two giant reed clones (109 and 85mg/g). Variability existed among Bermuda grass cultivars for conversion to ethanol after esterase and cellulase treatments, with Tifton 85 (289mg/g) and Coastcross II (284mg/g) being superior to Coastal (247mg/g) and Tifton 44 (245mg/g). Results suggest that ethanol yields vary significantly for feedstocks by species and within species and that genetic breeding for improved feedstocks should be possible (Anderson et al, 2007).

Internet resources

www.hort.purdue.edu/newcrop/indices/
http://forages.oregonstate.edu/fi
http://news.bbc.co.uk/2/hi/uk_news/wales/mid_/7852 906.stm
www.ips.org

TIMOTHY (*Phleum pratense* L.)

Description

Timothy grass (*Phleum pratense*), is an abundant perennial grass native to most of Europe except for the

Source: Rasbak 2005 commons.wikimedia.org

Figure 10.233 *Phleum pratense*

Mediterranean region. It grows to 50–150cm tall, with leaves up to 45cm long and 1cm broad. The flower head is 7–15cm long and 8–10mm broad, with densely packed spikelets (wikipedia.org).

Timothy is a short-lived perennial plant most commonly found in meadows, pastures, ditches and along trails. The leaves are ungrooved with a matt underside, and are light green to mid-green. The inflorescence is a dense cylindrical false ear which is usually about 6–7mm in diameter and 10cm in length. Flowering usually starts at the beginning of June (Siebert, 1975; Kaltofen and Schrader, 1991).

Plant material from timothy crops has mineral concentrations of approximately 5 per cent ash, 1 per cent silicon dioxide (SiO_2), 1.0–1.5 per cent nitrogen, 50mg/kg iron, 30–40mg/kg manganese and 4–5mg/kg copper (Pahkala et al, 1994).

Ecological requirements

Timothy readily adapts to most types of soil, especially to fertile soils where it responds by producing a high yield. Nutrient rich soils in locations ranging from fresh to damp are most appropriate for timothy crop cultivation. The plant is unable to tolerate long periods of dry weather, though it can tolerate winters and high levels of precipitation (Kaltofen and Schrader, 1991).

Propagation

Timothy is propagated by seed.

Crop management

Timothy is usually sown in the beginning to middle of summer (June to August). The crop has a high demand for fertilizers, especiallynitrogen. Table 10.114 summarizes the fertilizer rates obtained from field tests at sites in Germany. A large percentage of the fertilizer was supplied in the form of liquid manure – in some cases as much as 100 per cent. The fertilizer rates vary with soil composition.

Because timothy is a perennial plant that has most commonly been used for fodder purposes, timothy-legume mixtures in hay and pasture seedings have been used in crop rotations in the USA. Crop rotation with respect to timothy as a source of biomass for biofuels has not received as much attention.

Production

In Germany, field tests using timothy, harvested five times a year, produced in 1994 (three years after sowing) fresh matter yields averaging 66.32t/ha and dry

Table 10.114 Fertilizer rates used for timothy at several locations in Germany

Fertilizer	Amount (kg/ha)
Nitrogen (N)	214–411
Phosphorus (P_2O_5)	42–130
Potassium (K_2O)	166–544
Magnesium (MgO)	35–55

Source: (Kusterer and Wurth, 1992, 1995; Wurth, 1994).

matter yields averaging 12.28t/ha. The variety 'Tiller' produced the highest fresh matter yield, and the variety 'Barnee' produced the highest dry matter yield (Kusterer and Wurth, 1995).

Mediavilla et al (1993, 1994, 1995) provide the dry matter yields (t/ha) from timothy field tests that were conducted using liquid manure fertilizer at two locations in Switzerland, Reckenholz and Anwil. The yields in 1993, 1994 and 1995 were, respectively, 15.7, 12.5 and 14.5t/ha at Reckenholz, and 18.5, 17.9 and 11.2t/ha at Anwil. The crop was harvested three times in 1993 and twice times in each of 1994 and 1995. In Reckenholz between 70 and 90kg/ha/year NH_4-N liquid manure was used, while at Anwil the rate was up to 180kg/ha/year NH_4-N.

It is relatively high in fibre, especially when cut late. Seed yields of 600 kg/ha are reported. Seed yields in New Zealand about 250 kg/ha; in Europe, 300–600 kg/ha; in Portugal, 120–180 kg/ha. 'It gives a large yield of about 980 kg/ha WM.' (C.S.I.R., 1948–1976), which is not very large. With 60 kg N/ha, timothy and clover mixtures have yielded 9.4 MT DM/ha in Russia. Sown alone, it yielded 4.9 MT/DM/ha in Italy. It is an important crop in Scandinavia, Canada, north eastern and the Pacific Coast United States where it is a common constituent of pastures on moist heavy soils.

According to the Phytomass files, annual productivity ranges from 2 to 15 MT/ha, with 10–15 MT/ha resorted in Belgium, 2 in Bulgaria, 6–11 in Czechoslovakia, 13 in France, 14 in Germany, 11 in Jamaica, 11 in Poland, 5 in US, and 2 in USSR. (Duke 1998)

Processing and utilization

Timothy is harvested by mowing, with the number of harvests per year varying from one to five for mature crops. Harvests of five times per year involve harvesting approximately every 3–4 weeks beginning in the middle of May through until the middle of October. The harvested material then needs to be dried relatively quickly if it is to be used for fodder or stored. If it is to be used as biomass for conversion into a biofuel the quality of the harvested material is not important, but drying is necessary for its economic utilization.

Traditionally, timothy has been one of the most important hay grasses for cooler temperate humid regions. Through the application of present and developing technologies timothy will be used as biomass for conversion into a biofuel. This most commonly involves burning the crop matter for the direct production of heat and indirect production of electricity. The specific methane yields of crops, determined in 100–200 d methane potential assays, varied from 0.17 to 0.49m^3 CH$_4$ kg^{-1} VS$_{added}$ (volatile solids added) and

Source: Rasbak 2009 wikipedia.org

Figure 10.234 *Phleum pratense* ssp. *pratense*

from 25 to 260 m^3 CH$_4$ t$_{ww}$$^{-1}$ (tonnes of wet weight). Jerusalem artichoke, timothy-clover grass and reed canary grass gave the highest potential methane yields of 2900–5400 m^3 CH$_4$ ha^{-1}, corresponding to a gross energy yield of 28–53 MWh ha^{-1} and ca. 40,000–60,000 km ha^{-1} in passenger car transport. The effect of harvest time on specific methane yields per VS of crops varied a lot, whereas the specific methane yields per t$_{ww}$ increased with most crops as the crops matured (Lehtomäki et al 2007).

Internet resources

www.hort.purdue.edu/newcrop/duke_energy/Phleum_pratense.html
http://en.wikipedia.org/wiki/Timothy_hay 2009

TOPINAMBUR (JERUSALEM ARTICHOKE) (*Helianthus tuberosus* L.)

Description

Topinambur is a short-day perennial C$_3$ plant that originated in North America. A relative of the sunflower, topinambur grows to a height varying from 1 to 4m. Opposing one another on the stalk, grow rough, heart-shaped to lancet-shaped leaves. Flowering begins in late summer with yellow flowers of 4–8cm diameter (Franke, 1985). In central Europe, the plant's flowers scarcely produce seeds (Hoffmann, 1993).

Underground stolons form irregularly shaped tubers that are similar to potatoes except that topinambur tubers have a higher water content and produce adventitious roots. Depending on the variety, the tuber has a yellow, brown or red peel with white flesh. The tubers consist of 75–79 per cent water, 2–3 per cent protein and 15–16 per cent carbohydrates, of which the fructose polymer inulin can constitute up to 7–8 per cent or more (Franke, 1985). By the end of the growing cycle the carbohydrates, which are initially in high levels in the stems, have been transferred to the tubers. Topinambur can be placed in the group of plants producing the highest total dry matter yield per hectare (Dambroth, 1984).

Ecological requirements

Topinambur does not make special demands on the environment. Most soils have been shown to be suitable for topinambur cultivation, though the soil should be siftable so excessively clayey soils should be avoided. In addition, the soil should have a low number of stones to facilitate tuber harvesting (Hoffmann, 1993). The pH of the soil should be between 5.5 and 7.0 (Bramm and Bätz, 1988). The plant is very sensitive to frost, but the tubers can tolerate a ground temperature as low as −30°C.

Because of topinambur's efficient ability to hold water and nutrients, areas with low precipitation or periods of low precipitation can be endured by the crop and may even have little effect on the yield, though drought results in a halt in growth (Hoffmann, 1993).

Propagation

According to Hoffmann (1993), topinambur can be propagated using seed, tuber sprouts or rhizome pieces.

Source: DarKone 2004 wikipedia.org

Figure 10.235 Topinambur crop

Crop management

Topinambur planting begins in the middle of April using a potato planter to minimize field preparation. The rows should be 75cm apart and the plants should be 33cm apart in their row. The desired crop density is 38,000 to 40,000 plants to the hectare (Bramm and Bätz, 1988).

Correct fertilizer amounts are important for the success of the crop, but organic fertilizers should not be used during tuber formation. Table 10.115 shows suggested fertilizer levels, as given by Bramm and Bätz (1988).

The crop can be mechanically weeded, but usually topinambur's rapid growth and leaf formation quickly crowd out most weeds (Dambroth, 1984). Diseases that are most likely to attack topinambur crops include the stem rot *Sclerotinia sclerotiarum,* the sunflower rust *Puccinia helianthi* and the mildew *Erysiphe cichoracearum.* The crop is also susceptible to damage by rabbits, deer, wild boar and mice.

The crop planted before topinambur is not of importance in establishing a crop rotation, but appropriate crops to follow it are maize and summer wheat. Because topinambur is susceptible to *Sclerotinia*, a break in cultivation of 4 to 5 years is recommended (Bramm and Bätz, 1988). In addition, topinambur crops have a higher yield when planted as an annual crop than when planted as a perennial crop (Dambroth, 1984). It should be noted that because topinambur is a perennial plant, it may also reappear in subsequent crops.

Production

The lack of a market for energy crops means that it is difficult to determine their value, but there are estimates for topinambur, based on the selling price for the crop (in Italy) that produces a gross margin equal to that from a traditional crop. Taking into account labour

Table 10.115 Recommended fertilizer levels for topinambur

Fertilizer	Amount (kg/ha)
Nitrogen (N)	40–80
Potassium (K_2O)	240–300
Phosphorus (P_2O_5)	90–140
Magnesium (MgO)	60–90

Table 10.116 Characteristics of three topinambur varieties

Characteristic	'Bianka'	'Waldspindel'	'Medius'
Growth height	low	very high	very high
Ripeness	very early	middle	late
Tuber size	very large	large	middle
Dry matter (%)	19.27	26.59	24.33
Fructose (%)	10.67	17.90	16.24
Total sugar (%)	13.7 I	20.80	19.50
Tuber yield (t/ha)	49.667	61.417	62.333
Dry matter (t/ha)	9.57 I	16.33 I	15.166
Fructose (t/ha)	5.299	10.994	10.123
Ethanol (L/ha)	3969.85	7447.67	7086.33

costs, topinambur's variable costs, production and yield, marketable co-products and traditional crops' gross margin, the farmer would need to receive from €18 to €47 per tonne of tubers from a topinambur biomass crop to gain an income comparable to that from a traditional crop (Bartolelli et al, 1991).

The topinambur clone 'Violet de Rennes' has been shown to be very productive in Spain. Dry matter yields for tubers have been approximately 16t/ha/year (Fernandez et al, 1991).

Portugal has shown tuber yields of up to 30–43t/ha for some clones tested on both sandy and clay soils (Rosa et al, 1991).

Breeding research in The Netherlands using selected clones of 'Bianka', 'Yellow Perfect', 'Columbia' and 'Précoce' on light sandy soil produced in 1989 fresh matter tuber yields of 92–105t/ha, inulin percentages of 15–18 per cent, and inulin yields of approximately 16t/ha. The following year produced tuber yields of 50–57t/ha, inulin percentages of 13–15 per cent, and inulin yields of approximately 7t/ha. The reductions in 1990 are the result of an extensive period of low precipitation during the tuber-filling stage (van Soest et al, 1993). Table 10.116 shows research results from Germany (Schittenhelm, 1987).

Processing and utilization

Tuber harvesting can begin in November. The tubers' winter hardiness means that the harvest can be extended into the early part of the following year, which has the advantage of permitting a more continuous harvest based on demand. Topinambur harvesting techniques are quite similar to those of potato harvesting, so the tubers are most easily harvested using a potato harvester. Because of their poor storage ability – worse than that of potato – topinambur tubers should be processed within 14 days of being harvested.

Inulin is the product of greatest value obtained from the tubers. This D-fructose polymer can be used to produce ethanol or, through hydrolysis separation and isomerization, fructose and glucose (Delmas and Gaset, 1991). The green mass of the plant, which has a heating value of 17.6MJ/kg dry matter, can also be processed as paper pulp or used as fodder.

The tubers are crushed and the pulp pressed to obtain an extract. The inulin can then be obtained from the extract through physical, chemical or microbiological processes. Physically, tangential ultrafiltration can be used to recover a high sugar yield. Chemically, ethanol precipitation can be used to recover the sugars, but this produces a significantly lower sugar yield. Biologically, yeast fermentation can be used to obtain both sugar and ethanol (Fontana et al, 1992). Approximately 8–10L of ethanol can be produced from 100kg of tubers (Franke, 1985). Topinambur stalks also show potential for industrial and energy uses after processing. Using catalytic dehydration, 5-hydroxymethyl-2-furfural (HMF) can be obtained from fructans. HMF can be used to produce agrochemicals, detergents, pharmaceuticals, solvents, etc.

Unlike most tubers, but characteristic of members of Asteraceae (sunflower family to which it belongs), the tubers store the carbohydrate inulin instead of starch. The inulin is isolated on the basis of its high solubility in hot water; by boiling the tuber and allowing it to cool, polysaccharides can be extracted. Yields tend to vary with soil conditions, cultivar and season, but fresh weights in excess of 100t/ha have been recorded, which is around 8t/ha of sugar. For this reason, Jerusalem artichoke tubers are an important source of energy (PostCarbon Institute, 2008).

The dry matter yield (tuber, stems and leaves) can reach 30t/ha. Biogas production could reach 8120m³/ha from green matter, plus 2125m³ from the tubers (wikipedia.org/wiki/Topinambur).

Internet resources

http://de.wikipedia.org/wiki/Topinambur
www.topis.de

Source: Paul Fenwick, 19th March 2005, http://commons.wikimedia.
org/wiki/File:Sunroot_Flowers.jpg

Figure 10.236 Topinambur flowers

Source: Dr Hagen Graebner 2006 de.wikipedia.org

Figure 10.237 Topinambur tubers

WATER HYACINTH (*Eichhornia crassipes* (Mart.) Solms)

Description

The water hyacinth is a perennial aquatic herb. The rhizomes and stems are normally floating. The plant is native to Brazil, but now it is growing in most tropical and subtropical countries. Its habitat is estimated to range from tropical desert to rain forest, through subtropical or warm temperate desert to rain forest life zones. The leaves are killed by frost, and plants cannot tolerate water temperatures greater than 34°C. Often it grows as a weed. Rafts of water hyacinth can block channels and rivers. Water hyacinth has been widely introduced throughout North America, Asia, Australia and Africa. It can be found in large water areas such as Louisiana, or in the Kerala Backwaters in India.

Water hyacinth is a free-floating perennial aquatic plant native to tropical South America. With broad, thick, glossy, ovate leaves, water hyacinth may rise above the surface of the water as much as 1m in height. The leaves are 10–20 cm across, and float above the water surface. They have long, spongy and bulbous stalks. The feathery, freely hanging roots are purple-black. An erect stalk supports a single spike of 8–15 conspicuously attractive flowers, mostly lavender to pink in colour with six petals (Duke, 1984).

One of the fastest-growing plants known, water hyacinth reproduces primarily by way of runners or stolons, which eventually form daughter plants. It also produces large quantities of seeds, and these are viable for up to 30 years. The common water hyacinth (*Eichhornia crassipes*) is a vigorous grower known to double its population in two weeks. In many areas it is an important and pernicious invasive species. First introduced to North America in 1884, an estimated 50kg/m² of hyacinth once choked Florida's waterways, although the problem there has since been mitigated. When not controlled, water hyacinth will cover lakes and ponds entirely; this dramatically impacts water flow and blocks sunlight from reaching native aquatic plants (wikipedia.org).

Production

Harvesting appears to be more critical than cultivation. Rafts of floating plants may be harvested with specially equipped dredgers and rakes. A floating mat of medium-sized plants may contain 2 million plants per hectare and weigh 270–400t (wet). The annual productivity is estimated from 15–30t/ha to 88t/ha. In Florida, compared with other plant species representing many life forms, the water hyacinth is very productive. Once harvested and dried, the dry matter of the water hyacinth is roughly equivalent to those of other species in terms of energy (Table 10.117).

Source: Katherine Parys, Louisiana State University, Bugwood.org, www.invasive.org/browse/detail.cfm?imgnum=5212026

Figure 10.238 Common water hyacinth in flower

Source: Valerius Tygart, http://en.wikipedia.org/wiki/Water_Hyacinth

Figure 10.239 Water hyacinth-choked lakeshore at Ndere Island, Lake Victoria, Kenya

Table 10.117 Biomass production of various plant species in Florida

Plant	Dry matter (t/ha/year)
Azolla	10
Beta	4.4–11.7
Brassica	3–10
Casuarina equisetifolia	8.3
Cichorium intybus	5.5–7.9
Colocasi esculenta	9–19
Cynodon dactylon	23.5–24.6
Daucus carota	2.5–5.5
Eichhornia crassipes	30–88
Elodea	3
Eucalyptus	5.6–20
Helianthus tuberosus	2.2–9.5
Hydrilia	15
Hydrocotyle umbellata	20–58
Ipomoea batatus	7–23
Lemna	12
Melaleuca quinquenervia	28.5
Paspalum notatum	22.4
Pennisetum spp.	57.3
Pinus clausa	9.0
Pinus elliottii	9.4
Saccharum	32–54
Sorghum	16–37
Sorghum 'Sordan'	22.4
Typha spp.	20–40

Processing and utilization

There are several ways to use the weed for energy production. Bengali farmers dry water hyacinths for fuel. Estimations of biogas production vary: 1kg of dry matter yields 0.2–0.37m^3 biogas. In India 1t of dried water hyacinth yields about 50L of ethanol and 200kg of residual fibre. Bacterial fermentation of 1t yields about 740m^3 of gas with 51.6 per cent methane, 25.4 per cent hydrogen, 22.1 per cent CO_2 and 1.2 per cent oxygen. Gasification of 1t by air and steam at high temperatures (800°C) gives about 1100m^3 gas containing 16.6 per cent hydrogen, 4.8 per cent methane, 21.7 per cent CO, 4.1 per cent CO_2 and 52.8 per cent N_2.

Brankas water hyacinth is made from the water hyacinth plant. The usable material is separated from the flower of the plant and dried in natural sunlight until the colour is light brown or yellow. This material is delicate and needs to be treated and maintained. The advantage is a longer life due in part to the care process and the fact that there is no twisting or braiding.

Flat water hyacinth is similar to the Brankas style. The main difference is what happens after the drying process. Once the material is dried in natural sunlight it is twisted and braided for added durability. It is very good for living room and dining room furniture. The perfect colour is dark brown with a black wash or brown wash (rattanland.com).

Since the plant has abundant nitrogen content, it can be used a substrate for biogas production and the sludge obtained from the biogas.

Biogas production from manure and from sewage and agricultural waste streams has been studied well. For water hyacinth, the literature is scarce, but some exists and that suggests the weed makes for a good biogas feedstock.

(a)

(b)

(c)

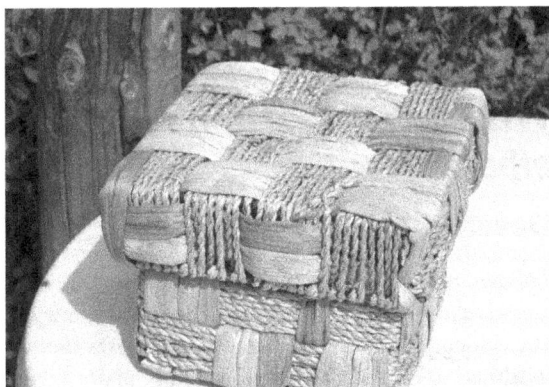

Source: rattanland.com

Figure 10.240 (a) Brankas water hyacinth; (b) water hyacinth shopping bag; and (c) water hyacinth box

Some conclusions:

- The total amount of gas produced from water hyacinth is about one-and- a-half times higher than cow dung per gram of volatile solid.
- A blend of water hyacinth and cow dung in the ratio of 2:3 by weight is most suitable for biogas production.
- Addition of very small amount of lower volatile fatty acid, particularly acetic acid, facilitates the gas production. This finding is very helpful for projects at the village level, where farmers often use biogas plants. In many villages all over the tropics, farmers produce sugar (from many different plants such as palm sugar, coconut, sugar cane or other local plants). If the residue from the process of making sugar juice is kept for fermentation for a few days the content will be highly rich in acetic acid. The addition of this residue would circumvent the problem of lower gas production during colder nights, and biogas plants could run successfully during all the seasons.
- The rate of production of biogas from water hyacinth is higher than from cow dung slurry. However, the fermentation process takes a longer time in the case of water hyacinth. Kinetic studies performed with water hyacinth and inoculum show that the gas production rate increases 12 times in a very short period of five days in comparison to cow dung and water hyacinth systems (20–40 days).
- The digested slurry can be used as a useful chemical-free eco-friendly fertilizer.

The Kottappuram Integrated Development Society (KIDS) of India which promotes 'bottom-up' environmental and bioenergy services for thousands of poor people, is taking on its most challenging project to date: using the hyacinth as a major resource to enhance access to energy, drinking water and mobility. With the support of the India-Canada Environment Facility (ICEF), KIDS will implement the Water Hyacinth Project through 'self-help groups'. The project will provide livelihoods for poor rural women. It is expected to become a model for eco-restoration of water bodies infested with water hyacinth (Rademakers, 2006).

Fermentation modes and microorganisms related to two typical free-floating aquatic plants, water hyacinth and

Figure 10.241 Harvesting water hyacinth

water lettuce, were investigated for their use in ethanol production. Except for arabinose, sugar contents in water lettuce resembled those in water hyacinth leaves. Water lettuce had slightly higher starch contents and lower contents of cellulose and hemicellulose. A traditional strain, *Saccharomyces cerevisiae* NBRC 2346, produced 14.4 and 14.9g/L ethanol, respectively, from water hyacinth and water lettuce. Moreover, a recombinant strain, *Escherichia coli* KO11, produced 16.9 and 16.2g/L ethanol in the simultaneous saccharification and fermentation mode (SSF), which was more effective than the separated hydrolysis and fermentation mode (SHF). The ethanol yield per unit biomass was comparable to those reported for other agricultural biomasses: 0.14–0.17g/g dry matter for water hyacinth and 0.15–0.16g/g dry matter for water lettuce. (Mishima et al, 2008)

The huge biomass potential of water hyacinth, up to 500t/ha, should lead to an intensive consideration of its use for producing biofuels such as biodiesel and biogas.

Internet resources

http://news.mongabay.com/bioenergy/2006/06/turning-pest-into-profit-bioenergy.html
http://en.wikipedia.org/wiki/Water_hyacinth
www.rattanland.com

WHITE FOAM (*Limnanthes alba* Hartw.)

Description

Limnanthes alba is an annual herb. The stems are erect or ascending and 10–30cm tall. The leaves are up to 10cm long. The plant is native to the Sierra Nevada foothills and the adjacent rolling plains from Sacramento to Chico, California, USA. The seeds contain 25–30 per cent oil with 1.56 per cent volatile isocyanates, and 20 per cent protein. The high concentration of C-20 fatty acids is unique. More than

Source: Marsha's Flat Road, Tuolumne County, CA
www.rangenet.org/directory/christie/plants/limnanthaceae/index.htm

Figure 10.242 *Limnanthes alba* Hartw.

90 per cent of the fatty acids have a chain length greater than C-18. Table 10.118 shows various characteristics of white foam seeds (Duke, 1983).

Ecological requirements

The range extends across warm temperate moist through subtropical dry to moist forest life zones. White foam is reported to tolerate slopes and waterlogging. The plant is also reported to tolerated an annual precipitation of 700–1100mm and an annual temperature of 12–19°C (Duke, 1983). The killing temperature is 0°C. The pH value should be 6–6.5 (FAO, 1996a).

White foam is commonly found on banks and the gravelly bars of small intermittent streams in the Sierra

Table 10.118 Characteristics of white foam seeds

Seed characteristic	Content
Oil (%)	17–29
20:1 (%)	50–65
22:1 (%)	10–29
22:2 (%)	15–30
Protein (%)	11–28
Glucosinolates (%)	3–10
Seed weight (g/1000)	4.2–9.8

Source: Kleiman (1990)

Nevada foothills. It grows on porous, quick-drying soils. Essentially, it is a xerophyte, flowering and setting seed on the last seasonal soil and stem moisture. It has about the same water requirement as dry-farmed winter grains, and seems to require less moisture than other species of this genus.

No pests or diseases have been reported for this species.

Propagation

Limnanthes alba is propagated by seed. Normally, it is sown in the autumn and harvested in early summer (FAO, 1996a). In experiments, seeds germinated at 4.5°C, gave poor germination at 21°C, and went dormant at 26.5°C. Growth and seedling periods closely match those of oats and barley. Good weed control is essential for good yields (achieved using propachlor and diclofog). As small streams dry in late spring, growth is terminated and plants rapidly mature their seeds. Seed dates range from 5 to 30 May. Seed shattering may cause 16–54 per cent of seed loss in *Limnanthes alba*, and 71–93 per cent in *Limnanthes douglasii*. With improved cultivars and earlier harvests a seed recovery greater than 95 per cent has been achieved with direct combine harvesting or windrowing and then combining.

Production

Experimental plantings with a seed rate of 2.9kg/ha yielded 1650kg/ha; 24.7kg/ha yielded about 2000kg/ha. With respect to the two seed rates, applying 48–50kg/ha N following a non-legume crop in early March gave seed yields of 790kg/ha and 700kg/ha respectively; without any fertilizer, yields of 530kg/ha were obtained. Yields of 400kg/ha are reported for *Limnanthes bakeri* and 900kg/ha for *Limnanthes alba*, with 1900kg/ha for *Limnanthes douglasii*, the most promising species of this genus.

Researchers hope to increase the yields, but for now, oil yields of 500kg/ha are difficult to obtain. Oregon could already produce 1100kg/ha seed, yielding 275kg oil. At Cornwallis, seed yields have ranged from 900–1800kg/ha. White foam is grown as a winter annual, principally in the Willamette valley, Oregon, where yields over 1.1t/ha are routine.

Figure 10.243 White flowers are *Limnanthes alba*, Lifornia Richvale Vernal Pool, Butte County, California

Processing and utilization

The oil composition of white foam is unique in several ways. Over 95 per cent of the fatty acids are longer than C-18, and about 90 per cent of these fatty acids have double bonds in delta-5 position. The 22:25 fatty acids react essentially like a monoenoic fatty acid in terms of oxidative stability. The oil should be oxidatively stable because of the absence of polyunsaturated fatty acids and the long-chain nature, and because the delta-5 double bond is more stable than olefins with the double bond in the centre of the molecule. The oil can be used as a fuel source and the stems offer various possibilities as solid biofuels for combustion, gasification or pyrolysis.

An oil obtained from the seed has similar properties to whale sperm oil and to jojoba (*Simmondsia chinensis*). It has specialized industrial applications. The seed contains about 20 per cent protein, 25–30 per cent oil, with 1.56 per cent volatile isothiocyanates. The high concentration of C20 and fatty acids in the seed oil is unique. No other seed oil is known to have as high concentration (>90 per cent) of total fatty acids of chain length greater than C18 (pfaf.org).

Internet resources

www.pfaf.org/database/plants.php?Limnanthes+alba

WILLOW (*Salix* spp.)

Description

Willow is grown mainly in continental climate zones including Europe, western Siberia and central Asia, as well as the Caucasus, northern Persia and Asia Minor (Schutt et al, 1994), but it is most successful in northern Europe.

(a)

(b)

Figure 10.244 (a) Weeping willow *Salix sepulcralis*; and (b) white willow *Salix alba*

Willows, sallows and osiers form the genus *Salix*, around 400 species of deciduous trees and shrubs, found primarily on moist soils in cold and temperate regions of the northern hemisphere. Most species are known as willow, but some narrow-leaved shrub species are called osier, and some broader-leaved species are called sallow (the latter name is derived from the Latin word salix, willow). Some willows (particularly arctic and alpine species), are low-growing or creeping shrubs; for example the dwarf willow (*Salix herbacea*) rarely exceeds 6cm in height, though spreading widely across the ground. Willows are very cross-fertile, and

numerous hybrids occur, both naturally and in cultivation. A well-known example is the weeping willow (*Salix × sepulcralis*), a widely planted ornamental tree, which is a hybrid of a Chinese species, the Peking willow, and a European species, the white willow (wikipedia.org/wiki/Willow).

Some deposits of willow plantations can be found at altitudes as high as 2000m in mountainous regions, but willow is a typical plant of plains and hilly areas. Most *Salix* spp. are compact shrubs with numerous thick branches and grow to a height between 0.3 and 4m. Others are true trees. *S. fragilis* is able to reach a height of 15m and *S. alba* a height of 20m (Schmeil and Fitschen, 1982). Some species of *Salix* produce leaves before blossoms in spring (Schütt et al, 1994). Normally, the shrubs develop an extensive root system. In Sweden, willow is especially used as a short rotation coppice (SRC) crop. Willow species such as *Salix viminalis* (L.) and *Salix dasyclados* (Wimm.) are cultivated on biomass plantations for use as energy sources.

Ecological requirements

Willow can be grown on a wide range of soil types, on light soils as well as on loamy soils, with a pH range from 6.0 to 7.5 and an optimum pH of 6.5. *Salix* can tolerate salinity up to 4dS/m (FAO, 1996a). *Salix* species are more dependent on water than other agricultural crops so dry locations should be avoided (Johansson et al, 1992). Willow's water consumption can reach 4.8mm/m^2 per day in June and July. More than 500L water is needed to produce 1kg of dry matter (Pohjonen, 1980). Irrigation is necessary if the yearly rainfall is less than 600L/m^2 (Parfitt and Royle, 1996). Most of the types selected for SRC could not produce economically viable yields on very dry or alkaline soils. To a certain degree they are tolerant of waterlogging. The optimal growing temperature of *Salix* spp. ranges between 15°C and 26°C, with a minimum growing temperature for the different *Salix* species of between 5°C and 10°C and a maximum growing temperature between 30°C and 40°C. Some species are frost tolerant, but not below –30°C.

Propagation

Willow cuttings from one-year-old shoots can be planted in quite coarsely prepared soil. For ease of planting, successful rooting and subsequent management, the site should be subsoiled if necessary, deep ploughed (25–30cm) in autumn and power harrowed to produce a level and uncompacted tilth (Parfitt and Royle, 1996). Planting is usually delayed until the spring, when about 18,000 cuttings 20cm long and more than 0.8cm thick are planted per hectare, after storage at –4°C during the winter and subsequent watering for a couple of days (Johansson et al, 1992).

The *Salix* cuttings are placed in double rows with a row spacing of 75cm, and 125cm between the double rows; the plant spacing in the row is 55cm (Johansson et al, 1992). In the first year only a few shoots on each cutting will grow, which will be cut back to the ground level in the first winter and can be used for making cuttings to establish new plantations. The regrowth of cut-back SRC-*Salix* is vigorous and dense, reaching up to 4m high in the next vegetation period.

Several laboratories have developed micropropagation techniques, and this is now a satisfactory propagation method.

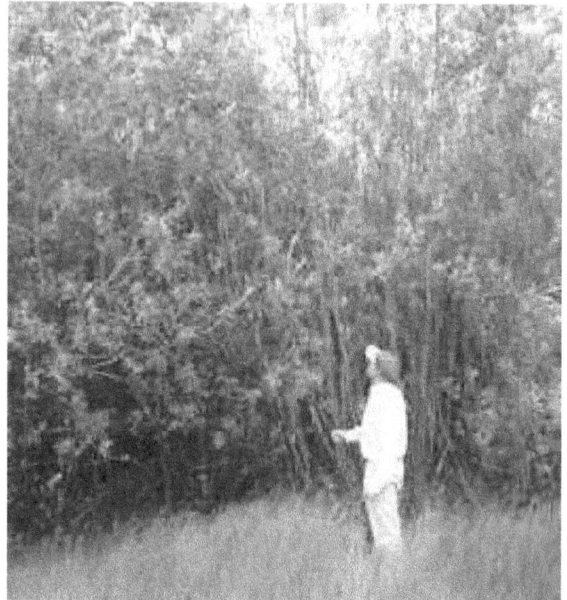

Source: www.agroforestry.ac.uk/images/willow_2002.jpg

Figure 10.245 Traditional willow coppice

Crop management

During the establishment phase, weeds are the worst enemy of willow. Weed control is necessary during the first year of establishment and the fields must be kept free from perennial weed rhizomes before planting. After the first year of establishing, willows are able to suppress weeds.

An average application of 60–80kg N, 10kg P and 35kg K is suitable from the second year onwards (Johansson et al, 1992), because in the planting year (first year) the application of fertilizers could enhance weed growth. The fertilizer regime should be adapted to the soil type and natural mineralization, and to the nitrogen entry and deposition via rainwater. In relation to its mass production, willow can be considered a low-input plant (El Bassam, 1996).

Willow may be attacked by a wide range of leaf-consuming, stem-sucking and wood-boring insects (Hunter et al, 1988). None of these insects has so far been more than an occasional problem. A *Salix* plantation provides a good habitat for many different insects, birds and animals. Diseases are more serious, and numerous pathogens are described, especially different species of rust. Different clones have varying

susceptibilities to fungi and insects, and can be selected for better resistance. Generally, there is little reason to introduce inputs for insects, and disease control. Considerable damage may be caused by mice and deer during the establishment phase of a *Salix* plantation.

Production

Willow can produce its first yield three years after planting. Current expectations are that there are eight or more cycles of harvests during the *Salix* plantation life period of three years, each as SRC (Figure 10.246). Estimated annual yields per hectare, in oven dried tonnes (ODT), are 5 (Ireland), 8–10 (Sweden), 8–20 (UK) and 15–20 (Italy). Increased yields are expected if the full benefits from improved breeding materials have been obtained and when agricultural practices have been optimized.

Processing and utilization

Willow is grown for biomass or biofuel, in energy forestry systems, as a consequence of its high energy in–energy out ratio, large carbon mitigation potential

Source: Abrahamson et al (2008)

Figure 10.246 New CNH short rotation coppice header being tested in the UK in March

and fast growth. Large-scale projects to support willow's development as an energy crop have been established, such as the Willow Biomass Project in the US and the Energy Coppice Project in the UK. Willow is also grown for charcoal.

Harvesting of stems take places in winter when the leaves have fallen off. There are two different approaches to harvesting: direct chipping and whole shoot harvesting. Direct chipping is practicable if chips can be combusted at the harvest moisture content (50 per cent) in large heating plants. Chips with a high moisture content will quickly deteriorate through microbial activity, so must be burned immediately or ventilated for storage. Drier chips are needed in smaller plants and stationary ovens on farms. In this case shoots are harvested whole and dried in piles during summer. This material reaches a dry matter content of about 70 per cent.

Dry wood chips for energy production need no further processing and can be burned or gasified (Parfitt and Royle, 1996). Willow chips can be utilized in two ways. Direct combustion in special automated boilers will give low-grade space heating of 50 to 400kWh. High-pressure steam techniques can be used to drive steam turbines of up to 10MW installed capacity (Graef, 1997). Another possibility is the gasification (pyrolysis) of wood chips to drive an engine to power a generator and also produce heat, a system

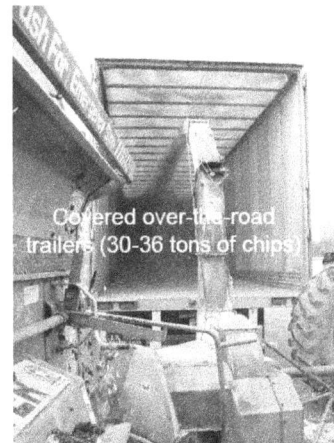

Source: Abrahamson et al (2008)

Figure 10.247 Getting the willow chips to the end user

known as combined heat and power (CHP). A growth rate of 150DT/ha/year is equivalent to 7000L of oil.

On the Danish island of Funen, a group of local players headed by the development council plan to establish a biogas plant which uses slurry and other organic waste from farming, as well as growing willow on 500ha of land for production of biomass that will be used as an energy source at the local Assens district heating plant. Growing willow for energy is an alternative to woodchip production. The energy from one hectare of willow corresponds to the energy from 5000 litres of oil. The 500ha will cover 20 per cent of district needs. There is a market for 100,000 hectares of energy willow for Danish district heating plants alone. Production of electricity and heat at Assens district heating plant is currently generated from woodchips, most of which is imported. Woodchip prices are, however, constantly rising as biofuel is increasingly used for energy production. The 500ha energy willow plot will not replace cornfields since corn and willow require two different types of soil, an important factor at a time when the connection between production of bioenergy and the global shortage of food is being discussed (*Børsen*, 2008).

Willow has been cultivated by Swedish farmers since the 1970s, and there are now over 16,000ha in production. The predominant variety is salix, which grows to 5–7m in height and has a productive lifespan of 25–30 years. The willow is processed into woodchips which are then used as a fuel in CHP and district heating plants (including CHP). The crop now contributes around 1 per cent of Sweden's fuel requirements. Plantations expanded rapidly in the 1990s with the availability of subsidies and the support of the Federation of Swedish Farmers Co-operatives.

Despite a short period of decline when EU set-aside rules were changed, confidence in the crop is growing. Whilst most of the woodchip produced is used by large municipal CHP stations, farmers have also established cooperative enterprises to run community heating plants. There are 15 farmer-owned heating plants, and around 40 larger plants in which farmers have a stake, alongside farmer support bodies and service cooperatives (Agrobränsle AB, 2004).

E.ON has stepped up its push into the biomass energy space, announcing that it has submitted a planning application to build a new £60 million biomass power station in Sheffield. The new facility is designed to provide power for around 40,000 homes through the burning of recycled wood, forestry waste and specially grown crops such as willow and elephant grass. E.ON said that by burning biomass as opposed to traditional fossil fuels the plant would save around 80,000 tonnes of carbon dioxide a year from being emitted. Should the project receive planning approval E.ON start construction in 2009 and deliver its first energy from the plant in 2011. The move is the latest in a line of biomass energy investments from E.ON, which in 2008 opened the UK's largest wood-fired power station near Lokerbie in Scotland. The £90 million plant is expected to provide enough power to meet the demands of around 70,000 homes (Murray, 2008).

Internet resources

www.cinram.umn.edu/srwc/docs/Powerpoints/L.Abrah
amson_New%20Holland%20forage%20harvester.pdf
www.cooperatives-uk.coop/live/images/cme_resources/
Users/Nick/Farmar-Energi.pdf
http://en.wikipedia.org/wiki/Willow

Part III

11
Ethanol Crops

Grains from cereals should continue to serve as a major source for food and feed. Cereal straw combined with rice straw and maize husks constitutes one of the principal biomass sources arising from present agricultural activities in Europe. This represents a large potential resource for utilization as energy feedstocks. Although there are some plans to use cereals as an energy source, the main justifications are overproduction of cereals and advanced technology in seed production, tillage, sowing, harvesting, baling and storage. Cereal crops that are whole crop harvested may be used under certain circumstances as energy crops.

In recent decades, plant-breeding activities have made good progress in increasing grain yields and improving the genetic make-up of crops to increase nutrient utilization efficiency and to improve their adaptability for wider environmental conditions. This has been achieved by improving the harvest index (the ratio of grain to straw) to almost 0.5 or even higher. The total biomass has been influenced very little by breeding activities. Cereals as energy crops on set-aside areas might be grown according to the following guidelines:

- Suitable genotypes will have a low harvest index – that is, they should produce considerably more straw than grain. Landraces and wild types are most suitable.
- The nitrogen fertilization rate could be 50 per cent of the normal rate for grain cereals because most of the nitrogen needed by cereals is for grain production and to improve the protein content of the grain. This is not needed in whole-crop harvesting for energy feedstocks.
- The whole-crop-harvesting procedure will involve cutting and collecting the unfractionated crop from the field.

- Harvesting should take place earlier than for grain cereals.

This system will prevent conflicts with the production of food cereal grains. Breeders should try to produce cultivars with a higher straw proportion as energy cereal crops. This would minimize the inputs and decrease environmental damage. There are already attempts to breed cereals, especially barley, with higher enzyme contents that enable efficient fermentation and ethanol production from the grain.

Worldwide average cereal production rates (t/ha of grain) in 1994 were: barley 2.18, maize 4.33, oats 1.70, rye 2.05, sorghum 1.39 and wheat 2.22 (FAO, 1995a). The average total biomass to be expected by following the above (environmentally consistent) guidelines might be calculated as follows: wheat 10t/ha, barley 6t/ha, rye 6t/ha, oats 4t/ha, triticale 12t/ha, sorghum 12t/ha and maize 12t/ha.

Significantly less total biomass should be expected from non-European Mediterranean countries. From the ecological point of view, rye and triticale possess some advantages because these two types can be considered as relatively low-input genotypes. No attempt has been made here to describe any of these cereal crops – they are not new crops, and breeding and cultivation procedures are already well established.

Depending on the potential of specific varieties of wheat and triticale, a considerable amount of ethanol can be produced.

Cereal grain yield and biomass production are affected by fertilizer application strategies. In order to quantify the performance of wheat, rye and triticale cultivars for use as energy crops, field experiments with either modified phosphorus–potassium or potassium applications were designed at two locations in Denmark over a 3-year period.

Table 11.1 Different crops in their production of ethanol per tonne or per acre

Estimated ethanol yield per tonne of cereal grain	
Winter wheat	392L/tonne
CPS & SWS wheat	382L/tonne
Triticale	382L/tonne
Durum	377L/tonne
CWRS & rye	364L/tonne
Corn – ethanol	400L/tonne
Barley (hull-less)	380L/tonne
Oats (hull-less)	353L/tonne
Oats (regular)	317L/tonne
Estimate ethanol yield per acre from survey of literature	
Sugar beets	2700L/acre
Sugar cane	2500L/acre
Corn	1500L/acre
Winter wheat	800L/acre
Wheat, barley	600L/acre
One bushel wheat = 10 litres ethanol	

Source: This information was mostly compiled by Solulski and Tarasoff in 1997 and was gleaned off the Alberta Ag Website (Day, 2008).

Table 11.2 Crop yields in t/ha in Belgium, 1997

Crop species	Harvested product	Crop yield (t/ha) National mean
Maize grain – wet – dried	Grain	10.14
	Grain	7.61
Spring wheat	Grain	6.57
	Straw	4.06
Winter wheat	Grain	9.13
	Straw	4.68
Spelt	Grain	6.25
	Straw	4.84
Rye	Grain	5.28
	Straw	4.31
Spring barley	Grain	4.82
	Straw	3.10
Winter barley	Grain	7.77
	Straw	4.27
Oats	Grain	4.98
	Straw	3.69
Triticale	Grain	7.53
	Straw	5.30
Potatoes for plant	Tuber	21.24
Early		30.07
Bintje variety		42.72
Other varieties		40.11
Sugar beet	Roots	54.53
Chicory for inulin	Roots	34.13
Linseed	Grain	1.66
Flax	Straw	6.50
Spring oilseed rape	Grain	1.62
Winter	Grain	3.67
Dry peas	Grain	5.10

Source: Report from the state of Belgium forming part of the Ienica project, from Institut National de Statistique, Recensement Agricole 1997, Belgium.

Five wheat cultivars ('Astron', 'Herzog', 'Kosack', 'Kraka' and 'Ure'), two winter rye cultivars (the population cultivar 'Motto' and the hybrid cultivar 'Marder') and the triticale cultivar 'Alamo' were selected. The grain and straw fractions were analysed for biomass, ash and contents of nitrogen (N), K, Cl, sulphur (S) and Na. Dry matter yields varied between 11.5 and 15.9t/ha at the two locations. Triticale and rye had a higher total dry matter yield than wheat, even at lower inputs of N fertilizer. Thus, the constant high yield of rye and triticale is an advantage for biomass for energy purposes. The mineral content of the grain fraction changed only a little between years and locations. By contrast, large variations in the analysed ions in the straw fraction between years and locations were observed. The use of K fertilizers resulted in a significantly increased concentration of K in the straw. However, this increased concentration was eliminated in years with high precipitation in the final 3 weeks before harvest, where substantial amounts of K, Cl and S were removed. (Jørgensen et al, 2007)

Triticale (*Triticosecale*) is a crop species resulting from a plant breeder's cross between a wheat (*Triticum* sp.) 'mother' and rye (*Secale* sp.) 'father'. Today, it has been cultivated in more than 50 countries worldwide.

Rye, triticale, and barley were evaluated as starch feedstock to replace wheat for ethanol production. Preprocessing of grain by abrasion on a Satake mill reduced fiber and increased starch concentrations in feedstock for fermentations. Higher concentrations of starch in flours from preprocessed cereal grains would increase plant throughput by 8–23 per cent since more starch is processed in the same weight of feedstock. Increased concentrations of starch for fermentation resulted in higher concentrations of ethanol in beer. Energy requirements to produce one litre of ethanol from preprocessed grains were reduced, the natural gas by 3.5–11.4 per cent, whereas power consumption was reduced by 5.2–15.6 per cent. (Sosulski et al, 1997)

In experiments carried out by Vuåuroviä and Pejin (2007) on triticale fermentation, the results obtained for the content of fermentable starch and the ethanol yield showed that native amylolytic enzymes of triticale can degrade 80–90 per cent of the available starch. The addition of glucoamylase, during the second preparation mode, increased the content of fermentable starch and ethanol yield. The best results were achieved applying the third mode of preparation. Comparing the preparation modes, it could be concluded that the application of both α-amylase and glucoamylase in the preparation step, increased the content of fermentable starch and ethanol yield by 7–13 per cent.

Ethanol yields of triticale samples prepared under different temperature regimes were between 29.76 and 39.30 (g ethanol/100g sample).

Barley (*Hordeum vulgare*)

Barley (*Hordeum vulgare*) is an annual cereal grain, which serves as a major animal feed crop, with smaller amounts used for malting and in health food. It is a member of the grass family, Poaceae. In 2005, barley ranked fourth in quantity produced and in area of cultivation of cereal crops in the world (560,000km², or an area larger than continental France). The domesticated form (*H. vulgare*) is descended from wild barley (*H. spontaneum*). Both forms are diploid (2n = 14 chromosomes) and are inter-fertile. The two forms are therefore often treated as one species, *Hordeum vulgare*, divided into subspecies *spontaneum* (wild) and subspecies *vulgare* (domesticated). The main difference between the two forms is the brittle rachis of the former, which enables seed dispersal in the wild.

Barley is more tolerant of soil salinity than wheat, which might explain the increase of barley cultivation on Mesopotamia from the second millennium BC onwards. Barley is not as cold tolerant as the winter wheats (*Triticum aestivum*), fall rye (*Secale cereale*) or winter triticale (× *Tricticale* Witt.), but may be sown as a winter crop in warmer areas of the world such as Australia.

Barley was grown in about 100 countries worldwide in 2005. The world production in 1974 was 148,818,870 tonnes, showing little change in the amount of barley produced worldwide.

Half of the United States' barley production is used as animal feed. A large part of the remainder is used for malting and is a key ingredient in beer and whisky

Table 11.3 Top ten barley producers in 2005 (million tonnes)

	Russia	16.7
	Canada	12.1
	Germany	11.7
	France	10.4
	Ukraine	9.3
	Turkey	9.0
	Australia	6.6
	United Kingdom	5.5
	United States	4.6
	Spain	4.4
World Total		**138**

Source: FAO

production. Two-row barley is traditionally used in German and English beers, and six-row barley was traditionally used in US beers. Both varieties are in common usage in the US now. Non-alcoholic drinks

Source: http://upload.wikimedia.org/wikipedia/commons/9/9a/Beer_wuerzburger_hofbraue.jpg

Figure 11.1 Beer

such as barley water and mugicha (popular in Korea and Japan) are also made from unhulled barley. Barley is also used in soups and stews, particularly in Eastern Europe. A small amount is used in health foods and coffee substitutes.

Researchers at the US Department of Agriculture's Eastern Regional Research Center (ERRC) are working to develop specialized variants of barley that could be used in ethanol production (greencarcongress.com).

In the US, most ethanol is corn-based – and hence most production facilities are located in the Corn Belt, not on either of the coasts. Barley, the reasoning goes, grows well in areas where corn does not, and so might become a financially cost-effective ethanol feedstock, once a number of severely limiting problems are solved.

Barley has an abrasive hull that causes expensive wear and tear on grain handling and milling equipment. Furthermore, it has a much lower starch content than corn; 50–55 per cent compared to corn's 72 per cent. As starch is the starting point of the production process, the production yield would be much lower.

The beta-glucan polysaccharide in barley also makes the mash too viscous to mix, ferment and distil economically. Accordingly, the Agricultural Research Service (ARS) researchers are working to develop:

- high-starch barley;
- hull-less barley varieties suitable for ethanol production;
- new milling processes that separate barley kernels into a starch-enriched fraction for ethanol production and a protein- and fibre-enriched fraction for food and feed co-products; and
- a process that uses beta-glucanase enzymes in fermentation to dramatically break down high-viscosity beta-glucans into low-viscosity oligosaccharides, solving the beta-glucan problem (greencarcongress.com).

Osage BioEnergy announced that it will break ground 2009 on its Appomattox Bio Energy plant, a barley ethanol plant that will be the largest in the US, using barley as a feedstock and producing 65 million gallons of ethanol per year. Barley produces a high-quality meal in addition to fuel ethanol. The Appomattox Bio Energy (ABE) facility is projected to use regionally grown barley as the primary raw material. Barley is a

Source: Hicks et al (2005)

Figure 11.2 Doyce barley field, 2005; first results indicate that Doyce barley produced higher ethanol yields (2.27 versus 1.64 gal/bushel)

moderate to high-yield winter crop and can be grown in double crop systems with other food crops such as soya beans (Shake, 2008).

Maize (Corn) (*Zea mays* L.)

(*Zea mays* ssp. *mays*), known as corn in some countries, is a cereal grain domesticated in Mesoamerica and subsequently spread throughout the American continents. After European contact with the Americas

Source: www.precisiongps.com/images/Manure%20Pics/corn%20field.jpg

Figure 11.3 Corn field

in the late fifteenth and early sixteenth century, maize spread to the rest of the world (wikipedia.org/Maize).

Maize is the most widely grown crop in the Americas (332 million tonnes annually in the United States alone). Hybrid maize, due to its high grain yield as a result of heterosis ('hybrid vigour'), is preferred by farmers over conventional varieties. While some maize varieties grow up to 7m tall, most commercially grown maize has been bred for a standardized height of 2.5m. Sweetcorn is usually shorter than field-corn varieties.

Maize is widely cultivated throughout the world, and a greater weight of maize is produced each year than any other grain. While the United States produces almost half of the world's harvest (~42.5 per cent), other top producing countries include China, Brazil, Mexico, Argentina, India and France. Worldwide production was around 800 million tonnes in 2007 – just slightly more than rice (~650 million tonnes) or wheat (~600 million tonnes). In 2007, over 150 million hectares of maize were planted worldwide, with a yield of 4970.9kg/ha (wikipedia.org/Maize).

Blue corn is another grain growing in demand for food-grade milling. It is a speciality dried corn that is used mainly for making tortillas and chips. Blue corn or maize is an open-pollinated flour corn and contains soft starch useful in the milling of speciality foods (Johnson and Jha, 1993).

Corn (*Zea mays* L.), the single largest US crop, is increasingly being used as a biomass fuel. It is currently harvested from 30 million hectares within the United States, which is almost a quarter of all the harvested cropland in the country. The average yield of moist corn grain is 8600kg/ha. According to the National Corn Growers Association, 1.3 billion bushels of corn were allocated to ethanol production in 2004. Corn residues, including the stalk and cob, may also prove useful in future energy production.

Maize is harvested at immature stages in October in Europe for whole plant (yields around 15t/ha dry matter) or cob silage. About 5000 to 15,000ha drilled in early varieties are harvested as grain (at about 70 per cent dry matter) which needs to be dried to 85 per cent DM and with yields of about 8–12t DM/ha. Seed rate aims to install about 100,000 plants per hectare, weed control is based upon treatments with triazines (rates limited for environmental reasons) often associated to

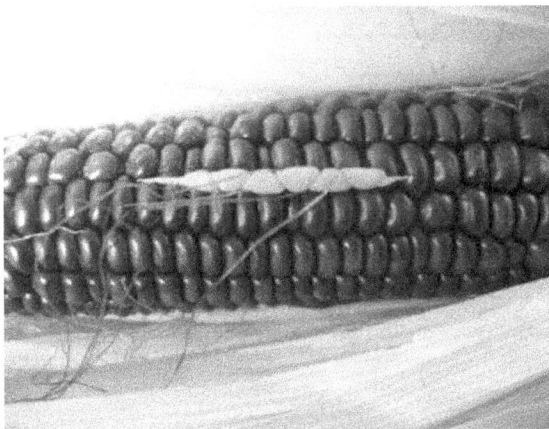

Source: Wayne Campbell, hila.webcentre.ca

Figure 11.4 Blue corn

Table 11.4 Top ten maize producers in 2007

Country	Production (Tonnes)
United States	332,092,180
People's Republic of China	151,970,000
Brazil	51,589,721
Mexico	22,500,000
Argentina	21,755,364
India	16,780,000
France	13,107,000
Indonesia	12,381,561
Canada	10,554,500
Italy	9,891,362
World	784,786,580

Source: FAO

other herbicides. Nitrogen needs of about 250kg N are partly provided by organic fertilizers (animal manures) and partly by mineral fertlizers.

Maize is increasingly used as a biomass fuel, such as ethanol, which as researchers search for innovative ways to reduce fuel costs, has unintentionally caused a rapid rise in food costs. This has led to the 2007 harvest being one of the most profitable corn crops in modern history for farmers. Maize is widely used in Germany as a feedstock for biogas plants. Here the maize is harvested, shredded, then placed in silage clamps from which it is fed into the biogas plants.

A biomass gasification power plant in Strem near Güssing, Burgenland, Austria was begun in 2005. Research is being done to make diesel out of the biogas by the Fischer-Tropsch method. Increasingly ethanol is being used at low concentrations (10 per cent or less) as an additive in gasoline (gasohol) for motor fuels to increase the octane rating, lower pollutants, and reduce petroleum use (what is nowadays also known as

'biofuels') and has been generating an intense debate regarding the necessity for new sources of energy, on the one hand, and the need to maintain, in regions such as Latin America, the food habits and culture which have been the essence of civilizations such as the one originated in Mesoamerica. The entry, in January 2008, of maize among the commercial agreements of NAFTA has increased this debate, considering the bad labour conditions of workers in the fields, and mainly the fact that NAFTA 'opened the doors to the import of corn from the United States, where the farmers who grow it receive multi-million dollar subsidies and other government supports'. According to OXFAM UK, after NAFTA went into effect, the price of maize in Mexico fell 70 per cent between 1994 and 2001. The number of farm jobs dropped as well: from 8.1 million in 1993 to 6.8 million in 2002. Many of those who found themselves without work were small-scale maize growers. However, introduction in the northern latitudes of the US of tropical maize for biofuels, and

Source: Alex Marshall, Clarke Energy Ltd, wikipedia.org/wiki/Maize#Biofuel

Figure 11.5 Farm-based maize silage digester located near Neumünster in Germany, 2007. Inflatable biogas holder is shown on top of the digester.

Source: Bill Hobbs

Figure 11.6 American ethanol plant using corn

not for human or animal consumption, may potentially alleviate this.

As a result of the US federal government announcing its production target of 35 billion gallons of biofuels by 2017, ethanol production was planned to grow to 7 billion gallons by 2010, up from 4.5 billion in 2006, boosting ethanol's share of corn demand in the US from 22.6 to 36.1 per cent.

'Feed corn' is being used increasingly for heating; specialized corn stoves (similar to wood stoves) are available and use either feed corn or wood pellets to generate heat. Corncobs are also used as a biomass fuel source. Maize is relatively cheap and home-heating furnaces have been developed which use maize kernels as a fuel. They feature a large hopper that feeds the uniformly sized corn kernels (or wood pellets or cherry pits) into the fire (wikipedia.org/ wiki/Maize#biofuel).

Potato

The potato is a starchy, tuberous crop from the perennial *Solanum tuberosum* of the Solanaceae family. The word potato may refer to the plant itself as well. In the region of the Andes, there are some other closely related cultivated potato species. Potatoes are the world's fourth largest food crop, following rice, wheat, and corn.

The FAO reports that the world production of potatoes in 2006 was 315 million tonnes. The largest producer, China, accounted for one quarter of the global output, followed by Russia and India.

In 2008, several international organizations began to give more emphasis to the potato as a key part of world food production, due to several developing economic problems. They cited the potato's potential for a beneficial role in world food production, owing to its status as a cheap and plentiful crop which can be raised in a wide variety of climates and locales.

In Europe, a large proportion of potatoes is planted in April and harvested in September–October yielding 30–50t/ha tubers at 20–24 per cent dry matter.

Potatoes are used to brew alcoholic beverages such as vodka and as food for domestic animals; potato starch is used to produce organic chemicals, in the textile industry, and in the manufacture of papers and boards. Maine companies are exploring the possibilities of using waste potatoes to obtain polylactic acid for use

Source: NightThree, wikipedia.org

Figure 11.7 Potato field, Fort Fairfield, Maine

Source: Vishalsh521, wikipedia.org

Figure 11.8 Potato farmer in India sitting beside the day's harvest

Table 11.5 Top ten potato producers in 2006 (million tonnes)

	People's Republic of China	70
	Russia	39
	India	24
	United States	20
	Ukraine	19
	Germany	10
	Poland	9
	Belgium	8
	Netherlands	7
	France	6
World Total		**315**

Source: FAO

Source: Oliver Spalt, wikipedia.org/Rice

Figure 11.9 Manual harvesting of rice field

in plastic products; other research projects seek ways to use the starch as a base for biodegradable packaging (wikipedia.org/Potato).

Bioethanol production from potatoes is based on the utilization of waste potatoes. Waste potatoes are produced from 5–20 per cent of crops as by-products in potato cultivation. At present, waste potatoes are used as feedstock only in one plant in Finland. Oy Shaman Spirits Ltd in Tyrnävä (near Oulu) uses 1.5 million kg of waste potatoes per year.

Alcohol yields varied significantly between cultivars. The highest alcohol yield was 9.5g/100g and the lowest yield 6.5g/100g. The average alcohol yield was 7.6g/100g (Liimatainen et al, 2004).

Rice (*Oryza sativa*)

Rice is a staple food for a large part of the world's human population, especially in tropical Latin America, and East, South and Southeast Asia, making it the second most consumed cereal grain. A traditional food plant in Africa, rice has the potential to improve nutrition, boost food security, foster rural development and support sustainable land care.

World production of rice has risen steadily from about 200 million tonnes of paddy rice in 1960 to 600 million tonnes in 2004. Milled rice is about 68 per cent of paddy rice by weight. In the year 2004, the top four producers were China (26 per cent of world production), India (20 per cent), Indonesia (9 per cent) and Bangladesh.

World trade figures are very different, as only about 5–6 per cent of rice produced is traded internationally. The largest three exporting countries are Thailand (26 per cent of world exports), Vietnam (15 per cent) and the United States (11 per cent), while the largest three importers are Indonesia (14 per cent), Bangladesh (4 per cent) and Brazil (3 per cent). Although China and India are the two largest producers of rice in the world, both of countries consume the majority of the rice produced (wikipedia.org/Rice).

Japan's first commercial plant to produce ethanol fuel for automobiles from locally grown rice was expected to reach full capacity of 1 million litres (220,000 gallons) a year by March 2009, a few months behind schedule. The project in Niigata, central Japan, for which the Japanese government is paying half the plant construction cost of 1.6 billion yen (US$15 million), is one of the nation's three such government-backed commercial production schemes. It is managed by the National Federation of Agriculture Co-operative Associations (Zen-Noh) and will use non-food rice.

Farmers were planting two types of super-harvest rice in more areas than planned this year after the 2007 harvest of one type gave a yield some 25 per cent less than expected.

This type is expected to consistently harvest 800kg per $1000m^2$ – 30–40 per cent more than that of ordinary rice. But the experience in 2007 showed that this is hard to achieve.

The Niigata project involves engineering company Mitsui Engineering and Shipbuilding Co and Satake Corp., a food processing machinery maker based in Hiroshima, western Japan, and is seen as more viable to apply across the country than the other two in the northern island of Hokkaido, which each have a capacity of 15 million litres.

Unlike Brazil, Japan lacks competitive locally grown farm produce to make ethanol to mix with gasoline and reduce greenhouse gas emissions. The project aims to use non-food rice planted on abandoned farmland, which now accounts for some 10 per cent of Japan's total farmlands, as cheaper imports drive farmers out of business and into the cities (reuters.com).

The farmers have a new reason to cultivate those plots – the government will be paying farmers to grow two types of super-harvest rice suitable for ethanol conversion (Levenstein, no date).

There are a number of technologies that can convert rice straw to ethanol including acid hydrolysis, enzymatic processes, gasification, steam explosion and solvent process technologies. Of these, gasification,

Source: Bernard Gagnon, wikipedia.org/Rice

Figure 11.10 Rice crop in Madagascar

acid hydrolysis and enzymatic processes appear most promising. All of these approaches have been shown to work in converting biomass residues to ethanol, but none are commercialized due to remaining technical and economic constraints.

Rice straw has been used, on an experimental basis in California, to produce power through direct combustion processes. There remain some technical challenges related to this potential use of rice straw, although straw-fuelled power stations are operating in Europe under more favourable economic conditions resulting from progressive environmental policies and government incentive programmes.

The most notable problem for using rice straw in combustion processes is the impact of ash melting and slagging. Pretreatments can address these challenges, but are generally too costly.

Gasification is a developing alternative to direct combustion for producing electricity from rice straw. In this process, gas is produced and then used in boilers, engines and gas turbines to generate power or manufacture liquid fuels and chemicals. All these areas have potential, provided adequate R&D funds are devoted to addressing the current economic disadvantages of using rice straw, rather than traditional fossil fuels, for the production of electricity and transportation (calrice.org).

Wheat

Wheat (*Triticum* spp.) is a worldwide cultivated grass from the Levant area of the Middle East. Globally, after maize, wheat is the second most produced food among the cereal crops; rice ranks third. Wheat grain is a staple food used to make flour for leavened, flat and steamed breads; cookies, cakes, breakfast cereal, pasta, noodles and couscous; and for fermentation to make beer, alcohol, vodka or biofuel. Wheat is planted to a limited extent as a forage crop for livestock, and the straw can be used as fodder for livestock or as a construction material for roofing thatch.

While winter wheat lies dormant during a winter freeze, wheat normally requires between 110 and 130 days between planting and harvest, depending upon climate, seed type and soil conditions. Crop management decisions require the knowledge of stage of development of the crop. In particular, spring fertilizer applications, herbicides, fungicides and growth regulators are typically applied at specific stages of plant development (wikipedia.org/Wheat).

Harvest takes place in Europe in August and yields vary between 7 and 10t/ha. The crop needs about 250kg N/ha (on average 3kg N per 100kg grain) of which about 160kg is from mineral fertilizers.

Due to the nature of European agriculture, high yield is necessary. In the last decades yields have increased at an annual rate of about 0.15t/ha. Straw is produced at an average yield of 4.5t/ha.

In 2006, the Wessex Grain subsidiary Green Spirit Fuels was given planning permission to create one of Britain's first bioethanol plants, which will eventually convert 340,000 tonnes of locally grown wheat per year into 131 million litres of ethanol (Madslien, 2006).

Surging demand for British grain around 2010 as major bioethanol plants come on line will wipe out the UK's wheat exports unless there is a big jump in output by domestic farmers.

In 2007, Britain's wheat outlook has been transformed by news that oil giant BP, Associated British Foods and US chemical company DuPont planned to build a bioethanol plant in Hull. The plant was expected to consume about one million tonnes of

Table 11.6 Top ten wheat producers in 2007 (million metric tonnes)

	European Union	124.7
	China	104
	India	69.3
	United States	49.3
	Russia	44.9
	Canada	25.2
	Pakistan	21.7
	Turkey	17.5
	Argentina	15.2
	Iran	14.8
World Total		**725**

Source: UN Food & Agriculture Organization (FAO)[28]

Source: Food Resource, Health and Human Sciences, Oregon State University, http://food.oregonstate.edu/gimages/images_wheat.html

Figure 11.11 Wheat harvester

wheat a year and followed the announcement of a similar plant from Ensus in Teesside, financed by two US private equity funds. Both plants were expected to come on line some time during 2009 and make their impact on Britain's wheat market in 2010. Demand for biofuels was expected to climb in 2010, driven by government rules that 5 per cent of motor fuel must come from renewable sources by that year. In 2007 Britain has an exportable wheat surplus of about 2.5 million tonnes but that total was expected to fall by about 750,000 tonnes later that year when a Cargill sweetener plant in Manchester, which uses wheat as its feedstock, comes on line (Hunt, 2007).

Selection of wheat and triticale genotypes as a source of biofuel

Contributed by Professor Dr R. K. Behl and Mr Molla Assefa, CCS Haryana Agricultural University, Hisar, India

Grains from cereals should continue to serve as a major source for food and feed. Cereal straw combined with rice straw and maize husks constitutes one of the principal biomass sources arising from present agricultural activities in Europe. This represents a large potential resource for utilization as energy feedstocks. Although there are some plans to use cereals as an energy source, the main justifications are overproduction of cereals and advanced technology in seed production, tillage, sowing, harvesting, baling and storage. Cereal crops that are whole crop harvested may be used under certain circumstances as energy crops. As per estimates provided by Food Corporation of India (FCI) huge quantities of cereal grains are getting spoiled every year due to unfavourable climatic conditions, and become unfit for human and animal consumption, and these are very cheap. There are about one million tonnes of damaged grains lying unutilized in FCI stores (Suresh et al, 1999). The damage includes discoloration, breakage, cracking, attack by fungi, insect damage, chalky grains, partial softening by being damp, dirty and bad smell etc. The damaged grains used for ethanol production are ten times cheaper than fine quality.

Triticale

Triticale production increased to about 3.5 million ha worldwide at the beginning of the new millennium.

This crop has the exceptional ability to give satisfactory yields under tough conditions and triticale is a promising species in sustainable production systems. In the Carpathian Basin region of Europe, winter triticale can be grown in place of wheat, barley, oats and rye and the spring variety can be grown in place of maize and spring barley. The share of winter triticale with its 90–95 per cent is dominant over spring triticale; however, there is a growing interest in the later variety. In a five-year period, triticale out-yielded wheat and rye except in one year where it had a lower yield than wheat at the sandy soil station in Hungary (Bona, 2007). On-farm data also proved that on the areas where triticale is grown, it has an advantage over wheat (in terms of yield stability). Growers appreciate a crop that can produce satisfactory yields without increasing the rate of fertilizer and pesticide use and intensive tillage. The average grain yield on the extremely poor areas varies between 2.5 and 4.5t/ha. However, fertilizer studies revealed that even in the above infertile acidic sandy soils in northeast Hungary, triticale can produce up to 8t/ha yield, if properly fertilized with N, P, K, Ca and Mg (Kádár et al, 1999). Small-grain variety performance tests have been conducted at the National Variety Field Tests network and proved the excellence of triticale within small-grain cereals.

The utilization of triticale has tremendous prospects both as animal feed and food for humans. Moreover, it will be one of the most promising non-food small-grain cereals for industrial production – i.e. biofuels (ethanol), organic and industrial chemicals, paper, building and plastics industry, and beverages. Triticale has yet to achieve its appropriate market position and image in Europe. In 1980, this crop is not on the EU list of cereal species that can benefit from government subsidy. Throughout the world – including mid-Europe – there is an extremely strong force on costs and benefits. Triticale has run a short period in its history: from the time of the first hexaploid triticales (Kiss, 1966), the modern hexaploid triticale cultivars today have the highest yield potentials within small-grain cereals. Prospects, however, are promising for triticale because of its strong yielding ability, high level of adaptation and also the possibility of using this species as an energy crop in the future (Green, 2002). In a field trial at Haryana Agricultural University, Hisar, India, it was observed that biological yield in elite tall-type triticale was 102g per plant as

against only 58g in standard check wheat variety. This was mainly due to plant height being 184cm in triticale and 106cm in wheat (Behl, 1980)

Energy grain (whole crop triticale, wheat)

In areas of cool weather and with a short growing season, where crops like maize cannot be produced in large quantity, alternative feedstocks need to be explored for fuel ethanol production. There is an abundant but underutilized supply of agricultural residues and herbaceous grasses available in different wheat- and barley-growing areas. In 2006, for example, Montana produced 5.2 million tonnes of wheat and 0.9 million tonnes of barley. Over 9 million tonnes of residues were left behind as a by-product of these crops. The annual and/or perennial grasses and cereal forage crops may serve both as livestock feed and lignocellulosic feedstock for fuel ethanol production. Suresh et al (1999) reported production of ethanol by utilizing gelatinized starch from damaged wheat grains. The simultaneous saccharification and fermentation was used to produce ethanol from raw starch of damaged wheat grains by utilizing crude amylase preparation from *Bacillus subtilis* VB2 and an amylolytic yeast strain *Saccharomyces cerevisiae* VSJ4. Various concentrations of damaged wheat starch were used and 25 per cent was found to be the optimum for damaged wheat starch yielding 4.40 per cent V/V ethanol.

The two main types of liquid biofuel are biodiesel and bioethanol. Biodiesel can be blended with diesel and bioethanol is primarily blended with petrol. Currently the majority of vehicle engines are designed to run on blends of at least 5 per cent biofuel. At present, the main vegetable oil used for biodiesel comes from oilseed rape, and bioethanol is produced from sugar beet or cereal grains. In future it may become possible to produce sugar, hence alcohol, from plant biomass, which is much cheaper and more plentiful. Using crops to produce fuel will help to meet targets for reducing greenhouse gas emissions such as carbon dioxide (CO_2). Cereals for combustion or for fermentation may be produced in the same manner as for food or fodder use. For fermentation use, high grain yields must be obtained, while for combustion a high total yield is aimed at, as both grain and straw are used.

Growing crops for biofuels

The best crops for producing bioethanol include wheat, triticale, barley, sugar beet, sugar cane and maize. Growers must maximize the economic production of ethanol per hectare. Ethanol production depends upon the yield of starch and the ease with which it can be processed. For wheat use, varieties with good starch/distilling properties are needed. Crop inputs, except nitrogen, should be applied at economically optimum amounts for yield. Nitrogen increases grain protein at the expense of starch, so N applications should be made earlier and at a slightly lower rate than for other markets. Processors may pay a premium for grain giving higher alcohol yields, which can be maximized through variety choice and nitrogen management.

- The nitrogen fertilization rate could be 50 per cent of the normal rate for grain cereals because most of the nitrogen needed by cereals is for grain production and to improve the protein content of the grain. This is not needed in whole-crop harvesting for energy feedstocks.
- The whole-crop-harvesting procedure will involve cutting and collecting the unfractionated crop from the field.
- Harvesting should take place earlier than for grain cereals.

This system will prevent conflicts with the production of food cereal grains.

Biofuel potential of wheat and triticale

In Germany, wheat, triticale and rye have been investigated. Wheat has the highest yield potential on good soils, while rye produces better on poor soils. Triticale performance is intermediate and attracts interest for energy purposes, as it is not used for food production. A mean total yield of 12 over dried tonnes (ODT) per ha (5.5ODT/ha of grain) is expected under German conditions and in Denmark total mean yields of 10.9ODT/ha at commercial conditions. In Austria total yields of about 10ODT/ha were obtained and in France yields of 10–14ODT/ha.

Cereals are highly developed crops due to their use for food production, and the knowledge of production is widespread among farmers. Therefore, energy grain production can easily be implemented in agriculture and high and stable yields can be expected.

Breeding wheat for biofuel production

To meet society's needs with renewable biopolymers, biochemicals and energy, we have to develop the potential

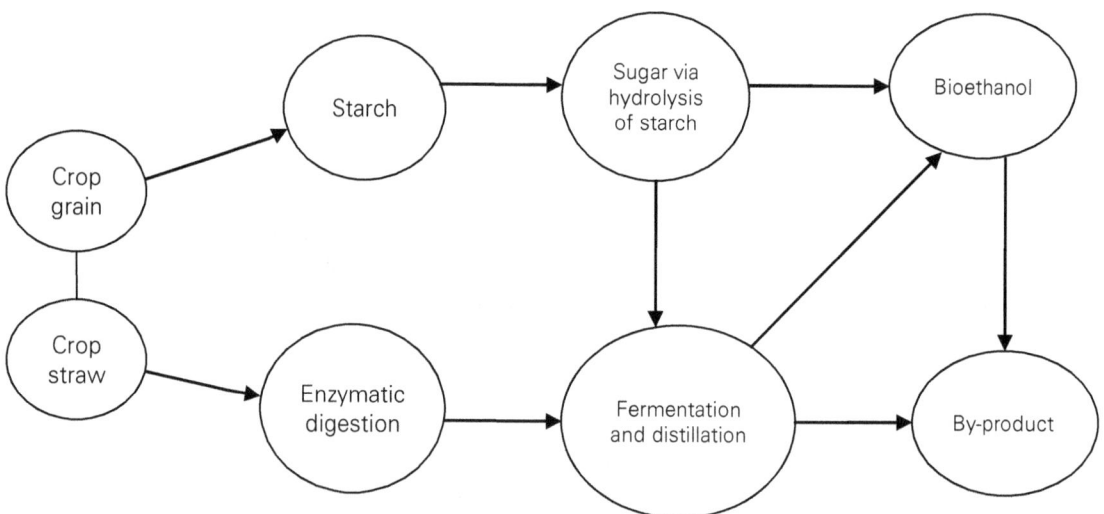

Figure 11.12 Biofuel production process

wheat and triticale genotypes. Known agronomic attributes including the highest grain and biomass yield, high starch content, and genotypes having a favourable net energy balance, and resistance to biotic and abiotic stresses. Some of these preferred characteristics make triticale and specific wheat genotypes crops of choice for industrial and energy end uses.

In recent decades, plant-breeding activities have made good progress in increasing grain yields and improving the genetic make-up of crops to increase their physiological efficiency to produce higher biomass through enhanced and prolonged photosynthesis, and improving their adaptability for wider environmental conditions and better input (nutrients and water) use efficiency. In a post-green revolution era the jump in grain yield in wheat has been achieved mainly through improving the harvest index (the ratio of grain to straw) to almost 0.5 or even higher and better agro-technology and agronomical production package. The total biomass has been influenced very little by breeding activities. However, a kind of yield plateau is now being witnessed. Therefore increasing biomass coupled with a high harvest index is needed. Breeders should try to produce cultivars with a higher straw proportion as energy cereal crops. This would minimize the inputs and decrease environmental damage. Attempts have already been made to breed cereals, especially barley, with higher enzyme contents that enable efficient fermentation and ethanol production from the grain.

Cereals as energy crops on set-aside or non-traditional stress-prone areas may need the attention of wheat/triticale breeders, to develop suitable genotypes which will have high biomass with a low harvest index to produce considerably more straw than grain, and/or high biomass coupled with higher harvest index to produce higher total tonnage of produce. Tall-type local wheat genotypes of the pre-green revolution era have been by and large replaced with semi-dwarf high-yielding varieties with re-tailored grain to straw ratio while total biomass remains almost the same. The breeders should use the tall types, landraces, and wild types which have a robust root system in a crossing programme so as to further enhance biomass through selection among recombinants. Crosses between winter and spring wheat generally exhibit luxuriance for biomass and resilience for adaptability in segregating generations. In such crosses, selection for high seedling vigour, fast and appropriate vegetative growth to support reproductive phase and development, higher plant height, more numerous tillers and broad leaves with stay-green trait would be worthwhile to develop the genotypes with a higher biomass and consequently a higher grain yield. There is therefore an urgent need to develop new and more efficient wheat-breeding methods to complement existing techniques, as well as to identify new traits to drive faster yield gains by exploiting the true biological yield potential of wheat. Various research activities in the world whose main aim was to capitalize on promising new techniques, based on morpho-physiological and biochemical criteria, which could be used for selecting high-yielding wheat varieties for breeding, have shown promising results.

Assessing the potential of genetic sources of variation in some physiological traits, such as high biomass production, longer rapid spike growth phase, or large kernel size, and stay-green characters is very important. In addition, the evaluation of germplasms should also focus on producing lines with high starch and more amylose content together with potential yield gains. Selection for high biomass yield should bring about positive improvements in biomass yield, grain yield, effective tiller number, and number of kernels per spike. As yield improvement due to increased partitioning of biomass to grain yield reaches its theoretical limit (Austin et al, 1980), breeding for larger total biomass becomes increasingly necessary if further genetic gains in yield potential are to be realized. Higher biomass may be achieved by:

- increased interception of radiation by the crop;
- greater intrinsic radiation use efficiency (RUE) throughout the crop cycle; and
- improved source–sink balance permitting higher sink demand and, therefore, higher RUE during grain filling.

Increased light interception could be achieved via early ground cover or improved 'stay-green' at the end of the cycle. Improved RUE may be achieved, for example, by decreasing photorespiration or photoinhibtion (Loomis and Amthor, 1999), or through improved canopy photosynthesis related to factors such as canopy architecture. Better source–sink balance may result from increased partitioning of assimilates during spike development so that grain number is increased; this improves RUE during grain filling as a consequence of increased demand for assimilates.

Some considerations in exploiting genetic resources

Traits have been identified in the germplasm bank of CIMMYT (the International Maize and Wheat Improvement Center) with potential to improve 'source' and 'sinks' to raise yield potential, and to improve stress tolerance.

- Traits are introgressed into good backgrounds to establish potential genetic gains.
- 'Source' and 'sink' type traits are crossed together to obtain synergy.

Traits to improve spike fertility ('sink')

- Large spikes. Good sources available but seed often shrivelled.
- Multi-ovary florets. Trait expressed in high-yield backgrounds.
- Branched spikelets. Introgressed with good results in Yugoslavia.
- Higher grain weight potential. Expressed when extra assimilates available in boot stage.
- Phenology. Genetic variation exists for duration of juvenile spike growth.

Traits to improve assimilate availability ('source')

- Green area duration. Rapid full light interception and stay-green sources identified.
- Stem reserves. Significant variation in accumulation and utilization exists.
- Erect leaf. Being introgressed into high-biomass Baviacora.

Traits to improve stress tolerance

Many traits have been postulated to confer stress tolerance in wheat, depending on specific environments. Germplasm is being screened for sources of these characters (Reynolds et al, 2004).

Wheat starch and wheat flour with high amylose content

Wheat starch having high amylose content may be prepared by isolating starch from wheat seeds lacking

Source: CIMMYT

Figure 11.13 Big spike wheat may improve 'sink' potential

Source: CIMMYT

Figure 11.14 Erect leaves and high chlorophyll content may improve 'source' potential

SGP-1 (starch granule protein-1) according to any appropriate method known in the field. Wheat starch and wheat flour of US invention are believed to be novel materials characterized by having a high level of apparent amylose content which has not been previously known. Such wheat starch and wheat flour may be useful in various industrial and food applications. Moreover, it may be useful as a breeding material for developing wheat which produces starch having an amylose content as high as that of maize (60–70 per cent).

Example: Production of wheat lacking SGP-1 (starch granule protein-1 null wheat)

To produce a wheat which lacks SGP-1 (SGP-1 null wheat), the following four parental wheat (*Triticum aestivum* L.) cultivars were used: Chousen 30 (C 30) and 57 (C 57) lacking SGP-A1; Kanto 79 (K 79) lacking SGP-B1 and Turkey 116 (T 116) lacking SGP-D1.

First, T 116 and K 79 were crossed to obtain F_1 seeds. F_1 plants which grew from the F_1 seeds were self-pollinated to obtain F_2 seeds. Starches were purified from the distal half of the F_2 seeds. SDS-polyacrylamide gel electrophoresis (SDS-PAGE) was performed using the purified starches so as to examine the presence or absence of SGP-D1 and SGP-B1. As a result, F_2 seeds lacking both SGP-D1 and SGP-B1 from cross K 79/T 116 were selected.

F_2 plants which grew from the selected F_2 seeds lacking both SGP-D1 and SGP-B1 were pollinated by either C 30 or C 57, both lacking SGP-A1, to obtain new F_1 seeds. New F_1 plants grown from the new F_1 seeds were self-pollinated to obtain new F_2 seeds. Starches were purified from the distal half of the new F_2 seeds. SDS-PAGE was performed using the purified starches so as to examine the presence or the absence of SGP-A1, SGP-B1 and SGP-D1. As a result, from the cross (K 79/T 116)F_2//C 30 or C 57, variant progeny (new F_2 plant) lacking SGP-1 was selected.

Conclusions

Currently, there exists little information on bioethanol yield advancement from wheat and triticale. In order to achieve this we need to set several objectives which include the development and optimization of analytical protocols for determining grain starch content and structure, fermentable sugar levels, and ethanol yield.

The establishment of correlations between grain ethanol yield and agronomic characteristics would be logical, to use these in order to quantify genetic variability of available germplasm, both local and international. A pre-breeding programme for genetic enhancement for desirable plant traits for biofuel production based on morpho-physiological and marker-assisted recurrent selection (MARS) should be established. Exploring use of molecular DNA for use in marker-assisted selection (MAS) deserves a priori consideration to complement conventional plant-breeding methods. Some research activities at molecular levels have already been initiated in some developed countries and promising results are awaited. However, a global network is needed in this direction so as to make the venture environmentally sustainable and economically profitable. Based on initial success, we can conclude that both possibilities and technical feasibilities exist for genetic and management options to produce agro-technology for cereal production for biofuels.

Pseudocereals: Amaranthus, buckwheat, quinoa

'Pseudocereals' is an expression used to refer to plant species cultivated for their starch-containing seeds. They have a very low harvest index and low cereal yields as a result of extensive breeding activities, though they have been cultivated for a very long time. Some of them have high biomass potential: amaranthus (*Amaranthus* spp.), quinoa (*Chenopodium quinoa* Willd.) and buckwheat (*Fagopyrum esculentum* Moench). These plant species can be grown in cool regions, and are the focus of research programmes in different parts of Europe.

Amaranthus is a herbaceous annual plant species with simple leaves. Only a few of the many amaranthus species are cultivated. Most cultivated species are hybrids. Investigations in 1995 showed that the total biomass achieved from amaranthus field trials was nearly 11t/ha dry matter, with a seed production of 2.4t/ha. Some genotypes delivered dry matter total biomass yields of up to 20t/ha.

Buckwheat is an annual plant species with branched stems growing from about 50 to 150cm high. The total biomass yield could reach 8.5t/ha dry matter and 3–4t/ha grain. But under common cultivation practices the grain yield lies between 1 and 2t/ha and the dry matter total biomass is approximately 5.5t/ha.

Quinoa possesses high genetic variability and ecological adaptability characteristics, and can be grown in different environments. There has been some breeding activity in Austria and Denmark and new varieties could be released in the near future.

Quinoa has been grown in northern Europe as an alternative crop for 10 years. Its total biomass yield lies between 4 and 10t/ha dry matter with a seed yield of between 1 and 2t/ha. Others have achieved higher figures – 10–13t/ha dry matter and 2–3t/ha grain (Aufhammer et al, 1995).

Internet resources

http://www.deq.state.mt.us/energy/bioenergy/June2005 EthanolConference/26Hicks_HullessBarley.pdf

http://uk.reuters.com/article/businessNews/idUKL299 1975820070629

http://inventorspot.com/articles/rice_ethanol_replaces_ oil_saves__10099, accessed February 2009, Figure 11.8

http://www.cimmyt.cgiar.org/

http://gas2.org/2008/09/26/osage-bioenergy-to-open-largest-ethanol-plant-in-us/

www.calrice.org/downloads/a5b5_ethanol.pdf

www.greencarcongress.com/2005/07/barley_ethanol_ .html

www.umanitoba.ca/afs/agronomists_conf/proceedings/ 2007/Scott_Day.pdf

http://en.wikipedia.org/wiki/Maize#Biofuel (2009)

http://hila.webcentre.ca/research/teosinte/

http://ecotality.com/life/author/bill/

http://en.wikipedia.org/wiki/Wheat

www.reuters.com/article/latestCrisis/idUST33749

12
Oil Crops

Calendula officinalis

The main problem is seed shattering at harvest. Yields varied from 1.5 to 2.5t/ha of seeds with oil contents around 20 per cent with 60 per cent calendic acid and 30 per cent linoleic acid. This oil displays important viscosity with high refraction index. Its ease of drying is of value in paints.

Camelina sativa

Spring or winter varieties are respectively sown in April or mid-October with a 7–8kg/ha seed density. Yields obtained have been 1.5–3t/ha for the spring varieties and about 3–4t/ha for winter varieties. The oil (30–40 per cent of the grain) is interesting for its high rate of gondoic acid (up to 19 per cent) and anti-oxidizing agents. Uses: paints, inks and cosmetics.

Carthamus tinctorius

Good pollination by insects is needed to obtain high seed yield up to 3t/ha with a maximum of 50 per cent oil (linoleic and oleic acids). A red-orange pigment extracted from the flowers can be used in textiles or cosmetics.

All these crops are grown for their seeds. Yield is linked to good crop establishment, dry weather at pollination notably for entomogameous crops and dry weather during ripening. It suffers highly from seedset spreading, which induces irregular ripening, and from seed shattering before harvest.

Crambe abyssinica

Sown at a density of 10kg/ha, this spring species is very sensitive to frost. The crop can be harvested about 100 days after sowing with a maximum yield of 2.5 to 3 tonnes of grains per ha and high inter-annual variability. Aside from these negative traits, an important brake to its development is the difficulty of extracting the oil (35 per cent of the grain with an interesting 55 per cent erucic acid). Uses: lubricants, plastics or nylon.

Dimorphoteca pluvialis

This species suffers from weak resistance to cold. Grain ripening is spread out and seeds shatter, which results in low yields at harvest (0.6–0.7t/ha).

Flax

The sowing density is about 140kg seeds per hectare corresponding to about 2000 plants per square metre. Yields are about 6.5t straw per ha.

Lesquerella grandiflora and L. gordonii

These species are not really adapted to the temperate climate of Belgium. Their seeds contain 25–35 per cent of oil with 55 per cent of lesquerolic acid (C20:1-OH). Uses: lubricants, nylon and plastics, gums which could be used as thickeners in food or cosmetics.

Limnanthes alba

Good climatic conditions are needed at the flowering stage in order to favour insect pollination which is necessary to obtain yields from 0.8 to 1.8t/ha. Seeds contain about 30 per cent oil of which 95 per cent is constituted by mono-unsaturated long-chain fatty acids showing high stability to heat and air.

Linseed

Linseed is known to be a very versatile crop with fluctuating yields (1.5–3.5t/ha) due to frequent difficulties with early sowing, late ripening, difficult seed desiccation after rain and risks of sprouting before harvest. The crop is sown from mid-March to the end of April at a density of 60–75kg seeds per hectare. Nutrient needs are limited.

Its economic effectiveness is low. Farmers would be interested in the development of winter varieties with earlier and easier harvest, and higher and more regular yield.

Table 12.2 shows the oil yields in gallons per acre (one gallon of oil = 7.3 pounds), keeping in mind that the yields will vary in different agro climatic zones.

Table 12.1 Vegetable oil yields

Ascending order			Alphabetical order		
Crop	litres oil/ha	US gal/acre	Crop	litres oil/ha	US gal/acre
corn (maize)	172	18	avocado	2638	282
cashew nut	176	19	brazil nut	2392	255
oats	217	23	calendula	305	33
lupin	232	25	camelina	583	62
kenaf	273	29	cashew nut	176	19
calendula	305	33	castor bean	1413	151
cotton	325	35	cocoa (cacao)	1026	110
hemp	363	39	coconut	2689	287
soya bean	446	48	coffee	459	49
coffee	459	49	coriander	536	57
linseed (flax)	478	51	corn (maize)	172	18
hazelnut	482	51	cotton	325	35
euphorbia	524	56	euphorbia	524	56
pumpkin seed	534	57	hazelnut	482	51
coriander	536	57	hemp	363	39
mustard seed	572	61	jatropha	1892	202
camelina	583	62	jojoba	1818	194
sesame	696	74	kenaf	273	29
safflower	779	83	linseed (flax)	478	51
rice	828	88	lupin	232	25
tung oil	940	100	macadamia nut	2246	240
sunflower	952	102	mustard seed	572	61
cocoa (cacao)	1026	110	oats	217	23
peanut	1059	113	oil palm	5950	635
opium poppy	1163	124	olive	1212	129
rapeseed	1190	127	opium poppy	1163	124
olive	1212	129	peanut	1059	113
castor bean	1413	151	pecan nut	1791	191
pecan nut	1791	191	pumpkin seed	534	57
jojoba	1818	194	rapeseed	1190	127
jatropha	1892	202	rice	828	88
macadamia nut	2246	240	safflower	779	83
brazil nut	2392	255	sesame	696	74
avocado	2638	282	soya bean	446	48
coconut	2689	287	sunflower	952	102
oil palm	5950	635	tung oil	940	100

Note: This table shows conservative estimates as crop yields vary widely. This data is compiled from a variety of sources. Where sources vary averages are given. The yield figures are most useful as comparative estimates: a high-yielding crop may not be 'better' (more suitable) than a lower-yielding crop, it depends on the particular situation.

Source: Addison (2001)

Table 12.2 Oil-producing crops

Plant	Latin Name	Gal Oil/Acre	Plant	Latin Name	Gal Oil/Acre
Oil palm	*Elaeis guineensis*	610	Rice	*Oriza sativa L.*	85
Macauba palm	*Acrocomia aculeata*	461	Buffalo gourd	*Cucurbita foetidissima*	81
Pequi	*Caryocar brasiliense*	383	Safflower	*Carthamus tinctorius*	80
Buriti palm	*Mauritia flexuosa*	335	Crambe	*Crambe abyssinica*	72
Oiticia	*Licania rigida*	307	Sesame	*Sesamum indicum*	71
Coconut	*Cocos nucifera*	276	Camelina	*Camelina sativa*	60
Avocado	*Persea americana*	270	Mustard	*Brassica alba*	59
Brazil nut	*Bertholletia excelsa*	245	Coriander	*Coriandrum sativum*	55
Macadamia nut	*Macadamia terniflora*	230	Pumpkin seed	*Cucurbita pepo*	55
Jatropha	*Jatropha curcas*	194	Euphorbia	*Euphorbia lagascae*	54
Babassu palm	*Orbignya martiana*	188	Hazelnut	*Corylus avellana*	49
Jojoba	*Simmondsia chinensis*	186	Linseed	*Linum usitatissimum*	49
Pecan	*Carya illinoensis*	183	Coffee	*Coffea arabica*	47
Bacuri	*Platonia insignis*	146	Soya bean	*Glycine max*	46
Castor bean	*Ricinus communis*	145	Hemp	*Cannabis sativa*	37
Gopher plant	*Euphorbia lathyris*	137	Cotton	*Gossypium hirsutum*	33
Piassava	*Attalea funifera*	136	Calendula	*Calendula officinalis*	31
Olive tree	*Olea europaea*	124	Kenaf	*Hibiscus cannabinus L.*	28
Rapeseed	*Brassica napus*	122	Rubber seed	*Hevea brasiliensis*	26
Opium poppy	*Papaver somniferum*	119	Lupin	*Lupinus albus*	24
Peanut	*Ariachis hypogaea*	109	Palm	*Erythea salvadorensis*	23
Cocoa	*Theobroma cacao*	105	Oat	*Avena sativa*	22
Sunflower	*Helianthus annuus*	98	Cashew nut	*Anacardium occidentale*	18
Tung oil tree	*Aleurites fordii*	96	Corn	*Zea mays*	18

Source: Scurlock (2005)

Table 12.3 Typical oil extraction from 100kg of oilseeds

Crop	Oil/100kg seeds
Castor seed	50kg
Copra	62kg
Cotton seed	13kg
Groundnut kernel	42kg
Mustard	35kg
Palm kernel	36kg
Palm fruit	20kg
Rapeseed	37kg
Sesame	50kg
Soya bean	14kg
Sunflower	32kg

Source: journeytoforever.org

Table 12.4 Oils and esters characteristics

Type of oil	Melting range (degree C)			Iodine number	Cetane number
	Oil/fat	Methyl ester	Ethyl ester		
Rapeseed oil, h. eruc.	5	0	−2	97 to 105	55
Rapeseed oil, i. eruc.	−5	−10	−12	110 to 115	58
Sunflower oil	−18	−12	−14	125 to 135	52
Olive oil	−12	−6	−8	77 to 94	60
Soya bean oil	−12	−10	−12	125 to 140	53
Cotton seed oil	0	−5	−8	100 to 115	55
Corn oil	−5	−10	−12	115 to 124	53
Coconut oil	20 to 24	−9	−6	8 to 10	70
Palm kernel oil	20 to 26	−8	−8	12 to 18	70
Palm oil	30 to 38	14	10	44 to 58	65
Palm oleine	20 to 25	5	3	85 to 95	65
Palm stearine	35 to 40	21	18	20 to 45	85
Tallow	35 to 40	16	12	50 to 60	75
Lard	32 to 36	14	10	60 to 70	65

Source: journeytoforever.org

Table 12.5 Selected 'typical' properties of certain common bioenergy feedstocks and biofuels, compared with coal and oil

		Composition		
		cellulose (%)	hemicellulose (%)	lignin (%)
Bioenergy feedstocks	corn stover	35	28	16–21
	sweet sorghum	27	25	11
	sugar cane bagasse	32–48	19–24	23–32
	sugar cane leaves			
	hardwood	45	30	20
	softwood	42	21	26
	hybrid poplar	42–56	18–25	21–23
	bamboo	41–49	24–28	24–26
	switchgrass	44–51	42–50	13–20
	miscanthus	44	24	17
	Arundo donax	31	30	21
Liquid biofuels	bioethanol	N/A	N/A	N/A
	biodiesel	N/A	N/A	N/A
Fossil fuels	Coal (low rank; lignite/sub-bituminous)	N/A	N/A	N/A
	Coal (high rank; bituminous anthracite)	N/A	N/A	N/A
	Oil (typical distillate)	N/A	N/A	N/A

		Chemical characteristics				
		heating value (gross, unless specified: GJ/t)	ash (%)	sulphur (%)	potassium (%)	Ash melting temperature [some ash sintering observed] (C)
Bioenergy feedstocks	corn stover	17.6	5.6			
	sweet sorghum	15.4	5.5			
	sugar cane bagasse	18.1	3.2–5.5	0.10–0.15	0.73–0.97	
	sugar cane leaves	17.4	7.7			
	hardwood	20.5	0.45	0.009	0.04	[900]

Table 12.5 Selected 'typical' properties of certain common bioenergy feedstocks and biofuels, compared with coal and oil (*Cont'd*)

		Chemical characteristics				
		heating value (gross, unless specified: GJ/t)	ash (%)	sulphur (%)	potassium (%)	Ash melting temperature [some ash sintering observed] (C)
	softwood	19.6	0.3	0.01		
	hybrid poplar	19.0	0.5–1.5	0.03	0.3	1350
	bamboo	18.5–19.4	0.8–2.5	0.03–0.05	0.15–0.50	
	switchgrass	18.3	4.5–5.8	0.12		1016
	miscanthus	17.1–19.4	1.5–4.5	0.1	0.37–1.12	1090 [600]
	Arundo donax	17.1	5–6	0.07		
Liquid biofuels	bioethanol	28		<0.01		N/A
	biodiesel	40	<0.02	<0.05	<0.0001	N/A
Fossil fuels	Coal (low rank; lignite/sub-bituminous)	15–19	5–20	1.0–3.0	0.02–0.3	~1300
	Coal (high rank; bituminous/anthracite)	27–30	1–10	0.5–1.5	0.06–0.15	~1300
	Oil (typical distillate)	42–45	0.5–1.5	0.2–1.2		N/A

		Physical characteristics		
		Cellulose fibre length (mm)	Chopped density at harvest (kg/m³)	Baled density [compacted bales] (kg/m³)
Bioenergy feedstocks	corn stover	1.5		
	sweet sorghum			
	sugar cane bagasse	1.7	50–75	
	sugar cane leaves		25–40	
	hardwood	1.2		
	softwood			
	hybrid poplar	1–1.4	150 (chips)	
	bamboo	1.5–3.2		
	switchgrass		108	105–133
	miscanthus		70–100	130–150 [300]
	Arundo donax	1.2		
				(typical bulk densities or range given below)
Liquid biofuels	bioethanol	N/A	N/A	790
	biodiesel	N/A	N/A	875
Fossil fuels	Coal (low rank; lignite/sub-bituminous)	N/A	N/A	700
	Coal (high rank; bituminous/anthracite)	N/A	N/A	850
	Oil (typical distillate)	N/A	N/A	700–900

Source: Scurlock (2005)

Internet resources

http://journeytoforever.org/biodiesel_yield.html
http://cat.inist.fr/?aModele=afficheN&cpsidt=2743848

13

Biogas from Crops

Introduction

Anaerobic digestion is the breaking-down of biological material by micro-organisms in the absence of oxygen i.e., under anaerobic conditions. A number of families of bacteria, working together, transform biological material into biogas. Biogas consists of around two-thirds combustible methane and about one-third carbon dioxide and other gases. The anaerobic bacteria are some of the oldest 'inhabitants' of our planet. With a power output of 40GWh of heat and 20GWh of electricity biogas plants are a major source of renewable energy. Anaerobic digestion of organic waste is especially suited to the treatment and stabilization of liquid or solid biological material. The process takes place in a sealed vessel (the so-called fermenter or digester) under controlled conditions and has major advantages: up to 90 per cent of the carbon content will be converted into biogas, and biogas supplies valuable energy in the form of electricity, heat and transportation fuels.

Biogas is a well-established fuel and it is a gas mixture comprising around 60 per cent methane and 40 per cent carbon dioxide that is formed when organic materials, such as dung or vegetable matter, are broken down by microbiological activity in the absence of air, at slightly elevated temperatures (most effective at 30–40°C or 50–60°C).

Biogas in developing countries

China has over 7.5 million household biogas digesters, 750 large- and medium-scale industrial biogas plants, and a network of rural 'biogas service centres' to provide the infrastructure necessary to support dissemination, financing and maintenance. India has also had a large programme, with about three million

household-scale systems installed (Martinot, 2003). Other countries in the South with active programmes include Nepal, Sri Lanka, Kenya and several countries in Latin America. As carbon emission levels are becoming of greater concern and as people realize the benefits of developing integrated energy supply options, then biogas becomes an increasingly attractive option. The most widespread designs of digester are the Chinese fixed dome digester and the Indian floating cover biogas digester. The digestion process is the same in each digester but the gas collection method is different. In the floating cover type, the water-sealed cover of the digester is capable of rising as gas is produced, where it acts as a storage chamber, whereas the fixed dome type has a lower gas storage capacity and requires good sealing if gas leakage is to be prevented. Both have been designed for use with animal waste or dung. The production of methane for use as a fuel, which reduces the amount of woodfuel required, thus reduces desertification. The waste is reduced to slurry that has a high nutrient content, making an ideal fertilizer. During the digestion process, dangerous bacteria in the dung and other organic matter are killed, which reduces the pathogens dangerous to human health.

For those without cattle or within urban centers, a conventional digester may not be appropriate. The Indian Appropriate Rural Technology Institute (ARTI) has introduced a small biogas digester that uses starchy or sugary wastes as feedstock, including waste flour, vegetable residues, waste food, fruit peelings, rotten fruit, oil cake, rhizomes of banana, canna (a plant similar to a lily but rich in starch) and non-edible seeds. The compact plants are made from cut-down high-density polythene (HDPE) water tanks, which are adapted using a heat gun and standard HDPE piping. The standard plant uses two tanks, with volumes of

Table 13.1 Typical composition of biogas

Constituent	%
Methane, CH_4	50–75
Carbon dioxide, CO_2	25–50
Nitrogen, N_2	0–10
Hydrogen, H_2	0–1
Hydrogen sulphide, H_2S	0–3
Oxygen, O_2	0–2

Source: wikipedia.org/Biogas

typically 0.75 and $1m^3$. The smaller tank is the gas holder and is inverted over the larger one which holds the mixture of decomposing feedstock and water (slurry). The feedstock must be blended so that it is smooth using a blender powered by electricity or by hand. 2kg of such feedstock produces about 500g of methane, and the reaction is completed within 24 hours. An inlet is provided for adding feedstock, and an overflow for removing the digested residue. The digester is set up in a sunny place close to the kitchen, and a pipe takes the biogas to the kitchen (Karve, 2006). Biogas has a wide variety of applications. It can be used directly for cooking and lighting or for heat generation, and for electricity production and fuel for cars.

Biogas is a clean fuel, thus reducing the levels of indoor air pollution, a major cause of ill-health for those living in poverty. Lighting is a major social asset, and already there are estimated to be over 10 million households with lighting from biogas (Martinot, 2003). Improved lighting is associated with longer periods for work or study. Where biogas is substituted for woodfuel, there are two benefits: a reduction in the pressures on the forest, and a time-saving for those who have to collect wood – usually women and children. If a biogas plant is linked to latrines in a sanitation programme, it is a positive way of reducing pathogens and converting the waste into safe fertilizer. Where biogas is linked with sales of the resultant fertilizer, it is an excellent source of additional income. Fertilizer can be used on crops to increase their yield. Biogas can be used to generate electricity, bringing with it the possibilities of improved communications, telephone, computer, radio and television for remote communities. Fuel produced locally is not as vulnerable to disruption as, for example, grid electricity or imported bottled gas (Bates, 2007).

Biogas from crops and their residues

In Europe, biogas is being made more and more often from energy crops that are used as a single substrate, instead of manure which is traditionally used. Several research efforts and trials are under way, analysing the potential of specially bred biogas maize, exotic grass species such as Sudan grass and sorghum, or new hybrid grass types.

Compared to making liquid biofuels, biogas has the advantage that the entire crop can be utilized and not merely the starch-, sugar- or oil-rich parts, which is the case with first-generation ethanol or biodiesel

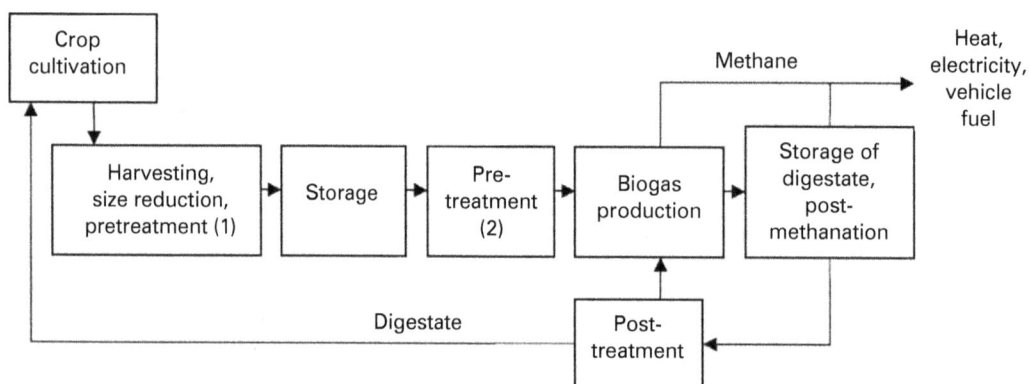

Source: Lehtomäki (2006) modified from Weiland (2003)

Figure 13.1 Biogas from energy crops – production chain

production. A biomethane digester can ferment a much wider range of biomass sources. This is why biogas has a large potential (as was recently illustrated by a report showing that biogas can replace all Russian gas imports in Europe).

Once the biogas is produced, it can be used either directly in gas engines and generators or in more efficient cogeneration plants for the production of power and heat. It can be purified to natural gas grade standards, after which it can be fed into the natural gas grid and utilized like ordinary fossil methane, by households, industries, or in cars and fuel cells. As an automotive fuel, used in compressed natural gas (CNG)-capable vehicles, biogas has the highest well-to-wheel efficiency and the lowest carbon dioxide footprint of all biofuels.

This is why the large-scale production of biogas in the tropics and subtropics is seen as a promising bioenergy sector, because the technology is well understood and is already used in most developing countries (on a micro-scale, at the household level); it yields more energy per hectare than liquid biofuels, which implies a better use of resources and less land needed; and the variety of suitable feedstocks is much larger.

A very important study by the Institut für Energetik und Umwelt, based in Leipzig, and by the

Source: zorg-biogas.com

Figure 13.2 Preparation of green biomass and screw charger

Table 13.2 Energy balance of various biofuels

Crop	Biofuel	Energy GJ/ha	Balance Energy Ratio (input:output)
Wheat	Bioethanol	34.67	1:2.3
Wheat	Biogas	68.48	1:3
Oilseed rape	Biodiesel	18.25	1:1.8

Source: photobucket.com

Öko-Instituts Darmstadt outlined in 2007 that the EU has a biogas production potential large enough to replace all natural gas imports from Russia by 2020. It shows that if current production trends continue, all of Europe's natural gas imports from Russia will be covered by locally produced biogas within two decades (Biopact, 2007f). The EU currently imports some 40 per cent of all its natural gas from Russia. In 2030, this dependency will have increased to 60 per cent. This outlook worries many, as it opens obvious questions about energy security. Last year, gas disputes between Russia and Belarus and the Ukraine, affected energy supplies to the EU. The Leipzig report on biogas, entitled 'Möglichkeiten einer europäischen Biogaseinspeisungsstrategie' ('The opportunities of a European strategy to feed biogas into the natural gas grid') puts this geopolitical question into an entirely different perspective.

The main findings of the study are:

- Europe's potential for the sustainable production of biomethane is 500 billion m³ of natural gas equivalent per year. This is roughly the total amount of natural gas currently consumed by the entire European Union.

- The entire EU's natural gas needs for the medium-term future (2020) can be met by biogas; all imports from Russia can be replaced, while the excess can substitute petroleum and coal.

- The production of 500 billion m³ of biogas, fed into the grid, will result in a reduction of 15 per cent of Europe's CO_2 emissions. The Kyoto Protocol demands a reduction of 10 per cent.

- An efficient biogas feed-in strategy will be built around the concept of 'biogas corridors': such corridors consist of biomass plantations established alongside the pipelines, so that the biogas can be fed into Europe's main natural gas grid without the need for new pipelines and infrastructures.

- A Europe-wide biogas feed-in strategy will result in the creation of 2.7 million new jobs within the EU. Employment will be generated mainly in agriculture, in the manufacture, construction and management of biogas plants and biogas purification plants.

The study says that in Europe, many different energy crops are being developed for the production of the gaseous biofuel. The researchers have found that dedicated maize varieties, as well as new grass hybrids, yield such high amounts of energy per hectare that Europe has enough land available for the biogas strategy (Biopact, 2007f).

Industrialized countries commonly use biogas digesters where animal dung, and increasingly fuel crops, are used as feedstock for large-scale biogas digesters. Brazil and the Philippines lead the world in crop-based digesters using sugar-cane residues as feedstock.

Interest and public support for biogas has been growing in most of the European countries. After a period of stagnation caused by technical and

Source: wikipedia.org/Biogas

Figure 13.3 Biogas holder and flare

Source: Biopact (2007f)

Figure 13.4 Biogas injection in the gas pipeline

Table 13.3 Comparison of various biofuels in Germany

Biofuels	Annual yield L/ha	Fuel equivalent	Prices at fuel station per litre in Germany
Rape oil	1.480L	1L = 0.96L diesel	€1.18 (May 2008)
Biodiesel (rape methyl ester)	1.550L	1L = 0.91L diesel	€1.40 (June 2008)
Bioethanol	2.560L	1L = 0.65L benzin	€1.21 (E85, May 2008)
BTL Fuel (biomass-to-liquid)	Up to 4.030L	1L = 0.97L diesel	
Biomethane	3.540kg	1kg = 1.40L gasoline	€0.93 (June 2008)

Source: Fachagentur Nachwachsende Rohstoffe e.V. (FNR) (2008)

Table 13.4 Biogas production of various biofuels

Material	Biogas yield (FM = Fresh matter)	Methane content
Maize silage	202m³/t FM	52%
Grass silage	172m³/t FM	54%
Rye (whole plant)	163m³/t FM	52%
Fodder beet	111m³/t FM	51%
Bio residues	100m³/t FM	61%
Poultry slurry	80m³/t FM	60%
Sugar beet slices	67m³/t FM	72%
Pig slurry	60m³/t FM	60%
Cattle manure	45m³/t FM	60%
Cereal swill	40m³/t FM	61%
Pigs liquid manure	28m³/t FM	65%
Cattle liquid manure	25m³/t FM	60%

Source: Fachagentur Nachwachsende Rohstoffe e.V. (FNR) (2008)

economical difficulties, the environmental benefits and increasing price of fossil fuel have improved the competitiveness of biogas as an energy fuel. This has been seen in both small- and large-scale plants in Denmark, Germany (with over 3000 plants producing 500MW electricity and 1000MW of heat) and Switzerland, and as a transport fuel in Sweden (where vehicles using biomass were voted environmental cars of the year in 2005). There have been interesting biogas projects in the UK, Ireland and The Netherlands. Despite this, the use of biogas in Europe is modest in relation to the raw-material potential, and biogas produces only a very small share of the total energy supply.

Several countries are experimenting with dedicated biogas energy crops, such as newly bred grass varieties (Sudan grass and tropical grass hybrids) or biogas 'super maize' developed in France. The crops are developed in such a way that they ferment easily and yield enough gas when used as a single substrate. Biogas crops can be used whole, which allows for the use of far more biomass per hectare.

When produced on a large scale, biogas can be fed into the natural gas grid and enter the energy-mix without consumers being aware of the change. A select number of European firms have already begun doing so, while farmers who generate excess biogas on their farms make use of incentives to sell the electricity they generate from it to the main power grid. In Germany, electricity from biogas is an integral part of the energy market. In 2005, biogas units produced 2.9 billion kWh of electricity (Fachagentur Nachwachsende Rohstoffe e.V. (FNR), 2008).

India is planning to deal with one of its major problems – air pollution from transport, through the use of compressed biogas (CBG). Since over 70 per cent of the world's long-term (2030) growth in demand for automotive fuels will come from rapidly developing countries like India this is highly relevant and is currently in the research phase (Biopact, 2007g).

A feasibility study concerning utilizing energy crops and crop residues in methane production through anaerobic digestion in boreal conditions was evaluated by Lehtomäki (2006):

Potential boreal energy crops and crop residues were screened for their suitability for methane production, and the effects of harvest time and storage on the methane

Table 13.5 Biogas yield

Substrate	Biogas yield m³/t
Corn silage	400
Fresh grass	500
Vegetables	400
Grain	560

Source: zorg-biogas.com

potential of crops was evaluated. Co-digestion of energy crops and crop residues with cow manure, as well as digestion of energy crops alone in batch leach bed reactors with and without a second stage upflow anaerobic sludge blanket reactor (UASB) or methanogenic filter (MF) were evaluated. The methane potentials of crops, as determined in laboratory methane potential assays, varied from 0.17 to 0.49m³ CH_4 per kilogram VS_{added} (volatile solids added) and from 25 to 260m³ CH_4 per tonne ww (tonnes of wet weight). Jerusalem artichoke, timothy-clover and reed canary grass gave the highest methane potentials of 2900–5400m³ CH_4 per hectare, corresponding to a gross energy potential of 28–53MWh/ha and 40,000–60,000km/ha in passenger car transport. The methane potentials increased with most crops as the crops matured. Ensiling without additives resulted in minor losses (0–13 per cent) in the methane potential of sugar beet tops but more substantial losses (17–39 per cent) in the methane potential of grass, while ensiling with additives was shown to have potential in improving the methane potentials of these substrates by up to 19–22 per cent. In semi-continuously fed laboratory continuously stirred tank reactors (CSTRs) co-digestion of manure and crops was shown feasible with feedstock VS containing up to 40 per cent of crops. The highest specific methane yields of 0.268, 0.229 and 0.213m³ CH_4 per kilogram VS_{added} in co-digestion of cow manure with grass, sugar beet tops and straw, respectively, were obtained with 30 per cent of crop in the feedstock, corresponding to 85–105 per cent of the methane potential in the substrates as determined by batch assays. Including 30 per cent of crop in the feedstock increased methane production per digester volume by 16–65 per cent above that obtained from digestion of manure alone. In anaerobic digestion of energy crops in batch leach bed reactors, with and without a second stage methanogenic reactor, the highest methane yields were obtained in the two-stage process without pH adjustment. This process was well suited for anaerobic digestion of the highly degradable sugar beet and grass-clover silage, yielding 0.382–0.390m³ CH₄ per kilogram VS$_{added}$ within the 50–55 day solids retention time, corresponding to 85–105 per cent of the methane potential in the substrates. With the more recalcitrant substrates, first year shoots of willow and clover-free grass silage, the methane yields in this process remained at 59–66 per cent of the methane potential in substrates. Only 20 per cent of the methane potential in grass silage was extracted in the one-stage leach bed process, while up to 98 per cent of the total methane yield in the two-stage process originated from the second stage methanogenic reactor. (Lehtomäki, 2006)

Internet resources

http://practicalaction.org/practicalanswers/product_info.php?products_id=42

http://news.mongabay.com/bioenergy/2007/02/study-biogas-can-replace-all-eu-imports.html

www.arti-india.org/content/view/45/52/

http://en.wikipedia.org/wiki/Biogas

http://i76.photobucket.com/albums/j14/biopact/comparison_biogas_biofuels-1.jpg?t=1174075940

http://zorg-biogas.com/biogas-plants/industrial-solutions/biogas-plant-for-energy-crops

Table 13.6 Examples of the methane and gross energy potentials of energy crops and crop residues as reported in the literature

Substrate	Methane potential				Gross energy potential	Ref.
	(m³ CH₄/ kg VS$_{added}$)	(m³ CH₄/ kg TS$_{added}$)	(m³ CH₄/ t ww)	(m³ CH₄/ ha/yr)	(MWh/ha/yr)	
Forage beet	0.46	n.r.	n.r.	5 800[a]	56[ac]	1
"	0.36	0.32[c]	55[c]	3 240[b]	34[b]	2
Alfalfa	0.41	n.r.	n.r.	3 965[a]	38[ac]	1
"	0.32	0.28[c]	56[c]	2 304[b]	24[b]	2
Potato	0.28	n.r.	n.r.	2 280[a]	22[ac]	1
Maize	0.41	n.r.	n.r.	5 780[a]	56[ac]	1
Wheat	0.39	n.r.	n.r.	2 960[a]	28[ac]	1
Barley	0.36	n.r.	n.r.	2 030[a]	20[ac]	1
Rape	0.34	n.r.	n.r.	1 190[a]	12[ac]	1
Grass	0.41	n.r.	n.r.	4 060[a]	39[ac]	1
"	0.27	0.24[c]	46[c]	1 908[c]	20[b]	2
"	0.27–0.35	0.25–0.32	64–83	n.r.	n.r.	3
Clover	0.35	n.r.	n.r.	2 530[a]	25[ac]	1
"	0.14–0.21	0.12–0.19	24–36	n.r.	n.r.	3
Marrow	0.26	n.r.	n.r.	1 680[a]	16[ac]	1
kale	0.32	0.28[c]	42[c]	2 304[b]	24[b]	2
Jerusalem artichoke	0.27	0.24[c]	49[c]	2 862[b]	30[b]	2
Sugar beet	0.23	0.19[c]	n.r.	n.r.	n.r.	4
tops	0.36–0.38	0.29–0.31[c]	36–38[c]	n.r.	n.r.	5
Straw	0.25–0.26	0.23–0.24	139–145	n.r.	n.r.	3
"	0.30[c]	0.25[c]	n.r.	n.r.	n.r.	6

[a] in Germany, [b] in Sweden, [c] Values calculated from the data reported, VS = volatile solids, TS = total solids, t ww = tonnes of wet weight, n.r. = not reported. 1: Weiland, 2003; 2: Brolin et al, 1988; 3: Kaparaju et al, 2002; 4: Gunaseelan, 2004; 5: Zubr, 1986; 6: Badger et al, 1979.

Source: Lehtomäki (2006)

Table 13.7 Examples of the effect of harvest time on methane potentials of energy crops as reported in the literature

Substrate	Stage of growth at harvest	Methane potential			Ref.
		(m³ CH₄/ kg VS_added)	(m³ CH₄/ kg TS_added)	(m³ CH₄/ t ww)	
Clover	vegetative	0.14–0.21	0.13–0.19	24–36	1
	flowering	0.14	0.12	17	1
Clover	vegetative	0.38	n.r.	n.r.	2
	budding	0.55	n.r.	n.r.	2
	flowering	0.56	n.r.	n.r.	2
Ryegrass	vegetative	0.42	n.r.	n.r.	2
	earing	0.62	n.r.	n.r.	2
	flowering	0.63	n.r.	n.r.	2
Wheat	flowering	0.42	n.r.	n.r.	2
	milky	0.39	n.r.	n.r.	2
	pasty	0.38	n.r.	n.r.	2
Barley	flowering	0.44	0.40[a]	74[a]	2
	milky	0.50	0.47[a]	129[a]	2
	pasty	0.35	0.33[a]	155[a]	2
Rye	flowering	0.37	0.34[a]	85[a]	3
	milky	0.41	0.38[a]	112[a]	3
	pasty	0.28	0.27[a]	164[a]	3
Triticale	flowering	0.53	0.50[a]	177[a]	3
	milky	0.46	0.44[a]	148[a]	3
	pasty	0.34	0.33[a]	215[a]	3

[a] Values calculated from the data reported, n.r. = not reported. 1: Kaparaju et al, 2002; 2: Pouech et al, 1998; 3: Heiermann et al, 2002.

Source: Lehtomäki (2006)

Table 13.8 Durations of the batch assays, total methane potentials of the substrates (methane potential of inoculum subtracted) and short-term (30 and 50 day) methane potentials expressed as proportion of the total methane potential (averages of replicates ± standard deviations where applicable)

Substrate	Harvest	Duration of the batch assay	Total methane potential			Short-term methane potential	
		(d)	$(m^3\ CH_4/$ kg $VS_{added})$	$(m^3\ CH_4/$ kg $TS_{added})$	$(m^3\ CH_4/$ t ww)	(% of total on day 30)	(% of total on day 50)
GRASSES:							
Timothy clover grass	1	146	0.37 ± 0.02	0.34 ± 0.02	85 ± 5	70	90
	2	125	0.38 ± 0.00	0.36 ± 0.00	72 ± 0	49	87
Reed canarygrass	1	125	0.34 ± 0.00	0.33 ± 0.00	97 ± 0	76	88
	2	194	0.43 ± 0.02	0.42 ± 0.02	167 ± 8	54	71
Lawn	1	124	0.30 ± 0.04	0.27 ± 0.04	58 ± 8	87	96
LEGUMES:							
Red clover	1	125	0.30 ± 0.06	0.27 ± 0.05	41 ± 8	61	89
	2	124	0.28 ± 0.06	0.26 ± 0.06	68 ± 15	66	89
Vetch–oat mixture	1	130	0.41 ± 0.02	0.37 ± 0.02	57 ± 3	56	89
	2	124	0.40 ± 0.04	0.37 ± 0.04	95 ± 10	72	89
Lupin	1	140	0.36 ± 0.04	0.33 ± 0.04	40 ± 4	71	89
	2	125	0.31 ± 0.06	0.29 ± 0.06	41 ± 8	74	92
LEAFY CROPS:							
Jerusalem artichoke	1	150	0.37 ± 0.06	0.34 ± 0.06	93 ± 15	54	90
	2	140	0.36 ± 0.04	0.34 ± 0.04	110 ± 12	46	86
Giant knotweed	1	125	0.17 ± 0.08	0.16 ± 0.08	32 ± 15	78	80
	2	164	0.27 ± 0.00	0.25 ± 0.00	76 ± 0	73	78
Nettle	1	130	0.21 ± 0.00	0.17 ± 0.00	25 ± 0	93	99
	2	125	0.42 ± 0.06	0.36 ± 0.05	60 ± 9	75	91
Rhubarb	1	107	0.49 ± 0.03	0.42 ± 0.03	40 ± 2	85	94
	2	107	0.32 ± 0.02	0.28 ± 0.02	25 ± 2	89	99
Marrow kale	1	150	0.31 ± 0.02	0.28 ± 0.02	37 ± 2	21	73
	2	140	0.32 ± 0.02	0.29 ± 0.02	38 ± 2	20	77
Tops of sugar beet		139	0.34 ± 0.00	0.29 ± 0.00	34 ± 0	47	88
STRAWS:							
Straw of oats		150	0.32 ± 0.02	0.29 ± 0.02	260 ± 16	44	76
Straw of rapeseed		154	0.24 ± 0.02	0.22 ± 0.02	199 ± 17	60	69

Source: Lehtomäki (2006)

Table 13.9 Annual dry matter yields of crops per hectare in boreal growing conditions, methane and gross energy potentials per hectare, and corresponding passenger car transport in km/ha

Substrate	Yield (t TS/ha)	Methane potential (m³ CH₄/ha/yr)	Gross energy potential (MWh/ha/yr)	Passenger car transport (1000 km/ha/yr)
GRASSES:				
Timothy clover grass	8–11[a]	2900 – 4000	28–38	36–50
Reed canarygrass	9–10[b]	3800 – 4200	37–41	47–53
Lawn	2[c]	500	5	7
LEGUMES:				
Red clover	5–7[a]	1400 – 1900	13–18	17–24
Vetch–oat mixture	5–7[d]	1900 – 2600	18–25	23–32
Lupin	4–7[e]	1300 – 2300	13–22	17–29
LEAFY CROPS:				
Jerusalem artichoke	9–16[f]	3100 – 5400	30–53	38–68
Giant knotweed	15[g]	3 800	36	47
Nettle	6–10[h]	2200 – 3600	21–35	27–45
Rhubarb	2–4[i]	800 – 1700	8–16	11–21
Marrow kale	6–8[j]	1700 – 2300	17–23	22–29
Tops of sugar beet	3–5[j]	900 – 1500	8–14	11–18
STRAWS:				
Straw of oats	2[b,j]	600	6	7
Straw of rapeseed	2[b]	400	4	6

[a] Kangas et al, 2004; [b] Sankari, 1993; [c] Tenhunen and Pelkonen, 1987; [d] Kiljala and Isolahti, 2003; [e] Aniszewski, 1993; [f] Häggblom, 1988; [g] yield in United Kindgom, Callaghan et al, 1985a; [h] Galambosi, 1995; [i] Nissi, 2003; [j] Hyytiäinen et al, 1999.

Source: Lehtomäki (2006)

14

Hydrogen and Methanol Crops

Hydrogen is viewed by many as the most promising fuel for light-duty vehicles (LDVs) for the future. Hydrogen can be produced through a large number of pathways and from many feedstocks (both fossil and renewable). A key issue in evaluating the sustainability of a hydrogen fuel cell vehicle (FCV) is an analysis of the processes employed to produce the hydrogen and the efficiency of its use in the vehicle, the 'well-to-wheel' (WTW) activities.

Several recent WTW studies, which include conventional as well as alternative fuel/propulsion system LDVs, are examined and compared. One potentially attractive renewable feedstock for hydrogen is biomass. A biomass to hydrogen pathway and its use in a FCV has only recently been included in WTW studies. The analysis is based on those studies which include biomass-derived hydrogen, comparing it to gasoline/diesel internal combustion engine vehicles (ICEVs) and FCV which utilize hydrogen from other feedstocks (natural gas and wind-generated electricity).

Since hydrogen is not commercially produced from biomass, all of these studies utilize process and emissions data based on research results and extrapolations to commercial scale. We find that direct comparison of results between studies is challenging due to the differences in study methodologies and assumptions concerning feedstocks, production processes, and vehicles. Overall however, WTW results indicate that hydrogen produced via the gasification of biomass and its use in a FCV has the potential to reduce greenhouse gas (GHG) emissions by between 75 and 100 per cent and utilize little fossil energy compared to conventional gasoline ICEVs and hydrogen from natural gas FCVs. (Fleming et al, 2005)

Bio-feedstocks

The large role biomass is expected to play in future energy supply can be explained by the fact that biomass fuels can substitute more or less directly for fossil fuels in the existing energy supply infrastructure. Intermittent renewables such as wind and solar energy are more challenging to the ways we distribute and consume energy, especially in the transport sector.

The literature distinguishes between first-generation and second-generation biofuels according to production technology and feedstock. The production of first-generation biofuels (bioethanol, biodiesel) employs specially cultivated, often food crops. Second-generation technologies are based on lignocellulosic crops for ethanol and diesel production, i.e. they are based on a feedstock in (potentially) significant supply worldwide.

The plant materials can be categorized into two groups according to their characteristics.

- Cereals, grains, sugar crops and other starches can easily be fermented to produce ethanol, which might be used pure or as a blend with normal petrol.
- Cellulosic materials like grasses, trees, different types of waste products and residues from crops, wood processing, and municipal solid waste, can also be converted into alcohol or synthesis gas, but these processes are much more complex. The drawback of these processes is the rather immature technology.

Conversion pathways

Gasification to provide fuel gas

Biomass gasification with turbine or engine to power production has an efficiency of about 40–50 per cent. This technology is also commercially available. The biomass gasification process is also referred to as 'pyrolysis by partial oxidation'. It intends to maximize the gaseous product, and generally takes place between 800 and 1000°C. The product is fuel gas, which can be upgraded to methanol by synthesis, combusted to generate heat, or can be used in engines, high-temperature turbines or fuel cells to generate power. The gas is very costly to store or transport because of the low energy density so it has to be used locally.

Biological gasification to produce hydrogen

The main objective of this process is to produce hydrogen from crops and wastes employing anaerobic, thermophilic or hyperthermophilic micro-organisms in order to supply the fuel cell industry with clean hydrogen gas derived from renewable resources. The final product is hydrogen.

Gasification of biomass into synthesis gas

Gasification is the process of gaseous fuel production by partial oxidation of a solid fuel. This means in common terms to burn with an oxygen deficit. The gasification of coal is well known, and has a history back to the year 1800. The oil shortage of World War II imposed an introduction of almost a million gasifiers to fuel cars, trucks and buses. One major advantage with gasification is the wide range of biomass resources available, ranging from agricultural crops and dedicated energy crops to residues and organic wastes. The feedstock might have a highly variable quality, but still the produced gas is quite standardized and produces a homogeneous product. This makes it possible to choose the feedstock that is the most available and economic at all times.

Gasification occurs in a number of sequential steps:

- drying to evaporate moisture;
- pyrolysis to give gas, vaporized tars or oils and a solid char residue; and
- gasification or partial oxidation of the solid char, pyrolysis tars and pyrolysis gases.

Not all the liquids from the pyrolysis are converted to syngas, due to physical limitations of the reactor and chemical limitations of the reactions. These residues form contaminant tars in the product gas, and have to be removed prior to e.g. a Fischer-Tropsch reactor. Other impurities in the producer gas are the organic BTX (benzene, toluene and xylene – benzene components with one or two methyl groups attached), and inorganic impurities as NH_3, HCN, H_2S, COS and HCl.

There are also volatile metals, dust and soot. The tars have to be cracked or removed first, to enable the use of conventional dry gas cleaning or advanced wet gas cleaning of the remaining impurities. There are three main ways of tar removal/cracking: thermal cracking, catalytic cracking or scrubbing.

Despite the long experience with gasification of biomass, there are some problems with large-scale reliable operations. No manufacturer of gasifiers is willing to give a full guarantee for technical performance of their gasification technology. Though they are sold commercially, they are not delivered with the same kind of operational guarantee as, for example, a gas turbine. This shows the limited operational experience and lack of confidence in the technology, but in comparison with alternative routes to utilize cellulosic biomass, gasification is well proven and one of the possible technologies to be introduced commercially as a major part of the energy route to biofuel (Adam, 2006).

Fischer-Tropsch synthesis

The Fischer-Tropsch (FT) reaction produces hydrocarbons of variable chain length from syngas. This process is operated commercially at Sasol South Africa (from coal-derived syngas) and Shell Malaysia (from natural gas-derived syngas). However, the application of biomass-derived syngas offers several challenges. There are no commercial processes yet, but in April 2008 Choren Fuel Freiberg GmbH & Co. KG completed the building phase of the first scaled-up production facility. The planned production capacity of this facility is 15,000t/yr.

Ethanol from syngas

Another application of syngas is bacterium conversion to ethanol done by the firm Bioengineering Resources Inc. (BRI). The anaerobic bacterium *Clostridium ljungdahlii* converts H_2 and CO into ethanol. The reactor fermentation vessels are constructed with the aim of short retention time, even at atmospheric pressure, thereby keeping equipment costs low. A distinct advantage of the syngas fermentation route is its ability to process nearly any biomass resource, and the high selectivity of the modified bacteria culture. Expected yields from a biomass syngas-to-ethanol facility with no external fuel source provided to the gasifier, are 264–397 litres of ethanol per tonne of dry biomass fed. This gives an energy efficiency of 39 per cent. The syngas fermentation approach has received very modest levels of support in the past.

Currently, there are only a handful of academic groups working in this area. More resources are needed if the pace of progress is to increase.

Pyrolysis

Pyrolysis is the thermal degradation of biomass in the absence of oxidizing agents at 200–500°C. Depending on the method used, the process leads to a mixture of tar vapours, gases and highly reactive carbonaceous char of different proportions. Using high heating rates, moderate temperatures (500°C) and very short residence times the tar compound can be maximized (bio-oil production), while using low temperatures and long residence times the char yield can be maximized.

The char produced can be upgraded to activated carbon, domestic cooking fuel or barbecue charcoal.

There are several ways to produce fuels from biomass. Examples of liquid biofuels with potential are biodiesel, bioethanol, bio-oil and Fischer-Tropsch liquids.

Pyrolysis liquid – bio-oil

Pyrolysis liquid is referred to by many names including pyrolysis oil, bio-oil, bio-crude oil, biofuel oil, wood liquids, wood oil, liquid smoke, wood distillates, pyroligneous tar, pyroligneous liquid and liquid wood. The crude bio-oil is dark brown and approximates to biomass in elemental composition. It cannot be used in conventional gasoline or diesel engines because of corrosion and plugging.

Biomass-to-liquid (BTL) fuels

The synthesis gas produced from the gasification process can be delivered to a Fischer-Tropsch (FT) reactor, where long hydrocarbon chains are produced and it is possible to obtain large amounts of e.g. diesel. There are many types of fuels which are obtainable from syngas (Figure 14.2).

The diesel produced in this way is very similar to conventional diesel and can be used without any adjustment to existing infrastructure or engine systems. It has a high cetane number and therefore has much better ignition performance than conventional diesel fuel. It is a cleaner fuel as it has no aromatics or sulphur, which significantly reduces pollutants from exhaust emissions (choren.com).

New developments

Shell Hydrogen LLC and Virent Energy Systems, Inc. announced in 2007 a five-year joint agreement to further develop and commercialize Virent's BioForming technology platform for hydrogen production. Virent's technology enables the economic production of hydrogen, among other fuels and chemicals, from renewable glycerol and sugar-based feedstocks. The vast majority of hydrogen today is produced using fossil fuels, including natural gas and coal (renewableenergyworld.com)

Scientists, in a report published in the new journal *ChemSusChem* (Idriss et al, 2008), claim they have created an entirely natural and renewable

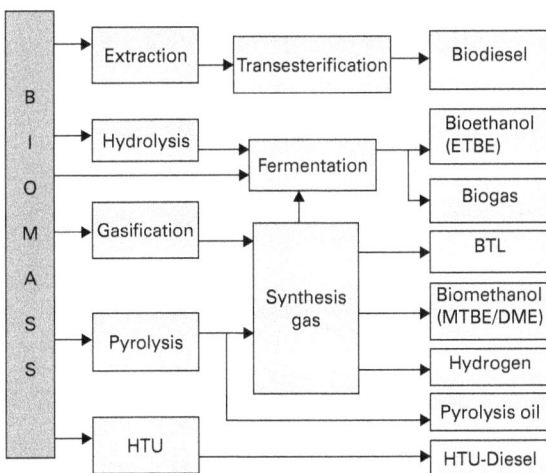

Source: Adam (2006)

Figure 14.1 Liquid biofuels

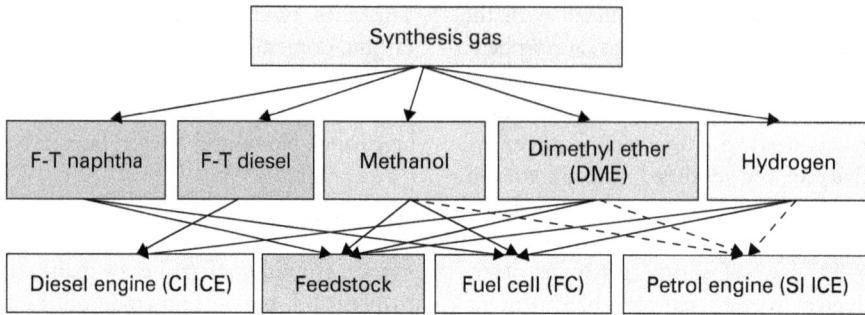

Source: EU (2005)

Figure 14.2 Conversion routes and possible applications of syngas (BTL fuels)

method for producing hydrogen to generate electricity which could drastically reduce the dependency on fossil fuels in the future. The breakthrough means ethanol which comes from the fermentation of crops can be completely converted to hydrogen and carbon dioxide for the first time. They created the first stable catalyst which can generate hydrogen using ethanol produced from crop fermentation at realistic conditions.

The catalyst is made of very small nanoparticles of metals deposited on larger nanoparticles of a support called cerium oxide which is also used in catalytic converters in cars. At present the generation of hydrogen needed to power a medium-sized fuel cell can be achieved using 1kg of this catalyst. As with traditional methods of hydrogen production, carbon dioxide is still created during the process they have developed. However unlike fossil fuels which are underground they

are using ethanol generated from an above-ground source – plants or crops. This means that any carbon dioxide created during the process is assimilated back into the environment and is then used by plants as part of their natural cycle of growth (rdmag.com).

Internet resources

www.senternovem.nl/mmfiles/Status_ perspectives_ biofuels_EU_2005_tcm24-152475.pdf
www.choren.com/en
www.rdmag.com/ShowPR.aspx?PUBCODE=014&ACCT=1400000100&ISSUE=0901&RELTYPE=MS&PRODCODE=00000000&PRODLETT=O&CommonCount=0
www.renewableenergyworld.com/rea/news/article/2007/08/shell-invests-in-hydrogen-from-biomass-production-49525

15
Underutilized and Unexploited Crops

Unexploited and underutilized crops represent a huge potential which exists in some regions of the world with specific climatic conditions. Their exploitation and improvement would lead to ensuring the supply of food, fodder, fuels and fibre to people in rural areas. The utilization of these crops could contribute to preserving these crops worldwide (Duke, 1984).

Internet resources

www.hort.purdue.edu/newcrop/indices/
http://en.wikipedia.org/wiki/File:Starr_020617-0019_Syagrus_romanzoffiana.jpg
http://en.wikipedia.org/wiki/File:Queenfruit.JPG

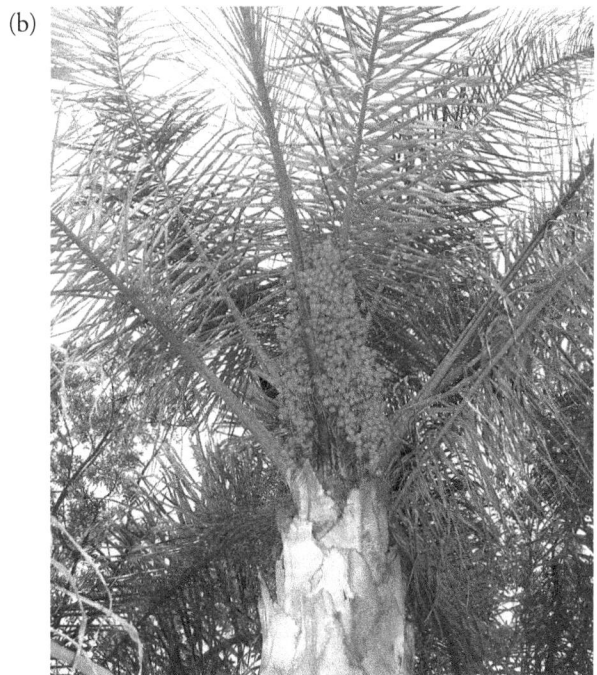

(a)

(b)

Source: (a) Forest & Kim Starr; (b) Mmcknight4, Wikipedia.org

Figure 15.1 Palmeira jerivá (*Syagrus romanzoffiana*): (a) tree; and (b) ripe fruit

Table 15.1 Various potential oil- and wax-producing crops

Botanical name	Common name	Notes/region
Anacardiaceae		
Joxicodendron succedaneum (L.) O. Kuntze (*Rhus succedanea* L.)	Japan tallow, wax tree, sumac cirier	Fatty oil in kernel; fruit mesocarp contains 65% wax, used for polishes, candles, matches. Subtropics
Arecoideae		
Euterpe edulis Mart.	Juçara, jiçara, assai	Oil in pulp. Brazil
Jessenia bataua (Mart.) Burret.	Ungurahui, seje, patauá, kumbu	Oil in pulp, good tasting, similar to olive oil. The palm is planted near the dwellings of native people. South America
J. polycarpa Karst.	Seje grande, coroba, milpesos, jagua	Oil in pulp, cooking oil of the Choco Indians. South America
J. repanda Engel	Aricagui, aricacua	Oil in pulp. The oil content of the mesocarp and the taste of the oil are similar to seje. Central America
Manicaria saccifera Gaertn.	Sleeve palm, monkey cap palm, temiche, ubussu, guagara, bassu.	Oil in seeds, locally used as food. Northern South America
Oenocarpus bacaba Mart.	Turu palm, bacaba	Oil in pulp, good tasting edible oil. The palm is planted near Indian settlements. Northern South America
Betulaceae		
Corylus avellana L.	Hazelnut, hazel, noisetier, coudrier	Nuts contain 60–68% fat
Caryocaraceae		
Caryocar villosum (Aubl.) Pers.	Piquia, pequia	Mesocarp and seeds contain 70% fat. Tropics
Chrysobalanaceae		
Licania rigida	Oiticicabaum, oiticica	Evergreen tree, occurs mostly wild; seeds contain 55–63% fat, like tung oil. South America
Licania rigida Benth.	Oiticia	Oil contains 73–83% licanic acid, uses similar to tung oil; production of 50,000t seeds (wild grown). Tropics
Cocoidae		
Acrocomia aculeata (jacj.) Lodd. ex Mart.	Paraguay palm, macaya, micauba, gru-gru	Oil in seeds and pulp, 50–55% oil in the kernels; fruit oil for soap, kernel oil is edible. Tropics
A. mexicana Karw.	Coyol, babosa, Mexican gru-gru palm, wine palm	40% fat in the seeds, sold in local markets. Mexico to Guatemala
A totai Mart.	Mbocayá palm, totai palm, gru-gru palm	Oil in seeds and pulp; industrial extraction in Paraguay, small amount exported. Latin America
Astrocaryum jauary Mart.	Awarra palm, jauary palm	Edible oil in pulp and seeds. South America
A murumuru Mart.	Murumuru	43% oil in the seeds, high melting point (33–36°C). Amazon region
A. tucuma Mart.	Tucuma palm, tucum	Oil in pulp and seeds, pulp oil for soap; 30–50% fat in kernels; also exported. Tropical America
A. vulgare Mart.	Awarra palm, cumara palm, aoura palm	Oil in pulp and seeds; grows well on sandy ground. South America
Attalea funifera Mart, ex Spreng	Bahia piasawa, coquilla nut	Oil in seeds, used for margarine and chocolate industry; main product: piassava fibre. South America
Bactris gasipaës H. B. K. (*Guilielma gasipaes*)	Peribaye, peach palm, paréton piritu, gacipaes, masato, uvito, chonta ruru, pupunha, amana	Oil in seeds, important food plant because of its high-starch fruit flesh; seeds provide macanilla fat. Central and South America

Table 15.1 Various potential oil- and wax-producing crops (*Cont'd*)

Botanical name	Common name	Notes/region
Maximiliana maripa Drude	Maripa, naxáribo	Oil in seeds (maripa fat), fruit flesh tastes of apricot. Guyana
M. regia Mart.	Cocorite palm, cucurita palm, inajá palm, jaguá palm	Oil in seeds, similar to *M. maripa*. South America
Orbygnia cohune (Mart.) Dahlgr. ex Standl	Cohune palm, manaca	Oil in seeds; the oil is used as salad oil or for technical purposes. Central America
Scheelea martiana Burret (*Attalea excelsa* Mart, ex Spreng)	Ouricoury, urucuri	Oil in seeds; the kernels are pressed; the underside of the leaves produces wax. Latin America
Syagrus coronata (Mart.) Becc.	Ouricoury, licuri	70% oil in seeds; the leaves provide ouricury wax; the fruit flesh is edible. South America

Compositae

Guizotia abyssinica (L. f.) Cass.	Nigerseed, niger, tilangi, ram-till, nook, nug, guja	Valuable vegetable oil. Tropics, tropical highlands and subtropics
Madia sativa Mol.	Chilean tarweed, coast tarweed, pitchweed, madi. melosa	Fruit contains 35–40% oil, undemanding plant; ancient oil plant of the Indians. Subtropics

Coryphoideae

Erythea salvadorensis (H. Wendl. ex Becc.) H. E. Moore	Palma	Oil in seeds, used locally. Central America

Cruciferae

Berteroa incana (L.) DC.	Grey cress	Oil in seeds, used locally
Brassica carinata A. Braun	Abyssinian mustard, Ethiopian cabbage	Oil and vegetable plant. Tropical highlands
B. junkea Czern. et Coss	Chinese mustard, Indian brown mustard, raya, rai	Seeds are used for oil extraction or seasoning. Tropics, subtropics
B. napus L. emend. Metzger var. *napus*	Rapeseed, colza, colsat. navette	Oil for nutrition and technical purposes. Subtropics, temperate zones
B. rapa L. emend. Metzger (*B. carnpestris*)	Various varieties	Oil of all varieties edible or for technical purposes
var. *dichotoma* Watt	Brown sarson, Indian rape	Temperate zones, subtropics, tropics
var. *sarson* Prain	Indian colza, yellow sarson, sarisa	Temperate zones, subtropics, tropics
var. *silvestris* (Lam) Briggs (var. *oleifera* DC)	Turnip rape, Chinese colza	Cultivation limited to Europe. Temperate zones, subtropics
var. *toria* Duthie & Fuller	Indian rape, toria	Mostly grown with irrigation. Temperate zones, subtropics, tropics
Camelina sativa (L.) Crantz	Leindotter, false flax, gold of pleasure	30–35% drying oil in seeds; originally a weed in flax fields. Temperate regions
Crambe abyssinica	Abyssinian kale, crambe, cole wort	The oil is only utilized technically. Tropics, subtropics, tropical highlands
Eruca vesicaria (L.) Cav. ssp. *sativa*	Olrauke, rocket	26–33% oil in seeds, for nutrition and as fuel. Russia, India
Raphanus sativus L. var. *oleiformis* Pers.	Chinesischer, Ölrettich, oilseed radish, oil radish, fodder radish	Oil in seeds, for nutrition and as fuel. East Asia
Raphanus raphanistrum L.	Hederich	Oil in seeds, used locally

Cucurbitaceae

Telfairia pedata (Sm. ex Sims) Hook.	Kweme, oyster nut tree	Perennial, climbing on trees; fruits, 36% oil in seeds, 60% in shelled seeds; cultivated plants productive for 30 years

Cyperaceae

Cyperus esculentus	Earth almond, yellow nutsedge, chufa	20–25% oil in tubers, mostly ester of oleic acid, yield 0.5–1.2t/ha. Tropics of East Africa

Table 15.1 Various potential oil- and wax-producing crops (*Cont'd*)

Botanical name	Common name	Notes/region
Dipterocarpaceae		
Shorea stenoptera Burck	Borneo tallow, pontianak kernels, illipe, engkabang	62% saturated fatty acids, 43% stearic acid, 37% oleic acid; several species, but only wild trees, are used. Tropics
Euphorbiaceae		
Aleurites fordii Hemsl., *A. montana, A. cordatus*	Tung oil tree, alévrite	Related to castor, up to 12m high, seed in capsules; 42–53% drying oil in seed, oil purgative and emetic. Tropics
Sopium sebiferum (L.) Roxb.	Chinese tallow tree, tallow berry, páu de sebo, chii an shu	Mesocarp 62% palmitic acid, 27% oleic acid; kernels 53% linoleic acid, 30% linolenic acid. Subtropics, tropical highlands
Tetracarpidium conophorum Hutchin. et Dalz (syn. *Plukenetia conophora*)	Conophor, awusa nut, African walnut	Seeds contain 54–60% drying oil. Western Africa
Euphorbia antisyphilitica Zucc.	Candelilla	Wax covers shoots, obtained by boiling, gathered from wild plants, used for polishes, painting materials, waxing of fruits. Tropics
Pedilanthus pavonis Boiss.	Candelilla	Other Euphorbiaceae species are used similarly. Tropics
Fagaceae		
Fagus sylvatica L.	European beech, hêtre	Seeds contain up to 46% oil, mostly glycerides of oleic and linoleic acid; trees only bloom and form fruit every 5–10 years
Gramineae		
Zea mays L.	Mais, maize, corn, maïs	Embryos, divided during starch production from the endosperm, contain 33–36% oil, by-product of starch production
Labiatae		
Hyptis spicigera Lam.	Moño, nino, kindi, andoka, hard simsim	Seeds eaten or pressed, 24–37% of drying oil; cultivation similar to sesame. Tropics
Perilla frutescens (L.) Britt. (*P. ocymoides* L.)	Perilla, tzu-su, hsiang-sui	Seeds contain 30–51% oil, 63–70% linolenic acid, for cooking oil and technical purposes. Subtropics, tropical highlands
Lauraceae		
Persea americana	Avocado	Pulp contains up to 30% oil used for oil production locally in Guatemala
Lecythidaceae		
Bertholletia excelsa Humb. et Bonpl.	Brazil nut tree, paranut tree	Seeds contain about 66% oil, rich in oleic and linoleic acid, used for nutrition
Lepidocaryoorideae		
Mauritia flexuosa L. f.	Burity do brejo, aguaje, bâche, rniriti, ita. rnoriche	Oil in seeds. Northern South America
Linaceae		
Linum usitatissimum L.	Linseed, flax, lin, tétard, lino	Main use for painting materials, biofuels; yields 800–2200kg/ha, linseeds 765–1300kg/ha. Temperate zones, subtropics, tropical highlands

Table 15.1 Various potential oil- and wax-producing crops (*Cont'd*)

Botanical name	Common name	Notes/region
Malvaceae		
Gossypium hirsutum L.	Baumwolle, cotton, coton	16–24% oil in seeds, by-product of fibre production; oil and press cake contain poisonous phenolic dialdehyde; press cake contains 23–44% protein, used as fodder
Myriaceae		
Myrica cerifera L	Wax myrtle, bayberry, arrayán, cera vegetal	Obtained from wild plants; several other Myrica species also produce bayberry wax. Subtropics, tropical highlands
Myristicaceae		
Virola surinamensis	Ucuhuba	Seeds contain 65% fat, 63–72% is myristic acid; collected from wild trees. Tropics.
Palmae		
Ceroxylon alpinum Bonpl. ex DC. (*C. andicala* Humb. et Bonpl.)	South American wax palm, palma Cera	Stem is covered by wax, used locally for production of candles and matches. Tropical highlands
Capernicia alba Morong (*C. australis* Becc.)	Caranday, carandá	Large stands in Mato Grosso, as yet little used. Tropical highlands
C. prunifera (Mill.) H. E. Moore (*C. cerifera* (Arr. da Cam. ex Koster) Mart.)	Carnauba wax palm	Large stands in northeast Brazil, 12,000t annual production; wax covers leaves, high melting point (83–86°C). Tropical highlands
Syagrus coronata (Mart.) Becc.	Ouricury palm, ouricuru, uricuri, Iicuri, nicuri	Mainly from Bahia, on dry sites; wax from leaves, obtained like carnauba wax. Tropical highlands
Papaveraceae		
Papaver somniferum L.	Opium poppy, pavot somnifere, dormideira	Vegetable oil, also used technically. Temperate zones, subtropics, tropical highlands
Pedaliaceae		
Sesamum indicum L.	Sesam, oriental sesame, sesame	Annual herb; seed in capsules, old cultivars with opening capsules; short-day plant; non-drying oil, 42–70% protein in the press cake
Polygalaceae		
Polygala butyracea Heckel	Black beniseed, malukang, ankalaki, cheyi	Seeds contain 30% fat (55–60% palmitic acid, 30% oleic acid). Tropics
Madhuca longifolia (J. G. Koenig) Macbr. var. *longifolia,* var. *latifolia* Roxb.) Cheval. (*M. indica* J. F. Gmel.)	Illipé, mahua, moa tree, butter tree, mahawa, mowara	Seeds contain 55–60% fat (23% palmitic acid, 19% stearic acid, 43% oleic acid, 13% linoleic acid), mostly used for technical purposes. Tropics
Theaceae		
Camellia sasanqua Thunb.	Teaseed, sasanqua camellia	Seeds contain 40% oil with 72–87% oleic acid, mostly used for nutritional purposes. Tropics

Source: Diercke (1981); Franke (1985); Rehm and Espig (1991)

Table 15.2 Various sugar- and starch-producing plant species with potential for ethanol production

Botanical name	Common name	Notes/region
Alismataceae		
Sagittaria sagittifolia L. ssp. *leucopetala* (Miq.) Hartog ex van Steenis	Chinese arrowhead, muya muya, ubi keladi	Cultivated in ponds; not edible raw, used for fodder and starch production. Subtropics, tropics
Andropogonoideae		
Coix lacryma-jobi	Adlay, capim rosarie, larmes de Job, Iágrima de San pedro, Job's tears	Perennial; thin-shelled forms are used as a cereal, thick-shelled as pearls. Tropics
Araceae		
Amorphophallus konjak K. Koch	Elephant yam, konjaku (Japan)	Flour used for the production of vermicelli and cakes. Subtropics, tropics
A. paeoniifolius (Dennst.) Nicolson (*A. campanulatus* Bl. ex Deene.)	Oroy, elephant foot, whitespot arum, telinga potato, suran, pongapong	Old tubers weigh up to 25kg; some forms contain large amounts of oxalate. Tropics
Colocasia esculenta (L.) Schott	Dasheen, taro, 'old' cocoyam, eddo	About 1000 cultivars; single large tuber, cultivated in water (dasheen) or many small daughter tubers, cultivated on dry land (eddoes); 6–15 month growing period, withstands high soil salt contents. Tropics and subtropics
Cyrtosperma chamissonis (Schott) Merr.	Giant swamp taro, maota (Tahiti), gallan, opeves, palauan	Tubers weigh up to 50kg after several years; grows well on coral islands, also on swampy saline soil. Tropics
Xanthosoma sagittifolium (L.) Schott	Yautia, 'new' cocoyam, tannia, tarias, Chinese taro, taye, ocumo, mangarito, taioba	Many cultivated forms. Grown worldwide in the tropics
Basellaceae		
Ullucus tuberosus Lozano	Ulluco, melloco, papa lisa, kipa uljuco, chugua	Production in Peru about 30,000t/yr. Tropical highlands
Cannaceae		
Canna edulis Ker-Gawl.	Edible canna, gruya, achira	Yields up to 50t/ha, important fodder. Subtropics, tropics
Chenopodiaceae		
Beta vulgaris var. *altissima* Döll	Sugar beet	Under favourable conditions: yield 57.4t/ha; sugar content 16% of fresh matter; sugar yield 9.18t/ha; ethanol production 5600L/ha. Temperate climates
B. vulgaris var. *rapacea* W.D.J. Koch	Fodder beet, mangel-wurzel	Good yields approach 98.5t/ha; starch content 8.2% of fresh matter; starch yield 8.08t/ha; ethanol production 4923L/ha. Temperate climates
Cruciferae		
Lepidum meyennii Walp.	Maca, chijura	Cultivated at heights of 3800–4200m. Tropical highlands
Cyperaceae		
Eleocharis dulcis (Burm. f.) Trin. ex Henschel (*Etuberosa* Schult.)	Chinese water chestnut, waternut, pi-tsi, teker, matei (China)	Yields of 20–40t/ha, cultivated in water. Tropics

Table 15.2 Various sugar- and starch-producing plant species with potential for ethanol production (*Cont'd*)

Botanical name	Common name	Notes/region
Dioscorea		
D. alata L.	Water yam, greater yam, winged yam, white yam	Important species; large tubers, 8–10 month growing period
D. bulbifera L.	Aerial yam, potato yam, igname bulbifere, batata de aire	Mostly only aerial tubers edible, tubers store well
D. cayenensis L.	Yellow Guinea yam, igname jaune, ñame amarillo	10-month growing period
D. dumetorum (Kunth) Pax	Cluster yam, African bitter yam, three-leaved yam	Growing period 8–10 months; both poisonous and non-poisonous forms cultivated
D. esculenta (Lour.) Burk	Asiatic yam, lesser yam, potato yam	Only known in cultivation, no bitter forms; growing period 7–10 months
D. hispida Dennst.	Asiatic bitter yam, intoxicating yam	Mostly gathered wild, all forms poisonous
D. japonica Thunb.	Japanese yam, shan-yu-tsai (China)	Food plant in Japan, medicinal plant in China
D. nummularia Lam.	Kerung (Java), ubing basol (Philippines)	Tubers large, deep lying; usually harvested 2–3 years after planting
D. opposita Thunb. (*D. batatas* Decne.)	Chinese yam, cinnamon yam, igname de Chine	Growing period 6 months; tolerant of cold
D. pentaphyllia	Sand yam, buck yam, ubi passir (Java)	Tubers slightly poisonous
D. rotundata	White Guinea yam, eboe yam, ñame blanco	Growing period 6–10 months; only known in cultivation; large tubers
D. trifida	Cush-cush yam, couche-couche, yampi, mapuey, aja	Growing period 9–10 months
Eragrostoideae		
Eleusine corocan (L) Gaertn.	Finger millet, dagusa (Ethiopia), ragi (India), kurakkan	Stores well, used for beer production, flat breads. Cultivated in central and eastern Africa from India to Japan
Eragrostis tef (Zuccagni), Trotter (*E. abyssinica* (Jacq.) Link)	Tef (Amhari), tafi (Galinya), paturin de Abessinie	Beer production, flat breads, fodder. Ethiopia
Labiatae		
Plectranthus edulis (Vatke) Agnew (*Coleus edulis* Vatke)	Gala dinich, oromo diniche	Tropical highlands
P. esculentus N. E. Br. (*Coleus esculentus* (N. E. Br) G. Tayl.)	Kafir potato, dazo, ndazu, rizga	Widely cultivated in the drier parts of tropical Africa
Solenostemon rotundifolius (Poir.) J. K. Morton (*Coleus rotundifolius* (Poir.) A. Chev. et Perr., *C. parviflorus* Benth.)	Hausa potato, Chinese potato, county potato, fra-fra-salaga, koorkan	Cultivars, generally small tubers; vegetation period 3–4 months. Tropics
Leguminosae		
Flemingia vestita Benth. Ex Barker (*Maughania vestita*)	Sòh-phlong	Yields up to 10t/ha, 7 month vegetation period. Tropical highlands
Pachyrhizus ahipa (Wedd.) Parodi	Yam bean, ahipa, jiquima, jicama, ajipa	Old cultivated plant of the Indians. Tropics
P. erosus L. Urb.	Potato bean, yam bean, patate cochon, jicama, sin kama, fan-ko	Young tubers are sweet; tubers over 1 year old are mainly used for starch production; yields up to 95t/ha, propagated by seed. Tropics
P. tuberosus (Lam.) Spreng.	Yam bean, jicama, fejião yacatupé	Similar to *P. erosus*, but the tubers are larger. Tropics
Sphenostylis stenocarpa (Hochst.) Harms	African yam bean, kutonoso, roya,	Growing period 8 months, mostly propagated by seed. Tropics

Table 15.2 Various sugar- and starch-producing plant species with potential for ethanol production (*Cont'd*)

Botanical name	Common name	Notes/region
Marantaceae		
Calathea allouia (Aubl.) Lindl.	Guinea arrowroot, sweet corn root, Ilerén, allouya	Small tubers, similar to potato. Tropics
Maranta arundinacea L.	St Vincent arrowroot, herb aux fléches, araruta	Only cultivated for starch production. Tropics
Nyctaginaceae		
Mirabilis expansa Ruiz et Pav.	Marvel of Peru, Mauka	Cultivated in a limited region for its edible roots. Tropical highlands
Oxalidaceae		
Oxalis tuberosa Mol.	Oca, oka, ibia, apillia	Tubers with high sugar and starch content, yields up to 20t/ha; only edible when cooked. Tropical
Panicoideae		
Digitaria cruciata (Nees) A. Camus	Raishan	Only cultivated locally in India (Khasi hills)
Echinachloa frumentacea (Roxb.) Link	Barnyard millet, billion dollar grass, sawa, sanwa	Quick ripening, salt tolerant. Cultivated in India, Southeast Asia
E. utilis Ohwi et Yabuno	Japanese millet, Japanese barnyard millet	Properties as for *Echinochloa frumentocea* but suitable for cooler climates. Cultivated in Japan, China
Panicum miliaceum L.	Common millet, true millet, mijo común, proso	Beer production, unleavened breads, bird seed. Cultivated in Mediterranean countries to East Asia
P. sumatrense Roth ex Roem. et Schult. (*P. miliare* Lam)	Little millet, samo, samai, kutki (India)	Very undemanding; drought tolerant, also tolerant of wetness. Cultivated in India, Sri Lanka
Paspalum scrobiculatum L.	Bastard millet, ditch millet, koda (India)	Undemanding, drought tolerant; fodder plant. Southeast Asia
Pennisetum americanum (L.) Leeke	Pearl millet, bulrush millet, millet á chandelle, mijo perla, dochan, kala sat, bajra (India)	Salt tolerant, drought tolerant. In the Sahel, cultivars are grown with a vegetation period of 60 days. Africa, India
Phalaris canariensis L.	Canarygrass, alpiste	Used as bird seed, also green fodder. Mediterranean countries (Morocco), Argentina
Setaria itolica (L) Beauv.	Italian millet, foxtail millet, millet des oiseaux, moha, painço	Less need for warmth than other millets. Mediterranean countries to Japan, particularly China, central Asia, India
Solanaceae		
Solanum tuberosum	Potato	Cultivation in the tropics increasing; may yield up to 20t/ha, often below 10t/ha
Tropaeolaceae		
Tropaeolum tuberosum	Añu, cubio, isañu, mashua	Yields of 20–30t/ha. Tropical Highlands
Zingiberaceae		
Curcuma angustifolia Roxb.	East Indian arrowroot, tikhur, tavakhira	Provides East Indian arrowroot starch. Tropics

Source: Diercke (1981); El Bassam (1996); Franke (1985); Rehm and Espig (1991)

Table 15.3 Potential energy plant species in arid and semi-arid regions

Botanical name	Common name	Notes
Amaranthaceae		
Aerva		Undershrubs
A. javanica	Eigaab, ghell, toorf, shagaret el-ghazal	Perennial plant, propagated by seed, grows in sandy places; erect, densely stellate, hairy undershrub
A. lanata	Eigaab gebeli	Perennial plant, propagated by seed; grows in sandy and stony places, especially lower parts of mountains; erect woolly undershrub, small leaves
Chenopodiaceae		
Hammada		Shrub with jointed stem
H. elegans	Rimth, remeh, balbal, belbel	Perennial plant, propagated by seed; grows on sandy and stony ground; stout shrub with thick glaucous branches, leaves rudimentary
Kachia		Annual herbs with flat leaves
K. indica	Amm shoor, kokia	Annual plant, propagated by seed, grows in moist conditions; highly branched herb, under favourable conditions reaching a height of 2m
Compositae		
Conyza		Mostly shrubby plants
C. aegyptiaca	Nashash ed-dibbaan	Annual plant, propagated by seed; grows along canals; densely villous herb,
C. bovei	Qessaniya, belleikh	Perennial plant, propagated by seed; grows on sandy and stony ground; glabrous shrub
C. dioscoridis	Barnoof	Perennial plant, propagated by seed; grows in loamy and sandy places; highly branched, hairy shrub, often 2–3m high
C. linifolia	Hasheesh el-gebl, ain el-katkoot	Annual plant, propagated by seed; grows on canal banks and wasteland; densely pubescent herb, numerous leaves
C. triloba	Heleba, habaq atshaan	Perennial plant, propagated by seed; grows on sandy and stony ground; shrublet
Cyperaceae		
Cyperus		Sometimes with creeping rhizome
C. alopecuroides	Samaar helw, mat sedge	Perennial plant, propagated by rhizomes; grows on moist soils; 1m or more high, up to 2cm broad leaves, nut plano-convex
C. articulatus	Boot, dees medawar	Perennial plant, propagated by rhizomes; grows in marshy places and along canals; stout culms, 1–2m high
C. capitatus	Se'd	Perennial plant, propagated by rhizomes; grows on maritime sand; 10–15cm high, stout pale green stems
C. conglomeratus	Se'di, oshob	Propagated by rhizomes; grows on coastal sand dunes; 10–30cm, tufted
C. laevigatus	Barbeit, okreish	Perennial plant, propagated by rhizomes; grows on salty lands; stem leafless or with short fleshy leaves
C. michelianus ssp. *pygmaeu*	Se'ed, sa'ad	Annual plant, propagated by rhizomes; grows on muddy and dry clay soils; small, densely tufted, mostly 15cm high
C. mundtii	Qateefa, qateef	Perennial plant, propagated by rhizomes; grows in moist places; stems creeping below, leafing to the middle with short, broad, flat leaves
C. papyrus	Fafeer, bordi, bardi, papyrus, paper reed	Perennial plant, propagated by rhizomes; grows around lakes; leafless, giant plant, usually over 1.5m high, could be cultivated

Table 15.3 Potential energy plant species in arid and semi-arid regions (*Cont'd*)

Botanical name	Common name	Notes/region
C. rotundus	Sa'd el-homar, nutsedge, purple nutsedge, purple nutgrass	Perennial plant, propagated by rhizomes; grows in sandy places as well as on moist ground; stem only leafy at the ground, stolons sometimes swollen into tubers that are collected as a drug
Gramineae		
Agrostis		Perennial grasses
A. stolonifera	Creeping bentgrass	Perennial plant, propagated by seed; grows in moist places; procumbent, branched weed
Arundo		Robust perennial reeds
A. donax	Ghaab baladi, ghaab farisi, giant reed, Spanish reed	Perennial plant, propagated by rhizomes; mesophytic, robust reed
Avena		Annuals
A. barbata	Zommeyr, slender oat, barbed oat	Annual plant, propagated by seed; grows on sandy fields; glabrous culm, up to 1m high
A. fatua	Zommeyr, wild oat	Annual plant, propagated by seed; grows on clay and sandy fields; common weed, associated with winter cereals
A. sativa	Zommeyr, oat, oats	Annual herb, propagated by seed; grows on clay and sandy fields
A. sterilis	Zommeyr, sterile oat, winter wild oat, wild red oat	Annual herb, propagated by seed; grows on clay and sandy fields
Bromus		Annual grasses
B. adoensis	Yadaab, yadaab gabal	Annual grass, propagated by seed; grows in sandy places; fine, slender grass, culms glabrous, 30cm tall
B. aegyptiacus	Abufakhour, abu keneitla	Annual grass, propagated by seed; grows on clay and sandy ground; 25–50cm high
B. diandrus	Abufakhour, abu keneitla	Annual grass, propagated by seed; grows on clay and sandy ground; up to 80cm high, softly villous leaves
B. fasciculatus	Sabal abu el-hossein	Annual grass, propagated by seed; grows in sandy places; only 5–10cm high
B. fasciculatus var. alexandrinus	Sabal abu el-hossein	Annual grass, propagated by seed; grows on clay and sandy ground; culm upwards, panicle branches, softly pubescent
B. japonicus	Safsoof	Annual grass, propagated by seed; grows in sandy places; culm up to 10cm, hairy leaves
B. lanceolatus	Khafoor, abufakhour	Annual grass, propagated by seed; grows on clay ground; culm 10–15cm
B. madritensis	Khafoor, abufakhour	Annual grass, propagated by seed; grows on sandy and stony ground; medium-sized grass, erect contracted panicle, culm glabrous below the panicle
B. rigidus	Khafoor, abufakhour	Annual grass, propagated by seed; grows in sandy places, close in characteristics to the preceding
B. rubens	Imendi, deil et-ta'lab	Annual grass, propagated by seed; grows in sandy fields; small, tufted, softly pubescent grass
B. scoparius	Sabal el-faar, sabat	Annual grass, propagated by seed; grows on clay and sandy ground; culm glabrous, 20–50cm high
B. sinaicus	Sabal abu el-hossein, sasoof	Annual grass, propagated by seed; grows in desert conditions
B. tectorum	Sabal abu el-hossein, sasoof, downy brome, drooping brome	Annual grass, propagated by seed; grows in sandy places; fine, slender grass, culm up to 60cm high
B. unioloides	Khafoor, abu fakhour, rescue grass, prairie grass	Annual grass, propagated by seed; suitable for several habitats; tall grass

Table 15.3 Potential energy plant species in arid and semi-arid regions (*Cont'd*)

Botanical name	Common name	Notes/region
Cymbopogon		Perennial, aromatic grasses
C. schoenanthus ssp. *proximus*	Halfa barr, camel grass	Perennial plant, propagated by seed; grows in dry places; stout densely tufted grass with narrow leaves
C. schoenanthus ssp. *schoenanthus*	Hashma, camel grass	Perennial plant, propagated by seed; grows in dry and stony places; stout, densely tufted grass, very slight fragrance
Cynodon		Creeping rhizomes or stolons
C. dactylon	Nigeel baladi, Bermuda grass, Bahama grass, couch grass, dhubgrass	Perennial plant, propagated by creeping rhizome; grows on clay and sandy ground; herb with extensively creeping rhizomes, producing rows of leafy culms
Desmostachya		Perennial rigid grasses
D. bipinnata	Halfa, halfa'a	Perennial plant, propagated by rhizomes; grows on canal banks and waste ground; rigid grass with rosetted very long leaves
Hordeum		Annual plants
H. leporinum	Abu shtirt, reesh (abu) el hossein, hare barley	Annual plant, propagated by seed; grows in rocky places and on dry calcareous sandy or loamy ground; flat leaf blades, small culm
H. marinum	Reesh (abu) el hossein (el-hosny)	Annual plant, propagated by seed; grows on sandy soil, Nile alluvium and borders of canals; similar to preceding, but smaller
H. spontaneum	Bahma, bohma, shaeeryia	Annual plant, propagated by seed; grows on desert pastures; tall desert grass
Imperata		Medium-sized grasses
I. cylindrica	Halfa, halfaa, alang-alang, blady grass, cogon grass	Perennial plant, propagated by rhizome; grows in dry or moist sandy places along channels and on wasteland; long leaves in a dense basilar rosette
Panicum		Annuals or perennials
P. turgidum	Abu rokba, turgid panic grass, desert grass	Perennial plant, propagated by rhizomes; grows on sandy and stony deserts; desert grass, growing in dense bushes, up to 1m high or more
Pennisetum		Annuals or perennials
P. divisum	Thommaam	Perennial plant, propagated by rootstocks; grows in desert conditions; bushy desert grass, growing in large thickets
P. orientale	Sabat, sasoof	Perennial plant, propagated by rootstocks; grows on stony and rocky ground; usually tufted slender grass with flat leaves
Phragmites		Perennial robust reeds
P. australis	Hagna, ghaab hagna, common reed grass	Perennial reed, propagated by rhizomes; grows in dry or moist places and along channels; robust plant, culm erect and tall, in undisturbed marsh rising some 5m above the water, much shorter in dry areas
P. mauritianus	Hagna, ghaab hagna, heesh-meddaad	Perennial plant, propagated by rhizomes; grows in dry or moist places and along channels; usually tall and stout
Polypogon		Annual grasses
P. monspliensis	Deil el-qott	Annual grass, propagated by seed; grows in moist sandy places, on canal banks and in sandy fields
Saccharum		Tall perennial grasses
S. spontaneum var. *aegyptiacum*	Heesh, boos, boos boos, farisi, wild cane, wild sugar cane, kans grass	Perennial plant, propagated by seed; grows on canal banks, sandy islets, waste ground and also in rock fissures; up to 3–5m long, grows in dense clumps, very long leaves

Table 15.3 Potential energy plant species in arid and semi-arid regions (*Cont'd*)

Botanical name	Common name	Notes/region
Sorghum		Tall annuals or rarely perennials
S. halepense	Gierraaoo, djarraaoo, Johnson grass	Perennial plant, propagated by rhizomes; grows in sandy and clay places; culm from stout scaly rhizome, up to 150cm tall
S. sudanense	Garawa, Sudangrass	Annual plant, propagated by rhizomes; grows in sandy and clay places; culm 1.5–3m high
S. verticilliflorum	Garawa	Perennial and annual plant, propagated by rhizomes; grows in sandy and clay places; culm 3.5cm high
S. virgatum	Garawa, hasheesh el-fa ras	Perennial and annual plant, propagated by rhizomes; grows along channels and in fields
Juncaceae		
Juncus		Marsh herbs
J. acutus	Sammar morr, asal	Perennial plant, propagated by rhizomes; grows on salty land, in moist places and dry sandy localities; stout tufted culm 1m long
J. littoralis	Sammaar	Perennial plant, propagated by rhizomes; grows on salty land, in moist places and dry sandy localities; tall, pungent, like preceding
J. punctorius var. *exallatus*	Shemoor, dees	Perennial plant, propagated by rhizomes; grows in moist places and in stagnant water; up to 2m high, stem rather thick
J. rigidus	Sammaar morr, sammar hoser el-gibn, sea hard-rush	Perennial plant, propagated by rhizomes; grows on salt marshes and salty land; culms up to 1m long
Leguminosae		
Alhagi		Spiny shrublet
A. maurorum	Aqool	Perennial plant, propagated by rhizomes; grows on canal banks and in waste places; richly branched, spiny shrublet
Liliaceae		
Asparagus		Perennial plants
A. africanus	Aqool gabal, shoak	Perennial plant, propagated by rootstocks; grows in sandy places; tall, woody, prickly shrub
A. aphyllus	Aqool gabal, shoak	Perennial plant, propagated by rootstocks; grows in sandy places; intricately branched plant, 50–100cm high
A. stipularis	Aqool gabal, shoak, aqool berri	Perennial plant, propagated by rootstocks; grows on sandy and stony ground; stem climbing or trailing over the ground
Asphodelus		Annual or perennial herbs
A. fistulosus var. *tenuifolius*	Baroos, basal ibles, basal esh-sheitaan, fistulate asphodel	Perennial plant, propagated by tubers; grows on sandy and stony ground; small or medium-sized herb
A. microcarpus	Basal el-onsol	Perennial plant, propagated by tubers; grows on sandy or stony ground; stout robust herb with root of several spindle-shaped tubers, 1–2cm broad leaves, scape richly branched over 1m high
Pontederiaceae		
Eichhornia		Water weeds
E. crassipes	Yasint el-mayya, ward en-nil, water hyacinth	Perennial plant, propagated by rhizomes and stolons; grows along watery canals; free-floating plant with rosetted glabrous leaves, producing richly branched roots at the nodes; native in Florida, infesting waterways in Egypt

Table 15.3 Potential energy plant species in arid and semi-arid regions (*Cont'd*)

Botanical name	Common name	Notes/region
Tiliaceae		
Corchorus		Annual, sometimes perennial, herbs
C. depressus		Perennial plant, propagated by seed; grows in sandy and stony places; mat-shaped plant with thick tortuous branches
C. olitorius	Melokhia, long-fruited jute	Annual plant, propagated by seed; grows on loamy and sandy ground; plant glabrous except for petioles
C. trilocularis	Melokhia sheitaani	Annual plant, propagated by seed; grows on loamy and sandy ground; herb with pubescent stem
Typhaceae		
Typha		Tall marsh herbs
T. domingensis	Bordi, bardi, birdi, berdi, dees	Perennial plant, propagated by rhizomes; grows in ditches and marshy places; tall marsh herb, often more than 1.5m high
T. elephantina	Bordi, bardi, birdi, berdi, dees, timeyn, elephant grass	Perennial plant, propagated by rhizomes; grows in ditches and marshy places; taller and more robust than the preceding
Umbelliferae		
Ammi		Annual weeds
A. majus	Khilla sheitaani, greater ammi, toothpick ammi, bishop's weed	Annual plant, propagated by seed; grows in loamy clay places; slender herb
A. visnaga	Khilla, khella, toothpick ammi, lesser bishop's weed	Annual plant, propagated by seed; grows on loamy clay ground; stout tall plant with thick stem

Note: All plants listed in this table are harvested by mowing the aerial vegetative shoots of the plant and combining them into piles for storage.
Source: Khalifa (no date)

Table 15.4 Fast-growing trees and shrubs for fuelwood production

Scientific name	Common name	Description/region	Yields
Casuarinaceae			
Sesbania bispinosa (Jacq.) W. F. Wight	Canicha, danchi, dunchi fibre	Provides a strong fibre, grown as green manure and for fuelwood production. Annual subshrub, up to 7m tall, native to northern India, China and Sri Lanka; adapted to wet areas and heavy soils. Ranging from subtropical moist through tropical dry to moist forest life zones	The wood has a density of 0.3g/cm³. In Italy 15t/ha dry matter yields are reported; in the tropics yield may be higher
Casuarina cunninghamiana Miq.	River she-oak	Ranging from warm temperate dry to moist through tropical thorn to dry forest life zones; native to eastern and northern Australia, medium-sized tree, 15–20m high; has survived 8°C frost	*Casuarina* spp. have dense wood of 0.8–1.2g/cm³, a heating value of about 5000kcal/kg. The wood burns slowly with little smoke or ash, can be burned green, and is good for charcoal, losing only two-thirds of its weight
Casuarina glauca Sieber	Swamp she-oak	Native to eastern Australia. Grows on estuarine plains that are flooded with brackish tidal water; thrives on seaside dunes, often in the path of ocean spray. In fuelwood plantations trees rapidly regenerate from root sprouts	In Israel reaching 20m in 12–14 years, even on saline water tables. From the fourth year trees shed annually about 4t cones that make a good pellet-sized fuel
Casuarina equisetifolia J. R. & G. Forst.	Common ironwood, beefwood, Australian pine, Polynesian ironwood	Evergreen tree, up to 30m high, indigenous from Indonesia and Malaysia to India and Sri Lanka, and northeast Australia; ranging from subtropical woodland to wet through tropical thorn to wet forest life zones. Does not coppice readily, works under a clear-felling system, with a rotation of 7–35 years	With spacing of 2m and 7–10-year rotation, trees may yield 75–200t/ha wood, i.e. 10–20t/ha/year. In China litterfall is 4t/ha/year. Wood burns with immense heat, even when green, heating a good charcoal; calorific values: wood 4956kcal/kg, charcoal 7181kcal/kg
Casuarina junghuhniana Miq.	Jemara	Up to 35m high, native to highlands of eastern Indonesia to East Java, Bali, lesser Sunda Islands; pioneers natural revegetation of deforested grassland. There are commercial plantings in salt marsh areas, sometimes inundated with saline water. Propagated by cuttings, or by coppice or root sprouts	As other *Casuarina* spp.
Mimosaceae			
Acacia seyal	Acacia	Easy to regenerate, main energy species in Sudan; occurs on cracking clay soils	Principal species providing energy, gum and timber in Sudan
Acacia nilotica		Easy to regenerate, can withstand flooding	
Acacia senegal		High coppicing power; produces gum	
Acacia auriculiformis A. Cunn.	Darwin black wattle	Native to the savannahs of New Guinea, the islands of the Torres Strait, and northern Australia; widely introduced, e.g. to Africa, India, Fiji, Indonesia, Philippines; from subtropical moist to wet through tropical dry to wet forest life zones. Can produce good fuelwood on poor soils, even where there are extended dry seasons. Tolerates high and low pH, poor soil, sand dunes and savannah	Yields reported to run higher than 20m³/ha/year on a 10–20-year rotation; dropping to 8–12m³/ha/year on poor soils. Density of wood 0.6–0.75g/cm³, heating value 4800–4900kcal/kg. Wood yields excellent charcoal. Litter, branches and leaves may also be used as fuel. Litter amounts to 4.5–6t/ha/year

Table 15.4 Fast-growing trees and shrubs for fuelwood production (*Cont'd*)

Scientific name	Common name	Description/region	Yields
Acacia mangium Willd.	Forest mangrove, mange	Up to 30m high, native to Australia, with small stands in New Guinea and the Moluccas. Probably capable of growing in tropical very dry to moist through subtropical dry to wet forest life zones	Fast growing, attaining 15m in 5 years, 23m in 9 years. Reported yields 30m³/ha/year, or 20m³/ha/year on poor sites. Heating value 4800–4900kcal/kg; Untended 9-year-old stands have yielded 415m³/ha timber
Albizia lebbek (L.) Benth.	East Indian walnut, siris tree, kokko	Up to 30m high tree, native to tropical Africa, Asia and northern Australia, widely planted and naturalized. Ranging from tropical thorn to tropical wet through subtropical thorn to wet forest life zones	Trees coppice well. Annual wood yields of 5m³/ha and 18–28m³/ha reported. Ripe fruits contain 15% moisture and 17% reducing sugar. Possible fruit yields of 10t/ha suggest an ethanol yield of 1700L/ha/year
Leucaena leucocep hala (Lam.) de Wit	Wild tamarind, lead tree, jumbie bean, hediondilla	Fast-growing tree, up to 20m high, forming dense stands. Native throughout the West Indies, from the Bahamas to Cuba and from southern Mexico to northern South America. Requires long, warm, wet growing seasons, doing best in full sun. Trees, propagated by seed or cuttings, coppice well	On 3–8-year-old trees, annual wood increment varies from 24 to 100m³/ha, averaging 30–40m³/ha. Dry leucaena wood has 39% of the heating value of fuel oil. Leucaenas are multipurpose trees that withstand almost any type/ frequency of pruning or coppicing. Used for fodder, green manure and food.
Leucaena diversifolia		Better than *L. leucocephala* for rainfall of 500–2000m and above	
Albizia falcatarina (L.) Fosberg	Molucca albizia	Native to the eastern islands of the Indonesian archipelago and New Guinea; up to 30m high. Reported to tolerate poor soils. Trees can be closely spaced at 1000–2000 plants per ha	Harvested in the Philippines after 7–8 years, then every 8 years from coppice. Wood soft, with a density of 0.30–0.35g/cm³. In a 9-year-old stand the above-ground biomass was 102t/ha. Annual net productivity was about 20t/ha
Prosopis juliflora DC	Velvet mesquite	Small tree, up to 12m high, originated from Central or South America, now pantropically introduced; tolerates drought, grazing, heavy soils and sand as well as saline dry flats; probably ranging from tropical thorn to dry through subtropical thorn to dry forest life zones	Yields of a 15-year rotation 75–100t/ha, of a 10-year rotation 50–60t/ha. The specific gravity of the wood is 0.70g/cm³ or higher, the wood burns slowly and evenly, holding the heat well
Samanea saman (Jacq.) Merr.	Rain tree	Large tree, up to 60m high, 80m crown diameter, native from Yucatan and Guatemala to Peru, Bolivia and Brazil, widely introduced elsewhere. Ranging from subtropical very dry to moist through tropical dry to moist forest life zones. Easy to propagate from seed and cuttings	A 15-year-old tree yields about 200–275kg pods per season; 10t pods can yield 1150L of ethanol. Branches and stem may be used as fuelwood
Dalbergio sissoo Roxb. ex D C	Sisu, sissoo, Indian rosewood	Indigenous to India, Nepal and Pakistan, now widely grown in the tropics; from sea level to over 1500m. Ranging from subtropical thorn to moist through tropical dry to moist forest life zones. Can stand temperatures from below freezing to nearly 50°C. Propagation may be by direct seeding, seedling transplantation, or with root suckers	Young trees coppice vigorously and reproduce from suckers. A height of 7m has been reported in 20 months. Yields (m³/ha): 20-year-old stand 100, 30-year stand 210, 40-year stand 280, 50-year stand 370, 60-year stand 460. Heating values: sapwood 4908kcal/kg, heartwood 5191 kcal/kg. The wood is an excellent fuel and gives good charcoal

Table 15.4 Fast-growing trees and shrubs for fuelwood production (*Cont'd*)

Scientific name	Common name	Description/region	Yields
Sesbania grandiflora (L.) Pers.	Agati, corkwood tree, West Indian Pea	Propagated readily by seeding or cuttings, cultivated for fuelwood, fodder and reforestation. After harvest, shoots resprout with vigour. Rapid growth rate, particularly during first 3–4 years. The wood is white, soft, with a density of only 0.42g/cm³. The tree is frost sensitive, ranging from tropical dry through tropical moist forest life zones; native to many Asian countries	If planted along field edges, 3m³/ha of stacked fuelwood in a 2-year rotation are yielded. In Indonesia it is grown in 5-year rotation and yields 20–25m³/ha
Myrtaceae			
Melaleuca quinquenervia (Cav.) S. T. Blake	Cajeput	Ranging from subtropical dry to wet through tropical dry forest life zones. Evergreen tree, up to 30m high, native from eastern Australia through Malaysia and Burma, now widely introduced. Forms brackish swamp forests immediately behind mangroves. Propagated by seed or cuttings of immature wood. Source of tea tree oil.	Annual wood production 10–16m³/ha air dried wood. Provides an excellent fuel; average heating values (kJ/kg): wood 18,422, bark 25,791, terminal branches 19,301, foliage 20,139. Seven-year rotations were suggested in Malaysia. The heat of combustion of the bark is unique. Moderately heavy wood: 0.740–0.785g/cm³

Source: Duke (1984); Brewbaker (1995)

References

Abbas, W. A. (1999) 'Les lectines dans les dattes'

Abrahamson, L. P., Volk, T. A., Priepke, E., Posselius, J., Aneshansley, D. J. and Smart, L. B. (2008) 'Development of a willow biomass crop harvesting system in New York', SRWCOWG Meeting, Minneapolis, 19 August, www.cinram.umn.edu/srwc/docs/Powerpoints/L.Abrahamson_New%20Holland%20forage%20harvester.pdf

Abuelsamid, S. (2007) 'First sugar beet ethanol plant opens in the UK', 23 November, http://green.autoblog.com/2007/11/23/first-sugar-beet-ethanol-plant-opens-in-the-uk/

Academie des Sciences (2007) 'Principaux enjeux et verrous scientifiques au debut du XXI siecle', *Syntheses des Rapports sur la Science et la Technolgie*, April

Adam, J. (2006) 'Possibilities for conversion of biomass to liquid fuels in Norway: Background information for a roadmap, NTNU Biomass Seminar 09, May, SINTEF/Oslo, www.pfi.no/Biodrivstoff/Bakgrunnsdokument%20Biodrivstoffveikart.pdf

Adams, C., Peter, J. F., Rand, M. C., Schroer, B. J. and Ziemke, K. C. (1983) 'Investigation of soybean oil as a diesel fuel extender: Endurance tests', *Journal of the American Oil Chemists' Society*, vol 60, pp1574–1579

Adams, D. M. (1992) 'Long-term timber supply in the United States: An analysis of resources and policies in transition', *Journal of Business Administration*, vol 20, pp131–156

Adams, D. M., Alig, R. J., Callaway, J. M., McCarl, B. A. and Winnett, S. M. (1995) *The Forest and Agricultural Sector Optmization Model (FASOM): Model Structure and Policy Implications*, Research paper PNW-RP USDA-Forest Service, Pacific Northwest Research Station, USA

Addison, K. (2001) 'Handmade projects', http://journeyto forever.org/biodiesel_yield.html

ADEME (1996) *Rape Methyl Ester: Energy, Ecological and Economic Assessments*, TELLES Report, EUREC Network on Biomass (Bio-electricity)

Ae, N., Arihara, J., Okada, K., Yoshihara, T. and Johansen, C. (1990) 'Phosphorus uptake by pigeonpea and its role in cropping systems of the Indian subcontinent', *Science*, vol 248, no 4954, pp477–480

Agrobränsle AB (2004) 'Recycling of wastewater and sludge in salix plantations', www.cooperatives-uk.coop/live/images/cme_resources/Users/Nick/Farmar-Energi.pdf 2008

AGTD (Allison Gas Turbine Division) of General Motors Corporation (1994) 'Research and development of proton-exchange membrane (PEM) fuel cell system for transportation applications', initial conceptual design report prepared for the Chemical Energy Division of Argonne National Laboratory, US Department of Energy

Ahmad, S. (1954) *Grasses and Sedges of Lahore District*, Department of Botany, The University of Punjab, Lahore, Publication No 12

Ahmad, S. (1977) *Grasses and Sedges of Pakistan*, Department of Botany, The University of Punjab, Lahore

Ahmad, S. (1980) *Flora of the Punjab: Key to Genera and Species, Part 1. Monocotyledons*, Biological Society of Pakistan Monograph No 9, Government College, Lahore

Ahmed, A. H. H., Khalil, M. K. and Farrag, A. M., (2002) 'Nitrate accumulation, growth, yield and chemical composition of rocket (Eruca sativa) plant as affected by NPK fertilisation, kinetin and salicylic acid', *Annals of Agricultural Science* (Cairo), vol 47, no 1, pp1–26

Akhtaruzzaman, A. F. M., Siddique, A. B. and Chowdhury, A. R. (1986) 'Potentiality of pigeonpea (arhar) plant for pulping', *Bano Biggyan Patrika*, vol 15, nos 1–2, pp31–36

Akhter, J. (2001) 'Effects on some physico-chemical and mineralogical characteristics of salt-affected soil by growing Kallar grass using saline water', PhD thesis, Institute of Geology, University of the Punjab, Lahore

Akhter, J., Mahmood, K., Malik, K. A., Ahmed, S. and Murray, R. (2003) 'Amelioration of a saline sodic soil through cultivation of a salt-tolerant grass *Leptochloa fusca*', *Environmental Conservation*, vol 30, no 2, pp168–174

Akinola, J. O. and Whiteman, P. C. (1972) 'A numerical classification of *Cajanus cajan* (L.) Millsp. accessions based on morphological and agronomic attributes', *Australian Journal of Agricultural Research*, vol 23, no 23, pp995–1005.

Akobundu, I. O., Ekeleme, F. Chikoye D. (1999) 'Influence of fallow management systems and frequency of cropping on weed growth and crop yield', *Weed Research*, vol 39, no 3, pp241–256

Al Ain (2006) Workshop on Methyl Bromide substitute for date fumigation, Date Palm Global Network, 17 November

Alcamo, J. and Swart, R. (1998) 'Future trends of land use emissions of major greenhouse gases', *Mitigation and Adaptation Strategies to Global Change*, vol 3, pp343–381

Ali, M. (1990) 'Pigeonpea: Cropping systems', in Y. L. Nene, et al (eds) *The Pigeonpea*, CAB International, UK, pp279–301

Alig, R., Adams, D., McCarl, B., Callaway, J. M. and Winnett, S. (1997) 'Assessing effects of mitigation strategies for global climate change with an intertemporal model of the US forest and agricultural sectors', in R. A. Sedjo, R. N. Sampson and J. Wisniewski (eds) *Economics of Carbon Sequestration in Forestry*, Lewis Publishers, Boca Raton, pp185–193

Alig, R. J., Hohenstein, W. G., Murray, B. C. and Haight, R. G. (1990) *Changes in Area of Timberland in the United States, 1952–2040, by Ownership, Forest Type, Region and State*, General Technical Report SE 64, USDA-Forest Service, Southeastern Forest Experiment Station

Allen, R. K., Platt, K. and Wiser, S. (1995) 'Biodiversity in New Zealand plantation', *New Zealand Forestry*, February, pp26–29

Allen, A. (2000). *Kudzu in Appalachia*, ASPI Technical Series TP 55, Appalachia – Science in the Public Interest

Althaus, H.-J., Chudacoff, M., Hischier, R., Jungbluth, N., Osses, M. and Primas A. (2004) *Life Cycle Inventories of Chemicals*, final report Ecoinvent 2000 No 8

Alvim, P. de T. and Kozlowski, T. T. (eds) (1977) *Ecophysiology of Tropical Crops*, Academic Press, New York

Amey, M. A. (1987) 'Some traditional methods of cassava conservation and processing in Uganda', paper presented at the Third East and Southern Africa Crops Workshop, 7–11 December, Muzuzu, Malawi

Anders, M. M., Potdar, M. V., Pathak, P. and Laryea, K. (1992) 'Ongoing production agronomy studies with groundnut at ICRISAT Center', in *Proceedings of the Fifth Regional Groundnut Workshop for South Africa*, pp111–120

Anderson, D. W. (1995) 'Decomposition of organic matter and carbon emissions from soils', in R. Lal, J. Kimble, E. Levine and B. A. Stewart (eds) *Soils and Global Change*, CRC Lewis Publishers, Boca Raton, pp165–175

Anderson, W. F., Dien, B. S., Brandon, S. K. and Peterson, J. D. (2007) Assessment of bermudagrass and bunch grasses as feedstock for conversion to ethanol', *Applied Biochemistry and Biotechnology*, vol 145, nos 1–3, pp13–21

Anitei, S. (2007) 'Coconut oil replacing diesel in the Solomon Islands', Softpedia, 9 May, http://news.softpedia.com/news/Coconut-Oil-Replacing-Diesel-54294.shtml

Anon (1993) 'Inulin Gewinnung bei Suiker-Unie', *Starch/Stärke*, vol 45, p158

Anon (1967) *Regional Pulse Improvement Project, Annual Report, No 5*, RPIP, New Delhi, India.

Anon (1999) *Land-use, Land-use Change and Forestry in Canada and the Kyoto Protocol*, Canada's National Climate Change Process, Sinks Table Options Paper, www.nccp.ca/html/index.htm

Anon (2006) 'Ethanol: l'industrie mondiale de l'ethanol en pleine ascension', *Focus*, vol 30, no 2

Apps, M. J. and Kurz, W. A. (1993) 'The role of Canadian forests in the global carbon balance', in M. Kannien (ed) *Carbon Balance on World's Forested Ecosystems: Towards a Global Assessment. Proceedings Intergovernmental Panel on Climate Change Workshop*, Joensuu, Finland, 11–15 May 1992, Academy of Finland, Helsinki, pp14–28

Apps, M. J., Kurz, W. A., Beukema, S. J. and Bhatti, J. S. (1999) 'Carbon budget of the Canadian forest product sector', *Environmental Science and Policy*, vol 2, pp25–41

Apps, M. J., Bhatti, J. S., Halliwell, D., Jiang, H. and Peng, C. (2000) 'Simulated carbon dynamics in the boreal forest of central Canada under uniform and random disturbance regimes', in R. Lal, J. M. Kimble and B. A. Stewart (eds) *Global Climate Change and Cold Regions Ecosystems: Advances in Soil Science*, Lewis Publishers, Boca Raton, pp107–121

Arnosti, D., Abbas, D. and Demchik, M. (2008) *Harvesting Fuel*, IATP, USA

Arnoux, M. (1974) 'Recherches sur la canne de Provence (*Arundo donax* L.) en vue de sa production et de sa transformation en pate a papier', *Ann. Amelior. Plantes Paris*, vol 24, no 4, pp349–376

Ash, A. J., Howden, S. M. and McIvor, J. G. 1996) 'Improved rangeland management and its implications for carbon sequestration', in *Proceedings of the Fifth International Rangeland Congress*, Salt Lake City, Utah, 23–28 July 1995, vol 1, pp19–20

Ashburner, G. R., Faure, M. G., Tomlinson, D. R. and Thompson, W. K. (1996) 'Collection of coconut (*Cocos nucifera*) embryos from remote locations', *Seed, Science and Technology*, vol 24, no l, pp159–69

Ashley, J. M. (1984) 'Groundnut', in P. R. Goldsworthy and N. M. Fisher (eds) *The Physiology of Tropical Field Crops*, John Wiley & Sons, Chichester, pp453–494

Aso, T. (1977) 'Studies on the germination of seeds of *Miscanthus sinensis* Anderss.', *Science Reports of the Yokohama National University, Section II, Biological & Geological Science*, vol 23, pp27–37

ASU (Arizona State University), (no date) 'Algal-based biofuels & biomaterial', http://biofuels.asu.edu/biomaterials.shtml

Atienza, S. G., Satovic, Z., Petersen, K. K., Dolstra, O. and Martín, A. (2003) 'Influencing combustion quality in *Miscanthus sinensis* Anderss.: Identification of QTLs for calcium, phosphorus and sulphur content', *Plant Breeding*, vol 122, no 2, pp141–145

Aufhammer, W., Lee, J. H., Kübler, E., Kuhn, M. and Wagner, S. (1995) 'Anbau und Nutzung der Pseudocerealien Buchweizen (*Fagopyrum esculentum* Moench), Reismelde (*Chenopodium quinoa* Willd.) und Amarant (*Amaranthus* ssp. L.) als Kornerfruchtarten', *Mitteilungen der Gesellschaft für Pflanzenbauwissenschaften*, Band 8, Wissenschaftlicher Verlag Giessen, pp125–139

Austin, R. B., Bingham, J., Blackwell, R. D., Evans, L. T., Ford, M. A., Morgan, C. L. and Taylor, M. (1980) 'Genetic improvement in winter wheat yields since 1900 and associated physiological changes', *Journal of Agricultural Science*, vol 94, pp675–689

Avouampo, E. (1996) 'Contribution á l'étude de la valorisation alimentaire du safou: transport, stockage, procédés de séchage des pulpes et d'extraction d'huile', Séminaire d'Information sur la Recherche en Alimentation et Nutrition, 6–8 October, Brazzaville, Congo

Ayala, F. and O'Leary, J. W. (1995) 'Growth and physiology of *Salicornia bigelovii* Torr. at suboptimal salinity', *International Journal of Plant Sciences*, vol 156, no 2, pp197–205

Ayerza, R (1990) 'The potential of jojoba (*Simmondsia chinensis*) in the arid chaco: Rooting capacity and seed and wax yield', in H. H. Naqvi, A. Estilai and I. P Ting (eds) *Proceedings of the First International Conference on New Industrial Crops and Products*, Riverside, California

Badiane, O. (2008) 'Sustaining and accelerating Africa's agricultural growth', IFPRI website, www.ifpri.org

Baert, J. R. A. (1991) 'Cultivation and breeding of root chicory for inulin production', in *Proceedings of the 1st European Symposium on Industrial Crops and Products*, Maastricht, Netherlands

Baert, J. R. A. (1993) 'The potential of inulin chicory as an alternative high income crop', in *Proceedings of EC Workshop on the Production and Impact of Specialist Minor Crops in the Rural Community*, Brussels, Belgium.

Bailey, W. W. and Bailey, R. Jr. (1996) 'Clean heat, steam, and electricity from rice hull gasification', in *Bioenergy '96, Proceedings of the Seventh National Bioenergy Conference*, 15–20 September 1996, pp284–287

Bajracharya, R. M., Lal, R. and Kimble, J. M. (1997) 'Long-term tillage effects on soil organic carbon distribution in aggregates and primary particle fractions of two Ohio soils', in R. Lal, et al (eds) *Management of Carbon Sequestration in Soil*, CRC Press, Boca Raton, pp113–123

Bala, A., Murphy, P. and Giller, K. E. (2003) 'Distribution and diversity of rhizobia nodulating agroforestry legumes in soils from three continents in the tropics', *Molecular Ecology*, vol 12, no 4, pp917–929

Baldelli, C. (1987a) 'Robinia and ginestra from weed plants to an ecological fuel', *Proceedings of the 4th EC Conference on 'Biomass for Energy and Industry'*, Orlèans, France, 11–15 May 1987, pp355–368

Baldelli, C. (1987b) *Agricultural Raw Materials for Energy and Industry on Marginal and Poor Lands*, UNITAR/UNDP, Rome, Italy, Newsletter No 9

Baldelli, C. (1988) 'Energy crops on marginal lands (biomass perugia)', in *Proceedings of the International Congress, 'Euroforum new energies'*, Saarbrücken, Germany, 24–28 October 1988, pp465–467.

Baldelli, C. (1989) 'Cropping bio-fuels', in *Proceedings of the International Conference on 'Biomass for Energy and Industry'*, Lisbon, Portugal, 9–13 October 1989, Vol I, pp499–502

Baldelli, C. (1992) 'Black locust as a source of energy', in J. W. Hanover, K. Miller and S. Plesko (eds) *Proceedings of the International Conference on Black Locust: Biology, Culture and Utilization*, Department of Forestry, Michigan State University, East Lansing, Michigan, pp237–243

Baldini, S. (1987) 'Forest biomass for energy in EEC countries from harvesting to storage', in G. L. Ferrero, G. Grassi and H. E. Williams (eds) *Biomass Energy: From Harvesting to Storage*, Elsevier Applied Science, London, pp22–40

Baldini, S. (1988) 'Potenzialità e possibilità di recupero della biomassa forestale in Italia', *Accademia Economico-Agraria dei Georgofili*, vol 35, pp3–19

Balesdent, J., Besnard, E., Arrouays, D. and Chenu, C. (1998) 'The dynamics of carbon in particle-size fractions of soil in a forest-cultivation sequence', *Plant and Soil*, vol 201, pp49–57

Ball, B. C., Scott, A. and Parker, J. P. (1999) 'Field N_2O, CO_2 and CH_4 fluxes in relation to tillage, compaction and soil quality in Scotland', *Soil and Tillage Research*, vol 53, pp29–39

Bambara, S. (1994) http://sunsite.unc.edu/london/orgfarm/biocontrol/

Bara, S. (1970) *Estudio sobre* Eucalyptus globulus. I. *Composición mineral de las hojas en relacion con su posicion en el arbol, la composición del suelo y la edad. Evolución del suelo por el cultivo del Eucalyptus*, Comunicacion no 39, IFIE, Madrid

Bara, S., Rigueiro, A., Gil, M. C., Mansilla, P. and Alonso, M. (1985) *Efectos ecologicos del* Eucalyptus globulus *en Galicia: Estudio comparativo con* Pinus pinaster *y* Quercus robur, Monografias INIA no 50, Madrid

Barnes, D. K. and Sheaffer, C. C. (1995) 'Alfalfa', in R. F. Barnes, D. A. Miller and C. J. Nelson (eds) *Forages, Vol. 1: An Introduction to Grassland Agriculture*, Iowa State University Press, Ames, Iowa, pp205–216

Barnett, P. E. and Curtin, D. T. (1986) *Development of Forest Harvesting Technology: Application in Short Rotation Intensive Culture (SRIC) Woody Biomass*, Technical Note B58, Tennessee Valley Authority, Muscle Shoals, Alabama

Barrett, D. J. and Gifford, R. M. (1999) 'Increased C-gain by an endemic Australian pasture grass at elevated atmospheric CO_2 concentration when supplied with non-labile inorganic phosphorus', *Australian Journal of Plant Physiology*, vol 26, pp443–451

Barrett, R. P., Mebrahtu, T. and Hanover, J. W. (1990) 'Black locust: A multi-purpose tree species for temperate climates', in J. Janick and J. E. Simon (eds) *Advances in New Crops*, Timber Press, Portland, OR, pp278–283

Bartolelli, V. and Mutinati, G. (1991) 'Economic convenience of cotton, jojoba and broom cultivation', in *Proceedings of the International Conference on 'Biomass for Energy, Industry and Environment'*, Athens, Greece, 22–26 April 1991, pp317–320

Bartolelli, V. and Mutinati, G. (1992) 'Economic convenience of vegetable oils as combustible', in *Biomass for Energy and Industry, 7th EC Conference*, Florence

Bartolelli, V., Mutinati, G. and Pisani, F. (1991) 'Microeconomic aspects of energy crops cultivation', in *Biomass for Energy, Industry and Environment, 6th EC Conference*, Athens, Elsevier Applied Science, pp233–237

Bass, S., Ford, J., Duboois, O., Moura-Costa, P., Wilson, C., Pinard, M. and Tipper, R. (2000) *Rural Livelihoods and Carbon Management*, an issues paper, International Institute for Environment and Development, Forestry and Land Use, London, UK

Bassham, J. A. (1980) 'Energy crops (energy farming)', in A. G. San Pietro (ed) *Biochemical and Photosynthetic Aspects of Energy Production*, Academic Press, New York, pp147–73

Bates, L. (2007) 'Biogas, practical answers: Technical information online', March 2007, http://practicalaction.org/practical answers/product_info.php?products_id=42

Batjes, N. H. (1998) 'Mitigation of atmospheric CO_2 concentrations by increased carbon sequestration in the soil', *Biology and Fertility of Soils*, vol 27, pp230–235

Batjes, N. H. (1999) *Management Options for Reducing CO_2-Concentrations in the Atmosphere by Increasing Carbon Sequestration in the Soil*, Report 410 200 031, Dutch National Research Programme on Global Air Pollution and Climate Change, Project executed by the International Soil Reference and Information Centre, Wageningen, The Netherlands

Bauer, C. (2003) 'Holzenergie', in R. Dones (ed) *Sachbilanzen von Energiesystemen: Grundlagen für den ökologischen Vergleich von Energiesystemen und den Einbezug von Energiesystemen in Ökobilanzen für die Schweiz*, Paul Scherrer Institut Villigen, Swiss Centre for Life Cycle Inventories, Dübendorf, www.ecoinvent.org

Bawa, K. S. and Dayanandan, S. (1997) 'Socioeconomic factors and tropical deforestation', *Nature*, vol 386, pp562–563

BBC News (2004) 'Bananas could power Aussie homes', http://news.bbc.co.uk/go/pr/fr/-/2/hi/science/nature/3604666.stm (accessed September 2009)

BBC News (2008) 'Airline in first biofuel flight', 24 February, http://news.bbc.co.uk/1/hi/uk/7261214.stm (accessed September 2009)

Beale, C. V. and Long, S. P. (1995) 'Can perennial C4 grasses attain high efficiencies of radiant energy conversion in cool climates?', *Plant, Cell and Environment*, vol 18, pp641–650

Behl, R. K. (1980) 'Studies on genetic divergence, heterosis and combining ability in hexaploid triticale', PhD dissertation submitted to Haryana Agricultural University, Hisar, India; Assefa, Molla

Belkhadem, M. A. (2006) 'Republique Algerienne Democratique et Polulaire', *Pacte National Economique et Social*, November

Bellido, L. P. (1984) 'World report on lupin', *Proceedings of the 3rd International Lupine Conference*, La Rochelle, France, pp466–487

Benzioni, A. (1997) 'Jojoba', www.hort.purdue.edu/newcrop/CropFactSheets/jojoba.html

Berardino, L. D., Berardino, F. D., Castelli, A. and Torre, F. D. (2006) 'A case of contact dermatitis from jojoba', *Contact Dermatitis*, vol 55, no 1, pp57–58

Bercy (2006) 'Flex Fuel: apercus internationaux', round table on 7 June 2006

Berg, C. (2004) 'World fuel ethanol analysis and outlook', www.distill.com/World-Fuel-Ethanol-A&O-2004.html

Berglund, D. R., Riveland, N. and Bergman, J. (2007) 'Safflower production A-870 (Revised)', August, www.ag.ndsu.edu/pubs/plantsci/crops/a870w.htm

Bergman, J. W. and Flynn, C. R. (2009) 'High oleic safflower as a diesel fuel extender: A potential new market for montana safflower', www.sidney.ars.usda.gov/state/saffcon/abstracts/SafflowerProducts/bergman.htm

Berndes, G., Hoogwijk, M. and van den Broek, R. (2003) 'The contribution of biomass in the future global energy supply: A review of 17 studies', *Biomass and Bioenergy*, vol 25, no 1, pp1–28

Berti, M., Johnson, B., Gesch, R. and Forcella, F. (2007) 'Cuphea plant nitrate content and seed yield response to nitrogen fertilizer', in J. Janick and A. Whipkey (eds) *Issues in New Crops and New Uses*, ASHS Press, Alexandria, VA

Beuch, S., Boelcke, B. and Belau, L. (2000) 'Effect of the organic residues of *Miscanthus* × *giganteus* on the soil organic matter level of arable soils', *Journal of Agronomy and Crop Science*, vol 184, no 2, pp111–120

Bezard, J., Silou, Th., Kiakouama, S. and Sempore, G. (1991) 'Variation de la fraction glycéridique de l'huile de la pulpe de safou avec l'état de maturité du fruit', *Rev. franç. Corps Gras*, vol 38, nos 7/8, pp233–241

BFE (2000) 'Schweizerische Gesamtenergiestatistik 1999'. Bundesamt für Energie, Bern, CH, www.energieschweiz.ch/bfe/de/statistik/gesamtenergie/

BFE/EWG (2004) 'Potenziale zur energetischen Nutzung von Biomasse in der Schweiz: überarbeitetes und ergänztes zweites Inputpapier', Bearbeitet von der Arbeitsgemeinschaft INFRAS (Martina Blum, Bernhard Oettli, Othmar Schwank, INFRAS, Denis Bedniaguine, Edgard Gnansounou, François Golay)

BfS/BUWAL (2003) 'Wald und Holz in der Schweiz: Jahrbuch 2002', www.umweltschweiz.ch/buwal/de/fachgebiete/fg_wald/rubrik2/holzinfos/#sprungmarke5

Bhatti, J. S., Apps M. J. and Jiang, H. (2001) 'Examining the carbon stocks of boreal forest ecosystems at stand and regional scales', in R. Lal, M. Kimble, R. F. Follett and B. A. Stewart (eds) *Assessment of Methods for Soil C Pools: Advances in Soil Science*, Lewis Publishers, Boca Raton, pp513–532

Bilbro, J. D. (2007) 'Castor: An old crop with new potentials', http://waynesword.palomar.edu/plmar99.htm#castor

Binggeli, D. and Guggisberg, B. (2003) 'Biomasse: Überblicksbericht zum Forschungsprogramm 2003', BfE

Binggeli, D. and Guggisberg, B. (2004) 'Biomasse: Überblicksbericht zum Forschungsprogramm 2004', BfE

Binkley, C. S., Apps, M. J., Dixon, R. K., Kauppi, P. E. and Nilsson, L. O. (1997) 'Sequestering carbon in natural forests', *Critical Reviews in Environmental Science and Technology*, vol 27, pp23–45

Biopact (2007a) 'US scientists develop drought tolerant sorghum', http://news.mongabay.com/bioenergy/2007/05/us-scientists-develop-drought-tolerant.html

Biopact (2007b) 'Sweet super sorghum yield data for the Icrisat hybrid', http://news.mongabay.com/bioenergy/2007/02/sweet-super-sorghum-yield-data-for.html

Biopact (2007c) 'Icrisat launches pro poor biofuels initiative in dry lands', http://news.mongabay.com/bioenergy/2007/03/icrisat-launches-pro-poor-biofuels.html

Biopact (2007d) 'Green steel made from tropical biomass: European project', http://news.mongabay.com/bioenergy/2007/02/tropical-biomass-for-production-of.html

Biopact (2007e) 'Joint Genome Institute announces 2008 genome sequencing targets with focus on bioenergy and carbon cycle', http://news.mongabay.com/bioenergy/2007/06/joint-genome-institute-announces-2008.html

Biopact (2007f) 'Study: Biogas can replace all EU imports of Russian gas by 2020', http://news.mongabay.com/bioenergy/2007/02/study-biogas-can-replace-all-eu-imports.html

Biopact (2007g) 'India's bright green idea: Compressed biogas for cars', http://news.mongabay.com/bioenergy/2006/10/indias-bright-green-idea-compressed.html

Birdsey, R. A. and Heath, L. S. (1995) 'Carbon changes in US forests', *Productivity of America's Forests and Climate Change*, vol 271, pp56–70

Birdsey, R. A., Alig, R. and Adams, D. (2000) 'Mitigation activities in the forest sector to reduce emissions and enhance sinks of greenhouse gases', in L. Joyce and R. Birdsey (eds) *The Impact of Climate Change on America's Forests*, US Department of Agriculture, Rocky Mountain Research Station, Fort Collins, CO

Björndahl, G. (1983) 'Structure and biomass of *Phragmites* stands', dissertation, Goteborg

Blackwell, A., Evans, K. A., Strang, R. H. C. and Cole, M. (2004) 'Toward development of neem-based repellents against the Scottish Highland biting midge *Culicoides impunctatus*', *Medical and Veterinary Entomology*, vol 18, no 4, pp449–452

Blaschek, H. P. (1995) 'Recent developments in the ABE fermentation', *Workshop on Energy from Biomass and Wastes*, Dublin Castle, Ireland, p28

Blatter, E. S. J., McCann, C. and Sabnis, T. S. (1929) *The Flora of the Indus Delta*, printed and published for the Indian Botanical Society by the Methodis Publishing House, Madras, India

Bludau, D. (1987) 'Landwirtcchftliche Kendaten und Angepasste Produktiontechnik der Zuckerhirse als Energierpflanze', in *Der Bundesrepublik Deutschland*, Diplomarbeit, Tu-Munchen-Weihenstephan.

Blümke, S. (1960) 'Die Robinie in Deutschland', *Forstwirtschaft, Holzwirtschaft*, vol 4, pp8–16

Boateng, A. A., Jung, H. G. and Adler, P. R. (2006) 'Pyrolysis of energy crops including alfalfa stems, reed canarygrass, and eastern gamagrass', *Fuel*, vol 85, nos 17–18, pp2450–2457

Bodansky, D. (1996) 'May we engineer the climate?', *Climatic Change*, vol 33, pp309–321

Boe, A. and Lee, D. K. (2007) 'Genetic variation for biomass production in prairie cordgrass and switchgrass', *Crop Science*, vol 47, pp929–934

Boland, D. J., Brophy, J. J. and House, A. P. N. (1991) *Eucalyptus Leaf Oils: Use, Chemistry, Distillation and Marketing*, Inkata Press, Sydney

Bolin, B., Sukumar, R., Ciais, P., Cramer, W., Jarvis, P., Kheshgi, H., Nobre, C., Semenov, S. and Steffen, W. (2000) 'Global perspective', in R. T. Watson, I. R. Noble, B. Bolin, N. H. Ravindranath, D. J. Verardo and D. J. Dokken (eds) *Land Use, Land Use Change and Forestry*, a special report of the IPCC, Cambridge University Press, pp23–51

Bombelli, V., Piccolo, A., Marceddu, E., Parrini, F. and Schenone, G. (1995) 'Biomass for energy in Sardinia (*Eucalyptus globulus, Arundo donax, Cynara cardunculus, Sorghum bicolor, Hibiscus cannabinus*)', in *Biomass for Energy, Environment, Agriculture and Industry, Proceedings of the 8th EC Conference*, Vienna, 1994, Pergamon, Vol 1, pp423–430

Bona, L. (2007) 'Triticale, an escalating stress tolerant crop in the carpathian basin region of Europe', in D. P. Singh, V. S. Tomar, R. K. Behl, S. D. Upadhyaya, M. S. Bhale and D. Khare (eds) *Sustainable Crop Production in Stress Environments*, Agribios (International) Publishers, Jodhpur, India, pp43–48

Bonan, G. B. and Shugart, H. H. (1992) 'Soil temperature, nitrogen mineralization and carbon source sink relationships in boreal forests', *Canadian Journal of Forest Research*, vol 22, pp629–639

Bonari, E., Mazzoncini, M. and Peruzzi, A. (1995) 'Effects of conventional and minimum tillage on winter oilseed rape (*Brassica napwus* L.) in a sandy soil', *Soil and Tillage Research*, vol 33, no 2, pp91–108

Bonduki, Y. and J. N. Swisher (1995) 'Options for mitigating greenhouse gas emissions in Venezuela's forest sector: A general overview', *Interciencia*, vol 20, no 6, pp380–387

Bongarten, B. C., Huber, D. A. and Apsley, D. K. (1992) 'Environmental and genetic influences on short-rotation biomass production of black locust (*Robinia pseudoacacia* L.) in the Georgia Piedmont', *Forest Ecology and Management Netherlands*, vol 55, nos 1–4, pp315–331

Booth, F. E. M. (1983) *Survey of Economic Plants for Arid and Semi-arid Tropics (SEPASAT)*, Royal Botanical Gardens, Kew, Richmond, England

Borough, C. J., Incoll, W. D., May, J. R. and Bird, T. (1984) 'Yield statistics', in W. E. Hillis and M. T. Brown (eds) *Eucalypts for Wood Production*, Academic Press, New York

Borralho, N. M. G. and Coterril, P. P. (1992) 'Genetic improvement of *Eucalyptus globulus* for pulp production', in J. S. Pereira and H. Pereira (eds) *Eucalyptus for Biomass Production*, Commission of the European Communities.

Børsen (2008) 'Funen plans to grow willow for energy production', *Børsen* financial daily newspaper

Bouteflika, A. (2007) 'Les changements climatiques en Afrique', speech by the President in Addis Ababa, 30 January

Boyce, S. G. (1995) *Landscape Forestry*, John Wiley & Sons, Inc., New York, pp67–166

Bradford, J. (2005) 'Threshold models applied to seed germination ecology', *New Phytologist*, vol 165, no 2, pp338–341

Bramm, A. and Bätz, W. (1988) *Industriepflanzenbau, Produktions- und Verwendungsalternativen*, BML, Bonn

Braswell, B. H., Schimel, D. S. Linder, E. and Moore, B., III (1997) 'The response of global terrestrial ecosystems to interannual temperature variability', *Science*, vol 278, pp870–872

Brewbaker, J. L. (1995) '*Leucaena*, new crop fact sheet', www.hort.purdue.edu/newcrop/CropFactSheets/leucaena.html

Broadhead, D. M. and Freeman, K. C. (1980) 'Stalk and sugar yield of sweet sorghum as affected by spacing', *Agronomy Journal*, vol 72, pp523–524

Brooker, M. I. H. and Kleinig, D. A. (1991) *Field Guide to Eucalypts. South-Eastern Australia*, Inkarta Press, Sydney

Brown, L. M. (1996) 'Uptake of carbon dioxide from flue gas by microalgae', Energy Conversion and Management, vol 37, nos 6–8, pp1363–1367

Brown, S., Sathaye, J., Cannell, M. and Kauppi, P. E. (1996a) 'Mitigation of carbon emissions to the atmosphere by forest management', *Commonwealth Forestry Review*, vol 75, pp80–91

Brown, S., Sathaye, J., Cannell, M. and Kauppi, P. E. (1996b) 'Management of forests for mitigation of greenhouse gas emissions', in R. T. Watson, M. C. Zinyowera, R. H. Moss and D. J. Dokken (eds) *Climate Change 1995 – Impacts, Adaptations and Mitigation of Climate Change: Scientific-Technical Analyses*, Contribution of Working Group II to the Second Assessment Report of the Intergovernmental Panel on Climate Change, Cambridge University Press, Cambridge, UK, pp773–797

Brown, S., Masera, O. and Sathaye, J. (2000) 'Project based activities', in R. T. Watson, I. R. Noble, B. Bolin, N. H. Ravindranath, D. J. Verardo and D. J. Dokken (eds) *Land Use, Land Use Change and Forestry*, a special report of the IPCC, Cambridge University Press, pp283–338

Bruce, J. P., Frome, M., Haites, E., Janzen, H., Lal, R. and Paustian, K. (1999) 'Carbon sequestration in soils', *Journal of Soil and Water Conservation*, vol 54, pp381–389

Brunner, J., Talbot, K. and Elkin, C. (1998) *Logging Burma's Frontier Forests: Resources and the Regime*, World Resources Institute, Washington, DC

Bryan, W. L., Monroe, G. E., Nichols, R. L. and Bascho, G. J. (1981) *Evaluation of Sweet Sorghum for Fuel Alcohol*, ASAE Paper No 81-3571, ASAE, St Joseph, Missouri

Buchanan, A. H. and Levine, S. B. (1999) 'Wood based building materials and atmospheric carbon emissions', *Environmental Science and Policy*, vol 2, pp427–437

Bullard, M. J., Heath, M. C., Nixon, P. M. I., Speller, C. S. and Kilpatrick, J. B. (1994) 'The comparative physiology of *Miscanthus sinensis, Tritium aestivum* and *Zea mays* grown under UK conditions', COST 814 Workshop

Bullard, M. J., Christian, D. G. and Wilkins, C. (1995) *Quantifying Biomass Production in Crops Grown for Energy*, interim report, ETSU Contract B/CR/003 87/00/00, Energy Technology Support Unit, Harwell, UK

Bunting, E. S. (1978) 'Agronomic and physiological factors affecting forage maize production', in E. S. Bunting, B. F. Pain, R. H. Phipps, J. M. Wilkinson and R. E. Gunn (eds) *Forage Maize, Production and Utilisation*, Agricultural Research Council, London, pp5745

Burke, I. C., Lauenroth, W. K. and Milchunas, D. G. (1997) 'Biogeochemistry of managed grasslands in central North America', in E. A. Paul, E. T. Elliott, K. Paustian and C. V. Cole (eds) *Soil Organic Matter in Temperate Agroecosystems: Long-Term Experiments in North America*, CRC Press, Boca Raton, pp85–102

Burke, I. C., Lauenroth, W. K., Vinton, M. A., Hook, P. B., Kelly, R. H., Epstein, H. E., Aguiar, M. R., Robles, M. D., Aguilera, M. O., Murphy, K. L. and Gill, R. A. (1998) 'Plant-soil interactions in temperate grasslands', *Biogeochemistry*, vol 42, pp121–143

Burschel, P., Kuersten, E. and Larson, B. C. (1993) *Die Rolle von Wald und Forstwirtschaft im Kohlenstoffhaushalt. Eine Betrachtung für die Bundesrepublik Deutschland [Role of Forests and Forestry in the Carbon Cycle: A Try-out for Germany]*, Forstliche Forschungsberichte München, Schriftenreihe der Forstwissenschaftlichen Fakultät der Universität München und der Bayerischen Forstlichen Versuchs- und Forschungsanstalt, 126

Burton, G. W. (1993) 'African grasses', in J. Janick and J. E. Simon (eds) *New Crops*, Wiley, New York, pp294–298

Burtraw, D. (2000) *Innovation Under the Tradable Sulfur Dioxide Emission Permits Program in the US Electricity Sector*, RFF Discussion Paper 00–38, Resources for the Future, Washington, DC

Burvall, J. (1996) Personal communication

Business Wire (2008) 'Pakistan's First Sugar cane Biogas Plant Powered by GE Energy's Jenbacher Gas Engines', 31 July, BNET Business Network

Cailhiez (1990) 'Planations industrielles d'Eucalyptus en République populaire du Congo', International Workshop on Large Scale Reforesation, Corvallis, Oregon, 8–10 May 1990.

Callaghan, T. V., Scott, R., Lawson, G. J. and Mainwaring, A. M. (1984) *An Experimental Assessment of Native and Naturalized Species of Plants as Renewable Sources of Energy in Great Britain: II. Cordgrass* – Spartina anglica, Report No ETSU-B-1086/P-2, NERC, Cambridge Institute of Terrestrial Ecology

Callaway, J. M. and McCarl, B. A. (1996) 'The economic consequences of substituting carbon payments for crop subsidies in US agriculture', *Environmental and Resource Economics*, vol 7, pp15–43

Cameron, L. (2008) 'The potential of sunflower biodiesel', Africa News Network, 20 November, http://africanagriculture. blogspot.com/2008/11/potential-of-sunflower-biodiesel.html

Campbell, K. (1996) 'The Minnesota agri-power project', in *Bioenergy '96: Proceedings of the Seventh National Bioenergy Conference*, 15–20 September 1996

Campbell, I., Campbell, C., Yu, Z., Vitt, D. and Apps, M. J. (2000) 'Millennial-scale rhythms in peatlands in the western interior of Canada and in the global carbon cycle', *Quaternary Research*, vol 54, pp155–158

Candolle, A. de (1967) *Origin of Cultivated Plants*, Hafner, New York and London, pp411–15

Cannell, M. G. (1988) 'The scientific background in biomass forestry in Europe: A strategy for the future', in F. C. Hummel, W. Palz and G. Grassi (eds) *Energy from Biomass*, 3, Elsevier Applied Science, pp83–140

Cao, M. and Woodward, F. I. (1998) 'Dynamic reponses of terrestrial ecosystem carbon cyling to global climate change', *Nature*, vol 393, pp249–252

Cara, C., Ruiz, E., Oliva, J. M., Sáez, F. and Castro, E. (2008) 'Conversion of olive tree biomass into fermentable sugars by dilute acid pretreatment and enzymatic saccharification', *Bioresource Technology*, vol 99, no 6, pp1869–1876

Carter, J. K. (ed) (1978) *Sunflower Science and Technology*, American Society of Agronomy, Madison, WI

Carter, M. R., Gregorich, E. G., Anderson, D. W., Doran, J. W., Janzen, H. H. and Pierce, F. J. (1997) 'Concepts of soil quality and their significance', in E. G. Gregorich and M. R. Carter (eds) *Soil Quality for Crop Production and Ecosystem Health*, Elsevier, Amsterdam, The Netherlands, pp1–19

Carter, M. R., Gregorich, E. G., Angers, D. A., Donald, R. G. and Bolinder, M. A. (1998) 'Organic C and N storage, and organic C fractions, in adjacent cultivated and forested soils of eastern Canada', *Soil and Tillage Research*, vol 47, pp253–261

CBC (Columbia Basin College) (no date) 'Research and development', www.columbiabasin.edu/home/index.asp?page=2360

Chandra, S., Asthana, A. N., Ali, M., Sachan, J. N., Lal, S. S., Singh, R. A. and Gangal, L. K. (1983) *Pigeonpea Cultivation in India: 'A Package of Practices'*, All India Coordinated Pulse Improvement Project, Kanpur, Extension Bulletin No 2

Chapelle, J. (1996) Personal communication, Sorghal, Belgium

Chaudhry, S. A. (1969) *Flora in Lyallpur and the Adjacent Canal Colony Districts*, West Pakistan Agricultural University, Lyallpur, Faisalabad

Chen, S. L. and Renvoize, S. A. (2005) 'A new species and a new combination of *Miscanthus* (Poaceae) from China', *Kew Bulletin*, vol 60, pp605–607

Chen, Y., Qin, W., Li, X., Gong, J. and Gong, N. (1985) 'The chemical composition of ten bamboo species', in A. N. Rao, G. Dhanarajan and C. B. Sastry (eds) *Recent Research on Bamboos: Proceedings of the International Bamboo Workshop*, 6–14 October, Hangzhou, China, pp110–113

Chianese, L. (1940) 'La ginestra e le sue specie nel sistema dell'agricoltura italiana', *Cellulosa*, vol 2, pp63–69

Chiaramonti, D., Grimm, P., Cendagorta, M. and El Bassam, N. (1998) 'Small energy farm scheme implementation for rescuing deserting land in small Mediterranean islands, coastal areas, having water and agricultural land constraints: Feasibility study', *Proceedings of the 10th International Conference 'Biomass for Energy and Industry'*, Würzburg, pp1259–1262

Chin, L. (1979) 'Fuels by hydrogenation', PhD thesis, Pennsylvania State University

Chomitz, K. M. and Kumari, K. (1996) *The Domestic Benefits of Tropical Forests: A Critical Review Emphasizing Hydrological Functions*, Policy Research Working Paper No WPS1601, World Bank

Chou, C.-H., Chiang, T.-Y. and Chiang, Y.-C. (2001) 'Towards an integrative biology research: A case study on adaptive and evolutionary trends of *Miscanthus* populations in Taiwan', *Weed Biology and Management*, vol 1, no 2, pp81–88

Christen, O. and Sieling, K. (1995) 'Effect of different preceding crops and crop rotations on yield of winter oil-seed rape *(Brassica napwus* L.)', *Journal of Agronomy and Crop Science [Zeitschrift für Acker und Pflanzenbau]*, vol 174, no 4, pp265–271

Christensen, B. T. and Johnston, A. E. (1997) 'Soil organic matter and soil quality: Lessons learned from long-term experiments at Askov and Rothamsted', in E. G. Gregorich and M. R. Carter (eds) *Soil Quality for Crop Production and Ecosystem Health*, Elsevier, Amsterdam, The Netherlands, pp399–430

Christersson, L. (2008) 'Poplar plantations for paper and energy in the south of Sweden', *Biomass and Bioenergy*, vol 32, no 11, pp997–1000

Christian, D. G. (1994) 'Quantifying the yield of perennial grasses grown as a biofuel for energy generation', *Renewable Energy*, vol 5, no 2, pp762–766

Christian, D. G. and Riche, A. B. (1998) 'Nitrate leaching losses under *Miscanthus* grass planted on a silty clay loam soil', *Soil Use and Management*, vol 14, no 3, pp131–135

Christiansen, R. C. (2008) 'Sea asparagus can be oil feedstock', *Biodiesel Magazine*, August, www.biodieselmagazine.com/article.jsp?article_id=2600

Chung, C.-H. (1993) 'Thirty years of ecological engineering with *Spartina* plantations in China', *Ecological Engineering*, vol 2, pp261–289

CIAT (Centro International de Agriculture Tropical) (1976) *Annual Report, 1975*, CIAT, Cali, Colombia

CIAT (1997) 'Cassava boom in Southeast Asia', www.worldbank.org/html/cgiar/newsletter/june97/9ciat.html (accessed September 2009)

Clark, A. (1994) *Aramco World,* November/December

Clawson, M. (1979) 'Forests in the long sweep of American history', *Science*, vol 204, pp1168–1174

Clayton, W. D. and Renovize, S. A. (1986) 'Genera graminum, grasses of the world', *Kew Bulletin*, Additional Series XIII

Clifton-Brown, J. C. and Lewandowski, I. (2000) 'Overwintering problems of newly established *Miscanthus* plantations can be overcome by identifying genotypes with improved rhizome cold tolerance', *New Phytologist*, vol 148, no 2, pp287–294

Clifton-Brown, J. C., Lewandowski, I. et al (2001) 'Performance of 15 Miscanthus genotypes at five sites in Europe', *Agronomy Journal*, vol 93, pp1013–1019

Clifton-Brown, J. C., Lewandowski, I., Bangerth, F. and Jones, M. B. (2002) 'Comparative responses to water stress in stay-green, rapid- and slow senescing genotypes of the biomass crop, *Miscanthus*', *New Phytologist*, vol 154, no 2, pp335–345

Clifton-Brown, J. C., Stampfl, P. F. and Jones, M. B. (2004) '*Miscanthus* biomass production for energy in Europe and its potential contribution to decreasing fossil fuel carbon emissions', *Global Change Biology*, vol 10, no 4, pp509–518

Clifton-Brown, J. C., Breuer, J. and Jones, M. B. (2007) 'Carbon mitigation by the energy crop, *Miscanthus*', *Global Change Biology*, vol 13, no 11, pp2296–2307

Coale, K. H., Johnson, K. S., Fitzwater, S. E., Gordon, R. M., Tanner, S., Chavez, F. P., Ferioli, L., Sakamoto, C., Rogers, P., Millero, F., Steinberg, P., Nightingale, P., Cooper, D., Cochran, W. P., Landry, M. R., Constantinou, J., Rollwagen, R., Trasvina, A. and Kudela, R. (1996) 'A massive phytoplankton bloom induced by an ecosystem-scale iron fertilization experiment in the equatorial Pacific Ocean', *Nature*, vol 383, pp495–501

Cohen, J. D. (2007) *Petrobras and the Biofuels*, Petrobras Europe Limited

Cohen, W. B., Harmon, M. E., Wallin, D. O. and Fiorella, M. (1996) 'Two decades of carbon flux from forests of the Pacific Northwest', *BioScience*, vol 46, no 11, 836–844

Cole, C. V., Flach, K., Lee, J., Sauerbeck, D. and Stewart, B. (1993) 'Agricultural sources and sinks of carbon', *Water, Air, and Soil Pollution*, vol 70, pp111–122

Cole, C. V., Duxbury, J., Freney, J., Heinemeyer, O., Minami, K., Mosier, A., Paustian, K., Rosenberg, N., Sampson, N., Sauerbeck, D. and Zhao, Q. (1996) 'Agricultural options for mitigation of greenhouse gas emissions', in R. T. Watson, M. C. Zinyowera, R. H. Moss and D. J. Dokken (eds) *Climate Change 1995 – Impacts, Adaptations, and Mitigation of Climate Change: Scientific-Technical Analyses*, Cambridge University Press, pp745–771

Cole, C. V., Duxbury, J., Freney, J., Heinemeyer, O., Minami, K., Mosier, A., Paustian, K., Rosenberg, N., Sampson, N., Sauerbeck, D. and Zhao, Q. (1997) 'Global estimates of potential mitigation of greenhouse gas emissions by agriculture', *Nutrient Cycling in Agroecosystems*, vol 49, pp221–228

Collins, R. P. and Jones, M. B. (1985) 'The influence of climatic factors on the distribution of C4 species in Europe', *Vegetatio*, vol 64, pp121–129

Columbus Foods (2008) 'Babassu palm tree', www.columbusfood.net

Conant, R. T., Paustian, K. and Elliot, E. T. (2001) 'Grassland management and conversion into grassland: Effects on soil carbon', *Ecological Applications*, vol 11, no 2, pp343–355

Conceição, M. M., Candeia, R. A., Silva, F. C., Bezerra, A. F., Fernandes V. J. and Souza A. G. (2007) 'Thermoanalytical characterization of castor oil biodiesel', *Renewable and Sustainable Energy Reviews*, vol 11, no 5, pp964–975

Cony, M. A. (1995) 'Rational afforestation of arid and semiarid lands with multipurpose trees', *Interciencia*, vol 2, no 5, pp249–253

Cook, C. D. K. (1974) *Water Plants of the World*, Dr W. Junk Publishers, The Hague

Cooper, D. J., Watson, A. J. and Nightingale, P. D. (1996) 'Large decrease in ocean surface CO_2 fugacity in response to in situ iron fertilization', *Nature*, vol 382, pp511–513

Corley, W. L. (1989) 'Propagation of ornamental grasses adapted to Georgia and the US southeast', *International Propagators' Society Combined Proceedings*, vol 39, pp322–337

Cosentino, S. (1996) 'Crop physiology of sweet sorghum (*Sorghum bicolor* [L.] Moench) in relation to water and nitrogen stress', in *First European Seminar on Sorghum for Energy and Industry*, Institut National de la Recherche Agronomique, France, pp30–41

Couto, L. and Betters, D. R. (1995) *Short Rotation Eucalypt Plantations in Brazil: Social and Enviromental Issues,* Oak Ridge National Laboratory, Tennessee

Cromie, W. J., 'Kudzu cuts alcohol consumption', *Harvard University Gazette*, http://news.harvard.edu/gazette/2005.05.19/09-kudzu.html (accessed August 2009)

Curt, M. D., Fernandez, J. and Martinez, M. (1994) 'Potential of sweet sorghum crop for biomass and sugars production in Madrid', in Hall et al (eds) *Biomass for Energy and Industry, Proceedings of the 7th EC Conference*, Ponte Press, Bochum, Germany, pp632–635

Curt, M. D., Fernandez, J. and Martinez, M. (1996) 'Effect of nitrogen fertilization on biomass production and macronutrients extraction of sweet sorghum cv keller', presented at the 9th EU Conference on Biomass for Energy, Agriculture and Environment, Copenhagen, 24–27 June

Da Motta, S., Young, C. and Ferraz, C. (1999) *Clean Development Mechanism and Climate Change: Cost Effectiveness and Welfare Maximization in Brazil*, report to the World Resources Institute, Institute of Economics, Federal University of Rio de Janeiro, Brazil

da Rosa, A. (2005) *Fundamentals of Renewable Energy Processes*, Elsevier, pp501–502

Da Silva, F. Jr., Rubio, S. and de Souza, F. (1995) 'Regeneration of an Atlantic forest formation in the understory of a Eucalyptus grandis plantation in south-eastern Brazil', *Journal of Tropical Ecology*, vol 11, pp147–152

Daggett, D., Hadaller, O., Hendricks, R. and Walther, R. (2006) 'Alternative fuels and their potential impact on aviation', *25th Congress of the International Council of the Aeronautical Sciences (ICAS 2006)*, Hamberg, Germany, 3–8 September

Daggett, D., Hendricks, R., Robert, C., Walther, R. and Corporan, E. (2007) *Alternate Fuels for Use in Commercial Aircraft*, American Institute of Aeronautics and Astronautics, Report No ISABE-2007-1196

Dajue, L. (2008) 'Sweet sorghum: A new sugar crop with bright future', Beijing Green Energy Institute, www.sustainable-agro.com

Dajue, L. and Mündel, H.-H. (1996) *Safflower Carthamus tinctoris L. Promoting the Conservation and Use of Underutilized and Neglected Crops, 7*, International Plant Genetic Resources Institute, Gatersleben/International Plant Genetic Resources Institute, Rome, Italy

Dajue, L. and Yonggang, L. (1996) 'Research and production of sweet sorghum in China', in *First European Seminar on Sorghum for Energy and Industry*, Institut National de la Recherche Agronomique, France, pp342–350

Dajue, L. and Yunzhou, H. (1993) 'The development and exploitation of safflower tea', in *Proceedings of the Third International Safflower Conference*, 9–13 June 1993, Beijing, pp837–843

Dalianis, C. (1995) *Improvement of Eucalypts Management*, report submitted to EU in the framework of AIR-CT93-1678 project

Dalianis, C., Sooter, C. and Christou, M. (1995a) 'Growth, biomass productivity of *Arundo donax* and *Miscanthus sinensis* "giganteus"', in Chartier et al (eds) *Biomass for Energy, Environment, Agriculture and, Industry: Proceedings of the 8th EU Biomass Conference*, Pergamon Press, UK, vol 1, pp575–582

Dalianis, C. D., Sooter, C. and Christou, M. (1995b) 'Sweet sorghum (*Sorghum bicolor* (L) Moench) biomass and sugar yields potential in Greece', in Chartier et al (eds) *Biomass for Energy, Environment, Agriculture and Industry: Proceedings of the 8th EU Biomass Conference*, Pergamon Press, Oxford, pp622–628

Dalianis, C., Alexopoulou, E., Dercas, N. and Sooter, C. (1996) 'Effect of plant density on growth, productivity and sugar yields of sweet sorghum in Greece', presented at the 9th EU Conference on Biomass for Energy, Agriculture and Environment, Copenhagen, Denmark

Dalianis, C., Christou, M., Sooter, C., Kyritsis, S., Zafiris, C. and Samiotakis, G. (1994) 'Productivity and energy potential of densely planted eucalypts in a two year short rotation', in D. O. Hall, et al (eds), *Biomass for Energy and Industry: Proceedings of the 7th Biomass Conference*, Ponte Press, Bochum, Germany

Dambroth, M. (1984) 'Topinambur-eine Konkurrenz fiir den Industriekartoffelanbau?', *Der Kartoffelbau*, vol 35, no 11, pp450–453

Dambroth, M. and El Bassam, N. (1990) 'Genotypic variation in plant productivity and consequences for breeding of "low-input cultivars"', in *Genetic Aspects of Plant Mineral Nutrition, Developments in Plant and Soil Sciences*, Kluwer, Dordrecht, pp1–7

Damor, K. (2008) 'Guj cotton yield may cross world average', *Business Standard*, 7 October, www.business-standard.com/india/storypage.php?autono=336512

Danfors, B. (1994) 'Harvesting *Salix*: Technical data on harvesting machines, harvesting capacity and further development', in *Harvesting Techniques for Energy Forestry*, Swedish University of Agricultural Sciences, pp46–49

Dannenmann, B. M. E., Choocharoen, C., Spreer, W., Naglea, M., Leisa, H., Neef, A. and Mueller, J. (2007) 'The potential of bamboo as a source of renewable energy in Northern Laos Tropentag', University of Kassel-Witzenhausen and University of Göttingen, 9–11 October 2007

Darwin, R., Tsigas, M., Lewandrowski, J. and Raneses, A. (1995) *World Agriculture and Climate Change: Economic Adaptations*, Agricultural Economic Report No 703, US Department of Agriculture, Economic Research Service, Washington, DC

Darwin, R., Tsigas, M., Lewandrowski, J. and Raneses, A. (1996) 'Land use and cover in ecological economics', *Ecological Economics*, vol 17, no 3, pp157–181

Date Palm Global Network (2005) The international workshop true-to-typeness of date palm tissue culture-derived plants, 23 May

Day, S. (2008) *The Crops that Will Power Biofuels (Ethanol in Particular)*, Manitoba Agriculture, Food and Rural Initiatives, 25 January

Davidson, E. A. and Ackerman, I. L. (1993) 'Changes in soil carbon inventories following cultivation of previously untilled soils', *Biogeochemistry*, vol 20, pp161–164

Davidson, E. A., Nepstad, D. C., Klink, C. and Trumbore, S. E. (1995) 'Pasture soils as carbon sink', *Nature*, vol 376, pp472–473

de Brujin, G. H. (1971) *Etude da caractere cyanogenetique du manioc (Manihot esculenta Crantz)*, Mededelingen Landbowhogeschool, Wageningen, The Netherlands, pp17–18

Decker, J. (2007a) 'Corn to run: Can ethanol be used as a clean alternative?', *Flight International*, 16 January, pp40–42

Decker, J. (2007b) 'Flying green: The use of alternative fuels for aviation', *Renewable Energy World*, 1 March, pp118–124

Decker, J. (2008a) 'Environmental special: Are alternative fuels really cleaner?', *Flight International*, 4 February, pp45–47

Decker, J. (2008b) 'Power players', *Flight International*, 17 June, pp41–44

de Jabrun, P. L. M., Byth, D. E. and Wallis, E. S. (1981) 'Imbibition by and effects of temperature on germination of mature seed of pigeonpea', in *Proceedings of the International Workshop on Pigeonpeas*, 15–19 December 1980, Patancheru, India, vol 2, pp181–187

Delesalle, L. and Dhellemmes, C. (1984) 'Semences de betteraves et de chicorée industrielles. Betterave porte-graine le repiquage: une pratique exigeante', *Bulletin Semences*, vol 86, pp39–40

Delmas, M. and Gaset, A. (1991) 'A new industrial crop network for the production of animal food, sugars and derivatives, and paper pulps', in *Biomass for Energy, Industry and Environment, 6th EC Conference*, Athens, Elsevier Applied Science, pp1262–1268

De Los Santos, J. A. (1987) *Aprovechamiento celulóosico del cardo*, Investigación Tècnica del Papel, Madrid, pp568–606

Deng, T. X. and Liu, G. F. (1991) 'Mathematical model of the relationship between nitrogen fixation by black locust and soil conditions', *Soil Biology and Biochemistry*, vol 23, no 1, pp1–7

Dercas, N., Dalianis, C., Panoutsou, C. and Sooter, C. (1995) 'Sweet sorghum (*Sorghum bicolor* (L) Moench) response to four irrigation and two nitrogen fertilization rates', in Chartier et al (eds) *Biomass for Energy, Environment, Agriculture and Industry: Proceedings of the 8th EU Biomass Conference*, Pergamon Press, Oxford, vol 1, pp629–639

Dercas, N., Panoutsou, C. and Dalianis, C. (1996) 'Water and nitrogen effects on sweet sorghum and productivity', presented at the 9th EU Conference on Biomass for Energy, Agriculture and Environment, Copenhagen, Denmark

Deuter, M. (1996) Personal communication

Dhellemmes, C. (1987) 'Agronomy of chicory', in R. J. Clarke and R. MacRae (eds) *Coffee, Vol. 5: Related Beverages,* Elsevier, London, pp179–191

Dick, W. A., Blevins, R. L., Frye, W. W., Peters, S. E., Christenson, D. R., Pierce, F. J. and Vitosh, M. L. (1998) 'Impacts of agricultural management practices on C sequestration in forest-derived soils of the eastern Corn Belt', *Soil and Tillage Research*, vol 47, pp235–244

Dickinson, R. E. (1996) 'Climate engineering: A review of aerosol approaches to changing the global energy balance', *Climatic Change*, vol 33, pp279–290

Diercke (1981) *Weltwirtschaftsatlas* l. Deutscher Taschenbuchverlag/ Westermann, Germany

Dippon, D. R., Rockwood, D. L. and Comer, C. W. (1985) 'Cost sensitivity analysis of *E. grandis* woody biomass systems', in W. H. Smith (ed) *Biomass Energy, Development*, Plenum Press, London

Dixon, R. K., Winjum, J. K. and Schroeder, P. E. (1993) 'Conservation and sequestration of carbon: The potential of forest and agroforest management practices', *Global Environmental Change*, vol 3, no 2, pp159–173

Dixon, R. K., Brown, S., Houghton, R. A., Solomon, A. M., Trexler, M. C. and Wisniewski, J. (1994) 'Carbon pools and flux of global forest ecosystems', *Science*, vol 263, pp185–190

DOB (Department of Biotechnology) (2008) 'Bio-control strategies for eco-friendly pest management', Ministry of Science & Technology, Government of India, www.dbtbiopesticides.nic.in/ event/EventDetails.asp?EventId=251 (accessed September 2009)

DOE (US Department of Energy) (1994) *The American Farm*, document prepared by the National Renewable Energy Laboratory, NREL/SP-420-5877

DOE (1997) *Biomass Power Program*, document prepared by the National Renewable Energy Laboratory, DOE/GO-10097-412

DOE (1998) *Renewable Energy Technology Characterizations*, Office of Utility Technologies, Washington, DC

Dogget, H. (1988) *Sorghum*, Longman, UK

Doka, G. (2003) *Life Cycle Inventories of Waste Treatment Services*, final report Ecoinvent 2000 No 13, EMPA St. Gallen, Swiss Centre for Life Cycle Inventories, Dübendorf, www.ecoinvent.org

Dolciotti, I., Mambelli, S., Grandi, S. and Venturi, G. (1996) 'A comparative analysis of the growth and yield performances of sweet and non sweet sorghum genotypes', in *First European Seminar on Sorghum for Energy and Industry*, Institute National de la Recherche Agronomique, France, pp207–212

Dornburg, V. and Faaij, A. (2001) 'Efficiency and economy of woodfired biomass energy systems in relation to scale regarding heat and power generation using combustion and gasification technologies', *Biomass and Bioenergy*, vol 21, no 2, pp91–108

Dornburg, V. and Faaij, A. (2005) 'Cost and CO_2-emission reduction of biomass cascading: Methodological aspects and case study of SRF poplar', *Climatic Change*, vol 71, no 3, pp373–408

Dossier pour la Science (2007) Edition francaise de Scientific American, 'Climat Coment eviter la surchauffe?' January–March

Douglas, J. (1994) 'A rich harvest from halophytes', *Resource*, May, pp15–18

Downing, M., Bain, R. and Overend, R. (1996) 'Economic development through biomass systems integration', in *Bioenergy '96: Proceedings of the Seventh National Bioenergy Conference*, 15–20 September 1996

DTI (Department of Trade and Industry) (1994) *Short Rotation Coppice Production and the Environment*, Technology Status Report 014

Duan, Z., Liu., X. and Qu, J. (1995) 'Effect of land desertification on CO_2 content of atmosphere in China', *Agricultural Input and Environment*, pp279–302

Duke, J. A. (1978) 'The quest for tolerant germplasm', in *ASA Special Symposium 32, Crop Tolerance to Suboptimal Land Conditions, American Society of Agronomy*, Madison, WI, pp1–61

Duke, J. A. (1979) 'Ecosystematic data on economic plants', *Quarterly Journal of Crude Drug Research*, vol 17, nos 3–4, pp91–110

Duke, J. A. (1983) *Handbook of Energy Crops*, published via Internet

Duke, J. A. (1984) *Handbook of Energy Crops*, published via Internet

Duke, S. H., Collins, M. and Soberalske, R. M. (1980) 'Effects of potassium fertilization on nitrogen fixation and nodule enzymes of nitrogen metabolism in alfalfa', *Crop Science*, vol 20, pp213–219

Dumbleton, F. (1997) 'Biomass conversion technologies: An overview', *Aspects of Applied Biology, 49: Biomass and Energy Crops*, pp341–347

Dupuy, B. and Kanga, A. N'G. (1991) 'Utilisation des acacias pour règènèrer les anciennes cocoteraies', *Revue Bois et Forets des Tropiques*, vol 230, 4th quarter

Dyson, F. J. (1977) 'Can we control the carbon dioxide in the atmosphere ?', *Energy*, vol 2, pp287–291

Dzondo-Gadet, M., Nzikou, J. M., Etoumongo, A., Linder, M. and Desobry, S. (2005) 'Encapsulation and storage of safou pulp oil in 6DE maltodextrins', *Process Biochemistry*, vol 40, no 1, pp265–271

Eaglesham, A., Brown, W. F. and Hardy, R. W. F. (eds) (2000) *National Agricultural Biotechnology Council Report*, National Agricultural Biotechnology Council, New York

Early, J. T. (1989) 'Space-based solar screen to offset the greenhouse effect', *Journal of the British Interplanetary Society*, vol 42, pp567–569

The Economist (2006) 'Steady as she goes', *The Economist*, 20 April, pp65–67

EEA (European Environment Agency) (2006) *How Much Bioenergy can Europe Produce Without Harming the Environment?*, EEA Report No 7

Eganathan, P., Subramanian, H. M. Sr., Latha, R. and Srinivasa Rao, C. (2006) 'Oil analysis in seeds of *Salicornia brachiata*', *Industrial Crops and Products*, vol 23, no 2, pp177–179

El Baker, A. and Namrod, B. (2005) 'Integrated dates processing project', January

El Bassam, N. (1990) 'Requirements for biomass production with higher plants under artificial ecosystems', *Proceedings of 'Artificial Ecological Systems' Workshop*, Marseille, pp59–65

El Bassam, N. (1993) 'Möglichkeiten und Grenzen der Bereitstellung von Energie aus Biomasse', *Landbauforschung Völkenrode*, vol 43, nos 2/3, pp101–111

El Bassam, N. (1994a) *New Hybrid Plant: Fiber Sorghum*, Annual Report, FAIR 1-CT92-0071

El Bassam, N. (1994b) 'Miscanthus – Stand und Perspektiven in Europa', *Forum für Zukunfts-energien e.V. – Energetische Nutzung von Biomasse im Konsenz mit Osteuropa*, International Meeting, March 1994, Jena, pp201–212

El Bassam, N. (1995) 'Auswirkungen einer Klimaveränderung auf den Anbau von C4-Pflanzenarten zur Erzeugung von Energieträgern', *Mitteilungen der Gesellschaft für Pflanzenbauwissenschaften*, vol 8, pp169–172

El Bassam, N. (1996) *Renewable Energy: Potential Eenergy Crops for Europe and the Mediterranean Region*, REU Technical Series 46, Food and Agriculture Organization of the United Nations, Rome

El Bassam, N. (1997) unpublished data

El Bassam, N. (1998a) 'Biological life support systems under controlled environments', in N. El Bassam et al (eds) *Sustainable Agriculture for Food, Energy and Industry*, James & James (Science Publishers) Ltd, London

El Bassam, N. (1998b) *Energy Plant Species: Their Use and Impact on Environment and Development*

El Bassam, N. (1999) *Integrated Energy Farm Feasibility Study*, SREN-FAO

El Bassam, N. (2001a) 'Renewable energy for rural communities', *Renewable Energy*, vol 24, pp401–408

El Bassam, N. (2001b) *The Productivity of* manually Echinochloa *as a Potential Energy Crop*, IFEED Publication No 12

El Bassam N. (2002a) 'Innovative fuels and biomass resources', *3rd LAMNET Workshop*, Brazil, 2–4 December 2002

El Bassam, N. (2002b) 'The concept of an integrated energy farm', *Latin American Thematic Network on Bioenergy, LAMNET Workshop*, Brazil, www.bioenergy-lamnet.org

El Bassam, N. (2004a) *LAMNET Workshop*, Ribeirao Preto, Brazil, September

El Bassam, N. (2004b) *Integrated Renewable Energy for Rural Communities*, Elsevier Science Publishers, Amsterdam and London

El Bassam, N. (2007a) 'Clean transportation and sustainable mobility for development', 19 December

El Bassam, N. (2007b) 'Future automotive fuels in regional and global context. Growth and security: Energy and energy transportation', *The Transatlantic Market Conference*, Dräger Foundation, Washington, DC, www.ifeed.org/links/From%20 Fields%20to% 20Wheels2007-05.pdf

El Bassam, N. (2008a) 'Sustainable mobility', *World Future Energy Summit 08*, Almasdar, UAE

El Bassam, N. (2008b) *Biomass from Miscanthus for Heat, Electricity and Transportation Fuel Generation*, IFEED Publication No 338, www.ifeed.org

El Bassam, N. and Dambroth, M. (1991) *A Concept of Energy Plant Farms*, Institute of Crop Science, FAL Bundesallee 50, D-18116 Braunschweig, Germany

El Bassam, N. and Jakob, K. (1996) 'Bambus – eine neue Rohstoffquelle – Erstevaluierung', *Landbauforschung Völkenrode*, vol 2, pp76–83

El Bassam, N., Dambroth, M. and Jacks, I. (1994) 'The utilization of *Miscanthus sinensis* (china grass) as an energy and an industrial raw material', *Natural Resources and Development*, vol 39, pp85–93

El Bassam, N., Sauerbeck, G., Bacher, W. and Korte, A. (2000) *Bamboo for Europe: Determination of Water Requirement and Water Use Efficiency as Well as Plant Adaptation in Northern Germany*, Federal Agricultural Research Centre, Institute of Crop and Grassland Science, Braunschweig, Germany

El Bassam, N., Meier, D., Gerdes, C. and Korte, A.-M. (2001) 'Energy from bamboo', Research Project No. III-234, Braunschweig, Germany

El Moudjahid (2006) 'Allocation du President de la Republique a l'occasion de la signature de la loi des finances 2007', 26 December

El Moudjahid (2006) 'Pour l'amelioration de la gouvernance locale des territoires ruraux, Interview du President de la Republique', 17 September

El Moudjahid (2007) 'Renouveau rural: l'Algerie en mouvement', document, 12 February

Elbersen, H. W. (1998) 'Alfalfa (*Medicago sativa* L.)', in N. El Bassam (ed) *Energy Plant Species*, James & James (Science Publishers) Ltd, London

Eldridge, K. G., Davidson, J., Harwood, C. and van Wyck, G. (1994) *Eucalypt Domestication and Breeding*, Clarendon Press, Oxford

Elliott, P. and Booth, R. (1993) *Brazilian Biomass Power Demonstration Project*, Special Brief, Shell Centre, London

El Watan (2006) 'Production de dattes: nouvelles formes d'aide pour l'exportation', 11 November

El Watan (2006) 'Le desert pour quoi faire?', 29 August

Emerging Markets Online (2008) 'Biodiesel 2020: A global market survey', www.emerging-markets.com

Engle, D. L., Melack, J. M., Doyle, R. D. and Fisher, T. R. (2008) 'High rates of net primary production and turnover of floating grasses on the Amazon floodplain: Implications for aquatic respiration and regional CO_2 flux', *Global Change Biology*, vol 14, no 2, pp369–381

Enzenwa, I., Abrbisala, O. A. and Akenova, M.E. (1996) *Tropical Grasslands*, vol 30, pp414–417

EPA/USIJI (US Environmental Protection Agency/US Initiative on Joint Implementation) (1998) *Activities Implemented Jointly: Third Report to the Secretariat of the UN Framework Convention on Climate Change*, EPA-report 236-R-98-004, US Environmental Protection Agency, Washington, DC, vol I, p19 and vol II, p607

EREC (2007) 'Renewable energy technology roadmap: 20% by 2020', www.erec.org

ERM (1995) *Effectiveness of Tax Incentives: Bioethanol (as ETBE) for Gasoline*, research project summary, Environmental Resources Management

Esler, D. (2008) 'Out of the cane: The Ipanema's latest iteration helps to grow its own fuel', *Aviation Week and Space Technology*, 6 October, pp78–79

EPFL, Jean-Louis Hersener, Ingenieurbüro HERSENER, Urs Meier, MERITEC GmbH, Konrad Schleiss, Umwelt- und Kompostberatung). Bundesamt für Energie BFE, Ittigen, retrieved from www.energieschweiz.ch/internet/03262/index.html?lang=de.

EU (2005) 'Status and perspectives of biomass-to-liquid fuels in the European Union', www.senternovem.nl/mmfiles/Status_ perspectives_biofuels_EU_2005_tcm24-152475.pdf

European Biomass Industry Association (2007) 'Other energy crops', www.eubia.org/193.0.html

Everest, J., Miller, J., Ball, D. and Patterson, M. (1999) 'Kudzu in Alabama: History, uses, and control', Alabama Cooperative Extension System

Ewald, D., Naujkos, G., Hertel, H. and Eich, J. (1992) 'Hat die Robinie in Brandenburg eine Zukunft?', *Allg. Forstzeitschrift*, vol 14, pp738–740

Faaij, A. (2006) 'Modern biomass conversion technologies', *Mitigation and Adaptation Strategies for Global Change*, vol 11, no. 2, pp335–367

Faaij, A. (2008) 'Bioenergy and global food security', WBGU-Hauptgutachten 'Welt im Wandel: Zukunftsfähige Bioenergie und nachhaltige Landnutzung', Berlin

Faaij, A. P. C. and Domac, J. (2006) 'Emerging international bioenergy markets and opportunities for socio-economic development', *Energy for Sustainable Development*, vol X, no 1, pp7–19

Faaij, A., Wagener, M., Junginger, M., Best, G., Bradley, D., Fritsche, U., Grassi, A., Hektor, B., Heinimö, J., Klokk, S., Kwant, K., Ling, E., Ranta, T., Risnes, H., Peksa, M., Rosillo-Calle, F., Ryckmanns, Y., Utria, B., Walter, A. and Woods, J. (2005) 'Opportunities and barriers for sustainable international bio-energy trade; Strategic advice of IEA Bioenergy Task 40', 14th European Biomass Conference and Exhibition, October, Paris, France

Facciola, S. (1990) *Cornucopia: A Source Book of Edible Plants*, Kampong Publications, www.ibiblio.org/pfaf/cgi-bin/arr_html? Eruca+vesicaria+sativa

FACE Foundation (1997) *Annual Report 1996*, Foundation FACE, Arnhem, The Netherlands

Fachagentur Nachwachsende Rohstoffe e.V. (FNR) (2008) *Biokraftstoffe Basisdaten Deutschland* Stand

FAI (Fertiliser Association of India) (1993) *Fertiliser Statistics*, FAI, New Delhi

Fairless, D. (2007) 'Biofuel: The little shrub that could – maybe', *Nature*, vol 449, pp652–655

Faist, E. M., Heck, T. and Jungbluth, N. (2003) 'Erdgas', in R. Dones (ed) *Sachbilanzen von Energiesystemen: Grundlagen für den ökologischen Vergleich von Energiesystemen und den Einbezug von Energiesystemen in Ökobilanzen für die Schweiz*, Paul Scherrer Institut Villigen, Swiss Centre for Life Cycle Inventories, Dübendorf, www.ecoinvent.org

Faix, O., Meier, D. and Beinhof, O. (1989) 'Analysis of lignocelluloses and lignines from *Arundo donax* L. and *Miscanthus sinensis* Andress., and hydroliquefaction of *Miscanthus*', *Biomass*, vol 18, pp109–126

Fallon, P. D., Smith, P., Smith, J. U., Szabo, J., Coleman, K. and Marshall, S. (1998) 'Regional estimates of carbon sequestration potential: Linking the Rothamsted Carbon Model to GIS databases', *Biology and Fertility of Soils*, vol 27, pp236–241

Fan, S., Gloor, M., Mahlman, J., Pacala, S., Sarmiento, J., Takahashi, T. and Tans, P. (1998) 'A large terrestrial carbon sink in North America implied by atmospheric and oceanic carbon dioxide data and models', *Science*, vol 282, pp442–446

FAO (United Nations Food and Agriculture Organization) (1980) *Report on the Agro-ecological Zones Project*, FAO, Rome, vol 1, pp24–48

FAO (1988) *Traditional Food plants*, FAO Food and Nutrition Paper No 42, Rome

FAO (1993a) *Date Palm Products*, FAO Agricultural Services Bulletin No 101, FAO, Rome, www.fao.org/docrep/t0681E/t0681e02.htm

FAO (1993b) *Forestry Statistics Today and Tomorrow*, FAO, Rome

FAO (1994) *Production Yearbook 1993*, FAO, Rome, pp108–109

FAO (1995a) *Production Yearbook 1994*, FAO, Rome, pp15–87

FAO (1995b) *Future Energy Requirements for Africa's Agriculture*, FAO, Rome

FAO (1996a) *Forest Resources Assessment 1990: Survey of Tropical Forest Cover and Study of Change Processes*, FAO Forestry Paper No 130, FAO, Rome

FAO (1996b) 'Ecocrop 1 database', FAO, Rome

FAO (1997) *State of the World's Forests*, FAO, Rome

FAO (2003), FAO stat database', FAO, Rome, http://apps. fao.org/page/collections

FAO (2008a) 'Biofuels: Prospects, risks and opportunities – The state of food and agriculture', www.fao.org

FAO (2008b) 'World food situation: Food price indices', FAO, Rome, www.fao.org

FAO (2008c) 'The State of Food and Agriculture', FAO, Rome, http://farm2.static.flickr.com/1221/1094121494_d26b58 dd77.jpg?v=0

Farre, I., Robertson, M. J., Asseng, S., French, R. J. and Dracup, M. (2003) 'Variability in lupin yield due to climate in Western Australia', in *Proceedings of the Australian Agronomy Conference*, Australian Society of Agronomy

Farrell, E. P., Führer, E., Ryan, D., Andersson, F., Hüttl, R. and Piussi, P. (2000) 'European forest ecosystems: Building the future on the legacy of the past', *Forest Ecology and Management*, vol 132, pp5–20

Fearnside, P. M. (1997) 'Greenhouse gases from deforestation in Brazilian Amazonia: Net committed emissions', *Climatic Change*, vol 35, no 3, pp321–360

Fearnside, P. M. (1998) 'Plantation forestry in Brazil: Projections to 2050', *Biomass and Bioenergy*, vol 15, pp437–450

Fearnside, P. M. (1999) 'Forests and global warming mitigation in Brazil: Opportunities in the Brazilian forest sector for responses to global warming under the "clean development mechanism"', *Biomass and Bioenergy*, vol 16, pp171–189

Feller, C. and Beare, M. H. (1997) 'Physical control of soil organic matter dynamics in the tropics', *Geoderma*, vol 79, pp69–116

Fernández, J. (1992) *Production and Utilization of* Cynara Cardunculus *L. Biomass for Energy, Paper Pulp and Food Industry*, Project JOUB0030-E C.C.E (DGXII), May 1990 to August 1992, final report, Brussels

Fernández, J. and Curt, M. D. (1995) 'Estimated cost of thermal power from *Cynara cardunculus* biomass in Spanish conditions: Application to electricity production', *Proceedings of the 8th European Conference on Biomass for Energy, Environment, Agriculture and Industry*, Vienna 1994, Pergamon, vol 1, pp342–350

Fernández, J. and Manzanares, P. (1989) '*Cynara cardunculus* L., a new crop for oil, paper pulp and energy', *Proceedings of the 5th European Conference on Biomass for Energy and Industry*, Lisbon, 9–13 October 1989, Elsevier, pp1184–1189

Fernández, J. and Marquez, L. (1995) 'Energetisches Potential von *Cynara cardunculus* L. aus landwirtschaftlichem Anbau in Trockengebieten des Mittelmeerraumes', *Der Tropenlandwirt [Journal of Agriculture in the Tropics and Subtropics]*, vol 53, pp121–129

Fernández, J., Curt, M. D. and Martinez, M. (1991) 'Water use efficiency of *Helianthus tuberosus* L. "Violet de Rennes", grown in drainage lysimeter', in *Biomass for Energy, Industry and Environment, 6th EC Conference*, Athens, Elsevier Applied Science, pp297–301

Fernándes, E. C. M., Motavalli, P. P., Castilla, C. and Mukurumbira, L.(1997) 'Management control of soil organic matter dynamics in tropical land-use systems', *Geoderma*, vol 79, pp49–67

Festerling, O. (1993) Personal communication

Feuerstein, U. (1995) *Erzeugung standortgerechter zur Ganzpflanzenverbrennung geeigneter Gräser für die Nutzung als nachwachsende Rohstoffe*, GFP-Project F 46/91 NR-90 NR 026, Deutsche Saatveredelung Lippstadt – Bremen GmbH

Fisher, M. J., Rao, I. M., Ayarza, M. A., Lascano, C. E., Sanz, J. I., Thomas, R. J. and Vera, R. R. (1994) 'Carbon storage by introduced deep-rooted grasses in the South American savannas', *Nature*, vol 371, pp236–238

Fisher, M. J., Thomas, R. J. and Rao, I. M. (1997) 'Management of tropical pastures in acid-soil savannas of South America for carbon sequestration in the soil', in R. Lal et al (eds) *Management of Carbon Sequestration in Soil*, CRC Press, Boca Raton, pp405–420

Flach, K. W., Barnwell, T. O. Jr. and Crosson, P. (1997) 'Impacts of agriculture on atmospheric carbon dioxide', in E. A. Paul, E. T. Elliott, K. Paustian and C. V. Cole (eds) *Soil Organic Matter in Temperate Agroecosystems: Long-Term Experiments in North America*, CRC Press, Boca Raton, pp3–13

Flannery, B. P., Kheshgi, H., Marland, G., MacCracken, M. C., Komiyama, H., Broecker, W., Ishitani, H., Rosenberg, N., Steinberg, M., Wigley, T. and Morantine, M. (1997) 'Geoengineering Climate', in R. G. Watts (ed) *Engineering Response to Global Climate Change*, CRC/Lewis Publishers, Boca Raton, pp379–427

Fleming, J. S., Habibi, S., MacLean, H. L. and Brinkman, N. (2005) 'Evaluating the sustainability of producing hydrogen from biomass through well-to-wheel analyses', *SAE Transactions*, vol 114, no 5, pp729-745

Fobil, J. N. (2008) 'Research and development of the shea tree and its products', Horizon Solutions Site, www.solutions-site.org/artman/publish/article_10.shtml (accessed September 2009)

Foidl, N., Foidl, G., Sanchez, M., Mittelbach, M. and Hackel, S. (1996) '*Jatropha curcas* L. as a source for the production of biofuel in Nicaragua', *Bioresource Technology*, vol 58, no 1, pp77–82

Fontana, A., Schorr-Galindo, S. and Guiraud, J. P. (1992) 'Inulin-containing crops: Improvement of fermentation economics through co-products valorization', in *Biomass for Energy and Industry, 7th EC Conference*, Florence, Ponte Press, pp1060–1065

Forgrave, A. (2008) 'Ryegrass could solve Wales' energy woes', www.dailypost.co.uk/farming-north-wales/farming-news/2008/04/17/ryegrass-could-solve-wales-energy-woes-55578-20775631/#

Forman, R. T. T. (1995) *Land Mosaics: The Ecology of Landscapes and Regions*, Cambridge University Press, New York

Forster, K. E. and Wright, N. G. (2002) 'Constraints to Arizona Agriculture and Possible Alternatives', Office of Arid Land Studies, University of Tuscon, Arizona

Francis, F. and Falisse, A. (1997) *IENICA Project*, Faculty of Agricultural Sciences Crops Production Unit, Gembloux, Belgium, p70

Francis, G., Edinger, R. and Becker, K. (2005) 'A concept for simultaneous wasteland reclamation, fuel production, and socio-economic development in degraded areas in India: Need, potential and perspectives of *Jatropha* plantations', *Natural Resources Forum*, vol 29, no 1, pp12–24

Franke, G., Hammer, K., Hanelt, P, Ketz, H.-A., Natho, G. and Reinbothe, H. (1988) *Früchte der Erde*, 3rd edn, Urania Verlag, Leipzig, Jena and Berlin

Franke, W. (1985) 'Inulin liefernde Pflanzen', in *Nutzpflanzenkunde*, 3rd edn, Georg Thieme Verlag, Stuttgart and New York

Franklin, J. F. and Spies, T. A. (1991) *Composition, Function, and Structure of Old-Growth Douglas-Fir Forests: Wildlife and Vegetation of Unmanaged Douglas-Fir Forests*, USDA Forest Service General Technical Report, PNW-GTR-285, pp71–80

Frese, L., Damroth, M. and Bramm, A. (1991) 'Breeding potential of root chicory', *Plant Breeding*, vol 106, pp107–113

Frischknecht, R., Jungbluth, N., Althaus, H.-J., Doka, G., Dones, R., Heck, T., Hellweg, S., Hischier, R., Nemecek, T., Rebitzer, G. and Spielmann, M. (2004a) *Overview and Methodology*, final report Ecoinvent 2000 No 1, Swiss Centre for Life Cycle Inventories, Dübendorf, www.ecoinvent.org

Frischknecht, R., Jungbluth, N., Althaus, H.-J., Doka, G., Dones, R., Hellweg, S., Hischier, R., Humbert, S., Margni, M., Nemecek, T. and Spielmann, M. (2004b) *Implementation of Life Cycle Impact Assessment Methods*, final report Ecoinvent 2000 No 3, Swiss Centre for Life Cycle Inventories, Dübendorf, retrieved from www.ecoinvent.org

Frost, B. W. (1996) 'Phytoplankton bloom on iron rations', *Nature*, vol 383, pp475–476

Frumhoff, P. C. and Losos, E. C. (1998) *Setting Priorities for Conserving Biological Diversity in Tropical Timber Production Forests*, Union of Concerned Scientists, Cambridge, MA

Frumhoff, P. C., Goetze, D. C. and Hardner, J. J. (1998) *Linking Solutions to Climate Change and Biodiversity Loss Through the Kyoto Protocol's Clean Development Mechanism*, Union of Concerned Scientists, Cambridge, MA

Fujimori, T. (1997) 'Overview of forest resources and forestation activities in Japan: Environmentally friendly tree products and their processing technology', in *Proceedings of the Japanese/Australian Workshop on Environmental Management*, Japan Science and Technology Agency and Forestry and Forest Products Research Institute, pp21–27

Fukushima, Y. (1987) 'Influence of forestation on mountainside at granite highlands', *Water Science*, vol 177, pp17–34

Garbe, C. (1995) 'Strohheizwerk Schkölen – nur ein Strohfeuer in Deutschland?', *Agronomical*, vol 1, no 95, pp10–13

Garratt, J. R. (1993) 'Sensitivity of climate simulations to land-surface and atmospheric boundary-layer treatments: A review', *Journal of Climate*, vol 6, pp419–449

Gawel, N. J., Robacker, C. D. and Corley, W. L. (1987) 'Propagation of *Miscanthus sinensis* through tissue culture', *Hortscience*, vol 22, p1137 (abstract)

GEIE Eurosorgho (1996) *New Hybrid Plant (Fiber Sorghum) as a Component in Mixtures of Pulps Standardly Used in the Paper Industry*, Periodic Progress Report 1995/96, FAIR l-CT92-0071

Geller, D. P. (1998) 'Alternative feedstocks for the production of biodiesel', www.google.com/search?q=CUPHEA% 2Ffuel&rls= com.microsoft:de:IE-SearchBox&ie=UTF-8&oe=UTF-8&sourceid=ie7&rlz=1I7SUNA

Geller, D. P., Goodrum, J. W. and Knapp, S. J. (1999) 'Fuel properties of oil from genetically altered *Cuphea viscosissima*', *Industrial Crops and Products*, vol 9, no 2, pp85–91

Germer, J. and Sauerborn, J. (1997) 'The oil palm (*Elais guinesis*) as a source of renewable energy', in *Proceedings of the International Conference on Sustainable Agriculture for Food, Energy and Industry*

Ghosh, S. P., Kumar, B. M., Kabeerathumma, S. and Nair, G. M. (1996) 'Productivity, soil fertility and soil erosion under cassava based agroforestry systems', *Agroforesty Systems*, vol 8, pp67–82

Giardina, C. P. and Ryan, M. G. (2000) 'Evidence that decomposition rates of organic carbon in mineral soils do not vary with temperature', *Nature*, vol 404, pp858–861

Gibson, J. G. (1986) 'Hints for propagating ornamental grasses, azaleas and junipers', *American Nurseryman*, vol 164, pp90–97

Gielis, J. (1995) 'Bamboo and biotechnology', *European Bamboo Society Journal*, May, pp27–39

Glencross, B. and Hawkins, W. (2004) 'A comparison of the digestibility of lupin (*Lupinus* sp.) kernel meals as dietary protein resources when fed to either, rainbow trout, *Oncorhynchus mykiss* or red seabream, *Pagrus auratus*', *Aquaculture Nutrition*, vol 10, no 2, pp65–73

Glenn, E. P. and O'Leary, J. W. (1985) 'Productivity and irrigation requirements of halophytes grown with hypersaline seawater in the Sonoran Desert', *Journal of Arid Environments*, vol 9, pp81–91

Glenn, E. P. and Watson, M. C. (1993) 'Halophyte crops for direct salt water irrigation', in H. Lieth and A. Al Masoom (eds) *Towards the Rational Use of High Salinity Tolerant Plants*, vol 1, pp379–85

Glenn, E. P., O'Leary, J. W., Watson, M. C., Thompson, T. L. and Khuel, R. (1991) '*Salicornia bigelovii*: an oilseed halophyte for seawater irrigation', *Science*, vol 251, pp1065–1067

Gliese, J. and Eitner, A. (1995) *Jatropha Oil as Fuel*, DNHE-GTZ, Project Pourghere

Global Industry Analysts, Inc. (2008) 'Ethanol: A global strategic business report', www.strategyr.com

Glück, P. and Weiss, G. (1996) 'Forestry in the context of rural development: Future research needs', in *Proceedings of the COST Seminar 'Forestry in the Context of Rural Development'*, Vienna, Austria, 15–17 April 1996, EFI Proceedings No 15, European Forest Institute Joensuu, Finland

Goering, C. E. and Daugherty, M. J. (1982) 'Energy accounting for eleven vegetable oil fuels', *Transactions of the ASAE*, vol 25, no 5, pp1209–1215

Göhre, K. (1952) *Die Robinie und ihr Holz*, Deutscher Bauernverlag, Berlin

Goldemberg, J. (2000) *World Energy Assessment: Energy and the Challenge of Sustainability*, UNDP/UN-DESA/World Energy Council

Goldemberg, J. (2004) *World Energy Assessment Overview: 2004 Update*, UNDP/UN-DESA/World Energy Council

Goldemberg, J., Teixeira Coelho, S., Mário Nastari, P. and Lucon, O. (2004) 'Ethanol learning curve: The Brazilian experience', *Biomass and Bioenergy*, vol 26, no 3, pp301–304

Goldstein, B. (2001) *Technical and Economical Feasibility of Buffalo Gourd as a Novel Energy Crop: Final Report*, New Mexico Solar Energy Inst., Las Cruces, USA

Gooding, H. J. (1962) 'The agronomic aspects of pigeonpeas', *Field Crop Abstracts*, vol 15, no 1, pp1–5

Gorham, E. (1991) 'Northern peatlands: Role in the carbon cycle and probable response to climate warming', *Ecological Applications*, vol 1, no 2, pp182–195

Gorham, E. (1995) 'The biogeochemistry of northern peatlands and its possible responses to global warming', in G. M. Woodwell and F. T. McKenzie (eds) *Biotic Feedback in the Global Climatic System: Will the Warming Feed the Warming?*, Oxford University Press, New York, pp169–187

Govindasamy, B. and Caldeira, K. (2000) 'Geoengineering Earth's radiation balance to mitigate CO_2-induced climate change', *Geophysical Research Letters*, vol 27, pp2141–2144

Grace, M. R. (1977) *Cassava Processing*, Plant Production Series No 3, FAO, Rome

Graef, M. (1997) Personal communication

Graef, M., Vellguth, G., Krahl, J. and Munack, A. (1994) 'Fuel from sugar beet and rape seed oil – mass and energy balances for evaluation', in *Proceedings of the 7th European Conference on Biomass for Energy, Environment, Agriculture and Industry*, Vienna, pp1–4

Grassi, G. (1997) 'Operational employment in the energy sector', presentation at the Tagung des SPD – Umweltforums, 5 July

Grassi, G. and Bridgewater, T. (1992) *Biomass for Energy and Environment, Agriculture and Industry in Europe: A Strategy for the Future*, Commission of the European Communities, Directorate General for Science Research and Development, Luxembourg

Gray, A. J., Benham, P. E. M. and Raybould, A. F. (1990) '*Spartina anglica*: The evolutionary and ecological background', in A. J. Gray and P. E. M. Benham (eds) Spartina anglica: *A Research Review*, ITE publication No 2, NERC, Swindon, UK

Greef, J. M. and Deuter, M. (1993) 'Syntaxonomy of *Miscanthus x giganteus* Greef et Deu.', *Angew. Bot.*, vol 67, pp87–90

Green Car congress (2007) 'China Planning Massive Jatropha Planting For Biodiesel', 7 February, www.greencarcongress.com/ 2007/02/china_planning_.html (accessed September 2009)

Green, C. (2002) 'The competitive position of triticale in Europe', in *Proceedings of the 5th International Triticale Symposium*, Radzikow, Poland, 30 June–5 July, 2002, vol I, pp22–26

Green, D. (1991) 'The new North American garden', *The Garden*, Royal Horticultural Society, London, vol 116 (Part I), January, pp18–22

Greenland, D. J. (1995) 'Land use and soil carbon in different agroecological zones: Soils and global change', in R. Lal, J. Kimble, E. Levine and B. A. Stewart (eds) *Soil Management and the Greenhouse Effect*, Lewis Publishers, pp9–24

Greenway, H. and Munns, R. (1980) 'Mechanisms of salt tolerance in nonhalophytes', *Annual Review of Plant Physiology*, vol 31, pp149–90

Gregory, W. C. and Gregory, M. P. (1976) 'Groundnut', in N. W. Simmonds (ed) *Evolution of Crop Plants*, Longman, London, pp151–154

Grimm, H. P. (1996) Personal communication

Gross, L. (1993) *Warum der 00-raps so interessant für uns ist... Arbeitsheft*, Verband Deutscher Oelmühlen, Bonn.

Guibet, J. C. (1988) 'L'utilisation des huiles vegetables et de derives comme carburants diesel', in *Liquid Fuels from Biomass, Report and Proceedings of CNRE Technical Consultation*, Saint-Remy-les-Chevreuse, France, 2–4 December 1988, FAO CNRE Bulletin, No 20

Guiraud, J. P., Bajou, A. M., Chautard, P. and Galzy, P. (1983) 'Inulin hydrolysis by an immobilized yeast-cell reactor', *Enzyme and Microbial Technology*, vol 5, pp185–190

Gumpon, P. and Teerawat, A. (2003) 'Palm oil as fuel for agricultural diesel engines: Comparative testing against diesel oil', *Songklanakarin Journal of Science and Technology*, vol 25, no 3, pp1–8

Gupta, A. K. and Agarwal, H. R. (1997) 'Taramira: A potential oilseed crop for marginal lands of north-west India', in *Proceedings of the International Conference on Sustainable agriculture for Food, Energy and Industry*, 22–28 June 1997, FAL, Braunschweig, Germany

Guretzky J. A. and Norton, S. (2008) 'Forage yields from 2007–2008 annual ryegrass variety trial', www.noble.org/ag/Forage/0708Ryegrass/index.html (accessed September 2009)

Haase, E. (1988) 'Pflanzen reinigen Schwermetallboden', *Umwelt*, vols 7–8, p342–344

Habibi, S., Maclean, H. and Brinkman, N. (2005) *Evaluating the Sustainability of Producing Hydrogen from Biomass through Well-to-Wheel Analyses*, University of Toronto, Document Number 2005-01-1552

Hahn, S. K. and Keyser, J. (1985) 'Cassava: A basic food of Africa', *Outlook on Agriculture*, vol 4, pp95–100

Hahn, S. K., Mahungu, N. M., Ottoo, J. A., Msabaha, M. A. M., Lutaladio, N. B. and Dahniya M. T. (1987) 'Cassava and African food crisis', in *Proceedings of the 3rd International Symposium for Tropical Root Crops*, August 1986, pp24–29

Hahn, S. K., Bai, K. V., Chukwuma, A. R., Dixon, A. and Ng, S. Y. (1994) 'Polyploid breeding of cassava', *Acta Horticulturae*, vol 380, pp102–109

Hall, D. O. (1994) 'Biomass energy in industrialized countries: A view from Europe', in *Agroforestry and Land Use Change in Industrialized Nations, 7th International Symposium of CIEC*, Berlin, pp287–329

Hall, D. O. and House, J. (1995) 'Biomass: An environmentally acceptable fuel for the future', *Proceedings of the Institution of Mechanical Engineers*, vol 209, pp203–213

Hall, D. O., Mynick, H. E. and Williams, R. H. (1991) 'Cooling the greenhouse with bioenergy', *Nature*, vol 353, pp11–12

Hall, D. O., Rosillo-Calle, F., Williams, R. W. and Woods, J. (1993) 'Biomass for energy: Supply prospects', in T. B. Johansson et al (eds) *Renewable Energy: Sources of Fuels and Electricity*, Island Press, Washington, DC, pp593–651

Hall, M. H. (1994) *Tall Fescue*, Agronomy Facts 28, The Pennsylvania State University, http://cropsoil.psu.edu/extension/facts/agfact28.pdf (accessed September 2009)

Hall, P. J. and Moody, B. (1994) *Forest Depletions Caused by Insects and Diseases in Canada 1982–1987*, Report ST-X-8, Natural Resources Canada, Canadian Forest Service, Ottawa

Hamelinck, C. and Faaij, A. (2006) 'Outlook for advanced biofuels', *Energy Policy*, vol 34, no 17, pp3268–3283

Hamid, S. and Haider, A. (2001) 'Physico-chemical characteristics of oil from buffalo gourd (*Cucurbita foetidissima*)', *Journal of Food Science and Technology*, vol 38, no 6, pp598–600

Hanover, J. W. (1993) 'Black locust: An excellent fiber crop', in J. Janick and J. E. Simon (eds) *New Crops*, Wiley, New York, pp432–435

Hanover, J. W., Mebrathu, T. and Bloese, P. (1991) 'Genetic improvement of black locust: A prime agroforestry species', *Forestry Chronicle*, vol 67, pp227–231

Harden, J. W., Sundquist, E. T., Stallard, R. F. and Mark, R. K. (1992) 'Dynamics of soil carbon during deglaciation of the Laurentide ice sheet', *Science*, vol 258, pp1921–1924

Hardin, G. (1968) 'The tragedy of the commons', *Science*, vol 162, pp1243–1248

Harmon, M. E., Ferrel, W. K. and Franklin, J. F. (1990) 'Effects on carbon storage of conversion of old-growth forests to young forests', *Science*, vol 247, no 4943, pp699–703

Harmon, M. E., Harmon, J. M., Ferell, W. K. and Brooks, D. (1996) 'Modelling carbon stores in Oregon and Washington forest products: 1900–1992', *Climatic Change*, vol 33, pp521–550

Harris, L. D. (1984) *The Fragmented Forest: Island Biogeography Theory and the Preservation of Biotic Diversity*, The University of Chicago

Harrison, F. (1992) 'Genista', *American Nurseryman*, December, pp57–61

Harrison, K. G., Broecker, W. S. and Bonani, G. (1993) 'The effect of changing land use on soil radiocarbon', *Science*, vol 262, pp725–726

Harrison, W. (2006) 'The role of Fischer Tropsch fuels for the US military', *National Aerospace Fuels Research Complex*, US Air Force, 30 August, www.purdue.edu/dp/energy/pdf/Harrison08-30-06.pdf

Hartmann, H. (1994) 'Systems for harvesting and compaction of solid herbaceous biofuels', in *Environmental Aspects of Production and Conversion of Biomass for Energy*, REUR Technical Series No 38, FAO, Rome, pp86–93

Hartwell, J. L. (1967–1971) 'Plants used against cancer: A survey', *Lloydia*, vols 30–34

Hayward, H. E. and Bernstein, L. (1958) 'Plant growth relationships on salt affected soils', *The Botanical Review*, vol 24, p584

Heath, L. S. and Birdsey, R. A. (1993) 'Carbon trends of productive temperate forests of the coterminous United States', *Water, Air, and Soil Pollution*, vol 70, pp279–293

Heath, L. S., Birdsey, R. A., Row, C. and Plantinga, A. J. (1996) 'Carbon pools and fluxes in US forest products', in M. J. Apps and D. T. Price (eds) *Forest Ecosystems, Forest Management, and the Carbon Cycle*, NATO ASI Series I, vol 40, pp271–278

Heath, M. C., Bullard, M. J., Kilpatrick, J. B. and Speller, C. S. (1994) 'A comparison of the production and economics of biomass crops for use in agricultural or set-aside land', *Aspects of Applied Biology*, vol 40, pp505–515

Heinsdorf, D. (1987) 'Ergebnisse eines Nährstoffmangelversuchs zur Robinie (*Robinia pseudoacacia* L.) auf Kipprohböden [Results of a nutrient deficiency study on robinia (*Robinia pseudoacaeia* L.) on spoil mound soil]', *Beitrage f. Forstwirtschaft*, vol 21, no 1, pp13–17

Heller, J. (1992) *Untersuchungen über genotypische Eigenschaften und Vermehrungs- und Anbauerfahren bei der Purgiernu?* (Jatropha curcas L.), Verlag Dr Kovač, Hamburg

Henderson, L. (2001) *Alien Weeds and Invasive Plants: A Complete Guide to Declared Weeds and Invaders in South Africa*, Plant Protection Research Institute, Handbook 12 (Weeds SAfr 2001)

Henning, R. K. (1995) 'Combating desertification: Fuel oil from jatropha plants in Africa', information packet from the GTZ Projet Pourghère, Bamako, Mali

Henning, R. K. and Mitzlaff, K. V. (1995) 'Nachwachsende Rohstoffe, *Del- Tropenlandwirt*', Beiheft 53, Universitat Gesamthochschule Kassel

Henrikson, R. (2008) 'How spirulina is ecologically grown', Spirulina Source.com, www.spirulinasource.com/earthfoodch6c.html (accessed September 2009)

Herger, G., Klingauf, F., Mangold, D., Pommer, E. H. and Scherer, M. (1988) 'Die Wirkung von Auszugen aus dem Sachalin-Staudenknoterich, *Reynoutria sacbalinensis* (F. Schmidt) Nakai, gegen Pilzkrankheiten, insbesondere Echte Mehltau-Pilze', *Nachrichtenbl. Deut. Pflanzenschutzd.*, vol 40, no 4, pp56–60

Herger, G., Kowalewski, A. and Guttler, J. (1990) *Untersuchungen über die Moglichkeiten eines feldmäbigen Anbaus von Knöterich-Arten mit fungiziden und insektiziden Eigenschaften und Entwicklung von Anbau-, Pflege-, Ernte- und Aufbereitungs-verfahren*, report, Institut fur biologische Schädlingsbekämpfung der Biologischen Bundesanstalt fur Land- und Forstwirtschaft, Darmstadt

Hernandez, C. E. and Witter, S. G. (1996) 'Evaluating and managing the environmental impact of banana production in Costa Rica: A systems approach', *Ambio*, vol 25, no 3, pp171–178

Herrick, J. E. and Wander, M. M. (1997) 'Relationships between soil organic carbon and soil quality in cropped and rangeland soils: The importance of distribution, composition, and soil biological activity', in R. Lal, J. M. Kimble, R. F. Follett and B. A. Stewart (eds) *Soil Processes and the Carbon Cycle*, CRC Press, Boca Raton, pp405–425

Hersener, J.-L. (no date) *Bundesamt für Energie BFE*, EPFL, retrieved from www.energieschweiz.ch/internet/03262/index.html?lang=de

Hersener, J.-L. and Meier, U. (1999) 'Energetisch nutzbares Biomassepotential in der Schweiz sowie Stand der Nutzung in ausgewählten EU-Staaten und den USA', in *Auftrag des Bundesamtes für Energie*, Bundesamt für Energie, Bern

Hicks, K. B., Flores, R. A., Taylor, F., McAloon, A. J., Moreau, R. A., Johnston, D. B., Senske, G. E., Brooks, W. S. and Griffey, C. A. (2005) 'Hulless barley: A new feedstock for fuel ethanol?', 15th EPAC Meeting, Cody, WY, 12–14 June 2005, www.deq.state.mt.us/energy/bioenergy/June2005EthanolConference/26Hicks_HullessBarley.pdf (accessed September 2009)

Hillis, W. E. (1990) 'Fast growing eucalypts and some of their characteristics', in D. Werner and P. Müller (eds) *Fast Growing Trees and Nitrogen Fixing Trees*, Gustav Fischer Verlag, Stuttgart and New York

Hirose, T., Miyazaki, A., Hashimoto, K., Yamamoto, Y., Yoshida, T. and Song, X. F. (2003) 'Specific differences in matter production and water purification efficiency in plants grown by the floating culture system', *Japanese Journal of Crop Science*, vol 72, no 4, pp424–430

Hitchcock, A. S. (1935) *Manual of the Grasses of the United States*, US Department of Agriculture miscellaneous publication No 200, Washington, DC

Hladick, A. (1993) 'Perspectives de développement pour l'Agroforesterie', in *Antropologie alimentaire et développement en afrique intetropicale: du biologique au social*, proceedings of conference, Yaoundé, 27–29 April, pp483–492

Hoen, H. F. and Solberg, B. (1994) 'Potential and economic efficiency of carbon sequestration in forest biomass through silvicultural management', *Forest Science*, vol 40, pp429–451

Hoffmann, G. (1960) 'Untersuchungen über die symbiontische Stickstoffbindung der Robinie *(Robinia pseudoacacia* L.)', dissertation, Eberswalde

Hoffmann, G. (1993) 'Der Anbau von Topinambur: Alternative oder botanische "Kuriosat"?' *Neue Landwirtschaft*, 12/93.

Hoffmann, W., Mudra, A. and Plarre, W. (1985) 'Lupinen *(Lupinus* spec.)', in *Lehrbuch der Züchtung landwirtschaftlicher Kulturpflanzen*, Parey, Berlin, vol 2, pp185–96

Höges, C. (2009) 'Ethanolsprit aus Brasilien: Blut im Tank', 23 January, *Speigel Online*, www.spiegel.de/wirtschaft/0,1518, 602457,00.html (accessed September 2009)

Hohenstein, W. G. and Wright, L. L. (1994) 'Biomass energy production in the United States: An overview', *Biomass and Bioenergy*, vol 6, no 3, pp161–173

Hohmeyer, O. and Ottinger, R. L. (eds) (1994) *Social Costs of Energy: Present Status and Future Trends*, Springer-Verlag, Berlin, pp61–68

Hol, P., Sikkema, R., Blom, E., Barendsen, P. and Veening, W. (1999) *Private Investments in Sustainable Forest Management, I: Final Report*, Form Ecology Consultants and Netherlands Committee for IUCN, The Netherlands

Hondelmann, W. (1984) 'Lupin: Ancient and modern crop plant', *Theoretical and Applied Genetics*, vol 68, pp1–9

Hondelmann, W. (1996) 'Die Lupine: Geschichte und Evolution einer Kulturpflanze', *Landbauforschung Völkenrode*, Sonderheft 162

Honermeier, B., Adam, L., Gottwald, R., Hanff, H., Hoffmann, G, Krüger, K, Patschke, K. and Zimmermann, K. H. (1993) *Ölfrücte – Empfehlungen zum Anbau in Brandenburg*, Druckhaus Schmergow GmbH, Potsdam

Hoogwijk, M., Faaij, A., van den Broek, R., Berndes, G., Gielen, D. and Turkenburg, W. (2003) 'Exploration of the ranges of the global potential of biomass for energy', *Biomass and Bioenergy*, vol 25, no 2, pp119–133

Hoogwijk, M., Faaij, A., Eickhout, B., de Vries, B. and Turkenburg, W. (2005a) 'Potential of biomass energy out to 2100, for four IPCC SRES land-use scenarios', *Biomass and Bioenergy*, vol 29, no 4, pp225–257

Hoogwijk, M., Faaij, A., de Vries, B. and Turkenburg, W. (2005b) 'Global potential of biomass for energy from energy crops under four GHG emission scenarios Part B: The economic potential', Bioenergy Trade, www.bioenergytrade.org/t40reportspapers/other reportspublications/potentialofbiomassbymhoogwijk032004/

Hope, H. T. and McElroy, A. (1990) 'Low temperature tolerance of switchgrass', *Canadian Journal of Plant Science*, vol 70, pp1091–1096

Hoppner, F. and Menge-Harmann, U. (1994) 'Anbauversuche zur Stickstoffdiingung und Bestandesdichte von Faserhanf', *Landbauforschung Völkenrode*, vol 44, no 4, pp314–324

Hotz, A., Kolb, W. and Schwarz, T. (1989) 'Nachwachsende Rohstoffe- eine Chance für die Landwirtschaft?', *Bayerisches Landwirtschaftliches Jahrbuch*, vol 66, no 1

Houghton, R. A. (1995a) 'Changes in the storage of terrestrial carbon since 1850', in R. Lal, J. Kimble, E. Levine and B. A. Stewart (eds) *Soils and Global Change*, Lewis Publishers, Boca Raton, pp45–65

Houghton, R. A. (1995b) 'Effects of land-use change, surface temperature, and CO_2 concentration on terrestrial stores of carbon', in G. M. Woodwell and E. MacKenzie (eds) *Biotic Feedbacks in the Global Climate Systems*, Oxford University Press, New York, pp333–350

Houghton, R. A., Davidson, E. A. and Woodwell, G. M. (1998) 'Missing sinks, feedbacks, and understanding the role of terrestrial ecosystems in the global carbon balance', *Global Biogeochemical Cycles*, vol 12, pp25–34

Houghton, R. A., Hackler, J. L. and Lawrence, K. T. (1999) 'The US carbon budget: Contributions from land-use change', *Science*, vol 285, pp574–578

Houghton, R. A., Skole, D. L., Nobre, C. A., Hackler, J. L., Lawrence, K. T. and Chomentowski, W. H. (2000) 'Annual fluxes of carbon from deforestation and regrowth in the Brazilian Amazon', *Nature*, vol 403, no 6767, pp301–304

Hsu, F. H. (1988) 'Effects of water stress on germination and seedling growth of *Miscanthus*', *Journal of Taiwan Livestock Research*, vol 21, pp37–52

Hsu, F. H. (2000) 'Seed longevity of *Miscanthus* species', *Journal of Taiwan Livestock Research*, vol 33, pp145–153

Hsu, F. H., Nelson, C. J. and Matches, A. G. (1985a) 'Temperature effects on germination of perennial warm-season forage grasses', *Crop Science*, vol 25, pp215–220

Hsu, F. H., Nelson, C. J. and Matches, A. G. (1985b) 'Temperature effects on seedling development of perennial warm-season forage grasses', *Crop Science*, vol 25, pp249–255

Hubbard, H. M. (1991) 'The real cost of energy', *Scientific American*, vol 264, pp18–23

Huber, H. P. (1984) *Jojoba*, Fortuna Finanz Verlag, CH-8172 Niederglatt ZH

Huffaker, M. (1994) http://sunsite.unc.edu/london/organfarm/cover-crops/

Huguenin-Elie, O., Stutz, C. J., Gago, R. and Lüscher, A. (2008) 'Sustainable management of foxtail meadows through hay making at seed maturity', http://orgprints.org/11958/01/Huguenin-Elie_11958_ed.doc (accessed September 2009)

Huggins, D. R., Buyanovsky, G. A., Wagner, G. H., Brown, J. R., Darmody, R. G., Peck, T. R., Lesoing, G. W., Vanotti, M. B. and Bundy, L. G. (1998) 'Soil organic C in the tallgrass prairie-derived region of the corn belt: Effects of long-term crop management', *Soil and Tillage Research*, vol 47, pp219–234

Huisman, W. and Kortleve, W. J. (1994) 'Mechanization of harvest and conservation of *Miscanthus sinensis giganteus*', COST 814 Workshop

Hulscher, W. S., Luo, Z. and Koopmans, A. (1999) 'Stoves on the carbon market', *Wood Energy News*, vol 14, no 3, pp20–21

Hungate, B. A., Holland, E. A., Jackson, R. B., Chapin, F. S., Mooney, H. A. and Field, C. B. (1997) 'The fate of carbon in grasslands under carbon dioxide enrichment', *Nature*, vol 388, pp576–579

Hunt, N. (2007) 'Biofuels may wipe out UK wheat exports', 29 June, http://uk.reuters.com/article/businessNews/idUKL29919758 20070629 (accessed September 2009)

Hunter, T., Royle, D. J. and Stott, K. G. (1988) 'Diseases of Salix biomass plantations: Results of an international survey in 1987', *Proceedings of the International Energy Agency/Bioenergy Agreement Task II Workshop Biotechnology Development*, Uppsala, Sweden, pp37–48

Hussain, M. and Hussain, M. A. (1970) '*Tolerance of Diplachne fusca Beauv. (Australian Grass) to Salt and Alkali*', Research Publication 11(25), Directorate of Land Reclamation, Lahore, West Pakistan

Husz, G. St. (1972) *Sugar Cane: Cultivation and Fertilization*, Ruhr-Stickstoff, Bochum

Husz, G. St. (1989) 'Zuckerrohr', in S. Rehm (ed) *Spezieller Pflanzenbau in den Tropen und Subtropen. Handbuch der Landwirtschaft und Ernährung in den Entwicklungs-ländern*, 2nd edn, Ulmer, Stuttgart, vol 4

Huzayyin, A. S., Bawady, A. H., Rady, M. A. and Dawood, A. (2004) 'Experimental evaluation of Diesel engine performance and emission using blends of jojoba oil and diesel fuel', *Energy Conversion and Management*, vol 45, nos 13–14, pp2093–2112

H.W. (1995) 'Neuer Fixstern oder nur Sternschnuppe?' *DLG-Mitteilungen*, vol 4, p50

ICRA (International Centre for Development Oriented Research in Agriculture) (2003) *Valorisation des savoirs et savoir-faire: perspectives d'implication des acteurs, dont la femme, dans la conservation in-situ de la biodiversité du palmier dattier dans les oasis du Djerid, Tunisie*, ICRA, Montpelier, France

ICRAF (International Centre for Research in Agroforestry) (1997) via Internet: World Bank Newsletter, June 1997.

Idriss, H., Scott, M., Llorca, J., Chan, S. C., Chiu, W., Sheng, P.-Y., Yee, A., Blackford, M. A., Pas, S. J., Hill, A. J., Alamgir, F. M., Rettew, R., Petersburg, C., Senanayake, S. D. and Barteau, M. A. (2008) 'A phenomenological study of the metal–oxide interface: The role of catalysis in hydrogen production from renewable resources', *ChemSusChem*, vol 1, no 11, pp905–910

IEA (International Energy Agency) (2004) *Biofuels for Transport – An International Perspective*, Office of Energy Efficiency, Technology and R&D, OECD/IEA, Paris

IEA (2006a) *World Energy Outlook 2006*, OECD/IEA, Paris

IEA (2006b) *Energy Technology Perspectives – Scenarios and Strategies to 2050*, OECD/IEA, Paris

IEA (2007) *Potential Contribution of Bioenergy to the World's Future Energy Demand*, www.ieabioenergy.com

IEA (2008) *Bioenergy Annual Report 2007*, www.ieabioenergy.com

IFAP (International Federation of Agricultural Producers) (2008) 'Biofuels: A new opportunity for family agriculture', www.ifap.org

IFFS (1995) 'Alfalfa (*Medicago sativa* L.)', International Forage Factsheet Series

IFPRI (International Food Policy Research Institute) (2001) 'Global food projections to 2020: Emerging trends and alternative futures', IFPRI, Washington, DC

IFPRI (2006) 'Bioenergy and Agricultural Promises and Challenges', FOCUS14, www.ifpri.org, p23

IGBP (International Geosphere-Biosphere Programme) (1998) 'The terrestrial carbon cycle: Implications for the Kyoto Protocol', *Science*, vol 280, pp1393–1394

IITA (International Institute of Tropical Agriculture) (1974) *Annual Report*, IITA, Ibadan, Nigeria.

Ingram, L. O., Bothast, R. J., Doran, J. B., Beall, D. S., Brooks, T. A., Wood, B. E., Lai, X., Asghari, K. and Yomano, L. P. (1995) 'Genetic engineering of bacteria for the conversion of lignocellulose to ethanol', *Proceedings of the Workshop on Energy from Biomass and Wastes*, Dublin Castle, Ireland, p17

Institut für Energetik und Umwelt, and Öko-Instituts Darmstadt (2007) *Möglichkeiten einer europäischen Biogaseinspeisungsstrategie [The Opportunities of a European Strategy to Feed Biogas into the Natural*

Gas Grid], (The Leipzig report on biogas), Institut für Energetik und Umwelt, Leipzig and Öko-Instituts Darmstadt.

International Centre for Underutilized Crops (2001) 'Fruits for the future: Safou', Factsheet No 3, www.icuc-iwmi.org/files/News/Resources/Factsheets/dacryodes.pdf (accessed September 2009)

International Trade Centre (2001) 'Exporting Groundnuts', *International Trade Forum*, no 1, www.tradeforum.org/news/fullstory.php/aid/135/Exporting_Groundnuts.html

IPCC (Intergovernmental Panel on Climate Change) (1996) *Climate Change 1995: Impacts, Adaptations and Mitigation of Climate Change: Scientific-Technical Analyses*, R. Watson, M. C. Zinyowera and R. Moss (eds), Contribution of Working Group II to the Second Assessment Report of the Intergovernmental Panel on Climate Change, Cambridge University Press, Cambridge, UK

IPCC (2000) *Special Report on Emission Scenarios*, IPCC, Cambridge University Press, Cambridge, UK

IPCC (2000a) *Land Use, Land Use Change and Forestry*, R. T. Watson, I. R. Noble, B. Bolin, N. H. Ravindranath, D. J. Verardo and D. J. Dokken (eds), A Special Report of the IPCC, Cambridge University Press, Cambridge, UK

IPCC (2000b) *Methodological and Technological Issues in Technology Transfer*, B. Metz, O. R. Davidson, J.-W. Martens, S. N. M. van Rooijen and L. van Wie-McGrory (eds), A Special Report of the IPCC, Cambridge University Press, Cambridge, UK

IPCC (2007) *IPCC Fourth Assessment Report*, Contribution of Working Group III, available from www.ipcc.ch

IPS (Inter Press Service International Association) (2008) 'Energy-Brazil: Elephant grass for biomass', www.ips.org

Irawati (1994) 'Selection for cold hardiness of sweet potatoes from Baliem Valley by *in vitro* culture', *Acta Horticulturae*, vol 369, pp220–225

Isensee, A. R. and Sadeghi, A. M. (1996) 'Effect of tillage reversal on herbicide leaching to groundwater', *Soil Science*, vol 161, pp382–389

ITTA (International Tropical Timber Agreement) (1983) 1983 United Nation Conference on Trade and Development, Geneva

ITTA (1994) 1994 United Nations Conference on Trade and Development, Geneva

Iwaki, H., Midorikawa, B. and Hogetsu, H. (1964) 'Studies on the productivity and nutrient element circulation in Kirigamine grassland, central Japan: II. Seasonal change in standing crop', *Botanical Magazine* (Tokyo) vol 77, p918

Izac, A.-M. N. (1997) 'Developing policies for soil carbon management in tropical regions, *Geoderma*, vol 79, pp261–276

Izaurralde, R. C., Nyborg, M., Solberg, E. D., Janzen, H. H., Arshad, M. A., Malhi, S. S. and Molina-Ayala, M. (1997) 'Carbon storage in eroded soils after five years of reclamation techniques', in R. Lal, J. M. Kimble, R. F. Follett and B. A. Stewart (eds) *Soil Processes and the Carbon Cycle*, CRC Press, Boca Raton, pp369–385

IZNE (Interdisziplinäres Zentrum für Nachhaltige Entwicklung) (2005) 'The bioenergy village: Self-sufficient heating and electricity supply using biomass', www.bioenergiedorf.de

IZNE (2007) 'Wärme- und Stromversorgung durch heimische Biomasse', Interdisziplinäres Zentrum für Nachhaltige Entwicklung (IZNE) der Universität Göttingen – Projektgruppe Bioenergiedörfer, April, www.gar-bw.de/fileadmin/gar/pdf/Energie_und_ Klima/Juehnde.pdf (accessed September 2009)

Jacinthe, P. A. and Dick, W. A. (1997) 'Soil management and nitrous oxide emissions from cultivated fields in southern Ohio', *Soil and Tillage Research*, vol 41, pp221–235

Jacobs, M. R. (1979) *Eucalypts for Planting*, FAO Forestry Series No 11, Rome

Jamieson, D. (1996) 'Ethics and intentional climate change', *Climatic Change*, vol 33, pp323–336

Janiak, B. (1994) Personal communication

Janick, J. and Whipkey, A. (eds) (2007) *Issues in New Crops and New Uses*, ASHS Press, Alexandria, VA

Janvier (2000) 'Etude des principaux marches europeens de la datte et du potentiel commercial des varietes non traditionnelles', FAO

Janzen, H. H., Campbell, C. A., Izaurralde, R. C., Ellert, B. H., Juma, N., McGill, W. B. and Zentner, R. P. (1998) 'Management effects on soil C storage on the Canadian prairies', *Soil and Tillage Research*, vol 47, pp181195

Jeltsch, P. J. M. (2005) 'Production industrielle du bioethanol', May

Jepma, C. J., Nilsson, S., Amano, M., Bonduki, Y., Lonnstedt, L., Sathaye, J. and Wilson, T. (1997) 'Carbon sequestration and sustainable forest management: Common aspects and assessment goals', in R. A. Sedjo, R. N. Sampson and J. Wisniewski (eds) 'Economics of carbon sequestration in forestry', *Critical Reviews in Environmental Science and Technology*, vol 27, pp83–96

Jodice, R., Vecchiet, M., Parrini, F. and Schenone, G. (1995a) 'Giant reed multiplication and cultivation experiences', in Chartier et al (eds) *Biomass for Energy, Environment, Agriculture and Industry: Proceedings of the 9th EU Biomass Conference*, Pergamon Press, UK, vol 1, pp686–691

Jodice, R., Vecchiet, M., Parrini, F. and Schenone, G. (1995b) 'Giant reed as energy crop: Initial economic evaluation of production costs in Italy', in Chartier et al (eds) *Biomass for Energy, Environment, Agriculture and Industry: Proceedings of the 8th EU Biomass Conference*, Pergamon Press, UK, vol 1, pp695–699

Johansson, H., Ledin, S. arid Forsse, L. S. (1992) 'Practical energy forestry in Sweden: A commercial alternative for farmers', in Hall, D. O. et al (eds) *Biomass for Energy and Industry*, Ponte Press, Bochum, pp117–126

Johnson, D. L. and Jha, M. N. (1993) 'Blue corn', in J. Janick and J. E. Simon (eds) *New Crops*, Wiley, New York, pp228–230

Johnson, J. D. and Hinman, C. W. (1980) 'Oil and rubber from arid land plants', *Science*, vol 208, pp460–464.

Johnson, M. G. (1995) 'The role of soil management in sequestering soil carbon', in R. Lal, J. Kimble, E. Levine and B. A. Stewart (eds) *Soil Management and Greenhouse Effect*, CRC Lewis Publishers, Boca Raton, pp351–363

Joint Genome Institute (2009) 'Scientists publish complete genetic blueprint of key biofuels crop', *Science News*, http://esciencenews.com/articles/2009/01/28/scientists.publish.complete.genetic.blueprint.key.biofuels.crop (accessed September 2009)

Jones, I. S. F. and Young, H. E. (1998) 'Enhanced oceanic uptake of carbon dioxide: An AIJ candidate', in P. W. F. Riemer, A. Y. Smith and K. V. Thambimuthu (eds) *Greenhouse Gas Mitigation: Technologies for Activities Implemented Jointly*, Elsevier Science Ltd, Oxford, UK, pp267–272

Jones, L. and Miller, J. H. (1991) Jatropha curcas: *A Multipurpose Species for Problematic Sites*, Astag Technical Paper, Land Resource Series No 1, The World Bank Asia Technical Department Agriculture Division

Jones, M. B. and Gravett, A. (1988) 'The productivity of C4 perennials grown for biomass in North West Europe', *Proceedings of the Euroforum, New Energies Congress*, Saarbrücken, Germany, vol 3, pp494–496

Jones, M. B., Long, S. P. and McNally, S. F. (1987) 'The potential productivity of C_4 cordgrasses and galingale for low input biomass production in Europe: Plant establishment', in G. Grassi, B. Delman, J. F. Molle and H. Zibetta, *Biomass for Energy and Industry: Proceedings of the 4th EC Conference*, Orlèans, France, Elsevier Applied Science, pp106–110

Jong, B. H. J. de, Montoya-Gomez, G., Nelson, K. and Soto-Pinto, L. (1995) 'Community forest management and carbon sequestration: A feasibility study from Chiapas, Mexico', *Interciencia*, vol 20, pp409–416

Joos, F., Sarmiento, J. L. and Siegenthaler, U. (1991) 'Estimates of the effect of Southern Ocean iron fertilization on atmospheric CO_2 concentrations', *Nature*, vol 349, pp772–774

Jørgensen, J. R., Deleuran, L. C. and Wollenweber, B. (2007) 'Prospects of whole grain crops of wheat, rye and triticale under different fertilizer regimes for energy production', *Biomass and Bioenergy*, vol 31, no 5, pp308–317

Jørgensen, U. and Kjeldsen, J. B. (1992) *Production of elefantgræs*, Danish Research Service for Plant and Soil Science, Grøn Viden 110

Jørgensen, U., Mortensen, J. and Ohlsson, C. (2003) 'Light interception and dry matter conversion efficiency of miscanthus genotypes estimated from spectral reflectance measurements', *New Phytologist*, vol 157, no 2, pp263–270

Jung, G. A., Shaffer, J. A. and Stout, W. L. (1988) 'Switchgrass and big bluestem responses to amendments on strongly acid soil', *Agronomy Journal*, vol 80, pp669–676

Jung, G. A., Shaffer, J. A., Stout, W. L. and Panciera, M. T. (1990) 'Warm-season grass diversity in yield, plant morphology and nitrogen concentration and removal in Northeastern USA', *Agronomy Journal*, vol 82, pp21–26

Jungbluth, N. (2004) 'Erdöl', in R. Dones (ed) *Sachbilanzen von Energiesystemen: Grundlagen für den ökologischen Vergleich von Energiesystemen und den Einbezug von Energiesystemen in Ökobilanzen für die Schweiz*, Paul Scherrer Institut Villigen, Swiss Centre for Life Cycle Inventories, Dübendorf, www.ecoinvent.org.

Jungbluth, N. and Frischknecht, R. (2004) 'Vorstudie "LCA von Energieprodukten"', Project No 100428, ESU-services for Bundesamt für Energie und Bundesamt für Umwelt, Wald und Landschaft (BUWAL), Uster

Jvanjukovic, L. K. (1981) 'Specific and intraspecific classification of sorghum species', *Kulturpflanze*, vol XXIX, pp273–280

Kádár, I., Németh, T. and Szemes, I. (1999) 'Fertilizer response of triticale in long-term experiment in Nyirlugos', *Novenytermeles*, vol 48, pp647–661 (in Hungarian with English Summary)

Kägi T., Freiermuth-Knuchel, R., Nemecek, T. and Gaillard, G. (2007) 'Ökobilanz von Energieprodukten: Bewertung der landwirtschaftlichen Biomasse Produktion(in Vorbereitung)', *GAIA*

Kahl, A. (1987) *Pelleting Plants for All Grades of Powdered or Granular Raw Materials Either Dry or in Paste Form*, Tech. Report 1326, Amandus Kahl, Hamburg

Kaimowitz, D. and Anglesen, A. (1998) *Economic Models of Tropical Deforestation: A Review*, Center for International Forestry Research (CIFOR), Bogor, Indonesia

Kaffka, S. R., Kearney, T. E., Knowles, P. D. and Miller, M. D. (2000) 'Safflower production in California', http://agric. ucdavis.edu/crops/oilseed/saff1intro.htm (accessed September 2009)

Kallivroussis, L., Natsis, A. and Papadakis, G. (2002) *The Energy Balance of Sunflower Production for Biodiesel in Greece*, Silsoe Research Institute, published by Elsevier Science Ltd

Kaltofen, H. and Schrader, A. (1991) *Gräser*, 3rd edn, Deutscher Landwirtschaftsverlag, Berlin

Kaltschmitt, M. and Bridgwater, A. V. (eds) (1997) *Proceedings of the International. Conference 'Gasification and Pyrolysis of Biomass'*, Stuttgart, 9–11 April 1997, CPL Press, Newbury

Kampen, J. (1982) *An Approach to Improved Productivity on Deep Vertisols*, ICRISAT Bulletin No 11, p14

Kanowski, P. J., Savill, P. S., Adlard, P. G., Burley, J., Evans, J., Palmer, J. R. and Wood, P. J. (1992) 'Plantation forestry', in N. P. Sharma (ed) *Managing the World's Forests*, Kendall-Hunt, Dubuque, Iowa, pp375–401

Kanwar, J. S. and Rego, T. J. (1983) 'Fertiliser use and watershed management in rainfed areas for increasing crop production', *Fertiliser News*, vol 28, no 9, pp33–43

Karjalainen, T. (1996) 'Dynamics and potentials of carbon sequestration in managed stands and wood products in Finland under changing climatic conditions', *Forest Ecology and Management*, vol 80, pp113–132

Karjalainen, T., Kellomaeki, S. and Pussinen, A. (1994) 'Role of wood based products in absorbing atmospheric carbon', *Silva Fennica*, vol 28, no 2, pp67–80

Karjalainen, T., Pussinen, A., Kellomäki, S. and Mäkipää, R. (1998) 'The history and future carbon dynamics of carbon sequestration in Finland's forest sector', in G. H. Kohlmaier, M. Weber and R.A. Houghton (eds) *Carbon Mitigation in Forestry and Wood Industry*, Springer-Verlag, Berlin, pp25–42

Karschon, R. (1961) 'Soil evolution as affected by eucalyptus', in *Relatoriose documentos, sequnda conferencia mundial do Eucalypto*, Food and Agriculture Organization, São Paulo, vol 2, pp897–904

Karus, M. and Leson, G. (1995) Nova-Institute, Cologne Germany, cited in *Textilforum*, vol 2, no 95, p26

Karve, A. (2006) *ARTI Biogas Plant: A Compact Digester for Producing Biogas from Food Waste*, Appropiate Rural Technology Institute (ARTI), www.arti-india.org/content/view/45/52/ (accessed September 2009)

Kasimir-Klemedtsson, A., Klemedtsson, L., Berglund, K., Martikainen, P., Silvola, J. and Oenema, O. (1997) 'Greenhouse gas emissions from farmed organic soils: A review', *Soil Use and Management*, vol 13, pp245–250

Katri, P., Forsman, K., Isolahti M. and Lötjönen, T. (2003–2005) 'Production of reed canary grass for bioenergy plants', Duration: 01.02.2003–31.12.2005, Phase: Conclusion

Katsumi, K. and Yota, Y. (2003) 'Spatiotemporal patterns of shoots within an isolated *Miscanthus sinensis* patch in the warm-temperate region of Japan', *Ecological Research*, vol 18, no 1, pp41–51

Katyal, S. K. and Iyer, P. V. R. (2000) 'Thermochemical characterization of pigeon pea stalk for its efficient utilization as an energy source', *Energy Sources*, vol 22, no 4, pp363–375

Kauffman, J. B., Cummings, D. L. and Ward, D. E. (1998) 'Fire in the Brazilian Amazon: 2. Biomass, nutrient pools and losses in cattle pastures', *Oecologia*, vol 113, pp415–427

Kauppi, P. E., Mielkainen, K. and Kuusela, K. (1992) 'Biomass and carbon budget of European forests, 1971–1990', *Science*, vol 256, pp70–74

Kauppi, P. E., Posch, M., Hänninen, P., Henttonen, H. M., Ihalainen, A., Lappalainen, E., Starr, M. and Tamminen, P. (1997) 'Carbon reservoirs in peatlands and forests in the boreal regions of Finland', *Silva Fennica*, vol 31, no 1, pp13–25

Keenan, R., Lamb, D., Woldring, O., Irvine, T. and Jensen, R. (1997) 'Restoration of plant biodiversity beneath tropical tree plantations in Northern Australia', *Forest Ecology and Management*, vol 99, pp117–131

Keeve, R., Krüger, G. H. J., Loubser, H. L. and Van Der Mey, J. A. M. (1999) 'Effect of temperature and photoperiod on the development of *Lupinus albus* L. in a controlled environment', *Journal of Agronomy and Crop Science*, vol 183, no 4, pp217–223

Keith, D. W. (2000) 'Geoengineering the climate: History and prospect', *Annual Reviews Energy and the Environment*, vol 25, pp245–284

Keith, D. W. (2001) 'Geoengineering', in A. S. Goudie (ed) *Oxford Encyclopedia of Global Change: Environmental Change and Human Society*, Oxford University Press, New York

Kellner, O. and Becker, M. (1966) *Grundzüge der Fütterungslehre*, 14th edn, Parey, Berlin

Kengue, J. and Nya Ngatchou, J. (1991) 'Perspectives d'amélioration du safoutier', in *Actes du seminaire sous régional sur la valorisation du safoutier*, 25–28 November, Brazzaville, Congo

Kengue, J. and Tchio, F. (1994) 'Essai de bouturage et de marcottage de safoutier (*Dacryodes edulis*)', in J. Nya Ngatcchou and J. Kengué (eds) *Seminaire sur la valorisation du Safoutier*, 4–6 October, Douala, Cameroun, pp80–98

Keresztesi, B. (1983) 'Breeding and cultivation of black locust (*Robinia pseudoacacia* L.) in Hungary', *Forest Ecology and Management*, vol 6, pp217–244

Keresztesi, B. (1988) 'Black locust: The tree of agriculture', *Outlook on Agriculture* (UK), vol 17, no 2, pp77–85

Kern, J. S. and Johnson, M. G. (1993) 'Conservation tillage impacts on national soil and atmospheric carbon levels', *Soil Science Society America Journal*, vol 57, pp200–210

Kern, J. S., Zitong, G., Ganlin, Z., Huizhen, Z. and Guobao, L. (1997) 'Spatial analysis of methane emissions from paddy soils in China and the potential for emissions reduction', *Nutrient Cycling in Agroecosystems*, vol 49, pp181–195

Kernick, M. D. (1986) 'Forage plants for salt affected areas in developing countries', *Reclamation and Revegetation Research*, vol 5, pp451–549

Kertz, G. (2009) Valcent Products Inc., http://blog.valcent.net/?p=375

Khalifa, S. F. (no date) *Report on Potential Energy Crops of the Mediterranean Region in Egypt*, Ain Shams University, Abbassia, Cairo, Egypt

Kharin, N. (1996) 'Strategy to combat desertification in Central Asia', *Desertification Control Bulletin*, vol 29, pp29–34

Khellil, C. (2004) Renewable 2004, International Conference for Renewable Energies, 1 June 2004

Kheshgi, H. (1995) 'Sequestering carbon dioxide by increasing ocean alkalinity', *Energy*, vol 20, pp915–922

Kiakouama, S. and Silou, T. (1990) 'Evolution des lipides de la pulpe de safou *Dacryodes edulis* en fonction de l'état de maturité du fruit', *Fruits*, vol 45, no 4, pp403–408

Kilpatrick, J. B., Heath, M. C., Speller, C. S., Nixon, P. M. I., Bullard, M. J., Spink, J. G. and Cromack, H. T. H. (1994) 'Establishment, growth and dry matter yield of *Miscanthus sacchariflorus* over two years under UK conditions', COST 814 Workshop

Kinkéla, T., Kama-Niamayoua, R., Mampouya, D. and Silou, T. (2006) 'Variations in morphological characteristics, lipid content and chemical composition of safou (*Dacryodes edulis* (G. Don) H. J. Lam.) according to fruit distribution: A case study', *African Journal of Biotechnology*, vol 5, no 12, pp1233–1238

Kirchhof, G. (2006) 'Analysis of biophysical and socio-economic constraints to soil fertility management in the PNG Highlands', Australian Centre for International Agricultural Research, www.aciar.gov.au/project/SMCN/2005/043

Kišgeci, J. et al (eds) (1989) *Agricultural Biomass for Energy*, CNRE Study No 4, University of Novi Sad, FAO, vol 21, nos 58–59

Kishimoto, M. (1997) 'Oil production', in K. Miyamoto (ed) *Renewable Biological Systems for Alternative Sustainable Energy Production*, FAO Agricultural Services Bulletin, pp81–95

Kiss, Á. (1966) 'Experiments with hexaploid triticale', *Novenytermeles*, vol 15, pp311–328 (in Hungarian with English summary)

Klasnja B., Orlovic, S., Galic, Z. and Drekic, M. (2006) 'Poplar biomass of short rotation plantations as renewable energy raw material', in M. D. Brenes (ed) *Biomass and Bioenergy Research*, Nova Science Publisher, Inc.

Kleiman, R. (1990) 'Chemistry of new industrial oilseed crops', in J. Janick and J. E. Simon (eds) *Advances in New Crops*, Timber Press, Portland, OR, pp196–203

Knapp, S. J. (1990) 'New temperate oilseed crops', in J. Janick and J. E. Simon (eds) *Advances in New Crops*, Timber Press, Portland, OR, pp203–210

Knapp, S. J. (1993) 'Breakthroughs towards the domestication of *Cuphea*', in J. Janick and J. E. Simon (eds) N*ew Crops*, Wiley, New York, pp372–379

Knoblauch, F., Tychsen, K. and Kjeldsen, J. B. (1991) Miscanthus sinensis *'giganteus' (elefantgræs)*, Landbrug Grøn Viden 85, [English version: *Manual for Growing* Miscanthus sinensis *'Giganteus'*, Danish Research Service for Plant and Soil Science, Institute of Landscape Plants, Hornum, Denmark].

Knoef, H. (ed) (2005) *Handbook Biomass Gasification*, Biomass Technology Group, European Commission, PyNe and GasNet, Enschede, the Netherlands

Kobayashi, K. (1987) 'Hydrologic effects of rehabilitation treatment for bare mountain slopes', *Bulletin of the Forestry and Forest Products Research Institute*, vol 300, pp151–185

Kolb, W., Hotz, A. and Kuhn, W. (1990) 'Untersuchungen zur Leistungsfähigkeit aus-dauernder Gräser für die Energie- und Rohstoffgewinnung [Investigations relating to the productivity of perennial grasses for the production of energy and raw materials]', *Rasen-Turf-Gazon*, Institute of Viticulture and Horticulture, Würzburg, vol 4, pp75–79

Kolton, I. and Ingber, Y. (2003) 'Solutions d'irrigations goutte a goutte pour plantations de mangue'

Kolchugina, T. P., Vinson, T. S., Gaston, G. G., Rozhkov, V. A. and Shwidenko, A. Z. (1995) 'Carbon pools, fluxes, and sequestration potential in soils of the former Soviet Union', in R. Lal, J. Kimble, E. Levine and B. A. Stewart (eds) *Soil Management and Greenhouse Effect*, CRC Lewis Publishers, Boca Raton, pp25–40

Kowalewski, A. and Herger, G. (1992) 'Investigations about the occurrence and chemical nature of the resistance inducing factor in the extract of *Reynoutria sachalinensis*', *Med. Fac. Landbouww*, (University of Gent), vol 57, no 2b, pp449–456

Krahl, J. (1993) 'Bestimmung der Schadstoffemissionen von landwirtschaftlichen Schleppern beim Betrieb mit Rapsölmethylester in Vergleich zu Dieselkraftstoff', dissertation, TU Braunschweig

Krapovickas, A. (1968) 'The origin, variability and spread of the groundnut (*Arachis hypogaea*)', in P. J. Ucko and G. W. Dimbleby (eds) *The Domestication and Exploitation of Plants and Animals*, Duckworth, London, pp427–441

Kremen, C., Niles, J. O., Dalton, M. G., Daily, G. C., Ehrlich, P. R., Fray, J. P., Gerwal, D. and Guillery, R. P. (2000) 'Economic incentives for rain forest conservation across scales', *Science*, vol 288, pp1828–1832

Krochmal, A. (1969) 'Propagation of cassava', *World Crops*, vol 21, pp193–195

Kuliev, A. (1996) 'Forests: An important factor in combatting desertification', *Problems of Desert Development*, vol 4, pp29–31

Kumar, A. (1996) 'Use of *Leptochloa fusca* for the improvement of salt-affected soils', *Experimental Agriculture*, vol 32, pp143–149

Kumar, D. and Yadav, I. S. (1992) 'Taramira (*Eruca sativa* Mill) research in India: Present status and future programme on yield enhancement', in D. Kumar and M. Rai (eds) *Advances in Oilseeds Research*, Scientific Publishers, Jodhpur, India, vol 1, pp329–358

Kumar, V., Ghosh, B. C. and Bhat, R. (1999) 'Recycling of crop wastes and green manure and their impact on yield and nutrient uptake of wetland rice', *Journal of Agricultural Science*, vol 132, pp149–154

Kurz, W. A. and Apps, M. J. (1999) 'A 70-year retrospective analysis of carbon fluxes in the Canadian forest sector', *Ecological Applications*, vol 9, pp526–547

Kurz, W. A., Apps, M. J., Webb, T. and MacNamee, P. (1992) *The Carbon Budget of the Canadian Forest Sector: Phase 1*, ENFOR Information Report NOR-X-326, Forestry Canada Northwest Region, Edmonton, Alberta, Canada

Kurz, W. A., Apps, M. J., Stocks, B. J. and Volney, W. J. A. (1995a) 'Global climatic change: Disturbance regimes and biospheric feedbacks of temperate and boreal forests', in G. M. Woodwell and F. T. Mackenzie (eds) *Biotic Feedbacks in the Global Climatic System: Will the Warming Speed the Warming?*, Oxford University Press, New York, pp119–133

Kurz, W. A., Apps, M. J., Beukema, S. J. and Lekstrum, T. (1995b) 'Twentieth century carbon budget of Canadian forests', *Tellus*, vol 47B, pp170–177

Kurz, W. A., Beukema, S. J. and Apps, M. J. (1997) 'Carbon budget implications of the transition from natural to managed disturbance regimes in forest landscapes', *Mitigation and Adaptation Strategies for Global Change*, vol 2, pp405–421

Kusterer, G. and Wurth, W. (1992) 'Ergebnisse der Landessortenversuche mit ausdauern-den Gräsern 1985–1991', *Information für die Pflanzenproduktion*, Landesanstalt fur Pflanzenbau Forchheim, vol 1

Kusterer, B. and Wurth, W. (1995) 'Ergebnisse der Landessortenversuche mit ausdauern-den Grisern 1994', *Information fuir die Pflanzenproduktion*, Landesanstalt fur Pflanzenbau Forchheim, vol 3

Kuusela, K. (1994) *Forest Resources in Europe 1950–1990*, Research Report 1, European Forest Institute, Joensuu, Finland

Lähde, E., Laiho, O. and Norokorpi, Y. (1999) 'Diversity-oriented silviculture in the boreal zone of Europe', *Forest Ecology and Management*, vol 118, pp223–243

Lal, R. and Bruce, J. P. (1999) 'The potential of world cropland soils to sequester C and mitigate the greenhouse effect', *Environmental Science and Policy*, vol 2, pp177–185

Lal, R., Kimble, J. and Follett, R. (1997) 'Land use and soil C pools in terrestrial ecosystems', in R. Lal, J. M. Kimble, R. F. Follett and B. A. Stewart (eds) *Management of Carbon Sequestration in Soil*, CRC Press, Boca Raton, pp1–10

Lal, R., Kimble, J. M., Follett, R. F. and Cole, C. V. (1998) *The Potential of US Cropland to Sequester Carbon and Mitigate the Greenhouse Effect*, Sleeping Bear Press, Inc., Ann Arbor Press, Chelsea, MI

Lal, R., Follett, R. F., Kimble, J. and Cole, C. V. (1999) 'Managing US cropland to sequester carbon in soil', *Journal of Soil and Water Conservation*, vol 54, pp374–381

Lamb, J. F., Jung, H. G., Sheaffer, C. C. and Samac, D. A. (2007) 'Alfalfa leaf protein and stem cell wall polysaccharide yields under hay and biomass management systems', *Crop Science*, vol 47, pp1407–1415

Lapeyrie, F. (1990) 'Controlled mycorrhizal inoculation of eucalyptus: Results and perspectives of applied and basic research', in D. Werner and P. Müller (eds) *Fast Growing Trees and Nitrogen Fixing Trees*, Gustav Fischer Verlag, Stuttgart and New York

Largeau, C., Casadevall, E., Berkaloff, C. and Dhamelincourt, P. (1980) 'Sites of accumulation and composition of hydrocarbons in *Botryococcus braunii*', *Phytochemistry*, vol 19, pp1043–1051

Larionova, A. A., Yermolayev, A. M., Blagodatsky, S. A., Rozanova, L. N., Yevdokimov, I. V. and Orlinsky, D. B. (1998) 'Soil respiration and carbon balance of gray forest soils as affected by land use', *Biology and Fertility of Soils*, vol 27, pp251–257

Larson, E. D. and Williams, R. H. (1995) 'Biomass plantation energy systems and sustainable development', in J. Goldemberg and T. B. Johansson (eds) *Energy as an Instrument for Socio-Economic Development*, United Nations Development Programme, New York, pp91–106

Lechtenberg, V. L., Johnson, K. D., Moore, K. J. and Hertel, J. M. (1981) 'Management of cool-season grasses for biomass production', *Agronomy Abstracts*, 73rd Annual Meeting, American Society of Agronomy, Madison, Wisconsin

Lederer, E. (2007) 'Production of biofuels "is a crime"', *The Independent*, 27 October, www.independent.co.uk/environment/green-living/production-of-biofuels-is-a-crime-398066.html

Lee, C. W., Glenn, E. P. and O'Leary, J. W. (1992) 'In vitro propagation of *Salicornia bigelovii* by shoot-tip cultures', *HortScience*, vol 27, no 5, p472

Lee, S. C., Lee, J. S. and Jeong, S. J. (2005) 'Germination and seedling growth of *Miscanthus sacchariflorus* as influenced by different plug cells and medium composition', *Korean Journal of Horticultural Science and Technology*, vol 23, no 3, pp315–318

Lee, Y. K., Ding, S. Y, Low, C. S., Chang, Y. C., Forday, W. L. and Chew, P. C. (1995) 'Design of an α-type tubular photobioreactor for mass cultivation of microalgae', *Journal of Applied Phycology*, vol 7, no 1, pp47–51

Leer, S. (2004) 'Agronomists: Long-term corn, soybean yield trends acres apart', Purdue News, 15 July, http://news.uns.purdue.edu/UNS/html4ever/2004/040715.Volenec.yields.html (accessed September 2009)

Le Maitre, D. C. and Versfeld, D. B. (1997) 'Forest evaporation models: Relationships between stand growth and evaporation', *Journal of Hydrology*, vol 193, pp240–257

Le Monde (2005) 'Avec les biocarburants, une nouvelle ere s'ouvre pour l'agriculture', 11 June

Le Monde (2006) 'La fievre du biodiesel s'est emparee du Bresil', 12 August

Le Monde (2006) 'La fonte des sols geles de Siberie s'accelere et renforce l'effet de serre', 7 September

Lehtomäki, A. (2006) 'Biogas production from energy crops and crop residues', thesis, University of Jyväskylä

Lehtomäki, A., Viinikainen, T. A. and Rintala, J. A. (2007) 'Screening boreal energy crops and crop residues for methane biofuel production', *Biomass and Bioenergy*, vol 32, no 6, pp541–550

Lemke, R. L., Izaurralde, R. C., Nyborg, M. and Solberg, E. D. (1999) 'Tillage and N source influence soil-emitted nitrous oxide in the Alberta Parkland region', *Canadian Journal of Soil Science*, vol 79, pp15–24

Lennerts, L. (1984) *Ölschrote, Ölkuchen, pflanzliche Öle und Fette*, Verlag Alfred Strothe Hannover, Germany

Leppick, E. E. (1971) 'Assumed gene centre of peanuts and soybean', *Economic Botany*, vol 25, pp199–294

Levenstein, S. (no date) 'Rice ethanol replaces oil, saves farm jobs', http://inventorspot.com/articles/rice_ethanol_replaces_oil_saves__10099 (accessed February 2009)

Lewandowski, I. and Kicherer, A. (1995) 'CO$_2$-balance for the cultivation and combustion of *Miscanthus*', *Biomass and Bioenergy*, vol 8, no 2, pp81–90

Lewandowski et al (2002) www.eubia.org/193.0.html

Lewandowski, I., Clifton-Brown, J. C., Andersson, B., Basch, G., Christian, D. G., Jørgensen, U., Jones, M. B., Riche, A. B., Schwarz ,K. U., Tayebi, K. and Teixeira, F. (2003) 'Environment and harvest time affects the combustion qualities of *Miscanthus* genotypes', *Agronomy Journal*, vol 95, pp1274–1280

Lewington, A. and Parker, E. (2002) *Ancient Trees: Trees that Live for a Thousand Years*, Collins and Brown, London

Li, G. and Tay, W. K. (2006) *Case Study of Palm Oil Waste Fuelled Power Plant*, report, China EnerSave Ltd

Li, G. J., Benoit, F., Ceustermans, N. and Xu Z. H. (2002) 'The possibilities of Chinese reed fibres as an environmentally sound organic substrate', *Acta Agriculturae Zhejiangensis*

Li, S.-G, Lai, C.-T., Tomoko, Y. and Takehisa, O. (2003) 'Carbon dioxide and water vapor exchange over a *Miscanthus*-type grassland: Effects of development of the canopy', *Ecological Research*, vol 18, no 6, pp661–675

Lieberei, R., Reisdorff, C. and Machado, D. M. (1996) *Interdisciplinary Research on the Conservation and Sustainable Use of the Amazonian Rain Forest and its Information Requirements*, GKSS-Forschungszentrum Geesthacht GmbH, Germany

Liese, W. (1985) *Bamboos: Biology, Silvics, Properties, Utilization*, Deutsche Gesellschaft für Technische Zusammenarbeit (GTZ), Schiiftenreihe No 180, TZ verlagsges, Rossdorf

Liese, W. (1995) 'Anatomy and utilization of bamboos', *European Bamboo Society Journal*, May, pp5–12

Lima, W. P. (1993) *Impacto Ambiental da Eacalypto*, 2nd edn, Editora da Universidade de Sao Paulo

Liimatainen, H., Kuokkanen, T. and Kääriäinen, J. (2004) 'Development of bio-ethanol production from waste potatoes', in E. Pongrácz (ed) *Proceedings of the Waste Minimization and Resources Use Optimization Conference*, University of Oulu, Finland, Oulu University Press, Oulu, pp123–129

Lipinsky, E. S. (1978) *Sugar Crops as a Source of Fuels, Vol II: Processing and Conversion*, final report, Research Department of Energy, Battelle Columbus Labs, OH

Liquid Biofuels Newsletter (1995) 'Information from the IEA-BA "Liquid Biofuels Activity"', Wieselburg, vol 2, pp1–20

Liski, J., Ilvesniemi, H., Makela, A. and Starr, M. (1998) 'Model analysis of the effects of soil age, fires and harvesting on the carbon storage of boreal forest soils', *European Journal of Soil Science*, vol 49, pp397–406

Liski, J., Ilvesniemi, H., Makela, A. and Westman, C. J. (1999) 'CO$_2$ emissions from soil in response to climatic warming are overestimated: The decomposition of old soil organic matter is tolerant of temperature', *Ambio*, vol 28, no 2, pp171–174

List, P. H. and Horhammer, L. (1969–1979) *Hager's handbuch der pharmazeutischen praxis*, Springer-Verlag, Berlin, vols 2–6

Liu, L. (1989) Plant resources of Gramineae: *Triarrhena*', Unpublished manuscript

Löhner, K. (1963) *Die Brennkraftmaschine*, VDI Verlag, Düsseldorf

London Free Press (2005) 'CSEPA to build sweet potato ethanol plant in Ontario', 21 October, www.allbusiness.com/operations/shipping/585121-1.html (accessed September 2009)

Long, E. (1995) 'Unravelling new fine fibre markets for hemp', *Arable Farming*, 14 November, pp29 and 32

Long, S. P. and Woolhouse, H. W. (1979) 'Primary production in *Spartina* marshes', in R. L. Jefferies and A. J. Davy (eds) *Ecological Processes in Coastal Environments*, Blackwell Scientific Publications, Oxford, pp333–352

Long, S. P., Dunn, R., Jackson, D, Othman, S. B. and Yaakub, M. H. (1990) 'The primary productivity of *Spartina anglica* on an East Anglian estuary', in A. J. Gray and P. E. M. Benham (eds) *Spartina anglica: A Research Review*, ITE publication No 2, NERC, Swindon, UK

Long, W. H. and Hensley, S. D. (1972) 'Insect pests of sugar cane', *Ann. Reo. Entom.*, vol 17, pp149–176

Loomis, R. S. and Amthor, J. S. (1999) 'Yield potential and plant assimilatory capacity and metabolic efficiencies', *Crop Science*, vol 39, pp1584–1596

Lötschert, W. and Beese, G. (1984) *Pflanzen der Tropen*, 2nd edn, BLV Verlagsgesellschaft, München, Wien and Zürich

Lowe D. C. (2006) 'Global change: A green source of surprise', *Nature*, vol 439, pp148–149

Lugo, A. (1997) 'The apparent paradox of reestablishing species richness on degraded lands with tree monocultures', *Forest Ecology and Management*, vol 99, pp9–19

Lunnan, A., Navrud, S., Rorstad, P. K., Simensen, K. and Solberg, B. (1991) *Skog og skogproduksjon i norge som virkemiddel mot CO$_2$-opphopning i atmosfaeren [Forest and Wood Products in Norway as a Means to Reduce CO$_2$-Accumulation in the Atmosphere]*, Skogforsk 6-1991, Norsk Institut for skogforskning og institut fro skogfag, Norges Landbrukshogskole, Norway

Luo, Z. and Hulscher, W. (1999) 'Woodfuel emissions and costs', *Wood Energy News*, vol 14, no 3, pp13–15

Lutz, A. (1992) 'Vegetable oil as fuel', *GTZ Gate*, vol 4

Lybbert T. J. (2007) 'Patent disclosure requirements and benefit sharing: A counterfactual case of Morocco's argan oil', *Ecological Economics*, vol 64, pp12–18

Lynd, L. R. (1996) 'Overview and evaluation of fuel ethanol from lignocellulosic biomass: Technology, economics, the environment and policy', *Annual Review Energy Environment*, vol 21, pp403–465

MacDicken, K. G. and Vergara, N. T. (1990) 'Introduction to agroforestry', in K. G. Macdicken and N. T. Vergara (eds) *Agroforestry: Classification and Management*, John Wiley and Sons, New York, pp1–30

MacKenzie, A. F., Fan, M. X. and Cadrin, F. (1997) 'Nitrous oxide emission as affected by tillage, corn-soybean-alfalfa rotations and nitrogen fertilization', *Canadian Journal of Soil Science*, vol 77, pp145–152

Mackenzie, A. (2004) 'Giant reed', in C. Harrington and A. Hayes (eds) *The Weed Workers' Handbook*, www.cal-ipc.org/file_library/19646.pdf

MacLaren, J. P. (1996) 'Plantation forestry – its role as a carbon sink: Conclusions from calculations based on New Zealand's planted forest estate', in M. J. Apps and D. T. Price (eds) *Forest Ecosystems, Forest Management and the Global Carbon Cycle*, NATO ASI Series I (Global Environmental Change), vol 40, Springer-Verlag Academic Publishers, Heidelberg, Germany, pp257–270

Madslien, J. (2006) 'Biofuel raises global dilemmas', http://news.bbc.co.uk/2/hi/business/4603272.stm

Magness, J. R., Markle, G. M. and Compton, C. C. (1971) *Food and Feed Crops of the United States*, Interregional Research Project IR-4, IR Bul. 1 (Bul. 828, New Jersey Agr. Exp. Sta.)

Mahmood, K. (1995) 'Salinity effects on seed germination, growth and chemical composition of *Atriplex lentiformis* (Torr.)', *Wats. Acta Sci.*, vol 5, pp59–66

Mahmood, K. (1997) 'Competitive superiority of *Kochia indica* over *Leptochloa fusca* (Kallar grass) under varying levels of soil moisture and salinity', *Pakistan Journal of Botany*, vol 29, pp289–297

Mahmood, K., Malik, K. A., Sheikh, K. H. and Lodhi, M. A. K. (1989) 'Allelopathy in saline agricultural land: Vegetation successional changes and patch dynamics', *Journal of Chemical Ecology*, vol 15, pp565–579

Mahmood, K., Malik, K. A., Lodhi, M. A. K. and Sheikh, K.H. (1993) 'Competitive interference by some invader species against Kallar grass (*Leptochloa fusca*) under different salinity and watering regimes', *Pakistan Journal of Botany*, vol 25, pp145–155

Mahmood, K., Malik, K. A., Sheikh, K. H., Hussain, A. and Lodhi, M. A. K. (1999) 'Allelopathic potential of weed species invading Kallar grass (*Leptochloa fusca*) in saline agricultural land', *Pakistan Journal of Botany*, vol 31, pp137–149

Makipaa, R., Karjalainen, T., Pussinen, A. and Kellomaki, S. (1999) 'Effects of climate change and nitrogen deposition on the carbon sequestration of a forest ecosystem in the boreal zone', *Canadian Journal of Forest Research*, vol 29, pp1490–1501

Makundi, W. R. (1997) 'Global climate change mitigation and sustainable forest management: The challenge of monitoring and verification', *Journal of Mitigation and Adaptation Strategies for Global Change*, vol 2, pp133–155

Makundi, W. R. (1998) 'Mitigation options in forestry, land-use change and biomass burning in Africa', in G. Mackenzie, J. Turkson and O. Davidson (eds) *Climate Change Mitigation in Africa, Proceedings of UNEP/SCEE Workshop*, Harare, May 1998

Makundi, W. R., Razali, W., Jones, D. and Pinso, C. (1998) 'Tropical forests in the Kyoto Protocol: Prospects for carbon offset projects after Buenos Aires', *Tropical Forestry Update*, vol 8, no 4, pp5–8

Malik, K. A., Zafar, Y. and Hussain, A. (1981) 'Associative dinitrogen fixation in *Diplachne fusca* (Kallar grass)', in P. H. Graham and S. C. Haris (eds) *BNF Technology for Tropical Agriculture*, UAT, Coli Columbia, pp503–507

Malik, K. A., Aslam, Z. and Naqvi, M. (1986) *Kallar Grass: A Plant for Saline Land*, Ghulamali Printers, Lahore

Mamma, D., Koulias, D., Fountoukidis, G., Kekos, D. and Macris, B. J. (1996) 'Bioethanol from sweet sorghum: Simultaneous saccharification and fermentation of carbohydrates by a mixed microbial culture', *Process Biochemistry*, vol 31, no 4, pp377–381

Mampouya, P. C. (1991) 'Marcottage du safoutier', in *Actes du seminaire sous regional sur la valorisation du safoutier*, 25–28 November, Brazzaville, Congo, pp32–39

Mangales, R. J. G. and Rezende, M. (1989) 'Charcoal and by-products from planted forests: A Brazilian experience', paper presented in the first workshop of CNRE on charcoal production and pyrolysis technologies held in Poros, Norway, 23–25 October 1989

Manitoba Agriculture, Food and Rural Initiatives, (no date) 'Industrial hemp production', www.gov.mb.ca/agriculture/crops/hemp/bko05s00.html (accessed September 2009)

Manley, J. T., Schuman, G. E., Reeder, J. D. and Hart, R. H. (1995) 'Rangeland soil carbon and nitrogen responses to grazing', *Journal of Soil and Water Conservation*, vol 50, pp294–298

Manzanares, M., Tenorio, J. L. and Ayerbe, L. (1995) 'Some aspects of kenaf (*Hibiscus cannabinus* L.) agronomy in the central area of the Iberian peninsula', in *Biomass for Energy, Environment, Agriculture and Industry, Proceedings of the 8th EC Conference*, Vienna, Pergamon, vol 1, pp717–722

Marchetti, C. (1976) *On Geoengineering and the CO_2 Problem*, Research Memorandum RM-76-17, International Institute for Applied Systems Analysis, Vienna, Austria

Mariau, D., Dery, S. K., Sangarè, A., N'Cho Yavo, P. and Phillipe, R. (1996) 'Coconut lethal yellowing disease in Ghana and planting material tolerance', *Plantations, Recherche, Developpement*, vol 3, no 2, pp105–112

Marland, G. (1996) 'Geoengineering'. in S. H. Schneider (ed) *Encyclopedia of Climate and Weather*, Oxford University Press, New York, vol 1, pp338–339

Marland, G. and Schlamadinger, B. (1997) 'Forests for carbon sequestration or fossil fuel substitution? A sensitivity analysis', *Biomass and Bioenergy*, vol 13, no 6, pp389–397

Marland, G. and Schlamadinger, B. (1999) 'The Kyoto Protocol could make a difference for the optimal forest-based CO_2 mitigation strategy: Some results from GORCAM', *Environmental Science and Policy*, vol 2, pp111–124

Martin, J. H. (1941) 'Climate and sorghum', in *Climate and Man*, USDA Yearbook, pp343–347.

Martin, J. H. (1990) 'A new iron age, or a ferric fantasy', *US Joint Global Ocean Flux Study Newsletter*, vol 1, no 4, pp5–6

Martin, J. H. (1991) 'Iron, Liebig's Law, and the greenhouse', *Oceanography*, vol 4, pp52–55

Martin, J. H., Coale, K. H., Johnson, K. S. and Fitzwater, S. E. (1994) 'Testing the iron hypothesis in ecosystems of the equatorial Pacific Ocean', *Nature*, vol 371, pp123–130

Martin, P. H., Valentini, R., Kennedy, P. and Folving, S. (1998) 'New estimate of the carbon sink strength of EU forests integrating flux measurements, field surveys, and space observations: 0.17–0.35 Gt(C)', *Ambio*, vol 27, no 7, pp582–584

Masera, O. R. (1995) 'Carbon mitigation scenarios for Mexican forests: Methodological considerations and results', *Interciencia*, vol 20, pp388–395

Masera, O. R. and Ordóñez, A. (1997) 'Forest management mitigation options', In C. Sheinbaum (Coord.) *Final Report to the USAID-Support to the National Climate Change Plan for Mexico*, Instituto de ingeniería, National University of Mexico (UNAM), Report 6133, UNAM, Mexico City, pp77–93

Masera, O. R., Bellon, M. and Segura, G. (1995) 'Forest management options for sequestering carbon in Mexico', *Biomass and Bioenergy*, vol 8, no 5, pp357–368

Masera, O. R., Ordoñez, M. J. and Dirzo, R. (1997a) 'Carbon emissions from Mexican forests: Current situation and long-term scenario', *Climatic Change*, vol 35, pp265–295

Masera, O. R., Bellon, M. R. and Segura, G. (1997b) 'Forestry options for sequestering carbon in Mexico: Comparative economic analysis of three case studies', *Critical Reviews in Environmental Science and Technology*, vol 27, pp227–244

Mastrorilli, M., Katerji, N. and Rana, G. (1996) 'Evaluation of the risk derived from introducing sweet sorghum in Mediterranean zone', in *First European Seminar on Sorghum for Energy and Industry*, Institut National de la Recherche Agronomique, France, pp323–328

Mather, A. S. (1990) 'Historical perspectives on forest resource use', in *Global Forest Resources*, Timber Press, Portland, OR, pp30–57

Matthews, R., Nabuurs, G. J., Alexeyev, V., Birdsey, R. A., Fischlin, A., MacLaren, J. P., Marland, G. and Price, D. (1996) 'WG3 summary: Evaluating the role of forest management and forest products in the carbon cycle', in M. J. Apps and D. T. Price (eds) *Forest Ecosystems, Forest Management and the Global Carbon Cycle*, Proceedings of a workshop held in September 1994 in Banff, Canada, NATO Advanced Science Institute Series, NATO-ASI Vol I 40, Berlin, Heidelberg, pp293–301

Matzke, W. (1988) 'The suitability of arundo reed *(Arundo donax* L.) as raw material for the paper industry', in *Escher Wyss Ltd, Manufacturing Programme*, pp40–45

Mayet, M. (2007) 'Africa's biodiesel: Going nuts!', African Centre for Biosafety, www.biosafetyafrica.net/portal/DOCS/GROUND NUT_BIODIESEL_BRIEFING.pdf

Mayeux, A. (1992) 'Effect of sowing time on groundnut yield in Botswana', in *Proceedings of the Fifth Regional Groundnut Workshop for South Africa*, pp72–79.

Mayton, H., Hansen, J., Neally, J. and Viands, D. (2008) *Accelerated Evaluation of Perennial Grass and Legume Feedstocks for Biofuel Production in New York State*, Department of Plant Breeding and Genetics, Cornell University, Ithaca, NY

McCormick, R. L. (2009) *Biodiesel Handling and Use Guide*, 4th edn

McDermott, M. (2008) 'Biofuel crops grown in saltwater could supply 35% of US's liquid fuel needs', www.treehugger.com/files/2008/12/biofuels-grown-in-saltwater-could-be-35-percent-united-states-fuels.php

McDonald, D. (1984) 'The ICRISAT groundnut programme', *Proceedings of the Regional Groundnut Workshop, Southern Africa*, pp17–30

McIvor, J. G. (2001) 'Litterfall from trees in semiarid woodlands of north-east Queensland', *Austral Ecology*, vol 26, no 2, pp150–155

McLaughlin, S. B., Samson, R., Bransby, D. and Wiselogel, A. (1996) 'Evaluating physical, chemical, and energetic properties of perennial grasses as biofuels', in *Bioenergy '96, Proceedings of the Seventh National Bioenergy Conference*, 15–20 September, Nashville, Tennessee, vol 1, pp1–8

Mediavilla, V., Lehmann, J. and Meister, E. (1991) *Energiegras/ Feldholz – Teilprojekt A: Energiegras, Jahresbericht 1991*, Bundesamt für Energiewirtschaft, Berne

Mediavilla, V., Lehmann, J. and Meister, E. (1993) *Energiegras/Feldholz – Teilprojekt A: Energiegras, Jahresbericht 1993*, Bundesamt für Energiewirtschaft, Berne

Mediavilla, V., Lehmann, J., Meister, E., Stünzi, H. and Serafin, F. (1994) *Energiegras/Feldholz – Energiegras, Jahresbericht 1994*, Bundesamt für Energiewirtschaft, Berne

Mediavilla, V., Lehmann, J., Meister, E., Stünzi, H. and Serafin, F. (1995) *Energiegras/Feldholz – Energiegras, Jahresbericht 1995*, Bundesamt für Energiewirtschaft, Berne

Meijer, W. J. M. and Mathijssen, E. W. J. M. (1991) 'Inulineproduktie via aardpeer of cichorei?', in W. J. M. Meijer and M. Vertregt (eds) *Agrobiologische Thema's CABO-DLO, Vol 4: Gewasdiversifikatie en Agrificatie,* Wageningen, Netherlands

Meijer, W. J. M., Vanderwerf, H. M. G., Mathijssen, E. W. J. M. and Vandenbrink, P. W. M. (1995) 'Constraints to dry matter production in fibre hemp *(Cannabis sativa* L.)', *European Journal of Agronomy*, vol 4, no 1, pp109–17

Melillo, J. R., McGuire, A. D., Kicklighter, D. W., Moore, B., Vorosmarty, C. J. and Schloss, A. L. (1993) 'Global climate change and terrestrial net primary production', *Nature*, vol 363, pp234–240

Mennessier, M. (2006) 'La chimie non polluante en plein essor', *Le Figaro*, 5 August

Meyer, W. B. and Turner, B. L. II (1992) 'Human population growth and global land-use/cover change', *Annual Review Ecology Systematics*, vol 23, pp39–61

Micales, J. A. and Skog, K. E. (1997) 'The decomposition of forest products in landfills', *International Biodeterioration and Biodegradation*, vol 39, pp145–158

Middleton, C. (1995) *Leucaena for Beef Production in Queensland*, http://leaky.rock.lap.csiro.au/facts/leucext2-txt.html

Miguel, A. (1988) 'Short rotation forest biomass plantations in Spain', in F. C. Hummel et al (eds) *Biomass Forestry in Europe: A Strategy for the Future*, Elsevier Applied Science, London.

Miller, E. C. (1916) 'Comparative study of the root systems and leaf areas of corn and sorghums', *Journal of Agricultural Research*, vol 6, p311

Miller, E. C. (1938) *Plant Physiology*, McGraw-Hill

Miller, G. T. (1992) *Living in the Environment*, 7th edn, Wadsworth, Belmont, CA

Miller, J. H. and Edwards, B. (1982) 'Kudzu: Where did it come from? And how can we stop it?', *Southern Journal of Applied Forestry*, vol 7, pp165–169

Millot, B. and Wheeler, F. (2006) 'Le bioethanol a partir des dattes, produit de choix environnemental pour l'Algerie?', November

Minami, K. (1995) 'The effect of nitrogen fertilizer use and other practices on methane emission from flooded rice', *Fertilizer Research*, vol 40, pp71–84

Mishima, D., Kuniki, M., Sei, K., Soda, S., Ike, M. and Fujita, M. (2008) 'Ethanol production from candidate energy crops: Water hyacinth (*Eichhornia crassipes*) and water lettuce (*Pistia stratiotes* L.)', *Bioresource Technology*, vol 99, no 7, pp2495–2500

Mitsch, W. J. and Wu, X. (1995) 'Wetlands and global change', in R. Lal, J. Kimble, E. Levine and B. A. Stewart (eds) *Soil Management and Greenhouse Effect*, CRC Lewis Publishers, Boca Raton, pp205–230

Miyagi, E. (1980) 'The effect of planting density on yield of napier grass (*Pennisetum purpureum* Schumach)', *Sci. Bul. Coll. Agr., U. Ryukus*, vol 27, pp293–301

Miyamoto, T., Kawahara, M. and Minamisawa, K. (2004) 'Novel endophytic nitrogen-fixing clostridia from the grass *Miscanthus sinensis* as revealed by terminal restriction fragment length polymorphism analysis', *Applied and Environmental Microbiology*, vol 70, no 11, pp6580–6586

Mobberley, D. G. (1956) 'Taxonomy and distribution of the genus *Spartina*', *Iowa State College Journal of Science*, vol 30, pp471–574

Mogilaer, I., Orioli, G. A. and Bletter, C. M. (1967) 'Trial to study topophysis and photo-periodism in cassava', *Bouplandia*, vol 2, pp265–272

Mohanan, C. and Liese, W. (1990) 'Diseases of bamboos', *International Journal of Tropical Plant Diseases*, vol 8, pp1–20

Monastersky, R. (1995) 'Iron versus the greenhouse', *Science News*, vol 148, pp220–222

Monfort, J. (2008) *World Watch Magazine*, July/August, vol 21, no 4, www.worldwatch.org/node/5777

Monteith, J. L. (1978) 'Reassessment of maximum growth rates for C_3 and C_4 crops', *Experimental Agriculture*, vol 14, pp1–5

Moon, S. (1997) 'Sowing seed, planting trees, producing power', *Solar Today*, pp16–19

Morgan, J. A., Knight, W. G., Dudley, L. M. and Hunt, H. W. (1994) 'Enhanced root system C-sink activity, water relations and aspects of nutrient acquisition in mycotrophic *Bouteloua gracilis* subjected to CO_2 enrichment', *Plant and Soil*, vol 165, pp139–146

Morgana, B. and Sardo, V. (1995) 'Giant reeds and C_4 grasses as a source of biomass', in Chartier et al (eds) *Biomass for Energy, Environment, Agriculture and Industry. Proceedings of the 9th EU Biomass Conference*, Pergamon Press, UK, vol 1, pp700–707

Morozov, V. L. (1979) 'Productivity of tall herbaceous vegetation in the Far East', *Izuest. Sibir. Otd. Akad. Nauk SSSR*, Ser. Biol. Nauk, vol 2, pp32–39; cited in Walter, H. (1981) 'Über hochstwerte der produktion von naturlichen Pflanzenbestanden in N. O. Asien', *Vegetatio*, vol 44, pp37–41

Moser, L. E. and Vogel, K. P. (1995) 'Switchgrass, big bluestem, and Indiangrass', in R. F. Barnes, D. A. Miller and C. J. Nelson (eds) *An Introduction to Grassland Agriculture: Forages*, 5th edn, vol I, Iowa State University Press, Ames, pp409–420

Mosier, A. R. (1998) 'Soil processes and global change', *Biology and Fertility of Soils*, vol 27, pp221–229

Moulton, R. and Richards, K. (1990) *Costs of Sequestrating Carbon Through Tree Planting and Forest Management in the United States*, GTR WO-58, USDA Forest Service, Washington, DC

Moura-Costa, P. and Stuart, M. (1998) 'Forestry based greenhouse gas mitigation: A story of market evolution', *Commonwealth Forestry Review*, vol 77, pp191–202

Moussouris, Y. and Pierce, A. (2000) 'Biodiversity links to cultural identity in southwest Morocco: The situation, the problems and proposed solutions', *Arid Lands Newsletter*, no 48

Moyer, J. L., Sweeney, D. W. and Lamond, R. E. (1995) 'Response of tall fescue to fertilizer placement at different levels of phosphorus, potassium, and soil pH', *Journal of Plant Nutrition*, vol 18, no 4, pp729–746

Moyo, C. C., Benesi, I. R. M., Chipungu, F. P., Mwale, C. H. L., Sandifolo, V. S. and Mahungu, N. M. (2006) 'Africa: Cassava and sweetpotato yield assessment in Malawi', *African Crop Science Journal* (Uganda), vol 12, no 3, special issue, pp295–303, http://knowledge.cta.int/en/content/view/full/2784

Mueller-Warrant, G. W., Young, W. C. and Mellbye, M. E. (1995) 'Residue removal method and herbicides for tall fescue seed production. 2: Crop tolerance', *Agronomy Journal*, vol 87, no 3, pp558–562

Müller, F. (1990) 'Die Robinie als Biomasseproduzent in Kurzumtriebsplantagen', *Österreichische Forstzeitung*, vol 5, p221

Münch, E. and Kiefer, J. (1989) 'Die Purgiernuß', *Schriftenreihe der GTZ*, No 209

Muñoz-Valenzuela S., Ibarra-López, A. A., Rubio-Silva, L. M., Valdez-Dávila, H. and Borboa-Flores, J. (2007) 'Neem tree morphology and oil content', in J. Janick and A. Whipkey (eds) *New Crops and New Uses*, ASHS Press, Alexandria, VA, p126

Munns, R., Greenway, H. and Kirst, G. O. (1983) 'Halotolerant eukaryotes', in O. L. Lange, P. S. Nobel, C. B. Osmond and H. Ziegler (eds) *Physiological Plant Ecology, Vol. 3: Responses to the Chemical and Biological Environment*, Springer, Berlin

Murray, J. (2008) 'Willow and elephant grass to provide power for 40,000 homes', BusinessGreen, Incisive Media Ltd

Musangi, R. S. and Soneji, S. C. (1967) 'Feeding groundnut (*Arachis hypogaea* L.) hulms to dairy cows in Uganda', *E. Afri. Forest Journal*, vol 33, pp170–174

Myllylä, I. (2007) 'Use of reed canary grass growing', *Vapoview*, no 1, 16 August, http://www.vapoviesti.fi/vapoview/index.php?id=1368&articleId=86&type=9 (accessed September 2009)

NABC (National Agricultural Biotechnology Council) (2000) *NABC 12: The Biobased Economy of the 21st Century: Agriculture Expanding into Health, Energy, Chemicals, and Materials*, http://nabc.cals.cornell.edu/pubs/pubs_reports.cfm#nabc12

Nabuurs, G. J. (1996) 'Significance of wood products in forest sector carbon balances', in M. J. Apps and D. T. Price (eds) *Forest Ecosystems, Forest Management and the Global Carbon Cycle*, Proceedings of the workshop held in September 1994 in Banff, Canada, NATO Advanced Science Institute Series, Vol I 40, Berlin, Heidelberg, pp245–256

Nabuurs, G. J. (1998) 'Bos wordt meer geld waard [Dutch forests become more valuable]', *Nederlands Bosbouwtijdschrift*, vol 70, no 2, p69

Nabuurs, G. J. and Sikkema, R. (1998) *The Role of Harvested Wood Products in National Carbon Balances: An Evaluation of Alternatives for IPCC Guidelines*, IBN Research Report 98/3, Institute for Forestry and Nature Research, Institute for Forest and Forest Products

Nabuurs, G. J., Paeivinen, R., Sikkema, R. and Mohren, G. M. J. (1997) 'The role of European forests in the global carbon cycle: A review', *Biomass and Bioenergy*, vol 13, no 6, pp345–358

Nabuurs, G. J., Dolman, A. V., Verkaik, E., Kuikman, P. J., van Diepen, C. A., Whitmore, A., Daamen, W., Oenema, O., Kabat, P. and Mohren, G. M. J. (2000) 'Article 3.3 and 3.4 of the Kyoto Protocol: Consequences for industrialised countries' commitment, the monitoring needs and possible side effect', *Environmental Science and Policy*, vol 3, nos 2/3, pp123–134

Nadelhoffer, K. J., Emmett, B. A., Gundersen, P., Kjønaas, O. J., Koopmans, C. J., Schleppi, P., Tietema, A. and Wright, R. F. (1999) 'Nitrogen deposition makes a minor contribution to carbon sequestration in temperate forests', *Nature*, vol 398, pp145–148

Nadgir, A. L., Phadke, C. H., Gupta, P. K., Parasharami, V. A., Nair, S. and Mascarenhas, A. F. (1984) 'Rapid multiplication of bamboo by tissue culture' *Silvae Genetica*, vol 33, pp219–223

Nagao, Fs. Ohigashi, M. Kuwal and H. Hisam (1948). *Japan Soc. Mech. Eng. Transactions* vol 51, no 354, p92

Nair, P. K. R. (1989) 'The role of trees in soil productivity and protection', in P. K. R. Nair (ed) *Agroforestry Systems in the Tropics*, Kluwer Academic Publishers, Dordrecht, The Netherlands, pp567–589

Nam, L. and Ma, J. (1989) 'Research on sweet sorghum and its synthetic application', in Grassi et al (eds) *Biomass for Energy and Industry, Proceedings of the 5th EC Conference*, Elsevier Applied Science, London, vol 1

NAS (National Academy of Sciences) (1975) *Underexploited Tropical Plants with Promising Economic Value*, Washington, DC

NAS (1992) *Policy Implications of Greenhouse Warming: Mitigation, Adaptation, and the Science Base*, Panel on Policy Implications of Greenhouse Warming, US National Academy of Sciences, National Academy Press, Washington, DC

NAS (National Academy of Sciences) (2008) *Lost Crops of Africa: Volume III: Fruits*, National Academy of Sciences, Washington, DC

Nasrin, A. B., Ma, A. N., Choo, Y. M., Mohamad, S., Rohaya, M. H., Azali A. and Zainal, Z. (2008) 'Oil palm biomass as potential substitution raw materials for commercial biomass briquettes production', *American Journal of Applied Sciences*, vol 5, no 3, pp179–183, http://findarticles.com/p/articles/mi_7109/is_/ai_n28458738

Nations Unies (1998) *Protocole de Kyoto a la convention cadre des Nations Unies sur les changements climatiques*

Ndamba, J. (1989) 'Analyse bromatologique du tourteau du safou en vue de son utilisation en alimentation animale. Résultats préliminaires', thése de doctorat vétérinaire, EISMV, Dakar

Negro, M. J., Manzanares, P., Ballesteros, I., Oliva, J., Araceli Cabañas, M. and Ballesteros, M. (2007) 'Hydrothermal pretreatment conditions to enhance ethanol production from poplar biomass', *Applied Biochemistry and Biotechnology*, vol 105, nos 1–3, pp87–100

Nelson, P. (1993) 'The development of the lupin industry in Western Australia and its role in farming systems', in *Proceedings VII International Conference*, Evora, Portugal

Nemecek, T., Heil, A., Huguenin, O., Meier, S., Erzinger, S., Blaser, S., Dux, D. and Zimmermann, A. (2004) *Life Cycle Inventories of Agricultural Production Systems*, final report Ecoinvent 2000 No 15, Agroscope FAL Reckenholz and FAT Taenikon, Swiss Centre for Life Cycle Inventories, Dübendorf, www.ecoinvent.org

Nene, Y. L. and Sheila, V. K. (1990) 'Pigeonpea: Geography and importance', in Y. L. Nene et al (eds) *The Pigeonpea*, CAB International, UK, pp1–14

Neue, H. U. (1997) 'Fluxes of methane from rice fields and potential for mitigation', *Soil Use and Management*, vol 13, pp258–267

Ng, W.-K. (2002) 'Potential of palm oil utilisation in aquaculture feeds', *Asia Pacific Journal of Clinical Nutrition*, vol 11, no s7, ppS473–S476

Nguyen, T. L. T., Gheewala, S. H. and Garivait, S. (2007) Full chain energy analysis of fuel ethanol from cassava in Thailand', *Environmental Science and Technology*, vol 41, no 11, pp4135–4142

Nielsen, H. K., Kunze, M. and Ahlhaus, M. (2006) 'Lignocellulosic energy crops: Four years' experiences from National Centre for Renewable Energy in Grimstad, Norway', *Use of Bioenergy in the Baltic Sea Region – Proceedings of the 2nd IBBC*, Stralsund, Germany

Nielson, P. N. (1987) 'Vegetative propagation of *Miscanthus sinensis* "giganteus"', *Tidssk Planteavl*, vol 91, pp361–368

Nilsson, S. and Shvidenko, A. (1998) *Is Sustainable Development of the Russian Forest Sector Possible?*, IUFRO Occasional Paper No 11

Nitske, W. R. and Wilson, C. M. (1965) *Rudolph Diesel, Pioneer of the Age of Power*, University of Oklahoma Press, pp122–123

Noble, I., Apps, M., Houghton, R., Lashof, D., Makundi, W., Murdiyarso, D., Murray, B., Sombroek, W. and Valentini, R. (2000) 'Implications of different definitions and generic issues', in R. T. Watson, I. R. Noble, B. Bolin, N. H. Ravindranath, D. J. Verardo and D. J. Dokken (eds) *Land Use, Land Use Change and Forestry*, A Special Report of the IPCC, Cambridge University Press, UK, pp55–126

Nobre, C. A., Sellers, P. J. and Shukla, J. (1991) 'Amazonian deforestation and regional climate change', *Journal of Climate*, vol 4, pp957–988

Nogueira, L. A. H., Trossero, M. A. and Couto, L. (1998) 'A discussion of the relationship between wood fibre for energy supply and overall supply of wood fibre for industry', *Unasylva*, vol 49, pp51–56

Norby, R. J. and Cotrufo, M. F. (1998) 'A question of litter quality', *Nature*, vol 396, pp17–18

Norden, A. J. (1980) 'Peanut', in W. R. Fehr and H. Hadley (eds) *Hybridization of Crop Plants*, American Society of Agronomy and Crop Science

Nouaim, R. (2005) *L'arganier au Maroc: entre mythes et réalités. Une civilisation née d'un arbre, une espèce fruitière-forestière à usages multiples*, L'Harmattan, Paris

Numata, M. (ed) (1975) *Ecological Studies in Japanese Grasslands with Special Reference to the IBP Areas: Productivity of Terrestrial Communities*, Japanese Committee for the International Biological Program (JIBP Synthesis), 13, University of Tokyo Press

Nyina-wamwiza, L., Wathelet, B. and Kestemont, P. (2007) 'Potential of local agricultural by-products for the rearing of African catfish *Clarias gariepinus* in Rwanda: Effects on growth, feed utilization and body composition', *Aquaculture Research*, vol 38, no 2, pp206–214

Odunfa, S. A. (1985) 'African fermented foods', in B. J. B Wood (ed) *Microbiology of Fermented Foods*, Elsevier Applied Science, Amsterdam, vol 2, pp155–161

OECD (1991) *Energy Policies of IEA Countries, 1990 Review*, OECD, Paris

OECD (2008) *OECD Environmental Outlook to 2030 Report*, OECD, Paris

OECD and FAO (2007) *Agricultural Outlook 2007–2016*, OECD, Paris

Ogden, J. M., Steinbugler, M. M. and Kreutz, T. G. (1999) 'A comparison of hydrogen, methanol and gasoline as fuels for fuel cell vehicles: Implications for vehicle design and infrastructure development', *Journal of Power Sources*, vol 79, pp143–168

Okafor, J. C. (1983) 'Varietal delimitation in *Dacryodes edulis* (G. Don) H. J. Lam (Burseraceae), *International Tree Crops Journal*, vol 2, pp255–265

Oldeman, L. R. (1994) 'The global extent of soil degradation', in D. J. Greenland and I. Szaboles (eds) *Soil Resilience and Sustainable Land Use*, CAB International, Wallingford, UK, pp99–118

Oliviera, P. (1996) UNL, Lisbon, Personal Communication

Olsson, R. (1993) 'Production methods and costs for reed canarygrass as an energy crop', in *Bioenergy Research Programme*, Publication 2, Bioenergy 93 Conference, Finland

Olsson, R. (1994) 'A new concept for reed canary grass production and its combined processing to energy and pulp', in *Non-wood Fibres for Industry*, Pira Int./Silsoe Research Institute Joint Conference, Bedfordshire, UK

Olsson, R. (2005) *Reed Canarygrass Development in Sweden*, Swedish University of Agricultural Sciences, Department of Agricultural Research for Northern Sweden, Laboratory for Chemistry and Biomass, www.p2pays.org/ref/17/16274/ollson.pdf

Omoti, U. and Okiy, A. D. (1987) 'Characteristics and composition of pulp oil and cake of African pear *Dacryodes edulis* (G. Don) H. J. Lam', *Journal of the Science of Food and Agriculture*, vol 38, pp67–72

Onwueme, I. C. (ed) (1982) *The Tropical Tuber Crops: Yams, Cassava, Sweet Potato and Cocoyrus*, English Language Book Society/John Wiley, Chichester

Organ, J. (1960) *Rare Vegetables for Garden and Table*, Faber

Ortmaier, E. and Schmittinger, B. (1988) 'Possibilities of tapping the natural yield potential of the babassu forests in northeastern Brazil by environment-orientated resource management', *Quarterly Journal of International Agriculturei*, vol 27, no 2, pp170–185

Ortiz-Cañavate, J. (1994) 'Technical applications of existing biofuels', in *Application of Biologically Derived Products as Fuels or Additives in Combustion Engines*, Directorate-General XII, Science, Research and Development, EUR 15647 EN, pp1–20, 52–67

Osman, S. M. and Ahmad, F. (1981) 'Forest oilseeds', in E. H. Pryde, L. H. Princen and K. D. Mukherjee (eds) *New Sources of Fats and Oils*, American Oil Chemist's Society, IN

Osuntokum, B. O. (1981) 'Cassava diet, chronic cyanide intoxication and neuropathy in the Nigerian Africans', *World Rev. Nutro Diet*, vol 26, pp141–73

Otani, T. and Ae, N. (1996) 'Phosphorus (P) uptake mechanisms of crops crops in soils with low P status. I: Screening crops for efficient P uptake', *Soil Science and Plant Nutrition*, vol 42, no 1, pp155–63

Ott, A., Strasser, H. R., Ammon, H. U., Mediavilla, V., Zürcher, N. and Grether, T. (1995) *Anbau und Verwertung von Kenaf*, Landwirtschaftliche Forschung Beratung, Merkblatt, UFA-Revue, Bern

Overend, R. P. (1995) 'Production of electricity from biomass crops: US perspective', in *Proceedings, Workshop on Energy from Biomass and Wastes*, Dublin Castle, Ireland, p2

Owensby, C. E. (1993) 'Potential impacts of elevated CO_2 and above- and below-ground litter quality of a tallgrass prairie', *Water, Air, and Soil Pollution*, vol 70, pp413–424

Paavola, T., Lehtomäki, A., Seppälä, M. and Rintala, J. (2007) 'Methane production from reed canary grass', University of Jyväskylä, Department of Biological and Environmental Science, University of Jyväskylä, Finland, www.cropgen.soton.ac.uk/publications/8%20Other/Oth_27_FCES2007%20paper_Paavola.pdf (accessed September 2009)

Padulosi, S. and Pignone, D. (eds) (1997) 'Rocket: A Mediterranean crop for the world', Report of a workshop, 13–14 December 1996, Legnaro (Padova), Italy, International Plant Genetic Resources Institute, Rome, Italy

Pahkala, K. A., Mela, T. J. N. and Laamanen, L. (1994) 'Mineral composition and pulping characteristics of several field crops cultivated in Finland', in *Biomass for Energy, Environment, Agriculture and Industry, Proceedings of the 8th EC Conference*, Vienna, Pergamon, pp395–400

Palo, M. and Uusivuori, J. (1999) 'Globalization of forests, societies and environments', in M. Palo and J. Uusivuori (eds) *World Forests, Society and Environment*, Kluwer, pp3–14

Palz, W. (1995) 'Future options for biomass in Europe', in *Proceedings, Workshop on Energy from Biomass and Wastes*, Dublin Castle, Ireland, p2

Palz, W. and Chartier, P. (eds) (1981) *Energy from Biomass in Europe*, Applied Science Publishers Ltd, London

Panwar, K. S. and Yadav, H. L. (1981) 'Response of short-duration pigeonpea to early planting and phosphorus levels in different cropping systems', in *Proceedings of the International Workshop on pigeonpeas*, 15–19 December 1980, Patancheru, India, vol 1, pp37–44

Parente, E. (2007) *Babassu Palms*, BioPact, Belgium

Parfitt, R. J. and Royle, D. J. (1996) '*Populas*, poplar', in *Renewable Energy: Potential Energy Crops for Europe and the Mediterranean Region*, REU Technical Series No 46, FAO

Parfitt, R. J., Clay, D. V., Arnold, G. M. and Foulkes, A. (1991) 'Weed control in new plantations of short rotation willow and poplar coppice', *Aspects of Applied Biology*, vol 29, pp419–423

Pari, D. and Ragno, I. (1998) 'Biomass crop energy balance', in H. Kopetz, T. Weber, W. Palz, P. Chartier and G. L. Ferrero (eds), *Proceedings of the 10th European Conference on Biomass for Energy and Industry*, CARMEN, Wurzburg, Germany, 8–11 June, pp819–823

Parker, T. D., Adams, D. A., Zhou, K., Harris, M. and Yu, L. (2003) 'Fatty acid composition and oxidative stability of cold-pressed edible seed oils', *Journal of Food Science*, vol 68, no 4, pp1240–1243

Parks, P. J. and Hardie, I. W. (1995) 'Least-cost forest carbon reserves: Cost effective subsidies to convert marginal agricultural land to forests', *Land Economics*, vol 71, pp122–136

Parlati, M. V., Bellini, E., Pandolfi, S., Giordani, E. and Martelli, S. (1994) 'Genetic improvement of olive: Initial observations on selections made in Florence', *Acta Horticulturae*, vol 356, pp87–90

Parrish, D. J., Wolf, D. D. and Daniels, W. L. (1993) *Perennial Species for Optimum Production of Herbaceous Biomass in the Piedmont*, Management Study 1987–1991, ORNL/Sub/85-

27413/7, National Technical Information Service, US Department of Commerce, Springfield, VA

Parrish, D. J., Wolf, D. D. and Daniels, W. L. (1997) *Switchgrass as a Biofuel Crop for the Upper Southeast: Variety Trials and Cultural Improvements*, Final Report 1992–1997, Biofuels Feedstock Development Program Environmental Sciences Division, Oak Ridge National Laboratory, Oak Ridge, Tennessee

Patil, V. C., Patil, S. V. and Hanamashetti, S. I. (1991) 'Bamboo farming: An economic alternative on marginal lands', in *Proceedings of the 4th International Bamboo Workshop*, Chiangmai, Thailand, FORSPA Publication 6, pp133–135

Paustian, K., Elliot, E. T. and Killian, K. (1997a) 'Modeling soil carbon in relation to management and climate change in some agroecosystems in central North America', in R. Lal, J. M. Kimble, R. F. Follett and B. A. Stewart (eds) *Soil Processes and the Carbon Cycle*, CRC Press, Boca Raton, pp459–471

Paustian, K., Andren, O., Janzen, H. H., Lal, R., Smith, P., Tian, G., Tiessen, H., van Noordwijk, M. and Woomer, P. L. (1997b) 'Agricultural soils as a sink to mitigate CO_2 emissions', *Soil Use and Management*, vol 13, pp230–244

Paustian, K., Collins, H. P. and Paul, E. A. (1997c) 'Management controls on soil carbon', in E. A. Paul, E. T. Elliott, K. Paustian and C. V. Cole (eds) *Soil Organic Matter in Temperate Agroecosystems: Long-Term Experiments in North America*, CRC Press, Boca Raton, pp15–49

Paustian, K., Cole, C. V., Sauerbeck, D. and Sampson, N. (1998) 'CO_2 mitigation by agriculture: An overview', *Climatic Change*, vol 40, pp135–162

Paustian, K., Six, J., Elliott, E. T. and Hunt, H. W. (2000) 'Management options for reducing CO_2 emissions from agricultural soils', *Biogeochemistry*, vol 48, pp147–163

Peck, A. J. and Williamson, D. R. (1987) 'Effects of forest cleaning on groundwater', *Journal of Hydrology*, vol 94, pp47–65

Peksel, A. and Kubicek, C. (2003) 'Effects of sucrose concentration during citric acid accumulation by *Aspergillus niger*', *Turkish Journal of Chemistry*, vol 27, pp581–590

Peng, T. H. and Broecker, W. S. (1991) 'Dynamical limitations on the Antarctic iron fertilization strategy', *Nature*, vol 349, pp227–229

Pereira, H. (1992) 'The raw material quality of *Eucalyptus globulus*', in J. S. Pereira and H. Pereira (eds) *Eucalyptus for Biomass Production*, Commission of the European Communities.

Perez-Garcia, J., Joyce, L. A., Binkley, C. S. and McGuire, A. D. (1997) 'Economic impacts of climate change on the global forest sector', in R. S. Sedjo (ed) *Economics of Carbon Sequestration in Forestry*, Lewis Publishers, Boca Raton

Pernkopf, J. (1984) 'The commercial and practical aspects of utilizing vegetable oils as diesel fuel substitute', Bio-Energy 84 World Conferences, 18–21 June, Gothenburg, Sweden

Peter, J. F., Rand, M. C. and Ziemke, M. C. (1982) *Investigation of a Soybean Oil as Diesel Fuel Extender*, SAE Paper No 823615

Petersen, K. K., Hagberg, P. and Kristiansen, K. (2002) '*In vitro* chromosome doubling of *Miscanthus sinensis*', *Plant Breeding*, vol 121, no 5, pp445–450

Peterson, C. L. (1985) 'Vegetable oil as a diesel fuel: Status and research priorities', American Society of Agricultural Engineers Summer Meeting, Michigan State University, East Lansing, 23–26 June, Paper No 85-3069

Petrikovi, V., Vina, J. and Ustjak, S. (1996) *Growing and Using of Technical and Energy Crops on Reclaimed Soils*, Metodika UZPI, Praha, No 17

Petrini, C. (1994) 'Response of fibre and sweet sorghum varieties to plant density', in *8th EC Biomass Conference for Industry Energy and Environment*, Vienna, Austria, 9–11 October

Petrini, C. and Belletti, A. (1991) 'Kenaf adaptability and productive potentialities in the north centre of Italy', in *Proceedings, Biomass for Energy, Industry and Environment, 6th EC Conference*, Athens, pp292–296

Phatak, S. C., Nadimpalli, R. G., Tiwari, S. C. and Bhardwaj, H. L. (1999) 'Pigeonpeas: Potential new crop for the southeastern United States', in J. Janick and J. E. Simon (eds) *New Crops*, Wiley, New York pp597–599

Philippine Information Agency (2006) 'Cong. Suarez proposes power plant fueled by biomass', PIA Press Release 2006/05/16, www.pia.gov.ph/?m=12&sec=reader&rp=3&fi=p060516.htm&no=26&date=05/16/2006

Philliphis, A. (1956) 'Protecion des los cultivos y defensa del suelo', Primeira Conferencia Mundial do Eucalypto, Food and Agriculture Organization, Rome

Phillips, S. J. and Wentworth Comus, P. (eds) (2000) *A Natural History of the Sonoran Desert*, University of California Press, pp256–257

Pichard, A. (2003) 'Ether de methyle et de butyle tertiaire', INERIS, www.ineris.fr/chimie/fr/LesPDF/MetodExpChron/mtbe.pdf (accessed September 2009)

Pickett, S. T. A. and White, P. S. (1985) *The Ecology of Natural Disturbance and Patch Dynamics*, Academic Press Inc., San Diego

Piedade, M. T. F., Junk, W. J. and Long, S. P. (1991) 'The productivity of the C4 grass *Echinochloa polystachya* on the Amazon Floodplain', *Ecology*, vol 72, no 4, pp1456–1463

Piedade M. T. F., Long, S. P. and Junk, W. J. (1994) 'Leaf and canopy photosynthetic CO_2 uptake of a stand of Echinocloa polystachya on the Central Amazon floodplain', *Oecologia*, vol 97, no 2, pp 193–201

Pielke, R. A. and Avissar, R. (1990) 'Influence of landscape structure on local and regional climate', *Landscape Ecology*, vol 4, pp133–155

Plantinga, A. J., Mauldin, T. and Miller, D. J. (1999) 'An econometric analysis of the costs of sequestering carbon in forests', *American Journal of Agricultural Economics*, vol 81, no 4, pp812–824

Plarre, W. (1989) '*Lupinus* spp.', in S. Rehm (ed) *Spezieller Pflanzenbau in den Tropen und Subtropen. Handbuch der Landwirtschaft und Ernährung in den Entwicklungs-ländern*, 2nd edn, Ulmer, Stuttgart, vol 4

Pignatti, S. (1982) *Flora d'Italia*, Edagricole, Bologna

Pohjonen, V. (1980) 'Energiaviljely sitoo auringon energgiaa', *Tyotehoseuran Metsatiedotus*, vol 3

Poitrat, E. (1995) 'Energy balance of bioethanol and rapeseed methyl ester', in *Biomass for Energy, Environment, Agriculture and Industry, Proceedings of the 8th EC Conference*, Vienna, Pergamon, pp1156–1158

Poore, D., Burges, P., Palmer, J., Rietbergen, S. and Synnott, T. (1989) *No Timber Without Trees: Sustainability in the Tropical Forest*, Earthscan Publications, London

Pope, P. E. and Anderson, C. P. (1982) 'Biomass yields and nutrient removal in short rotation black locust plantations', in

R. N. Müller (ed) *Proceedings of the Fourth Central Hardwood Forest Conference*, University of Kentucky, Lexington, pp244–259

Popular Mechanics (1941) 'Pinch hitters for defense', *Popular Mechanics*, December, vol 76, no 6

Post, W. M. and Kwon, K. C. (2000) 'Soil carbon sequestration and land-use change: Processes and potential', *Global Change Biology*, vol 6, pp317–327

PostCarbon Institute (2008) 'Jerusalem artichoke', Sebastopol, CA

Postel, S. and Heise, L. (1988) *Reforesting the Earth*, Worldwatch Institute, Washington, DC

Potter, L., Bingham, M. J., Baker, M. G. and Long, S. P. (1995) 'The potential of two perennial C_4 grasses and a perennial C_4 sedge as lignocellulosic fuel crops in NW Europe: Crop establishment and yields in E England,' *Annals of Botany*, vol 76, pp513–520

Potter, K. N., Torbert, H. A., Johnson, H. B. and Tischler, C. R. (1999) 'Carbon storage after long-term grass establishment on degraded soils', *Soil Science*, vol 164, pp718–725

Prabu, M. J. (2007) 'Biofuel: Meeting the needs of dryland farmers', http://bioenergy.checkbiotech.org/news/biofuel_meeting_needs_dryland_farmers (accessed September 2009)

Prentice, I. C., Farquhar, G. D., Fasham, M. J. R., Goulden, M. L., Heimann, M., Jaramillo, V. J., Kheshgi, H. S., Le Quéré, C., Scholes, R. J. and Wallace, D. W. R. (2001) 'The carbon cycle and atmospheric CO_2', in *Climate Change 2001: The Scientific Basis*, Contribution of Working Group I to the IPCC Third Assessment Report, Cambridge University Press, Cambridge, UK

Price, D. T., Apps, M. J. and Kurz, W. A. (1998) 'Past and possible future dynamics of Canada's boreal forest ecosystems', in G. H. Kohlmaier, M. Weber and R. A. Houghton (eds) *Carbon Dioxide Mitigation in Forestry and Wood Industry*, Springer-Verlag, Berlin, Germany, pp63–88

Prine, G. M. and McConnel, W. V. (1996) 'Growing tall-grass energy crops on sewage effluent spray field at Tallahassee, FL', in *Bioenergy '96, Proceedings of the Seventh National Bioenergy Conference*, 15–20 September, Nashville, Tennessee

Prine, G. M., Stricker, J. A. and McConnel, W. V. (1997) 'Opportunities for bioenergy development in the lower south USA', in *Making Business From Biomass, Proceedings of the Third Biomass Conference of the Americas*, Montreal, Quebec, Canada, 24–29 August, pp227–235

Prutpongse, P. and Gavinlertvatana, P. (1992) '*In vitro* micropropagation of 54 species from 15 genera of bamboo', *HortScience*, vol 27, no 5, pp453–454

Purseglove, J. W. (1968) *Tropical Crops: Dicotyledons*, Longmans, Green, London, vol 1

Quinby, J. R., Kramer, N. W., Stephens, J. C., Lahr, K. A. and Karper, R. E. (1958) 'Grain sorghum production in Texas', *Texas Agricultural Experiment Station Bulletin*, no 912

Rabenhorst, M. C. (1995) 'Carbon storage in tidal marsh soils', in R. Lal, J. Kimble, E. Levine and B. A. Stewart (eds) *Soils and Global Change*, CRC Lewis Publishers, Boca Raton, pp93–103

Rachie, K. O. and Roberts, L. M. (1974) 'Grain legumes of the lowland tropics', in N. C. Brady (ed) *Advances in Agronomy*, Academic Press, New York, pp1–132

Rademakers, L. (2006) 'Turning pest into profit: Bioenergy from water hyacinth', Biopact, 18 June, http://news.mongabay.com/bioenergy/2006/06/turning-pest-into-profit-bioenergy.html (accessed September 2009)

Radlein, D. and Piskorz, J. (1997) 'Production of chemicals from bio-oil', in M. Kaltschmitt and A. V. Bridgwater (eds) *Proceedings of the International Conference 'Gasification and Pyrolysis of Biomass'*, Stuttgart, Germany, 9–11 April 1997, CPL Press, Newbury, UK, 471–480

Ragland, K. W., Aerts, D. J. and Weiss, C. (1996) 'Co-firing switchgrass in a 50MW pulverized coal burner', in *Bioenergy 96, Proceedings of the Seventh National Bioenergy Conference*, 15–20 September, Nashville, Tennessee, vol 1, pp113–120

Rajoka, M. I., Tabassum, R., Ahmad, M., Latif, F., Parvez, S. and Malik, K. A. (1996) 'Production of biofuels from renewable biomass', in *Proceedings of the Asia-Pacific Solar Experts' Meeting*, Islamabad, Pakistan, 18–21 December 1995, pp44–67

Rahmani, M., Hodges, A. W., Stricker, J. A. and Kiker, C. F. (1997) 'Economic analysis of biomass crop production in Florida', in *Making Business from Biomass, Proceedings of the Third Biomass Conference of the Americas*, Montreal, Quebec, Canada, 24–29 August, pp91–99

Raghu, S., Wiltshire, C. and Dhileepan, K. (2005) 'Intensity of pre-dispersal seed predation in the invasive legume *Leucaena leucocephala* is limited by the duration of pod retention', *Austral Ecology*, vol 30, no 3, pp310–318

Ramakrishna, A., Wani, S. P., Rao, C. S. and Reddy, U. S. (2005) 'Effect of improved crop production technology on pigeonpea yield in resource poor rainfed areas', in *Global Theme – Agroecosystems*, International Crops Research Institute for the Semi-Arid Tropics (ICRISAT), Patancheru, India, vol 1, no 1

Ramesh, K. and Padhya, A. (1990) '*In vitro* propagation of neem, *Azadirachta indica* (A. Juss), from leaf discs', *Indian Journal of Experimental Biology*, vol 28, no 10, pp932–935

Randall, J. M. and Marinelli, J. (1996) *Invasive Plants: Weeds of the Global Garden*, Brooklyn Botanic Garden Club, Handbook No 149

Rangaprasad, R. (2003) 'Wood plastic composite: An overview', July

Rao, J. V. (1974) 'Studies on fertiliser management of wheat in maize-wheat and arhar-wheat cropping systems', PhD thesis, Indian Agricultural Research Institute, New Delhi, India

Rao, I. V., Ramanuja, I. V. and Narang, V. (1992) 'Rapid propagation of bamboos through tissue culture', in F. W. G. Baker (ed) *Rapid Propagation of Fast-Growing Woody Species*, CAB International for CASAFA, pp57–70

Ranwell, D. S. (1972) *Ecology of Salt Marshes and Sand Dunes*, Chapman and Hall, London

Rasmussen, P. E. and Albrecht, S. L. (1997) 'Crop management effects on organic carbon in semi-arid Pacific Northwest soils', in R. Lal, J. M. Kimble, R. F. Follett and B. A. Stewart (eds) *Management of Carbon Sequestration in Soil*, CRC Press, Boca Raton, pp209–219

Raun, W. R., Johnson, G. V., Phillips, S. B. and Westerman, R. L. (1998) 'Effect of long-term N fertilization on soil organic C and total N in continuous wheat under conventional tillage in Oklahoma', *Soil and Tillage Research*, vol 47, pp323–330

Raveendran T. S. (no date) 'Sweet sorghum', Centre for Plant Breeding & Genetics, Tamil Nadu University, India

Ravindranath, N. H. and Hall, D. O. (1994) 'Indian forest conservation and tropical deforestation', *Ambio*, vol 23, no 8, pp521–523

Ravindranath, N. H. and Hall, D. O. (1995) *Biomass, Energy and Environment: A Developing Country Perspective from India*, Oxford University Press, Oxford, UK

Ravindranath, N. H. and Somashekhar, B. S. (1995) 'Potential and economics of forestry options for carbon sequestration in India', *Biomass and Bioenergy*, vol 8, pp323–336

Reddy, B. V. S., Ramesh, S., Kumar, A. A., Wani, S. P., Ortiz, R., Ceballos, H. and Sreedevi, T. K. (2008) 'Bio-fuel crops research for energy security and rural development in developing countries', *Bioenergy Research*, vol 1, nos 3–4, pp248–258

Reeder, J. D., Schuman, G. E. and Bowman, R. A. (1998) 'Soil C and N changes on conservation reserve program lands in the Central Great Plains', *Soil and Tillage Research*, vol 47, pp339–349

Rehm, S. and Espig, G. (1991) *The Cultivated Plants of the Tropics and Subtropics*, Verlag Joseph Margraf, Priese GmbH, Berlin

Reinsvold, R. J. and Pope, P. E. (1987) 'Combined effect of soil nitrogen and phosphorus on nudulation and growth *of Robinia pseudoacacia*', *Canadian Journal of Forest Research*, vol 17, no 8, pp964–969

Reis, M. G., Kimmins, J. P., Rezende, G. C. and Barros, N. E. (1995) 'Acumulo de biomassa em una sequencia de idade de *E. grandis* plantado no cerrado em duas areas com deiferents produtividades', *Rivista Aruore*, vol 9, pp149–162

REN21 (2007) 'Renewables 2007 Global Status Report', www.ren21.net

Reuters (2008) 'Global energy finalizes first stage of Ethiopian castor farming project', www.reuters.com/article/pressRelease/idUS123302+28-

Reynolds, M., Skovmand, B., Trethowan, R., Singh, R. and van Ginkel, M. (published 2001, accepted for Internet publication 2004) *Applications of Physiology to Wheat Breeding, Wheat Program*, CIMMYT, www.cimmyt.cgiar.org/

Ricardo, R. P. and Madeira, M. A. V. (1985) *Relacoes Solo-Eucalipto*, Universidade Tecnica de Lisboa

Roberts, L. (1999) *World Resources: 1998–99*, World Resources Institute, United Nations Environment Programme, United Nations Development Programme and the World Bank, Oxford University Press, New York

Rockwood, D. L., Comer., C. D., Dippon, D. R. and Huffrnan, J. B. (1985) 'Wood biomass production options for Florida', *Florida Agricultural Experiment Station Bulletin*, no 856

Rocha, L. B. (2005) 'The future of fuel ethanol in Thailand and India and its impact on the world sugar market', F. O. Licht, 10 August

Römpp, H. (1974) *Römpps Chemie Lexikon*, 7th edn, Franckh'sche Verlagshandlung, Stuttgart.

Rosa, M. F., Bartolomeu, M. L., Novias, J. M., Sá-Correia, I., Barradas, M. C., Romano, M. C., Antunes, M. P. and Sampaio, T. M. (1991) 'The Portuguese experience on the direct ethanolic fermentation of Jerusalem artichoke tubers', in *Biomass for Energy, Industry and Environment, 6th EC Conference*, Athens, Elsevier Applied Science, pp546–555

Rosenzweig, C. and Hillel, D. (2000) 'Soils and global climate change: Challenges and opportunities', *Soil Science*, vol 165, pp47–56

Rossillo-Calle, F. (1987) 'Brazil: A biomass society', in D. O. Hall and R. P. Overend (eds) *Biomass Regenerable Energy*, John Wiley and Sons, New York

Rottmann-Meyer, S., Rose, H. and Martens, R. (eds) (1995) *Nachwachsende Rohstoffe – Möglichkeiten und Chancen für den Industrie- und Energiepflanzenanbau*, 2nd edn, Landwirtschaftskammer, Hannover

Rousseau, S. and Loiseau, P. (1982) 'Structure et cycle de developpement des peulements a *Cytisus scoparius* dans la claine des domes', *Acta Oecologie Applica*, vol V, no 2, pp121–125

Row, C. (1996) 'Effects of selected forest management options on carbon storage', in N. Sampson and D. Hair (eds) *Forests and Global Change: Vol 2. Forest Management Opportunities for Mitigating Carbon Emissions, American Forests*, Washington, DC, pp27–58

Ruiz-Altisent, M. (ed) (1994) *Biofuels: Application of Biologically Derived Products as Fuels or Additives in Combustion Engines*, Directorate-General XII, Science, Research and Development, EUR 15647 EN

Rutherford, L. and Heath, M. C. (eds) (1992) *The Potential of Miscanthus as a Fuel Crop*, Energy Technology Support Unit (ETSU) B1354, Harwell, UK

Ruwisch, V. and Sauer, B. (2007) 'Bioenergy Village Jühnde: Experiences in rural self-sufficiency', IZNE, www.bioenergiedorf.de

Saeidy, E. El (2004) 'Technological fundamentals of briquetting cotton stalks as a biofuel', dissertation, Landwirtschaftlich-Gärtnerische Fakultät, Humboldt-Univ., Berlin

Sage, R. F., Coiner, H. A., Way, D. A., Brett Runion, G., Prior, S. A., Allen Torbert, H., Sicher, R. and Ziska, L. (2009) 'Kudzu [*Pueraria montana* (Lour.) Merr. Variety lobata]: A new source of carbohydrate for bioethanol production', *Biomass and Bioenergy*, vol 33, no 1, pp57-61

Saleh, N. and Hartojo, K. (2000) 'Present status and future research in sweetpotato in Indonesia', www.papuaweb.org/dlib/tema/ubi/psp-2003-saleh-hartojo.pdf

Salter, A. (2006) 'Selection of energy crops: Agroeconomic and environmental considerations', Cropgen dissemination BOKU – IFA – Tulln, 6 February, www.cropgen.soton.ac.uk/publications/5%20Vienna%20mini-symposium/VMS_02_Salter.pdf (accessed September 2009)

Samson, R. and Chen, Y. (1995) 'Short-rotation forestry and water problem', in *Proceedings, Canadian Energy Plantation Workshop*, pp43–49

Samson, R. A. and Omielan, J. A. (1994) 'Switchgrass: A potential biomass energy crop for ethanol production', in *Proceedings of the Thirteenth North American Prairie Conference*, Canada, pp253–258

Sampson, R. N., Apps, M., Brown, S., Cole, C. V., Downing, J., Heath, L. S., Ojima, D. S., Smith, T. M., Solomon, A. M. and Wisniewski, J. (1993) Workshop summary statement – Terrestrial biospheric carbon fluxes: Quantification of sinks and sources of CO_2', *Water, Air, and Soil Pollution*, vol 70, nos 1–4, pp3–15

Sampson, R. N., Scholes, R. J., Cerri, C., Erda, L., Hall, D. O., Handa, M., Hill, P., Howden, M., Janzen, H., Kimble, J., Lal, R., Marland, G., Minami, K., Paustian, K., Read, P., Sanchez, P. A., Scoppa, C., Solberg, B., Trossero, M. A., Trumbore, S., Van Cleemput, O., Whitmore, A. and Xu, D. (2000) 'Article 3.4: Additional human-induced activities', in R. T. Watson, I. R. Noble, B. Bolin, N. H. Ravindranath, D. J. Verardo and D. J. Dokken (eds) *Land Use, Land-use Change, and Forestry*, A Special Report of the Intergovernmental Panel on Climate Change, Cambridge University Press, Cambridge, UK

Sanderson, M. A. and Wolf, D. D. (1995) 'Switchgrass biomass composition during morphological development in diverse environments', *Crop Science*, vol 35, pp1432–1438

Sanderson, M. A., Reed, R. L., McLaughlin, S. B., Wullschleger, S. D., Conger, B. V, Parrish, D. J., Wolf, D. D., Taliaferro, C., Hopkins, A. A., Ocumpaugh, W. R., Hussey, M. A., Read, J. R. and Tischler, C. R. (1996) 'Switchgrass as a sustainable bioenergy crop', *Bioresource Technology*, vol 56, pp83–93

Sandhu, G. R. and Malik, K. A. (1975) 'Plant succession: A key to the utilization of saline soils', *Nucleus* (Karachi), vol 12, pp35–38

Sathaye, J. and Ravindranath, N. H. (1998) 'Climate change mitigation in the energy and forestry sectors of developing countries', *Annual Review of Energy and Environment*, vol 23, pp387–437

Sator, C. (1979) 'Lupinenanbau zur Kornerproduktion unter besonderer Berücksichtigung der Gelben Lupine *(Lupinus luteus* L.)', *Garten Organisch*, vol 3, no 79, pp74–77

Sauti, R. F. N., Saka, J. O. K. and Kumsiya, E. G. (1987) 'Preliminary communication research note on composition and nutritional value of cassava maize composite flour', in *The 3rd East and Southern Africa Root Crops Workshop*, 7–11 December 1987, Mzuzu, Malawi, p6

Saynor, B., Bauen, A. and Leach, M. (2003) 'The potential for renewable energy sources in aviation', Imperial College Centre for Technology and Policy, www3.imperial.ac.uk/pls/portallive/docs/1/7294712.PDF (accessed September 2009)

Scharpenseel, H. W. (1993) 'Major carbon reservoirs of the pedosphere; source-sink relations; potential of d14C and d13C as supporting methodologies', *Water, Air, and Soil Pollution*, vol 70, pp431–442

Scharpenseel, H. W. and Becker-Heidmann, P. (1994) 'Sustainable land use in the light of resilience/elasticity to soil organic matter fluctuations', in D. J. Greenland and I. Szabolcs (eds) *Soil Resilience and Sustainable Land Use*, CAB International, Wallingford, pp249–264

Scheer, H. (1993) *Sonnenstrategie – Politik ohne Alternative*, Piper Verlag, München

Scheffer, K. (1998) 'Use of wet biomass for thermal power generation, oil and fiber production', in D. E. Leihner and T. A. Mitschein (eds) *A Third Millenium for Humanity? The Search for Paths of Sustainable Development, Proceedings of the Conference 'Forum Belém I'*, 26–29 November 1996, Belém do Pará, Brazil, Europäischer Verlag der Wissenschaften, Frankfurt/Main

Schelling, T. C. (1996) 'The economic diplomacy of geoengineering', *Climatic Change*, vol 33, pp303–307

Scheurer, K. and Baier, U. (2001) *Biogene Güter in der Schweiz: Massen- und Energieflüsse*, Schlussbericht, Im Auftrag des Bundesamtes für Energie BFE, Hochschule Wädenswil, Wädenswil

Schimel, D. S. (1995) 'Terrestrial ecosystems and the carbon cycle', *Global Change Biology*, vol 1, pp77–91

Schimel, D., Melillo, J., Tian, H., McGuire, A. D., Kicklighter, D., Kittel, T., Rosenbloom, N., Running, S., Thornton, P., Ojima, D., Parton, W., Kelly, R., Sykes, M., Neilson, R. and Rizzo, B. (2000) 'Contribution of increasing CO_2 and climate to carbon storage by ecosystems in the United States', *Science*, vol 287, pp2004–2006

Schindler, D. W. and Bayley, S. E. (1993) 'The biosphere as an increasing sink for atmospheric carbon: Estimates from increased nitrogen deposition', *Global Biogeochemical Cycles*, vol 7, pp717–733

Schittenhelm, S. (1987) 'Topinambur – eine Pflanze mit Zukunft', *Lohnunternehmer fahrbuch*, pp169–174

Schittenhelm, S. (1999) 'Agronomic performance of root chicory, Jerusalem artichoke, and sugar beet in stress and nonstress environments', *Crop Science*, vol 39, pp1815–1823

Schlamadinger, B. and Marland, G. (1996) 'The role of forest and bioenergy strategies in the global carbon cycle', *Biomass and Bioenergy*, vol 10, pp275–300

Schlamadinger, B., Faaij, A., Junginger, M., Daugherty, E. and Schlesinger, W. H. (1999) 'Carbon sequestration in soils', *Science*, vol 284, p2095

Schmeil, O. and Fitschen, J. (1982) *Flora von Deutschland undseinen angrenzenden Gebieten*, 87, Auflage, Quelle & Mayer, Heidelberg

Schmitz, N. (2005) 'Innovations in the production of bioethanol and their implications for energy and greenhouse gas balances', F. O. Licht, 12 August

Schnepf, R. (2006) *Agriculture Based Renewable Energy Production*, CRS Report for Congress, 4 August

Scholes, R. J. and van Breemen, N. (1997) 'The effects of global change on tropical ecosystems', *Geoderma*, vol 79, pp9–24

Scholes, R. J., Schulze, E. D., Pitelka, L. F. and Hall, D. O. (1999) 'Biogeochemistry of terrestrial ecosystems', in B. Walker, W. Steffen, J. Canadell and J. Ingram (eds) *The Terrestrial Biosphere and Global Change: Implications for Natural and Managed Ecosystems*, Cambridge University Press, Cambridge, pp271–303

Schön, H. and Strehler, A. (1992) 'Wirtschaftlich sinnvolle Erzeugung von Biomasse – Kostentrends, Umweltentlastung', in *Nachwachsende Rohstoffe*, Proceedings of the International CLAAS Symposium, Harsewinkel, pp1–16

Schrottmaier, J., Pernkopf, J. and Wöirgetter, M. (1988) 'Plant oil as fuel: A preliminary evaluation', *Proceedings, OKL Colloqium*, pp41–48

Schuhmacher, K.-D. (1991) 'Analyse des Weltmarktes für Ölsaaten, Ölfrüchte und pflanzliche Öle', dissertation, Institut für Agrarpolitik und Marktforschung der Justus Liebig Universität, Gießen

Schulze, E.-D., Lloyd J., Kelliher, F. M., Wirth, C., Rebmann, C., Luhker, B., Mund, M., Knohl, A., Milyukova, I. M., Schulze, W., Ziegler, W., Varlagin, A. B., Sogachev, A. F., Valentini, R., Dore S., Grigoriev, S., Kolle, O., Panfyorov, M. I., Tchebakova, N., Vygodskaya, N. N. (1999) 'Productivity of forests in the Eurosiberian boreal region and their potential to act as a carbon sink: A synthesis', *Global Change Biology*, vol 5, no 6, pp703–722

Schulze, E. D. (2000) *Carbon and Nitrogen Cycling in European Forest Ecosystems*, Ecological Studies Vol 142, Springer, Heidelberg, Germany

Schuster, W. H. (1992) *Olpflanzen in Europa*, DLG Verlag, Frankfurt am Main, Germany

Schuster, W. H. (1989) 'Saflor', in S. Rehm (ed) *Spezieller Pflanzenbau in den Tropen und Subtropen. Handbuch der Landwirtschaft und Ernährung in den Entwicklungsländern*, 2nd edn, Ulmer, Stuttgart, vol 4

Schuster, W. H. (1989) 'Sojabohne', in S. Rehm (ed) *Spezieller Pflanzenbau in den Tropen und Subtropen. Handbuch der*

Landwirtschaft und Ernährung in den Entwicklungsländern, 2nd edn, Ulmer, Stuttgart, vol 4

Schütt, P., Schuck, H. J., Aas, G. and Lang, U. M. (eds) (1994) *Enzyklopädie der Holzgewächse, Handbuch und Atlas der Dendrologie*, ecomed Verlagsgesellschaft Landsberg am Lech

Schwarz, K. U., Murphy, D. P. L. and Schnug, E. (1994) 'Studies on the growth and yield of *Miscanthus* x *giganteus* in Germany', *Aspects of Applied Biology*, vol 40, pp533–540

Schwartz, L. (2008) 'China fuels ethanol industry with yams, sweet potatoes and cassava', 16 May, www.renewableenergyworld.com/rea/news/story?id=52450 (accessed September 2009)

Scott, R., Callaghan, T. U. and Jackson, G. J. (1990) '*Spartina* as a biofuel', in A. J. Gray and P. E. M. Benham (eds) Spartina anglica: *A Research Review*, ITE publication No 2, NERC, Swindon, UK

Scurlock J. (2005) *Bioenergy Feedstock Characteristics*, Oak Ridge National Laboratory, Bioenergy Feedstock Development Programs

Sedjo, R. A. (1983) *The Comparative Economics of Plantation Forestry*, Johns Hopkins Press, Resources for the Future, Baltimore, MD

Sedjo, R. A. (1992) 'Forest ecosystem in the global carbon cycle', *Ambio*, vol 21, no 4, pp274–277

Sedjo, R. A. (1997) 'The economics of forest-based biomass supply', *Energy Policy*, vol 25, no 6, pp559–566

Sedjo, R. A. (1999a) 'Land use change and innovation in US forestry', in D. Simpson (ed) *Productivity in Natural Resources Industries: Improvement through Innovation*, Resources for the Future, Washington, DC, pp141–174

Sedjo, R. A. (1999b) *Potential for Carbon Forest Plantations in Marginal Timber Forests: The Case of Patagonia, Argentina*, RFF Discussion Paper 99-27, Resources for the Future, Washington, DC

Sedjo, R. A. and Botkin, D. (1997) 'Using forest plantations to spare natural forests', *Environment*, vol 39, pp14–20

Sedjo, R. A. and Lyon, K. S. (1990) *The Long-Term Adequacy of World Timber Supply*, Resources for the Future, Washington, DC

Sedjo, R. A. and Sohngen, B. (2000) *Forestry Sequestration of CO₂ and Markets for Timber*, RFF Discussion Paper 00-35, Resources for the Future, Washington, DC

Sedjo, R. A. and Solomon, A. M. (1989) 'Climate and forests', in N. J. Rosenberg, W. E. Easterling, P. R. Crosson and J. Darmstadter (eds) *Greenhouse Warming: Abatement and Adaptation*, Resources for the Future, Washington, DC, pp105–120

Sedjo, R. A., Wisniewski, J., Sample, A. V. and Kinsman, J. D. (1995) 'The economics of managing carbon via forestry: Assessment of existing studies', *Environmental and Resource Economics*, vol 6, no 2, pp139–165

Selim, M. Y. E., Radwan, M. S. and Elfeky, S. M. S. (2003) 'Combustion of jojoba methyl ester in an indirect injection diesel engine', *Renewable Energy*, vol 28, no 9, pp1401–1420

Shah, P. (1997) 'Biomass for industry: Primary productivity of pigeonpea as affected by plant population and phosphorus', in *Abstract International Conference on Sustainable Agriculture for Food, Energy and Industry*, Braunschweig, 22–28 June 1997, p288

Shake, A. (2008), 'Osage BioEnergy to open largest barley ethanol plant in US', http://gas2.org/2008/09/26/osage-bioenergy-to-open-largest-ethanol-plant-in-us/

Shakir, S. (1996) 'Flash pyrolysis: A technology to produce biofuels from miscanthus', *REAIIEC – Technische Mitteilung*, pp2–14

Shantz, H. L. and Piemeisel, L. N. (1927) 'The water requirements of plants at Akron, Colo', *Journal of Agricultural Research*, vol 12, pp1093–1191

Shauck, M. and Zanin, G. (2001) *The Present and Future Potential of Biomass Fuels in Aviation*, Renewable Aviation Fuels Development Center at Baylor University

Sheaffer, C. C., Tanner, C. B. and Kirkham, M. B. (1988) 'Alfalfa water relations and irrigation', in A. A. Hanson, D. K. Barnes and Hill, R. R. Jr. (eds) *Alfalfa and Alfalfa Improvement*, American Society of Agronomy, Monograph 29, Madison, Wisconsin, pp373–409

Sheaffer, C. C. Marten, G. C., Rabas, D. L., Martin, N. P. and Miller, D. W. (1990) *Reed Canarygrass*, Station Bulletin 595, Minnesota Agricultural Experiment Station

Sheinbaum, C. and Masera, O. R. (2000) 'Mitigating carbon emissions while advancing national development priorities: The case of Mexico', *Climatic Change*, vol 47, no 3, pp259–282

Shen, S. Y., Yvas, A. D. and Jones, P. C. (1984) 'Economic analysis of short rotation and ultra-short rotation forestry', *Resources and Conservation*, vol 10, pp255–270

Shepashenko, D., Shvidenko, A. and Nilsson, S. (1998) 'Phytomass (live biomass) and carbon of Siberian forests', *Biomass and Bioenergy*, vol 14, no 1, pp21–31

Siebert, K. (1975) *Kriterien der Futterpflanzen einschließlich Rasengraser und ihre Bewertung zur Sortenidentifizierung*, Bundesverband Deutscher Pflanzenzüchter e.V. Bonn

Sikander, A., Ikram-ul-Haq, Qadeer, M. A. and Iqbal, J. (2002) 'Production of citric acid by *Aspergillus niger* using cane molasses in a stirred fermentor', *Electronic Journal of Biotechnology*, vol 5, no 3, www.ejbiotechnology.cl/content/vol5/issue3/full/3/index.html (accessed September 2009)

Silva Dias, J. C. (1997) 'Rocket in Portugal: Botany, cultivation, uses and potential', in Padulosi, S. and Pignone, D. (eds) *Rocket: A Mediterranean Crop for the World: Report of a Workshop*, 13–14 December 1996, Legnaro (Padova), Italy, International Plant Genetic Resources Institute, Rome, Italy, pp81–85

Silou, T. (1996) 'Le safoutier (*Dacryodes edulis*): un arbre mal connu', *Fruits*, vol 51, pp47–60

Silou, T. and Avouampo, E. (1997) 'Perspectives de production de l'huile de safou', in *Bulletin Africain bioressources et energies pour le développement et l'environnement*, Enda tm Dakar no 8

Silou, T., Kiakouama, S., Bezard, J. and Sempore, G. (1991) 'Note sur la composition en acides gras et en triglycerides de l'huile de safou en relation avec la solidification partielle de cette huile', *Fruits*, vol 46, no 3, pp26–27

Silou, T., Kiakouama, S. and Koucka-Gokana, B. (1994) 'Evaluation de la production et etude de la variabilité morphologique et physico-chimique du safou, *Dacryodes edulis*. Mise au point méthodologique et résultats préliminaires', in J. Nya Ngatcchou and J. Kengué (eds) *Séminaire sur la valorisation du safoutier*, Douala, Cameroun, 4–6 October, pp30–44

Singh, R. P. and Das, S. K. (1985) 'Management of chickpea and pigeonpea under stress conditions, with particular reference to drought', in *Proceedings of the Consultant Workshop on Adaptation of Chickpea and Pigeonpea to Abiotic Stresses*, 19–21 December 1984, pp51–61

Singh, B. R., Borresen, T., Uhlen, G. and Ekeberg, E. (1997a) 'Long-term effects of crop rotation, cultivation practices, and fertilizers on carbon sequestration in soils in Norway', in R. Lal, J. M. Kimble, R. F. Follett and B. A. Stewart (eds) *Management of Carbon Sequestration in Soil*, CRC Press, Boca Raton, pp195–208

Singh, B., Singh Y., Maskina M. S. and Meelu, O. P. (1997b) 'The value of poultry manure for wetland rice grown in rotation with wheat', *Nutrient Cycling in Agroecosystems*, vol 47, pp243–250

Sipilä, K. (1995) 'Research into thermochemical conversion of biomass into fuels, chemicals and fibres', in *Biomass for Energy, Environment, Agriculture and Industry, Proceedings of the 8th EC Conference*, Vienna, Pergamon, pp156–167

Skerman, P. J. and Riveros, F. (1990) *Tropical Grasses*, FAO, Rome

Skoulou, V., Zabaniotou, A., Stavropoulos, G. and Sakelaropoulos, G. (2007) *Syngas Production from Olive Tree Cuttings and Olive Kernels in a Downdraft Fixed-Bed Gasifier*, International Association for Hydrogen Energy, Elsevier

Sladden, S. E., Bransby, D. I. and Aiken, G. E. (1991) 'Biomass yield, composition and production costs for eight switchgrass varieties in Alabama', *Biomass and Bioenergy*, vol 1, no 2, pp119–122

Slatyer, R. O. (1955) 'Studies of the water relations of crop plants grown under natural rainfall in northern Australia', *Australian Journal of Agricultural Research*, vol 6, pp365–377

Smart, J. (1994) *The Groundnut Crop: Scientific Basis for Improvement*, Chapman and Hall, London

Smeets, E. and Faaij, A. (2007) 'Bioenergy potentials from forestry to 2050: An assessment of the drivers that determine the potentials', *Climatic Change*, vol 81, pp 353–390

Smeets, E., Faaij, A., Lewandowski, I. and Turkenburg, W. (2007) 'A quickscan of global bioenergy potentials to 2050', *Progress in Energy and Combustion Science*, vol 33, no 1, pp56–106

Smith, P., Powlson, D. S., Glendining, M. J. and Smith, J. U. (1997) 'Opportunities and limitations for C sequestration in European agricultural soils through change in management', in R. Lal, J. M. Kimble, R. F. Follett and B. A. Stewart (eds) *Management of Carbon Sequestration in Soil*, CRC Press, Boca Raton, pp143–152

Sohngen, B. and Mendelsohn, R. (1998) 'Valuing the impact of large-scale ecological change in a market: The effect of climate change on US timber', *American Economic Review*, vol 88, no 4, pp686–710

Sohngen, B. and Sedjo, R. A. (1999) *Carbon Sequestration by Forestry: Effects of Timber Markets*, Report No PH3/10, IEA Greenhouse Gas R&D Programme, UK

Sohngen, B., Mendelsohn, R. and Sedjo, R. A. (1999) 'Forest management, conservation, and global timber markets', *American Journal of Agricultural Economics*, vol 81, pp1–13

Soileau J. M. and Brandford, B. N. (1985) 'Biomass and sugar yield response of sweet sorghum to lime and fertilizer', *Agronomy Journal*, vol 77, pp471–475

Solberg, B. (1997) 'Forest biomass as carbon sink: Economic value and forest management/policy implications', *Critical Reviews in Environmental Science and Technology*, vol 27, ppS323–333

Solberg, B., Brooks, D., Pajuoja, H., Peck, T. J. and Wardle, P. A. (1996) *Long-Term Trends and Prospects in World Supply and Demand for Wood and Implications for Sustainable Forest Management*, EFI Research Report No 6, European Forest Institute/Norwegian Forest Research Institute, a contribution to the CSD Ad Hoc Intergovernmental Panel on Forests

Soldatos, P. and Chatzidaki, M. (1999) 'Economic evaluation of biofuel production in Greece: The case of ethanol', in *Proceedings of AgEnergy 99*, vol 2. pp973–980

Solomon, D., Lehmann, J. and Zech, W. (2000) 'Land use effects on soil organic matter properties of chromic luvisols in semi-arid northern Tanzania: Carbon, nitrogen, lignin and carbohydrates', *Agriculture, Ecosystems and Environment*, vol 78, pp203–213

Sosulski, K., Wang, S., Ingledeww, M., Sosulski, F. W. and Tang, J. (1997) 'Preprocessed barley, rye, and triticale as a feedstock for an integrated fuel ethanol-feedlot plant', *Applied Biochemistry and Biotechnology*, vol 63–65, no 1, pp59–70

Spiecker, H., Mielikainen, K., Kohl, M. and Skovsgaard, J. P. (1996) *Growth Trends in European Forests*, Springer-Verlag, Heidelberg, Germany

Spielmann, M., Kägi, T., Stadler, P. and Tietje, O. (2004) *Life Cycle Inventories of Transport Services*, final report Ecoinvent 2000 No 14, UNS, ETH-Zurich, Swiss Centre for Life Cycle Inventories, Dübendorf, www.ecoinvent.org

Suresh, K., Kiransree, N. and Rao, V. (1999) 'Production of ethanol by raw starch hydrolysis and fermentation of damaged grains of wheat and sorghum', *Biosystems Engineering*, vol 21, pp165–168

Stallard, R. F. (1998) 'Terrestrial sedimentation and the carbon cycle: Coupling weather and erosion to carbon burial', *Global Biogeochemical Cycles*, vol 12, pp231–257

Stampfl, P. F., Clifton-Brown, J. C. and Jones, M. B. (2007) 'European-wide GIS-based modelling system for quantifying the feedstock from *Miscanthus* and the potential contribution to renewable energy targets', *Global Change Biology*, vol 13, no 11, pp2283–2295

Stanghellini, E., Mihail, J. D., Rasmussen, S. L. and Turner, B. C. (1992) '*Macrophomina phaseolina*: A soilborne pathogen of *Salicornia bigelovii* in a marine habitat', *Plant Disease*, vol 76, pp751–752

Stanghellini, E., Werner, F. G., Turner, B. C. and Watson, M. C. (1988) 'Seedling death of *Salicornia* attributed to *Metachroma* larvae', *The Southwest Entomologist*, vol 13, no 4, p305

Stavins, R. (1999) 'The costs of carbon sequestration: A revealed preference approach', *The American Economic Review*, vol 89, pp994–1009

Stockfisch, N., Forstreuter, T. and Ehlers, W. (1999) 'Ploughing effects on soil organic matter after twenty years of conservation tillage in Lower Saxony, Germany', *Soil and Tillage Research*, vol 52, pp91–101

Stover, R. H. and Simmonds, N. W. (1987) *Bananas*, Longman Group/John Wiley

Strezov, V., Evans, T. J. and Hayman, C. (2008) 'Thermal conversion of elephant grass (*Pennisetum purpureum* Schum) to bio-gas, bio-oil and charcoal', *Bioresource Technology*, vol 99, no 17, pp8394–8399

Stringer, J. W. and Carpenter, S. B. (1986) 'Energy yield of black locust biomass fuel', *Forest Science*, vol 32, no 4, pp1049–1057

Stuart, M. D. and Moura-Costa, P. H. (1998) 'Greenhouse gas mitigation: A review of international policies and initiatives', in *Policies that Work for People*, Series No 8, International Institute of Environment and Development, London, UK, pp27–32

Sujatha, M. and Mukta, N. (1996) 'Morphogenesis and plant regeneration from tissue cultures of *Jatropha curcas*', *Plant Cell, Tissue and Organ Culture*, vol 44, no 2, pp125–141

Sunpower Inc. (1997) *Status of Sunpower Technology*, information brochure, April

Suporn, K. (1987) *Using of Palm Oil in Diesel Engine*, Senior Project ME17/1987, Department of Mechanical Engineering, Faculty of Engineering, Prince of Songkla University, Hat Yai

Suszkiw, J. (2006) *Scientists Gear Up to Decode Cassava Genome*, ARS, US Department of Agriculture

Sutor, P., Sturm, M., Hotz, A., Kolb, W. and Kuhn, W. (1991) 'Anbau von *Miscanthus sinensis* "giganteus"', Bayerische Landesanstalt für Bodenkultur und Pflanzenbau, Technical University, Munich, 8/91, ppIII 5–III 10

Swisher, J. N. and Renner, F. P. (1996) 'Carbon offsets from biomass energy projects', in *Bioenergy '96, Proceedings of the Seventh National Bioenergy Conference*, 15–20 September, Nashville, Tennessee

Tabard, P. (1985) 'Une plante energetique a cycle court. La genet: *Cytisus soparius*', in *Proceedings of 3rd EC Conference on 'Energy from Biomass'*, Venice, Italy, 25–29 March 1985, pp283–287

Tabuna, H. (1993) *La commercialisation du Safou a Brazzaville*, CIRAD-SAR Internal Report No 82/93, Montpellier, France

Takyi, S. K. (1974) 'Effects of nitrogen, planting method and seed bed type on yield of cassava (*Manihot esculenta* Crantz)', *Ghana Journal of Agricultural Science*, vol 7, pp69–73

Tchebokova, N. and Vygodskaya, N. N. (1999) 'Productivity of forests in the Eurosiberian boreal region and their potential to act as a carbon sink: A synthesis', *Global Change Biology*, vol 5, pp703–722

Tchendji, C., Severin, M., Wathelet, J. P. and Deroanne, C. (1987) 'Composition de la graisse de *Dacryodes edulis*', *Rev. Franc. Corps Gras*, vol 28, no 3, p123

Teixeira, M. A. (2005) 'Heat and power demands in babassu palm oil extraction industry in Brazil', *Energy Conversion and Management*, vol 46, pp2068–2074

Teller, E., Wood, L. and Hyde, R. (1997) *Global Warming and Ice Ages: I. Prospects for Physics-Based Modulation of Global Change*, UCRL-JC-128715, Lawrence Livermore National Laboratory, Livermore, California

Thiagarajan, M. and Murali, P. M. (1994) 'Optimum conditions for embryo culture of *Azadirachta indica* (A. Juss)', *Indian Forester*, vol 120, no 6, pp500–503

Thurmond, W. (2008) *Biodiesel 2020: A Global Market Survey, 2nd Edition*, February, www.emerging-markets.com/biodiesel/ (accessed September 2009)

Tian, G., Kolawole, G. O., Salako, F. K. and Kang, B. T. (1999) 'An improved cover crop-fallow system for sustainable management of low activity clay soils of the tropics', *Soil Science*, vol 164, pp671–682

Tiessen, H., Feller, C., Sampaio, E. V. S. B. and Garin, P. (1998) 'Carbon sequestration and turnover in semiarid savannas and dry forest', *Climatic Change*, vol 40, pp105–117

Tiki Manga, T. and Kengue, J. (1994) 'Strategie d'amélioration du safoutier (*Dacryodes edulis*)', in J. Nya Ngatcchou and J. Kengué (eds) *Séminaire sur la valorisation du safoutier*, Douala, Cameroun, 4–6 October, pp12–16

Toblez, F. (1940) '*Arundo donax* (Pfahlrohr) als Zellstoffquelle', *Faserforschung*, vol 15, no 1, pp41–42

Tole, L. (1998) 'Source of deforestation in tropical developing countries', *Environmental Management*, vol 22, pp19–33

Torbert, H. A., Rogers, H. H., Prior, S. A., Schlesinger, W. H. and Runions, G. B. (1997) 'Effects of elevated atmospheric CO_2 in agro-ecosystems on soil carbon storage', *Global Change Biology*, vol 3, pp513–521

Tothill, J. C. and Hacker, J. B. (1973) *The Grasses of Southeast Queensland*, University of Queensland Press, St Lucia

Townsend, A. R. and Rastetter, E. B. (1996) 'Nutrient constraints on carbon storage in forested ecosystems', in M. J. Apps and D. T. Price (eds) *Forest Ecosystems, Forest Management and the Global Carbon Cycle*, NATO ASI Series I (Global Environmental Change), Springer-Verlag Academic publishers, Heidelberg, Germany, vol 40, pp35–46

Tremper, G. (2004) 'The history and promise of jojoba', Armchair World, www.armchair.com/warp/jojoba1.html (accessed September 2009)

Trexler and Associates, Inc. (1998) *Final Report of the Biotic Offsets Assessment Workshop*, Baltimore, USA, Prepared for the US Environmental Protection Agency, Washington, DC

Turnbull, J. W. and Pryor L. D. (1984) 'Choice of species and seed sources', in W. E. Hillis and M. T. Brown (eds) *Eucalypts for Wood Production*, Academic Press, New York

Turner, D. P., Koerper, G. J., Harmon, M. E. and Lee, J. L. (1995) 'A carbon budget for forests of the coterminous United States', *Ecological Applications*, vol 5, pp421–436

Turner, S. M., Nightingale, P. D., Spokes, L. J., Liddicoat, M. I. and Liss, P. S. (1996) 'Increased dimethyl sulphide concentrations in sea water from in situ iron enrichment', *Nature*, vol 383, pp513–517

Turner, D. P., Winjum, J. K., Kolchugina, T. P. and Cairns, M. A. (1997) 'Accounting for biological and anthropogenic factors in national land-based carbon budgets', *Ambio*, vol 26, pp220–226

Turner, A. (2008) 'Green journey', *Flight International*, 17 June, pp39–40

Uma, D. M., Santaiah, V., Rama, R. S., Prasada, R. A. and Singa, R. M. (1990) 'Interaction of conversion and nitrogen fertilization on growth and yield of rainfed castor', *Journal of Oilseeds Research*, vol 7, pp98–105

UNDP (United Nations Development Programme) (1997) *Energy after Rio: Prospects and Challenges*, Executive Summary

UN-ECE/FAO (UN Economic Committee for Europe/Food and Agricultural Organisation) (2000) *Forest resources of Europe, CIS, North America, Australia, Japan and New Zealand*, Geneva Timber and Forest Study Papers No 17, UN-ECE/FAO, Geneva, Switzerland

UN Energy (2005) 'The Energy Challenge for Achieving the Millennium Development Goals', UN Energy Paper, 22 June

UNFCCC (UN Framework Convention on Climate Change) (1997) *The Kyoto Protocol to the United Nations Framework Convention on Climate Change*, Document FCCC/CP/1997/7/ Add.1, http://www.unfccc.de/

Untitled (1994) *Journal of Hunan Agricultural College*, 1994-4, p160

USDA (United States Department of Agriculture) (2004) *USDA Agricultural Baseline Projections to 2013*, USDA, Washington, DC

USDA/Agricultural Research Service (2008) 'Sweet potato out-yields corn in ethanol production study', *ScienceDaily*, 28 August, www.sciencedaily.com/releases/2008/08/080825200752.htm (accessed September 2009)

Ustjak, S. and Ustjakova, M., *Potential for Agricultural Biomass to Produce Bioenergy in the Czech Republic*, Research Institute of Crop Production Prague, Czech Republic, Google books online, http://books.google.com/books?id=DcIk34hjV7wC&pg= PA239&lpg=PA239&dq=SORREL/bioenergy&source=bl&ots= z1JHCqepHm&sig=woy8jLDeLoDfLLRMnKJXo6hjVUI&hl= de&sa=X&oi=book_result&resnum=4&ct=result#PPA230,M1

Ustjak, S., Honzik, R. and Malírová, J. (1996) 'Production of biomass as a source of energy', in *Annual Report*, Research Institute of Crop Production, Prague, p8

Vaidya, S. (2007) 'Tapping green alternative', http://gulfnews.com/news/gulf/oman/tapping-green-alternative-1.185379

Valentini, R., Matteucci, A., Dolman, A. J. and Schulze, E. D. (2000) 'Respiration as the main determinant of carbon balance of European forests', *Nature*, vol 404, pp861–865

van Basshuysen, R., Steinwart, J., Stähle, H. and Bauder, A. (1989) 'Audi Turbodieselmotor mit Direkteinspritzung', *Motortechnische Zeitschrift*, vol 50, no 10, pp58–65

Vance, C. P., Heichel, G. H. and Phillips, D. A. (1988) 'Nodulation and symbiotic dinitrogen fixation', in A. A. Hanson, D. K. Barnes and R. R. Hill Jr. (eds) *Alfalfa and Alfalfa Improvement*, American Society of Agronomy, Monograph 29, Madison, Wisconsin, pp229–257

Van der Gon, H. D. (2000) 'Changes in CH_4 emission from rice fields from 1960 to 1990s: 1. Impacts of modern rice technology', *Global Biogeochemical Cycles*, vol 14, pp61–72

van der Grift, R. M., Eboug, C. and Essers, A. S. A. (1995) 'Solid substrate fermentation and sundrying: Two popular cassava processing techniques in Uganda', *Roots Newsletter* of the Southern African Root Crops Research Network (SARRNET) and the East Africa Root Crops Research Network (EARRNET), vol 2, no l, September, pp17–20

van der Palen, J. (1995) 'Musa basjo', *Groei & Bloei*, June

Van der Werf, H. M. G., Meijer, W. J. M., Mathijssen, W. E. J. M. and Darwinkel, A. (1993) 'Potential dry matter production of *Miscanthus sinensis* in the Netherlands', *Industrial Crops and Products*, vol 1, pp203–210

Van Ginkel, J. H., Gorissen, A. and van Veen, J. A. (1996) 'Long-term decomposition of grass roots as affected by elevated atmospheric carbon dioxide', *Journal of Environmental Quality*, vol 25, pp1122–1128

van Loo, S. and Koppejan, J. (eds) (2002) *Handbook of Biomass Combustion and Co-firing*, IEA Bioenergy Task 32, Twente University Press

Van Noordwijk, M., Cerri, C., Woomer, P. L., Nugroho, K. and Bernoux, M. (1997) 'Soil carbon dynamics in the humid tropical forest zone', *Geoderma*, vol 79, pp187–225

van Soest, L. J. M., Mastebroek, H. D. and de Meijer, E. P. M. (1993) 'Genetic resources and breeding: A necessity for the success of industrial crops', *Industrial Crops and Products*, vol 1, pp283–288

VEB (1986) *Durchwachsene Silphie* (Silphium perforliatum *L.*) *Anbau und Verwendung*, Saat-und Pflanzgut Erfurt, Saatzuchtstation Bendeleben

Vellguth, G. (1991) 'Energetische Nutzung von Rapsöl und Rapsrömethylester, Documentation', *Nachwachsende Rohstoff*, FAL, pp17–21

Venkataratnam, N. and Sheldrake, A. R. (1985) 'Second harvest yields of medium-duration pigeonpea (*Cajanus cajan*) in peninsular India', *Field Crops Research*, vol 10, pp323–332

Venkateswarlu, J., Vittal, K. P. R. and Das, S. K. (1986) 'Evaluation of two land configurations in production of pearl millet and castor beans on shallow red soils of semi-arid tropics', *International Agrophysics*, vol 2, no 2, pp145–152

Verdeil J. L., Buffard-Morel, J., Magnaval, C., Huet, C. and Grosdemagne, F. (1993) 'Vegetative propagation of coconut (*Cocos nucifera* L.): Towards coordinated European research in conjunction with producer countries', in *La recherche europèenne au service du cocotier*, Proceedings of the Seminar held 8–10 September, Montpellier, France

Vetter, A. and Wurl, G. (1994) 'Moglichkeiten der Nutzung von Topinambur und weiteren Pflanzenarten als Festbrennstoff', in *Workshop on Environmental Aspects of Production and Conversion of Biomass for Energy*, REUR Technical Series No 38, pp43–53

Viator, R. P., Garrison, D. D., Dufrene, E. O. Jr., Tew, T. L. and Richard, E. P. Jr. (2005) 'Planting method and timing effects on sugar cane yield', Plant Management Network, www.plantmanagementnetwork.org/pub/cm/research/2005/sugarcane/ (accessed September 2009)

Vickery, M. L. (1976) *Plant Products of Tropical Africa*, Macmillan

Viet Meyer (1990) *Butter Fruit, 'africado'*, draft, US National Academy of Sciences, Washington, DC

Villax, E. J. (1963) *La Culture des Plantes Fourragères dans la Région Méditerranéenne Occidentale*, Inst. Nat. Rech. Agron., Rabat

Vitousek, P. M., Mooney, H. A., Lubchenco, J. and Melillo, J. M. (1997) 'Human domination of Earth's ecosystems', *Science*, vol 277, pp494–499

Vogel, K. P. (1987) 'Seeding rates for establishing big bluestem and switchgrass with pre-emergence Atrazine applications', *Agronomy Journal*, vol 79, pp509–512

Vourdoubas, J. (2005) 'Possibilities of using olive kernel wood for power generation on Crete – Greece', TEI of Crete, Department of Natural Resources and Environment Greece, http://ape.chania.teicrete.gr/gr/files/synedrio3.doc (accessed September 2009)

Vuǎuroviǎ, V. M. and Pejin, D. J. (2007) 'Zbornik Matice srpske za prirodne nauke', *Proc. Nat. Sci., Matica Srpska Novi Sad*, vol 113, pp285–291

Vymazal, J. (1995) 'Constructed wetlands for wastewater treatment in the Czech Republic: State of the art', *Water Science and Technology*, vol 32, no 3, pp357–364

Wachsmann, N. and Jochinke, D. (2004) 'Safflower in Australia', www.safflower.jochinke.com.au/ (accessed September 2009)

Waggoner, P., 1994: *How Much Land Can Ten Billion People Spare for Nature?* Task Force Report No 121, Council for Agricultural Science and Technology, Ames, IA

Walsh D. P. (2008*)* 'Banana waste to produce fuel in Australia', www.nextautos.com/plantsmanufacturing/banana-waste-to-produce-fuel-in-australia

Wan Razali, W. M. and Tay, J. (2000) 'Forestry carbon emission offset and carbon sinks project: Examples of opportunity in carbon sequestration and its implications from Kyoto Protocol', International Workshop on the Response of Tropical Forest Ecosystems to Long Term Cyclic Climate Change, Science and Technology Agency, Japan, National Research Council of Thailand, 24–27 January 2000, Kanchanaburi, Thailand

Wang, X. and Feng, Z. (1995) 'Atmospheric carbon sequestration through agroforestry in China', *Energy*, vol 20, no 2, pp117–121

Wang, M., Saricks, C. and Santini, D. (1999) *Effects of Fuel Ethanol Use on Fuel-Cycle Energy and Greenhouse Gas Emission*, research report, Center for Transportation Research, Energy Systems Division, Argonne National Laboratory, Argonne, IL

Watson, A. J., Law, C. S., Van Scoy, K. A., Millero, F. L., Yao, W., Friedrich, G. E., Liddcoat, M. I., Wanninkhof, R. H., Barber, R. T.

and Coale, K. H. (1994) 'Minimal effect of iron fertilization on sea-surface carbon dioxide concentrations', *Nature*, vol 371, pp143–145

WBGU (Wissenschaftlicher Beirat der Bundesregierung Globale Umweltveränderungen) (1998) 'Die Anrechnung biologischer Quellen und Senken im Kyoto-Protokoll: Fortschritt oder Rückschlag für den globalen Umweltschutz?' Sondergutachten 1998, WBGU, Bremerhaven, Germany (available in English)

Weber, E. (2003) *Invasive Plant Species of the World: A Reference Guide to Environmental Weeds*, CABI Publishing

Weber, L. (1993) 'Yield components of five kenaf cultivars', *Agronomy Journal*, vol 85, pp533–535

Weber, M. G. and Flanningan, M. D. (1997) 'Canadian boreal forest ecosystem structure and function in a changing climate: Impact on fire regimes', *Environmental Reviews*, vol 5, pp145–166

Weber, R. (1987) 'Stirling-motor treibt Warmepumpe an', VDI-*Nachrichtung*, No 19

WEC (World Energy Council) (2007) *Energy and Climate Change Study*, World Energy Council, London

Wedin, D. A. and Tilman, D. (1996) 'Influence of nitrogen loading and species composition on the carbon balance of grasslands', *Science*, vol 274, pp1720–1723

Weiland, P. (1997) Personal communication

Weiland, P. (2003) 'Production and energetic use of biogas from energy crops and wastes in Germany', *Applied Biochemistry and Biotechnology*, vol 109, pp263–274

Weiss, E. A. (1983) *Oilseed Crops*, Longman, London

Wells, M. L. (1994) 'Pumping iron in the Pacific', *Nature*, vol 368, pp295–296

Werner, F., Althaus, H.-J., Künniger, T., Richter, K. and Jungbluth, N. (2003) *Life Cycle Inventories of Wood as Fuel and Construction Material*, final report Ecoinvent 2000 No 9, EMPA Dübendorf, Swiss Centre for Life Cycle Inventories, Dübendorf, www.ecoinvent.org

Wernick, I. K., Waggoner, P. E. and Ausubel, J. H. (1998) Searching for leverage to conserve forests: The industrial ecology of wood products in the United States', *Journal of Industrial Ecology*, vol 1, pp125–145

Whiteman, P. C., Byth, D. E and Wallis, E. S. (1985) 'Pigeonpea (*Cajanuscajan* (L.) Millsp)', in R. J. Summerfield and E. H. Roberts (eds) *Grain Legume Crops*, Collins Professional and Technical Books, London, p658

Wiggins, I. L. (1980) *Flora of Baja California*, Stanford University Press, Stanford, CA

Wiklund, A. (1992) 'The genus *Cynara* L. (Asteraceae-Carduaceae)', *Botanical Journal of the Linnean Society, London*, vol 109, pp75–123

Willey, R. W., Rao, M. R. and Natarajan, M. (1981) 'Traditional cropping systems with pigeonpea and their improvement', in *Proceedings of the International Workshop on Pigeonpeas*, Patancheru, India, 15–19 December 1980, vol 1, pp11–25

Williams, B. C. and Campbell, P. E. (1994) 'Application of fuel cells in "clean" energy systems', *Transactions of the Institution of Chemical Engineers B: Process Safety and Environmental Protection*, vol 72(B), pp252–256

Williams, R. H. (1993) 'Fuel cells, their fuels and the US automobile', prepared for the First Annual World Car 2001 Conference, University of California, Riverside, CL, 20–24 June 1993

Williams, R. H. (1994) 'The clean machine: Fuel cells and their fuels for cars', *Technology Review*, April, pp20–30

Williams, R. H. (1995) 'The prospects for renewable energy', *Siemens Review*, pp1–16

Williams, R. H. and Larson, E. D. (1993) 'Advanced gasification-based biomass power generation', in T. B. Johansson, H. Kelly, A. K. N. Reddy and R. H. Williams (eds) *Renewable Energy: Sources for Fuels and Electricity*, Island Press, Washington, DC, pp729–785

Williams, R. H. and Larson, E. D. (1996) 'Biomass gasifier gas turbine power generating technology', *Biomass and Bioenergy*, vol 10, pp149–166

Williams, R. H., Larson, E. D., Katofsky, R. E. and Chen, J. (1995) *Methanol and Hydrogen from Biomass for Transportation, with Comparisons to Methanol and Hydrogen from Natural Gas and Coal*, The Centre for Energy and Environmental Studies, Princeton University

Willison, T. W., Baker, J. C. and Murphy, D. V. (1998) 'Methane fluxes and nitrogen dynamics from a drained fenland peat', *Biology and Fertility of Soils*, vol 27, pp279–283

Winjum, J. K., Brown, S. and Schlamadinger, B. (1998) 'Forest harvests and wood products: Sources and sinks of atmospheric carbon dioxide', *Forest Science*, vol 44, no 2, pp272–284

Woess-Gallasch, S. (2006) 'Should we trade biomass, bio-electricity, green certificates or CO_2 credits?', in *IEA Bioenergy Agreement Annual Report 2005*, IEA Bio-energy, ExCo:2006:01, January, pp4–19

Woodwell, G. M., Mackenzie, F. T., Houghton, R. A., Apps, M., Gorham, E. and Davidson, E. (1998) 'Biotic feedbacks in the warming of the Earth', *Climatic Change*, vol 40, pp495–518

Woomer, P. L., Palm, C. A., Qureshi, J. N. and Kotto-Same, J. (1997) 'Carbon sequestration and organic resource management in African smallholder agriculture', in R. Lal, J. M. Kimble, R. F. Follett and B. A. Stewart (eds) *Management of Carbon Sequestration in Soil*, CRC Press, Boca Raton, pp153–173

World Bank (1992) *World Development Report 1992*, World Bank, Washington, DC

World Bank (2009) *Global Economic Prospects: Commodities at the Crossroads*, The International Bank for Reconstruction and Development, Washington, DC, http://go.worldbank.org/Y3FILKN180, p22

WEA (World Energy Assessment) (2000) *World Energy Assessment of the United Nations*, UNDP, UNDESA and the World Energy Council, New York

WEA (2004) *Overview: 2004 Update*, UNDP, UNDESA and the World Energy Council

Worldwatch Institute (2007) *Biofuels for Transport: Global Potential and Implications for Energy and Agriculture*, Earthscan/James & James, London, p336

WRI (World Resources Institute) (1987) *Tropical Forestry Action Plan: Recent Developments*, World Resources Institute, Washington, DC

Wright, R. M. (1996) *Jamaica's Energy: Old Prospects New Resources*, Twin Guinep Ltd/Stephenson's Litho Press Ltd

Wurth, W. and Kusterer, B. (1995) *Ergebnisse der Landessortenaersuche mit Futterrüben und Welschem Weidelgrass 1994 Information für die Pflanzenproduktion*, Landesanstalt für Pflanzenbau Forchheim

Xi, Q. (2000) 'Investigation on the distribution and potential of giant grasses in China', PhD dissertation, University of Kiel, Cuvillier Verlag, Goettingen

Xi, Q. G. (2008) 'Description of an Introduced Plant *Miscanthus x giganteus* Greef et Deu.', *Pratacultural Science*, 2008-2 (in Chinese with English abstract)

Yadav, R. L., Yadav, D. S., Singh, R. M. and Kumar, A. (1998) 'Long term effects of inorganic fertilizer inputs on crop productivity in a rice-wheat cropping system', *Nutrient Cycling in Agroecosystems*, vol 51, pp193–200

Yagi, K., Tsuruta, H. and Minami, K. (1997) 'Possible options for mitigating methane emission from rice cultivation', *Nutrient Cycling in Agroecosystems*, vol 49, pp213–220

Yamasaki, S. (1981) 'Effect of water level on the development of rhizomes of three hygrophytes', *Japanese Journal of Ecology*, vol 31, pp353–359

Yazaki, Y., Mariko, S. and Koizumi, H. (2004) 'Carbon dynamics and budget in a *Miscanthus sinensis* grassland in Japan', *Ecological Research*, vol 19, no 5, pp511–520

Ye, B., Saito, A. and Minamisawa, K. (2005) 'Effect of inoculation with anaerobic nitrogen-fixing consortium on salt tolerance of *Miscanthus sinensis*', *Soil Science and Plant Nutrition*, vol 51, no 2, pp243–249

Yoshida, K. (1983) 'Heterogeneous environmental structure in a moth community of Tomakomai Experiment Forest, Hokkaido University', *Japan Journal of Ecology*, vol 33, pp445–451

Youmbi, E., Clair-Maczulajtys, D. and Bory, G. (1989) 'Variation de la composition des fruits de *Dacryodes edulis* (Don) Lam', *Fruits*, vol 44, no 3, pp149–157

Young, A. (1997) *Agroforestry for Soil Management*, 2nd edn, CAB International, Oxford, UK

Zah, R., Böni, H., Gauch, M., Hischier, R., Lehmann, M. and Wäger, P. (2007) 'Ökobilanzierung von Energieprodukten: Ökologische Bewertung von Biotreibstoffen', Schlussbericht, Entwurf, Abteilung Technologie und Gesellschaft, Empa im Auftrag des Bundesamtes für Energie, des Bundesamtes für Umwelt und des Bundesamtes für Landwirtschaft, Bern

Zafar, Y. and Malik, K. A. (1984) 'Photosynthetic system of *Leptochloa fusca* (L.) Kunth', *Pakistan Journal of Botany*, vol 16, pp109–116

Zafar, Y., Ashraf, M. and Malik, K. A. (1986) 'Nitrogen fixation associated with roots of kallar grass (*Leptochloa fusca* L. Kunth)', *Plant and Soil*, vol 90, pp93–105

Zaman, A. and Maiti, A. (1990) 'Cost of safflower cultivation and benefit-cost ratio at different nitrogen levels and limited moisture supply conditions', *Environment and Ecology*, vol 8, no 1, pp232–235

Zauner, E. and Küntzel, U. (1986) 'Methane production from ensiled plant material', *Biomass*, vol 10, pp207–223

Zhang, X., Peterson, C. L., Reece, D., Möller, G. and Haws, R. (1995) 'Biodegradability of biodiesel in the aquatic environment', paper presented to ASAE meeting, p14

Zhang, Z. X. (1996) 'Some economic aspects of climate change', *International Journal Environment and Pollution*, vol 6, pp185–195

Zimmerman, G., *Potential of Reed Canary Grass as a Biofuel in Michigan's Eastern Upper Peninsula*, Lake Superior State University, Sault Sainte Marie, MI

Zimmermann, R. W., Roberts, D. R. and Carpenter, S. B. (1982) 'Nitrogen fertilization effects on black locust seedlings examined using the acetylene reduction assay', in B. A. Thielges (ed) *Proceedings of Seventh North American Forest Biology Workshop*, Lexington, pp305–309

Zoltai, S. T. and Martikainen, P. J. (1996) 'Estimated extent of forested peatlands and their role in the global carbon cycle', in M. J. Apps and D. T. Price (eds) *Forest Ecosystems, Forest Management and the Global Carbon Cycle*, NATO ASI Series, Series I: Global Environmental Change, vol 40, pp47–58

Zsuffa, L. and Barkley, B. (1984) 'The commercial and practical aspects of short rotation forestry in temperate regions: A state-of-the-art review', *Proceedings of Conference on 'Bio Energy'*, Goteborg, 15–21 June 1984, vol l, pp39–57

Index of Botanical Names

Index of Common Names

Index of General Terms

For Product Safety Concerns and Information please contact our EU
representative GPSR@taylorandfrancis.com
Taylor & Francis Verlag GmbH, Kaufingerstraße 24, 80331 München, Germany

www.ingramcontent.com/pod-product-compliance
Lightning Source LLC
Chambersburg PA
CBHW060952210326
41598CB00031B/4800

9 781138 975712